Extensive and Varied Exercise Sets	An abundant collection of exercises is included in an exercise set at the end of each section. Exercises are organized within categories. Your instructor will usually provide guidance on which exercises to work. The exercises in the first category, Practice Exercises, follow the same order as the section's worked examples.	worked examples and use them as models for solving these problems.
Practice Plus Problems	This category of exercises contains more challenging problems that often require you to combine several skills or concepts.	It is important to dig in and develop your problem-solving skills. Practice Plus Exercises provide you with ample opportunity to do so.

3 Review for Quizzes and Tests

Feature	Description	Benefit
Mid-Chapter Check Points	At approximately the midway point in the chapter, an integrated set of review exercises allows you to review the skills and concepts you learned separately over several sections.	By combining exercises from the first half of the chapter, the Mid-Chapter Check Points give a comprehensive review before you move on to the material in the remainder of the chapter.
Chapter Review Grids	Each chapter contains a review chart that summarizes the definitions and concepts in every section of the chapter. Examples that illustrate these key concepts are also included in the chart.	Review this chart and you'll know the most important material in the chapter!
Chapter Review Exercises	A comprehensive collection of review exercises for each of the chapter's sections follows the review grid.	Practice makes perfect. These exercises contain the most significant problems for each of the chapter's sections.
Chapter Tests	Each chapter contains a practice test with approximately 25 problems that cover the important concepts in the chapter. Take the practice test, check your answers, and then watch the Chapter Test Prep Video CD to see worked-out solutions for any exercises you miss.	You can use the chapter test to determine whether you have mastered the material covered in the chapter.
Chapter Test Prep Video CD	This video CD found at the back of your text contains worked-out solutions to every exercise in each chapter test.	The video lets you review any exercises you miss on the chapter test.
Cumulative Review Exercises	Beginning with Chapter 2, each chapter concludes with a comprehensive collection of mixed cumulative review exercises. These exercises combine problems from previous chapters and the present chapter, providing an ongoing cumulative review.	Ever forget what you've learned? These exercises ensure that you are not forgetting anything as you move forward.

More Tools for Your Mathematics Success

Student Study Pack

Get math help when YOU need it! The **Student Study Pack** provides you with the ultimate set of support resources to go along with your text. Packaged at no charge with a new textbook, the **Student Study Pack** contains these invaluable tools:

☑ **Student Solutions Manual**

A printed manual containing full solutions to odd-numbered textbook exercises.

☑ **Prentice Hall Math Tutor Center**

Tutors provide one-on-one tutoring for any problem with an answer at the back of the book. You can contact the Tutor Center via a toll-free phone number, fax, or email.

☑ **CD-ROM Lecture Series**

A comprehensive set of textbook-specific CD-ROMs containing short video clips of the textbook objectives being reviewed and key examples being solved.

Tutorial and Homework Options

MYMATHLAB

MyMathLab, packaged at no charge with a new textbook, is a complete online multimedia resource to help you succeed in learning. **MyMathLab** features:

☑ The entire textbook online.

☑ Problem-solving video clips and practice exercises, correlated to the examples and exercises in the text.

☑ Online tutorial exercises with guided examples.

☑ Online homework and tests.

☑ Generates a personalized study plan based on your test results.

☑ Tracks all online homework, tests, and tutorial work you do in **MyMathLab** gradebook.

Introductory Algebra
for College Students

Fourth Edition

Introductory Algebra
for College Students

Robert Blitzer
Miami-Dade College

PEARSON

Prentice
Hall

Upper Saddle River, New Jersey 07458

Library of Congress Cataloging-in-Publication Data

Blitzer, Robert.
   Introductory algebra for college students / Robert Blitzer.—4th ed.
     p. cm.
   Includes indexes.
   ISBN 0-13-149262-4
    1. Algebra—Textbooks. I. Title.

   QA152.3.B648 2006
   512.9—dc22   2004058736

Senior Acquisitions Editor: *Paul Murphy*
Editor in Chief: *Christine Hoag*
Project Manager: *Liz Covello*
Production Editor: *Barbara Mack*
Assistant Managing Editor: *Bayani Mendoza de Leon*
Senior Managing Editor: *Linda Mihatov Behrens*
Executive Managing Editor: *Kathleen Schiaparelli*
Media Project Manager: *Audra J. Walsh*
Media Production Editor: *Allyson Kloss*
Managing Editor, Digital Supplements: *Nicole M. Jackson*
Assistant Manufacturing Manager/Buyer: *Michael Bell*
Executive Marketing Manager: *Eilish Collins Main*
Marketing Assistant: *Rebecca Alimena*
Director of Marketing: *Patrice Jones*
Development Editor: *Don Gecewicz*
Editor in Chief, Development: *Carol Trueheart*
Editorial Assistant: *Mary Burket*
Assistant Editor/Print Supplements Editor: *Christina Simoneau*
Art Director: *Maureen Eide*
Interior Design: *Running River Design*
Cover Art and Design: *Koala Bear Design*
Art Editor: *Thomas Benfatti*
Director of Creative Services: *Paul Belfanti*
Director, Image Resource Center: *Melinda Reo*
Manager, Rights and Permissions: *Zina Arabia*
Manager, Visual Research: *Beth Brenzel*
Image Permission Coordinator: *Debbie Hewitson*
Photo Researcher: *Elaine Soares*
Art Studio: *Scientific Illustrators*
Compositor: *Prepare, Inc.*

© 2006, 2002, 1998, 1995 Pearson Education, Inc.
Pearson Prentice Hall
Pearson Education, Inc.
Upper Saddle River, New Jersey 07458

Printed in the United States of America
10 9 8 7 6 5 4 3 2 1

ISBN 0-13-149262-4

Pearson Education LTD., *London*
Pearson Education Australia PTY, Limited, *Sydney*
Pearson Education Singapore, Pte. Ltd
Pearson Education North Asia Ltd, *Hong Kong*
Pearson Education Canada, Ltd., *Toronto*
Pearson Educación de Mexico, S.A. de C.V.
Pearson Education—Japan, *Tokyo*
Pearson Education Malaysia, Pte. Ltd

CONTENTS

A BRIEF GUIDE TO GETTING THE MOST FROM THIS BOOK
Inside Front Cover

V

PREFACE

Introductory Algebra for College Students, Fourth Edition, provides comprehensive, in-depth coverage of the topics required in a one-term course in beginning or introductory algebra. The book is written for college students who have no previous experience in algebra and for those who need a review of basic algebraic concepts. I wrote the book to help diverse students, with different backgrounds and career plans, to succeed in beginning algebra. *Introductory Algebra for College Students*, Fourth Edition, has two primary goals:

1. To help students acquire a solid foundation in the basic skills of algebra.
2. To show students how algebra can model and solve authentic real-world problems.

One major obstacle in the way of achieving these goals is the fact that very few students actually read their textbook. This has been a regular source of frustration for me and for my colleagues in the classroom. Anecdotal evidence gathered over years highlights two basic reasons why students do not take advantage of their textbook:

- "I'll never use this information."
- "I can't follow the explanations."

I've written every page of the Fourth Edition with the intent of eliminating these two objections. The ideas and tools I've used to do so are described in the features that follow. These features and their benefits are highlighted for the student in "A Brief Guide to Getting the Most from This Book" that appears inside the front cover.

What's New in the Fourth Edition?

I believe students and instructors will welcome the following new features:

- **Practice Plus Exercises.** More challenging practice exercises that often require students to combine several skills or concepts have been added to the exercise sets. The 452 Practice Plus Exercises in the Fourth Edition, averaging 8 of these exercises per exercise set, provide instructors with the option of creating assignments that take practice exercises to a more challenging level than in the previous edition.

- **Mid-Chapter Check Points.** At approximately the midway point in each chapter, an integrated set of review exercises allows students to review and assimilate the skills and concepts they learned separately over several sections. The 177 exercises that make up the Mid-Chapter Check Points, averaging 18 exercises per check point, are of a mixed nature, requiring students to discriminate which concepts or skills to apply. The Mid-Chapter Check Points should help students bring together the different objectives covered in the first half of the chapter before they move on to the material in the remainder of the chapter.

- **New Applications and Real-World Data.** I researched hundreds of books, magazines, almanacs, and online data sites to prepare the Fourth Edition. As a result, many new, innovative applications, supported by data that extend as far up to the present as possible, appear in 165 new application exercises and examples.

- **Over 1400 New Examples and Exercises.** In addition to the 452 Practice Plus Exercises, the 177 Mid-Chapter Check Points, and the 165 new application exercises, the Fourth Edition contains 644 new exercises that appear in the various categories of the exercise sets.

- **Increased Study Tip Boxes.** The book's Study Tip boxes offer suggestions for problem solving, point out common student errors, and provide informal tips and suggestions. These invaluable hints appear in greater abundance in the Fourth Edition.

- **Chapter Test Prep Video CD.** Packaged at the back of the text, this video CD provides students with step-by-step solutions for each of the exercises in the book's chapter tests.

- **Graphical and Numerical Approaches to Problems.** Although the use of graphing utilities is optional, students can look at the new side-by-side features in Using Technology boxes and begin to understand both graphical and numerical approaches to problems even if they are not using a graphing utility in the course.

What Content and Organizational Changes Have Been Made to the Fourth Edition?

- Section 1.1 (Fractions) contains new discussions on mixed numbers and improper fractions, as well as the use of prime factorizations in reducing fractions and finding LCDs.

- Section 2.4 (Formulas and Percents) includes new discussions on percent increase, percent decrease, and ways percents are frequently used incorrectly.

- Section 10.1 (Solving Quadratic Equations by the Square Root Property) now includes a discussion of the distance between two points in rectangular coordinates.

- Section 10.6 (Introduction to Functions) provides a new introduction to relations and functions using data for calories burned in various activities.

What Familiar Features Have Been Retained in the Fourth Edition?

The features described below that helped make the Third Edition so popular continue in the Fourth Edition.

- **Detailed Worked-Out Examples.** Each worked example is titled, making clear the purpose of the example. Examples are clearly written and provide students with detailed step-by-step solutions. No steps are omitted and each step is thoroughly explained to the right of the mathematics.

- **Check Point Examples.** Each example is followed by a similar matched problem, called a Check Point, offering students the opportunity to test their understanding of the example by working a similar exercise. The answers to the Check Points are provided in the answer section.

- **Explanatory Voice Balloons.** Voice balloons are used in a variety of ways to demystify mathematics. They translate algebraic ideas into everyday English, help clarify problem-solving procedures, present alternative ways of understanding concepts, and connect problem solving to concepts students have already learned.

- **Extensive and Varied Exercise Sets.** An abundant collection of exercises is included in an exercise set at the end of each section. Exercises are organized within seven category types: Practice Exercises, Practice Plus Exercises, Application Exercises, Writing in Mathematics, Critical Thinking Exercises, Technology Exercises, and Review Exercises. This format makes it easy to create well-rounded homework assignments. The order of the practice exercises is exactly the same as the order of the section's worked examples. This parallel order enables students to refer to the titled examples and their detailed explanations to achieve success working the practice exercises.

- **Chapter-Opening and Section-Opening Scenarios.** Every chapter and every section open with a scenario presenting a unique application of algebra in students' lives outside the classroom. These scenarios are revisited in the course of the chapter or section in an example, discussion, or exercise.

- **Section Objectives.** Learning objectives open every section. These objectives help students recognize and focus on the section's most important ideas. The objectives are stated in the margin at their point of use.

- **Chapter Review Grids.** Each chapter contains a review chart that summarizes the definitions and concepts in every section of the chapter. Examples that illustrate these key concepts are also included in the chart.

- **End-of-Chapter Materials.** A comprehensive collection of review exercises for each of the chapter's sections follows the review grid. This is followed by a chapter test that enables students to test their understanding of the material covered in the chapter. Beginning with Chapter 2, each chapter concludes with a comprehensive collection of mixed cumulative review exercises.

- **Graphing.** Chapter 1 contains an introduction to graphing, a topic that is integrated throughout the book. Line, bar, circle, and rectangular coordinate graphs that use real data appear in nearly every section and exercise set. Many examples and exercises use graphs to explore relationships between data and to provide ways of visualizing a problem's solution.

- **Geometric Problem Solving.** Chapter 3 on problem solving contains a section that teaches geometric concepts that are important to a student's understanding of algebra. There is a frequent emphasis on problem solving in geometric situations, as well as on geometric models that allow students to visualize algebraic formulas.

- **Thorough, Yet Optional Technology.** Although the use of graphing utilities is optional, they are utilized in Using Technology boxes to enable students to visualize and gain numerical insight into algebraic concepts. The use of graphing utilities is also reinforced in the technology exercises appearing in the exercise sets for those who want this option. With the book's early introduction to graphing, students can look at the calculator screens in the Using Technology boxes and gain an increased understanding of an example's solution even if they are not using a graphing utility in the course.

- **Study Tips.** Study Tip boxes appear throughout the book.

- **Enrichment Essays.** These discussions provide historical, interdisciplinary, and otherwise interesting connections to the algebra under study, showing students that math is an interesting and dynamic discipline.

- **Discovery.** Discover for Yourself boxes, found throughout the text, encourage students to further explore algebraic concepts. These explorations are optional and their omission does not interfere with the continuity of the topic under consideration.

- **Chapter Projects.** At the end of each chapter is a collaborative activity that gives students the opportunity to work cooperatively as they think and talk about mathematics. Additional group projects can be found in the *Instructor's Resource Manual*. Many of these exercises should result in interesting group discussions.

Resources for the Instructor

Print

Annotated Instructor's Edition (ISBN: 0-13-149263-2)

- Answers in place on the same text page as exercises, or in the Graphing Answer Section

- Answers to all exercises in the exercise sets, Mid-Chapter Check Points, Chapter Reviews, Chapter Tests, and Cumulative Reviews

Instructor's Solutions Manual (ISBN: 0-13-192184-3)

- Detailed step-by-step solutions to the even-numbered section exercises

- Solutions to every exercise (odd and even) in the Mid-Chapter Check Points, Chapter Reviews, Chapter Tests, and Cumulative Reviews

- Solution methods reflect those emphasized in the text.

Instructor's Resource Manual with Tests (ISBN: 0-13-192182-7)

- Six test forms per chapter—3 free response, 3 multiple choice

- Two *Cumulative Tests* for all even-numbered chapters

- Two *Final Exams*

- Answers to all test items

- *Mini-Lectures* for each section with brief lectures including key learning objectives, classroom examples, and teaching notes

- Additional *Activities*, two per chapter, providing short group activities in a convenient ready-to-use handout format

- *Skill Builders* providing an enhanced worksheet for each text section, including concept rules, explained examples, and extra problems for students

- Twenty *Additional Exercises* per section for added test exercises or worksheets

Media

Lab Pack CD Lecture Series (ISBN: 0-13-133077-2)

- Organized by section, *Lab Pack CD Lecture Series* contains problem-solving techniques and examples from the textbook.

- Step-by-step solutions to selected exercises from each textbook section marked with a video icon

TestGen (ISBN: 0-13-192183-5)

- Windows and Macintosh compatible

- Algorithmically driven, text-specific testing program covering all objectives of the text

- Chapter Test file for each chapter provides algorithms specific to exercises in each *Chapter Test* from the text.

- Edit and add your own questions with the built-in question editor, which allows you to create graphs, import graphics, and insert math notation.

- Create a nearly unlimited number of tests and worksheets, as well as assorted reports and summaries.

- Networkable for administering tests and capturing grades online, or on a local area network

MyMathLab
www.mymathlab.com

An all-in-one, online tutorial, homework, assessment, and course management tool with the following features:

- Rich and flexible set of course materials, featuring free-response exercises algorithmically generated for unlimited practice and mastery

- Entire textbook online with links to multimedia resources including video clips, practice exercises, and animations that are correlated to the textbook examples and exercises

- Homework and test managers to select and assign online exercises correlated directly to the text

- A personalized Study Plan generated based on student test results. The Study Plan links directly to unlimited tutorial exercises for the areas students need to study and retest, so they can practice until they have mastered the skills and concepts.

- *MyMathLab Gradebook* allows you to track all of the online homework, tests, and tutorial work while providing grade control.

- Import *TestGen* tests

MathXL®
www.mathxl.com

A powerful online homework, tutorial, and assessment system that allows instructors to:

- Create, edit, and assign online homework and tests using algorithmically generated exercises correlated at the objective level to the textbook

- Track student work in *MathXL*'s online gradebook

Resources for the Student

Student Solutions Manual   CD/Video   PH Math/Tutor Center   MathXL Tutorials on CD   MathXL®   MyMathLab   Interactmath.com

Student Study Pack (ISBN: 0-13-154938-3)

Get math help when YOU need it! Available at no charge when packaged with a new text, the *Student Study Pack* provides the ultimate set of support resources to go along with the text. The *Student Study Pack* includes the *Student Solutions Manual*, access to the *Prentice Hall Math Tutor Center*, and the *CD Lecture Videos*.

Print

Student Solutions Manual (ISBN: 0-13-192185-1)

- Solutions to all odd-numbered section exercises
- Solutions to every exercise (odd and even) in the Mid-Chapter Check Points, Chapter Reviews, Chapter Tests, and Cumulative Reviews
- Solution methods reflect those emphasized in the text.

Media

MyMathLab

An all-in-one, online tutorial, homework, assessment, course management tool with the following student features:

- Entire textbook online with links to multimedia resources including video clips, practice exercises, and animations correlated to the textbook examples and exercises
- Online tutorial, homework, and tests
- A personalized Study Plan based on student test results. The Study Plan links directly to unlimited tutorial exercises for the areas students need to study and retest, so they can practice until they have mastered the skills and concepts..

MathXL®

A powerful online homework, tutorial, and assessment system that allows students to:

- Take chapter tests and receive a personalized study plan based on their test results
- See diagnosed weaknesses and link directly to tutorial exercises for the objectives they need to study and retest
- Access supplemental animations and video clips directly from selected exercises

New! www.InterActMath.com

- The power of the *MathXL®* text-specific tutorial exercises available for unlimited practice online, without an access code

MathXL® Tutorials on CD (ISBN: 0-13-133069-1)
An interactive tutorial that provides:

- Algorithmically generated practice exercises correlated at the objective level
- Practice exercises accompanied by an example and a guided solution
- Tutorial video clips within the exercise to help students visualize concepts
- Easy-to-use tracking of student activity and scores and printed summaries of students' progress

Chapter Test Prep Video CD (ISBN: 0-13-133076-4)

- Provides step-by-step video solutions to each exercise in each Chapter Test in the textbook
- Packaged at no charge with the text, inside the back cover

PH Tutor Center (ISBN: 0-13-064604-0)

- Tutorial support via phone, fax, or email bundled at no charge with a new text, or purchased separately with a used book
- Staffed by developmental math faculty
- Available 5 days a week, 7 hours a day

Acknowledgments

An enormous benefit of authoring a successful series is the broad-based feedback I receive from the students, dedicated users, and reviewers. Every change to this edition is the result of their thoughtful comments and suggestions. I would like to express my appreciation to all the reviewers, whose collective insights form the backbone of this revision. In particular, I would like to thank:

Howard Anderson	*Skagit Valley College*
John Anderson	*Illinois Valley Community College*
Michael H. Andreoli	*Miami Dade College – North Campus*
Jana Barnard	*Angelo State University*
Gale Brewer	*Amarillo College*
Warren J. Burch	*Brevard Community College*
Alice Burstein	*Middlesex Community College*
Edie Carter	*Amarillo College*
Sandra Pryor Clarkson	*Hunter College*
Sally Copeland	*Johnson County Community College*
Carol Curtis	*Fresno City College*
Robert A. Davies	*Cuyahoga Community College*
Ben Divers, Jr.	*Ferrum College*
Irene Doo	*Austin Community College*
Charles C. Edgar	*Onondaga Community College*
Karen Edwards	*Diablo Valley College*
Susan Forman	*Bronx Community College*
Wendy Fresh	*Portland Community College*
Gary Glaze	*Eastern Washington University*
Jay Graening	*University of Arkansas*
Robert B. Hafer	*Brevard Community College*
Mary Lou Hammond	*Spokane Community College*
Andrea Hendricks	*Georgia Perimeter College*
Donald Herrick	*Northern Illinois University*
Beth Hooper	*Golden West College*
Sandee House	*Georgia Perimeter College*
Tracy Hoy	*College of Lake County*
Laura Hoye	*Trident Community College*
Margaret Huddleston	*Schreiner University*
Marcella Jones	*Minneapolis Community and Technical College*
Shelbra B. Jones	*Wake Technical Community College*
Sharon Keenee	*Georgia Perimeter College*
Gary Kersting	*North Central Michigan College*
Gary Knippenberg	*Lansing Community College*
Mary Kochler	*Cuyahoga Community College*
Robert Leibman	*Austin Community College*
Jennifer Lempke	*North Central Michigan College*
Ann M. Loving	*J. Sargent Reynolds Community College*

Kent MacDougall	*Temple College*
Hank Martel	*Broward Community College*
Kim Martin	*Southeastern Illinois College*
John Robert Martin	*Tarrant County Junior College*
Irwin Metviner	*State University of New York at Old Westbury*
Jean P. Millen	*Georgia Perimeter College*
Lawrence Morales	*Seattle Central Community College*
Lois Jean Niemi	*Minneapolis Community and Technical College*
Allen R. Newhart	*Parkersburg Community College*
Peg Pankowski	*Community College of Allegheny County – South Campus*
Robert Patenaude	*College of the Canyons*
Christopher Reisch	*The State University of New York at Buffalo*
Nancy Ressler	*Oakton Community College*
Haazim Sabree	*Georgia Perimeter College*
Gayle Smith	*Lane Community College*
Dick Spangler	*Tacoma Community College*
Janette Summers	*University of Arkansas*
Robert Thornton	*Loyola University*
Lucy C. Thrower	*Francis Marion College*
Andrew Walker	*North Seattle Community College*

Additional acknowledgments are extended to Paul Murphy, senior acquisitions editor; Liz Covello, project manager; Barbara Mack, production editor; and Bayani Mendoza de Leon, assistant managing editor. Lastly, my thanks to the staff at Prentice Hall for their continuing support: Eilish Main, executive marketing manager; Patrice Jones, director of marketing; Chris Hoag, editor-in-chief, and to the entire Prentice Hall sales force for their confidence and enthusiasm about my books.

I hope that my love for learning, as well as my respect for the diversity of students I have taught and learned from over the years, is apparent throughout this new edition. By connecting algebra to the whole spectrum of learning, it is my intent to show students that their world is profoundly mathematical, and indeed, π is in the sky.

Robert Blitzer

TO THE STUDENT

I've written this book so that you can learn about the power of algebra and how it relates directly to your life outside the classroom. All concepts are carefully explained, important definitions and procedures are set off in boxes, and worked-out examples that present solutions in a step-by-step manner appear in every section. Each example is followed by a similar matched problem, called a Check Point, for you to try so that you can actively participate in the learning process as you read the book. (Answers to all Check Points appear in the back of the book.) Study Tips offer hints and suggestions and often point out common errors to avoid. A great deal of attention has been given to applying algebra to your life to make your learning experience both interesting and relevant.

As you begin your studies, I would like to offer some specific suggestions for using this book and for being successful in this course:

1. Read the book. Read each section with pen (or pencil) in hand. Move through the worked-out examples with great care. These examples provide a model for doing exercises in the exercise sets. As you proceed through the reading, do not give up if you do not understand every single word. Things will become clearer as you read on and see how various procedures are applied to specific worked-out examples.

2. Work problems every day and check your answers. The way to learn mathematics is by doing mathematics, which means working the Check Points and assigned exercises in the exercise sets. The more exercises you work, the better you will understand the material.

3. Review for quizzes and tests. After completing a chapter, study the chapter review chart, work the exercises in the Chapter Review, and work the exercises in the Chapter Test. Answers to all these exercises are given in the back of the book.

> The methods that I've used to help you read the book, work the problems, and review for tests are described in "A Brief Guide to Getting the Most from This Book" that appears inside the front cover. Spend a few minutes reviewing the guide to familiarize yourself with the book's features and their benefits.

4. Use the resources available with this book. Additional resources to aid your study are described following the guide to getting the most from your book. These resources include a Solutions Manual, a Chapter Test Prep Video CD, MyMathLab, an online version of the book with links to multimedia resources, MathXL®, an online homework, tutorial, and assessment system of the text, and tutorial support at no charge at the PH Tutor Center.

5. Attend all lectures. No book is intended to be a substitute for valuable insights and interactions that occur in the classroom. In addition to arriving for lecture on time and being prepared, you will find it useful to read the section before it is covered in lecture. This will give you a clear idea of the new material that will be discussed.

I wrote this book in Point Reyes National Seashore, 40 miles north of San Francisco. The park consists of 75,000 acres with miles of pristine surf-washed beaches, forested ridges, and bays bordered by white cliffs. It was my hope to convey the beauty and excitement of mathematics using nature's unspoiled beauty as a source of inspiration and creativity. Enjoy the pages that follow as you empower yourself with the algebra needed to succeed in college, your career, and in your life.

Regards,
Bob
Robert Blitzer

ABOUT THE AUTHOR

Bob Blitzer is a native of Manhattan and received a Bachelor of Arts degree with dual majors in mathematics and psychology (minor: English literature) from the City College of New York. His unusual combination of academic interests led him toward a Master of Arts in mathematics from the University of Miami and a doctorate in behavioral sciences from Nova University. Bob is most energized by teaching mathematics and has taught a variety of mathematics courses at Miami-Dade College for nearly 30 years.

He has received numerous teaching awards, including Innovator of the Year from the League for Innovations in the Community College, and was among the first group of recipients at Miami-Dade College for an endowed chair based on excellence in the classroom. In addition to *Introductory Algebra for College Students*, Bob has written *Intermediate Algebra for College Students*, *Essentials of Intermediate Algebra for College Students*, *Introductory and Intermediate Algebra for College Students*, *Essentials of Introductory and Intermediate Algebra for College Students*, *Algebra for College Students*, *Thinking Mathematically*, *College Algebra*, *Algebra and Trigonometry*, and *Precalculus*, all published by Prentice Hall.

Introductory Algebra
for College Students

Sitting in the biology department office, you overhear two of the professors discussing the possible adult heights of their respective children. Looking at the blackboard that they've been writing on, you see that there are formulas that can estimate the height a child will attain as an adult. If the child is x years old and h inches tall, that child's adult height, H, in inches, is approximated by one of the following formulas:

Girls:

$$H = \frac{h}{0.00028x^3 - 0.0071x^2 + 0.0926x + 0.3524}$$

Boys:

$$H = \frac{h}{0.00011x^3 - 0.0032x^2 + 0.0604x + 0.3796}$$

You will use one of these formulas in Exercise 118 on page 91.

CHAPTER 1

The Real Number System

Wh<!-- -->en you encounter formulas with symbols such as those that appear in the estimation of a child's adult height, don't panic! You are already familiar with many symbols—the smiley face, the peace symbol, the heart symbol, the dollar sign, and even symbols on your calculator or computer. In this chapter, you will become familiar with the special symbolic notation of algebra. You will see that the language of algebra describes your world and holds the power to solve many of its problems.

SECTION 1.1

Objectives

1 Convert between mixed numbers and improper fractions.

2 Write the prime factorization of a composite number.

3 Reduce or simplify fractions.

4 Multiply fractions.

5 Divide fractions.

6 Add and subtract fractions with identical denominators.

7 Add and subtract fractions with unlike denominators.

8 Solve problems involving fractions.

FRACTIONS

The Nine Justices of the U.S. Supreme Court

"If I were asked where I place the American aristocracy, I should reply without hesitation that it occupies the judicial bench and bar. Scarcely any political question arises in the United States that is not resolved, sooner or later, into a judicial question."

Alexis de Tocqueville, *Democracy in America*

President Jimmy Carter had no opportunity to make an appointment to the Supreme Court. However, he selected more African Americans, Hispanics, and women for the lower federal courts than all other prior presidents combined—40 women, 37 African Americans, and 16 Hispanics. The graph in Figure 1.1 shows the number of female and minority appointments to federal judgeships for Presidents Carter, Reagan, Bush Senior, and Clinton.

We can use the numbers shown in Figure 1.1 to form various fractions. For example, the graph indicates that President Carter appointed a total of 258 federal judges. Of the 258 judges, there were 40 women. We can say that the *fraction* of women appointed to federal judgeships under Carter was $\frac{40}{258}$. Can you see how this fraction is used to refer to part (40 women) of a whole (258 total appointments)?

In a fraction, the number that is written above the fraction bar is called the **numerator**. The number below the fraction bar is called the **denominator**.

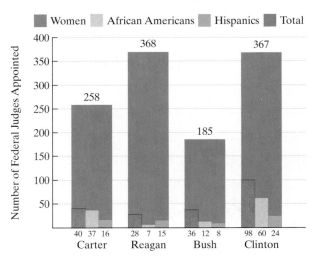

Female and Minority Appointments to Federal Judgeships

FIGURE 1.1

Source: James Burns et al., *Government by the People*, Prentice Hall, 2002

Fraction bar $\dfrac{40}{258}$

40 is the numerator.

258 is the denominator.

The numbers 40 and 258 are examples of *natural numbers*. The **natural numbers** are the numbers that we use for counting:

$$1, 2, 3, 4, 5, \ldots.$$

The three dots after the 5 indicate that the list continues in the same manner without ending.

Fractions appear throughout algebra. We are frequently given numerical information that involves fractions. In this section, we present a brief review of operations with fractions that we will use in algebra.

1 Convert between mixed numbers and improper fractions.

Mixed Numbers and Improper Fractions A **mixed number** consists of the addition of a natural number and a fraction, expressed without the use of an addition sign. Here is an example of a mixed number:

$$3\frac{4}{5}.$$ The natural number is 3 and the fraction is $\frac{4}{5}$. $3\frac{4}{5}$ means $3 + \frac{4}{5}$.

An **improper fraction** is a fraction whose numerator is greater than its denominator. An example of an improper fraction is $\frac{19}{5}$.

The mixed number $3\frac{4}{5}$ can be converted to the improper fraction $\frac{19}{5}$ using the following procedure:

CONVERTING A MIXED NUMBER TO AN IMPROPER FRACTION

1. Multiply the denominator of the fraction by the natural number and add the numerator to this product.

2. Place the result from step 1 over the denominator in the mixed number.

EXAMPLE 1 Converting from Mixed Number to Improper Fraction

Convert $3\frac{4}{5}$ to an improper fraction.

SOLUTION

$$3\frac{4}{5} = \frac{5 \cdot 3 + 4}{5}$$ Multiply the denominator by the natural number and add the numerator.

$$= \frac{15 + 4}{5} = \frac{19}{5}$$ Place the result over the mixed number's denominator.

■

✔ **CHECK POINT 1** Convert $2\frac{5}{8}$ to an improper fraction.

An improper fraction can be converted to a mixed number using the following procedure:

CONVERTING AN IMPROPER FRACTION TO A MIXED NUMBER

1. Divide the denominator into the numerator. Record the quotient (the result of the division) and the remainder.

2. Write the mixed number using the following form:

$$\text{quotient}\ \frac{\text{remainder}}{\text{original denominator}}.$$

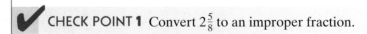

natural number part   fraction part

EXAMPLE 2   Converting from an Improper Fraction to a Mixed Number

Convert $\frac{42}{5}$ to a mixed number.

SOLUTION

Step 1.   Divide the denominator into the numerator.

$$
\begin{array}{r}
8 \quad \text{quotient}\\
5\overline{)42}\\
40\\
\hline
2 \quad \text{remainder}
\end{array}
$$

Step 2.   Write the mixed number using quotient $\dfrac{\text{remainder}}{\text{original denominator}}$. Thus,

$$\frac{42}{5} = 8\frac{2}{5}.$$

quotient — original denominator — remainder — original denominator

 CHECK POINT 2 Convert $\frac{5}{3}$ to a mixed number.

Using Variables to Express Fractions   Algebra uses letters, such as a and b, to represent numbers. Such letters are called **variables**. Throughout this section, we will use variables to express fractions. These fractions can be represented by the *variable expression*

$$\frac{a}{b}.$$

In this expression, the variable a represents any natural number and the variable b also represents any natural number. Can you see why $\frac{a}{b}$ is called a variable expression? Its value varies with the choice of a and the choice of b. Here are some examples:

Choose $a = 5$ and $b = 3$. $\dfrac{a}{b} = \dfrac{5}{3}$ Choose $a = 3$ and $b = 5$. $\dfrac{a}{b} = \dfrac{3}{5}$ Choose $a = 1$ and $b = 100$. $\dfrac{a}{b} = \dfrac{1}{100}.$

2 Write the prime factorization of a composite number.

Factors and Prime Factorizations   In the section opener, we saw that the fraction of women appointed to federal judgeships under President Carter was $\frac{40}{258}$. As you looked at this fraction, did you attempt to *simplify* it by *reducing it to its lowest terms*? Before we simplify fractions, let's first review how to factor the natural numbers that make up the numerator and the denominator.

To **factor** a natural number means to write it as a multiplication of two or more natural numbers. For example, 21 can be factored as $7 \cdot 3$. In the statement $7 \cdot 3 = 21$, 7 and 3 are called the **factors** and 21 is the **product**.

7 is a factor of 21. $7 \cdot 3 = 21$ The product of 7 and 3 is 21.

3 is a factor of 21.

Are 7 and 3 the only factors of 21? The answer is no because 21 can also be factored as $1 \cdot 21$. Thus, 1 and 21 are also factors of 21. The factors of 21 are 1, 3, 7, and 21.

Unlike the number 21, some natural numbers have only two factors: the number itself and 1. For example, the number 7 has only two factors: 7 (the number itself) and 1. The only way to factor 7 is $1 \cdot 7$ or, equivalently, $7 \cdot 1$. For this reason, 7 is called a *prime number.*

> **PRIME NUMBERS** A **prime number** is a natural number greater than 1 that has only itself and 1 as factors.

The first ten prime numbers are

$$2, 3, 5, 7, 11, 13, 17, 19, 23, \text{ and } 29.$$

Can you see why the natural number 15 is not in this list? In addition to having 15 and 1 as factors ($15 = 1 \cdot 15$), it also has factors of 3 and 5 ($15 = 3 \cdot 5$). The number 15 is an example of a *composite number.*

> **COMPOSITE NUMBERS** A **composite number** is a natural number greater than 1 that is not a prime number.

Every composite number can be expressed as the product of prime numbers. For example, the composite number 45 can be expressed as

$$45 = 3 \cdot 3 \cdot 5.$$

This product contains only prime numbers: 3 and 5.

Expressing a composite number as the product of prime numbers is called the **prime factorization** of that composite number. The prime factorization of 45 is $3 \cdot 3 \cdot 5$. The order in which we write these factors does not matter. This means that

$$45 = 3 \cdot 3 \cdot 5 \quad \text{or} \quad 45 = 5 \cdot 3 \cdot 3 \quad \text{or} \quad 45 = 3 \cdot 5 \cdot 3.$$

To find the prime factorization of a composite number, begin by selecting any two numbers whose product is the number to be factored. If one or both of the factors are not prime numbers, continue to factor each composite number. Stop when all numbers are prime.

EXAMPLE 3 Prime Factorization of a Composite Number

Find the prime factorization of 100.

SOLUTION Begin by selecting any two numbers whose product is 100. Here is one possibility:

$$100 = 4 \cdot 25.$$

Because the factors 4 and 25 are not prime, we factor each of these composite numbers.

$$100 = 4 \cdot 25 \qquad \text{This is our first factorization.}$$
$$= 2 \cdot 2 \cdot 5 \cdot 5 \qquad \text{Factor 4 and 25.}$$

Notice that 2 and 5 are both prime. The prime factorization of 100 is $2 \cdot 2 \cdot 5 \cdot 5$. ∎

 CHECK POINT 3 Find the prime factorization of 36.

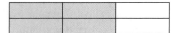

3 Reduce or simplify fractions.

FIGURE 1.2

Reducing Fractions   Two fractions are **equivalent** if they represent the same value. Writing a fraction as an equivalent fraction with a smaller denominator is called **reducing a fraction**. A fraction is **reduced to its lowest terms** when the numerator and denominator have no common factors other than 1.

Look at the rectangle in Figure 1.2. Can you see that it is divided into 6 equal parts? Of these 6 parts, 4 of the parts are red. Thus, $\frac{4}{6}$ of the rectangle is red.

The rectangle in Figure 1.2 is also divided into 3 equal stacks and 2 of the stacks are red. Thus, $\frac{2}{3}$ of the rectangle is red. Because both $\frac{4}{6}$ and $\frac{2}{3}$ of the rectangle are red, we can conclude that $\frac{4}{6}$ and $\frac{2}{3}$ are equivalent fractions.

How can we show that $\frac{4}{6} = \frac{2}{3}$ without using Figure 1.2? Prime factorizations of 4 and 6 play an important role in the process. So does the **Fundamental Principle of Fractions**.

FUNDAMENTAL PRINCIPLE OF FRACTIONS   In words: The value of a fraction does not change if the numerator and denominator are divided (or multiplied) by the same nonzero number.

In algebraic language: If $\frac{a}{b}$ is a fraction and c is a nonzero number, then

$$\frac{a \cdot c}{b \cdot c} = \frac{a}{b}.$$

We use prime factorizations and the Fundamental Principle to reduce $\frac{4}{6}$ to its lowest terms as follows:

$$\frac{4}{6} \;=\; \frac{2 \cdot 2}{3 \cdot 2} \;=\; \frac{2}{3}$$

Write prime factorizations of 4 and 6.

Divide the numerator and the denominator by the common prime factor, 2.

Here is a procedure for writing a fraction in lowest terms:

REDUCING A FRACTION TO ITS LOWEST TERMS

1. Write the prime factorization of the numerator and the denominator.
2. Divide the numerator and the denominator by the greatest common factor, the product of all factors common to both.

Division lines can be used to show dividing out a fraction's numerator and denominator by common factors:

$$\frac{4}{6} = \frac{2 \cdot \cancel{2}}{3 \cdot \cancel{2}} = \frac{2}{3}.$$

STUDY TIP

When reducing a fraction to its lowest terms, only factors that are common to the numerator and the denominator can be divided out. **If you have not factored** and expressed the numerator and denominator in terms of multiplication, **do not divide out**.

Correct:

$$\frac{2 \cdot \cancel{2}}{3 \cdot \cancel{2}} = \frac{2}{3}$$

Incorrect:

$$\frac{2 + \cancel{2}}{3 + \cancel{2}} = \frac{2}{3}$$

Note that $\frac{2+2}{3+2}$ $= \frac{4}{5}$, not $\frac{2}{3}$.

EXAMPLE 4 Reducing Fractions

Reduce each fraction to its lowest terms:

a. $\dfrac{6}{14}$ **b.** $\dfrac{15}{75}$ **c.** $\dfrac{25}{11}$ **d.** $\dfrac{11}{33}$.

SOLUTION For each fraction, begin with the prime factorization of the numerator and the denominator.

a. $\dfrac{6}{14} = \dfrac{3 \cdot 2}{7 \cdot 2} = \dfrac{3}{7}$ 2 is the greatest common factor of 6 and 14. Divide the numerator and the denominator by 2.

> Including 1 as a factor is helpful when all other factors can be divided out.

b. $\dfrac{15}{75} = \dfrac{3 \cdot 5}{3 \cdot 25} = \dfrac{1 \cdot 3 \cdot 5}{3 \cdot 5 \cdot 5} = \dfrac{1}{5}$ 3 · 5, or 15, is the greatest common factor of 15 and 75. Divide the numerator and the denominator by 3 · 5.

c. $\dfrac{25}{11} = \dfrac{5 \cdot 5}{1 \cdot 11}$

Because 11 and 25 share no common factor (other than 1), $\frac{11}{25}$ is already reduced to its lowest terms.

d. $\dfrac{11}{33} = \dfrac{1 \cdot 11}{3 \cdot 11} = \dfrac{1}{3}$ 11 is the greatest common factor of 11 and 33. Divide the numerator and denominator by 11.

When reducing fractions, it may not always be necessary to write prime factorizations. In some cases, you can use inspection to find the greatest common factor of the numerator and the denominator. For example, when reducing $\frac{15}{75}$, you can use 15 rather than 3 · 5:

$$\frac{15}{75} = \frac{1 \cdot 15}{5 \cdot 15} = \frac{1}{5}.$$

 CHECK POINT 4 Reduce each fraction to its lowest terms:

a. $\dfrac{10}{15}$ **b.** $\dfrac{42}{24}$ **c.** $\dfrac{13}{15}$ **d.** $\dfrac{9}{45}$.

4 Multiply fractions.

Multiplying Fractions The result of multiplying two fractions is called their **product**.

MULTIPLYING FRACTIONS In words: The product of two or more fractions is the product of their numerators divided by the product of their denominators.

In algebraic language: If $\frac{a}{b}$ and $\frac{c}{d}$ are fractions, then

$$\frac{a}{b} \cdot \frac{c}{d} = \frac{a \cdot c}{b \cdot d}.$$

Here is an example that illustrates the rule in the previous box:

$$\frac{3}{8} \cdot \frac{5}{11} = \frac{3 \cdot 5}{8 \cdot 11} = \frac{15}{88}.$$

> The product of $\frac{3}{8}$ and $\frac{5}{11}$ is $\frac{15}{88}$.

> Multiply numerators and multiply denominators.

EXAMPLE 5 Multiplying Fractions

Multiply. If possible, reduce the product to its lowest terms:

a. $\frac{3}{7} \cdot \frac{2}{5}$ **b.** $5 \cdot \frac{7}{12}$ **c.** $\frac{2}{3} \cdot \frac{9}{4}$ **d.** $\left(3\frac{2}{3}\right)\left(1\frac{1}{4}\right)$.

SOLUTION

a. $\frac{3}{7} \cdot \frac{2}{5} = \frac{3 \cdot 2}{7 \cdot 5} = \frac{6}{35}$
   > Multiply numerators and multiply denominators.

b. $5 \cdot \frac{7}{12} = \frac{5}{1} \cdot \frac{7}{12} = \frac{5 \cdot 7}{1 \cdot 12} = \frac{35}{12}$ or $2\frac{11}{12}$
   > Write 5 as $\frac{5}{1}$. Then multiply numerators and multiply denominators.

c. $\frac{2}{3} \cdot \frac{9}{4} = \frac{2 \cdot 9}{3 \cdot 4} = \frac{18}{12} = \frac{3 \cdot \cancel{6}}{2 \cdot \cancel{6}} = \frac{3}{2}$ or $1\frac{1}{2}$

   > Simplify $\frac{18}{12}$; 6 is the greatest common factor of 18 and 12.

d. $\left(3\frac{2}{3}\right)\left(1\frac{1}{4}\right) = \frac{11}{3} \cdot \frac{5}{4} = \frac{11 \cdot 5}{3 \cdot 4} = \frac{55}{12}$ or $4\frac{7}{12}$ ∎

✔ **CHECK POINT 5** Multiply. If possible, reduce the product to its lowest terms:

a. $\frac{4}{11} \cdot \frac{2}{3}$ **b.** $6 \cdot \frac{3}{5}$

c. $\frac{3}{7} \cdot \frac{2}{3}$ **d.** $\left(3\frac{2}{5}\right)\left(1\frac{1}{2}\right)$.

STUDY TIP

You can divide numerators and denominators by common factors before performing multiplication. Then multiply the remaining factors in the numerators and multiply the remaining factors in the denominators. For example,

$$\frac{7}{15} \cdot \frac{20}{21} = \frac{\cancel{7} \cdot 1}{\cancel{5} \cdot 3} \cdot \frac{\cancel{5} \cdot 4}{\cancel{7} \cdot 3} = \frac{1 \cdot 4}{3 \cdot 3} = \frac{4}{9}.$$

> 7 is the greatest common factor of 7 and 21.

> 5 is the greatest common factor of 15 and 20.

Study Tip (continued)

The divisions involving the common factors, 7 and 5, are often shown as follows:

$$\frac{7}{15} \cdot \frac{20}{21} = \frac{\overset{1}{\cancel{7}}}{\underset{3}{\cancel{15}}} \cdot \frac{\overset{4}{\cancel{20}}}{\underset{3}{\cancel{21}}} = \frac{1 \cdot 4}{3 \cdot 3} = \frac{4}{9}.$$

| Divide by 7. | Divide by 5. |

---

5 Divide fractions.

Dividing Fractions The result of dividing two fractions is called their **quotient**. A geometric figure is useful for developing a process for determining the quotient of two fractions.

Consider the division

$$\frac{4}{5} \div \frac{1}{10}.$$

We want to know how many $\frac{1}{10}$'s are in $\frac{4}{5}$. We can use Figure 1.3 to find this quotient. The rectangle is divided into fifths. The dashed lines further divide the rectangle into tenths.

Figure 1.3 shows that $\frac{4}{5}$ of the rectangle is red. How many $\frac{1}{10}$'s of the rectangle does this include? Can you see that this includes eight of the $\frac{1}{10}$ pieces? Thus, there are eight $\frac{1}{10}$'s in $\frac{4}{5}$:

$$\frac{4}{5} \div \frac{1}{10} = 8.$$

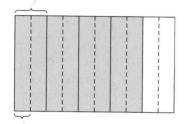

This is $\frac{1}{5}$ of the figure.

This is $\frac{1}{10}$ of the figure.

FIGURE 1.3

We can obtain the quotient 8 in the following way:

$$\frac{4}{5} \div \frac{1}{10} = \frac{4}{5} \cdot \frac{10}{1} = \frac{4 \cdot 10}{5 \cdot 1} = \frac{40}{5} = 8.$$

| Change the division to multiplication. | Invert the divisor, $\frac{1}{10}$. |

By inverting the divisor, $\frac{1}{10}$, and obtaining $\frac{10}{1}$, we are writing the divisor's *reciprocal*. Two fractions are **reciprocals** of each other if their product is 1. Thus, $\frac{1}{10}$ and $\frac{10}{1}$ are reciprocals because $\frac{1}{10} \cdot \frac{10}{1} = 1$.

Generalizing from the result shown above and using the word *reciprocal*, we obtain the following rule for dividing fractions:

DIVIDING FRACTIONS In words: The quotient of two fractions is the first fraction multiplied by the reciprocal of the second fraction.

In algebraic language: If $\frac{a}{b}$ and $\frac{c}{d}$ are fractions and $\frac{c}{d}$ is not 0, then

$$\frac{a}{b} \div \frac{c}{d} = \frac{a}{b} \cdot \frac{d}{c}.$$

| Change division to multiplication. | Invert $\frac{c}{d}$ and write its reciprocal. |

EXAMPLE 6 Dividing Fractions

Divide: **a.** $\dfrac{2}{3} \div \dfrac{7}{15}$ **b.** $\dfrac{3}{4} \div 5$ **c.** $4\dfrac{3}{4} \div 1\dfrac{1}{2}.$

SOLUTION

a. $\dfrac{2}{3} \div \dfrac{7}{15} = \dfrac{2}{3} \cdot \dfrac{15}{7} = \dfrac{2 \cdot 15}{3 \cdot 7} = \dfrac{30}{21} = \dfrac{10 \cdot \cancel{3}}{7 \cdot \cancel{3}} = \dfrac{10}{7}$ or $1\dfrac{3}{7}$

| Change division to multiplication. | Invert $\frac{7}{15}$ and write its reciprocal. | Simplify: 3 is the greatest common factor of 30 and 21. |

b. $\dfrac{3}{4} \div 5 = \dfrac{3}{4} \div \dfrac{5}{1} = \dfrac{3}{4} \cdot \dfrac{1}{5} = \dfrac{3 \cdot 1}{4 \cdot 5} = \dfrac{3}{20}$

| Change division to multiplication. | Invert $\frac{5}{1}$ and write its reciprocal. |

c. $4\dfrac{3}{4} \div 1\dfrac{1}{2} = \dfrac{19}{4} \div \dfrac{3}{2} = \dfrac{19}{4} \cdot \dfrac{2}{3} = \dfrac{19 \cdot 2}{4 \cdot 3} = \dfrac{38}{12} = \dfrac{19 \cdot \cancel{2}}{6 \cdot \cancel{2}} = \dfrac{19}{6}$ or $3\dfrac{1}{6}$ ■

✔ **CHECK POINT 6** Divide:

a. $\dfrac{5}{4} \div \dfrac{3}{8}$ **b.** $\dfrac{2}{3} \div 3$ **c.** $3\dfrac{3}{8} \div 2\dfrac{1}{4}$

6 Add and subtract fractions with identical denominators.

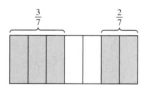

FIGURE 1.4

Adding and Subtracting Fractions with Identical Denominators The result of adding two fractions is called their **sum**. The result of subtracting two fractions is called their **difference**. A geometric figure is useful for developing a process for determining the sum or difference of two fractions with identical denominators.

Consider the addition

$$\frac{3}{7} + \frac{2}{7}.$$

We can use Figure 1.4 to find this sum. The rectangle is divided into sevenths. On the left, $\frac{3}{7}$ of the rectangle is red. On the right, $\frac{2}{7}$ of the rectangle is red. Including both the left and the right, a total of $\frac{5}{7}$ of the rectangle is red. Thus,

$$\frac{3}{7} + \frac{2}{7} = \frac{5}{7}.$$

We can obtain the sum $\frac{5}{7}$ in the following way:

$$\frac{3}{7} + \frac{2}{7} = \frac{3 + 2}{7} = \frac{5}{7}.$$

| Add numerators and put this result over the common denominator. |

Generalizing from this result gives us the following rule:

ADDING AND SUBTRACTING FRACTIONS WITH IDENTICAL DENOMINATORS
In words: The sum or difference of two fractions with identical denominators is the sum or difference of their numerators over the common denominator.
In algebraic language: If $\frac{a}{b}$ and $\frac{c}{b}$ are fractions, then

$$\frac{a}{b} + \frac{c}{b} = \frac{a + c}{b} \quad \text{and} \quad \frac{a}{b} - \frac{c}{b} = \frac{a - c}{b}.$$

EXAMPLE 7 Adding and Subtracting Fractions with Identical Denominators

Perform the indicated operations:

a. $\dfrac{3}{11} + \dfrac{4}{11}$　　　　**b.** $\dfrac{11}{12} - \dfrac{5}{12}$　　　　**c.** $5\dfrac{1}{4} - 2\dfrac{3}{4}.$

SOLUTION

a. $\dfrac{3}{11} + \dfrac{4}{11} = \dfrac{3+4}{11} = \dfrac{7}{11}$

b. $\dfrac{11}{12} - \dfrac{5}{12} = \dfrac{11-5}{12} = \dfrac{6}{12} = \dfrac{1 \cdot \cancel{6}}{2 \cdot \cancel{6}} = \dfrac{1}{2}$

c. $5\dfrac{1}{4} - 2\dfrac{3}{4} = \dfrac{21}{4} - \dfrac{11}{4} = \dfrac{21-11}{4} = \dfrac{10}{4} = \dfrac{\cancel{2} \cdot 5}{\cancel{2} \cdot 2} = \dfrac{5}{2}$ or $2\dfrac{1}{2}$ ∎

 CHECK POINT 7 Perform the indicated operations:

a. $\dfrac{2}{11} + \dfrac{3}{11}$　　　　**b.** $\dfrac{5}{6} - \dfrac{1}{6}$　　　　**c.** $3\dfrac{3}{8} - 1\dfrac{1}{8}.$

---

7 Add and subtract fractions with unlike denominators.

Adding and Subtracting Fractions with Unlike Denominators How do we add or subtract fractions with different denominators? We must first rewrite them as equivalent fractions with the same denominator. We do this by using the Fundamental Principle of Fractions: The value of a fraction does not change if the numerator and the denominator are multiplied by the same nonzero number. Thus, if $\frac{a}{b}$ is a fraction and c is a nonzero number, then

$$\frac{a}{b} = \frac{a \cdot c}{b \cdot c}.$$

EXAMPLE 8 Writing an Equivalent Fraction

Write $\frac{3}{4}$ as an equivalent fraction with a denominator of 16.

SOLUTION To obtain a denominator of 16, we must multiply the denominator of the given fraction, $\frac{3}{4}$, by 4. So that we do not change the value of the fraction, we also multiply the numerator by 4.

$$\frac{3}{4} = \frac{3 \cdot 4}{4 \cdot 4} = \frac{12}{16}$$ ∎

 CHECK POINT 8 Write $\frac{2}{3}$ as an equivalent fraction with a denominator of 21.

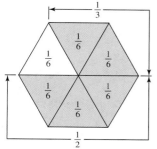

FIGURE 1.5 $\frac{1}{2} + \frac{1}{3} = \frac{5}{6}$

Equivalent fractions can be used to add fractions with different denominators, such as $\frac{1}{2}$ and $\frac{1}{3}$. Figure 1.5 indicates that the sum of half the whole figure and one-third of the whole figure results in 5 parts out of 6, or $\frac{5}{6}$, of the figure. Thus,

$$\frac{1}{2} + \frac{1}{3} = \frac{5}{6}.$$

We can obtain the sum $\frac{5}{6}$ if we rewrite each fraction as an equivalent fraction with a denominator of 6.

$$\frac{1}{2} + \frac{1}{3} = \frac{1 \cdot 3}{2 \cdot 3} + \frac{1 \cdot 2}{3 \cdot 2}$$ Rewrite each fraction as an equivalent fraction with a denominator of 6.

$$= \frac{3}{6} + \frac{2}{6}$$ Perform the multiplications. We now have a common denominator.

$$= \frac{3 + 2}{6}$$ Add the numerators and place this sum over the common denominator.

$$= \frac{5}{6}$$ Perform the addition.

When adding $\frac{1}{2}$ and $\frac{1}{3}$, there are many common denominators that we can use, such as 6, 12, 18, and so on. The given denominators, 2 and 3, divide into all of these numbers. However, the denominator 6 is the smallest number that 2 and 3 divide into. For this reason, 6 is called the *least common denominator*, abbreviated LCD.

ADDING AND SUBTRACTING FRACTIONS WITH UNLIKE DENOMINATORS

1. Rewrite the fractions as equivalent fractions with the least common denominator.
2. Add or subtract the numerators, putting this result over the common denominator.

EXAMPLE 9   Adding and Subtracting Fractions with Unlike Denominators

Perform the indicated operation: **a.** $\dfrac{1}{5} + \dfrac{3}{4}$ **b.** $\dfrac{3}{4} - \dfrac{1}{6}$ **c.** $2\dfrac{7}{15} - 1\dfrac{4}{5}$.

SOLUTION

a. Just by looking, can you can tell that the smallest number divisible by both 5 and 4 is 20? Thus, the least common denominator for the denominators 5 and 4 is 20. We rewrite both fractions as equivalent fractions with the least common denominator, 20.

$$\frac{1}{5} + \frac{3}{4} = \frac{1 \cdot 4}{5 \cdot 4} + \frac{3 \cdot 5}{4 \cdot 5}$$ To obtain denominators of 20, multiply the numerator and denominator of the first fraction by 4 and the second fraction by 5.

$$= \frac{4}{20} + \frac{15}{20}$$ Perform the multiplications.

$$= \frac{4 + 15}{20}$$ Add the numerators and put this sum over the least common denominator.

$$= \frac{19}{20}$$ Perform the addition.

b. By looking, can you tell that the smallest number divisible by both 4 and 6 is 12? Thus, the least common denominator for the denominators 4 and 6 is 12. We rewrite both fractions as equivalent fractions with the least common denominator, 12.

$$\frac{3}{4} - \frac{1}{6} = \frac{3 \cdot 3}{4 \cdot 3} - \frac{1 \cdot 2}{6 \cdot 2}$$ To obtain denominators of 12, multiply the numerator and denominator of the first fraction by 3 and the second fraction by 2.

$$= \frac{9}{12} - \frac{2}{12}$$ Perform the multiplications.

$$= \frac{9 - 2}{12}$$ Subtract the numerators and put this difference over the least common denominator.

$$= \frac{7}{12}$$ Perform the subtraction.

DISCOVER FOR YOURSELF

Try Example 9(a), $\frac{1}{5} + \frac{3}{4}$, using a common denominator of 40. Because both 5 and 4 divide into 40, 40 is a common denominator, although not the *least* common denominator. Describe what happens. What is the advantage of using the least common denominator?

c. $2\dfrac{7}{15} - 1\dfrac{4}{5} = \dfrac{37}{15} - \dfrac{9}{5}$ *Convert each mixed number to an improper fraction.*

> The smallest number divisible by both 15 and 5 is 15. Thus, the least common denominator for the denominators 15 and 5 is 15. Because the first fraction already has a denominator of 15, we only have to rewrite the second fraction.

$= \dfrac{37}{15} - \dfrac{9 \cdot 3}{5 \cdot 3}$ *To obtain denominators of 15, multiply the numerator and denominator of the second fraction by 3.*

$= \dfrac{37}{15} - \dfrac{27}{15}$ *Perform the multiplications.*

$= \dfrac{37 - 27}{15}$ *Subtract the numerators and put this difference over the common denominator.*

$= \dfrac{10}{15}$ *Perform the subtraction.*

$= \dfrac{2 \cdot \cancel{5}}{3 \cdot \cancel{5}} = \dfrac{2}{3}$ *Reduce to lowest terms.*   ■

 CHECK POINT 9 Perform the indicated operation:

a. $\dfrac{1}{2} + \dfrac{3}{5}$ **b.** $\dfrac{4}{3} - \dfrac{3}{4}$ **c.** $3\dfrac{1}{6} - 1\dfrac{11}{12}$.

EXAMPLE 10 Using Prime Factorizations to Find the LCD

Perform the indicated operation: $\dfrac{1}{15} + \dfrac{7}{24}$.

SOLUTION We need to first find the least common denominator. Using inspection, it is difficult to determine the smallest number divisible by both 15 and 24. We will use their prime factorizations to find the least common denominator:

$$15 = 5 \cdot 3 \quad \text{and} \quad 24 = 8 \cdot 3 = 2 \cdot 2 \cdot 2 \cdot 3.$$

The different prime factors are 5, 3, and 2. The least common denominator is obtained by using the greatest number of times each factor appears in any prime factorization. Because 5 and 3 appear as prime factors and 2 is a factor of 24 three times, the least common denominator is

$$5 \cdot 3 \cdot 2 \cdot 2 \cdot 2 = 5 \cdot 3 \cdot 8 = 120.$$

Now we can rewrite both fractions as equivalent fractions with the least common denominator, 120.

$\dfrac{1}{15} + \dfrac{7}{24} = \dfrac{1 \cdot 8}{15 \cdot 8} + \dfrac{7 \cdot 5}{24 \cdot 5}$ *To obtain denominators of 120, multiply the numerator and denominator of the first fraction by 8 and the second fraction by 5.*

$= \dfrac{8}{120} + \dfrac{35}{120}$ *Perform the multiplications.*

$= \dfrac{8 + 35}{120}$ *Add the numerators and put this sum over the least common denominator.*

$= \dfrac{43}{120}$ *Perform the addition.*   ■

 CHECK POINT 10 Perform the indicated operation: $\dfrac{3}{10} + \dfrac{7}{12}$.

8 Solve problems involving fractions.

Applications You can hardly pick up a newspaper without being bombarded with graphs that describe how we live, what we want, and even how we are likely to die. Understanding fractions and their operations is often helpful in interpreting the information given in these graphs.

Circle graphs, also called **pie charts**, show how a quantity is divided into parts. Circle graphs are divided into pieces, called **sectors**. The sizes of the sectors show the relative sizes of the categories. Figure 1.6 is an example of a typical circle graph. The graph shows the breakdown of the number of countries in the world that are free, partly free, or not free.

EXAMPLE 11 Using a Circle Graph to Interpret Information

Use Figure 1.6 to determine the fraction of the world's countries that are not free. Reduce the answer to its lowest terms.

SOLUTION The sector on the upper left shows that there are 48 countries that are not free. By adding the numbers in the three sectors, we see that there are a total of

$$48 + 58 + 86$$

or 192 countries.

Fraction of nonfree countries

$$= \frac{\text{number of countries that are not free}}{\text{total number of countries}}$$

$$= \frac{48}{192} = \frac{2 \cdot 24}{2 \cdot 96} = \frac{2 \cdot 2 \cdot 12}{2 \cdot 2 \cdot 48} = \frac{\cancel{2} \cdot \cancel{2} \cdot \cancel{12} \cdot 1}{\cancel{2} \cdot \cancel{2} \cdot \cancel{12} \cdot 4} = \frac{1}{4}$$

The fraction of the world's countries that are not free is $\frac{1}{4}$. ∎

World's Countries by Status of Freedom

| Not Free | Partly Free |
| 48 countries | 58 countries |

Free
86 countries

FIGURE 1.6

Source: Larry Berman and Bruce Murphy, *Approaching Democracy*, 4th edition, Prentice Hall, 2003

 CHECK POINT 11 Use Figure 1.6 to determine the fraction of the world's countries that are partly free. Reduce the answer to its lowest terms.

The graph in Figure 1.6 can show the fraction of countries, rather than the actual number of countries, in each of the three sectors. If a circle graph shows a fraction within each of its sectors, the sum of the fractional parts must be 1, representing one whole circle.

EXAMPLE 12 Miss America Hair Color

The Miss America title has been awarded, with some breaks, since 1921. The graph in Figure 1.7 might disprove the often-quoted claim that "gentlemen prefer blondes." What fraction of Miss America titleholders have been blondes?

SOLUTION We begin by finding the fraction of Miss Americas who have not been blondes. Add the fractions of brunettes, $\frac{7}{10}$, and redheads, $\frac{3}{50}$.

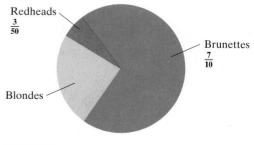

Miss America Hair Color 1921–2003

Redheads $\frac{3}{50}$

Brunettes $\frac{7}{10}$

Blondes

FIGURE 1.7

Source: Ben Schott, *Schott's Miscellany*, Bloomsbury, 2003

$$\frac{7}{10} + \frac{3}{50}$$ The least common denominator is 50.

$$= \frac{7 \cdot 5}{10 \cdot 5} + \frac{3}{50}$$ To obtain denominators of 50, multiply the numerator and denominator of the first fraction by 5.

$$= \frac{35}{50} + \frac{3}{50}$$ Perform the multiplications.

$$= \frac{35 + 3}{50} = \frac{38}{50}$$ Add the numerators and put this sum over the least common denominator.

$$= \frac{19 \cdot \cancel{2}}{25 \cdot \cancel{2}} = \frac{19}{25}$$ Reduce to lowest terms.

Thus, $\frac{19}{25}$ of the Miss Americas have not been blondes. The fractions representing all three hair colors in Figure 1.7 must add up to 1. Therefore, the fraction of Miss Americas who have been blondes can be found by subtracting $\frac{19}{25}$ from 1:

$$1 - \frac{19}{25} = \frac{25}{25} - \frac{19}{25} = \frac{25 - 19}{25} = \frac{6}{25}.$$

We see that $\frac{6}{25}$ of Miss America titleholders have been blondes. ◼

 CHECK POINT 12 At a workshop on enhancing creativity, $\frac{1}{4}$ of the participants are musicians, $\frac{2}{5}$ are artists, $\frac{1}{10}$ are actors, and the remaining participants are writers. Find the fraction of people at the workshop who are writers.

1.1 EXERCISE SET

Student Solutions Manual   CD/Video   PH Math/Tutor Center   MathXL Tutorials on CD   MathXL®   MyMathLab   Interactmath.com

Practice Exercises

In Exercises 1–6, convert each mixed number to an improper fraction.

1. $2\frac{3}{8}$ **2.** $2\frac{7}{9}$ **3.** $7\frac{3}{5}$

4. $6\frac{2}{5}$ **5.** $8\frac{7}{16}$ **6.** $9\frac{5}{16}$

In Exercises 7–12, convert each improper fraction to a mixed number.

7. $\frac{23}{5}$ **8.** $\frac{47}{8}$ **9.** $\frac{76}{9}$

10. $\frac{59}{9}$ **11.** $\frac{711}{20}$ **12.** $\frac{788}{25}$

In Exercises 13–28, identify each natural number as prime or composite. If the number is composite, find its prime factorization.

13. 22   **14.** 15   **15.** 20
16. 75   **17.** 37   **18.** 23
19. 36   **20.** 100   **21.** 140
22. 110   **23.** 79   **24.** 83
25. 81   **26.** 64
27. 240   **28.** 360

In Exercises 29–40, simplify each fraction by reducing it to its lowest terms.

29. $\frac{10}{16}$ **30.** $\frac{8}{14}$ **31.** $\frac{15}{18}$ **32.** $\frac{18}{45}$

33. $\frac{35}{50}$ **34.** $\frac{45}{50}$ **35.** $\frac{32}{80}$ **36.** $\frac{75}{80}$

37. $\frac{44}{50}$ **38.** $\frac{38}{50}$ **39.** $\frac{120}{86}$ **40.** $\frac{116}{86}$

In Exercises 41–90, perform the indicated operation. Where possible, reduce the answer to its lowest terms.

41. $\frac{2}{5} \cdot \frac{1}{3}$ **42.** $\frac{3}{7} \cdot \frac{1}{4}$ **43.** $\frac{3}{8} \cdot \frac{7}{11}$

44. $\frac{5}{8} \cdot \frac{3}{11}$ **45.** $9 \cdot \frac{4}{7}$ **46.** $8 \cdot \frac{3}{7}$

47. $\frac{1}{10} \cdot \frac{5}{6}$ **48.** $\frac{1}{8} \cdot \frac{2}{3}$ **49.** $\frac{5}{4} \cdot \frac{6}{7}$

50. $\frac{7}{4} \cdot \frac{6}{11}$ **51.** $\left(3\frac{3}{4}\right)\left(1\frac{3}{5}\right)$

52. $\left(2\frac{4}{5}\right)\left(1\frac{1}{4}\right)$ **53.** $\frac{5}{4} \div \frac{4}{3}$

54. $\frac{7}{8} \div \frac{2}{3}$ **55.** $\frac{18}{5} \div 2$ **56.** $\frac{12}{7} \div 3$

57. $2 \div \frac{18}{5}$ **58.** $3 \div \frac{12}{7}$ **59.** $\frac{3}{4} \div \frac{1}{4}$

60. $\frac{3}{7} \div \frac{1}{7}$ **61.** $\frac{7}{6} \div \frac{5}{3}$ **62.** $\frac{7}{4} \div \frac{3}{8}$

63. $\frac{1}{14} \div \frac{1}{7}$ **64.** $\frac{1}{8} \div \frac{1}{4}$ **65.** $6\frac{3}{5} \div 1\frac{1}{10}$

66. $1\frac{3}{4} \div 2\frac{5}{8}$ **67.** $\frac{2}{11} + \frac{4}{11}$ **68.** $\frac{5}{13} + \frac{2}{13}$

69. $\dfrac{7}{12} + \dfrac{1}{12}$

70. $\dfrac{5}{16} + \dfrac{1}{16}$

71. $\dfrac{5}{8} + \dfrac{5}{8}$

72. $\dfrac{3}{8} + \dfrac{3}{8}$

73. $\dfrac{7}{12} - \dfrac{5}{12}$

74. $\dfrac{13}{18} - \dfrac{5}{18}$

75. $\dfrac{16}{7} - \dfrac{2}{7}$

76. $\dfrac{17}{5} - \dfrac{2}{5}$

77. $\dfrac{1}{2} + \dfrac{1}{5}$

78. $\dfrac{1}{3} + \dfrac{1}{5}$

79. $\dfrac{3}{4} + \dfrac{3}{20}$

80. $\dfrac{2}{5} + \dfrac{2}{15}$

81. $\dfrac{3}{8} + \dfrac{5}{12}$

82. $\dfrac{3}{10} + \dfrac{2}{15}$

83. $\dfrac{11}{18} - \dfrac{2}{9}$

84. $\dfrac{17}{18} - \dfrac{4}{9}$

85. $\dfrac{4}{3} - \dfrac{3}{4}$

86. $\dfrac{3}{2} - \dfrac{2}{3}$

87. $\dfrac{7}{10} - \dfrac{3}{16}$

88. $\dfrac{7}{30} - \dfrac{5}{24}$

89. $3\dfrac{3}{4} - 2\dfrac{1}{3}$

90. $3\dfrac{2}{3} - 2\dfrac{1}{2}$

Practice Plus

In Exercises 91–94, perform the indicated operation. Write the answer as a variable expression.

91. $\dfrac{3}{4} \cdot \dfrac{a}{5}$

92. $\dfrac{2}{3} \div \dfrac{a}{7}$

93. $\dfrac{11}{x} + \dfrac{9}{x}$

94. $\dfrac{10}{y} - \dfrac{6}{y}$

In Exercises 95–96, perform the indicated operations. Begin by performing operations in parentheses.

95. $\left(\dfrac{1}{2} - \dfrac{1}{3} \right) \div \dfrac{5}{8}$

96. $\left(\dfrac{1}{2} + \dfrac{1}{4} \right) \div \left(\dfrac{1}{2} + \dfrac{1}{3} \right)$

Application Exercises

97. The circle graph shows the results of a survey in which American adults were asked the following question:

> If you got conflicting or different reports of the same news story from television, newspapers, radio, or magazines, which would you be most inclined to believe?

The circle graph shows the fraction of American adults finding each medium the most believable.

Most Believable News Source

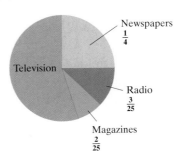

Newspapers $\frac{1}{4}$

Television

Radio $\frac{3}{25}$

Magazines $\frac{2}{25}$

Source: Thomas R. Dye, *Who's Running America? Seventh Edition*, Prentice Hall, 2002

a. What fraction of those surveyed found television the most believable news source?

b. If 2000 people were surveyed, how many found radio the most believable news source?

98. The circle graph shows the results of one of the questions in a telephone poll taken for *Time/CNN* during 2002's summer of corporate scandals.

Which of the following best describes your view of the recent accounting scandals involving major American corporations?

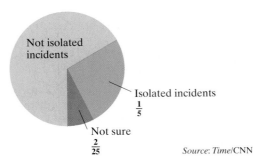

Not isolated incidents

Isolated incidents $\frac{1}{5}$

Not sure $\frac{2}{25}$

Source: *Time/CNN*

a. What fraction of those surveyed replied that the accounting scandals were not isolated incidents?

b. If 2000 people were surveyed, how many were not sure if the accounting scandals were isolated incidents?

The ratio of a to b can be expressed by the fraction $\frac{a}{b}$. In Exercises 99–100, use the natural numbers shown in Figure 1.1 on page 2 to find the indicated ratio. Reduce each fraction to its lowest terms.

99. the ratio of Hispanic judges to female judges appointed by Carter

100. the ratio of African-American judges to female judges appointed by Reagan

A common application of fractions involves preparing food for a different number of servings than what the recipe calls for. The amount of each ingredient can be found as follows:

Amount of ingredient needed

$$= \frac{\text{desired serving size}}{\text{recipe serving size}} \times \text{ingredient amount in the recipe.}$$

Use this information to solve Exercises 101–102. Give answers as mixed numbers.

101. A chocolate-chip cookie recipe for five dozen cookies requires $\frac{3}{4}$ cup sugar. If you want to make eight dozen cookies, how much sugar is needed?

102. A mix for eight servings of instant potatoes requires $2\frac{2}{3}$ cups of water. If you want to make six servings, how much water is needed?

103. If you walk $\frac{3}{4}$ mile and then jog $\frac{2}{5}$ mile, what is the total distance covered? How much farther did you walk than jog?

104. Some companies pay people extra when they work more than a regular 40-hour work week. The overtime pay is often $1\frac{1}{2}$ times the regular hourly rate. This is called *time and a half*. A summer job for students pays $12 per hour and offers time and a half for the hours worked over 40. If a student works 46 hours during one week, what is the student's total pay before taxes?

105. The legend of a map indicates that 1 inch = 16 miles. If the distance on the map between two cities is $2\frac{3}{8}$ inches, how far apart are the cities?

Writing in Mathematics

Writing about mathematics will help you to learn mathematics. For all writing exercises in this book, use complete sentences to respond to the questions. Some writing exercises can be answered in a sentence; others require a paragraph or two. You can decide how much you need to write as long as your writing clearly and directly answers the question in the exercise. Standard references such as a dictionary and a thesaurus should be helpful.

106. Explain how to convert a mixed number to an improper fraction and give an example.

107. Explain how to convert an improper fraction to a mixed number and give an example.

108. Describe the difference between a prime number and a composite number.

109. What is meant by the prime factorization of a composite number?

110. What is the Fundamental Principle of Fractions?

111. Explain how to reduce a fraction to its lowest terms. Give an example with your explanation.

112. Explain how to multiply fractions and give an example.

113. Explain how to divide fractions and give an example.

114. Describe how to add or subtract fractions with identical denominators. Provide an example with your description.

115. Explain how to add fractions with different denominators. Use $\frac{5}{6} + \frac{1}{2}$ as an example.

116. Explain what is wrong with this statement. "If you'd like to save some money, I'll be happy to sell you my computer system for only $\frac{3}{2}$ of the price I originally paid for it."

Critical Thinking Exercises

117. Which one of the following is true?

a. $\frac{1}{2} + \frac{1}{5} = \frac{2}{7}$ **b.** $\frac{2+6}{2} = \frac{\cancel{2}+6}{\cancel{2}} = 6$ **c.** $\frac{1}{2} \div 4 = 2$

d. Every fraction has infinitely many equivalent fractions.

118. Shown below is a short excerpt from "The Star-Spangled Banner." The time is $\frac{3}{4}$, which means that each measure must contain notes that add up to $\frac{3}{4}$. The values of the different notes tell musicians how long to hold each note.

$$\circ = 1 \quad \text{\textonehalfnote} = \frac{1}{2} \quad \text{\textquarternote} = \frac{1}{4} \quad \text{\texteighthnote} = \frac{1}{8}$$

Use vertical lines to divide this line of "The Star-Spangled Banner" into measures.

say does that   Star-span-gled   Ban-ner   yet   wave   O'er the

---

THE REAL NUMBERS

Objectives

1 Define the sets that make up the real numbers.

2 Graph numbers on a number line.

3 Express rational numbers as decimals.

4 Classify numbers as belonging to one or more sets of the real numbers.

5 Understand and use inequality symbols.

6 Find the absolute value of a real number.

The U.N. building is designed with three golden rectangles.

The United Nations Building in New York was designed to represent its mission of promoting world harmony. Viewed from the front, the building looks like three rectangles stacked upon each other. In each rectangle, the ratio of the width to height is $\sqrt{5} + 1$ to 2, approximately 1.618 to 1. The ancient Greeks believed that such a rectangle, called a **golden rectangle**, was the most visually pleasing of all rectangles.

The ratio 1.618 to 1 is approximate because $\sqrt{5}$ is an irrational number, a special kind of real number. Irrational? Real? Let's make sense of all this by describing the kinds of numbers you will encounter in this course.

1 Define the sets that make up the real numbers.

Natural Numbers and Whole Numbers Before we describe the set of real numbers, let's be sure you are familiar with some basic ideas about sets. A **set** is a collection of objects whose contents can be clearly determined. The objects in a set are called the **elements** of the set. For example, the set of numbers used for counting can be represented by

$$\{1, 2, 3, 4, 5, \ldots\}.$$

The braces, { }, indicate that we are representing a set. This form of representing a set uses commas to separate the elements of the set. Remember that the three dots after the 5 indicate that there is no final element and that the listing goes on forever.

We have seen that the set of numbers used for counting is called the set of **natural numbers**. When we combine the number 0 with the natural numbers, we obtain the set of **whole numbers**.

NATURAL NUMBERS AND WHOLE NUMBERS

The set of **natural numbers** is $\{1, 2, 3, 4, 5, \ldots\}$.

The set of **whole numbers** is $\{0, 1, 2, 3, 4, 5, \ldots\}$.

Integers and the Number Line The whole numbers do not allow us to describe certain everyday situations. For example, if the balance in your checking account is $30 and you write a check for $35, your checking account is overdrawn by $5. We can write this as -5, read *negative* 5. The set consisting of the natural numbers, 0, and the negatives of the natural numbers is called the set of **integers**.

INTEGERS The set of **integers** is

$$\{\ldots, \underbrace{-4, -3, -2, -1,}_{\text{Negative integers}}\ 0,\ \underbrace{1, 2, 3, 4, \ldots}_{\text{Positive integers}}\}.$$

Notice that the term **positive integers** is another name for the natural numbers. The positive integers can be written in two ways:

1. Use a "+" sign. For example, +4 is "positive four."
2. Do not write any sign. For example, 4 is assumed to be "positive four."

EXAMPLE 1 Practical Examples of Negative Integers

Write a negative integer that describes each of the following situations:

a. A debt of $10

b. The shore surrounding the Dead Sea is 1312 feet below sea level.

SOLUTION

a. A debt of $10 can be expressed by the negative integer -10 (negative ten).

b. The shore surrounding the Dead Sea is 1312 feet below sea level, expressed as -1312. ∎

 CHECK POINT 1 Write a negative integer that describes each of the following situations:

a. A debt of $500

b. Death Valley, the lowest point in North America, is 282 feet below sea level.

2 Graph numbers on a number line.

The **number line** is a graph we use to visualize the set of integers, as well as sets of other numbers. The number line is shown in Figure 1.8.

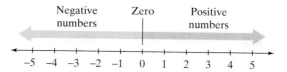

FIGURE 1.8 The number line

The number line extends indefinitely in both directions, shown by the arrows on the left and the right. Zero separates the positive numbers from the negative numbers on the number line. The positive integers are located to the right of 0, and the negative integers are located to the left of 0. Zero is neither positive nor negative. For every positive integer on a number line, there is a corresponding negative integer on the opposite side of 0.

Integers are graphed on a number line by placing a dot at the correct location for each number.

EXAMPLE 2 Graphing Integers on a Number Line

Graph: **a.** −3   **b.** 4   **c.** 0.

SOLUTION Place a dot at the correct location for each integer.

✔ CHECK POINT 2 Graph: **a.** −4   **b.** 0   **c.** 3.

Rational Numbers If two integers are added, subtracted, or multiplied, the result is always another integer. This, however, is not always the case with division. For example, 10 divided by 5 is the integer 2. By contrast, 5 divided by 10 is $\frac{1}{2}$, and $\frac{1}{2}$ is not an integer. To permit divisions such as $\frac{5}{10}$, we enlarge the set of integers, calling the new collection the *rational numbers*. The set of **rational numbers** consists of all the numbers that can be expressed as a quotient of two integers, with the denominator not 0.

> **THE RATIONAL NUMBERS** The set of **rational numbers** is the set of all numbers that can be expressed in the form $\frac{a}{b}$, where a and b are integers and b is not equal to 0, written $b \neq 0$. The integer a is called the **numerator** and the integer b is called the **denominator**.

STUDY TIP

In Section 1.7, you will learn that a negative number divided by a positive number gives a negative result. Thus, $\frac{-3}{4}$ can also be written as $-\frac{3}{4}$.

Here are two examples of rational numbers:

- $\frac{1}{2}$ *a*, the integer forming the numerator, is 1.
       b, the integer forming the denominator, is 2.

- $\frac{-3}{4}$ *a*, the integer forming the numerator, is −3.
       b, the integer forming the denominator, is 4.

Is the integer 5 another example of a rational number? Yes. The integer 5 can be written with a denominator of 1.

$$5 = \frac{5}{1}$$ *a*, the integer forming the numerator, is 5.
     b, the integer forming the denominator, is 1.

All integers are also rational numbers because they can be written with a denominator of 1.

How can we express a negative mixed number, such as $-2\frac{3}{4}$, in the form $\frac{a}{b}$? Copy the negative sign and then follow the procedure discussed in the previous section:

$$-2\frac{3}{4} = -\frac{4 \cdot 2 + 3}{4} = -\frac{8 + 3}{4} = -\frac{11}{4} = \frac{-11}{4}.$$

$a = -11$

$b = 4$

Copy the negative sign from step to step and convert $2\frac{3}{4}$ to an improper fraction.

Rational numbers are graphed on a number line by placing a dot at the correct location for each number.

EXAMPLE 3 Graphing Rational Numbers on a Number Line

Graph: **a.** $\frac{7}{2}$ **b.** -4.6.

SOLUTION Place a dot at the correct location for each rational number.

a. Because $\frac{7}{2} = 3\frac{1}{2}$, its graph is midway between 3 and 4.

b. Because $-4.6 = -4\frac{6}{10}$, its graph is $\frac{6}{10}$ of a unit to the left of -4.

$$(b) \qquad\qquad\qquad\qquad (a)$$
$$-5 \quad -4 \quad -3 \quad -2 \quad -1 \quad 0 \quad 1 \quad 2 \quad 3 \quad 4 \quad 5$$

3 Express rational numbers as decimals.

✔ **CHECK POINT 3** Graph: **a.** $\frac{9}{2}$ **b.** -1.2.

Every rational number can be expressed as a fraction or a decimal. To express the fraction $\frac{a}{b}$ as a decimal, divide the denominator, b, into the numerator, a.

EXAMPLE 4 Expressing Rational Numbers as Decimals

Express each rational number as a decimal: **a.** $\frac{5}{8}$ **b.** $\frac{7}{11}$.

SOLUTION In each case, divide the denominator into the numerator.

a.
```
    0.625
8)5.000
  48
  ──
   20
   16
   ──
    40
    40
    ──
     0
```

b.
```
     0.6363 ...
11)7.0000 ...
   66
   ──
    40
    33
    ──
     70
     66
     ──
      40
      33
      ──
       70
        ⋮
```

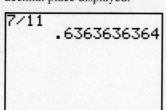
In Example 4, the decimal for $\frac{5}{8}$, namely 0.625, stops and is called a **terminating decimal**. Other examples of terminating decimals are

$$\frac{1}{4} = 0.25, \quad \frac{2}{5} = 0.4, \quad \text{and} \quad \frac{7}{8} = 0.875.$$

By contrast, the division process for $\frac{7}{11}$ results in $0.6363\ldots$, with the digits 63 repeating over and over indefinitely. To indicate this, write a bar over the digits that repeat. Thus,

$$\frac{7}{11} = 0.\overline{63}.$$

The decimal for $\frac{7}{11}$, $0.\overline{63}$, is called a **repeating decimal**. Other examples of repeating decimals are

$$\frac{1}{3} = 0.333\ldots = 0.\overline{3} \quad \text{and} \quad \frac{2}{3} = 0.666\ldots = 0.\overline{6}.$$

> **RATIONAL NUMBERS AND DECIMALS** Any rational number can be expressed as a decimal. The resulting decimal will either terminate (stop), or it will have a digit that repeats or a block of digits that repeat.

USING TECHNOLOGY

You can obtain decimal approximations for irrational numbers using a calculator. For example, to approximate $\sqrt{2}$, use the following keystrokes:

Many Scientific Calculators

$2\ \boxed{\sqrt{}}$ or $2\ \boxed{\substack{\text{2nd}\\ \text{INV}}}\ \boxed{x^2}$

Many Graphing Calculators

$\boxed{\sqrt{}}\ 2\ \boxed{\text{ENTER}}$

The display may read 1.41421356237, although your calculator may display more or fewer digits. Between which two integers would you graph $\sqrt{2}$ on a number line?

CHECK POINT 4 Express each rational number as a decimal:

a. $\dfrac{3}{8}$ **b.** $\dfrac{5}{11}$.

Irrational Numbers Can you think of a number that, when written in decimal form, neither terminates nor repeats? An example of such a number is $\sqrt{2}$ (read: "the square root of 2"). The number $\sqrt{2}$ is a number that can be multiplied by itself to obtain 2. No terminating or repeating decimal can be multiplied by itself to get 2. However, some approximations come close to 2.

- 1.4 is an approximation of $\sqrt{2}$:
$$1.4 \times 1.4 = 1.96.$$
- 1.41 is an approximation of $\sqrt{2}$:
$$1.41 \times 1.41 = 1.9881.$$
- 1.4142 is an approximation of $\sqrt{2}$:
$$1.4142 \times 1.4142 = 1.99996164.$$

Can you see how each approximation in the list is getting better? This is because the products are getting closer and closer to 2.

The number $\sqrt{2}$, whose decimal representation does not come to an end and does not have a block of repeating digits, is an example of an **irrational number**.

> **THE IRRATIONAL NUMBERS** Any number that can be represented on the number line that is not a rational number is called an **irrational number**. Thus, the set of irrational numbers is the set of numbers whose decimal representations are neither terminating nor repeating.

ENRICHMENT ESSAY

The Best and Worst of π

In 1999, two Japanese mathematicians used two different computer programs to calculate π to over 206 billion digits. The calculations took the computer 43 hours!

The most inaccurate version of π came from the 1897 General Assembly of Indiana. Bill No. 246 stated that "π was by law 4."

Perhaps the best known of all the irrational numbers is π (pi). This irrational number represents the distance around a circle (its circumference) divided by the diameter of the circle. In the *Star Trek* episode "Wolf in the Fold," Spock foils an evil computer by telling it to "compute the last digit in the value of π." Because π is an irrational number, there is no last digit in its decimal representation:

$$\pi = 3.1415926535897932384626433832795\ldots.$$

Because irrational numbers cannot be represented by decimals that come to an end, mathematicians use symbols such as $\sqrt{2}$, $\sqrt{3}$, and π to represent these numbers. However, **not all square roots are irrational**. For example, $\sqrt{25} = 5$ because 5 multiplied by itself is 25. Thus, $\sqrt{25}$ is a natural number, a whole number, an integer, and a rational number $\left(\sqrt{25} = \frac{5}{1}\right)$.

4 Classify numbers as belonging to one or more sets of the real numbers.

The Set of Real Numbers   All numbers that can be represented by points on the number line are called **real numbers**. Thus, the set of real numbers is formed by combining the rational numbers and the irrational numbers. Every real number is either rational or irrational.

The sets that make up the real numbers are summarized in Table 1.1. Notice the use of the symbol $\approx$ in the examples of irrational numbers. The symbol $\approx$ means "is approximately equal to."

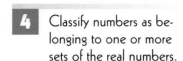

Real numbers

This diagram shows that every real number is rational or irrational.

Table 1.1	The Sets that Make Up the Real Numbers	
Name	**Description**	**Examples**
Natural numbers	$\{1, 2, 3, 4, 5, \ldots\}$ These numbers are used for counting.	$2, 3, 5, 17$
Whole numbers	$\{0, 1, 2, 3, 4, 5, \ldots\}$ The set of whole numbers is formed by adding 0 to the set of natural numbers.	$0, 2, 3, 5, 17$
Integers	$\{\ldots, -5, -4, -3, -2, -1, 0, 1, 2, 3, 4, 5, \ldots\}$ The set of integers is formed by adding negatives of the natural numbers to the set of whole numbers.	$-17, -5, -3, -2, 0,$ $2, 3, 5, 17$
Rational numbers	The set of rational numbers is the set of all numbers that can be expressed in the form $\frac{a}{b}$, where a and b are integers and b is not equal to 0, written $b \neq 0$. Rational numbers can be expressed as terminating or repeating decimals.	$-17 = \frac{-17}{1}, -5 = \frac{-5}{1}, -3, -2,$ $0, 2, 3, 5, 17,$ $\frac{2}{5} = 0.4,$ $\frac{-2}{3} = -0.6666\cdots = -0.\overline{6}$
Irrational numbers	The set of irrational numbers is the set of all numbers whose decimal representations are neither terminating nor repeating. Irrational numbers cannot be expressed as a quotient of integers.	$\sqrt{2} \approx 1.414214$ $-\sqrt{3} \approx -1.73205$ $\pi \approx 3.142$ $-\frac{\pi}{2} \approx -1.571$

EXAMPLE 5   Classifying Real Numbers

Consider the following set of numbers:
$$\left\{-7, -\frac{3}{4}, 0, 0.\overline{6}, \sqrt{5}, \pi, 7.3, \sqrt{81}\right\}.$$
List the numbers in the set that are
  a. natural numbers.   **b.** whole numbers.   **c.** integers.
  d. rational numbers.   **e.** irrational numbers.   **f.** real numbers.

SOLUTION
  a. Natural numbers: The natural numbers are the numbers used for counting. The only natural number in the set is $\sqrt{81}$ because $\sqrt{81} = 9$. (9 multiplied by itself is 81.)
  b. Whole numbers: The whole numbers consist of the natural numbers and 0. The elements of the set that are whole numbers are 0 and $\sqrt{81}$.

c. Integers: The integers consist of the natural numbers, 0, and the negatives of the natural numbers. The elements of the set that are integers are $\sqrt{81}$, 0, and −7.

d. Rational numbers: All numbers in the set that can be expressed as the quotient of integers are rational numbers. These include $-7\left(-7=\frac{-7}{1}\right)$, $-\frac{3}{4}$, $0\left(0=\frac{0}{1}\right)$, and $\sqrt{81}\left(\sqrt{81}=\frac{9}{1}\right)$. Furthermore, all numbers in the set that are terminating or repeating decimals are also rational numbers. These include $0.\overline{6}$ and 7.3.

e. Irrational numbers: The irrational numbers in the set are $\sqrt{5}\,(\sqrt{5}\approx 2.236)$ and $\pi\,(\pi\approx 3.14)$. Both $\sqrt{5}$ and π are only approximately equal to 2.236 and 3.14, respectively. In decimal form, $\sqrt{5}$ and π neither terminate nor have blocks of repeating digits.

f. Real numbers: All the numbers in the given set are real numbers. ∎

 CHECK POINT 5 Consider the following set of numbers:

$$\left\{-9, -1.3, 0, 0.\overline{3}, \frac{\pi}{2}, \sqrt{9}, \sqrt{10}\right\}.$$

List the numbers in the set that are

a. natural numbers.   **b.** whole numbers.

c. integers.   **d.** rational numbers.

e. irrational numbers.   **f.** real numbers.

---

5   Understand and use inequality symbols.

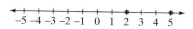

FIGURE 1.9

Ordering the Real Numbers On the real number line, the real numbers increase from left to right. The lesser of two real numbers is the one farther to the left on a number line. The greater of two real numbers is the one farther to the right on a number line.

Look at the number line in Figure 1.9. The integers 2 and 5 are graphed. Observe that 2 is to the left of 5 on the number line. This means that 2 is less than 5:

     $2 < 5$:   2 is less than 5 because 2 is to the *left* of 5 on the number line.

In Figure 1.9, we can also observe that 5 is to the right of 2 on the number line. This means that 5 is greater than 2.

     $5 > 2$:   5 is greater than 2 because 5 is to the *right* of 2 on the number line.

The symbols $<$ and $>$ are called **inequality symbols**. These symbols always point to the lesser of the two real numbers when the inequality is true.

> 2 is less than 5.
>
> 5 is greater than 2.

     $2 < 5$ The symbol points to 2, the lesser number.

     $5 > 2$ The symbol points to 2, the lesser number.

EXAMPLE 6   Using Inequality Symbols

Insert either $<$ or $>$ in the shaded area between each pair of numbers to make a true statement:

     a. 3   17   **b.** −4.5   1.2   **c.** −5   −83   **d.** $\dfrac{4}{5}$ $\dfrac{2}{3}$.

SOLUTION In each case, mentally compare the graph of the first number to the graph of the second number. If the first number is to the left of the second number, insert the symbol $<$ for "is less than." If the first number is to the right of the second number, insert the symbol $>$ for "is greater than."

     a. Compare the graphs of 3 and 17 on the number line. Because 3 is to the left of 17, this means that 3 is less than 17: $3 < 17$.

b. Compare the graphs of -4.5 and 1.2. Because -4.5 is to the left of 1.2, this means that -4.5 is less than 1.2: $\quad -4.5 < 1.2$.

c. Compare the graphs of -5 and -83. Because -5 is to the right of -83, this means that -5 is greater than -83: $\quad -5 > -83$.

d. Compare the graphs of $\frac{4}{5}$ and $\frac{2}{3}$. To do so, convert to decimal notation or use a common denominator. Using decimal notation, $\frac{4}{5} = 0.8$ and $\frac{2}{3} = 0.\overline{6}$. Because 0.8 is to the right of $0.\overline{6}$, this means that $\frac{4}{5}$ is greater than $\frac{2}{3}$: $\quad \frac{4}{5} > \frac{2}{3}$. ∎

 CHECK POINT 6 Insert either $<$ or $>$ in the shaded area between each pair of numbers to make a true statement:

a. $14 \quad 5$ **b.** $-5.4 \quad 2.3$ **c.** $-19 \quad -6$ **d.** $\dfrac{1}{4} \quad \dfrac{1}{2}$.

The symbols $<$ and $>$ may be combined with an equal sign, as shown in the table.

Symbols	Meaning	Example	Explanation
$a \leq b$	a is less than or equal to b.	$3 \leq 7$ $7 \leq 7$	Because $3 < 7$ Because $7 = 7$
$b \geq a$	b is greater than or equal to a.	$7 \geq 3$ $-5 \geq -5$	Because $7 > 3$ Because $-5 = -5$

When using the symbol $\leq$ (is less than or equal to), the inequality is a true statement if either the $<$ part or the $=$ part is true. When using the symbol $\geq$ (is greater than or equal to), the inequality is a true statement if either the $>$ part or the $=$ part is true.

EXAMPLE 7   Using Inequality Symbols

Determine whether each inequality is true or false:

a. $-7 \leq 4$ **b.** $-7 \leq -7$ **c.** $-9 \geq 6$.

SOLUTION

a. $-7 \leq 4$ is true because $-7 < 4$ is true.

b. $-7 \leq -7$ is true because $-7 = -7$ is true.

c. $-9 \geq 6$ is false because neither $-9 > 6$ nor $-9 = 6$ is true. ∎

 CHECK POINT 7 Determine whether each inequality is true or false:

a. $-2 \leq 3$ **b.** $-2 \geq -2$ **c.** $-4 \geq 1$.

6 Find the absolute value of a real number.

Absolute Value   Suppose you are feeling a bit lazy. Instead of your usual jog, you decide to use your finger to stroll along the number line. You start at 0 and end the finger walk at -5. You have covered a distance of 5 units.

    Absolute value describes distance from 0 on a number line. If a represents a real number, the symbol $|a|$ represents its absolute value, read "the absolute value of a." In terms of your walk, we can write

$$|-5| = 5.$$

> The absolute value of -5 is 5 because -5 is 5 units from 0 on the number line.

ABSOLUTE VALUE   The absolute value of a real number a, denoted by $|a|$, is the distance from 0 to a on the number line. Because absolute value describes a distance, it is never negative.

| EXAMPLE 8 | Finding Absolute Value |

Find the absolute value: **a.** $|-3|$ **b.** $|5|$ **c.** $|0|$.

SOLUTION The solution is illustrated in Figure 1.10.

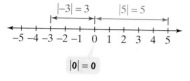

a. $|-3| = 3$ The absolute value of -3 is 3 because -3 is 3 units from 0.

b. $|5| = 5$ 5 is 5 units from 0.

c. $|0| = 0$ 0 is 0 units from itself.

FIGURE 1.10 Absolute value describes distance from 0 on a number line.

Can you see that the absolute value of a real number is either positive or zero? Zero is the only real number whose absolute value is 0: $|0| = 0$. **The absolute value of any other real number is always positive**.

✔ CHECK POINT 8 Find the absolute value:

a. $|-4|$ **b.** $|6|$ **c.** $|-\sqrt{2}|$.

1.2 EXERCISE SET

Student Solutions Manual   CD/Video   PH Math/Tutor Center   MathXL Tutorials on CD   MathXL®   MyMathLab   Interactmath.com

Practice Exercises

In Exercises 1–8, write a positive or negative integer that describes each situation.

1. Meteorology: 20° below zero
2. Navigation: 65 feet above sea level
3. Health: A gain of 8 pounds
4. Economics: A loss of $12,500.00
5. Banking: A withdrawal of $3000.00
6. Physics: An automobile slowing down at a rate of 3 meters per second each second.
7. Economics: A budget deficit of 4 billion dollars
8. Football: A 14-yard loss

In Exercises 9–20, start by drawing a number line that shows integers from −5 to 5. Then graph each real number on your number line.

9. 2   10. 5   11. −5   12. −2   13. $3\frac{1}{2}$ 14. $2\frac{1}{4}$

15. $\frac{11}{3}$ 16. $\frac{7}{3}$ 17. −1.8   18. −3.4   19. $-\frac{16}{5}$ 20. $-\frac{11}{5}$

In Exercises 21–32, express each rational number as a decimal.

21. $\frac{3}{4}$ 22. $\frac{3}{5}$ 23. $\frac{7}{20}$

24. $\frac{3}{20}$ 25. $\frac{7}{8}$ 26. $\frac{5}{16}$

27. $\frac{9}{11}$ 28. $\frac{3}{11}$ 29. $-\frac{1}{2}$

30. $-\frac{1}{4}$ 31. $-\frac{5}{6}$ 32. $-\frac{7}{6}$

*In Exercises 33–36, list all numbers from the given set that are: **a.** natural numbers, **b.** whole numbers, **c.** integers, **d.** rational numbers, **e.** irrational numbers, **f.** real numbers.*

33. $\left\{-9, -\frac{4}{5}, 0, 0.25, \sqrt{3}, 9.2, \sqrt{100}\right\}$

34. $\left\{-7, -0.\overline{6}, 0, \sqrt{49}, \sqrt{50}\right\}$

35. $\left\{-11, -\frac{5}{6}, 0, 0.75, \sqrt{5}, \pi, \sqrt{64}\right\}$

36. $\left\{-5, -0.\overline{3}, 0, \sqrt{2}, \sqrt{4}\right\}$

37. Give an example of a whole number that is not a natural number.

38. Give an example of an integer that is not a whole number.

39. Give an example of a rational number that is not an integer.

40. Give an example of a rational number that is not a natural number.

41. Give an example of a number that is an integer, a whole number, and a natural number.

42. Give an example of a number that is a rational number, an integer, and a real number.

43. Give an example of a number that is an irrational number and a real number.

44. Give an example of a number that is a real number, but not an irrational number.

In Exercises 45–62, insert either < or > in the shaded area between each pair of numbers to make a true statement.

45. $\dfrac{1}{2}$ 2

46. 4 -3

47. 3 $-\dfrac{5}{2}$

48. 3 $\dfrac{3}{2}$

49. -4 -6

50. $-\dfrac{5}{2}$ $-\dfrac{5}{3}$

51. -2.5 1.5

52. -1.25 -0.5

53. $-\dfrac{3}{4}$ $-\dfrac{5}{4}$

54. 0 $-\dfrac{1}{2}$

55. -4.5 3

56. -5.5 2.5

57. $\sqrt{2}$ 1.5

58. $\sqrt{3}$ 2

59. $0.\overline{3}$ 0.3

60. 0.6 $0.\overline{6}$

61. $-\pi$ -3.5

62. $-\dfrac{\pi}{2}$ -2.3

In Exercises 63–70, determine whether each inequality is true or false.

63. $-5 \geq -13$

64. $-5 \leq -8$

65. $-9 \geq -9$

66. $-14 \leq -14$

67. $0 \geq -6$

68. $0 \geq -13$

69. $-17 \geq 6$

70. $-14 \geq 8$

In Exercises 71–78, find each absolute value.

71. $|6|$

72. $|3|$

73. $|-7|$

74. $|-9|$

75. $\left|\dfrac{5}{6}\right|$

76. $\left|\dfrac{4}{5}\right|$

77. $|-\sqrt{11}|$

78. $|-\sqrt{29}|$

Practice Plus

In Exercises 79–86, insert either <, >, or = in the shaded area to make a true statement.

79. $|-6|$ $|-3|$

80. $|-20|$ $|-50|$

81. $\left|\dfrac{3}{5}\right|$ $|-0.6|$

82. $\left|\dfrac{5}{2}\right|$ $|-2.5|$

83. $\dfrac{30}{40} - \dfrac{3}{4}$ $\dfrac{14}{15} \cdot \dfrac{15}{14}$

84. $\dfrac{17}{18} \cdot \dfrac{18}{17}$ $\dfrac{50}{60} - \dfrac{5}{6}$

85. $\dfrac{8}{13} \div \dfrac{8}{13}$ $|-1|$

86. $|-2|$ $\dfrac{4}{17} \div \dfrac{4}{17}$

Application Exercises

Temperatures sometimes fall below zero. A combination of low temperature and wind makes it feel colder than the actual temperature. The table shows how cold it feels when low temperatures are combined with different wind speeds.

Windchill												
Wind	Temperature (°F)											
(mph)	35	30	25	20	15	10	5	0	−5	−10	−15	−20
5	31	25	19	13	7	−1	−5	−11	−16	−22	−28	−34
10	27	21	15	9	3	−4	−10	−16	−22	−28	−35	−41
15	25	19	13	6	0	−7	−13	−19	−26	−32	−39	−45
20	24	17	11	4	−2	−9	−15	−22	−29	−35	−42	−48
25	23	16	9	3	−4	−11	−17	−24	−31	−37	−44	−51

Source: National Weather Service

Use the information from the table to solve Exercises 87–88.

87. Write a negative integer that indicates how cold the temperature feels when the temperature is 15° Fahrenheit and the wind is blowing at 20 miles per hour.

88. Write a negative integer that indicates how cold the temperature feels when the temperature is 10° Fahrenheit and the wind is blowing at 15 miles per hour.

The following table shows the amount of money, in millions of dollars, collected and spent by the U.S. government from 1997 through 2002.

Year	Money Collected	Money Spent
1997	1,579,300	1,601,200
1998	1,721,800	1,652,600
1999	1,827,500	1,703,000
2000	2,025,200	1,788,800
2001	1,991,000	1,863,900
2002	1,946,100	2,052,300

Money is expressed in millions of dollars.
Source: Office of Management and Budget

Use the information from the table to solve Exercises 89–90.

89. List the years for which money collected < money spent. Was there a budget surplus or deficit in these years?

90. List the years for which money collected > money spent. Was there a budget surplus or deficit in these years?

Writing in Mathematics

Writing about mathematics will help you to learn mathematics. For all writing exercises in this book, use complete sentences to respond to the questions. Some writing exercises can be answered in a sentence; others require a paragraph or two. You can decide how much you need to write as long as your writing clearly and directly answers the question in the exercise. Standard references such as a dictionary and a thesaurus should be helpful.

91. What is a set?

92. What are the natural numbers?

93. What are the whole numbers?

94. What are the integers?

95. How does the set of integers differ from the set of whole numbers?

96. Describe how to graph a number on the number line.

97. What is a rational number?

98. Explain how to express $\frac{3}{8}$ as a decimal.

99. Describe the difference between a rational number and an irrational number.

100. If you are given two real numbers, explain how to determine which one is the lesser.

101. Describe what is meant by the absolute value of a number. Give an example with your explanation.

102. Give an example of an everyday situation that can be described using integers but not using whole numbers.

Critical Thinking Exercises

103. Which one of the following statements is true?
 a. Every rational number is an integer.
 b. Some whole numbers are not integers.
 c. Some rational numbers are not positive.
 d. Irrational numbers cannot be negative.

104. Which one of the following statements is true?
 a. $\sqrt{36}$ is an irrational number.
 b. Some real numbers are not rational numbers.
 c. Some integers are not rational numbers.
 d. All whole numbers are positive.

105. Answer this question without using a calculator. Between which two consecutive integers is $-\sqrt{47}$?

106. We have used a code to describe an activity that is important to your success in school and at work. Here is the coded message:

$$(20, 18) \quad (-8, -12) \quad (0, 9) \quad (-14, -17)$$
$$(0, 11) \quad (-17, 9) \quad (-16, 14) \quad (-22, 7).$$

Read across the two rows. Each number pair represents one letter of the coded message. The first number pair represents the first letter, the second pair, the second letter, and so on. Use the following table to decode the message.

First Decoding	Second Decoding				
Use the greater of the two numbers in each pair. If the greater number is negative, then use its absolute value.	0 = blank space 5 = E 6 = F 11 = K 12 = L 17 = Q 18 = R 23 = W 24 = X	1 = A 2 = B 7 = G 8 = H 13 = M 14 = N 19 = S 20 = T 25 = Y 26 = Z	3 = C 4 = D 9 = I 10 = J 15 = O 16 = P 21 = U 22 = V		

Technology Exercises

In Exercises 107–110, use a calculator to find a decimal approximation for each irrational number, correct to three decimal places. Between which two integers should you graph each of these numbers on the number line?

107. $\sqrt{3}$

108. $-\sqrt{12}$

109. $1 - \sqrt{2}$

110. $2 - \sqrt{5}$

SECTION 1.3

Objectives

1 Plot ordered pairs in the rectangular coordinate system.

2 Find coordinates of points in the rectangular coordinate system.

3 Interpret information given by line graphs.

4 Interpret information given by bar graphs.

ORDERED PAIRS AND GRAPHS

The beginning of the seventeenth century was a time of innovative ideas and enormous intellectual progress in Europe. English theatergoers enjoyed a succession of exciting new plays by Shakespeare. William Harvey proposed the radical notion that the heart was a pump for blood rather than the center of emotion. Galileo, with his new-fangled invention called the telescope, supported the theory of Polish astronomer Copernicus that the sun, not the Earth, was the center of the solar system. Monteverdi was writing the world's first grand operas. French mathematicians Pascal and Fermat invented a new field of mathematics called probability theory.

Into this arena of intellectual electricity stepped French aristocrat René Descartes (1596–1650). Descartes, propelled by the creativity surrounding him, developed a new branch of mathematics that brought together algebra and geometry in a unified way—a way that visualized numbers as points on a graph, equations as geometric figures, and geometric figures as equations. This new branch of mathematics, called *analytic geometry*, established Descartes as one of the founders of modern thought and among the most original mathematicians and philosophers of any age. We begin this section by looking at Descartes's deceptively simple idea, called the **rectangular coordinate system** or (in his honor) the **Cartesian coordinate system**.

Points and Ordered Pairs Descartes used two number lines that intersect at right angles at their zero points, as shown in Figure 1.11. The horizontal number line is the **x-axis**. The vertical number line is the **y-axis**. The point of intersection of these axes is their zero points, called the **origin**. Positive numbers are shown to the right and above the origin. Negative numbers are shown to the left and below the origin. The axes divide the plane into four quarters, called **quadrants**. The points located on the axes are not in any quadrant.

Each point in the rectangular coordinate system corresponds to an **ordered pair** of real numbers, (x, y). Examples of such pairs are $(4, 2)$ and $(-5, -3)$. The first number in each pair, called the **x-coordinate**, denotes the distance and direction from the origin along the x-axis. The second number, called the **y-coordinate**, denotes vertical distance and direction along a line parallel to the y-axis or along the y-axis itself.

1 Plot ordered pairs in the rectangular coordinate system.

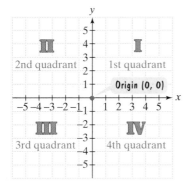

FIGURE 1.11 The rectangular coordinate system

Figure 1.12 shows how we **plot**, or locate, the points corresponding to the ordered pairs $(4, 2)$ and $(-5, -3)$. We plot $(4, 2)$ by going 4 units from 0 to the right along the x-axis. Then we go 2 units up parallel to the y-axis. We plot $(-5, -3)$ by going 5 units from 0 to the left along the x-axis and 3 units down parallel to the y-axis. The phrase "the point corresponding to the ordered pair $(-5, -3)$" is often abbreviated as "the point $(-5, -3)$."

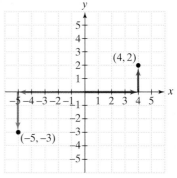

FIGURE 1.12 Plotting $(4, 2)$ and $(-5, -3)$

EXAMPLE 1 Plotting Points in the Rectangular Coordinate System

Plot the points: $A(-3, 5)$, $B(2 -4)$, $C(5, 0)$, $D(-5, -3)$, $E(0, 4)$ and $F(0, 0)$.

SOLUTION See Figure 1.13. We move from the origin and plot the points in the following way:

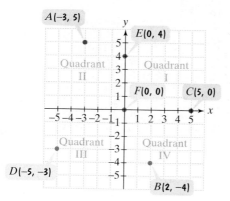

$A(-3, 5)$:	3 units left, 5 units up
$B(2, -4)$:	2 units right, 4 units down
$C(5, 0)$:	5 units right, 0 units up or down
$D(-5, -3)$:	5 units left, 3 units down
$E(0, 4)$:	0 units right or left, 4 units up
$F(0, 0)$:	0 units right or left, 0 units up or down

FIGURE 1.13 Plotting points

The phrase *ordered pair* is used because **order is important**. For example, the points $(2, 5)$ and $(5, 2)$ are not the same. To plot $(2, 5)$, move 2 units right and 5 units up. To plot $(5, 2)$, move 5 units right and 2 units up. The points $(2, 5)$ and $(5, 2)$ are in different locations. **The order in which coordinates appear makes a difference in a point's location.**

✔ **CHECK POINT 1** Plot the points: $A(-2, 4)$, $B(4, -2)$, $C(-3, 0)$, and $D(0, -3)$.

2 Find coordinates of points in the rectangular coordinate system.

In the rectangular coordinate system, each ordered pair corresponds to exactly one point. Example 2 illustrates that each point in the rectangular coordinate system corresponds to exactly one ordered pair.

EXAMPLE 2 Finding Coordinates of Points

Determine the coordinates of points A, B, C, and D shown in Figure 1.14.

SOLUTION

Point	Position from the Origin	Coordinates
A	6 units left, 0 units up or down	$(-6, 0)$
B	0 units right or left, 2 units up	$(0, 2)$
C	2 units right, 0 units up or down	$(2, 0)$
D	4 units right, 2 units down	$(4, -2)$

FIGURE 1.14 Finding coordinates of points

 CHECK POINT 2 Determine the coordinates of points E, F, and G shown in Figure 1.14.

The rectangular coordinate system lets us visualize relationships between two quantities, as shown in the next example.

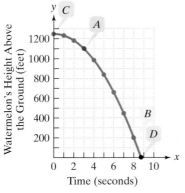

FIGURE 1.15

EXAMPLE 3 **An Application of the Rectangular Coordinate System**

Comedian David Letterman has been known to drop watermelons out of windows and watch them fall. Suppose that he drops a watermelon from the observation deck at the top of New York's Empire State Building. (Yes, the sidewalk on the ground below the building has been cleared!) Letterman drops the melon and watches it glide past the side of the building, smiling at the instant it splatters on the ground. The points in Figure 1.15 show the height of the melon above the ground at different times. Find the coordinates of point *A*. Then interpret the coordinates in terms of the information given.

SOLUTION   Let's take a few minutes to look at the rectangular coordinate system. The *x*-axis represents the time, in seconds, that the watermelon is in the air. Each mark on the *x*-axis represents one second. The *y*-axis represents the melon's height, in feet, above the ground. Each mark on the *y*-axis represents 100 feet. Because time and height are not negative, Figure 1.15 shows only the first quadrant of the rectangular coordinate system and its boundary.

   Now let us find the coordinates of point *A*. Point *A* is 3 units right and 1100 units up from the origin. Thus, the coordinates of point *A* are (3, 1100). This means that after 3 seconds, the watermelon is 1100 feet above the ground. ■

 CHECK POINT 3 Use Figure 1.15 to find the coordinates of point *B*. Then interpret the coordinates in terms of the information given.

Take another look at Figure 1.15. How can we find the coordinates of points *C* and *D*? It appears that we must estimate one or more of these coordinates and arrive at reasonable approximations.

EXAMPLE 4 **Estimating Coordinates of a Point**

Use Figure 1.15 to estimate the coordinates of point *C*. Then interpret the coordinates in terms of the information given.

SOLUTION   Point *C* is 0 units right from the origin. Thus, its *x*-coordinate is 0. How far up from the origin is point *C*? Its distance up is approximately midway between 1200 and 1300. Thus, its *y*-coordinate is approximately 1250. A reasonable estimate is that the coordinates of point *C* are (0, 1250). This means that after 0 seconds, or at the very instant Letterman dropped the watermelon, it was approximately 1250 feet above the ground. Equivalently, the melon was dropped from a height of approximately 1250 feet. ■

 CHECK POINT 4 Use Figure 1.15 to estimate the coordinates of point *D*. Then interpret the coordinates in terms of the information given.

STUDY TIP

Any point on the *x*-axis has a *y*-coordinate of 0. Any point on the *y*-axis has an *x*-coordinate of 0.

Throughout your study of algebra, you will see how the rectangular coordinate system is used to create graphs that show relationships between quantities. Magazines and newspapers also use graphs to display information. In the remainder of this section, we will discuss how to interpret line and bar graphs.

3 Interpret information given by line graphs.

Line Graphs **Line graphs** are often used to illustrate trends over time. Some measure of time, such as months or years, frequently appears on the horizontal axis. Amounts are generally listed on the vertical axis. Points are drawn to represent the given information. The graph is formed by connecting the points with line segments.

Figure 1.16 is an example of a typical line graph. The graph shows the average age at which women in the United States married for the first time from 1890 through 2002. The years are listed on the horizontal axis and the ages are listed on the vertical axis. The symbol ⌇ on the vertical axis shows that there is a break in values between 0 and 20. Thus, the first tick mark on the vertical axis represents an average age of 20.

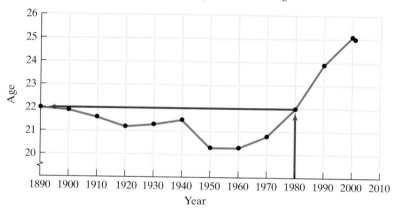

Women's Average Age of First Marriage

FIGURE 1.16
Source: U.S. Census Bureau

A line graph displays information in the first quadrant of a rectangular coordinate system. By identifying points on line graphs and their coordinates, you can interpret specific information given by the graph.

For example, the red lines in Figure 1.16 show how to find the average age at which women married for the first time in 1980.

STEP 1 Locate 1980 on the horizontal axis.

STEP 2 Locate the point above 1980.

STEP 3 Read across to the corresponding age on the vertical axis.

The age is 22. Thus, in 1980, women in the United States married for the first time at an average age of 22.

| **EXAMPLE 5** | Using a Line Graph |

The line graph in Figure 1.17 shows the fraction of federal prisoners in the United States sentenced for drug offenses from 1970 through 2001.

a. For the period shown, estimate the maximum fraction of federal prisoners sentenced for drug offenses. When did this occur?

b. Table 1.2 shows the number, in thousands, of federal prisoners in the United States for four selected years. Estimate the number of federal prisoners sentenced for drug offenses for the year in part (a).

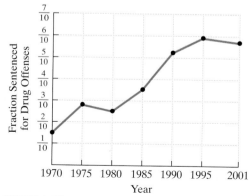

Fraction of U.S. Federal Prisoners Sentenced for Drug Offenses

FIGURE 1.17
Source: Frank Schmalleger, *Criminal Justice Today*, 7th Edition, Prentice Hall, 2003

Table 1.2

Number, in Thousands, of U.S. Federal Prisoners

Year	Federal Prisoners
1990	60
1995	90
2000	130
2001	140

Source: Bureau of Justice Statistics

SOLUTION

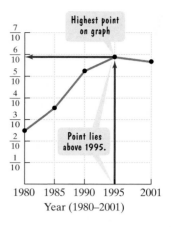

a. The maximum fraction of federal prisoners sentenced for drug offenses can be found by locating the highest point on the graph. This point lies above 1995 on the horizontal axis. Read across to the corresponding fraction on the vertical axis. The number falls slightly below $\frac{6}{10}$. It appears that $\frac{6}{10}$, or $\frac{3}{5}$, is a reasonable estimate. Thus, the maximum fraction of federal prisoners sentenced for drug offenses is approximately $\frac{3}{5}$. This occurred in 1995.

b. Table 1.2 shows that there were 90 thousand federal prisoners in 1995. The number sentenced for drug offenses can be determined as follows:

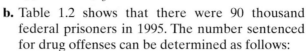

$$\text{Number sentenced for drug offenses} = \frac{3}{5} \cdot 90 = \frac{3 \cdot 90}{5} = \frac{270}{5} = 54.$$

In 1995, approximately 54 thousand federal prisoners were sentenced for drug offenses. ∎

 CHECK POINT 5 Use Figure 1.17 and Table 1.2 to estimate the number of federal prisoners sentenced for drug offenses in 1990.

4 Interpret information given by bar graphs.

Bar Graphs **Bar graphs** are convenient for showing comparisons among items. The bars may be vertical or horizontal, and their heights or lengths are used to show the amounts of different items. Figure 1.18 is an example of a typical bar graph. The graph shows the number of cars per 100 people in ten selected countries.

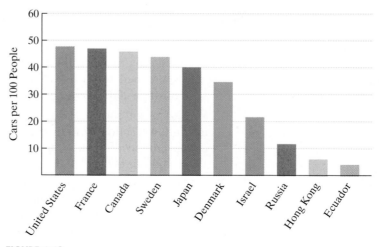

FIGURE 1.18

Source: The World Bank

EXAMPLE 6 Using a Bar Graph

Using Figure 1.18, answer the following:

a. Estimate the number of cars per 100 people in the United States.

b. Which countries have fewer than 20 cars per 100 people?

SOLUTION

a. To estimate the number of cars per 100 people in the United States, look at the top of the bar representing the United States and then read the cars-per-100-people scale on the vertical axis. The bar extends more than midway between 40 and 50, but is less than 50 by a few units. It appears that 48 is a reasonable estimate. Thus, there are approximately 48 cars per 100 people in the United States.

b. To find which countries have fewer than 20 cars per 100 people, we first locate the 20 mark on the vertical axis. Then we look for bars whose heights do not exceed 20. There are three such bars, namely, the three bars located in the right portion of the graph. The names of the countries below these bars show that Russia, Hong Kong, and Ecuador have fewer than 20 cars per 100 people. ∎

 CHECK POINT 6 Using Figure 1.18, answer the following:

a. Estimate the number of cars per 100 people in Israel.

b. Which countries have more than 40 cars per 100 people?

1.3 EXERCISE SET

 Student Solutions Manual   CD/Video   PH Math/Tutor Center   MathXL Tutorials on CD   MathXL®   MyMathLab   Interactmath.com

Practice Exercises

In Exercises 1–8, plot the given point in a rectangular coordinate system. Indicate in which quadrant each point lies.

1. $(3, 5)$ **2.** $(5, 3)$ **3.** $(-5, 1)$

4. $(1, -5)$ **5.** $(-3, -1)$ **6.** $(-1, -3)$

7. $(6, -3.5)$ **8.** $(-3.5, 6)$

In Exercises 9–24, plot the given point in a rectangular coordinate system.

9. $(-3, -3)$ **10.** $(-5, -5)$ **11.** $(-2, 0)$

12. $(-5, 0)$ **13.** $(0, 2)$ **14.** $(0, 5)$

15. $(0, -3)$ **16.** $(0, -5)$ **17.** $\left(\dfrac{5}{2}, \dfrac{7}{2}\right)$

18. $\left(\dfrac{7}{2}, \dfrac{5}{2}\right)$ **19.** $\left(-5, \dfrac{3}{2}\right)$ **20.** $\left(-\dfrac{9}{2}, -4\right)$

21. $(0, 0)$ **22.** $\left(-\dfrac{5}{2}, 0\right)$ **23.** $\left(0, -\dfrac{5}{2}\right)$

24. $\left(0, \dfrac{7}{2}\right)$

In Exercises 25–32, give the ordered pairs that correspond to the points labeled in the figure.

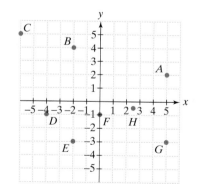

25. A **26.** B **27.** C

28. D **29.** E **30.** F

31. G **32.** H

33. In which quadrants are the y-coordinates positive?

34. In which quadrants are the x-coordinates negative?

35. In which quadrants do the x-coordinates and the y-coordinates have the same sign?

36. In which quadrants do the *x*-coordinates and the *y*-coordinates have opposite signs?

Practice Plus

37. Graph four points in at least two different quadrants such that the absolute value of the *x*-coordinate is the *y*-coordinate.

38. Graph four points in at least two different quadrants such that the absolute value of the *y*-coordinate is the *x*-coordinate.

39. The point $A(-2, y)$ lies on the line that connects $(-3, 5)$ and $(1, 3)$. Estimate the *y*-coordinate of point *A*.

40. The point $A(-1, y)$ lies on the line that connects $(-4, -1)$ and $(2, -4)$. Estimate the *y*-coordinate of point *A*.

Application Exercises

A football is thrown by a quarterback to a receiver. The points in the figure show the height of the football, in feet, above the ground in terms of its distance, in yards, from the quarterback. Use this information to solve Exercises 41–46.

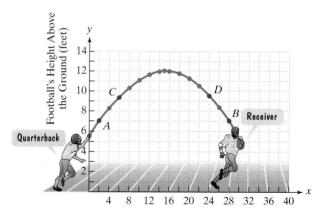

Distance of the Football from the Quarterback (yards)

41. Find the coordinates of point *A*. Then interpret the coordinates in terms of the information given.

42. Find the coordinates of point *B*. Then interpret the coordinates in terms of the information given.

43. Estimate the coordinates of point *C*.

44. Estimate the coordinates of point *D*.

45. What is the football's maximum height? What is its distance from the quarterback when it reaches its maximum height?

46. What is the football's height when it is caught by the receiver? What is the receiver's distance from the quarterback when he catches the football?

Afghanistan accounts for 76% of the world's illegal opium production. Opium-poppy cultivation nets big money in a country where most people earn less than $1 per day. (Source: Newsweek) The line graph shows opium-poppy cultivation, in thousands of acres, in Afghanistan from 1990 through 2002. Use the graph to solve Exercises 47–52.

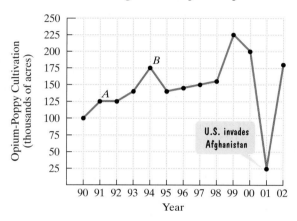

Afghanistan's Opium Crop

Source: U.N. Office on Drugs and Crime

47. What are the coordinates of point *A*? What does this mean in terms of the information given by the graph?

48. What are the coordinates of point *B*? What does this mean in terms of the information given by the graph?

49. For the period shown, when did opium cultivation reach a minimum? How many thousands of acres were used to cultivate the illegal crop?

50. For the period shown, when did opium cultivation reach a maximum? How many thousands of acres were used to cultivate the illegal crop?

51. Between which two years did opium cultivation not change?

52. Between which two years did opium cultivation increase at the greatest rate? What is a reasonable estimate of the increase, in thousands of acres, used to cultivate the illegal crop during this period?

The line graphs at the top of the next page show that oil consumption in the United States has continued to increase as domestic production of crude oil has been dropping. Use the information shown by the two line graphs to solve Exercises 53–60.

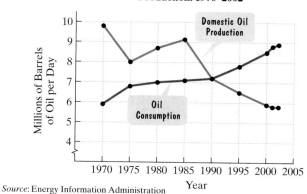

U.S. Oil Consumption and Domestic Production: 1970–2002

Source: Energy Information Administration

53. Find an estimate for oil consumption in 1985.

54. Find an estimate for domestic oil production in 1985.

55. In which year were oil consumption and domestic production the same? Estimate the millions of barrels of oil consumed and produced per day during that year.

56. For which years did oil consumption exceed domestic production?

57. For the period shown, when did domestic oil production reach a maximum? What is a reasonable estimate for oil production during that year?

58. For the period shown, when did oil consumption reach a maximum? What is a reasonable estimate for oil consumption during that year?

59. Find an estimate for the difference between production and consumption in 1980.

60. Find an estimate for the difference between consumption and production in 2000.

You can't judge the quality of a movie by its domestic box-office gross, but you can't keep making movies that don't make money. The bar graph shows that actor Kevin Costner needs a hit. Use the information in the graph to solve Exercises 61–64.

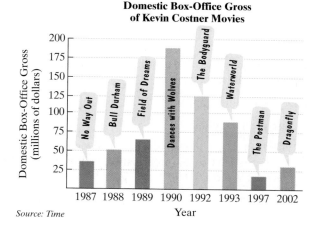

Domestic Box-Office Gross of Kevin Costner Movies

Source: Time

61. Estimate the box-office gross, in millions of dollars, of *The Bodyguard*.

62. Estimate the box-office gross, in millions of dollars, of *Waterworld*.

63. Which movies have a box-office gross of less than $50 million?

64. Which movies have a box-office gross of more than $100 million?

The bar graph shows life expectancy in the United States by year of birth. Use the graph to solve Exercises 65–68.

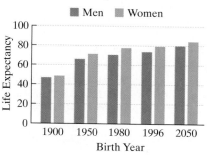

Life Expectancy in the U.S. by Birth Year

Source: U.S. Census Bureau

65. Estimate the life expectancy for men born in 1900.

66. Estimate the life expectancy for women born in 2050.

67. By approximately how many more years can women born in 1996 expect to live as compared to men born in 1950?

68. For which genders and for which birth years does life expectancy exceed 40 years but is at most 60 years?

The bar graph shows the annual median income (the income in the middle of the list when all of the groups' incomes are arranged in order) for six groups, organized by gender and race. Use the information in the graph to solve Exercises 69–72.

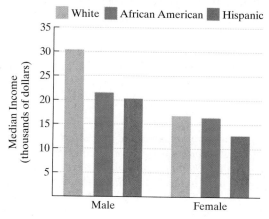

U.S. Median Income in 2001, by Gender and Race

Source: U.S. Census Bureau

(Refer to the graph on the previous page)

69. Estimate the median income, in thousands of dollars, for African-American men.

70. Estimate the median income, in thousands of dollars, for white women.

71. Estimate the difference, in thousands of dollars, between median income of Hispanic men and Hispanic women.

72. Estimate the difference, in thousands of dollars, between the median income of African-American men and African-American women.

Candidates for president of the United States keep jumping into the race earlier and earlier. The bar graph shows the average number of days before the nominating convention that candidates declared their intention to run for the office. Use the information in the graph to solve Exercises 73–74.

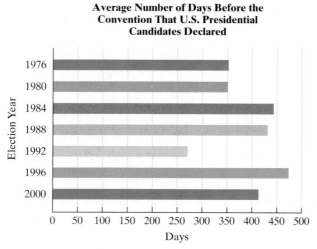

Average Number of Days Before the Convention That U.S. Presidential Candidates Declared

Source: Newsweek

73. In 1992, Clinton's 284 days, a short campaign, didn't cost him the win. Estimate the average number of days before the convention that candidates declared in 1992. By approximately how many days did Clinton's declaration exceed the 1992 average?

74. Typical of candidates today, in 2000, Bush declared 405 days prior to the convention. Estimate the average number of days before the convention that candidates declared in 2000. By approximately how many days was Bush's declaration less than the 2000 average?

Despite various programs over the past three decades, oil from other countries accounts for an ever-greater share of energy in the United States. The bar graph in the next column shows the fraction of oil used in the United States that was obtained from other countries for five selected years. In Exercises 75–78, use the information shown by the graph to write the year or years that fit the given description.

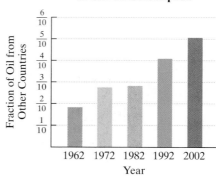

Oil Imports as a Fraction of U.S. Oil Consumption

Source: Energy Information Administration

75. The United States obtained more than $\frac{1}{2}$ of its oil from other countries.

76. The United States obtained less than $\frac{1}{5}$ of its oil from other countries.

77. The fraction of imported oil exceeded $\frac{1}{5}$ but was at most $\frac{1}{2}$.

78. The fraction of imported oil exceeded $\frac{2}{5}$ but was at most $\frac{1}{2}$.

79. In 2002, $\frac{1}{10}$ of U.S. imported oil was from Saudi Arabia. Use the information shown in the graph to estimate the fraction of oil used in the United States in 2002 that was imported from Saudi Arabia.

80. In 2002, $\frac{2}{25}$ of U.S. imported oil was from Venezuela. Use the information shown in the graph to estimate the fraction of oil used in the United States in 2002 that was imported from Venezuela.

Writing in Mathematics

81. What is the rectangular coordinate system?

82. Explain how to plot a point in the rectangular coordinate system. Give an example with your explanation.

83. Explain why $(5, -2)$ and $(-2, 5)$ do not represent the same point.

84. Explain how to find the coordinates of a point in the rectangular coordinate system.

85. Describe a line graph.

86. Describe a bar graph.

87. Describe how the information in the bar graph for Exercises 75–80 is related to the information in the line graphs for Exercises 53–60.

88. Find a graph in a newspaper, magazine, or almanac and describe what the graph illustrates.

Critical Thinking Exercises

89. a. Graph each of the following sets of points in a separate rectangular coordinate system:

$$A = \{(-3, 9), (-2, 4), (-1, 1),$$
$$(0, 0), (1, 1), (2, 4), (3, 9)\}$$
$$B = \{(9, -3), (4, -2), (1, -1),$$
$$(0, 0), (1, 1), (4, 2), (9, 3)\}.$$

   b. In which set (A or B) is each x-coordinate associated with exactly one y-coordinate? In which set are x-coordinates associated with more than one y-coordinate? Describe how these differences can be seen from your graphs in part (a).

90. a. Graph each of the following points:

$$\left(1, \frac{1}{2}\right), (2, 1), \left(3, \frac{3}{2}\right), (4, 2).$$

Parts (b)–(d) can be answered by changing the sign of one or both coordinates of the points in part (a).

   b. What must be done to the coordinates so that the resulting graph is a mirror-image reflection about the y-axis of your graph in part (a)?

   c. What must be done to the coordinates so that the resulting graph is a mirror-image reflection about the x-axis of your graph in part (a)?

   d. What must be done to the coordinates so that the resulting graph is a straight-line extension of your graph in part (a)?

Review Exercises

From here on, each exercise set will contain three review exercises. It is essential to review previously covered topics to improve your understanding of the topics and to help maintain your mastery of the material. If you are not certain how to solve a review exercise, turn to the section and the illustrative example given in parentheses at the end of each exercise.

91. Add: $\dfrac{3}{4} + \dfrac{2}{5}$. (Section 1.1, Example 9)

92. Insert $<$ or $>$ in the shaded area to make a true statement: $-\dfrac{1}{4}$ ▨ 0. (Section 1.2, Example 6)

93. Find the absolute value: $|-5.83|$. (Section 1.2, Example 8)

SECTION 1.4

Objectives

1 Evaluate algebraic expressions.

2 Understand and use the vocabulary of algebraic expressions.

3 Use commutative properties.

4 Use associative properties.

5 Use distributive properties.

6 Combine like terms.

7 Simplify algebraic expressions.

8 Use algebraic expressions that model reality.

BASIC RULES OF ALGEBRA

Algebraic Expressions Feeling attractive with a suntan that gives you a "healthy glow"? Think again. Direct sunlight is known to promote skin cancer. Although sunscreens protect you from burning, dermatologists are concerned with the long-term damage that results from the sun even without sunburn.

We have seen that algebra uses letters, such as x and y, to represent numbers. Recall that such letters are called **variables**. For example, we can let x represent the number of minutes that a person can stay in the sun without burning with no sunscreen. With a number 6 sunscreen, exposure time without burning is six times as long, or 6 times x. This can be written $6 \cdot x$, but it is usually expressed as $6x$. Placing a number and a letter next to one another indicates multiplication.

Notice that $6x$ combines the number 6 and the variable x using the operation of multiplication. A combination of variables and numbers using the operations of addition, subtraction, multiplication, or division, as well as powers or roots, is called an **algebraic expression**. Here are some examples of algebraic expressions:

$$x + 6, \quad x - 6, \quad 6x, \quad \frac{x}{6}, \quad 3x + 5, \quad \sqrt{x} + 7.$$

1 Evaluate algebraic expressions.

Evaluating Algebraic Expressions We can replace a variable that appears in an algebraic expression by a number. We are **substituting** the number for the variable. The process is called **evaluating the expression**. For example, we can evaluate $6x$ (from the sunscreen example) for $x = 15$. We substitute 15 for x. We obtain $6 \cdot 15$, or 90. This means if you can stay in the sun for 15 minutes without burning when you don't put on any lotion, then with a number 6 lotion, you can "cook" for 90 minutes without burning.

Many algebraic expressions involve more than one operation. The order in which we add, subtract, multiply, and divide is important. In Section 1.8, we will discuss the rules for the order in which operations should be done. For now, follow this order:

1. Perform calculations within parentheses first.
2. Perform multiplication before addition.

EXAMPLE 1 Evaluating an Algebraic Expression

The algebraic expression $2.35x + 179.5$ describes the population of the United States, in millions, x years after 1960. Evaluate the expression for $x = 40$. Describe what the answer means in practical terms.

SOLUTION We begin by substituting 40 for x. Because $x = 40$, we will be finding the U.S. population 40 years after 1960, in the year 2000.

$$2.35x + 179.5$$

Replace x with 40.

$= 2.35(40) + 179.5$
$= 94 + 179.5$ Perform the multiplication: $2.35(40) = 94$.
$= 273.5$ Perform the addition.

According to the given algebraic expression, in 2000, the population of the United States was 273.5 million. ∎

According to the U.S. Bureau of the Census, in 2000 the population of the United States was 281.4 million. Notice that the algebraic expression in Example 1 provides an approximate, but not an exact, description of the actual population. Many algebraic expressions approximately describe some aspect of reality, and we say that they *model* reality.

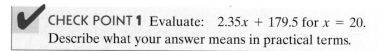 CHECK POINT 1 Evaluate: $2.35x + 179.5$ for $x = 20$.
Describe what your answer means in practical terms.

2 Understand and use the vocabulary of algebraic expressions.

The Vocabulary of Algebraic Expressions We have seen that an algebraic expression combines numbers and variables. Here is another example of an algebraic expression:

$$7x + 3.$$

The **terms** of an algebraic expression are those parts that are separated by addition. For example, the algebraic expression $7x + 3$ contains two terms, namely $7x$ and 3. Notice that a term is a number, a variable, or a number multiplied by one or more variables.

The numerical part of a term is called its **coefficient**. In the term $7x$, the 7 is the coefficient. If a term containing one or more variables is written without a coefficient, the coefficient is understood to be 1. Thus, x means $1x$ and ab means $1ab$.

A term that consists of just a number is called a **constant term**. The constant term of $7x + 3$ is 3.

The parts of each term that are multiplied are called the **factors of the term**. The factors of the term $7x$ are 7 and x.

Like terms are terms that have exactly the same variable factors. Here are two examples of like terms:

$7x$ and $3x$	These terms have the same variable factor, x.
$4y$ and $9y$.	These terms have the same variable factor, y.

By contrast, here are some examples of terms that are not like terms. These terms do not have the same variable factor.

$7x$ and 3	The variable factor of the first term is x. The second term has no variable factor.
$7x$ and $3y$	The variable factor of the first term is x. The variable factor of the second term is y.

Constant terms are like terms. Thus, the constant terms 7 and -12 are like terms.

EXAMPLE 2 Using the Vocabulary of Algebraic Expressions

Use the algebraic expression

$$4x + 7 + 5x$$

to answer the following questions:
   a. How many terms are there in the algebraic expression?
   b. What is the coefficient of the first term?
   c. What is the constant term?
   d. What are the like terms in the algebraic expression?

SOLUTION

   a. Because terms are separated by addition, the algebraic expression $4x + 7 + 5x$ contains three terms.

$$4x + 7 + 5x$$

First term	Second term	Third term

   b. The coefficient of the first term, $4x$, is 4.
   c. The constant term in $4x + 7 + 5x$ is 7.
   d. The like terms in $4x + 7 + 5x$ are $4x$ and $5x$. These terms have the same variable factor, x. ∎

 CHECK POINT **2** Use the algebraic expression $6x + 2x + 11$ and answer each of the four questions in Example 2.

Equivalent Algebraic Expressions In Example 2, we considered the algebraic expression

$$4x + 7 + 5x.$$

Let's compare this expression with a second algebraic expression

$$9x + 7.$$

Evaluate each expression for some choice of x. We will select $x = 2$.

$$
\begin{array}{ll}
4x + 7 + 5x & 9x + 7 \\
\text{Replace } x \text{ with 2.} & \text{Replace } x \text{ with 2.} \\
= 4 \cdot 2 + 7 + 5 \cdot 2 & = 9 \cdot 2 + 7 \\
= 8 + 7 + 10 & = 18 + 7 \\
= 25 & = 25
\end{array}
$$

Both algebraic expressions have the same value when $x = 2$. Regardless of what number you select for x, the algebraic expressions $4x + 7 + 5x$ and $9x + 7$ will have the same value. These expressions are called *equivalent algebraic expressions*. Two algebraic expressions that have the same value for all replacements are called **equivalent algebraic expressions**. Because $4x + 7 + 5x$ and $9x + 7$ are equivalent algebraic expressions, we write

$$4x + 7 + 5x = 9x + 7.$$

Properties of Real Numbers and Algebraic Expressions We now turn to basic properties, or rules, that you know from past experiences in working with whole numbers and fractions. These properties will be extended to include all real numbers and algebraic expressions. We will give each property a name so that we can refer to it throughout the study of algebra.

The Commutative Properties The addition or multiplication of two real numbers can be done in any order. For example, $3 + 5 = 5 + 3$ and $3 \cdot 5 = 5 \cdot 3$. Changing the order does not change the answer of a sum or a product. These facts are called **commutative properties**.

3 Use commutative properties.

> **THE COMMUTATIVE PROPERTIES** Let a, b, and c represent real numbers, variables, or algebraic expressions.
>
> **Commutative Property of Addition**
> $$a + b = b + a$$
> Changing order when adding does not affect the sum.
>
> **Commutative Property of Multiplication**
> $$ab = ba$$
> Changing order when multiplying does not affect the product.

EXAMPLE 3 Using the Commutative Properties

Use the commutative properties to write an algebraic expression equivalent to each of the following: **a.** $y + 6$ **b.** $5x$.

SOLUTION

a. By the commutative property of addition, an algebraic expression equivalent to $y + 6$ is $6 + y$. Thus,

$$y + 6 = 6 + y.$$

b. By the commutative property of multiplication, an algebraic expression equivalent to $5x$ is $x5$. Thus,

$$5x = x5.$$ ■

CHECK POINT 3 Use the commutative properties to write an algebraic expression equivalent to each of the following: **a.** $x + 14$ **b.** $7y$.

EXAMPLE 4   Using the Commutative Properties

Write an algebraic expression equivalent to $13x + 8$ using

   a. the commutative property of addition.
   b. the commutative property of multiplication.

SOLUTION

   a. By the commutative property of addition, we change the order of the terms being added. This means that an algebraic expression equivalent to $13x + 8$ is $8 + 13x$:
$$13x + 8 = 8 + 13x.$$

   b. By the commutative property of multiplication, we change the order of the factors being multiplied. This means that an algebraic expression equivalent to $13x + 8$ is $x13 + 8$:
$$13x + 8 = x13 + 8.$$ ∎

CHECK POINT 4 Write an algebraic expression equivalent to $5x + 17$ using
   a. the commutative property of addition.
   b. the commutative property of multiplication.

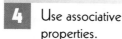 Use associative properties.

The Associative Properties   Parentheses indicate groupings. We perform operations within the parentheses first. For example,
$$(2 + 5) + 10 = 7 + 10 = 17$$
and
$$2 + (5 + 10) = 2 + 15 = 17.$$

   In general, the way in which three numbers are grouped does not change their sum. It also does not change their product. These facts are called the **associative properties**.

STUDY TIP

The associative property does not hold for subtraction or division.

$$(6 - 3) - 1 \neq 6 - (3 - 1)$$
$$(8 \div 4) \div 2 \neq 8 \div (4 \div 2)$$

THE ASSOCIATIVE PROPERTIES   Let a, b, and c represent real numbers, variables, or algebraic expressions.

Associative Property of Addition
$$(a + b) + c = a + (b + c)$$
Changing grouping when adding does not affect the sum.

Associative Property of Multiplication
$$(ab)c = a(bc)$$
Changing grouping when multiplying does not affect the product.

The associative properties can be used to simplify some algebraic expressions by removing the parentheses.

EXAMPLE 5   Simplifying Using the Associative Properties

Simplify: **a.** $3 + (8 + x)$ **b.** $8(4x)$.

SOLUTION

a. $3 + (8 + x)$ — This is the given algebraic expression.

$= (3 + 8) + x$ — Use the associative property of addition to group the first two numbers.

$= 11 + x$ — Add within parentheses.

Using the commutative property of addition, this simplified algebraic expression can also be written as $x + 11$.

b. $8(4x)$ — This is the given algebraic expression.

$= (8 \cdot 4)x$ — Use the associative property of multiplication to group the first two numbers.

$= 32x$ — Multiply within parentheses.

We can use the commutative property of multiplication and write this simplified algebraic expression as $x32$ or $x \cdot 32$. However, it is customary to express a term with its coefficient on the left. Thus, we use $32x$ as the simplified form of the algebraic expression. ■

CHECK POINT 5 Simplify:

a. $8 + (12 + x)$ **b.** $6(5x)$.

ENRICHMENT ESSAY

The Associative Property and the English Language

In the English language, sentences can take on different meanings depending on the way the words are associated with commas. Here are two examples.

• *Do not break your bread or roll in your soup.*   • *Woman, without her man, is nothing.*
  Do not break your bread, or roll in your soup.   *Woman, without her, man is nothing.*

The next example involves the use of both basic properties to simplify an algebraic expression.

EXAMPLE 6   Using the Commutative and Associative Properties

Simplify: $7 + (x + 2)$.

SOLUTION

$7 + (x + 2)$ — This is the given algebraic expression.

$= 7 + (2 + x)$ — Use the commutative property to change the order of the addition.

$= (7 + 2) + x$ — Use the associative property to group the first two numbers.

$= 9 + x$ — Add within parentheses.

Using the commutative property of addition, an equivalent algebraic expression is $x + 9$. ∎

✔ **CHECK POINT 6** Simplify: $8 + (x + 4)$.

5 Use distributive properties.

The Distributive Properties The **distributive property** involves both multiplication and addition. The property shows how to multiply the sum of two numbers by a third number. Consider, for example, $4(7 + 3)$, which can be calculated in two ways. One way is to perform the addition within the grouping symbols and then multiply:

$$4(7 + 3) = 4(10) = 40.$$

The other way is to *distribute* the multiplication by 4 over the addition by first multiplying each number within the parentheses by 4 and then adding:

$$4(\overset{\frown}{7 + 3}) = 4 \cdot 7 + 4 \cdot 3 = 28 + 12 = 40.$$

The result in both cases is 40. Thus,

$$4(7 + 3) = 4 \cdot 7 + 4 \cdot 3. \quad \text{Multiplication distributes over addition.}$$

The distributive property allows us to rewrite the product of a number and a sum as the sum of two products.

> **THE DISTRIBUTIVE PROPERTY** Let a, b, and c represent real numbers, variables, or algebraic expressions.
>
> $$a(\overset{\frown}{b + c}) = ab + ac$$
>
> Multiplication distributes over addition.

STUDY TIP

Do not confuse the distributive property with the associative property of multiplication.

Distributive:

$$4(5 + x) = 4 \cdot 5 + 4x$$
$$= 20 + 4x$$

Associative:

$$4(5 \cdot x) = (4 \cdot 5)x$$
$$= 20x$$

EXAMPLE 7 Using the Distributive Property

Multiply: $6(x + 4)$.

SOLUTION Multiply *each term* inside the parentheses, x and 4, by the multiplier outside, 6.

$$6(x + 4) = 6x + 6 \cdot 4 \quad \text{Use the distributive property to remove parentheses.}$$
$$= 6x + 24 \quad \text{Multiply: } 6 \cdot 4 = 24. \quad ∎$$

✔ **CHECK POINT 7** Multiply: $5(x + 3)$.

EXAMPLE 8 Using the Distributive Property

Multiply: $5(3y + 7)$.

SOLUTION Multiply *each term* inside the parentheses, $3y$ and 7, by the multiplier outside, 5.

$$5(3y + 7) = 5 \cdot 3y + 5 \cdot 7 \quad \text{Use the distributive property to remove parentheses.}$$
$$= 15y + 35 \quad \text{Multiply. Use the associative property of multiplication to find } 5 \cdot 3y: 5 \cdot 3y = (5 \cdot 3)y = 15y. \quad ∎$$

✔ **CHECK POINT 8** Multiply: $6(4y + 7)$.

When using the distributive property to remove parentheses, be sure to multiply *each term* inside the parentheses by the multiplier outside.

Incorrect!

$$5(3y + 7) = 5 \cdot 3y + 7$$

> 7 must also be multiplied by 5.

$$= 15y + 7$$

Table 1.3 shows a number of other forms of the distributive property.

Table 1.3	Other Forms of the Distributive Property	
Property	**Meaning**	**Example**
$a(b - c) = ab - ac$	Multiplication distributes over subtraction.	$5(4x - 3) = 5 \cdot 4x - 5 \cdot 3$ $= 20x - 15$
$a(b + c + d) = ab + ac + ad$	Multiplication distributes over three or more terms in parentheses.	$4(x + 10 + 3y)$ $= 4x + 4 \cdot 10 + 4 \cdot 3y$ $= 4x + 40 + 12y$
$(b + c)a = ba + ca$	Multiplication on the right distributes over addition (or subtraction).	$(x + 7)9 = x \cdot 9 + 7 \cdot 9$ $= 9x + 63$

6 Combine like terms.

Combining Like Terms The distributive property

$$a(b + c) = ab + ac$$

lets us add and subtract like terms. To do this, we will usually apply the property in the form

$$ax + bx = (a + b)x$$

and then combine a and b. For example,

$$3x + 7x = (3 + 7)x = 10x.$$

This process is called **combining like terms**.

EXAMPLE 9 Combining Like Terms

Combine like terms: **a.** $4x + 15x$ **b.** $7a - 2a$.

SOLUTION

a. $4x + 15x$ These are like terms because 4x and 15x have identical variable factors.

$= (4 + 15)x$ Apply the distributive property.

$= 19x$ Add within parentheses.

b. $7a - 2a$ These are like terms because 7a and 2a have identical variable factors.

$= (7 - 2)a$ Apply the distributive property.

$= 5a$ Subtract within parentheses.

DISCOVER FOR YOURSELF

Can you think of a fast method that will immediately give each result in Example 9? Describe the method.

✔ **CHECK POINT 9** Combine like terms: **a.** $7x + 3x$ **b.** $9a - 4a$.

When combining like terms, you may find yourself leaving out the details of the distributive property. For example, you may simply write

$$7x + 3x = 10x.$$

It might be useful to think along these lines: Seven things plus three of the (same) things give ten of those things. To add like terms, add the coefficients and copy the common variable.

COMBINING LIKE TERMS MENTALLY

1. Add or subtract the coefficients of the terms.
2. Use the result of step 1 as the coefficient of the term's variable factor.

When an expression contains three or more terms, use the commutative and associative properties to group like terms. Then combine the like terms.

EXAMPLE 10 Grouping and Combining Like Terms

Simplify: **a.** $7x + 5 + 3x + 8$ **b.** $4x + 7y + 2x + 3y$.

SOLUTION

a. $7x + 5 + 3x + 8$

$= (7x + 3x) + (5 + 8)$ Rearrange terms and group the like terms using the commutative and associative properties. This step is often done mentally.

$= 10x + 13$ Combine like terms: $7x + 3x = 10x$. Combine constant terms: $5 + 8 = 13$.

b. $4x + 7y + 2x + 3y$

$= (4x + 2x) + (7y + 3y)$ Group like terms.

$= 6x + 10y$ Combine like terms by adding coefficients and keeping the variable factor. ■

 CHECK POINT 10 Simplify:

a. $8x + 7 + 10x + 3$ **b.** $9x + 6y + 5x + 2y$.

---

7 Simplify algebraic expressions.

Simplifying Algebraic Expressions An algebraic expression is **simplified** when parentheses have been removed and like terms have been combined.

SIMPLIFYING ALGEBRAIC EXPRESSIONS

1. Use the distributive property to remove parentheses.
2. Rearrange terms and group like terms using commutative and associative properties. This step may be done mentally.
3. Combine like terms by combining the coefficients of the terms and keeping the same variable factor.

EXAMPLE 11 Simplifying an Algebraic Expression

Simplify: $5(3x + 7) + 6x$.

SOLUTION

$5(3x + 7) + 6x$

$= 5 \cdot 3x + 5 \cdot 7 + 6x$ Use the distributive property to remove the parentheses.

$= 15x + 35 + 6x$ Multiply.

$= (15x + 6x) + 35$ Group like terms.

$= 21x + 35$ Combine like terms.   ■

 CHECK POINT 11 Simplify: $7(2x + 3) + 11x$.

EXAMPLE 12 Simplifying an Algebraic Expression

Simplify: $6(2x + 4y) + 10(4x + 3y)$.

SOLUTION

$6(2x + 4y) + 10(4x + 3y)$

$= 6 \cdot 2x + 6 \cdot 4y + 10 \cdot 4x + 10 \cdot 3y$ Use the distributive property to remove the parentheses.

$= 12x + 24y + 40x + 30y$ Multiply.

$= (12x + 40x) + (24y + 30y)$ Group like terms.

$= 52x + 54y$ Combine like terms.   ■

CHECK POINT 12 Simplify: $7(4x + 3y) + 2(5x + y)$.

Applications Have you had a good workout lately? The next example involves an algebraic expression that approximately describes, or models, your optimum heart rate during exercise.

8   Use algebraic expressions that model reality.

EXAMPLE 13 Modeling Optimum Heart Rate

The optimum heart rate is the rate that a person should achieve during exercise for the exercise to be most beneficial. The algebraic expression

$$0.6(220 - a)$$

describes a person's optimum heart rate, in beats per minute, where a represents the age of the person in years.

a. Use the distributive property to rewrite the algebraic expression without parentheses.

b. Use each form of the algebraic expression to determine the optimum heart rate for a 20-year-old runner.

SOLUTION

a. $0.6(220 - a) = 0.6(220) - 0.6a$ Use the distributive property to remove parentheses.

$= 132 - 0.6a$ Multiply: $0.6(220) = 132$.

b. To determine the optimum heart rate for a 20-year-old runner, substitute 20 for a in each form of the algebraic expression.

Using $0.6(220 - a)$:

$$0.6(220 - 20)$$
$$= 0.6(200)$$
$$= 120$$

Using $132 - 0.6a$:

$$132 - 0.6(20)$$
$$= 132 - 12$$
$$= 120$$

Both forms of the algebraic expression indicate that the optimum heart rate for a 20-year-old runner is 120 beats per minute. ∎

 CHECK POINT 13 Use each form of the algebraic expression in Example 13 to determine the optimum heart rate for a 40-year-old runner.

1.4 EXERCISE SET

 Student Solutions Manual   CD/Video   PH Math/Tutor Center   MathXL Tutorials on CD   MathXL®   MyMathLab   Interactmath.com

Practice Exercises

In Exercises 1–10, evaluate each algebraic expression for the given value of the variable.

1. $x + 11; x = 4$

2. $x - 7; x = 11$

3. $8x; x = 10$

4. $12x; x = 5$

5. $5x + 6; x = 8$

6. $6x + 9; x = 4$

7. $7(x + 3); x = 2$

8. $8(x + 4); x = 3$

9. $\frac{5}{9}(F - 32); F = 77$

10. $\frac{5}{9}(F - 32); F = 50$

In Exercises 11–16, an algebraic expression is given. Use each expression to answer the following questions.

a. How many terms are there in the algebraic expression?

b. What is the numerical coefficient of the first term?

c. What is the constant term?

d. Does the algebraic expression contain like terms? If so, what are the like terms?

11. $3x + 5$

12. $9x + 4$

13. $x + 2 + 5x$

14. $x + 6 + 7x$

15. $4y + 1 + 3x$

16. $8y + 1 + 10x$

In Exercises 17–24, use the commutative property of addition to write an equivalent algebraic expression.

17. $y + 4$

18. $x + 7$

19. $5 + 3x$

20. $4 + 9x$

21. $4x + 5y$

22. $10x + 9y$

23. $5(x + 3)$

24. $6(x + 4)$

In Exercises 25–32, use the commutative property of multiplication to write an equivalent algebraic expression.

25. $9x$

26. $8x$

27. $x + y6$

28. $x + y7$

29. $7x + 23$

30. $13x + 11$

31. $5(x + 3)$

32. $6(x + 4)$

In Exercises 33–36, use an associative property to rewrite each algebraic expression. Once the grouping has been changed, simplify the resulting algebraic expression.

33. $7 + (5 + x)$

34. $9 + (3 + x)$

35. $7(4x)$

36. $8(5x)$

In Exercises 37–56, use a distributive property to rewrite each algebraic expression without parentheses.

37. $3(x + 5)$

38. $4(x + 6)$

39. $8(2x + 3)$

40. $9(2x + 5)$

41. $\frac{1}{3}(12 + 6r)$

42. $\frac{1}{4}(12 + 8r)$

43. $5(x + y)$

44. $7(x + y)$

45. $3(x - 2)$

46. $4(x - 5)$

47. $2(4x - 5)$

48. $6(3x - 2)$

49. $\frac{1}{2}(5x - 12)$

50. $\frac{1}{3}(7x - 21)$

51. $(2x + 7)4$

52. $(5x + 3)6$

53. $6(x + 3 + 2y)$

54. $7(2x + 4 + y)$

55. $5(3x - 2 + 4y)$

56. $4(5x - 3 + 7y)$

In Exercises 57–74, simplify each algebraic expression.

57. $7x + 10x$

58. $5x + 13x$

59. $11a - 3a$

60. $14b - 5b$

61. $3 + (x + 11)$

62. $7 + (x + 10)$

63. $5y + 3 + 6y$

64. $8y + 7 + 10y$

65. $2x + 5 + 7x - 4$

66. $7x + 8 + 2x - 3$

67. $11a + 12 + 3a + 2$

68. $13a + 15 + 2a + 11$

69. $5(3x + 2) - 4$

70. $2(5x + 4) - 3$

71. $12 + 5(3x - 2)$

72. $14 + 2(5x - 1)$

73. $7(3a + 2b) + 5(4a + 2b)$

74. $11(6a + 3b) + 4(12a + 5b)$

Practice Plus

In Exercises 75–78, name the property illustrated by each true statement.

75. $6x + (2y + 7y) = 6x + (7y + 2y)$

76. $6x + (2y + 7y) = (2y + 7y) + 6x$

77. $6x + (2y + 7y) = (6x + 2y) + 7y$

78. $(6x)4 = 4(6x)$

In Exercises 79–82, determine if each statement is true or false. Do not use a calculator.

79. $468(787 + 289) = 787 + 289(468)$

80. $468(787 + 289) = 787(468) + 289(468)$

81. $58 \cdot 9 + 32 \cdot 9 = (58 + 32) \cdot 9$

82. $58 \cdot 9 \cdot 32 \cdot 9 = (58 \cdot 32) \cdot 9$

Application Exercises

83. Suppose you can stay in the sun for x minutes without burning when you don't put on any lotion. The algebraic expression $15x$ describes how long you can tan without burning with a number 15 lotion. Evaluate the algebraic expression for $x = 20$. Describe what the answer means in practical terms.

84. Suppose that the cost of an item, excluding tax, is x dollars. The algebraic expression $0.06x$ describes the sales tax on that item. Evaluate the algebraic expression for $x = 400$. Describe what the answer means in practical terms.

The algebraic expression

$$405x + 5565$$

gives an approximate description of credit-card debt per U.S. household x years after 1994. The actual debt per household is shown in the bar graph at the top of the next column. Use this information to solve Exercises 85–86.

Credit-Card Debt per U.S. Household

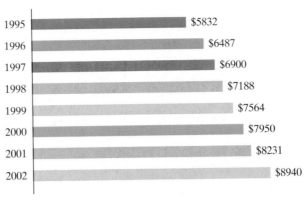

1995	$5832
1996	$6487
1997	$6900
1998	$7188
1999	$7564
2000	$7950
2001	$8231
2002	$8940

Source: Cardweb.com, Inc.

85. Evaluate the algebraic expression for $x = 8$. Describe what the answer means in practical terms. How well does the algebraic expression model the actual data for the appropriate year shown in the bar graph?

86. Evaluate the algebraic expression for $x = 7$. Describe what the answer means in practical terms. How well does the algebraic expression model the actual data for the appropriate year shown in the bar graph?

87. Testosterone is the hormone of choice for hundreds of thousands of middle-aged American men determined to stave off symptoms of andropause, the male version of menopause. The algebraic expression

$$2(0.18x + 0.01) + 0.02x$$

gives an approximate description of the millions of prescriptions for testosterone in the United States x years after 1997. The line graph indicates the actual number of testosterone prescriptions.

Testosterone Prescriptions in the U.S.

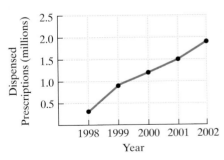

Source: NDC HEALTH

a. Simplify the algebraic expression.

b. Use the simplified algebraic expression to find the number of testosterone prescriptions in 2002. How well does the algebraic expression model the data shown in the line graph for 2002?

88. The equivalent algebraic expressions

$$\frac{DA + D}{24} \quad \text{and} \quad \frac{D(A + 1)}{24}$$

describe the drug dosage for children between the ages of 2 and 13. In each algebraic expression, D stands for an adult dose and A represents the child's age. If an adult dose of ibuprofen is 200 milligrams, what is the proper dose for a 12-year-old child? Use both forms of the algebraic expressions to answer the question. Which form is easier to use?

Writing in Mathematics

89. What is an algebraic expression? Provide an example with your description.

90. What does it mean to evaluate an algebraic expression? Provide an example with your description.

91. What is a term? Provide an example with your description.

92. What are like terms? Provide an example with your description.

93. What are equivalent algebraic expressions?

94. State a commutative property and give an example.

95. State an associative property and give an example.

96. State a distributive property and give an example.

97. Explain how to add like terms. Give an example.

98. What does it mean to simplify an algebraic expression?

99. An algebra student incorrectly used the distributive property and wrote $3(5x + 7) = 15x + 7$. If you were that student's teacher, what would you say to help the student avoid this kind of error?

100. You can transpose the letters in the word "conversation" to form the phrase "voices rant on." From "total abstainers" we can form "sit not at ale bars." What two algebraic properties do each of these transpositions (called anagrams) remind you of? Explain your answer.

Critical Thinking Exercises

101. Which one of the following statements is true?
   a. Subtraction is a commutative operation.
   b. $(24 \div 6) \div 2 = 24 \div (6 \div 2)$
   c. $7y + 3y = (7 + 3)y$ for any value of y
   d. $2x + 5 = 5x + 2$

102. Which one of the following statements is true?
   a. $a + (bc) = (a + b)(a + c)$ In words, addition can be distributed over multiplication.
   b. $4(x + 3) = 4x + 3$
   c. Not every algebraic expression can be simplified.
   d. Like terms contain the same coefficients.

103. A business that manufactures small alarm clocks has weekly fixed costs of $5000. The average cost per clock for the business to manufacture x clocks is described by

$$\frac{0.5x + 5000}{x}.$$

   a. Find the average cost when $x = 100$, 1000, and 10,000.

   b. Like all other businesses, the alarm clock manufacturer must make a profit. To do this, each clock must be sold for at least 50¢ more than what it costs to manufacture. Due to competition from a larger company, the clocks can be sold for $1.50 each and no more. Our small manufacturer can only produce 2000 clocks weekly. Does this business have much of a future? Explain.

Review Exercises

104. Express $\frac{4}{9}$ as a decimal. (Section 1.2, Example 4)

105. Plot $(-3, -1)$ in a rectangular coordinate system. (Section 1.3, Example 1)

106. Divide: $\frac{3}{7} \div \frac{15}{7}$. (Section 1.1, Example 6)

✔ MID-CHAPTER CHECK POINT

CHAPTER 1

What You Know: We reviewed operations with fractions. We defined the real numbers and represented them as points on a number line. We used the rectangular coordinate system to represent ordered pairs of real numbers. We saw how information is displayed in line and bar graphs. Finally, we introduced some basic rules of algebra and used the commutative, associative, and distributive properties to simplify algebraic expressions.

In Exercises 1–9, perform the indicated operation or simplify the given expression. Where possible, reduce fractional answers to lowest terms.

1. $15a + 14 + 9a - 13$

2. $\frac{7}{10} - \frac{8}{15}$

3. $\frac{2}{3} \cdot \frac{3}{4}$

4. $7(9x + 3) + \frac{1}{3}(6x - 15)$

5. $\dfrac{5}{22} + \dfrac{5}{33}$

6. $\dfrac{3}{5} \div \dfrac{9}{10}$

7. $\dfrac{23}{105} - \dfrac{2}{105}$

8. $2\dfrac{7}{9} \div 3$

9. $5\dfrac{2}{9} - 3\dfrac{1}{6}$

10. Plot $\left(3, -5\dfrac{1}{2}\right)$ in a rectangular coordinate system. In which quadrant does the point lie?

In Exercises 11–13, rewrite $5(x + 3)$ as an equivalent expression using the given property.

11. the commutative property of multiplication

12. the commutative property of addition

13. the distributive property

Use the bar graph to solve Exercises 14–15.

Percentage of Americans Ages 12 and Older Listening to Internet Radio

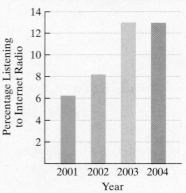

Source: Edison Media Research

14. Find an estimate for the difference between the percentage of Americans listening to Internet radio in 2004 and in 2001.

15. For which years shown did the percentage of Americans listening to Internet radio exceed 7%?

16. Insert either $<$ or $>$ in the shaded area to make a true statement:

$$-8000 \quad\quad -8\dfrac{1}{4}.$$

17. List all the rational numbers in this set:

$$\left\{-11, -\dfrac{3}{7}, 0, 0.45, \sqrt{23}, \sqrt{25}\right\}.$$

In Exercises 18–19, evaluate each algebraic expression for the given value of the variable.

18. $10x + 7; \; x = \dfrac{3}{5}$

19. $8(x - 2); \; x = \dfrac{5}{2}$

Use the graphs to solve Exercises 20–22.

Grades of U.S. Undergraduate College Students

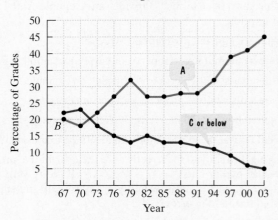

Source: UCLA Higher Education Research Institute

20. What are the coordinates of point B? What does this mean in terms of the information given by the graph?

21. For the period shown, when did the percentage of grades consisting of C or below reach a maximum? What is a reasonable estimate of the percentage of grades in this category for that year?

22. Find the difference between the percentage of A's in 2003 and the percentage of C's or below in 2003.

23. Find the absolute value: $|-19.3|$.

24. Express $\dfrac{1}{11}$ as a decimal.

25. Rewrite without parentheses:

$$8(7x - 10 + 3y).$$

SECTION

1.5

Objectives

1 Add numbers with a number line.

2 Find sums using identity and inverse properties.

3 Add numbers without a number line.

4 Use addition rules to simplify algebraic expressions.

5 Solve applied problems using a series of additions.

ADDITION OF REAL NUMBERS

It has not been a good day! First, you lost your wallet with $30 in it. Then, to get through the day, you borrowed $10, which you somehow misplaced. Your loss of $30 followed by a loss of $10 is an overall loss of $40. This can be written

$$-30 + (-10) = -40.$$

The result of adding two or more numbers is called the **sum** of the numbers. The sum of -30 and -10 is -40. You can think of gains and losses of money to find sums. For example, to find $17 + (-13)$, think of a gain of $17 followed by a loss of $13. There is an overall gain of $4. Thus, $17 + (-13) = 4$. In the same way, to find $-17 + 13$, think of a loss of $17 followed by a gain of $13. There is an overall loss of $4, so $-17 + 13 = -4$.

1 Add numbers with a number line.

Adding with a Number Line We use the number line to help picture the addition of real numbers. Here is the procedure for finding $a + b$, the sum of a and b, using the number line:

USING THE NUMBER LINE TO FIND A SUM Let a and b represent real numbers. To find $a + b$ using a number line,

1. Start at a.

2. a. If b is **positive**, move b units to the **right**.

   b. If b is **negative**, move b units to the **left**.

   c. If b is **0**, **stay** at a.

3. The number where we finish on the number line represents the sum of a and b.

This procedure is illustrated in Examples 1 and 2. Think of moving to the right as a gain and moving to the left as a loss.

EXAMPLE 1 Adding Real Numbers Using a Number Line

Find the sum using a number line:

$$3 + (-5).$$

SOLUTION   We find $3 + (-5)$ using the number line in Figure 1.19.

Step 1.   We consider 3 to be the first number, represented by a in the preceding box. We start at a, or 3.

Step 2.   We consider -5 to be the second number, represented by b. Because this number is negative, we move 5 units to the left.

Step 3.   We finish at -2 on the number line. The number where we finish represents the sum of 3 and -5. Thus,

$$3 + (-5) = -2.$$

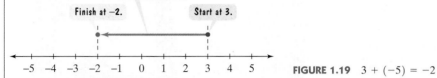

FIGURE 1.19 $3 + (-5) = -2$

Observe that if there is a gain of \$3 followed by a loss of \$5, there is an overall loss of \$2. ■

 CHECK POINT 1 Find the sum using a number line:

$$4 + (-7).$$

EXAMPLE 2   Adding Real Numbers Using a Number Line

Find each sum using a number line: **a.** $-3 + (-4)$ **b.** $-6 + 2$.

SOLUTION

a. To find $-3 + (-4)$, start at -3. Move 4 units to the left. We finish at -7. Thus,

$$-3 + (-4) = -7.$$

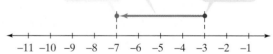

Observe that if there is a loss of \$3 followed by a loss of \$4, there is an overall loss of \$7.

b. To find $-6 + 2$, start at -6. Move 2 units to the right because 2 is positive. We finish at -4. Thus,

$$-6 + 2 = -4.$$

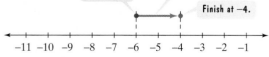

Observe that if there is a loss of \$6 followed by a gain of \$2, there is an overall loss of \$4. ■

 CHECK POINT 2 Find each sum using a number line:

a. $-1 + (-3)$ **b.** $-5 + 3$.

2 Find sums using identity and inverse properties.

The Number Line and Properties of Addition The number line can be used to picture some useful properties of addition. For example, let's see what happens if we add two numbers with different signs but the same absolute value. Two such numbers are 3 and −3. To find 3 + (−3) on a number line, we start at 3 and move 3 units to the left. We finish at 0. Thus,

$$3 + (-3) = 0.$$

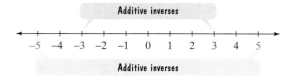

Numbers that are opposites, such as 3 and −3, are called *additive inverses*. **Additive inverses** are pairs of real numbers that are the same number of units from zero on the number line, but are on opposite sides of zero. Thus, −3 is the additive inverse of 3, and 5 is the additive inverse of −5. The additive inverse of 0 is 0. Other additive inverses come in pairs.

Additive inverses

-5 -4 -3 -2 -1 0 1 2 3 4 5

Additive inverses

In general, the sum of any real number, denoted by *a*, and its additive inverse, denoted by −*a*, is zero:

$$a + (-a) = 0.$$

This property is called the **inverse property of addition**. In Section 1.4, we discussed the commutative and associative properties of addition. We now add two additional properties to our previous list, shown in Table 1.4.

Table 1.4	**Identity and Inverse Properties of Addition**	

Let *a* be a real number, a variable, or an algebraic expression.

Property	Meaning	Examples
Identity Property of Addition	Zero can be deleted from a sum. $a + 0 = a$ $0 + a = a$	• $4 + 0 = 4$ • $-3x + 0 = -3x$ • $0 + (5a + b) = 5a + b$
Inverse Property of Addition	The sum of a real number and its additive inverse gives 0, the additive identity. $a + (-a) = 0$ $(-a) + a = 0$	• $6 + (-6) = 0$ • $3x + (-3x) = 0$ • $[-(2y + 1)] + (2y + 1) = 0$

3 Add numbers without a number line.

Adding without a Number Line Now that we can picture the addition of real numbers, we look at two rules for using absolute value to add signed numbers.

> ### ADDING TWO NUMBERS WITH THE SAME SIGN
> **1.** Add the absolute values.
> **2.** Use the common sign as the sign of the sum.

STUDY TIP

The sum of two positive numbers is always positive.
The sum of two negative numbers is always negative.

EXAMPLE 3 Adding Real Numbers

Add without using a number line:

a. $-11 + (-15)$ **b.** $-0.2 + (-0.8)$ **c.** $-\frac{3}{4} + \left(-\frac{1}{2}\right)$.

SOLUTION In each part of this example, we are adding numbers with the same sign.

a. $-11 + (-15) = -26$ — Add absolute values: $11 + 15 = 26$.

Use the common sign.

b. $-0.2 + (-0.8) = -1$ — Add absolute values: $0.2 + 0.8 = 1.0$ or 1.

Use the common sign.

c. $-\frac{3}{4} + \left(-\frac{1}{2}\right) = -\frac{5}{4}$ — Add absolute values: $\frac{3}{4} + \frac{1}{2} = \frac{3}{4} + \frac{2}{4} = \frac{5}{4}$.

Use the common sign.

✔ **CHECK POINT 3** Add without using a number line:

a. $-10 + (-25)$ **b.** $-0.3 + (-1.2)$ **c.** $-\frac{2}{3} + \left(-\frac{1}{6}\right)$.

We also use absolute value to add two real numbers with different signs.

ADDING TWO NUMBERS WITH DIFFERENT SIGNS
1. Subtract the smaller absolute value from the greater absolute value.
2. Use the sign of the number with the greater absolute value as the sign of the sum.

EXAMPLE 4 Adding Real Numbers

Add without using a number line: **a.** $-13 + 4$ **b.** $-0.2 + 0.8$ **c.** $-\frac{3}{4} + \frac{1}{2}$.

SOLUTION In each part of this example, we are adding numbers with different signs.

a. $-13 + 4 = -9$ — Subtract absolute values: $13 - 4 = 9$.

Use the sign of the number with the greater absolute value.

b. $-0.2 + 0.8 = 0.6$ — Subtract absolute values: $0.8 - 0.2 = 0.6$.

Use the sign of the number with the greater absolute value. The sign is assumed to be positive.

c. $-\frac{3}{4} + \frac{1}{2} = -\frac{1}{4}$ — Subtract absolute values: $\frac{3}{4} - \frac{1}{2} = \frac{3}{4} - \frac{2}{4} = \frac{1}{4}$.

Use the sign of the number with the greater absolute value.

USING TECHNOLOGY

You can use a calculator to add signed numbers. Here are the keystrokes for finding $-11 + (-15)$:

Scientific Calculator

11 $+/-$ + 15 $+/-$ =

Graphing Calculator

$(-)$ 11 + $(-)$ 15 ENTER

Here are the keystrokes for finding $-13 + 4$:

Scientific Calculator

13 $+/-$ + 4 =

Graphing Calculator

$(-)$ 13 + 4 ENTER

STUDY TIP

The sum of two numbers with different signs may be positive or negative. Keep in mind that the sign of the sum is the sign of the number with the greater absolute value.

✔ CHECK POINT 4 Add without using a number line:

  a. $-15 + 2$ **b.** $-0.4 + 1.6$ **c.** $-\dfrac{2}{3} + \dfrac{1}{6}$.

4 Use addition rules to simplify algebraic expressions.

Algebraic Expressions The rules for adding real numbers can be used to simplify certain algebraic expressions.

EXAMPLE 5 Simplifying Algebraic Expressions

Simplify:

  a. $-11x + 7x$

  b. $7y + (-12z) + (-9y) + 15z$

  c. $3(8 - 7x) + 5(2x - 4)$.

SOLUTION

  a. $-11x + 7x$ The given algebraic expression has two like terms. $-11x$ and $7x$ have identical variable factors.

  $= (-11 + 7)x$ Apply the distributive property.

  $= -4x$ Add within parentheses: $-11 + 7 = -4$.

  b. $7y + (-12z) + (-9y) + 15z$ The colors indicate that there are two pairs of like terms.

  $= 7y + (-9y) + (-12z) + 15z$ Arrange like terms so that they are next to one another.

  $= [7 + (-9)]y + [(-12) + 15]z$ Apply the distributive property.

  $= -2y + 3z$ Add within the grouping symbols: $7 + (-9) = -2$ and $-12 + 15 = 3$.

  c. $3(8 - 7x) + 5(2x - 4)$

  $= 3 \cdot 8 - 3 \cdot 7x + 5 \cdot 2x - 5 \cdot 4$ Use the distributive property to remove the parentheses.

  $= 24 - 21x + 10x - 20$ Multiply.

  $= (-21x + 10x) + (24 - 20)$ Group like terms.

  $= (-21 + 10)x + (24 - 20)$ Apply the distributive property.

  $= -11x + 4$ Perform operations within parentheses: $-21 + 10 = -11$ and $24 - 20 = 4$. ∎

✔ CHECK POINT 5 Simplify:

  a. $-20x + 3x$ **b.** $3y + (-10z) + (-10y) + 16z$

  c. $4(10 - 8x) + 4(3x - 5)$.

5 Solve applied problems using a series of additions.

Applications Positive and negative numbers are used in everyday life to represent such things as gains and losses in the stock market, rising and falling temperatures, deposits and withdrawals on bank statements, and ascending and descending motion. Positive and negative numbers are used to solve applied problems involving a series of additions.

One way to add a series of positive and negative numbers is to use the commutative and associative properties.

- Add all the positive numbers.
- Add all the negative numbers.
- Add the sums obtained in the first two steps.

The next example illustrates this idea.

EXAMPLE 6 An Application of Adding Signed Numbers

A glider was towed 1000 meters into the air and then let go. It descended 70 meters into a thermal (rising bubble of warm air), which took it up 2100 meters. At this point it dropped 230 meters into a second thermal. Then it rose 1200 meters. What was its altitude at that point?

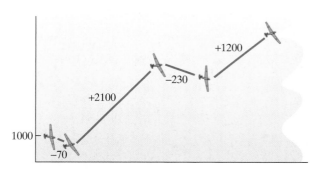

SOLUTION   We use the problem's conditions to write a sum. The altitude of the glider is expressed by the following sum:

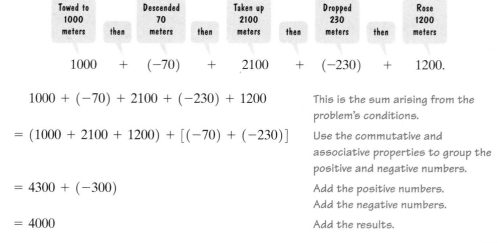

Towed to 1000 meters	then	Descended 70 meters	then	Taken up 2100 meters	then	Dropped 230 meters	then	Rose 1200 meters

$$1000 \ + \ (-70) \ + \ 2100 \ + \ (-230) \ + \ 1200.$$

$1000 + (-70) + 2100 + (-230) + 1200$ This is the sum arising from the problem's conditions.

$= (1000 + 2100 + 1200) + [(-70) + (-230)]$ Use the commutative and associative properties to group the positive and negative numbers.

$= 4300 + (-300)$ Add the positive numbers.
Add the negative numbers.

$= 4000$ Add the results.

The altitude of the glider is 4000 meters. ∎

DISCOVER FOR YOURSELF

Try working Example 6 by adding from left to right. You should still obtain 4000 for the sum. Which method do you find easier?

✔ **CHECK POINT 6** The water level of a reservoir is measured over a five-month period. During this time, the level rose 2 feet, then fell 4 feet, then rose 1 foot, then fell 5 feet, and then rose 3 feet. What was the change in the water level at the end of the five months?

1.5 EXERCISE SET

Student Solutions Manual   CD/Video   PH Math/Tutor Center   MathXL Tutorials on CD   Math XL MathXL®   MyMathLab MyMathLab   Interactmath.com

Practice Exercises

In Exercises 1–8, find each sum using a number line.

1. $7 + (-3)$ **2.** $7 + (-2)$ **3.** $-2 + (-5)$

4. $-1 + (-5)$ **5.** $-6 + 2$ **6.** $-8 + 3$

7. $3 + (-3)$ **8.** $5 + (-5)$

In Exercises 9–46, find each sum without the use of a number line.

9. $-7 + 0$ **10.** $-5 + 0$

11. $30 + (-30)$ **12.** $15 + (-15)$

13. $-30 + (-30)$ **14.** $-15 + (-15)$

15. $-8 + (-10)$ **16.** $-4 + (-6)$

17. $-0.4 + (-0.9)$ **18.** $-1.5 + (-5.3)$

19. $-\dfrac{7}{10} + \left(-\dfrac{3}{10}\right)$ **20.** $-\dfrac{7}{8} + \left(-\dfrac{1}{8}\right)$

21. $-9 + 4$ **22.** $-7 + 3$

23. $12 + (-8)$ **24.** $13 + (-5)$

25. $6 + (-9)$ **26.** $3 + (-11)$

27. $-3.6 + 2.1$ **28.** $-6.3 + 5.2$

29. $-3.6 + (-2.1)$ **30.** $-6.3 + (-5.2)$

31. $\dfrac{9}{10} + \left(-\dfrac{3}{5}\right)$ **32.** $\dfrac{7}{10} + \left(-\dfrac{2}{5}\right)$

33. $-\dfrac{5}{8} + \dfrac{3}{4}$ **34.** $-\dfrac{5}{6} + \dfrac{1}{3}$

35. $-\dfrac{3}{7} + \left(-\dfrac{4}{5}\right)$ **36.** $-\dfrac{3}{8} + \left(-\dfrac{2}{3}\right)$

37. $4 + (-7) + (-5)$ **38.** $10 + (-3) + (-8)$

39. $85 + (-15) + (-20) + 12$

40. $60 + (-50) + (-30) + 25$

41. $17 + (-4) + 2 + 3 + (-10)$

42. $19 + (-5) + 1 + 8 + (-13)$

43. $-45 + \left(-\dfrac{3}{7}\right) + 25 + \left(-\dfrac{4}{7}\right)$

44. $-50 + \left(-\dfrac{7}{9}\right) + 35 + \left(-\dfrac{11}{9}\right)$

45. $3.5 + (-45) + (-8.4) + 72$

46. $6.4 + (-35) + (-2.6) + 14$

In Exercises 47–64, simplify each algebraic expression.

47. $-10x + 2x$ **48.** $-19x + 10x$

49. $25y + (-12y)$ **50.** $26y + (-14y)$

51. $-8a + (-15a)$ **52.** $-9a + (-13a)$

53. $-4 + 7x + 5 + (-13x)$

54. $-5 + 8x + 3 + (-16x)$

55. $7b + 2 + (-b) + (-6)$

56. $10b + 7 + (-b) + (-15)$

57. $7x + (-5y) + (-9x) + 2y$

58. $13x + (-9y) + (-11x) + 3y$

59. $4(5x - 3) + 6$

60. $5(2x - 3) + 4$

61. $8(3 - 4y) + 35y$

62. $7(5 - 3y) + 25y$

63. $6(2 - 9a) + 7(3a + 5)$

64. $8(3 - 7a) + 4(2a + 3)$

Practice Plus

In Exercises 65–68, find each sum.

65. $|-3 + (-5)| + |2 + (-6)|$

66. $|4 + (-11)| + |-3 + (-4)|$

67. $-20 + [-|15 + (-25)|]$

68. $-25 + [-|18 + (-26)|]$

In Exercises 69–70, insert either $<, >,$ *or* $=$ *in the shaded area to make a true statement.*

69. $6 + [2 + (-13)]$ ▨ $-3 + [4 + (-8)]$

70. $[(-8) + (-6)] + 10$ ▨ $-8 + [9 + (-2)]$

Application Exercises

Solve Exercises 71–80 by writing a sum of signed numbers and adding.

71. The greatest temperature variation recorded in a day is 100 degrees in Browning, Montana, on January 23, 1916. The low temperature was $-56°F$. What was the high temperature?

72. In Spearfish, South Dakota, on January 22, 1943, the temperature rose 49 degrees in two minutes. If the initial temperature was $-4°F$, what was the high temperature?

73. The Dead Sea is the lowest elevation on earth, 1312 feet below sea level. What is the elevation of a person standing 712 feet above the Dead Sea?

74. Lake Assal in Africa is 512 feet below sea level. What is the elevation of a person standing 642 feet above Lake Assal?

75. The temperature at 8:00 A.M. was $-7°F$. By noon it had risen 15°F, but by 4:00 P.M. it had fallen 5°F. What was the temperature at 4:00 P.M.?

76. On three successive plays, a football team lost 15 yards, gained 13 yards, and then lost 4 yards. What was the team's total gain or loss for the three plays?

77. A football team started with the football at the 27-yard line, advancing toward the center of the field (the 50-yard line). Four successive plays resulted in a 4-yard gain, a 2-yard loss, an 8-yard gain, and a 12-yard loss. What was the location of the football at the end of the fourth play?

78. The water level of a reservoir is measured over a five-month period. At the beginning, the level is 20 feet. During this time, the level rose 3 feet, then fell 2 feet, then fell 1 foot, then fell 4 feet, and then rose 2 feet. What is the reservoir's water level at the end of the five months?

79. The bar graph shows the budget surplus or deficit for the United States government from 1999 through 2003.

U.S. Government Budget Surplus/Deficit

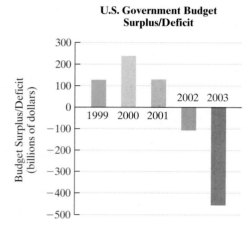

Source: Budget of the U.S. Government

In 1999, the U.S. government had a budget surplus of $126 billion. This surplus then changed as follows:

    2000: increased by $110 billion

    2001: decreased by $109 billion

    2002: decreased by $233 billion

    2003: decreased by $349 billion.

What was the government's budget surplus or deficit in 2003?

80. According to *Newsweek*, the Ford Motor Company "swung from most profitable carmaker to biggest loser." The bar graph in the next column shows Ford's net profits and losses (the sum of income and expenses) from 1998 through 2002. In 1998, Ford's net profit was $22 billion. This profit then changed as follows:

    1999: decreased by $16 billion

    2000: decreased by $3 billion

    2001: decreased by $8 billion

    2002: increased by $4 billion.

What was the company's net profit or loss in 2002?

Net Profits and Losses of the Ford Motor Company

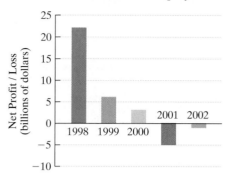

Source: Newsweek

Writing in Mathematics

81. Explain how to add two numbers with a number line. Provide an example with your explanation.

82. What are additive inverses?

83. Describe how the inverse property of addition

$$a + (-a) = 0$$

can be shown on a number line.

84. Without using a number line, describe how to add two numbers with the same sign. Give an example.

85. Without using a number line, describe how to add two numbers with different signs. Give an example.

86. Write a problem that can be solved by finding the sum of at least three numbers, some positive and some negative. Then explain how to solve the problem.

87. Without a calculator, you can add numbers using a number line, using absolute value, or using gains and losses. Which method do you find most helpful? Why is this so?

Critical Thinking Exercises

88. Which one of the following statements is true?

  a. The sum of a positive number and a negative number is a negative number.

  b. $|-9 + 2| = 9 + 2$

  c. If two numbers are both positive or both negative, then the absolute value of their sum equals the sum of their absolute values.

  d. $\dfrac{3}{4} + \left(-\dfrac{3}{5}\right) = -\dfrac{3}{20}$

89. Which one of the following statements is true?

  a. The sum of a positive number and a negative number is a positive number.

  b. If one number is positive and the other negative, then the absolute value of their sum equals the sum of their absolute values.

  c. $\dfrac{3}{4} + \left(-\dfrac{2}{3}\right) = -\dfrac{1}{12}$

  d. The sum of zero and a negative number is always a negative number.

In Exercises 90–91, find the missing term.

90. $5x +$ __ $+ (-11x) + (-6y) = -6x + 2y$

91. __ $+ 11x + (-3y) + 3x = 7(2x - 3y)$

Technology Exercises

92. Use a calculator to verify any five of the sums that you found in Exercises 17–46.

93. Use a calculator to verify any three of the answers that you obtained in Application Exercises 71–80.

Review Exercises

94. Determine whether this inequality is true or false: $19 \geq -18$. (Section 1.2, Example 7)

95. Consider the set

$$\{-6, -\pi, 0, 0.\overline{7}, \sqrt{3}, \sqrt{4}\}.$$

List all numbers from the set that are **a.** natural numbers, **b.** whole numbers, **c.** integers, **d.** rational numbers, **e.** irrational numbers, **f.** real numbers. (Section 1.2, Example 5)

96. Plot $\left(\frac{7}{2}, -\frac{5}{2}\right)$ in a rectangular coordinate system. In which quadrant does the point lie? (Section 1.3, Example 1)

SECTION **1.6**

SUBTRACTION OF REAL NUMBERS

Objectives

1 Subtract real numbers.

2 Simplify a series of additions and subtractions.

3 Use the definition of subtraction to identify terms.

4 Use the subtraction definition to simplify algebraic expressions.

5 Solve problems involving subtraction.

People are going to live longer in the 21st century. This will put added pressure on the Social Security and Medicare systems. How insecure is Social Security's future? In this section, we use subtraction of real numbers to numerically describe one aspect of the insecurity.

The Meaning of Subtraction Time for a new computer! Your favorite model, which normally sells for $1500, has an incredible price reduction of $600. The computer's reduced price, $900, can be expressed in two ways:

$$1500 - 600 = 900 \quad \text{or} \quad 1500 + (-600) = 900.$$

This means that

$$1500 - 600 = 1500 + (-600).$$

To subtract 600 from 1500, we add 1500 and the additive inverse of 600. Generalizing from this situation, we define subtraction as follows:

> **DEFINITION OF SUBTRACTION** For all real numbers a and b,
> $$a - b = a + (-b).$$
>
> In words: To subtract b from a, add the additive inverse of b to a. The result of subtraction is called the **difference**.

1 Subtract real numbers.

A Procedure for Subtracting Real Numbers The definition of subtraction gives us a procedure for subtracting real numbers.

> **SUBTRACTING REAL NUMBERS**
> **1.** Change the subtraction operation to addition.
> **2.** Change the sign of the number being subtracted.
> **3.** Add, using one of the rules for adding numbers with the same sign or different signs.

USING TECHNOLOGY

You can use a calculator to subtract signed numbers. Here are the keystrokes for finding 5 − (−6):

Scientific Calculator

5 − 6 +/− =

Graphing Calculator

5 − (−) 6 ENTER

Here are the keystrokes for finding −9 − (−3):

Scientific Calculator

9 +/− − 3 +/− =

Graphing Calculator

(−) 9 − (−) 3 ENTER

Don't confuse the subtraction key on a graphing calculator, −, with the sign change or additive inverse key, (−). What happens if you do?

EXAMPLE 1 Using the Definition of Subtraction

Subtract: **a.** $7 - 10$ **b.** $5 - (-6)$ **c.** $-9 - (-3)$.

SOLUTION

a. $7 - 10 = 7 + (-10) = -3$

> Change the subtraction to addition. Replace 10 with its additive inverse.

b. $5 - (-6) = 5 + 6 = 11$

> Change the subtraction to addition. Replace −6 with its additive inverse.

c. $-9 - (-3) = -9 + 3 = -6$

> Change the subtraction to addition. Replace −3 with its additive inverse.

 CHECK POINT 1 Subtract:

a. $3 - 11$ **b.** $4 - (-5)$ **c.** $-7 - (-2)$.

The definition of subtraction can be applied to real numbers that are not integers.

EXAMPLE 2 Using the Definition of Subtraction

Subtract: **a.** $-5.2 - (-11.4)$ **b.** $-\dfrac{3}{4} - \dfrac{2}{3}$ **c.** $4\pi - (-9\pi)$.

SOLUTION

a. $-5.2 - (-11.4) = -5.2 + 11.4 = 6.2$

> Change the subtraction to addition. Replace −11.4 with its additive inverse.

b. $-\dfrac{3}{4} - \dfrac{2}{3} = -\dfrac{3}{4} + \left(-\dfrac{2}{3}\right) = -\dfrac{9}{12} + \left(-\dfrac{8}{12}\right) = -\dfrac{17}{12}$

> Change the subtraction to addition. Replace $\frac{2}{3}$ with its additive inverse.

c. $4\pi - (-9\pi) = 4\pi + 9\pi = (4 + 9)\pi = 13\pi$

> Change the subtraction to addition. Replace −9π with its additive inverse.

Reading the symbol "−" can be a bit tricky. The way you read it depends on where it appears. For example,

$$-5.2 - (-11.4)$$

is read "negative five point two minus negative eleven point four." Read parts (b) and (c) of Example 2 aloud. When is "−" read "negative" and when is it read "minus"?

 CHECK POINT 2 Subtract:

a. $-3.4 - (-12.6)$ b. $-\dfrac{3}{5} - \dfrac{1}{3}$ c. $5\pi - (-2\pi)$.

2 Simplify a series of additions and subtractions.

Problems Containing a Series of Additions and Subtractions In some problems, several additions and subtractions occur together. We begin by converting all subtractions to additions of additive inverses, or opposites.

> **SIMPLIFYING A SERIES OF ADDITIONS AND SUBTRACTIONS**
>
> 1. Change all subtractions to additions of additive inverses.
> 2. Group and then add all the positive numbers.
> 3. Group and then add all the negative numbers.
> 4. Add the results of steps 2 and 3.

EXAMPLE 3 Simplifying a Series of Additions and Subtractions

Simplify: $7 - (-5) - 11 - (-6) - 19$.

SOLUTION

$$7 - (-5) - 11 - (-6) - 19$$

$= 7 + 5 + (-11) + 6 + (-19)$ Write subtractions as additions of additive inverses.

$= (7 + 5 + 6) + [(-11) + (-19)]$ Group the positive numbers. Group the negative numbers.

$= 18 + (-30)$ Add the positive numbers. Add the negative numbers.

$= -12$ Add the results. ■

 CHECK POINT 3 Simplify: $10 - (-12) - 4 - (-3) - 6$.

3 Use the definition of subtraction to identify terms.

Subtraction and Algebraic Expressions We know that the terms of an algebraic expression are separated by addition signs. Let's use this idea to identify the terms of the following algebraic expression:

$$9x - 4y - 5.$$

Because terms are separated by addition, we rewrite the algebraic expression as additions of additive inverses, or opposites. Thus,

$$9x - 4y - 5 = 9x + (-4y) + (-5).$$

The three terms of the algebraic expression are $9x$, $-4y$, and -5.

EXAMPLE 4 Using the Definition of Subtraction to Identify Terms

Identify the terms of the algebraic expression:

$$2xy - 13y - 6.$$

SOLUTION   Rewrite the algebraic expression as additions of additive inverses.

$$2xy - 13y - 6 = 2xy + (-13y) + (-6)$$

First term   Second term   Third term

Because terms are separated by addition, the terms are $2xy$, $-13y$, and -6. ■

CHECK POINT 4 Identify the terms of the algebraic expression:

$$-6 + 4a - 7ab.$$

 4 Use the subtraction definition to simplify algebraic expressions.

The procedure for subtracting real numbers can be used to simplify certain algebraic expressions that involve subtraction.

EXAMPLE 5 Simplifying Algebraic Expressions

Simplify: **a.** $2 + 3x - 8x$ **b.** $-4x - 9y - 2x + 12y.$

SOLUTION

a. $2 + 3x - 8x$ This is the given algebraic expression.

$= 2 + 3x + (-8x)$ Write the subtraction as the addition of an additive inverse.

$= 2 + [3 + (-8)]x$ Apply the distributive property.

$= 2 + (-5x)$ Add within the grouping symbols

$= 2 - 5x$ Be concise and express as subtraction.

b. $-4x - 9y - 2x + 12y$ This is the given algebraic expression.

$= -4x + (-9y) + (-2x) + 12y$ Write the subtractions as the additions of additive inverses.

$= -4x + (-2x) + (-9y) + 12y$ Arrange like terms so that they are next to one another.

$= [-4 + (-2)]x + (-9 + 12)y$ Apply the distributive property.

$= -6x + 3y$ Add within the grouping symbols. ■

STUDY TIP

You can think of gains and losses of money to work the distributive property mentally:

• $3x - 8x = -5x$ A gain of 3 dollars followed by a loss of 8 of those dollars is a net loss of 5 of those dollars.

• $-9y + 12y = 3y$ A loss of 9 dollars followed by a gain of 12 of those dollars is a net gain of 3 of those dollars.

 CHECK POINT 5 Simplify:

a. $4 + 2x - 9x$ **b.** $-3x - 10y - 6x + 14y.$

5 Solve problems involving subtraction.

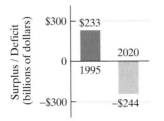

FIGURE 1.20 Social Security Annual Surplus/Deficit

Source: U.S. Office of Management and Budget

Applications Subtraction is used to solve problems in which the word "difference" appears. The difference between real numbers a and b is expressed as $a - b$.

EXAMPLE 6 An Application of Subtraction Using the Word "Difference"

The bar graph in Figure 1.20 shows that in 1995, Social Security had an annual cash surplus of $233 billion. By 2020, this amount is expected to be a negative number—a deficit of $244 billion. What is the difference between the 1995 surplus and the projected 2020 deficit?

SOLUTION

| The difference | is | the 1995 surplus | minus | the 2020 deficit. |

$$= \quad 233 \quad - \quad (-244)$$
$$= \quad 233 \quad + \quad 244 \quad = \quad 477$$

The difference between the 1995 surplus and the projected 2020 deficit is $477 billion.

 CHECK POINT 6 The peak of Mount Everest is 8848 meters above sea level. The Marianas Trench, on the floor of the Pacific Ocean, is 10,915 meters below sea level. What is the difference in elevation between the peak of Mount Everest and the Marianas Trench?

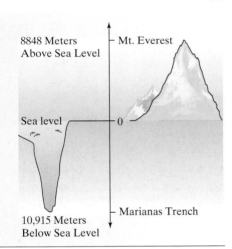

8848 Meters Above Sea Level — Mt. Everest

Sea level — 0

— Marianas Trench

10,915 Meters Below Sea Level

1.6 EXERCISE SET

 Student Solutions Manual   CD/Video   PH Math/Tutor Center   MathXL Tutorials on CD   MathXL® MyMathLab   Interactmath.com

Practice Exercises

1. Consider the subtraction $5 - 12$.
  a. Find the additive inverse, or opposite, of 12.
  b. Rewrite the subtraction as the addition of the additive inverse of 12.

2. Consider the subtraction $4 - 10$.
  a. Find the additive inverse, or opposite, of 10.
  b. Rewrite the subtraction as the addition of the additive inverse of 10.

3. Consider the subtraction $5 - (-7)$.
  a. Find the additive inverse, or opposite, of -7.
  b. Rewrite the subtraction as the addition of the additive inverse of -7.

4. Consider the subtraction $2 - (-8)$.
  a. Find the additive inverse, or opposite, of -8.
  b. Rewrite the subtraction as the addition of the additive inverse of -8.

In Exercises 5–50, perform the indicated subtraction.

5. $14 - 8$

6. $15 - 2$

7. $8 - 14$

8. $2 - 15$

9. $3 - (-20)$

10. $5 - (-17)$

11. $-7 - (-18)$

12. $-5 - (-19)$

13. $-13 - (-2)$

14. $-21 - (-3)$

15. $-21 - 17$

16. $-29 - 21$

17. $-45 - (-45)$

18. $-65 - (-65)$

19. $23 - 23$

20. $26 - 26$

21. $13 - (-13)$

22. $15 - (-15)$

23. $0 - 13$

24. $0 - 15$

25. $0 - (-13)$

26. $0 - (-15)$

27. $\dfrac{3}{7} - \dfrac{5}{7}$

28. $\dfrac{4}{9} - \dfrac{7}{9}$

29. $\dfrac{1}{5} - \left(-\dfrac{3}{5}\right)$

30. $\dfrac{1}{7} - \left(-\dfrac{3}{7}\right)$

31. $-\dfrac{4}{5} - \dfrac{1}{5}$

32. $-\dfrac{4}{9} - \dfrac{1}{9}$

33. $-\dfrac{4}{5} - \left(-\dfrac{1}{5}\right)$

34. $-\dfrac{4}{9} - \left(-\dfrac{1}{9}\right)$

35. $\dfrac{1}{2} - \left(-\dfrac{1}{4}\right)$

36. $\dfrac{2}{5} - \left(-\dfrac{1}{10}\right)$

37. $\dfrac{1}{2} - \dfrac{1}{4}$

38. $\dfrac{2}{5} - \dfrac{1}{10}$

39. $9.8 - 2.2$

40. $5.7 - 3.3$

41. $-3.1 - (-1.1)$

42. $-4.6 - (-1.1)$

43. $1.3 - (-1.3)$

44. $1.4 - (-1.4)$

45. $-2.06 - (-2.06)$

46. $-3.47 - (-3.47)$

47. $5\pi - 2\pi$

48. $9\pi - 7\pi$

49. $3\pi - (-10\pi)$

50. $4\pi - (-12\pi)$

In Exercises 51–68, simplify each series of additions and subtractions.

51. $13 - 2 - (-8)$

52. $14 - 3 - (-7)$

53. $9 - 8 + 3 - 7$

54. $8 - 2 + 5 - 13$

55. $-6 - 2 + 3 - 10$

56. $-9 - 5 + 4 - 17$

57. $-10 - (-5) + 7 - 2$

58. $-6 - (-3) + 8 - 11$

59. $-23 - 11 - (-7) + (-25)$

60. $-19 - 8 - (-6) + (-21)$

61. $-823 - 146 - 50 - (-832)$

62. $-726 - 422 - 921 - (-816)$

63. $1 - \dfrac{2}{3} - \left(-\dfrac{5}{6}\right)$

64. $2 - \dfrac{3}{4} - \left(-\dfrac{7}{8}\right)$

65. $-0.16 - 5.2 - (-0.87)$

66. $-1.9 - 3 - (-0.26)$

67. $-\dfrac{3}{4} - \dfrac{1}{4} - \left(-\dfrac{5}{8}\right)$

68. $-\dfrac{1}{2} - \dfrac{2}{3} - \left(-\dfrac{1}{3}\right)$

In Exercises 69–72, identify the terms in each algebraic expression.

69. $-3x - 8y$

70. $-9a - 4b$

71. $12x - 5xy - 4$

72. $8a - 7ab - 13$

In Exercises 73–84, simplify each algebraic expression.

73. $3x - 9x$

74. $2x - 10x$

75. $4 + 7y - 17y$

76. $5 + 9y - 29y$

77. $2a + 5 - 9a$

78. $3a + 7 - 11a$

79. $4 - 6b - 8 - 3b$

80. $5 - 7b - 13 - 4b$

81. $13 - (-7x) + 4x - (-11)$

82. $15 - (-3x) + 8x - (-10)$

83. $-5x - 10y - 3x + 13y$

84. $-6x - 9y - 4x + 15y$

Practice Plus

In Exercises 85–90, find the value of each expression.

85. $-|-9 - (-6)| - (-12)$

86. $-|-8 - (-2)| - (-6)$

87. $\dfrac{5}{8} - \left(\dfrac{1}{2} - \dfrac{3}{4}\right)$

88. $\dfrac{9}{10} - \left(\dfrac{1}{4} - \dfrac{7}{10}\right)$

89. $|-9 - (-3 + 7)| - |-17 - (-2)|$

90. $|24 - (-16)| - |-51 - (-31 + 2)|$

Application Exercises

91. The peak of Mount Kilimanjaro, the highest point in Africa, is 19,321 feet above sea level. Qattara Depression, Egypt, one of the lowest points in Africa, is 436 feet below sea level. What is the difference in elevation between the peak of Mount Kilimanjaro and the Qattara Depression?

92. The peak of Mount Whitney is 14,494 feet above sea level. Mount Whitney can be seen directly above Death Valley, which is 282 feet below sea level. What is the difference in elevation between these geographic locations?

The bar graph shows the occupations with the greatest projected growth and the greatest projected losses from 2000 through 2010. Use the graph to solve Exercises 93–96.

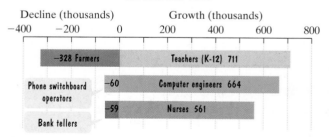

Projected Changes in the Number of Jobs: 2000–2010

Source: U.S. Bureau of Labor Statistics

93. What is the difference, in thousands of jobs, between the projected increase in teachers' jobs and the projected decrease in farmers' jobs?

94. What is the difference, in thousands of jobs, between the projected increase in computer engineering jobs and the projected decrease in jobs for phone switchboard operators?

95. By how many thousands of jobs does the decline for phone switchboard operators exceed the decline for farmers?

96. By how many thousands of jobs does the decline for bank tellers exceed the decline for farmers?

Do you enjoy cold weather? If so, try Fairbanks, Alaska. The average daily low temperature for each month in Fairbanks is shown in the bar graph. Use the graph to solve Exercises 97–100.

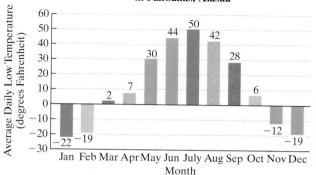

Each Month's Average Daily Low Temperature in Fairbanks, Alaska

Source: The Weather Channel Enterprises, Inc.

97. What is the difference between the average daily low temperatures for March and February?

98. What is the difference between the average daily low temperatures for October and November?

99. How many degrees warmer is February's average low temperature than January's average low temperature?

100. How many degrees warmer is November's average low temperature than December's average low temperature?

When a person receives a drug injected into a muscle, the concentration of the drug in the body depends on the time elapsed since the injection. The points in the rectangular system show the concentration of the drug, measured in milligrams per 100 milliliters, from the time the drug is injected until 13 hours later. Use this information to solve Exercises 101–106.

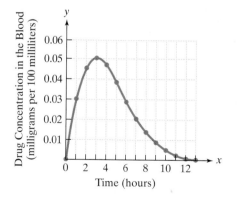

101. What is the drug's maximum concentration and when does this occur?

102. What happens by the end of 13 hours?

103. What is the approximate difference between the drug's concentration 4 hours after it was injected and 1 hour after it was injected?

104. What is the approximate difference between the drug's concentration 4 hours after it was injected and 7 hours after it was injected?

105. When is the drug's concentration increasing?

106. When is the drug's concentration decreasing?

Writing in Mathematics

107. Explain how to subtract real numbers.

108. How is $4 - (-2)$ read?

109. Explain how to simplify a series of additions and subtractions. Provide an example with your explanation.

110. Explain how to find the terms of the algebraic expression $5x - 2y - 7$.

111. Write a problem that can be solved by finding the difference between two numbers. At least one of the numbers should be negative. Then explain how to solve the problem.

Critical Thinking Exercises

112. Which one of the following statements is true?
   a. If a and b are negative numbers, then $a - b$ is a negative number.
   b. $7 - (-2) = 5$
   c. The difference between 0 and a negative number is always a positive number.
   d. None of the given statements is true.

113. The golden age of Athens culminated in 212 B.C. and the golden age of India culminated in A.D. 500. Determine the number of years that elapsed between these dates. (*Note:* When the calendar was reformed, the number 0 had not been invented. There was no year 0 and the year A.D. 1 followed the year 1 B.C. Calculate the difference between the years in the usual way and then use this added bit of information to modify your answer.)

114. Find the value:
$$-1 + 2 - 3 + 4 - 5 + 6 - \cdots - 99 + 100.$$

Technology Exercises

115. Use a calculator to verify any five of the differences that you found in Exercises 5–46.

116. Use a calculator to verify any three of the answers that you found in Exercises 51–68.

Review Exercises

117. Graph on a number line: -4.5. (Section 1.2; Example 3)

118. Use the commutative property of addition to write an equivalent algebraic expression: $10(a + 4)$. (Section 1.4; Example 4)

119. Give an example of an integer that is not a natural number (Section 1.2; Example 5)

SECTION 1.7

Objectives

1 Multiply real numbers.

2 Multiply more than two real numbers.

3 Find multiplicative inverses.

4 Use the definition of division.

5 Divide real numbers.

6 Simplify algebraic expressions involving multiplication.

7 Use algebraic expressions that model reality.

MULTIPLICATION AND DIVISION OF REAL NUMBERS

Technology is now promising to bring light, fast, and beautiful wheelchairs to millions of disabled people. The cost of manufacturing these radically different wheelchairs can be modeled by an algebraic expression containing division. In this section, we will see how this algebraic expression illustrates that low prices are possible with high production levels, urgently needed in this situation. There are more than half a billion people with disabilities in developing countries; an estimated 20 million need wheelchairs right now.

Multiplying Real Numbers   Suppose that things go from bad to worse for Social Security, and the projected $244 billion deficit in 2020 triples by the end of the twenty-first century. The new deficit is

$$3(-244) = (-244) + (-244) + (-244) = -732$$

or $732 billion. Thus,

$$3(-244) = -732.$$

The result of the multiplication, -732, is called the **product** of 3 and -244. The numbers being multiplied, 3 and -244, are called the **factors** of the product.

Rules for multiplying real numbers are described in terms of absolute value. For example, $3(-244) = -732$ illustrates that the product of numbers with different signs is found by multiplying their absolute values. The product is negative.

$$3(-244) \;=\; -732$$

> Factors have different signs and the product is negative.

> Multiply absolute values:
> $|3| \cdot |-244| = 3 \cdot 244 = 732.$

The following rules are used to determine the sign of the product of two numbers:

1 Multiply real numbers.

THE PRODUCT OF TWO REAL NUMBERS

- The product of two real numbers with **different signs** is found by multiplying their absolute values. The product is **negative**.
- The product of two real numbers with the **same sign** is found by multiplying their absolute values. The product is **positive**.
- The product of 0 and any real number is 0. Thus, for any real number a,

$$a \cdot 0 = 0 \quad \text{and} \quad 0 \cdot a = 0.$$

EXAMPLE 1   Multiplying Real Numbers

Multiply:

a. $6(-3)$ **b.** $-\dfrac{1}{5} \cdot \dfrac{2}{3}$ **c.** $(-9)(-10)$ **d.** $(-1.4)(-2)$ **e.** $(-372)(0)$.

SOLUTION

a. $6(-3) = -18$ Multiply absolute values: $6 \cdot 3 = 18$.

Different signs: negative product

b. $-\dfrac{1}{5} \cdot \dfrac{2}{3} = -\dfrac{2}{15}$ Multiply absolute values: $\frac{1}{5} \cdot \frac{2}{3} = \frac{1 \cdot 2}{5 \cdot 3} = \frac{2}{15}$.

Different signs: negative product

c. $(-9)(-10) = 90$ Multiply absolute values: $9 \cdot 10 = 90$.

Same sign: positive product

d. $(-1.4)(-2) = 2.8$ Multiply absolute values: $(1.4)(2) = 2.8$.

Same sign: positive product

e. $(-372)(0) = 0$ The product of 0 and any real number is 0: $a \cdot 0 = 0$.

■

✔ CHECK POINT 1 Multiply: **a.** $8(-5)$ **b.** $-\dfrac{1}{3} \cdot \dfrac{4}{7}$

c. $(-12)(-3)$ **d.** $(-1.1)(-5)$ **e.** $(-543)(0)$.

---

2 Multiply more than two real numbers.

Multiplying More Than Two Numbers How do we perform more than one multiplication, such as

$$-4(-3)(-2)?$$

Because of the associative and commutative properties, we can order and group the numbers in any manner. Each pair of negative numbers will produce a positive product. Thus, the product of an even number of negative numbers is always positive. By contrast, the product of an odd number of negative numbers is always negative.

$$-4(-3)(-2) = -24$$ Multiply absolute values: $4 \cdot 3 \cdot 2 = 24$.

Odd number of negative numbers (three): negative product

MULTIPLYING MORE THAN TWO NUMBERS

1. Assuming that no factor is zero,
  • The product of an **even** number of **negative numbers** is **positive**.
  • The product of an **odd** number of **negative numbers** is **negative**.
  The multiplication is performed by multiplying the absolute values of the given numbers.
2. If any factor is 0, the product is 0.

EXAMPLE 2 Multiplying More Than Two Numbers

Multiply: **a.** $(-3)(-1)(2)(-2)$ **b.** $(-1)(-2)(-2)(3)(-4)$.

SOLUTION

a. $(-3)(-1)(2)(-2) = -12$

Multiply absolute values: $3 \cdot 1 \cdot 2 \cdot 2 = 12.$

Odd number of negative numbers (three): negative product

b. $(-1)(-2)(-2)(3)(-4) = 48$

Multiply absolute values: $1 \cdot 2 \cdot 2 \cdot 3 \cdot 4 = 48.$

Even number of negative numbers (four): positive product

■

CHECK POINT 2 Multiply:

a. $(-2)(3)(-1)(4)$ **b.** $(-1)(-3)(2)(-1)(5)$.

Is it always necessary to count the number of negative factors when multiplying more than two numbers? No. If any factor is 0, you can immediately write 0 for the product. For example,

$$(-37)(423)(0)(-55)(-3.7) = 0.$$

If any factor is 0, the product is 0.

3 Find multiplicative inverses.

The Meaning of Division The result of dividing the real number a by the nonzero real number b is called the **quotient** of a and b. We can write this quotient as $a \div b$ or $\frac{a}{b}$.

We know that subtraction is defined in terms of addition of an additive inverse, or opposite:

$$a - b = a + (-b).$$

In a similar way, we can define division in terms of multiplication. For example, the quotient of 8 and 2 can be written as multiplication:

$$8 \div 2 = 8 \cdot \frac{1}{2}.$$

We call $\frac{1}{2}$ the *multiplicative inverse*, or *reciprocal*, of 2. Two numbers whose product is 1 are called **multiplicative inverses** or **reciprocals** of each other. Thus, the multiplicative inverse of 2 is $\frac{1}{2}$ and the multiplicative inverse of $\frac{1}{2}$ is 2 because $2 \cdot \frac{1}{2} = 1$.

EXAMPLE 3 Finding Multiplicative Inverses

Find the multiplicative inverse of each number:

a. 5   **b.** $\frac{1}{3}$ **c.** -4 **d.** $-\frac{4}{5}$.

SOLUTION

a. The multiplicative inverse of 5 is $\frac{1}{5}$ because $5 \cdot \frac{1}{5} = 1$.

b. The multiplicative inverse of $\frac{1}{3}$ is 3 because $\frac{1}{3} \cdot 3 = 1$.

c. The multiplicative inverse of -4 is $-\frac{1}{4}$ because $(-4)\left(-\frac{1}{4}\right) = 1$.

d. The multiplicative inverse of $-\frac{4}{5}$ is $-\frac{5}{4}$ because $\left(-\frac{4}{5}\right)\left(-\frac{5}{4}\right) = 1$. ■

 CHECK POINT 3 Find the multiplicative inverse of each number:

a. 7   **b.** $\frac{1}{8}$ **c.** -6 **d.** $-\frac{7}{13}$.

Can you think of a real number that has no multiplicative inverse? The number **0 has no multiplicative inverse** because 0 multiplied by any number is never 1, but always 0.

We now define division in terms of multiplication by a multiplicative inverse.

---

4 Use the definition of division.

DEFINITION OF DIVISION If a and b are real numbers and b is not 0, then the quotient of a and b is defined as

$$a \div b = a \cdot \frac{1}{b}.$$

In words: The quotient of two real numbers is the product of the first number and the multiplicative inverse of the second number.

USING TECHNOLOGY

You can use a calculator to multiply and divide signed numbers. Here are the keystrokes for finding

$$(-173)(-256):$$

Scientific Calculator

173 $\boxed{+/-}$ $\boxed{\times}$ 256 $\boxed{+/-}$ $\boxed{=}$

Graphing Calculator

$\boxed{(-)}$ 173 $\boxed{\times}$ $\boxed{(-)}$ 256 $\boxed{\text{ENTER}}$

The number 44288 should be displayed.

Division is performed in the same manner, using $\boxed{\div}$ instead of $\boxed{\times}$. What happens when you divide by 0? Try entering

$$8 \boxed{\div} 0$$

and pressing $\boxed{=}$ or $\boxed{\text{ENTER}}$.

EXAMPLE 4 Using the Definition of Division

Use the definition of division to find each quotient: **a.** $-15 \div 3$ **b.** $\dfrac{-20}{-4}$.

SOLUTION

a. $-15 \div 3 = -15 \cdot \dfrac{1}{3} = -5$

> Change the division to multiplication.   Replace 3 with its multiplicative inverse.

b. $\dfrac{-20}{-4} = -20 \cdot \left(-\dfrac{1}{4}\right) = 5$

> Change the division to multiplication.   Replace -4 with its multiplicative inverse.

■

 CHECK POINT 4 Use the definition of division to find each quotient:

a. $-28 \div 7$ **b.** $\dfrac{-16}{-2}$.

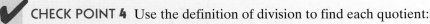

5 Divide real numbers.

A Procedure for Dividing Real Numbers Because the quotient $a \div b$ is defined as the product $a \cdot \frac{1}{b}$, the sign rules for dividing numbers are the same as the sign rules for multiplying them.

THE QUOTIENT OF TWO REAL NUMBERS
- The quotient of two real numbers with **different signs** is found by dividing their absolute values. The quotient is **negative**.
- The quotient of two real numbers with the **same sign** is found by dividing their absolute values. The quotient is **positive**.
- Division of a nonzero number by zero is undefined.
- Any nonzero number divided into 0 is 0.

EXAMPLE 5 Dividing Real Numbers

Divide: **a.** $\dfrac{8}{-2}$ **b.** $-\dfrac{3}{4} \div \left(-\dfrac{5}{9}\right)$ **c.** $\dfrac{-20.8}{4}$ **d.** $\dfrac{0}{-7}$.

SOLUTION

a. $\dfrac{8}{-2} = -4$ Divide absolute values: $\frac{8}{2} = 4$.

Different signs: negative quotient

b. $-\dfrac{3}{4} \div \left(-\dfrac{5}{9}\right) = \dfrac{27}{20}$ Divide absolute values: $\frac{3}{4} \div \frac{5}{9} = \frac{3}{4} \cdot \frac{9}{5} = \frac{27}{20}$.

Same sign: positive quotient

c. $\dfrac{-20.8}{4} = -5.2$ Divide absolute values: $4)\overline{20.8}^{\,5.2}$.

Different signs: negative quotient

d. $\dfrac{0}{-7} = 0$ Any nonzero number divided into 0 is 0.

Can you see why $\frac{0}{-7}$ must be 0? The definition of division tells us that

$$\frac{0}{-7} = 0 \cdot \left(-\frac{1}{7}\right)$$

and the product of 0 and any real number is 0. By contrast, the definition of division does not allow for division by 0 because 0 does not have a multiplicative inverse. It is incorrect to write

$$\frac{-7}{0} = -7 \cdot \frac{1}{0}.$$

0 does not have a multiplicative inverse.

Division by zero is not allowed or not defined. Thus, $\frac{-7}{0}$ does not represent a real number. A real number can never have a denominator of 0.

 CHECK POINT 5 Divide:

a. $\dfrac{-32}{-4}$ **b.** $-\dfrac{2}{3} \div \dfrac{5}{4}$ **c.** $\dfrac{21.9}{-3}$ **d.** $\dfrac{0}{-5}$.

6 Simplify algebraic expressions involving multiplication.

Multiplication and Algebraic Expressions In Section 1.4, we discussed the commutative and associative properties of multiplication. We also know that multiplication distributes over addition and subtraction. We now add some additional properties to our previous list (Table 1.5). These properties are frequently helpful in simplifying algebraic expressions.

Table 1.5 **Additional Properties of Multiplication**

Let a be a real number, a variable, or an algebraic expression.

Property	Meaning	Examples
Identity Property of Multiplication	1 can be deleted from a product. $a \cdot 1 = a$ $1 \cdot a = a$	• $\sqrt{3} \cdot 1 = \sqrt{3}$ • $1x = x$ • $1(2x + 3) = 2x + 3$
Inverse Property of Multiplication	If a is not 0: $a \cdot \dfrac{1}{a} = 1$ $\dfrac{1}{a} \cdot a = 1$ The product of a nonzero number and its multiplicative inverse, or reciprocal, gives 1, the multiplicative identity.	• $6 \cdot \dfrac{1}{6} = 1$ • $3x \cdot \dfrac{1}{3x} = 1$ (x is not 0.) • $\dfrac{1}{(y-2)} \cdot (y - 2) = 1$ (y is not 2.)
Multiplication Property of -1	Negative 1 times a is the additive inverse, or opposite, of a. $-1 \cdot a = -a$ $a(-1) = -a$	• $-1 \cdot \sqrt{3} = -\sqrt{3}$ • $-1\left(-\frac{3}{4}\right) = \frac{3}{4}$ • $-1x = -x$ • $-(x + 4) = -1(x + 4)$ $= -x - 4$
Double Negative Property	The additive inverse of $-a$ is a. $-(-a) = a$	• $-(-4) = 4$ • $-(-6y) = 6y$

In the preceding table, we used two steps to remove the parentheses from $-(x + 4)$. First, we used the multiplication property of -1.

$$-(x + 4) = -1(x + 4)$$

Then we used the distributive property, distributing -1 to each term in parentheses.

$$-1(x + 4) = (-1)x + (-1)4 = -x + (-4) = -x - 4$$

There is a fast way to obtain $-(x + 4) = -x - 4$ in just one step.

NEGATIVE SIGNS AND PARENTHESES If a negative sign precedes parentheses, remove the parentheses and change the sign of every term within the parentheses.

Here are some examples that illustrate this method.

$$-(11x + 5) = -11x - 5$$
$$-(11x - 5) = -11x + 5$$
$$-(-11x + 5) = 11x - 5$$
$$-(-11x - 5) = 11x + 5$$

EXAMPLE 6 — Simplifying Algebraic Expressions

Simplify: **a.** $-2(3x)$ **b.** $6x + x$ **c.** $8a - 9a$ **d.** $-3(2x - 5)$ **e.** $-(3y - 8)$.

SOLUTION We will show all steps in the solution process. However, you probably are working many of these steps mentally.

a. $-2(3x)$ This is the given algebraic expression.
$= (-2 \cdot 3)x$ Use the associative property and group the first two numbers.
$= -6x$ Numbers with opposite signs have a negative product.

b. $6x + x$ This is the given algebraic expression.
$= 6x + 1x$ Use the multiplication property of 1.
$= (6 + 1)x$ Apply the distributive property.
$= 7x$ Add within parentheses.

c. $8a - 9a$ This is the given algebraic expression.
$= (8 - 9)a$ Apply the distributive property.
$= -1a$ Subtract within parentheses: $8 - 9 = 8 + (-9) = -1$.
$= -a$ Apply the multiplication property of -1.

d. $-3(2x - 5)$ This is the given algebraic expression.
$= -3(2x) - (-3) \cdot (5)$ Apply the distributive property.
$= -6x - (-15)$ Multiply.
$= -6x + 15$ Subtraction is the addition of an additive inverse.

e. $-(3y - 8)$ This is the given algebraic expression.
$= -3y + 8$ Remove parentheses by changing the sign of every term inside the parentheses.

 CHECK POINT 6 Simplify: **a.** $-4(5x)$ **b.** $9x + x$
c. $13b - 14b$ **d.** $-7(3x - 4)$ **e.** $-(7y - 6)$.

Before turning to applications, let's try one additional example involving simplification.

EXAMPLE 7 — Simplifying an Algebraic Expression

Simplify: $5(2y - 9) - (9y - 8)$.

SOLUTION

$5(2y - 9) - (9y - 8)$ This is the given algebraic expression.
$= 5 \cdot 2y - 5 \cdot 9 - (9y - 8)$ Apply the distributive property over the first parentheses.
$= 10y - 45 - (9y - 8)$ Multiply.
$= 10y - 45 - 9y + 8$ Remove the second parentheses by changing the sign of each term within parentheses.
$= (10y - 9y) + (-45 + 8)$ Group like terms.
$= 1y + (-37)$ Combine like terms. For the variable terms, $10y - 9y = 10y + (-9y) = [10 + (-9)]y = 1y$.
$= y + (-37)$ Use the multiplication property of 1: $1y = y$.
$= y - 37$ Express addition of an additive inverse as subtraction.

✔ CHECK POINT **7** Simplify: $4(3y - 7) - (13y - 2)$

A Summary of Operations with Real Numbers Operations with real numbers are summarized in Table 1.6.

Table 1.6 **Summary of Operations with Real Numbers**

Signs of Numbers	Addition	Subtraction	Multiplication	Division
Both Numbers Are Positive Examples 8 and 2 2 and 8	Sum Is Always Positive $8 + 2 = 10$ $2 + 8 = 10$	Difference May Be Either Positive or Negative $8 - 2 = 6$ $2 - 8 = -6$	Product Is Always Positive $8 \cdot 2 = 16$ $2 \cdot 8 = 16$	Quotient Is Always Positive $8 \div 2 = 4$ $2 \div 8 = \frac{1}{4}$
One Number Is Positive and the Other Number Is Negative Examples 8 and −2 −8 and 2	Sum May Be Either Positive or Negative $8 + (-2) = 6$ $-8 + 2 = -6$	Difference May Be Either Positive or Negative $8 - (-2) = 10$ $-8 - 2 = -10$	Product Is Always Negative $8(-2) = -16$ $-8(2) = -16$	Quotient Is Always Negative $8 \div (-2) = -4$ $-8 \div 2 = -4$
Both Numbers Are Negative Examples −8 and −2 −2 and −8	Sum Is Always Negative $-8 + (-2) = -10$ $-2 + (-8) = -10$	Difference May Be Either Positive or Negative $-8 - (-2) = -6$ $-2 - (-8) = 6$	Product Is Always Positive $-8(-2) = 16$ $-2(-8) = 16$	Quotient Is Always Positive $-8 \div (-2) = 4$ $-2 \div (-8) = \frac{1}{4}$

7 Use algebraic expressions that model reality.

Applications Algebraic expressions that model reality frequently contain division.

EXAMPLE 8 Average Cost of Producing a Wheelchair

A company that manufactures wheelchairs has monthly fixed costs of $500,000. The average cost per wheelchair for the company to manufacture x wheelchairs per month is modeled by the algebraic expression

$$\frac{400x + 500{,}000}{x}.$$

Find the average cost per wheelchair for the company to manufacture

    a. 10,000 wheelchairs per month.   **b.** 50,000 wheelchairs per month.

    c. 100,000 wheelchairs per month.

What happens to the average cost per wheelchair as the production level increases?

SOLUTION

    a. We are interested in the average cost per wheelchair for the company if 10,000 wheelchairs are manufactured per month. Because x represents the number of wheelchairs manufactured per month, we substitute $10{,}000$ for x in the given algebraic expression.

$$\frac{400x + 500{,}000}{x} = \frac{400(10{,}000) + 500{,}000}{10{,}000} = \frac{4{,}000{,}000 + 500{,}000}{10{,}000}$$

$$= \frac{4{,}500{,}000}{10{,}000} = 450$$

The average cost per wheelchair of producing 10,000 wheelchairs per month is $450.

b. Now, 50,000 wheelchairs are manufactured per month. We find the average cost per wheelchair by substituting $50{,}000$ for x in the given algebraic expression.

$$\frac{400x + 500{,}000}{x} = \frac{400(50{,}000) + 500{,}000}{50{,}000} = \frac{20{,}000{,}000 + 500{,}000}{50{,}000}$$

$$= \frac{20{,}500{,}000}{50{,}000} = 410$$

The average cost per wheelchair of producing 50,000 wheelchairs per month is $410.

c. Finally, the production level has increased to 100,000 wheelchairs per month. We find the average cost per wheelchair for the company by substituting $100{,}000$ for x in the given algebraic expression.

$$\frac{400x + 500{,}000}{x} = \frac{400(100{,}000) + 500{,}000}{100{,}000} = \frac{40{,}000{,}000 + 500{,}000}{100{,}000}$$

$$= \frac{40{,}500{,}000}{100{,}000} = 405$$

The average cost per wheelchair of producing 100,000 wheelchairs per month is $405.

As the production level increases, the average cost of producing each wheelchair decreases. This illustrates the difficulty with small businesses. It is nearly impossible to have competitively low prices when production levels are low. ∎

The points in the rectangular coordinate system in Figure 1.21 show the relationship between production level and cost. The x-axis represents the number of wheelchairs produced per month. The y-axis represents the average cost per wheelchair for the company. The symbol ξ on the y-axis indicates that there is a break in the values of y between 0 and 400. Thus, the values of y begin at 400.

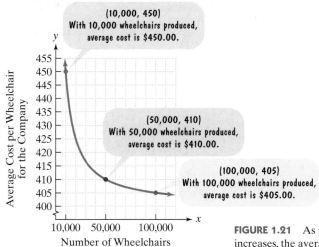

FIGURE 1.21 As production level increases, the average cost per wheelchair for the company decreases.

The three points with the voice balloons illustrate our computations in Example 8. The points in Figure 1.21 are falling from left to right. Can you see how this shows that the company's cost per wheelchair is decreasing as their production level increases?

 CHECK POINT 8 A company that manufactures running shoes has weekly fixed costs of $300,000. The average cost per pair of running shoes for the company to manufacture x pairs per week is modeled by the algebraic expression

$$\frac{30x + 300,000}{x}.$$

Find the average cost per pair of running shoes for the company to manufacture

a. 1000 pairs per week. **b.** 10,000 pairs per week.

c. 100,000 pairs per week.

1.7 EXERCISE SET

Student Solutions Manual CD/Video PH Math/Tutor Center MathXL Tutorials on CD MathXL® MyMathLab Interactmath.com

Practice Exercises

In Exercises 1–34, perform the indicated multiplication.

1. $5(-9)$ **2.** $10(-7)$

3. $(-8)(-3)$ **4.** $(-9)(-5)$

5. $(-3)(7)$ **6.** $(-4)(8)$

7. $(-19)(-1)$ **8.** $(-11)(-1)$

9. $0(-19)$ **10.** $0(-11)$

11. $\frac{1}{2}(-24)$ **12.** $\frac{1}{3}(-21)$

13. $\left(-\frac{3}{4}\right)(-12)$ **14.** $\left(-\frac{4}{5}\right)(-30)$

15. $-\frac{3}{5}\cdot\left(-\frac{4}{7}\right)$ **16.** $-\frac{5}{7}\cdot\left(-\frac{3}{8}\right)$

17. $-\frac{7}{9}\cdot\frac{2}{3}$ **18.** $-\frac{5}{11}\cdot\frac{2}{7}$

19. $3(-1.2)$ **20.** $4(-1.2)$

21. $-0.2(-0.6)$ **22.** $-0.3(-0.7)$

23. $(-5)(-2)(3)$ **24.** $(-6)(-3)(10)$

25. $(-4)(-3)(-1)(6)$ **26.** $(-2)(-7)(-1)(3)$

27. $-2(-3)(-4)(-1)$ **28.** $-3(-2)(-5)(-1)$

29. $(-3)(-3)(-3)$ **30.** $(-4)(-4)(-4)$

31. $5(-3)(-1)(2)(3)$ **32.** $2(-5)(-2)(3)(1)$

33. $(-8)(-4)(0)(-17)(-6)$

34. $(-9)(-12)(-18)(0)(-3)$

In Exercises 35–42, find the multiplicative inverse of each number.

35. 4   **36.** 3   **37.** $\frac{1}{5}$

38. $\frac{1}{7}$ **39.** -10 **40.** -12

41. $-\frac{2}{5}$ **42.** $-\frac{4}{9}$

In Exercises 43–46,

a. *Rewrite the division as multiplication involving a multiplicative inverse.*

b. *Use the multiplication from part (a) to find the given quotient.*

43. $-32 \div 4$ **44.** $-18 \div 6$

45. $\frac{-60}{-5}$

46. $\frac{-30}{-5}$

In Exercises 47–76, perform the indicated division or state that the expression is undefined.

47. $\frac{12}{-4}$ **48.** $\frac{40}{-5}$ **49.** $\frac{-21}{3}$

50. $\frac{-60}{6}$ **51.** $\frac{-90}{-3}$ **52.** $\frac{-66}{-6}$

53. $\frac{0}{-7}$ **54.** $\frac{0}{-8}$ **55.** $\frac{7}{0}$

56. $\frac{-8}{0}$ **57.** $-15 \div 3$ **58.** $-80 \div 8$

59. $120 \div (-10)$ **60.** $130 \div (-10)$

61. $(-180) \div (-30)$ **62.** $(-150) \div (-25)$

63. $0 \div (-4)$ **64.** $0 \div (-10)$

65. $-4 \div 0$ **66.** $-10 \div 0$

67. $\frac{-12.9}{3}$ **68.** $\frac{-21.6}{3}$

69. $-\dfrac{1}{2} \div \left(-\dfrac{3}{5}\right)$

70. $-\dfrac{1}{2} \div \left(-\dfrac{7}{9}\right)$

71. $-\dfrac{14}{9} \div \dfrac{7}{8}$

72. $-\dfrac{5}{16} \div \dfrac{25}{8}$

73. $\dfrac{1}{3} \div \left(-\dfrac{1}{3}\right)$

74. $\dfrac{1}{5} \div \left(-\dfrac{1}{5}\right)$

75. $6 \div \left(-\dfrac{2}{5}\right)$

76. $8 \div \left(-\dfrac{2}{9}\right)$

In Exercises 77–96, simplify each algebraic expression.

77. $-5(2x)$

78. $-9(3x)$

79. $-4\left(-\dfrac{3}{4}y\right)$

80. $-5\left(-\dfrac{3}{5}y\right)$

81. $8x + x$

82. $12x + x$

83. $-5x + x$

84. $-6x + x$

85. $6b - 7b$

86. $12b - 13b$

87. $-y + 4y$

88. $-y + 9y$

89. $-4(2x - 3)$

90. $-3(4x - 5)$

91. $-3(-2x + 4)$

92. $-4(-3x + 2)$

93. $-(2y - 5)$

94. $-(3y - 1)$

95. $4(2y - 3) - (7y + 2)$

96. $5(3y - 1) - (14y - 2)$

Practice Plus

In Exercises 97–104, write a numerical expression for each phrase. Then simplify the numerical expression by performing the given operations.

97. 8 added to the product of 4 and -10

98. 14 added to the product of 3 and -15

99. The product of -9 and -3, decreased by -2

100. The product of -6 and -4, decreased by -5

101. The quotient of -18 and the sum of -15 and 12

102. The quotient of -25 and the sum of -21 and 16

103. The difference between -6 and the quotient of 12 and -4

104. The difference between -11 and the quotient of 20 and -5

Application Exercises

The graph shows the millions of welfare recipients in the United States who received cash assistance from 1994 through 2001. The data can be modeled by the algebraic expression

$$-1.4t + 14.7,$$

which represents the number of welfare recipients, in millions, t years after 1994. Use this algebraic expression to solve Exercises 105–106.

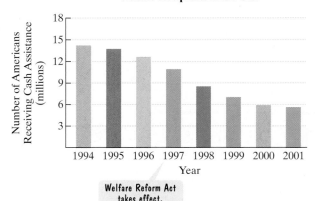

Welfare Recipients in the U.S.

Source: Department of Health and Human Services

105. According to the model, how many welfare recipients were there in 2000? How well does the algebraic expression model the actual data for 2000 shown in the bar graph?

106. According to the model, how many welfare recipients were there in 1997? How well does the algebraic expression model the actual data for 1997 shown in the bar graph?

In an experiment on memory, students in a language class are asked to memorize 40 vocabulary words in Latin, a language with which the students are not familiar. After studying the words for one day, students are tested each day after to see how many words they remember. The class average is taken and the results are graphed as points in the rectangular coordinate system. Use the points shown to solve Exercises 107–108.

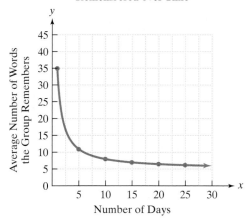

Average Number of Words Remembered over Time

107. a. Find a reasonable estimate of the number of Latin words remembered after 5 days.

  b. The algebraic expression

$$\frac{5x + 30}{x}$$

  models the number of Latin words remembered by the students after x days. Use this expression to find the number of Latin words remembered after 5 days. How does this compare with your estimate from part (a)?

108. a. Find a reasonable estimate of the number of Latin words remembered after 15 days.

  b. The algebraic expression

$$\frac{5x + 30}{x}$$

  models the number of Latin words remembered by the students after x days. Use this expression to find the number of Latin words remembered after 15 days. How does this compare with your estimate from part (a)?

In Palo Alto, California, a government agency ordered computer-related companies to contribute to a pool of money to clean up underground water supplies. (The companies had stored toxic chemicals in leaking underground containers.) The algebraic expression

$$\frac{200x}{100 - x}$$

models the cost, in tens of thousands of dollars, for removing x percent of the contaminants. Use this algebraic expression to solve Exercises 109–110.

109. a. Substitute 50 for x and find the cost, in tens of thousands of dollars, for removing 50% of the contaminants.

  b. Find the cost, in tens of thousands of dollars, for removing 80% of the contaminants.

  c. Describe what is happening to the cost of the cleanup as the percentage of contaminant removed increases.

110. a. Substitute 60 for x and find the cost, in tens of thousands of dollars, for removing 60% of the contaminants.

  b. Find the cost, in tens of thousands of dollars, for removing 90% of the contaminants.

  c. Describe what is happening to the cost of the cleanup as the percentage of contaminants removed increases.

Writing in Mathematics

111. Explain how to multiply two real numbers. Provide examples with your explanation.

112. Explain how to determine the sign of a product that involves more than two numbers.

113. Explain how to find the multiplicative inverse of a number.

114. Why is it that 0 has no multiplicative inverse?

115. Explain how to divide real numbers.

116. Why is division by zero undefined?

117. Explain how to simplify an algebraic expression in which a negative sign precedes parentheses.

118. A politician promises to do "whatever it takes" to seize "one hundred percent" of all illegal drugs that enter the country. Suppose that the cost, in millions of dollars, for seizing x percent of the illegal drugs entering the country is modeled by the algebraic expression

$$\frac{130x}{100 - x}.$$

According to this algebraic expression, can the politician keep his or her promise? Explain your answer.

Critical Thinking Exercises

119. Which one of the following statements is true?

  a. Multiplying a negative number by a nonnegative number will always give a negative number.

  b. The product of two negative numbers is always a positive number.

  c. The product of -3 and 4 is 12.

  d. The product of real numbers a and b is not always equal to the product of real numbers b and a.

120. Which one of the following statements is true?

  a. The product of two negative numbers is sometimes a negative number.

  b. Both the addition and the multiplication of two negative numbers result in a positive number.

  c. $\left(-\frac{1}{2}\right)\left(-\frac{1}{2}\right) = \frac{1}{4}$

  d. Reversing the order of the two factors in a product results in a different answer.

In Exercises 121–124, write an algebraic expression for the given English phrase.

121. The value, in cents, of x nickels

122. The distance covered by a car traveling at 50 miles per hour for x hours

123. The monthly salary, in dollars, for a person earning x dollars per year

124. The fraction of people in a room who are women if there are 40 women and x men in the room

Technology Exercises

125. Use a calculator to verify any five of the products that you found in Exercises 1–34.

126. Use a calculator to verify any five of the quotients that you found in Exercises 47–76.

127. Simplify using a calculator:

$$0.3(4.7x - 5.9) - 0.07(3.8x - 61).$$

128. Use your calculator to attempt to find the quotient of -3 and 0. Describe what happens. Does the same thing occur when finding the quotient of 0 and -3? Explain the difference. Finally, what happens when you enter the quotient of 0 and itself?

Review Exercises

In Exercises 129–131, perform the indicated operation.

129. $-6 + (-3)$ (Section 1.5, Example 3)

130. $-6 - (-3)$ (Section 1.6, Example 1)

131. $-6 \div (-3)$ (Section 1.7, Example 4)

SECTION 1.8

EXPONENTS, ORDER OF OPERATIONS, AND MATHEMATICAL MODELS

Objectives

1 Evaluate exponential expressions.

2 Simplify algebraic expressions with exponents.

3 Use the order of operations agreement.

4 Evaluate formulas.

It's been another one of those days! Traffic is really backed up on the highway. Finally, you see the source of the traffic jam—a minor fender-bender. Still stuck in traffic, you notice that the driver appears to be quite young. This might seem like a strange observation. After all, what does a driver's age have to do with his or her chance of getting into an accident? In this section, we see how algebra describes your world, including a relationship between age and numbers of car accidents.

1 Evaluate exponential expressions.

Natural Number Exponents   Although people do a great deal of talking, the total output since the beginning of gabble to the present day, including all baby talk, love songs, and congressional debates, only amounts to about 10 million billion words. This can be expressed as 16 factors of 10, or 10^{16} words.

Exponents such as 2, 3, 4, and so on are used to indicate repeated multiplication. For example,

$$10^2 = 10 \cdot 10 = 100,$$

$$10^3 = 10 \cdot 10 \cdot 10 = 1000, \quad 10^4 = 10 \cdot 10 \cdot 10 \cdot 10 = 10,000.$$

The 10 that is repeated when multiplying is called the **base**. The small numbers above and to the right of the base are called **exponents** or **powers**. The exponent tells the number of times the base is to be used when multiplying. In 10^3, the base is 10 and the exponent is 3.

Any number with an exponent of 1 is the number itself. Thus, $10^1 = 10$.

Multiplications that are expressed in exponential notation are read as follows:

10^1: "ten to the first power"
10^2: "ten to the second power" or "ten squared"
10^3: "ten to the third power" or "ten cubed"
10^4: "ten to the fourth power"
10^5: "ten to the fifth power"

etc.

Any real number can be used as the base. Thus,

$$7^2 = 7 \cdot 7 = 49 \quad \text{and} \quad (-3)^4 = (-3)(-3)(-3)(-3) = 81.$$

The bases are 7 and -3, respectively. Do not confuse $(-3)^4$ and -3^4.

$$-3^4 = -(3 \cdot 3 \cdot 3 \cdot 3) = -81$$

> The negative is not taken to the power because it is not inside parentheses.

An exponent applies only to a base. A negative sign is not part of a base unless it appears in parentheses.

EXAMPLE 1 Evaluating Exponential Expressions

Evaluate: **a.** 4^2 **b.** $(-5)^3$ **c.** $(-2)^4$ **d.** -2^4.

SOLUTION

> Exponent is 2.

a. $4^2 = 4 \cdot 4 = 16$ The exponent indicates that the base is used as a factor two times.

> Base is 4.

We read $4^2 = 16$ as "4 to the second power is 16" or "4 squared is 16."

> Exponent is 3.

b. $(-5)^3 = (-5)(-5)(-5)$ The exponent indicates that the base is used as a factor three times.

> Base is -5.

$= -125$ An odd number of negative factors yields a negative product.

We read $(-5)^3 = -125$ as "the number negative 5 to the third power is negative 125" or "negative 5 cubed is negative 125."

> Exponent is 4.

c. $(-2)^4 = (-2)(-2)(-2)(-2)$ The exponent indicates the base is used as a factor four times.

> Base is -2.

$= 16$ An even number of negative factors yields a positive product.

We read $(-2)^4 = 16$ as "the number negative 2 to the fourth power is 16."

Exponent is 4.

d. $-2^4 = -(2 \cdot 2 \cdot 2 \cdot 2)$ The negative is not inside parentheses and is not taken to the fourth power.

Base is 2.

$= -16$ Multiply the twos and copy the negative.

We read $-2^4 = -16$ as "the negative of 2 raised to the fourth power is negative 16" or "the opposite, or additive inverse, of 2 raised to the fourth power is negative 16."

 CHECK POINT 1 Evaluate:

a. 6^2 **b.** $(-4)^3$ **c.** $(-1)^4$ **d.** -1^4.

The formal algebraic definition of a natural number exponent summarizes our discussion:

DEFINITION OF A NATURAL NUMBER EXPONENT If b is a real number and n is a natural number,

Exponent

$$b^n = \underbrace{b \cdot b \cdot b \cdot \ldots \cdot b}_{\substack{b \text{ appears as a} \\ \text{factor } n \text{ times.}}}$$

Base

b^n is read "the nth power of b" or "b to the nth power." Thus, the nth power of b is defined as the product of n factors of b. The expression b^n is called an **exponential expression**.
Furthermore, $b^1 = b$.

2 Simplify algebraic expressions with exponents.

Exponents and Algebraic Expressions The distributive property can be used to simplify certain algebraic expressions that contain exponents. For example, we can use the distributive property to combine like terms in the algebraic expression $4x^2 + 6x^2$:

$$4x^2 + 6x^2 = (4 + 6)x^2 = 10x^2.$$

First term with variable factor x^2 Second term with variable factor x^2 The common variable factor is x^2.

EXAMPLE 2 Simplifying Algebraic Expressions

Simplify, if possible: **a.** $7x^3 + 2x^3$ **b.** $5x^2 + x^2$ **c.** $3x^2 + 4x^3$.

SOLUTION

a. $7x^3 + 2x^3$ There are two like terms with the same variable factor, namely x^3.

$= (7 + 2)x^3$ Apply the distributive property.

$= 9x^3$ Add within parentheses.

3 Use the order of operations agreement.

b. $5x^2 + x^2$ There are two like terms with the same variable factor, namely x^2.

$= 5x^2 + 1x^2$ Use the multiplication property of 1.

$= (5 + 1)x^2$ Apply the distributive property.

$= 6x^2$ Add within parentheses.

c. $3x^2 + 4x^3$ cannot be simplified. The terms $3x^2$ and $4x^3$ are not like terms because they have different variable factors, namely x^2 and x^3. ■

 CHECK POINT 2 Simplify, if possible:

a. $16x^2 + 5x^2$ **b.** $7x^3 + x^3$ **c.** $10x^2 + 8x^3$.

Order of Operations Suppose that you want to find the value of $3 + 7 \cdot 5$. Which procedure shown is correct?

$$3 + 7 \cdot 5 = 3 + 35 = 38 \quad \text{or} \quad 3 + 7 \cdot 5 = 10 \cdot 5 = 50$$

If you know the answer, you probably know certain rules, called the **order of operations**, to make sure that there is only one correct answer. One of these rules states that if a problem contains no parentheses or other grouping symbols, perform multiplication before addition. Thus, the procedure on the left is correct because the multiplication of 7 and 5 is done first. Then the addition is performed. The correct answer is 38.

Some problems contain grouping symbols, such as parentheses, ( ); brackets, [ ]; braces, { }; absolute value symbols, | |; or fraction bars. These grouping symbols tell us what to do first. Here are two examples:

• $(3 + 7) \cdot 5 = 10 \cdot 5 = 50$

> First, perform operations in grouping symbols.

• $8|6 - 16| = 8|{-10}| = 8 \cdot 10 = 80$.

Here are the rules for determining the order in which operations should be performed:

ORDER OF OPERATIONS

1. Perform all operations within grouping symbols.
2. Evaluate all exponential expressions.
3. Do all multiplications and divisions in the order in which they occur, working from left to right.
4. Finally, do all additions and subtractions in the order in which they occur, working from left to right.

In the third step, be sure to do all multiplications and divisions *as they occur* from left to right. For example,

$$8 \div 4 \cdot 2 = 2 \cdot 2 = 4$$ Do the division first because it occurs first.

$$8 \cdot 4 \div 2 = 32 \div 2 = 16.$$ Do the multiplication first because it occurs first.

EXAMPLE 3 Using the Order of Operations

Simplify: $18 + 2 \cdot 3 - 10$.

SOLUTION There are no grouping symbols or exponential expressions. In cases like this, we multiply and divide before adding and subtracting.

$$18 + 2 \cdot 3 - 10 = 18 + 6 - 10 \quad \text{Multiply: } 2 \cdot 3 = 6.$$
$$= 24 - 10 \quad \text{Add and subtract from left}$$
$$\text{to right: } 18 + 6 = 24.$$
$$= 14 \quad \text{Subtract: } 24 - 10 = 14. \quad \blacksquare$$

 CHECK POINT 3 Simplify: $20 + 4 \cdot 3 - 17$.

EXAMPLE 4 Using the Order of Operations

Simplify: $6^2 - 24 \div 2^2 \cdot 3 - 1$.

SOLUTION There are no grouping symbols. Thus, we begin by evaluating exponential expressions. Then we multiply or divide. Finally, we add or subtract.

$$6^2 - 24 \div 2^2 \cdot 3 - 1$$
$$= 36 - 24 \div 4 \cdot 3 - 1 \quad \text{Evaluate exponential expressions:}$$
$$6^2 = 6 \cdot 6 = 36 \text{ and } 2^2 = 2 \cdot 2 = 4.$$
$$= 36 - 6 \cdot 3 - 1 \quad \text{Perform the multiplications and divisions}$$
$$\text{from left to right. Start with } 24 \div 4 = 6.$$
$$= 36 - 18 - 1 \quad \text{Now do the multiplication: } 6 \cdot 3 = 18.$$
$$= 18 - 1 \quad \text{Finally, perform the subtraction}$$
$$\text{from left to right: } 36 - 18 = 18.$$
$$= 17 \quad \text{Complete the subtraction: } 18 - 1 = 17. \quad \blacksquare$$

 CHECK POINT 4 Simplify: $7^2 - 48 \div 4^2 \cdot 5 - 2$.

EXAMPLE 5 Using the Order of Operations

Simplify: **a.** $(2 \cdot 5)^2$ **b.** $2 \cdot 5^2$.

SOLUTION

a. Because $(2 \cdot 5)^2$ contains grouping symbols, namely parentheses, we perform the operation within parentheses first.

$$(2 \cdot 5)^2 = 10^2 \quad \text{Multiply within parentheses:}$$
$$2 \cdot 5 = 10.$$
$$= 100 \quad \text{Evaluate the exponential expression:}$$
$$10^2 = 10 \cdot 10 = 100.$$

b. Because $2 \cdot 5^2$ does not contain grouping symbols, we begin by evaluating the exponential expression.

$$2 \cdot 5^2 = 2 \cdot 25 \quad \text{Evaluate the exponential expression:}$$
$$5^2 = 5 \cdot 5 = 25.$$
$$= 50 \quad \text{Now do the multiplication:}$$
$$2 \cdot 25 = 50. \quad \blacksquare$$

 CHECK POINT 5 Simplify: **a.** $(3 \cdot 2)^2$ **b.** $3 \cdot 2^2$.

EXAMPLE 6 Using the Order of Operations

Simplify: $\left(\dfrac{1}{2}\right)^3 - \left(\dfrac{1}{2} - \dfrac{3}{4}\right)^2 (-4)$.

SOLUTION Because grouping symbols appear, we perform the operation within parentheses first.

$$\left(\dfrac{1}{2}\right)^3 - \left(\dfrac{1}{2} - \dfrac{3}{4}\right)^2 (-4)$$

$$= \left(\dfrac{1}{2}\right)^3 - \left(-\dfrac{1}{4}\right)^2 (-4) \qquad \text{Work inside parentheses first:}$$

$$\dfrac{1}{2} - \dfrac{3}{4} = \dfrac{2}{4} - \dfrac{3}{4} = \dfrac{2}{4} + \left(-\dfrac{3}{4}\right) = -\dfrac{1}{4}.$$

$$= \dfrac{1}{8} - \dfrac{1}{16}(-4) \qquad \text{Evaluate exponential expressions:}$$

$$\left(\dfrac{1}{2}\right)^3 = \dfrac{1}{2} \cdot \dfrac{1}{2} \cdot \dfrac{1}{2} = \dfrac{1}{8} \text{ and } \left(-\dfrac{1}{4}\right)^2 = \left(-\dfrac{1}{4}\right)\left(-\dfrac{1}{4}\right) = \dfrac{1}{16}.$$

$$= \dfrac{1}{8} - \left(-\dfrac{1}{4}\right) \qquad \text{Multiply: } \dfrac{1}{16} \cdot \left(\dfrac{-4}{1}\right) = -\dfrac{4}{16} = -\dfrac{1}{4}.$$

$$= \dfrac{3}{8} \qquad \text{Subtract: } \dfrac{1}{8} - \left(-\dfrac{1}{4}\right) = \dfrac{1}{8} + \dfrac{1}{4} = \dfrac{1}{8} + \dfrac{2}{8} = \dfrac{3}{8}. \qquad ■$$

✔ **CHECK POINT 6** Simplify: $\left(-\dfrac{1}{2}\right)^2 - \left(\dfrac{7}{10} - \dfrac{8}{15}\right)^2 (-18)$.

Some expressions contain many grouping symbols. An example of such an expression is $2[5(4 - 7) + 9]$. The grouping symbols are the parentheses and the brackets.

> The parentheses, the innermost grouping symbols, group $4 - 7$.

$$2[5(4 - 7) + 9]$$

> The brackets, the outermost grouping symbols, group $5(4 - 7) + 9$.

When combinations of grouping symbols appear, **perform operations within the innermost grouping symbols first**. Then work to the outside, performing operations within the outermost grouping symbols.

EXAMPLE 7 Using the Order of Operations

Simplify: $2[5(4 - 7) + 9]$.

SOLUTION

$$2[5(4 - 7) + 9]$$

$$= 2[5(-3) + 9] \qquad \text{Work inside parentheses first:}$$
$$4 - 7 = 4 + (-7) = -3.$$

$$= 2[-15 + 9] \qquad \text{Work inside brackets and multiply: } 5(-3) = -15.$$

$$= 2[-6] \qquad \text{Add inside brackets: } -15 + 9 = -6. \text{ The resulting}$$
$$\text{problem can also be expressed as } 2(-6).$$

$$= -12 \qquad \text{Multiply: } 2[-6] = -12. \qquad ■$$

Parentheses can be used for both innermost and outermost grouping symbols. For example, the expression $2[5(4 - 7) + 9]$ can also be written $2(5(4 - 7) + 9)$. However, too many parentheses can be confusing. The use of both parentheses and brackets makes it easier to identify inner and outer groupings.

 CHECK POINT 7 Simplify: $4[3(6 - 11) + 5]$.

EXAMPLE 8 Using the Order of Operations

Simplify: $18 \div 6 + 4[5 + 2(8 - 10)^3]$.

SOLUTION

$$18 \div 6 + 4[5 + 2(8 - 10)^3]$$

$= 18 \div 6 + 4[5 + 2(-2)^3]$ Work inside parentheses first: $8 - 10 = 8 + (-10) = -2$.

$= 18 \div 6 + 4[5 + 2(-8)]$ Work inside brackets and evaluate the exponential expression: $(-2)^3 = (-2)(-2)(-2) = -8$.

$= 18 \div 6 + 4[5 + (-16)]$ Work inside brackets and multiply: $2(-8) = -16$.

$= 18 \div 6 + 4[-11]$ Work inside brackets and add: $5 + (-16) = -11$.

$= 3 + 4[-11]$ Perform the multiplications and divisions from left to right. Start with $18 \div 6 = 3$.

$= 3 + (-44)$ Now do the multiplication: $4(-11) = -44$.

$= -41$ Finally, perform the addition: $3 + (-44) = -41$. ∎

 CHECK POINT 8 Simplify: $25 \div 5 + 3[4 + 2(7 - 9)^3]$.

Fraction bars are grouping symbols that separate expressions into two parts, the numerator and the denominator. Consider, for example,

The fraction bar is the grouping symbol.

The numerator is one part of the expression.

$$\frac{2(3 - 12) + 6 \cdot 4}{2^4 + 1}.$$

The denominator is the other part of the expression.

We can use brackets instead of the fraction bar. An equivalent expression is

$$[2(3 - 12) + 6 \cdot 4] \div [2^4 + 1].$$

The grouping suggests a method for simplifying expressions with fraction bars as grouping symbols:

- Simplify the numerator.
- Simplify the denominator.
- If possible, simplify the fraction.

EXAMPLE 9 Using the Order of Operations

Simplify: $\dfrac{2(3-12)+6\cdot 4}{2^4+1}$.

SOLUTION

$$\dfrac{2(3-12)+6\cdot 4}{2^4+1}$$

$$=\dfrac{2(-9)+6\cdot 4}{16+1}$$

Work inside parentheses in the numerator: $3-12=3+(-12)=-9$. Evaluate the exponential expression in the denominator: $2^4=2\cdot2\cdot2\cdot2=16$.

$$=\dfrac{-18+24}{16+1}$$

Multiply in the numerator: $2(-9)=-18$ and $6\cdot4=24$.

$$=\dfrac{6}{17}$$

Perform the addition in the numerator and the denominator. ■

 CHECK POINT 9 Simplify: $\dfrac{5(4-9)+10\cdot3}{2^3-1}$.

EXAMPLE 10 Using the Order of Operations

Evaluate: $-x^2-7x$ for $x=-2$.

SOLUTION We begin by substituting -2 for each occurrence of x in the algebraic expression. Then we use the order of operations to evaluate the expression.

$$-x^2-7x$$

Replace x with -2.

$$=-(-2)^2-7(-2)$$

$$=-4-7(-2)$$

Evaluate the exponential expression: $(-2)^2=(-2)(-2)=4$.

$$=-4-(-14)$$

Multiply: $7(-2)=-14$.

$$=10$$

Subtract: $-4-(-14)=-4+14=10$. ■

CHECK POINT 10 Evaluate: $-x^2-4x$ for $x=-5$.

Some algebraic expressions contain two sets of grouping symbols. Using the order of operations, grouping symbols are removed from innermost (parentheses) to outermost (brackets).

EXAMPLE 11 Simplifying an Algebraic Expression

Simplify: $18x^2+4-[6(x^2-2)+5]$.

SOLUTION

$18x^2 + 4 - [6(x^2 - 2) + 5]$

$= 18x^2 + 4 - [6x^2 - 12 + 5]$ Use the distributive property to remove parentheses:

$6(x^2 - 2) = 6x^2 - 6 \cdot 2 = 6x^2 - 12.$

$= 18x^2 + 4 - [6x^2 - 7]$ Add inside brackets: $-12 + 5 = -7.$

$= 18x^2 + 4 - 6x^2 + 7$ Remove brackets by changing the sign of each term within brackets.

$= (18x^2 - 6x^2) + 4 + 7$ Group like terms.

$= 12x^2 + 11$ Combine like terms. ∎

✔ **CHECK POINT 11** Simplify: $14x^2 + 5 - [7(x^2 - 2) + 4].$

---

4   Evaluate formulas.

Applications: Formulas and Mathematical Models One aim of algebra is to provide a compact, symbolic description of the world. These descriptions involve the use of *formulas*. A **formula** is a statement of equality that uses letters to express a relationship between two or more variables. For example, one variety of crickets chirps faster as the temperature rises. You can calculate the temperature by counting the number of times a cricket chirps per minute and applying the following formula:

$$T = 0.3n + 40.$$

In the formula, T is the temperature, in degrees Fahrenheit, and n is the number of cricket chirps per minute. We can use this formula to determine the temperature if you are sitting on your porch and count 80 chirps per minute. Here is how to do so:

$T = 0.3n + 40$ This is the given formula.

$T = 0.3(80) + 40$ Substitute 80 for n.

$T = 24 + 40$ Multiply: $0.3(80) = 24.$

$T = 64.$ Add.

When there are 80 cricket chirps per minute, the temperature is 64 degrees.

The process of finding formulas to describe real-world phenomena is called **mathematical modeling**. Such formulas, together with the meaning assigned to the variables, are called **mathematical models**. We often say that these formulas model, or describe, the relationship among the variables.

In creating mathematical models, we strive for both accuracy and simplicity. For example, the formula $T = 0.3n + 40$ is relatively simple to use. However, you should not get upset if you count 80 cricket chirps and the actual temperature is 62 degrees, rather than 64 degrees, as predicted by the formula. Many mathematical models give an approximate, rather than an exact, description of the relationship between variables.

Sometimes a mathematical model gives an estimate that is not a good approximation or is extended to include values of the variable that do not make sense. In these cases, we say that **model breakdown** has occurred. Here is an example:

Use the mathematical model $T = 0.3n + 40$ with $n = 1200$ (1200 cricket chirps per minute).

$$T = 0.3(1200) + 40 = 360 + 40 = 400$$

At 400° F, forget about 1200 chirps per minute! At this temperature, the cricket would "cook" and, alas, all chirping would cease.

EXAMPLE 12 Car Accidents and Age

The mathematical model

$$N = 0.4x^2 - 36x + 1000$$

approximates the number of accidents, N, per 50 million miles driven, for drivers who are x years old. The formula applies to drivers ages 16 through 74, inclusive. How many accidents, per 50 million miles driven, are there for 20-year-old drivers?

SOLUTION In the mathematical model, x represents the age of the driver. We are interested in 20-year-old drivers. Thus, we substitute 20 for each occurrence of x. Then we use the order of operations to find N, the number of accidents per 50 million miles driven.

$N = 0.4x^2 - 36x + 1000$ This is the given mathematical model.

$N = 0.4(20)^2 - 36(20) + 1000$ Replace each occurrence of x with 20.

$N = 0.4(400) - 36(20) + 1000$ Evaluate the exponential expression:
$(20)^2 = (20)(20) = 400.$

$N = 160 - 720 + 1000$ Multiply from left to right:
$0.4(400) = 160$ and $36(20) = 720.$

$N = -560 + 1000$ Perform additions and subtractions from left to right. Subtract:
$160 - 720 = 160 + (-720) = -560.$

$N = 440$ Add: $-560 + 1000 = 440.$

Thus, 20-year-old drivers have 440 accidents per 50 million miles driven. ∎

How do the number of accidents for 20-year-old drivers compare to, say, the number for 45-year-old drivers? The answer is given by the points in the rectangular coordinate system in Figure 1.22. The x-coordinate of each point represents the driver's age. The y-coordinate represents the number of accidents per 50 million miles driven.

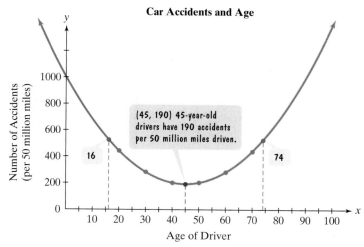

FIGURE 1.22

Can you see that the point (45, 190) is lower than any of the other points? In practical terms, this indicates that 45-year-olds have the least number of car accidents, 190 per 50 million miles driven. Drivers both younger and older than 45 have more accidents per 50 million miles driven.

> ✔ **CHECK POINT 12** Use the mathematical model described in Example 12, $N = 0.4x^2 - 36x + 1000$, to answer this question: How many accidents, per 50 million miles driven, are there for 40-year-old drivers?

Some formulas give an exact, rather than an approximate, relationship between variables. For example, Figure 1.23 shows temperatures on the Celsius scale and on the Fahrenheit scale. The formula

$$C = \frac{5}{9}(F - 32)$$

expresses an exact relationship between Fahrenheit temperature, F, and Celsius temperature, C.

The Formula	**What the Formula Tells Us**
$C = \dfrac{5}{9}(F - 32)$	If 32 is subtracted from the Fahrenheit temperature, $F - 32$, and this difference is multiplied by $\frac{5}{9}$, the resulting product, $\frac{5}{9}(F - 32)$, gives the Celsius temperature.

FIGURE 1.23 The Celsius scale is on the left and the Fahrenheit scale is on the right.

EXAMPLE 13   Converting from Fahrenheit to Celsius

The temperature on a warm spring day is 77°F. Use the formula $C = \frac{5}{9}(F - 32)$ to find the equivalent temperature on the Celsius scale.

SOLUTION   Because the temperature is 77°F, we substitute 77 for F in the given formula. Then we use the order of operations to find the value of C.

$$C = \frac{5}{9}(F - 32) \qquad \text{This is the given formula.}$$

$$C = \frac{5}{9}(77 - 32) \qquad \text{Replace F with 77.}$$

$$C = \frac{5}{9}(45) \qquad \text{Work inside parentheses first: } 77 - 32 = 45.$$

$$C = 25 \qquad \text{Multiply: } \frac{5}{9}(45) = \frac{5}{\cancel{9}} \cdot \frac{\cancel{45}^{\,5}}{1} = \frac{25}{1} = 25.$$

Thus, 77°F is equivalent to 25°C.   ∎

> ✔ **CHECK POINT 13** The temperature on a warm summer day is 86°F. Use the formula $C = \frac{5}{9}(F - 32)$ to find the equivalent temperature on the Celsius scale.

1.8 EXERCISE SET

Student Solutions Manual   CD/Video   PH Math/Tutor Center   MathXL Tutorials on CD   MathXL®   MyMathLab   Interactmath.com

Practice Exercises

In Exercises 1–14, evaluate each exponential expression.

1. 9^2 **2.** 3^2 **3.** 4^3

4. 6^3 **5.** $(-4)^2$ **6.** $(-10)^2$

7. $(-4)^3$ **8.** $(-10)^3$ **9.** $(-5)^4$

10. $(-1)^6$ **11.** -5^4 **12.** -1^6

13. -10^2 **14.** -8^2

In Exercises 15–28, simplify each algebraic expression, or explain why the expression cannot be simplified.

15. $7x^2 + 12x^2$ **16.** $6x^2 + 18x^2$

17. $10x^3 + 5x^3$ **18.** $14x^3 + 8x^3$

19. $8x^4 + x^4$ **20.** $14x^4 + x^4$

21. $26x^2 - 27x^2$

22. $29x^2 - 30x^2$

23. $27x^3 - 26x^3$

24. $30x^3 - 29x^3$

25. $5x^2 + 5x^3$

26. $8x^2 + 8x^3$

27. $16x^2 - 16x^2$

28. $34x^2 - x^2$

In Exercises 29–72, use the order of operations to simplify each expression.

29. $7 + 6 \cdot 3$

30. $3 + 4 \cdot 5$

31. $45 \div 5 \cdot 3$

32. $40 \div 4 \cdot 2$

33. $6 \cdot 8 \div 4$

34. $8 \cdot 6 \div 2$

35. $14 - 2 \cdot 6 + 3$

36. $36 - 12 \div 4 + 2$

37. $8^2 - 16 \div 2^2 \cdot 4 - 3$

38. $10^2 - 100 \div 5^2 \cdot 2 - 1$

39. $3(-2)^2 - 4(-3)^2$

40. $5(-3)^2 - 2(-4)^2$

41. $(4 \cdot 5)^2 - 4 \cdot 5^2$

42. $(3 \cdot 5)^2 - 3 \cdot 5^2$

43. $(2 - 6)^2 - (3 - 7)^2$

44. $(4 - 6)^2 - (5 - 9)^2$

45. $6(3 - 5)^3 - 2(1 - 3)^3$

46. $-3(-6 + 8)^3 - 5(-3 + 5)^3$

47. $[2(6 - 2)]^2$

48. $[3(4 - 6)]^3$

49. $2[5 + 2(9 - 4)]$

50. $3[4 + 3(10 - 8)]$

51. $[7 + 3(2^3 - 1)] \div 21$

52. $[11 - 4(2 - 3^3)] \div 37$

53. $\dfrac{10 + 8}{5^2 - 4^2}$

54. $\dfrac{6^2 - 4^2}{2 - (-8)}$

55. $\dfrac{37 + 15 \div (-3)}{2^4}$

56. $\dfrac{22 + 20 \div (-5)}{3^2}$

57. $\dfrac{(-11)(-4) + 2(-7)}{7 - (-3)}$

58. $\dfrac{-5(7 - 2) - 3(4 - 7)}{-13 - (-5)}$

59. $4|10 - (8 - 20)|$

60. $6|7 - 4 \cdot 3|$

61. $8(-10) + |4(-5)|$

62. $4(-15) + |3(-10)|$

63. $-2^2 + 4[16 \div (3 - 5)]$

64. $-3^2 + 2[20 \div (7 - 11)]$

65. $24 \div \dfrac{3^2}{8 - 5} - (-6)$

66. $30 \div \dfrac{5^2}{7 - 12} - (-9)$

67. $\dfrac{\dfrac{1}{4} - \dfrac{1}{2}}{\dfrac{1}{3}}$

68. $\dfrac{\dfrac{3}{5} - \dfrac{7}{10}}{\dfrac{1}{2}}$

69. $-\dfrac{9}{4}\left(\dfrac{1}{2}\right) + \dfrac{3}{4} \div \dfrac{5}{6}$

70. $\left[-\dfrac{4}{7} - \left(-\dfrac{2}{5}\right)\right]\left[-\dfrac{3}{8} + \left(-\dfrac{1}{9}\right)\right]$

71. $\dfrac{\dfrac{7}{9} - 3}{\dfrac{5}{6}} \div \dfrac{3}{2} + \dfrac{3}{4}$

72. $\dfrac{\dfrac{17}{25}}{\dfrac{3}{5} - 4} \div \dfrac{1}{5} + \dfrac{1}{2}$

In Exercises 73–80, evaluate each algebraic expression for the given value of the variable.

73. $x^2 + 5x; x = 3$

74. $x^2 - 2x; x = 6$

75. $3x^2 - 8x; x = -2$

76. $4x^2 - 2x; x = -3$

77. $-x^2 - 10x; x = -1$

78. $-x^2 - 14x; x = -1$

79. $\dfrac{6y - 4y^2}{y^2 - 15}; y = 5$

80. $\dfrac{3y - 2y^2}{y(y - 2)}; y = 5$

In Exercises 81–88, simplify each algebraic expression by removing parentheses and brackets.

81. $3[5(x - 2) + 1]$

82. $4[6(x - 3) + 1]$

83. $3[6 - (y + 1)]$

84. $5[2 - (y + 3)]$

85. $7 - 4[3 - (4y - 5)]$

86. $6 - 5[8 - (2y - 4)]$

87. $2(3x^2 - 5) - [4(2x^2 - 1) + 3]$

88. $4(6x^2 - 3) - [2(5x^2 - 1) + 1]$

Practice Plus

In Exercises 89–92, express each sentence as a single numerical expression. Then use the order of operations to simplify the expression.

89. Cube -2. Subtract this exponential expression from -10.

90. Cube -5. Subtract this exponential expression from -100.

91. Subtract 10 from 7. Multiply this difference by 2. Square this product.

92. Subtract 11 from 9. Multiply this difference by 2. Raise this product to the fourth power.

In Exercises 93–96, let x represent the number. Express each sentence as a single algebraic expression. Then simplify the expression.

93. Multiply a number by 5. Add 8 to this product. Subtract this sum from the number.

94. Multiply a number by 3. Add 9 to this product. Subtract this sum from the number.

95. Cube a number. Subtract 4 from this exponential expression. Multiply this difference by 5.

96. Cube a number. Subtract 6 from this exponential expression. Multiply this difference by 4.

Application Exercises

Medical researchers have found that the desirable heart rate, R, in beats per minute, for beneficial exercise is approximated by the mathematical models

$$R = 165 - 0.75A \qquad \text{for men}$$
$$R = 143 - 0.65A \qquad \text{for women}$$

where A is the person's age. Use these mathematical models to solve Exercises 97–98.

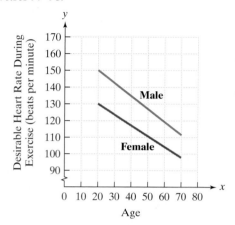

97. What is the desirable heart rate during exercise for a 40-year-old man? Identify your computation as an appropriate point in the rectangular coordinate system.

98. What is the desirable heart rate during exercise for a 40-year-old woman? Identify your computation as an appropriate point in the rectangular coordinate system.

The bar graph shows the cost of Medicare, in billions of dollars, through 2005. The data can be modeled by the formula

$$N = 1.2x^2 + 15.2x + 181.4,$$

where N represents Medicare spending, in billions of dollars, x years after 1995. Use this formula to solve Exercises 99–100.

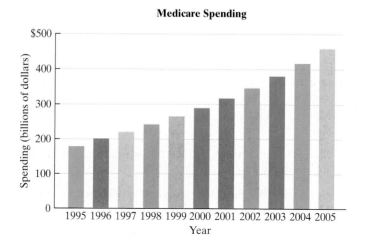

Medicare Spending

Source: Congressional Budget Office

99. According to the formula, what was the cost of Medicare, in billions of dollars, in 2000? How well does the formula describe the cost for that year shown by the bar graph?

100. According to the formula, what was the cost of Medicare, in billions of dollars, in 2005? How well does the formula describe the cost for that year shown by the bar graph?

Bariatrics is the field of medicine that deals with the overweight. Bariatric surgery closes off a large part of the stomach. As a result, patients eat less and have a diminished appetite. Celebrities like pop singer Carnie Wilson and the Today show's weatherman Al Roker have become no-longer-larger-than-life walking billboards for the operation. The bar graph shows the number of bariatric surgeries from 1992 through 2002.

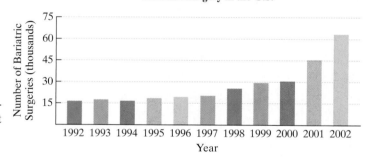

Bariatric Surgery in the U.S.

Source: American Society for Bariatric Surgery

The formula

$$N = 0.12x^3 - x^2 + 3x + 15$$

models the number of bariatric surgeries, N, in thousands, x years after 1992. Use this formula to solve Exercises 101–102.

101. According to the formula, how many bariatric surgeries were performed in 2002? How well does the formula model the data shown in the bar graph?

102. According to the formula, how many bariatric surgeries were performed in 1994? How well does the formula model the data shown in the bar graph?

The formula

$$C = \frac{5}{9}(F - 32)$$

expresses the relationship between Fahrenheit temperature, F, and Celsius temperature, C. In Exercises 103–106, use the formula to convert the given Fahrenheit temperature to its equivalent temperature on the Celsius scale.

103. 68°F

104. 41°F

105. −22°F

106. −31°F

Writing in Mathematics

107. Describe what it means to raise a number to a power. In your description, include a discussion of the difference between -5^2 and $(-5)^2$.

108. Explain how to simplify $4x^2 + 6x^2$. Why is the sum not equal to $10x^4$?

109. Why is the order of operations agreement needed?

110. What is a formula?

111. The formula $F = \frac{9}{5}C + 32$ expresses the relationship between Celsius temperature, C, and Fahrenheit temperature, F. You'll be leaving the cold of winter for a vacation to Hawaii. CNN International reports a temperature in Hawaii of 30°C. Should you pack a winter coat? Use the formula to explain your answer.

Critical Thinking Exercises

112. Which one of the following is true?

  a. If x is -3, then the value of $-3x - 9$ is -18.

  b. The algebraic expression $\dfrac{6x + 6}{x + 1}$ cannot have the same value when two different replacements are made for x such as $x = -3$ and $x = 2$.

  c. A miniature version of a space shuttle is an example of a mathematical model.

  d. The value of $\dfrac{|3 - 7|-2^3}{(-2)(-3)}$ is the fraction that results when $\frac{1}{3}$ is subtracted from $-\frac{1}{3}$.

113. Simplify: $\quad \dfrac{1}{4} - 6(2 + 8) \div \left(-\dfrac{1}{3}\right)\left(-\dfrac{1}{9}\right).$

Grouping symbols can be inserted into $4 + 3 \cdot 7 - 4$ so that the resulting value is 45. By placing parentheses around the addition we obtain

$$(4 + 3) \cdot 7 - 4 = 7 \cdot 7 - 4 = 49 - 4 = 45.$$

In Exercises 114–115, insert parentheses in each expression so that the resulting value is 45.

114. $2 \cdot 3 + 3 \cdot 5$

115. $2 \cdot 5 - \dfrac{1}{2} \cdot 10 \cdot 9$

Technology Exercises

The United States has more people in prison, as well as more people in prison per capita, than any other western industrialized nation. The bar graph shows the number of inmates in U.S. state and federal prisons in nine selected years from 1985 through 2001.

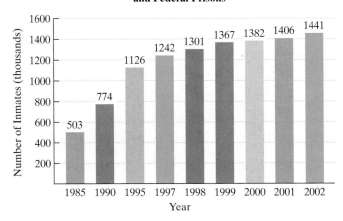

Number of Inmates in U.S. State and Federal Prisons

Source: U.S. Justice Department

The data in the graph can be modeled by each of the following formulas:

$$N = 59.5x + 505.12$$
$$N = -0.619x^2 + 69.52x + 485.28$$
$$N = -0.239x^3 + 5.288x^2 + 33.28x + 503.47.$$

In each formula, x represents the number of years after 1985 and N represents the number of inmates, in thousands. Use this information to solve Exercises 116–117.

116. Use a calculator to determine which of the three formulas serves as the best model for the inmate population in 2001.

117. Use a calculator to determine which, if any, of the three formulas serves as the best model for the inmate population for the period shown by the graph.

118. In this exercise, use a calculator and one of the formulas that estimates the adult height of a child given in the chapter introduction on page 1. Predict the adult height, H, in inches using the actual height of a child whose age, x, and height, h, in inches, you know.

Review Exercises

119. Simplify: $-8 - 2 - (-5) + 11$ (Section 1.6, Example 3)

120. Multiply: $-4(-1)(-3)(2)$. (Section 1.7, Example 2)

121. Give an example of a real number that is not an irrational number. (Section 1.2, Example 5).

GROUP PROJECT

CHAPTER 1

One measure of physical fitness is your *resting heart rate*. Generally speaking, the more fit you are, the lower your resting heart rate. The best time to take this measurement is when you first awaken in the morning, before you get out of bed. Lie on your back with no body parts crossed and take your pulse in your neck or wrist. Use your index and second fingers and count your pulse beat for one full minute to get your resting heart rate. A resting heart rate under 48 to 57 indicates high fitness, 58 to 62, above average fitness, 63 to 70, average fitness, 71 to 82, below average fitness, and 83 or more, low fitness.

Another measure of physical fitness is your percentage of body fat. You can estimate your body fat using the following formulas:

For men: Body fat $= -98.42 + 4.15w - 0.082b$

For women: Body fat $= -76.76 + 4.15w - 0.082b$

where w = waist measurement, in inches, and b = total body weight, in pounds. Then divide your body fat by your total weight to get your body fat percentage. For men, less than 15% is considered athletic, 25% about average. For women, less than 22% is considered athletic, 30% about average.

Each group member should bring his or her age, resting heart rate, and body fat percentage to the group. Using the data, the group should create three graphs.

a. Create a graph that shows age and resting heart rate for group members.

b. Create a graph that shows age and body fat percentage for group members.

c. Create a graph that shows resting heart rate and body fat percentage for group members.

For each graph, select the style (line or bar) that is most appropriate.

CHAPTER 1 SUMMARY

Definitions and Concepts	Examples

Section 1.1 Fractions

Mixed Numbers and Improper Fractions

A mixed number consists of the addition of a natural number $(1, 2, 3, \ldots)$ and a fraction, expressed without the use of an addition sign. An improper fraction has a numerator that is greater than its denominator. To convert a mixed number to an improper fraction, multiply the denominator by the natural number and add the numerator. Then place this result over the original denominator.

Convert $5\frac{3}{7}$ to an improper fraction.

$$5\frac{3}{7} = \frac{7 \cdot 5 + 3}{7} = \frac{35 + 3}{7} = \frac{38}{7}$$

To convert an improper fraction to a mixed number, divide the denominator into the numerator and write the mixed number using

$$\text{quotient} \frac{\text{remainder}}{\text{original denominator}}.$$

Convert $\frac{14}{3}$ to a mixed number.

$$\frac{14}{3} = 4\frac{2}{3} \qquad \begin{array}{r} 4 \\ 3\overline{)14} \\ \underline{12} \\ 2 \end{array}$$

A prime number is a natural number greater than 1 that has only itself and 1 as factors. A composite number is a natural number greater than 1 that is not a prime number.
The prime factorization of a composite number means to express the composite number as the product of prime numbers.

Find the prime factorization:

$$60 = 6 \cdot 10$$
$$= 2 \cdot 3 \cdot 2 \cdot 5$$

Definitions and Concepts	Examples

Section 1.1 Fractions (continued)

A fraction is reduced to its lowest terms when the numerator and denominator have no common factors other than 1. To reduce a fraction to its lowest terms, divide both the numerator and the denominator by their greatest common factor. The greatest common factor can be found by inspection or prime factorizations of the numerator and the denominator.	Reduce to lowest terms: $$\frac{8}{14} = \frac{2 \cdot 4}{2 \cdot 7} = \frac{4}{7}$$
Multiplying Fractions The product of two or more fractions is the product of their numerators divided by the product of their denominators.	Multiply: $$\frac{2}{7} \cdot \frac{5}{9} = \frac{2 \cdot 5}{7 \cdot 9} = \frac{10}{63}$$
Dividing Fractions The quotient of two fractions is the first multiplied by the reciprocal (or multiplicative inverse) of the second.	Divide: $$\frac{4}{9} \div \frac{3}{7} = \frac{4}{9} \cdot \frac{7}{3} = \frac{4 \cdot 7}{9 \cdot 3} = \frac{28}{27}$$
Adding and Subtracting Fractions with Identical Denominators Add or subtract numerators. Put this result over the common denominator.	Subtract: $$\frac{5}{8} - \frac{3}{8} = \frac{5 - 3}{8} = \frac{2}{8} = \frac{2 \cdot 1}{2 \cdot 4} = \frac{1}{4}$$
Adding and Subtracting Fractions with Unlike Denominators Rewrite the fractions as equivalent fractions with the least common denominator. Then add or subtract numerators, putting this result over the common denominator.	Add: $$\frac{3}{8} + \frac{5}{12} = \frac{3}{8} \cdot \frac{3}{3} + \frac{5}{12} \cdot \frac{2}{2}$$ **The LCD is 24.** $$= \frac{9}{24} + \frac{10}{24} = \frac{19}{24}$$

Section 1.2 The Real Numbers

A set is a collection of objects, called elements, whose contents can be clearly determined.	$$\{a, b, c\}$$
A line used to visualize numbers is called a number line.	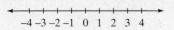 $-4\ -3\ -2\ -1\ \ 0\ \ 1\ \ 2\ \ 3\ \ 4$
Real Numbers: the set of all numbers that can be represented by points on the number line **The Sets That Make Up the Real Numbers** • Natural Numbers: $\{1, 2, 3, 4, \dots\}$ • Whole Numbers: $\{0, 1, 2, 3, 4, \dots\}$ • Integers: $\{\dots, -3, -2, -1, 0, 1, 2, 3, \dots\}$ • Rational Numbers: the set of numbers that can be expressed as the quotient of an integer and a nonzero integer; can be expressed as terminating or repeating decimals • Irrational Numbers: the set of numbers that cannot be expressed as the quotient of integers; decimal representations neither terminate nor repeat.	Given the set $$\left\{-1.4, 0, 0.\overline{7}, \frac{9}{10}, \sqrt{2}, \sqrt{4}\right\}$$ list the • natural numbers: $\sqrt{4}$, or 2 • whole numbers: $0, \sqrt{4}$ • rational numbers: $-1.4, 0, 0.\overline{7}, \frac{9}{10}, \sqrt{4}$ • irrational numbers: $\sqrt{2}$ • real numbers: $$-1.4, 0, 0.\overline{7}, \frac{9}{10}, \sqrt{2}, \sqrt{4}$$

Definitions and Concepts	Examples

Section 1.2 The Real Numbers (continued)

For any two real numbers, a and b, a is less than b if a is to the left of b on the number line.	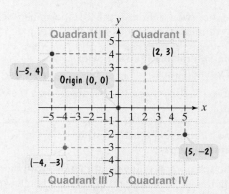

$-4\ -3\ -2\ -1\ \ 0\ \ 1\ \ 2\ \ 3\ \ 4$

Inequality Symbols

$<$: is less than	$>$: is greater than
$\leq$: is less than or equal to	$\geq$: is greater than or equal to

$$-2 < 0 \qquad 0 > -2$$
$$0 < 2.5 \qquad 2.5 > 0$$

The absolute value of a, written $|a|$, is the distance from 0 to a on the number line.	$|4| = 4 \quad |0| = 0 \quad |-6| = 6$

Section 1.3 Ordered Pairs and Graphs

The rectangular coordinate system consists of a horizontal number line, the x-axis, and a vertical number line, the y-axis, intersecting at their zero points, the origin. Each point in the system corresponds to an ordered pair of real numbers (x, y). The first number in the pair is the x-coordinate; the second number is the y-coordinate.

Plot: $(2, 3), (-5, 4), (-4, -3), (5, -2)$.

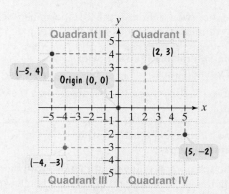

Information is often displayed using line graphs, bar graphs, and circle graphs. Line graphs are often used to illustrate trends over time. Bar graphs are convenient for showing comparisons among items.

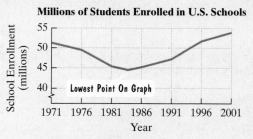

Millions of Students Enrolled in U.S. Schools

Source: National Education Association

In which year did enrollment reach a minimum? 1984
Estimate enrollment for that year: 44.5 million.

Section 1.4 Basic Rules of Algebra

A letter used to represent a number is called a variable. An algebraic expression is a combination of variables, numbers, and operation symbols. Terms are separated by addition. The parts of each term that are multiplied are its factors. Like terms have the same variable factors raised to the same powers. To evaluate an algebraic expression, substitute a given number for the variable and simplify.

Evaluate: $6(x + 3) + 4x$ when $x = 5$.

Replace x with 5.

$$= 6(5 + 3) + 4 \cdot 5$$

$$= 6(8) + 4 \cdot 5$$

$$= 48 + 20$$

$$= 68$$

Definitions and Concepts	Examples

Section 1.4 Basic Rules of Algebra (continued)

Properties of Real Numbers and Algebraic Expressions

- Commutative Properties:
$$a + b = b + a$$
$$ab = ba$$
- Associative Properties:
$$(a + b) + c = a + (b + c)$$
$$(ab)c = a(bc)$$
- Distributive Properties:
$$a(b + c) = ab + ac$$
$$(b + c)a = ba + ca$$
$$a(b - c) = ab - ac$$
$$(b - c)a = ba - ca$$
$$a(b + c + d) = ab + ac + ad$$

Commutative of Addition:
$$5x + 4 = 4 + 5x$$
Commutative of Multiplication:
$$5x + 4 = x5 + 4$$
Associative of Addition:
$$6 + (4 + x) = (6 + 4) + x = 10 + x$$
Associative of Multiplication:
$$7(10x) = (7 \cdot 10)x = 70x$$
Distributive:
$$8(x + 5 + 4y) = 8x + 40 + 32y$$
Distributive to Combine Like Terms:
$$8x + 12x = (8 + 12)x = 20x$$

Simplifying Algebraic Expressions

Use the distributive property to remove grouping symbols. Then combine like terms.

$$4(5x + 7) + 13x$$
$$= 20x + 28 + 13x$$
$$= (20x + 13x) + 28$$
$$= 33x + 28$$

Section 1.5 Addition of Real Numbers

Sums on a Number Line

To find $a + b$, the sum of a and b, on a number line, start at a. If b is positive, move b units to the right. If b is negative, move b units to the left. If b is 0, stay at a. The number where we finish on the number line represents $a + b$.

$$-7 + 5 = -2$$

Start at −7. Move 5 units to the right.

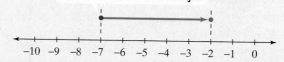

Additive inverses are pairs of real numbers that are the same number of units from zero on the number line, but on opposite sides of zero.

- Identity Property of Addition:
$$a + 0 = 0 \qquad 0 + a = a$$
- Inverse Property of Addition:
$$a + (-a) = 0 \qquad (-a) + a = 0$$

The additive inverse (or opposite) of 4 is −4. The additive inverse of −1.7 is 1.7.

Identity Property of Addition:
$$4x + 0 = 4x$$

Inverse Property of Addition:
$$4x + (-4x) = 0$$

Addition without a Number Line

To add two numbers with the same sign, add their absolute values and use their common sign. To add two numbers with different signs, subtract the smaller absolute value from the greater absolute value and use the sign of the number with the greater absolute value.

Add:
$$10 + 4 = 14$$
$$-4 + (-6) = -10$$
$$-30 + 5 = -25$$
$$12 + (-8) = 4$$

To add a series of positive and negative numbers, add all the positive numbers and add all the negative numbers. Then add the resulting positive and negative sums.

$$5 + (-3) + (-7) + 2$$
$$= (5 + 2) + [(-3) + (-7)]$$
$$= 7 + (-10)$$
$$= -3$$

Definitions and Concepts	Examples

Section 1.6 Subtraction of Real Numbers

To subtract b from a, add the additive inverse of b to a: $$a - b = a + (-b).$$ The result is called the difference between a and b.	Subtract: $$-7 - (-5) = -7 + 5 = -2$$ $$-\frac{3}{4} - \frac{1}{2} = -\frac{3}{4} + \left(-\frac{1}{2}\right)$$ $$= -\frac{3}{4} + \left(-\frac{2}{4}\right) = -\frac{5}{4}$$
To simplify a series of additions and subtractions, change all subtractions to additions of additive inverses. Then use the procedure for adding a series of positive and negative numbers.	Simplify: $$-6 - 2 - (-3) + 10$$ $$= -6 + (-2) + 3 + 10$$ $$= -8 + 13$$ $$= 5$$

Section 1.7 Multiplication and Division of Real Numbers

The result of multiplying a and b, ab, is called the product of a and b. If the two numbers have different signs, the product is negative. If the two numbers have the same sign, the product is positive. If either number is 0, the product is 0.	Multiply: $$-5(-10) = 50$$ $$\frac{3}{4}\left(-\frac{5}{7}\right) = -\frac{3}{4}\cdot\frac{5}{7} = -\frac{15}{28}$$
Assuming that no number is 0, the product of an even number of negative numbers is positive. The product of an odd number of negative numbers is negative. If any number is 0, the product is 0.	Multiply: $$(-3)(-2)(-1)(-4) = 24$$ $$(-3)(2)(-1)(-4) = -24$$
The result of dividing the real number a by the nonzero real number b is called the quotient of a and b. If two numbers have different signs, their quotient is negative. If two numbers have the same sign, their quotient is positive. Division by zero is undefined.	Divide: $$\frac{21}{-3} = -7$$ $$-\frac{1}{3} \div (-3) = \frac{1}{3}\cdot\frac{1}{3} = \frac{1}{9}$$
Two numbers whose product is 1 are called multiplicative inverses or reciprocals of each other. The number 0 has no multiplicative inverse. • Identity Property of Multiplication $$a \cdot 1 = a \qquad 1 \cdot a = a$$ • Inverse Property of Multiplication If a is not 0: $$a \cdot \frac{1}{a} = 1 \qquad \frac{1}{a} \cdot a = 1$$ • Multiplication Property of -1 $$-1a = -a \qquad a(-1) = -a$$ • Double Negative Property $$-(-a) = a$$	The multiplicative inverse of 4 is $\frac{1}{4}$. The multiplicative inverse of $-\frac{1}{3}$ is -3. Simplify: $$1x = x$$ $$7x \cdot \frac{1}{7x} = 1$$ $$4x - 5x = -1x = -x$$ $$-(-7y) = 7y$$
If a negative sign precedes parentheses, remove parentheses and change the sign of every term within parentheses.	Simplify: $$-(7x - 3y + 2) = -7x + 3y - 2$$

Definitions and Concepts | Examples

Section 1.8 Exponents, Order of Operations, and Mathematical Models

Definitions and Concepts	Examples
If b is a real number and n is a natural number, b^n, the nth power of b, is the product of n factors of b. Furthermore, $b^1 = b$.	Evaluate: $$8^2 = 8 \cdot 8 = 64$$ $$(-5)^3 = (-5)(-5)(-5) = -125$$
Order of Operations 1. Perform operations within grouping symbols, starting with the innermost grouping symbols. 2. Evaluate exponential expressions. 3. Multiply and divide in order from left to right. 4. Add and subtract in order from left to right.	Simplify: $$\begin{aligned} 5(4-6)^2 &- 2(1-3)^3 \\ &= 5(-2)^2 - 2(-2)^3 \\ &= 5(4) - 2(-8) \\ &= 20 - (-16) \\ &= 20 + 16 = 36 \end{aligned}$$
Some algebraic expressions contain two sets of grouping symbols: parentheses, the inner grouping symbols, and brackets, the outer grouping symbols. To simplify such expressions, use the order of operations and remove grouping symbols from innermost (parentheses) to outermost (brackets).	Simplify: $$\begin{aligned} 5 &- 3[2(x+1) - 7] \\ &= 5 - 3[2x + 2 - 7] \\ &= 5 - 3[2x - 5] \\ &= 5 - 6x + 15 \\ &= -6x + 20 \end{aligned}$$
A formula is a statement of equality that uses letters to express a relationship between two or more variables. Formulas that describe real-world phenomena are called mathematical models. Many mathematical models give an approximate, rather than an exact, description of the relationship between variables.	The formula $N = 0.4x^2 + 0.5$ models the millions of people, N, in the United States using cable modems x years after 1996.

CHAPTER 1 REVIEW EXERCISES

1.1 *In Exercises 1–2, convert each mixed number to an improper fraction.*

1. $3\dfrac{2}{7}$

2. $5\dfrac{9}{11}$

In Exercises 3–4, convert each improper fraction to a mixed number.

3. $\dfrac{17}{9}$

4. $\dfrac{27}{5}$

In Exercises 5–7, identify each natural number as prime or composite. If the number is composite, find its prime factorization.

5. 60

6. 63

7. 67

In Exercises 8–9, simplify each fraction by reducing it to its lowest terms.

8. $\dfrac{15}{33}$

9. $\dfrac{40}{75}$

In Exercises 10–15, perform the indicated operation. Where possible, reduce the answer to its lowest terms.

10. $\dfrac{3}{5} \cdot \dfrac{7}{10}$

11. $\dfrac{4}{5} \div \dfrac{3}{10}$

12. $1\dfrac{2}{3} \div 6\dfrac{2}{3}$

13. $\dfrac{2}{9} + \dfrac{4}{9}$

14. $\dfrac{5}{6} + \dfrac{7}{9}$

15. $\dfrac{3}{4} - \dfrac{2}{15}$

16. The gas tank of a car is filled to its capacity. The first day, $\frac{1}{4}$ of the tank's gas is used for travel. The second day, $\frac{1}{3}$ of the tank's original amount of gas is used for travel. What fraction of the tank is filled with gas at the end of the second day?

1.2 *In Exercises 17–18, graph each real number on a number line.*

17. -2.5

18. $4\dfrac{3}{4}$

In Exercises 19–20, express each rational number as a decimal.

19. $\dfrac{5}{8}$

20. $\dfrac{3}{11}$

21. Consider the set

$$\left\{ -17, -\frac{9}{13}, 0, 0.75, \sqrt{2}, \pi, \sqrt{81} \right\}.$$

List all numbers from the set that are: **a.** natural numbers, **b.** whole numbers, **c.** integers, **d.** rational numbers, **e.** irrational numbers, **f.** real numbers.

22. Give an example of an integer that is not a natural number.

23. Give an example of a rational number that is not an integer.

24. Give an example of a real number that is not a rational number.

In Exercises 25–28, insert either $<$ or $>$ in the shaded area between each pair of numbers to make a true statement.

25. -93 ▢ 17

26. -2 ▢ -200

27. 0 ▢ $-\frac{1}{3}$

28. $-\frac{1}{4}$ ▢ $-\frac{1}{5}$

In Exercises 29–30, determine whether each inequality is true or false.

29. $-13 \geq -11$

30. $-126 \leq -126$

In Exercises 31–32, find each absolute value.

31. $|-58|$

32. $|2.75|$

1.3 *In Exercises 33–36, plot the given point in a rectangular coordinate system. Indicate in which quadrant each point lies.*

33. $(1, -5)$

34. $(4, -3)$

35. $\left(\frac{7}{2}, \frac{3}{2} \right)$

36. $(-5, 2)$

37. Give the ordered pairs that correspond to the points labeled in the figure.

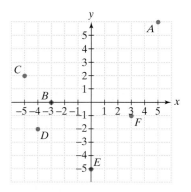

We live in an era of democratic aspiration. The number of democracies worldwide is on the rise. The line graph shows the number of democracies worldwide, in four-year periods, from 1973 through 2001, including 2002. Use the graph to solve Exercises 38–42.

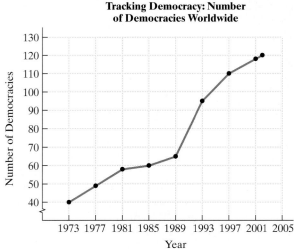

Tracking Democracy: Number of Democracies Worldwide

Source: The Freedom House

38. Find an estimate for the number of democracies in 1989.

39. How many more democracies were there in 2002 than in 1973?

40. In which four-year period did the number of democracies increase at the greatest rate?

41. In which four-year period did the number of democracies increase at the slowest rate?

42. In which year were there 49 democracies?

The United States has the highest rate of television ownership in the world. The bar graph shows the seven countries in the world with the greatest number of televisions per 100 people. Use the graph to solve Exercises 43–44.

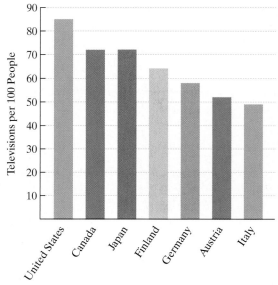

Television Ownership in Global Perspective

Source: The World Bank

43. Estimate the number of televisions per 100 people in the United States.

44. Which countries have more than 50 but fewer than 70 televisions per 100 people?

1.4 *In Exercises 45–46, evaluate each algebraic expression for the given value of the variable.*

45. $7x + 3$; $x = 10$ **46.** $5(x - 4)$; $x = 12$

47. Use the commutative property of addition to write an equivalent algebraic expression: $7 + 13y$.

48. Use the commutative property of multiplication to write an equivalent algebraic expression: $9(x + 7)$.

In Exercises 49–50, use an associative property to rewrite each algebraic expression. Then simplify the resulting algebraic expression.

49. $6 + (4 + y)$

50. $7(10x)$

51. Use the distributive property to rewrite without parentheses: $6(4x - 2 + 5y)$.

In Exercises 52–53, simplify each algebraic expression.

52. $4a + 9 + 3a - 7$

53. $6(3x + 4) + 5(2x - 1)$

54. Suppose that a store is selling all computers at 25% off the regular price. If x is the regular price, the algebraic expression $x - 0.25x$ describes the sale price. Evaluate the expression when $x = 2400$. Describe what the answer means in practical terms.

1.5

55. Use a number line to find the sum: $-6 + 8$.

In Exercises 56–58, find each sum without the use of a number line.

56. $8 + (-11)$ **57.** $-\dfrac{3}{4} + \dfrac{1}{5}$

58. $7 + (-5) + (-13) + 4$

In Exercises 59–60, simplify each algebraic expression.

59. $8x + (-6y) + (-12x) + 11y$

60. $10(4 - 3y) + 28y$

61. The Dead Sea is the lowest elevation on Earth, 1312 feet below sea level. If a person is standing 512 feet above the Dead Sea, what is that person's elevation?

62. The water level of a reservoir is measured over a five-month period. At the beginning, the level is 25 feet. During this time, the level fell 3 feet, then rose 2 feet, then rose 1 foot, then fell 4 feet, and then rose 2 feet. What is the reservoir's water level at the end of the five months?

1.6

63. Rewrite $9 - 13$ as the addition of an additive inverse.

In Exercises 64–66, perform the indicated subtraction.

64. $-9 - (-13)$ **65.** $-\dfrac{7}{10} - \dfrac{1}{2}$

66. $-3.6 - (-2.1)$

In Exercises 67–68, simplify each series of additions and subtractions.

67. $-7 - (-5) + 11 - 16$

68. $-25 - 4 - (-10) + 16$

69. Simplify: $3 - 6a - 8 - 2a$.

70. What is the difference in elevation between a plane flying 26,500 feet above sea level and a submarine traveling 650 feet below sea level?

1.7 *In Exercises 71–73, perform the indicated multiplication.*

71. $-7(-12)$ **72.** $\dfrac{3}{5}\left(-\dfrac{5}{11}\right)$

73. $5(-3)(-2)(-4)$

In Exercises 74–76, perform the indicated division or state that the expression is undefined.

74. $\dfrac{45}{-5}$ **75.** $-17 \div 0$

76. $-\dfrac{4}{5} \div \left(-\dfrac{2}{5}\right)$

In Exercises 77–78, simplify each algebraic expression.

77. $-4\left(-\dfrac{3}{4}x\right)$ **78.** $-3(2x - 1) - (4 - 5x)$

1.8 *In Exercises 79–81, evaluate each exponential expression.*

79. $(-6)^2$ **80.** -6^2 **81.** $(-2)^5$

In Exercises 82–83, simplify each algebraic expression, or explain why the expression cannot be simplified.

82. $4x^3 + 2x^3$ **83.** $4x^3 + 4x^2$

In Exercises 84–92, use the order of operations to simplify each expression.

84. $-40 \div 5 \cdot 2$ **85.** $-6 + (-2) \cdot 5$

86. $6 - 4(-3 + 2)$ **87.** $28 \div (2 - 4^2)$

88. $36 - 24 \div 4 \cdot 3 - 1$ **89.** $-8[-4 - 5(-3)]$

90. $\dfrac{6(-10 + 3)}{2(-15) - 9(-3)}$ **91.** $\left(\dfrac{1}{2} + \dfrac{1}{3}\right) \div \left(\dfrac{1}{4} - \dfrac{3}{8}\right)$

92. $\dfrac{1}{2} - \dfrac{2}{3} \div \dfrac{5}{9} + \dfrac{3}{10}$

In Exercises 93–94, evaluate each algebraic expression for the given value of the variable.

93. $x^2 - 2x + 3; x = -1$

94. $-x^2 - 7x; x = -2$

In Exercises 95–96, simplify each algebraic expression.

95. $4[7(a - 1) + 2]$

96. $-6[4 - (y + 2)]$

On the average, infant girls weigh 7 pounds at birth and gain 1.5 pounds for each month for the first six months. The formula

$$W = 1.5x + 7$$

models a baby girl's weight, W, in pounds, after x months, where x is less than or equal to 6. Use the formula to solve Exercises 97–98.

Average Weight for Infant Girls

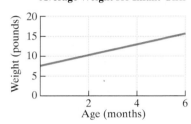

97. What does an infant girl weigh after four months? Identify your computation as an appropriate point on the line graph.

98. What does an infant girl weigh after six months? Identify your computation as an appropriate point on the line graph.

Among people under 40, opinions are split on downloading music: 45% think it's stealing, while 46% think it's not. (Source: Newsweek, September 22, 2003) Downloading and music CD prices that remained high through 2003 resulted in the decline of CD sales shown in the bar graph. The data can be modeled by the formula

$$N = -26x^2 + 143x + 740,$$

where N represents the number of music CD sales, in millions, x years after 1997. Use the formula to solve Exercises 99–100.

Music CD Sales in the U.S.

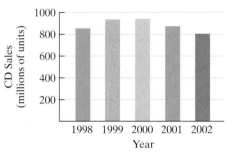

Source: RIAA

99. According to the formula, how many millions of CDs were sold in 2002? How well does the formula describe sales for that year shown by the bar graph?

100. What do you notice about CD sales from 1998 through 2002 from the bar graph that is not obvious by looking at the formula?

CHAPTER 1   TEST

Remember to use your Chapter Test Prep Video CD to see the worked-out solutions to the test questions you want to review.

In Exercises 1–10, perform the indicated operation or operations.

1. $1.4 - (-2.6)$

2. $-9 + 3 + (-11) + 6$

3. $3(-17)$

4. $\left(-\frac{3}{7}\right) \div \left(-\frac{15}{7}\right)$

5. $\left(3\frac{1}{3}\right)\left(-1\frac{3}{4}\right)$

6. $-50 \div 10$

7. $-6 - (5 - 12)$

8. $(-3)(-4) \div (7 - 10)$

9. $(6 - 8)^2(5 - 7)^3$

10. $\dfrac{3(-2) - 2(2)}{-2(8 - 3)}$

In Exercises 11–13, simplify each algebraic expression.

11. $11x - (7x - 4)$

12. $5(3x - 4y) - (2x - y)$

13. $6 - 2[3(x + 1) - 5]$

14. List all the rational numbers in this set.

$$\left\{-7, -\frac{4}{5}, 0, 0.25, \sqrt{3}, \sqrt{4}, \frac{22}{7}, \pi\right\}$$

15. Insert either $<$ or $>$ in the shaded area to make a true statement: -1 _____ -100.

16. Find the absolute value: $|-12.8|$.

17. Plot $(-4, 3)$ in a rectangular coordinate system. In which quadrant does the point lie?

18. Find the coordinates of point *A* in the figure.

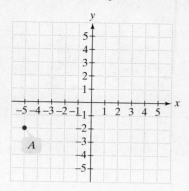

In Exercises 19–20, evaluate each algebraic expression for the given value of the variable.

19. $5(x - 7); x = 4$

20. $x^2 - 5x; x = -10$

21. Use the commutative property of addition to write an equivalent algebraic expression: $2(x + 3)$.

22. Use the associative property of multiplication to rewrite $-6(4x)$. Then simplify the expression.

23. Use the distributive property to rewrite without parentheses: $7(5x - 1 + 2y)$.

Tule elk are introduced into a newly acquired habitat. The points in the rectangular coordinate system show the elk population every ten years. Use this information to solve Exercises 24–25.

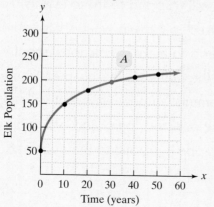

24. Find the coordinates of point *A*. Then interpret the coordinates in terms of the information given.

25. How many elk were introduced into the habitat?

26. Use the bar graph to find a reasonable estimate for the millions of DVDs sold in the United States in 2002.

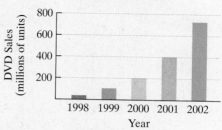

Source: International Recording Media Association

27. The formula

$$T = 3(A - 20)^2 \div 50 + 10$$

describes the average running time, *T*, in seconds, for a person who is *A* years old to run the 100-yard dash. How long does it take a 30-year-old runner to run the 100-yard dash?

28. What is the difference in elevation between a plane flying 16,200 feet above sea level and a submarine traveling 830 feet below sea level?

29. Use the graph to find a reasonable estimate, in thousands of dollars, for the average price of an existing single-family home in the United States in 1999.

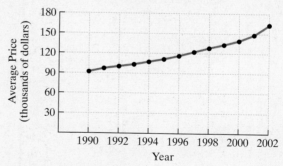

Source: National Association of Realtors

30. The data in Exercise 29 can be modeled by the formula

$$N = 0.3x^2 + 1.2x + 92.7,$$

where *N* represents the average price of an existing single-family home, in thousands of dollars, *x* years after 1989. Use the formula to find the average price of such a home in 1999.

Reading your psychology textbook, you were intrigued by a study that found undergraduate students with an average or high sense of humor reported few increases in depression over time. By contrast, students with a low sense of humor were more likely to become depressed in response to negative life events. Is there a way to model these variables and use formulas to predict how we will respond to difficult life events?

This problem appears as Exercises 75–76 in Exercise Set 2.3.

Linear Equations and Inequalities in One Variable

The belief that humor and laughter can have positive benefits on our lives is not new. The Bible tells us, "A merry heart doeth good like a medicine, but a broken spirit drieth the bones". (Proverbs 17:22). Algebra can be used to model the influence that humor plays in our responses to negative life events. The resulting formulas predict how low- and high-humor college students will respond to these events. Formulas can be used to explain what is happening in the present and to make predictions about what might occur in the future. In this chapter, you will learn to use formulas in new ways that will help you to recognize patterns, logic, and order in a world that can appear chaotic to the untrained eye.

SECTION

2.1

Objectives

1 Check whether a number is a solution to an equation.

2 Use the addition property of equality to solve equations.

3 Solve applied problems using formulas.

THE ADDITION PROPERTY OF EQUALITY

Unfortunately, many of us have been fined for driving over the speed limit. The amount of the fine depends on how fast we are speeding. Suppose that a highway has a speed limit of 60 miles per hour. The amount that speeders are fined, F, is described by the formula

$$F = 10x - 600,$$

where x is the speed in miles per hour. A friend, whom we shall call Leadfoot, borrows your car and returns a few hours later with a $400 speeding fine. Leadfoot is furious, protesting that the car was barely driven over the speed limit. Should you believe Leadfoot?

To decide if Leadfoot is telling the truth, use the formula

$$F = 10x - 600.$$

Leadfoot was fined $400, so substitute 400 for F:

$$400 = 10x - 600.$$

Now you need to find the value for x. This variable represents Leadfoot's speed, which resulted in the $400 fine.

The use of algebra in a variety of everyday applications often leads to an *equation*. An **equation** is a statement that two algebraic expressions are equal. Thus, $400 = 10x - 600$ is an example of an equation. The equal sign divides the equation into two parts, the left side and the right side:

$$\boxed{400} \quad = \quad \boxed{10x - 600}$$
$$\text{Left side} \qquad \text{Right side}$$

The two sides of an equation can be reversed, so we can express this equation as

$$10x - 600 = 400.$$

The form of this equation is $ax + b = c$, with $a = 10$, $b = -600$, and $c = 400$. Any equation in this form is called a **linear equation in one variable**. The exponent on the variable in such an equation is 1.

In the next three sections, we will study how to solve linear equations. **Solving an equation** is the process of finding the number (or numbers) that make the equation a true statement. These numbers are called the **solutions**, or **roots**, of the equation, and we say that they **satisfy** the equation.

Checking Whether a Number Is a Solution to an Equation   A proposed solution to an equation can be checked by substituting that number for each occurrence of the variable in the equation. If the substitution results in a true statement, the number is a solution. If the substitution results in a false statement, the number is not a solution.

1 Check whether a number is a solution to an equation.

EXAMPLE 1 Checking Proposed Solutions (Is Leadfoot Telling the Truth?)

Consider the equation

$$10x - 600 = 400.$$

(Remember that x represents Leadfoot's speed that resulted in the $400 fine.) Determine whether

a. 60 is a solution. **b.** 100 is a solution.

SOLUTION

a. To determine whether 60 is a solution to the equation, we substitute 60 for x.

$10x - 600 = 400$ This is the given equation.

$10(60) - 600 \overset{?}{=} 400$ Substitute 60 for x. The question mark over the equal sign indicates that we do not know yet if the statement is true.

$600 - 600 \overset{?}{=} 400$ Multiply: $10(60) = 600$.

This statement is false. $0 = 400$ Subtract: $600 - 600 = 0$.

Because the check results in a false statement, we conclude that 60 is not a solution to the given equation. (Leadfoot was not doing 60 miles per hour in your borrowed car.)

b. To determine whether 100 is a solution to the equation, we substitute 100 for x.

$10x - 600 = 400$ This is the given equation.

$10(100) - 600 \overset{?}{=} 400$ Substitute 100 for x.

$1000 - 600 \overset{?}{=} 400$ Multiply: $10(100) = 1000$.

This statement is true. $400 = 400$ Subtract: $1000 - 600 = 400$.

Because the check results in a true statement, we conclude that 100 is a solution to the given equation. Thus, 100 satisfies the equation. (Leadfoot was doing an outrageous 100 miles per hour, and lied with the claim that your car was barely driven over the speed limit.) ■

CHECK POINT 1 Consider the equation $5x - 3 = 17$. Determine whether

a. 3 is a solution. **b.** 4 is a solution.

---

2 Use the addition property of equality to solve equations.

Using the Addition Property of Equality to Solve Equations Consider the equation

$$x = 11.$$

By inspection, we can see that the solution to this equation is 11. If we substitute 11 for x, we obtain the true statement $11 = 11$.

Now consider the equation

$$x - 3 = 8.$$

If we substitute 11 for x, we obtain $11 - 3 \overset{?}{=} 8$. Subtracting on the left side, we get the true statement $8 = 8$.

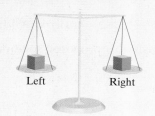

Left   Right

Consider $x - 3 = 8$.

$x - 3$ 8

The scale is balanced if the left and right sides are equal.

Add 3 to the left side.

$x - 3 + 3$ 8

Keep the scale balanced by adding 3 to the right side.

$x - 3 + 3$ $8 + 3$

Thus, $x = 11$.

The equations $x - 3 = 8$ and $x = 11$ both have the same solution, namely 11, and are called *equivalent equations*. **Equivalent equations** are equations that have the same solution.

The idea in solving a linear equation is to get an equivalent equation with the variable (the letter) by itself on one side of the equal sign and a number by itself on the other side. For example, consider the equation $x - 3 = 8$. To get x by itself on the left side, add 3 to the left side, because $x - 3 + 3$ gives $x + 0$, or just x. You must then add 3 to the right side also. By doing this, we are using the **addition property of equality**.

THE ADDITION PROPERTY OF EQUALITY   The same real number (or algebraic expression) may be added to both sides of an equation without changing the equation's solution. This can be expressed symbolically as follows:

$$\text{If } a = b, \text{ then } a + c = b + c.$$

EXAMPLE 2   Solving an Equation Using the Addition Property

Solve the equation: $x - 3 = 8$.

SOLUTION   We can isolate the variable, x, by adding 3 to both sides of the equation.

$$x - 3 = 8 \qquad \text{This is the given equation.}$$
$$x - 3 + 3 = 8 + 3 \qquad \text{Add 3 to both sides.}$$
$$x + 0 = 11 \qquad \text{This step is often done mentally and not listed.}$$
$$x = 11$$

By inspection, we can see that the solution to $x = 11$ is 11. To check this proposed solution, replace x with 11 in the original equation.

Check $x - 3 = 8$ This is the original equation.

$11 - 3 \overset{?}{=} 8$ Substitute 11 for x.

This statement is true. $8 = 8$ Subtract: $11 - 3 = 8$.

Because the check results in a true statement, we conclude that the solution to the given equation is 11.   ∎

The set of an equation's solutions is called its **solution set**. Thus, the solution set of the equation in Example 2 is $\{11\}$. The solution can be expressed as 11 or, using set notation, $\{11\}$. However, do not write the solution as $x = 11$. **The solution of an equation should not be given as an equivalent equation.**

 CHECK POINT **2** Solve the equation and check your proposed solution:

$$x - 5 = 12.$$

When we use the addition property of equality, we add the same number to both sides of an equation. We know that subtraction is the addition of an additive inverse. Thus, the addition property also lets us subtract the same number from both sides of an equation without changing the equation's solution.

EXAMPLE 3 Subtracting the Same Number from Both Sides

Solve and check: $z + 1.4 = 2.06$.

SOLUTION

$z + 1.4 = 2.06$	This is the given equation.
$z + 1.4 - 1.4 = 2.06 - 1.4$	Subtract 1.4 from both sides. This is equivalent to adding -1.4 to both sides.
$z = 0.66$	Subtracting 1.4 from both sides eliminates 1.4 on the left.

Can you see that the solution to $z = 0.66$ is 0.66? To check this proposed solution, replace z with 0.66 in the original equation.

Check	$z + 1.4 = 2.06$	This is the original equation.
	$0.66 + 1.4 \stackrel{?}{=} 2.06$	Substitute 0.66 for z.
	$2.06 = 2.06$	This statement is true.

This true statement indicates that the solution is 0.66. ∎

✔ **CHECK POINT 3** Solve and check: $z + 2.8 = 5.09$.

When isolating the variable, we can isolate it on either the left side or the right side of an equation.

EXAMPLE 4 Isolating the Variable on the Right

Solve and check: $-\dfrac{1}{2} = x - \dfrac{2}{3}$.

SOLUTION We can isolate the variable, x, on the right side by adding $\frac{2}{3}$ to both sides of the equation.

<table>
<tr><td>$-\dfrac{1}{2} = x - \dfrac{2}{3}$</td><td>This is the given equation.</td></tr>
<tr><td>$-\dfrac{1}{2} + \dfrac{2}{3} = x - \dfrac{2}{3} + \dfrac{2}{3}$</td><td>Add $\frac{2}{3}$ to both sides, isolating x on the right.</td></tr>
<tr><td>$-\dfrac{3}{6} + \dfrac{4}{6} = x$</td><td>Rewrite each fraction as an equivalent fraction with a denominator of 6: $-\dfrac{1}{2} + \dfrac{2}{3} = -\dfrac{1}{2} \cdot \dfrac{3}{3} + \dfrac{2}{3} \cdot \dfrac{2}{2} = -\dfrac{3}{6} + \dfrac{4}{6}$.</td></tr>
<tr><td>$\dfrac{1}{6} = x$</td><td>Add on the left side: $-\dfrac{3}{6} + \dfrac{4}{6} = \dfrac{-3+4}{6} = \dfrac{1}{6}$.</td></tr>
</table>

Take a moment to check the proposed solution, $\frac{1}{6}$. Substitute $\frac{1}{6}$ for x in the original equation. You should obtain $-\frac{1}{2} = -\frac{1}{2}$. This true statement indicates that the solution is $\frac{1}{6}$. ∎

✔ **CHECK POINT 4** Solve and check: $-\dfrac{1}{2} = x - \dfrac{3}{4}$.

STUDY TIP

The equations $a = b$ and $b = a$ have the same meaning. If you prefer, you can solve

$$-\frac{1}{2} = x - \frac{2}{3}$$

by reversing the two sides and solving

$$x - \frac{2}{3} = -\frac{1}{2}.$$

In Example 5, we combine like terms before using the addition property.

EXAMPLE 5 Combining Like Terms before Using the Addition Property

Solve and check: $5y + 3 - 4y - 8 = 6 + 9$.

SOLUTION

$$5y + 3 - 4y - 8 = 6 + 9 \qquad \text{This is the given equation.}$$

$$y - 5 = 15 \qquad \text{Combine like terms: } 5y - 4y = y, 3 - 8 = -5,$$
$$\text{and } 6 + 9 = 15.$$

$$y - 5 + 5 = 15 + 5 \qquad \text{Add 5 to both sides.}$$

$$y = 20$$

To check the proposed solution, 20, replace y with 20 in the original equation.

Check

$$5y + 3 - 4y - 8 = 6 + 9 \qquad \text{Be sure to use the original equation and not}$$
$$\text{the simplified form in the second step}$$
$$\text{above. (Why?)}$$

$$5(20) + 3 - 4(20) - 8 \overset{?}{=} 6 + 9 \qquad \text{Substitute 20 for y.}$$

$$100 + 3 - 80 - 8 \overset{?}{=} 6 + 9 \qquad \text{Multiply on the left.}$$

$$103 - 88 \overset{?}{=} 6 + 9 \qquad \text{Combine positive numbers and combine}$$
$$\text{negative numbers on the left.}$$

$$15 = 15 \qquad \text{This statment is true.}$$

This true statement verifies that the solution is 20. ■

 CHECK POINT 5 Solve and check: $8y + 7 - 7y - 10 = 6 + 4$.

Adding and Subtracting Variable Terms on Both Sides of an Equation In some equations, variable terms appear on both sides. Here is an example:

$$4x = 7 + 3x.$$

> A variable term, $4x$, is on the left side.

> A variable term, $3x$, is on the right side.

Our goal is to isolate all the variable terms on one side of the equation. We can use the addition property of equality to do this. The property allows us to add or subtract the same variable term on both sides of an equation without changing the solution. Let's see how we can use this idea to solve $4x = 7 + 3x$.

EXAMPLE 6 Using the Addition Property to Isolate Variable Terms

Solve and check: $4x = 7 + 3x$.

SOLUTION In the given equation, variable terms appear on both sides. We can isolate them on one side by subtracting $3x$ from both sides of the equation.

$$4x = 7 + 3x \qquad \text{This is the given equation.}$$

$$4x - 3x = 7 + 3x - 3x \qquad \text{Subtract 3x from both sides and isolate variable terms}$$
$$\text{on the left.}$$

$$x = 7 \qquad \text{Subtracting 3x from both sides eliminates 3x on the}$$
$$\text{right. On the left, } 4x - 3x = 1x = x.$$

To check the proposed solution, 7, replace x with 7 in the original equation.

Check

$$4x = 7 + 3x \qquad \text{Use the original equation.}$$
$$4(7) \overset{?}{=} 7 + 3(7) \qquad \text{Substitute 7 for x.}$$
$$28 \overset{?}{=} 7 + 21 \qquad \text{Multiply: } 4(7) = 28 \text{ and } 3(7) = 21.$$
$$28 = 28 \qquad \text{This statement is true.}$$

This true statement verifies that the solution is 7. ∎

 CHECK POINT 6 Solve and check: $7x = 12 + 6x$.

EXAMPLE 7 Solving an Equation by Isolating the Variable

Solve and check: $3y - 9 = 2y + 6$.

SOLUTION Our goal is to isolate variable terms on one side and constant terms on the other side. Let's begin by isolating the variable on the left.

$$3y - 9 = 2y + 6 \qquad \text{This is the given equation.}$$
$$3y - 2y - 9 = 2y - 2y + 6 \qquad \begin{array}{l}\text{Isolate the variable terms on the left by subtracting}\\ \text{2y from both sides.}\end{array}$$
$$y - 9 = 6 \qquad \begin{array}{l}\text{Subtracting 2y from both sides eliminates 2y on the}\\ \text{right. On the left, } 3y - 2y = 1y = y.\end{array}$$

Now we isolate the constant terms on the right by adding 9 to both sides.

$$y - 9 + 9 = 6 + 9 \qquad \text{Add 9 to both sides.}$$
$$y = 15$$

Check
$$3y - 9 = 2y + 6 \qquad \text{Use the original equation.}$$
$$3(15) - 9 \overset{?}{=} 2(15) + 6 \qquad \text{Substitute 15 for y.}$$
$$45 - 9 \overset{?}{=} 30 + 6 \qquad \text{Multiply: } 3(15) = 45 \text{ and } 2(15) = 30.$$
$$36 = 36 \qquad \text{This statement is true.}$$

The solution is 15. ∎

 CHECK POINT 7 Solve and check: $3x - 6 = 2x + 5$.

---

3 Solve applied problems using formulas.

Applications Our next example shows how the addition property of equality can be used to find the value of a variable in a mathematical model.

EXAMPLE 8 An Application: Vocabulary and Age

There is a relationship between the number of words in a child's vocabulary, V, and the child's age, A, in months, for ages between 15 and 50 months, inclusive. This relationship can be modeled by the formula

$$V + 900 = 60A.$$

Use the formula to find the vocabulary of a child at the age of 30 months.

SOLUTION In the formula, A represents the child's age, in months. We are interested in a 30-month-old child. Thus, we substitute 30 for A. Then we use the addition property of equality to find V, the number of words in the child's vocabulary.

$$V + 900 = 60A \qquad \text{This is the given formula.}$$
$$V + 900 = 60(30) \qquad \text{Substitute 30 for } A.$$
$$V + 900 = 1800 \qquad \text{Multiply: } 60(30) = 1800.$$
$$V + 900 - 900 = 1800 - 900 \qquad \text{Subtract 900 from both sides and solve for } V.$$
$$V = 900$$

At the age of 30 months, a child has a vocabulary of 900 words.

Vocabulary and Age

(30, 900)
At 30 months,
a child knows
900 words.

Vocabulary (number of words)

2400, 2000, 1600, 1200, 800, 400

0 10 20 30 40 50 60

Age (months)

FIGURE 2.1

The points in the rectangular coordinate system in Figure 2.1 allow us to "see" the formula $V + 900 = 60A$. The first coordinate of each point represents the child's age, in months. The second coordinate represents the child's vocabulary. The points are rising steadily from left to right. This shows that a typical child's vocabulary is steadily increasing with age.

✔ CHECK POINT 8 Use the formula $V + 900 = 60A$ to find the vocabulary of a child at the age of 50 months.

2.1 EXERCISE SET

Student Solutions Manual   CD/Video   PH Math/Tutor Center   MathXL Tutorials on CD   MathXL®   MyMathLab   Interactmath.com

Practice Exercises

Solve each equation in Exercises 1–44 using the addition property of equality. Be sure to check your proposed solutions.

1. $x - 4 = 19$
2. $y - 5 = -18$
3. $z + 8 = -12$
4. $z + 13 = -15$
5. $-2 = x + 14$
6. $-13 = x + 11$
7. $-17 = y - 5$
8. $-21 = y - 4$
9. $7 + z = 11$
10. $18 + z = 14$
11. $-6 + y = -17$
12. $-8 + y = -29$
13. $x + \dfrac{1}{3} = \dfrac{7}{3}$
14. $x + \dfrac{7}{8} = \dfrac{9}{8}$
15. $t + \dfrac{5}{6} = -\dfrac{7}{12}$
16. $t + \dfrac{2}{3} = -\dfrac{7}{6}$
17. $x - \dfrac{3}{4} = \dfrac{9}{2}$
18. $x - \dfrac{3}{5} = \dfrac{7}{10}$
19. $-\dfrac{1}{5} + y = -\dfrac{3}{4}$
20. $-\dfrac{1}{8} + y = -\dfrac{1}{4}$
21. $3.2 + x = 7.5$
22. $-2.7 + w = -5.3$
23. $x + \dfrac{3}{4} = -\dfrac{9}{2}$
24. $r + \dfrac{3}{5} = -\dfrac{7}{10}$
25. $5 = -13 + y$
26. $-11 = 8 + x$
27. $-\dfrac{3}{5} = -\dfrac{3}{2} + s$
28. $\dfrac{7}{3} = -\dfrac{5}{2} + z$
29. $830 + y = 520$
30. $-90 + t = -35$
31. $r + 3.7 = 8$
32. $x + 10.6 = -9$
33. $-3.7 + m = -3.7$
34. $y + \dfrac{7}{11} = \dfrac{7}{11}$

35. $6y + 3 - 5y = 14$
36. $-3x - 5 + 4x = 9$
37. $7 - 5x + 8 + 2x + 4x - 3 = 2 + 3 \cdot 5$
38. $13 - 3r + 2 + 6r - 2r - 1 = 3 + 2 \cdot 9$
39. $7y + 4 = 6y - 9$
40. $4r - 3 = 5 + 3r$
41. $12 - 6x = 18 - 7x$
42. $20 - 7s = 26 - 8s$
43. $4x + 2 = 3(x - 6) + 8$
44. $7x + 3 = 6(x - 1) + 9$

Practice Plus

The equations in Exercises 45–48 contain small geometric figures that represent real numbers. Use the addition property of equality to isolate x on one side of the equation and the geometric figures on the other side.

45. $x - \square = \triangle$
46. $x + \square = \triangle$
47. $2x + \triangle = 3x + \square$
48. $6x - \triangle = 7x - \square$

In Exercises 49–52, use the given information to write an equation. Let x represent the number described in each exercise. Then solve the equation and find the number.

49. If 12 is subtracted from a number, the result is -2. Find the number.
50. If 23 is subtracted from a number, the result is -8. Find the number.
51. The difference between $\dfrac{2}{5}$ of a number and 8 is $\dfrac{7}{5}$ of that number. Find the number.

52. The difference between 3 and $\frac{2}{7}$ of a number is $\frac{5}{7}$ of that number. Find the number.

Application Exercises

Formulas frequently appear in the business world. For example, the cost, C, of an item (the price paid by a retailer) plus the markup, M, on that item (the retailer's profit) equals the selling price, S, of the item. The formula is

$$C + M = S.$$

Use the formula to solve Exercises 53–54.

53. The selling price of a computer is $1850. If the markup on the computer is $150, find the cost to the retailer for the computer.

54. The selling price of a television is $650. If the cost to the retailer for the television is $520, find the markup.

The formula

$$d + 525{,}000 = 5000c$$

models the relationship between the annual number of deaths in the United States from heart disease, d, and the average cholesterol level, c, of blood. (Cholesterol level, c, is expressed in milligrams per deciliter of blood.) Use this formula to solve Exercises 55–56.

55. The average cholesterol level for people in the United States is 210. According to the formula, how many deaths per year from heart disease can be expected with this cholesterol level?

56. Suppose that the average cholesterol level for people in the United States could be reduced to 180. Determine the number of annual deaths from heart disease that can be expected with this reduced cholesterol level.

Statins, used to control high cholesterol, are the top-selling class of prescription drugs in the United States. The bar graph shows statin sales, in billions of dollars, from 1998 through 2002. The data can be modeled by the formula

$$S - 1.6x = 5.8,$$

where S is statin sales, in billions of dollars, x years after 1998. Use this formula to solve Exercises 57–58.

Statin Sales in the U.S.

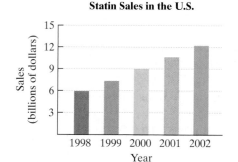

Source: Lawrence Berkeley
National Laboratory

57. According to the formula, what will Americans spend on statins in 2005?

58. According to the formula, what will Americans spend on statins in 2006?

Writing in Mathematics

59. What is an equation?

60. Explain how to determine whether a number is a solution to an equation.

61. State the addition property of equality and give an example.

62. Explain why $x + 2 = 9$ and $x + 2 = -6$ are not equivalent equations.

63. What is the difference between solving an equation such as

$$5y + 3 - 4y - 8 = 6 + 9$$

and simplifying an algebraic expression such as

$$5y + 3 - 4y - 8?$$

If there is a difference, which topic should be taught first? Why?

Critical Thinking Exercises

64. Which one of the following statements is true?
   a. If $y - a = -b$, then $y = a + b$.
   b. If $y + 7 = 0$, then $y = 7$.
   c. The solution to $4 - x = -3x$ is -2.
   d. If 7 is added on one side of an equation, then it should be subtracted on the other side.

65. Write an equation with a negative solution that can be solved by adding 100 to both sides.

Technology Exercises

Use a calculator to solve each equation in Exercises 66–67.

66. $x - 7.0463 = -9.2714$

67. $6.9825 = 4.2296 + y$

Review Exercises

68. Plot $(-3, 1)$ in rectangular coordinates. In which quadrant does the point lie? (Section 1.3, Example 1)

69. Simplify: $-16 - 8 \div 4 \cdot (-2)$. (Section 1.8, Example 4)

70. Simplify: $3[7x - 2(5x - 1)]$. (Section 1.8, Example 11)

SECTION

2.2

Objectives

1 Use the multiplication property of equality to solve equations.

2 Solve equations in the form $-x = c$.

3 Use the addition and multiplication properties to solve equations.

4 Solve applied problems using formulas.

---

1 Use the multiplication property of equality to solve equations.

THE MULTIPLICATION PROPERTY OF EQUALITY

Could you live to be 125? The number of Americans ages 100 or older could approach 850,000 by 2050. Some scientists predict that by 2100, our descendants could live to be 200 years of age. In this section, we will see how a formula can be used to make these kinds of predictions as we turn to a new property for solving linear equations.

Using the Multiplication Property of Equality to Solve Equations Can the addition property of equality be used to solve every linear equation in one variable? No. For example, consider the equation

$$\frac{x}{5} = 9.$$

We cannot isolate the variable x by adding or subtracting 5 on both sides. To get x by itself on the left side, multiply the left side by 5:

$$5 \cdot \frac{x}{5} = \left(5 \cdot \frac{1}{5}\right)x = 1x = x.$$

> 5 is the multiplicative inverse of $\frac{1}{5}$.

You must then multiply the right side by 5 also. By doing this, we are using the **multiplication property of equality**.

> **THE MULTIPLICATION PROPERTY OF EQUALITY** The same nonzero real number (or algebraic expression) may multiply both sides of an equation without changing the solution. This can be expressed symbolically as follows:
> If $a = b$ and $c \neq 0$, then $ac = bc$.

> **EXAMPLE 1** Solving an Equation Using the Multiplication Property
>
> Solve the equation: $\frac{x}{5} = 9$.
>
> **SOLUTION** We can isolate the variable, x, by multiplying both sides of the equation by 5.
>
> $$\frac{x}{5} = 9 \qquad \text{This is the given equation.}$$
> $$5 \cdot \frac{x}{5} = 5 \cdot 9 \qquad \text{Multiply both sides by 5.}$$
> $$1x = 45 \qquad \text{Simplify.}$$
> $$x = 45 \qquad 1x = x$$
>
> By substituting 45 for x in the original equation, we obtain the true statement $9 = 9$. This verifies that the solution is 45. ∎

✔ CHECK POINT 1 Solve the equation: $\dfrac{x}{3} = 12$.

When we use the multiplication property of equality, we multiply both sides of an equation by the same nonzero number. We know that division is multiplication by a multiplicative inverse. Thus, the multiplication property also lets us divide both sides of an equation by a nonzero number without changing the solution.

EXAMPLE 2 Dividing Both Sides by the Same Nonzero Number

Solve: **a.** $6x = 30$ **b.** $-7y = 56$ **c.** $-18.9 = 3z$.

SOLUTION   In each equation, the variable is multiplied by a number. We can isolate the variable by dividing both sides of the equation by that number.

a. $6x = 30$ This is the given equation.

$\dfrac{6x}{6} = \dfrac{30}{6}$ Divide both sides by 6.

$1x = 5$ Simplify.

$x = 5$ 1x = x

By substituting 5 for x in the original equation, we obtain the true statement $30 = 30$. The solution is 5.

b. $-7y = 56$ This is the given equation.

$\dfrac{-7y}{-7} = \dfrac{56}{-7}$ Divide both sides by -7.

$1y = -8$ Simplify.

$y = -8$ 1y = y

By substituting -8 for y in the original equation, we obtain the true statement $56 = 56$. The solution is -8.

c. $-18.9 = 3z$ This is the given equation.

$\dfrac{-18.9}{3} = \dfrac{3z}{3}$ Divide both sides by 3.

$-6.3 = 1z$ Simplify.

$-6.3 = z$ 1z = z

By substituting -6.3 for z in the original equation, we obtain the true statement $-18.9 = -18.9$. The solution is -6.3.   ■

✔ CHECK POINT 2 Solve:

a. $4x = 84$ **b.** $-11y = 44$ **c.** $-15.5 = 5z$.

Some equations have a variable term with a fractional coefficient. Here is an example:

$$\frac{3}{4}y = 12.$$

The coefficient of the term $\frac{3}{4}y$ is $\frac{3}{4}$.

To isolate the variable, multiply both sides of the equation by the multiplicative inverse of the fraction. For the equation $\frac{3}{4}y = 12$, the multiplicative inverse of $\frac{3}{4}$ is $\frac{4}{3}$. Thus, we solve $\frac{3}{4}y = 12$ by multiplying both sides by $\frac{4}{3}$.

EXAMPLE 3 Using the Multiplication Property to Eliminate a Fractional Coefficient

Solve: **a.** $\frac{3}{4}y = 12$ **b.** $9 = -\frac{3}{5}x$.

SOLUTION

a.

$$\frac{3}{4}y = 12$$ This is the given equation.

$$\frac{4}{3}\left(\frac{3}{4}y\right) = \frac{4}{3} \cdot 12$$ Multiply both sides by $\frac{4}{3}$, the multiplicative inverse of $\frac{3}{4}$.

$$1y = 16$$ On the left, $\frac{4}{3}\left(\frac{3}{4}y\right) = \left(\frac{4}{3} \cdot \frac{3}{4}\right)y = 1y$.

On the right, $\frac{4}{3} \cdot \frac{12}{1} = \frac{48}{3} = 16$.

$$y = 16$$ $1y = y$

By substituting 16 for y in the original equation, we obtain the true statement $12 = 12$. The solution is 16.

b.

$$9 = -\frac{3}{5}x$$ This is the given equation.

$$-\frac{5}{3} \cdot 9 = -\frac{5}{3}\left(-\frac{3}{5}x\right)$$ Multiply both sides by $-\frac{5}{3}$, the multiplicative inverse of $-\frac{3}{5}$.

$$-15 = 1x$$ Simplify.

$$-15 = x$$ $1x = x$

By substituting -15 for x in the original equation, we obtain the true statement $9 = 9$. The solution is -15. ∎

✔ **CHECK POINT 3** Solve: **a.** $\frac{2}{3}y = 16$ **b.** $28 = -\frac{7}{4}x$.

STUDY TIP

The equation

$$9 = -\frac{3}{5}x$$

can be expressed as

$$9 = -\frac{3x}{5}$$

or

$$9 = \frac{-3x}{5}.$$

2 Solve equations in the form $-x = c$.

Equations and Coefficients of -1 How do we solve an equation in the form $-x = c$, such as $-x = 4$? Because the equation means $-1x = 4$, we have not yet obtained a solution. The solution of an equation is obtained from the form $x = $ some number. The equation $-x = 4$ is not yet in this form. We still need to isolate x. We can do this by multiplying or dividing both sides of the equation by -1. We will multiply by -1.

EXAMPLE 4 Solving Equations in the Form $-x = c$

Solve: **a.** $-x = 4$ **b.** $-x = -7$.

SOLUTION We multiply both sides of each equation by -1. This will isolate x on the left side.

a.

$$-x = 4$$ This is the given equation.

$$-1x = 4$$ Rewrite $-x$ as $-1x$.

$$(-1)(-1x) = (-1)(4)$$ Multiply both sides by -1.

$$1x = -4$$ On the left, $(-1)(-1) = 1$. On the right, $(-1)(4) = -4$.

$$x = -4$$ $1x = x$

Check

$$-x = 4 \qquad \text{This is the original equation.}$$
$$-(-4) \stackrel{?}{=} 4 \qquad \text{Substitute } -4 \text{ for } x.$$
$$4 = 4 \qquad -(-a) = a, \text{ so } -(-4) = 4.$$

This true statement indicates that the solution is -4.

b.

$$-x = -7 \qquad \text{This is the given equation.}$$
$$-1x = -7 \qquad \text{Rewrite } -x \text{ as } -1x.$$
$$(-1)(-1x) = (-1)(-7) \qquad \text{Multiply both sides by } -1.$$
$$1x = 7 \qquad (-1)(-1) = 1 \text{ and } (-1)(-7) = 7.$$
$$x = 7 \qquad 1x = x$$

By substituting 7 for x in the original equation, we obtain the true statement $-7 = -7$. The solution is 7. ∎

STUDY TIP

If $-x = c$, then the equation's solution is the additive inverse (the opposite) of c. For example, the solution of $-x = -7$ is the additive inverse of -7, which is 7.

✔ **CHECK POINT 4** Solve: **a.** $-x = 5$ **b.** $-x = -3$.

---

3 Use the addition and multiplication properties to solve equations.

Equations Requiring Both the Addition and Multiplication Properties   When an equation does not contain fractions, we will often use the addition property of equality before the multiplication property of equality. Our overall goal is to isolate the variable with a coefficient of 1 on either the left or right side of the equation.

Here is the procedure that we will be using to solve the equations in the next three examples:

• Use the addition property of equality to isolate the variable term.
• Use the multiplication property of equality to isolate the variable.

EXAMPLE 5 Using Both the Addition and Multiplication Properties

Solve: $3x + 1 = 7$.

SOLUTION   We begin by isolating the variable term, $3x$, subtracting 1 from both sides. Then we isolate the variable, x, by dividing both sides by 3.

• **Use the addition property of equality to isolate the variable term.**

$$3x + 1 = 7 \qquad \text{This is the given equation.}$$
$$3x + 1 - 1 = 7 - 1 \qquad \text{Use the addition property, subtracting 1 from both sides.}$$
$$3x = 6 \qquad \text{Simplify.}$$

• **Use the multiplication property of equality to isolate the variable.**

$$\frac{3x}{3} = \frac{6}{3} \qquad \text{Divide both sides by 3.}$$
$$x = 2 \qquad \text{Simplify.}$$

By substituting 2 for x in the original equation, we obtain the true statement $7 = 7$. The solution is 2. ∎

✔ **CHECK POINT 5** Solve: $4x + 3 = 27$.

EXAMPLE 6 Using Both the Addition and Multiplication Properties

Solve: $-2y - 28 = 4$.

SOLUTION We begin by isolating the variable term, $-2y$, adding 28 to both sides. Then we isolate the variable, y, by dividing both sides by -2.

- **Use the addition property of equality to isolate the variable term.**

$$-2y - 28 = 4 \qquad \text{This is the given equation.}$$
$$-2y - 28 + 28 = 4 + 28 \qquad \text{Use the addition property, adding 28 to both sides.}$$
$$-2y = 32 \qquad \text{Simplify.}$$

- **Use the multiplication property of equality to isolate the variable.**

$$\frac{-2y}{-2} = \frac{32}{-2} \qquad \text{Divide both sides by } -2.$$
$$y = -16 \qquad \text{Simplify.}$$

Take a moment to substitute -16 for y in the given equation. Do you obtain the true statement $4 = 4$? The solution is -16. ∎

 CHECK POINT 6 Solve: $-4y - 15 = 25$.

EXAMPLE 7 Using Both the Addition and Multiplication Properties

Solve: $3x - 14 = -2x + 6$.

SOLUTION We will use the addition property to collect all terms involving x on the left and all numerical terms on the right. Then we will isolate the variable, x, by dividing both sides by its coefficient.

- **Use the addition property of equality to isolate the variable term.**

$$3x - 14 = -2x + 6 \qquad \text{This is the given equation.}$$
$$3x + 2x - 14 = -2x + 2x + 6 \qquad \text{Add 2x to both sides.}$$
$$5x - 14 = 6 \qquad \text{Simplify.}$$
$$5x - 14 + 14 = 6 + 14 \qquad \text{Add 14 to both sides.}$$
$$5x = 20 \qquad \text{Simplify. The variable term, 5x, is isolated on the left. The numerical term, 20, is isolated on the right.}$$

- **Use the multiplication property of equality to isolate the variable.**

$$\frac{5x}{5} = \frac{20}{5} \qquad \text{Divide both sides by 5.}$$
$$x = 4 \qquad \text{Simplify.}$$

Check
$$3x - 14 = -2x + 6 \qquad \text{Use the original equation.}$$
$$3(4) - 14 \stackrel{?}{=} -2(4) + 6 \qquad \text{Substitute the proposed solution, 4, for x.}$$
$$12 - 14 \stackrel{?}{=} -8 + 6 \qquad \text{Multiply.}$$
$$-2 = -2 \qquad \text{Simplify.}$$

The true statement $-2 = -2$ verifies that the solution is 4. ∎

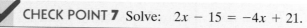 **CHECK POINT 7** Solve: $2x - 15 = -4x + 21$.

4 Solve applied problems using formulas.

Applications Your life expectancy is related to the year when you were born. The bar graph in Figure 2.2 shows life expectancy in the United States by year of birth. The data for U.S. women shown by the green bars can be modeled by the formula

$$E = 0.18t + 71,$$

where E is the life expectancy for women born t years after 1950.

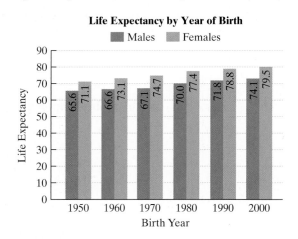

Life Expectancy by Year of Birth

FIGURE 2.2
Source: U.S. Bureau of the Census

EXAMPLE 8 Using the Formula For Life Expectancy

Use the formula

$$E = 0.18t + 71$$

to determine the year of birth for which U.S. women can expect to live 87.2 years.

SOLUTION We are interested in a life expectancy of 87.2 years. We substitute 87.2 for E in the formula and solve for t. Keep in mind that t represents birth years *after* 1950.

$E = 0.18t + 71$	This is the given formula.
$87.2 = 0.18t + 71$	Replace E with 87.2.

Our goal is to isolate t.

$87.2 - 71 = 0.18t + 71 - 71$	Isolate the term containing t by subtracting 71 from both sides.
$16.2 = 0.18t$	Simplify.
$\dfrac{16.2}{0.18} = \dfrac{0.18t}{0.18}$	Divide both sides by 0.18.
$90 = t$	Simplify: $0.18\overline{)16.20}$.

The formula indicates that U.S. women born 90 years after 1950, or in 2040, can expect to live 87.2 years. ∎

 CHECK POINT 8 The data for U.S. men shown by the blue bars in Figure 2.2 can be modeled by the formula $E = 0.16t + 65$, where E represents life expectancy for men born t years after 1950. Determine the year of birth for which U.S. men can expect to live 76.2 years.

2.2 EXERCISE SET

Student Solutions Manual CD/Video PH Math/Tutor Center MathXL Tutorials on CD MathXL® MyMathLab Interactmath.com

Practice Exercises

Solve each equation in Exercises 1–28 using the multiplication property of equality. Be sure to check your proposed solutions.

1. $\dfrac{x}{6} = 5$ **2.** $\dfrac{x}{7} = 4$

3. $\dfrac{x}{-3} = 11$ **4.** $\dfrac{x}{-5} = 8$

5. $5y = 35$ **6.** $6y = 42$

7. $-7y = 63$ **8.** $-4y = 32$

9. $-28 = 8z$ **10.** $-36 = 8z$

11. $-18 = -3z$ **12.** $-54 = -9z$

13. $-8x = 6$ **14.** $-8x = 4$

15. $17y = 0$ **16.** $-16y = 0$

17. $\dfrac{2}{3}y = 12$ **18.** $\dfrac{3}{4}y = 15$

19. $28 = -\dfrac{7}{2}x$ **20.** $20 = -\dfrac{5}{8}x$

21. $-x = 17$ **22.** $-x = 23$

23. $-47 = -y$ **24.** $-51 = -y$

25. $-\dfrac{x}{5} = -9$ **26.** $-\dfrac{x}{5} = -1$

27. $2x - 12x = 50$ **28.** $8x - 3x = -45$

Solve each equation in Exercises 29–54 using both the addition and multiplication properties of equality. Check proposed solutions.

29. $2x + 1 = 11$ **30.** $2x + 5 = 13$

31. $2x - 3 = 9$ **32.** $3x - 2 = 9$

33. $-2y + 5 = 7$ **34.** $-3y + 4 = 13$

35. $-3y - 7 = -1$ **36.** $-2y - 5 = 7$

37. $12 = 4z + 3$ **38.** $14 = 5z - 21$

39. $-x - 3 = 3$ **40.** $-x - 5 = 5$

41. $6y = 2y - 12$ **42.** $8y = 3y - 10$

43. $3z = -2z - 15$ **44.** $2z = -4z + 18$

45. $-5x = -2x - 12$ **46.** $-7x = -3x - 8$

47. $8y + 4 = 2y - 5$ **48.** $5y + 6 = 3y - 6$

49. $6z - 5 = z + 5$ **50.** $6z - 3 = z + 2$

51. $6x + 14 = 2x - 2$ **52.** $9x + 2 = 6x - 4$

53. $-3y - 1 = 5 - 2y$ **54.** $-3y - 2 = -5 - 4y$

Practice Plus

The equations in Exercises 55–58 contain small geometric figures that represent real numbers. Use the multiplication property of equality to isolate x on one side of the equation and the geometric figures on the other side.

55. $\dfrac{x}{\square} = \triangle$ **56.** $\triangle = \square x$

57. $\triangle = -x$ **58.** $\dfrac{-x}{\square} = \triangle$

In Exercises 59–62, use the given information to write an equation. Let x represent the number described in each exercise. Then solve the equation and find the number.

59. If a number is multiplied by 6, the result is 10. Find the number.

60. If a number is multiplied by −6, the result is 20. Find the number.

61. If a number is divided by −9, the result is 5. Find the number.

62. If a number is divided by −7, the result is 8. Find the number.

Application Exercises

The formula

$$M = \dfrac{n}{5}$$

models your distance, M, in miles, from a lightning strike in a thunderstorm if it takes n seconds to hear thunder after seeing the lightning. Use this formula to solve Exercises 63–64.

63. If you are 2 miles away from the lightning flash, how long will it take the sound of thunder to reach you?

64. If you are 3 miles away from the lightning flash, how long will it take the sound of thunder to reach you?

The Mach number is a measurement of speed, named after the man who suggested it, Ernst Mach (1838–1916). The formula

$$M = \dfrac{A}{740}$$

indicates that the speed of an aircraft, A, in miles per hour, divided by the speed of sound, approximately 740 miles per hour, results in the Mach number, M. Use the formula to determine the speed, in miles per hour, of the aircrafts in Exercises 65–66. (Note: When an aircraft's speed increases beyond Mach 1, it is said to have broken the sound barrier.)

65.   **66.**

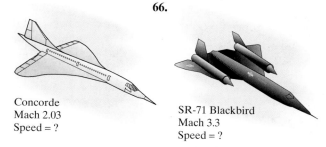

Concorde
Mach 2.03
Speed = ?

SR-71 Blackbird
Mach 3.3
Speed = ?

67. The formula $P = -0.5d + 100$ models the percentage, P, of lost hikers found in search and rescue missions when members of the search team walk parallel to one another separated by a distance of d yards. If a search and rescue team finds 70% of lost hikers, substitute 70 for P in the formula and find the parallel distance of separation between members of the search party.

68. The formula $M = 420x + 720$ models the data for the amount of money lost to credit card fraud worldwide, M, in millions of dollars, x years after 1989. In which year did losses amount to 4080 million dollars?

Writing in Mathematics

69. State the multiplication property of equality and give an example.

70. Explain how to solve the equation $-x = -50$.

71. Explain how to solve the equation $2x + 8 = 5x - 3$.

72. What might occur in the future to cause life expectancy to exceed the predictions made by the models in Example 8 and Check Point 8 on page 116?

Critical Thinking Exercises

73. Which one of the following statements is true?

    a. If $7x = 21$, then $x = 21 - 7$.

    b. If $3x - 4 = 16$, then $3x = 12$.

    c. If $3x + 7 = 0$, then $x = \dfrac{7}{3}$.

    d. The solution to $6x = 0$ is not a natural number.

In Exercises 74–75, write an equation with the given characteristics.

74. The solution is a positive integer and the equation can be solved by dividing both sides by -60.

75. The solution is a negative integer and the equation can be solved by multiplying both sides by $\dfrac{4}{5}$.

Technology Exercises

Solve each equation in Exercises 76–77. Use a calculator to help with the arithmetic. Check your solution using the calculator.

76. $3.7x - 19.46 = -9.988$

77. $-72.8y - 14.6 = -455.43 - 4.98y$

Review Exercises

78. Evaluate: $(-10)^2$. (Section 1.8, Example 1)

79. Evaluate: -10^2. (Section 1.8, Example 1)

80. Evaluate $x^3 - 4x$ for $x = -1$. (Section 1.8, Example 10)

2.3

Objectives

1 Solve linear equations.

2 Solve linear equations containing fractions.

3 Identify equations with no solution or infinitely many solutions.

4 Solve applied problems using formulas.

SOLVING LINEAR EQUATIONS

Yes, we overindulged, but it was delicious. Now if we could just grow a few inches taller, we'd be back in line with our recommended weights.

In this section, we will see how algebra models the relationship between weight and height. To use this mathematical model, it would be helpful to have a systematic procedure for solving linear equations. We open the section with such a procedure.

1 Solve linear equations.

A Step-By-Step Procedure for Solving Linear Equations Here is a step-by-step procedure for solving a linear equation in one variable. Not all of these steps are necessary to solve every equation.

SOLVING A LINEAR EQUATION
1. Simplify the algebraic expression on each side.
2. Collect all the variable terms on one side and all the constant terms on the other side.
3. Isolate the variable and solve.
4. Check the proposed solution in the original equation.

EXAMPLE 1   Solving a Linear Equation

Solve and check: $2x - 8x + 40 = 13 - 3x - 3$.

SOLUTION

Step 1.   **Simplify the algebraic expression on each side.**

$$2x - 8x + 40 = 13 - 3x - 3$$ This is the given equation.
$$-6x + 40 = 10 - 3x$$ Combine like terms: $2x - 8x = -6x$ and $13 - 3 = 10$.

Step 2.   **Collect variable terms on one side and constant terms on the other side.** We will collect variable terms on the left by adding $3x$ to both sides. We will collect the numbers on the right by subtracting 40 from both sides.

$$-6x + 40 + 3x = 10 - 3x + 3x$$ Add $3x$ to both sides.
$$-3x + 40 = 10$$ Simplify: $-6x + 3x = -3x$.
$$-3x + 40 - 40 = 10 - 40$$ Subtract 40 from both sides.
$$-3x = -30$$ Simplify.

Step 3.   **Isolate the variable and solve.** We isolate the variable, x, by dividing both sides by -3.

$$\frac{-3x}{-3} = \frac{-30}{-3}$$ Divide both sides by -3.
$$x = 10$$ Simplify.

Step 4.   **Check the proposed solution in the original equation.** Substitute 10 for x in the original equation.

$$2x - 8x + 40 = 13 - 3x - 3$$ This is the original equation.
$$2 \cdot 10 - 8 \cdot 10 + 40 \stackrel{?}{=} 13 - 3 \cdot 10 - 3$$ Substitute 10 for x.
$$20 - 80 + 40 \stackrel{?}{=} 13 - 30 - 3$$ Perform the indicated multiplications.
$$-60 + 40 \stackrel{?}{=} -17 - 3$$ Subtract: $20 - 80 = -60$ and $13 - 30 = -17$.
$$-20 = -20$$ Simplify.

By substituting 10 for x in the original equation, we obtain the true statement $-20 = -20$. This verifies that the solution is 10. ∎

✔ **CHECK POINT 1** Solve and check: $-7x + 25 + 3x = 16 - 2x - 3$.

EXAMPLE 2   Solving a Linear Equation

Solve and check: $5x = 8(x + 3)$.

SOLUTION

Step 1.   **Simplify the algebraic expression on each side.** Use the distributive property to remove parentheses on the right.

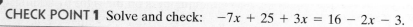

$$5x = 8(x + 3)$$ This is the given equation.
$$5x = 8x + 24$$ Use the distributive property.

DISCOVER FOR YOURSELF

Solve the equation in Example 1 by collecting terms with the variable on the right and numbers on the left. What do you observe?

Step 2. **Collect variable terms on one side and constant terms on the other side.** We will work with $5x = 8x + 24$ and collect variable terms on the left by subtracting $8x$ from both sides. The only constant term, 24, is already on the right.

$$5x - 8x = 8x + 24 - 8x \quad \text{Subtract 8x from both sides.}$$
$$-3x = 24 \quad \text{Simplify: } 5x - 8x = -3x.$$

Step 3. **Isolate the variable and solve.** We isolate the variable, x, by dividing both sides by -3.

$$\frac{-3x}{-3} = \frac{24}{-3} \quad \text{Divide both sides by } -3.$$
$$x = -8 \quad \text{Simplify.}$$

Step 4. **Check the proposed solution in the original equation.** Substitute -8 for x in the original equation.

$$5x = 8(x + 3) \quad \text{This is the original equation.}$$
$$5(-8) \stackrel{?}{=} 8(-8 + 3) \quad \text{Substitute } -8 \text{ for x.}$$
$$5(-8) \stackrel{?}{=} 8(-5) \quad \text{Perform the addition in parentheses:}$$
$$\qquad\qquad\qquad\qquad -8 + 3 = -5.$$
$$-40 = -40 \quad \text{Multiply.}$$

The true statement $-40 = -40$ verifies that -8 is the solution. ■

✔ **CHECK POINT 2** Solve and check: $8x = 2(x + 6)$.

EXAMPLE 3 Solving a Linear Equation

Solve and check: $2(x - 3) - 17 = 13 - 3(x + 2)$.

SOLUTION

Step 1. **Simplify the algebraic expression on each side.**

> Do not begin with 13 − 3. Multiplication (the distributive property) is applied before subtraction.

$$2(x - 3) - 17 = 13 - 3(x + 2) \quad \text{This is the given equation.}$$
$$2x - 6 - 17 = 13 - 3x - 6 \quad \text{Use the distributive property.}$$
$$2x - 23 = -3x + 7 \quad \text{Combine like terms.}$$

Step 2. **Collect variable terms on one side and constant terms on the other side.** We will collect variable terms on the left by adding $3x$ to both sides. We will collect the numbers on the right by adding 23 to both sides.

$$2x - 23 + 3x = -3x + 7 + 3x \quad \text{Add 3x to both sides.}$$
$$5x - 23 = 7 \quad \text{Simplify: } 2x + 3x = 5x.$$
$$5x - 23 + 23 = 7 + 23 \quad \text{Add 23 to both sides.}$$
$$5x = 30 \quad \text{Simplify.}$$

Step 3. **Isolate the variable and solve.** We isolate the variable, x, by dividing both sides by 5.

$$\frac{5x}{5} = \frac{30}{5} \quad \text{Divide both sides by 5.}$$
$$x = 6 \quad \text{Simplify.}$$

The compact, symbolic notation of algebra enables us to use a clear step-by-step method for solving equations, designed to avoid the confusion shown in the painting.

Squeak Carnwath *Equations* 1981, oil on cotton canvas 96 in. h × 72 in. w.

Step 4. **Check the proposed solution in the original equation.** Substitute 6 for x in the original equation.

$$2(x - 3) - 17 = 13 - 3(x + 2) \qquad \text{This is the original equation.}$$
$$2(6 - 3) - 17 \overset{?}{=} 13 - 3(6 + 2) \qquad \text{Substitute 6 for x.}$$
$$2(3) - 17 \overset{?}{=} 13 - 3(8) \qquad \text{Simplify inside parentheses.}$$
$$6 - 17 \overset{?}{=} 13 - 24 \qquad \text{Multiply.}$$
$$-11 = -11 \qquad \text{Subtract.}$$

The true statement $-11 = -11$ verifies that 6 is the solution. ∎

✔ **CHECK POINT 3** Solve and check: $4(2x + 1) - 29 = 3(2x - 5)$.

---

2 Solve linear equations containing fractions.

Linear Equations with Fractions Equations are easier to solve when they do not contain fractions. How do we remove fractions from an equation? We begin by multiplying both sides of the equation by the least common denominator of any fractions in the equation. The least common denominator is the smallest number that all denominators will divide into. Multiplying every term on both sides of the equation by the least common denominator will eliminate the fractions in the equation. Example 4 shows how we "clear an equation of fractions."

EXAMPLE 4 Solving a Linear Equation Involving Fractions

Solve and check: $\dfrac{3x}{2} = \dfrac{x}{5} - \dfrac{39}{5}$.

SOLUTION The denominators are 2, 5, and 5. The smallest number that is divisible by 2, 5, and 5 is 10. We begin by multiplying both sides of the equation by 10, the least common denominator.

$$\frac{3x}{2} = \frac{x}{5} - \frac{39}{5} \qquad \text{This is the given equation.}$$

$$10 \cdot \frac{3x}{2} = 10\left(\frac{x}{5} - \frac{39}{5}\right) \qquad \text{Multiply both sides by 10.}$$

$$10 \cdot \frac{3x}{2} = 10 \cdot \frac{x}{5} - 10 \cdot \frac{39}{5} \qquad \text{Use the distributive property. Be sure to multiply all terms by 10.}$$

$$\overset{5}{\cancel{10}} \cdot \frac{3x}{\underset{1}{\cancel{2}}} = \overset{2}{\cancel{10}} \cdot \frac{x}{\underset{1}{\cancel{5}}} - \overset{2}{\cancel{10}} \cdot \frac{39}{\underset{1}{\cancel{5}}} \qquad \text{Divide out common factors in the multiplications.}$$

$$15x = 2x - 78 \qquad \text{Complete the multiplications. The fractions are now cleared.}$$

At this point, we have an equation similar to those we previously have solved. Collect the variable terms on one side and the constant terms on the other side.

$$15x - 2x = 2x - 2x - 78 \qquad \text{Subtract 2x to get the variable terms on the left.}$$
$$13x = -78 \qquad \text{Simplify.}$$

Isolate x by dividing both sides by 13.

$$\frac{13x}{13} = \frac{-78}{13} \qquad \text{Divide both sides by 13.}$$
$$x = -6 \qquad \text{Simplify.}$$

Check the proposed solution. Substitute -6 for x in the original equation. You should obtain $-9 = -9$. This true statement verifies that the solution is -6. ∎

✔ **CHECK POINT 4** Solve and check: $\dfrac{x}{4} = \dfrac{2x}{3} + \dfrac{5}{6}$.

3 | Identify equations with no solution or infinitely many solutions.

Linear Equations with No Solution or Infinitely Many Solutions Thus far, each equation that we have solved has had a single solution. However, some equations are not true for even one real number. Such an equation is called an **inconsistent equation**, or a **contradiction**. Here is an example of such an equation:

$$x = x + 4.$$

There is no number that is equal to itself plus 4. This equation has no solution.

An equation that is true for all real numbers is called an **identity**. An example of an identity is

$$x + 3 = x + 2 + 1.$$

Every number plus 3 is equal to that number plus 2 plus 1. Every real number is a solution to this equation.

If you attempt to solve an equation with no solution, you will eliminate the variable and obtain a false statement, such as $2 = 5$. If you attempt to solve an equation that is true for every real number, you will eliminate the variable and obtain a true statement, such as $4 = 4$.

EXAMPLE 5 Solving a Linear Equation

Solve: $2x + 6 = 2(x + 4)$.

SOLUTION

$2x + 6 = 2(x + 4)$	This is the given equation.
$2x + 6 = 2x + 8$	Use the distributive property.
$2x + 6 - 2x = 2x + 8 - 2x$	Subtract 2x from both sides.
⟵ Keep reading. 6 = 8 is not the solution. $6 = 8$	Simplify.

The original equation is equivalent to the false statement $6 = 8$, which is false for every value of x. The equation is inconsistent and has no solution. ∎

✔ **CHECK POINT 5** Solve: $3x + 7 = 3(x + 1)$.

EXAMPLE 6 Solving a Linear Equation

Solve: $-3x + 5 + 5x = 4x - 2x + 5$.

SOLUTION

$-3x + 5 + 5x = 4x - 2x + 5$	This is the given equation.
$2x + 5 = 2x + 5$	Combine like terms: $-3x + 5x = 2x$ and $4x - 2x = 2x$.
$2x + 5 - 2x = 2x + 5 - 2x$	Subtract 2x from both sides.
⟵ Keep reading. 5 = 5 is not the solution. $5 = 5$	Simplify.

The original equation is equivalent to the true statement $5 = 5$, which is true for every value of x. The equation is an identity and all real numbers are solutions. Try substituting any number of your choice for x in the original equation. You will obtain a true statement. ∎

✔ **CHECK POINT 6** Solve: $3(x - 1) + 9 = 8x + 6 - 5x$.

4 Solve applied problems using formulas.

Applications The next example shows how our procedure for solving equations with fractions can be used to find the value of a variable in a mathematical model.

EXAMPLE 7 Modeling Weight and Height

The formula

$$\frac{W}{2} - 3H = 53$$

models the recommended weight, W, in pounds, for a male, where H represents the man's height, in inches, over 5 feet. What is the recommended weight for a man who is 6 feet, 3 inches tall?

SOLUTION Keep in mind that H represents height in inches *above 5 feet*. A man who is 6 feet, 3 inches tall is 1 foot, 3 inches above 5 feet. Because 12 inches = 1 foot, he is $12 + 3$, or 15 inches, above 5 feet tall. To find his recommended weight, we substitute 15 for H in the formula and solve for W:

$$\frac{W}{2} - 3H = 53 \qquad \text{This is the given formula.}$$

$$\frac{W}{2} - 3 \cdot 15 = 53 \qquad \text{Substitute 15 for } H.$$

$$\frac{W}{2} - 45 = 53 \qquad \text{Multiply: } 3 \cdot 15 = 45.$$

Multiply both sides of the equation by 2, the least common denominator:

$$2\left(\frac{W}{2} - 45\right) = 2 \cdot 53$$

$$2 \cdot \frac{W}{2} - 2 \cdot 45 = 2 \cdot 53 \qquad \text{Use the distributive property.}$$

$$W - 90 = 106 \qquad \text{Multiply: } 2 \cdot \frac{W}{2} = \frac{2}{1} \cdot \frac{W}{2} = \frac{2W}{2} = 1W = W.$$

$$W - 90 + 90 = 106 + 90 \qquad \text{Add 90 to both sides.}$$

$$W = 196 \qquad \text{Simplify.}$$

The recommended weight for a man whose height is 6 feet, 3 inches is 196 pounds. ∎

The points in the rectangular coordinate system in Figure 2.3 allow us to "see" the formula $\frac{W}{2} - 3H = 53$. The first coordinate of each point represents a man's height, in inches, above 5 feet. The second coordinate represents that man's recommended weight, in pounds. The points are rising steadily from left to right. This shows that the recommended weight increases as height increases.

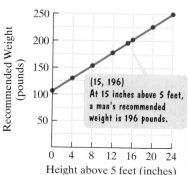

Recommended Weights for Men

(15, 196)
At 15 inches above 5 feet, a man's recommended weight is 196 pounds.

Height above 5 feet (inches)

FIGURE 2.3

✔ **CHECK POINT 7** Use the formula $\frac{W}{2} - 3H = 53$ to find the recommended weight for a man who is 5 feet, 3 inches tall.

2.3 EXERCISE SET

Student Solutions Manual   CD/Video   PH Math/Tutor Center   MathXL Tutorials on CD   MathXL®   MyMathLab   Interactmath.com

Practice Exercises

In Exercises 1–30, solve each equation. Be sure to check your proposed solution by substituting it for the variable in the original equation.

1. $5x + 3x - 4x = 10 + 2$
2. $4x + 8x - 2x = 20 - 15$
3. $4x - 9x + 22 = 3x + 30$
4. $3x + 2x + 64 = 40 - 7x$
5. $3x + 6 - x = 8 + 3x - 6$
6. $3x + 2 - x = 6 + 3x - 8$
7. $4(x + 1) = 20$
8. $3(x - 2) = -6$
9. $7(2x - 1) = 42$
10. $4(2x - 3) = 32$
11. $38 = 30 - 2(x - 1)$
12. $20 = 44 - 8(2 - x)$
13. $2(4z + 3) - 8 = 46$
14. $3(3z + 5) - 7 = 89$
15. $6x - (3x + 10) = 14$
16. $5x - (2x + 14) = 10$
17. $5(2x + 1) = 12x - 3$
18. $3(x + 2) = x + 30$
19. $3(5 - x) = 4(2x + 1)$
20. $3(3x - 1) = 4(3 + 3x)$
21. $8(y + 2) = 2(3y + 4)$
22. $8(y + 3) = 3(2y + 12)$
23. $3(x + 1) = 7(x - 2) - 3$
24. $5x - 4(x + 9) = 2x - 3$
25. $5(2x - 8) - 2 = 5(x - 3) + 3$
26. $7(3x - 2) + 5 = 6(2x - 1) + 24$
27. $6 = -4(1 - x) + 3(x + 1)$
28. $100 = -(x - 1) + 4(x - 6)$
29. $10(z + 4) - 4(z - 2) = 3(z - 1) + 2(z - 3)$
30. $-2(z - 4) - (3z - 2) = -2 - (6z - 2)$

Solve and check each equation in Exercises 31–46. Begin your work by rewriting each equation without fractions.

31. $\dfrac{x}{5} - 4 = -6$
32. $\dfrac{x}{2} + 13 = -22$
33. $\dfrac{2x}{3} - 5 = 7$
34. $\dfrac{3x}{4} - 9 = -6$
35. $\dfrac{2y}{3} - \dfrac{3}{4} = \dfrac{5}{12}$
36. $\dfrac{3y}{4} - \dfrac{2}{3} = \dfrac{7}{12}$
37. $\dfrac{x}{3} + \dfrac{x}{2} = \dfrac{5}{6}$
38. $\dfrac{x}{4} - \dfrac{x}{5} = 1$
39. $20 - \dfrac{z}{3} = \dfrac{z}{2}$
40. $\dfrac{z}{5} - \dfrac{1}{2} = \dfrac{z}{6}$

41. $\dfrac{y}{3} + \dfrac{2}{5} = \dfrac{y}{5} - \dfrac{2}{5}$
42. $\dfrac{y}{12} + \dfrac{1}{6} = \dfrac{y}{2} - \dfrac{1}{4}$
43. $\dfrac{3x}{4} - 3 = \dfrac{x}{2} + 2$
44. $\dfrac{3x}{5} - \dfrac{2}{5} = \dfrac{x}{3} + \dfrac{2}{5}$
45. $\dfrac{x - 3}{5} - 1 = \dfrac{x - 5}{4}$
46. $\dfrac{x - 2}{3} - 4 = \dfrac{x + 1}{4}$

In Exercises 47–64, solve each equation. Identify equations that have no solution, or equations that are true for all real numbers.

47. $3x - 7 = 3(x + 1)$
48. $2(x - 5) = 2x + 10$
49. $2(x + 4) = 4x + 5 - 2x + 3$
50. $3(x - 1) = 8x + 6 - 5x - 9$
51. $7 + 2(3x - 5) = 8 - 3(2x + 1)$
52. $2 + 3(2x - 7) = 9 - 4(3x + 1)$
53. $4x + 1 - 5x = 5 - (x + 4)$
54. $5x - 5 = 3x - 7 + 2(x + 1)$
55. $4(x + 2) + 1 = 7x - 3(x - 2)$
56. $5x - 3(x + 1) = 2(x + 3) - 5$
57. $3 - x = 2x + 3$
58. $5 - x = 4x + 5$
59. $\dfrac{x}{3} + 2 = \dfrac{x}{3}$
60. $\dfrac{x}{4} + 3 = \dfrac{x}{4}$
61. $\dfrac{x}{2} - \dfrac{x}{4} + 4 = x + 4$
62. $\dfrac{x}{2} + \dfrac{2x}{3} + 3 = x + 3$
63. $\dfrac{2}{3}x = 2 - \dfrac{5}{6}x$
64. $\dfrac{2}{3}x = \dfrac{1}{4}x - 8$

Practice Plus

The equations in Exercises 65–66 contain small figures ($\square$, $\triangle$, and $\$$) that represent real numbers. Use properties of equality to isolate x on one side of the equation and the small figures on the other side.

65. $\dfrac{x}{\square} + \triangle = \$$

66. $\dfrac{x}{\square} - \triangle = -\$$

67. If $\dfrac{x}{5} - 2 = \dfrac{x}{3}$, evaluate $x^2 - x$.

68. If $\dfrac{3x}{2} + \dfrac{3x}{4} = \dfrac{x}{4} - 4$, evaluate $x^2 - x$.

In Exercises 69–72, use the given information to write an equation. Let x represent the number described in each exercise. Then solve the equation and find the number.

69. When one-third of a number is added to one-fifth of the number, the sum is 16. What is the number?

70. When two-fifths of a number is added to one-fourth of the number, the sum is 13. What is the number?

71. When 3 is subtracted from three-fourths of a number, the result is equal to one-half of the number. What is the number?

72. When 30 is subtracted from seven-eighths of a number, the result is equal to one-half of the number. What is the number?

Application Exercises

In Massachusetts, speeding fines are determined by the formula

$$F = 10(x - 65) + 50,$$

where F is the cost, in dollars, of the fine if a person is caught driving x miles per hour. Use this formula to solve Exercises 73–74.

73. If a fine comes to $250, how fast was that person speeding?

74. If a fine comes to $400, how fast was that person speeding?

The graph indicates a relationship between sense of humor and depression. Persons with a low sense of humor have higher levels of depression in response to negative life events than those with a high sense of humor.

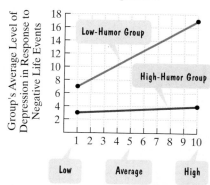

Sense of Humor and Depression

Low-Humor Group

High-Humor Group

Group's Average Level of Depression in Response to Negative Life Events

Low   Average   High

Intensity of Negative Life Event

Source: Steven Davis and Joseph Palladino, *Psychology Third Edition*, Prentice Hall, 2003.

The data can be modeled by the formulas

$$D = \frac{10}{9}N + \frac{53}{9} \quad \text{low humor}$$

$$D = \frac{1}{9}N + \frac{26}{9}, \quad \text{high humor}$$

where N is the intensity of a negative life event (from 1, low, to 10, high) and D is the level of depression in response to that event. Use these formulas to solve Exercises 75–76.

75. If the low-humor group averages a level of depression of 10 in response to a negative life event, what is the intensity of that event? How is the solution shown on the line graph?

76. If the high-humor group averages a level of depression of 3.5, or $\frac{7}{2}$, in response to a negative life event, what is the intensity of that event? How is the solution shown on the line graph?

The formula

$$p = 15 + \frac{5d}{11}$$

describes the pressure of sea water, p, in pounds per square foot, at a depth of d feet below the surface. Use the formula to solve Exercises 77–78.

77. The record depth for breath-held diving, by Francisco Ferreras (Cuba) off Grand Bahama Island, on November 14, 1993, involved pressure of 201 pounds per square foot. To what depth did Ferreras descend on this ill-advised venture? (He was underwater for 2 minutes and 9 seconds!)

78. At what depth is the pressure 20 pounds per square foot?

Writing in Mathematics

79. In your own words, describe how to solve a linear equation.

80. Explain how to solve a linear equation containing fractions.

81. Suppose that you solve $\frac{x}{5} - \frac{x}{2} = 1$ by multiplying both sides by 20, rather than the least common denominator of 5 and 2 (namely, 10). Describe what happens. If you get the correct solution, why do you think we clear the equation of fractions by multiplying by the *least* common denominator?

82. Suppose you are an algebra teacher grading the following solution on an examination:

Solve: $-3(x - 6) = 2 - x$
Solution: $-3x - 18 = 2 - x$
$-2x - 18 = 2$
$-2x = -16$
$x = 8.$

You should note that 8 checks, and the solution is 8. The student who worked the problem therefore wants full credit. Can you find any errors in the solution? If full credit is 10 points, how many points should you give the student? Justify your position.

Critical Thinking Exercises

83. Which one of the following statements is true?

a. The equation $3(x + 4) = 3(4 + x)$ has precisely one solution.

b. The equation $2y + 5 = 0$ is equivalent to $2y = 5$.

c. If $2 - 3y = 11$ and the solution to the equation is substituted into $y^2 + 2y - 3$, a number results that is neither positive nor negative.

d. The equation $x + \frac{1}{3} = \frac{1}{2}$ is equivalent to $x + 2 = 3$.

84. A woman's height, h, is related to the length of her femur, f (the bone from the knee to the hip socket), by the formula $f = 0.432h - 10.44$. Both h and f are measured in inches. A partial skeleton is found of a woman in which the femur is 16 inches long. Police find the skeleton in an area where a woman slightly over 5 feet tall has been missing for over a year. Can the partial skeleton be that of the missing woman? Explain.

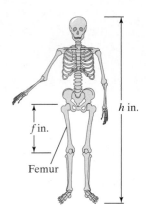

f in.

Femur

h in.

Solve each equation in Exercises 85–86.

85. $\dfrac{2x - 3}{9} + \dfrac{x - 3}{2} = \dfrac{x + 5}{6} - 1$

86. $2(3x + 4) = 3x + 2[3(x - 1) + 2]$

Technology Exercises

Solve each equation in Exercises 87–88. Use a calculator to help with the arithmetic. Check your solution using the calculator.

87. $2.24y - 9.28 = 5.74y + 5.42$

88. $4.8y + 32.5 = 124.8 - 9.4y$

Review Exercises

In Exercises 89–91, insert either < or > in the box between each pair of numbers to make a true statement.

89. $-24 \,\square\, -20$ (Section 1.2, Example 6)

90. $-\dfrac{1}{3} \,\square\, -\dfrac{1}{5}$ (Section 1.2, Example 6)

91. Simplify: $-9 - 11 + 7 - (-3)$. (Section 1.6, Example 3)

SECTION

2.4

FORMULAS AND PERCENTS

"And, if elected, it is my solemn pledge to cut your taxes by 10% for each of my first three years in office, for a total cut of 30%."

Objectives

1 Solve a formula for a variable.

2 Express a decimal as a percent.

3 Express a percent as a decimal.

4 Use the percent formula.

5 Solve applied problems involving percent change.

Did you know that one of the most common ways that you are given numerical information is with percents? In this section, you will learn to use a formula that will help you to understand percents, enabling you to make sense of the politician's promise.

1 Solve a formula for a variable.

Solving a Formula for One of Its Variables We know that solving an equation is the process of finding the number (or numbers) that make the equation a true statement. All of the equations we have solved contained only one letter, such as x or y.

By contrast, the formulas we have seen contain two or more letters, representing two or more variables. Here is an example:

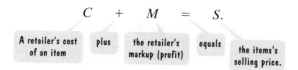

$$C \quad + \quad M \quad = \quad S.$$

A retailer's cost of an item — plus — the retailer's markup (profit) — equals — the items's selling price.

We say that this formula is solved for the variable S because S is alone on one side of the equation and the other side does not contain an S.

Solving a formula for a variable means rewriting the formula so that the variable is isolated on one side of the equation. It does not mean obtaining a numerical value for that variable.

The addition and multiplication properties of equality are used to solve a formula for one of its variables. Consider the retailer's formula, $C + M = S$. How do we solve this formula for C? Use the addition property to isolate C by subtracting M from both sides:

We need to isolate C.

$$C + M = S \qquad \text{This is the given formula.}$$
$$C + M - M = S - M \qquad \text{Subtract } M \text{ from both sides.}$$
$$C = S - M. \qquad \text{Simplify.}$$

Solved for C, the formula $C = S - M$ tells us that the cost of an item for a retailer is the item's selling price minus its markup.

To solve a formula for one of its variables, treat that variable as if it were the only variable in the equation. Think of the other variables as if they were numbers. Use the addition property of equality to isolate all terms with the specified variable on one side of the equation and all terms without the specified variable on the other side. Then use the multiplication property of equality to get the specified variable alone.

Our first example involves the formula for the area of a rectangle. The **area of a two-dimensional figure** is the number of square units it takes to fill the interior of the figure. A **square unit** is a square, each of whose sides is one unit in length, as illustrated in Figure 2.4. The figure shows that there are 12 square units contained within the rectangle. The area of the rectangle is 12 square units. Notice that the area can be determined in the following manner:

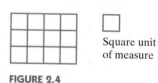

Square unit of measure

FIGURE 2.4

Across   Down

$$4 \text{ units} \cdot 3 \text{ units} = 4 \cdot 3 \text{ units} \cdot \text{units} = 12 \text{ square units}.$$

The area of a rectangle is the product of the distance across, its length, and the distance down, its width.

AREA OF A RECTANGLE The area, A, of a rectangle with length l and width w is given by the formula

$$A = lw.$$

EXAMPLE 1 Solving a Formula for a Variable

Solve the formula $A = lw$ for w.

SOLUTION Our goal is to get w by itself on one side of the formula. There is only one term with w, lw, and it is already isolated on the right side. We isolate w on the right by using the multiplication property of equality and dividing both sides by l.

 We need to isolate w.

$$A = lw \qquad \text{This is the given formula.}$$

$$\frac{A}{l} = \frac{lw}{l} \qquad \text{Isolate } w \text{ by dividing both sides by } l.$$

$$\frac{A}{l} = w \qquad \text{Simplify: } \frac{lw}{l} = 1w = w.$$

The formula solved for w is $\frac{A}{l} = w$ or $w = \frac{A}{l}$. Thus, the area of a rectangle divided by its length is equal to its width.   ∎

✓ **CHECK POINT 1** Solve the formula $A = lw$ for l.

The perimeter, P, of a two-dimensional figure is the sum of the lengths of its sides. Perimeter is measured in linear units, such as inches, feet, yards, meters, or kilometers.

Example 2 involves the perimeter of a rectangle. Because perimeter is the sum of the lengths of the sides, the perimeter of the rectangle shown in Figure 2.5 is $l + w + l + w$. This can be expressed as

$$P = 2l + 2w.$$

FIGURE 2.5 A rectangle with length l and width w

> **PERIMETER OF A RECTANGLE**   The perimeter, P, of a rectangle with length l and width w is given by the formula
>
> $$P = 2l + 2w.$$
>
> The perimeter of a rectangle is the sum of twice the length and twice the width.

EXAMPLE 2 Solving a Formula for a Variable

Solve the formula $2l + 2w = P$ for w.

SOLUTION First, isolate $2w$ on the left by subtracting $2l$ from both sides. Then solve for w by dividing both sides by 2.

We need to isolate w.

$$2l + 2w = P \qquad \text{This is the given formula.}$$

$$2l - 2l + 2w = P - 2l \qquad \text{Isolate } 2w \text{ by subtracting } 2l \text{ from both sides.}$$

$$2w = P - 2l \qquad \text{Simplify.}$$

$$\frac{2w}{2} = \frac{P - 2l}{2} \qquad \text{Isolate } w \text{ by dividing both sides by 2.}$$

$$w = \frac{P - 2l}{2} \qquad \text{Simplify.}$$

∎

✔ CHECK POINT **2** Solve the formula $2l + 2w = P$ for l.

EXAMPLE 3 Solving a Formula for a Variable

The total price of an article purchased on a monthly deferred payment plan is described by the following formula:

$$T = D + pm.$$

In this formula, T is the total price, D is the down payment, p is the amount of the monthly payment, and m is the number of payments. Solve the formula for p.

SOLUTION First, isolate pm on the right by subtracting D from both sides. Then isolate p from pm by dividing both sides of the formula by m.

> We need to isolate p.

$T = D + pm$	This is the given formula. We want p alone.
$T - D = D - D + pm$	Isolate pm by subtracting D from both sides.
$T - D = pm$	Simplify.
$\dfrac{T - D}{m} = \dfrac{pm}{m}$	Now isolate p by dividing both sides by m.
$\dfrac{T - D}{m} = p$	Simplify: $\dfrac{pm}{m} = \dfrac{p\overset{1}{\cancel{m}}}{\underset{1}{\cancel{m}}} = p \cdot 1 = p.$ ∎

✔ CHECK POINT **3** Solve the formula $T = D + pm$ for m.

The next example has a formula that contains a fraction. To solve for a variable in a formula involving fractions, we begin by multiplying both sides by the least common denominator of all fractions in the formula. This will eliminate the fractions. Then we solve for the specified variable.

EXAMPLE 4 Solving a Formula Containing a Fraction for a Variable

Solve the formula $\dfrac{W}{2} - 3H = 53$ for W.

SOLUTION Do you remember seeing this formula in the last section? It models the recommended weight, W, for a male, where H represents the man's height, in inches, over 5 feet. We begin by multiplying both sides of the formula by 2 to eliminate the fraction. Then we isolate the variable W.

$\dfrac{W}{2} - 3H = 53$	This is the given formula.
$2\left(\dfrac{W}{2} - 3H\right) = 2 \cdot 53$	Multiply both sides by 2.
$2 \cdot \dfrac{W}{2} - 2 \cdot 3H = 2 \cdot 53$	Use the distributive property.

> We need to isolate *W*.

$$W - 6H = 106 \qquad \text{Simplify.}$$
$$W - 6H + 6H = 106 + 6H \qquad \text{Isolate W by adding 6H to both sides.}$$
$$W = 106 + 6H \qquad \text{Simplify.}$$

This form of the formula makes it easy to find a man's recommended weight, *W*, if we know his height, *H*, in inches, over 5 feet. ■

 CHECK POINT 4 Solve for *x*: $\dfrac{x}{3} - 4y = 5$.

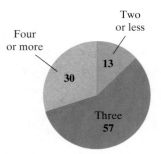

FIGURE 2.6 Number of bedrooms in privately owned single-family U.S. houses per 100 houses.
Source: U.S. Census Bureau and HUD

Before turning to a formula involving percent, let's review some of the basics of percent.

Basics of Percent **Percents** are the result of expressing numbers as a part of 100. The word *percent* means *per hundred*. For example, the circle graph in Figure 2.6 shows that 57 out of every 100 single-family homes have three bedrooms. Thus, $\frac{57}{100} = 57\%$, indicating that 57% of the houses have three bedrooms. The percent sign, %, is used to indicate the number of parts out of one hundred parts.

By definition, $57\% = \frac{57}{100}$. We can express the fraction $\frac{57}{100}$ in decimal notation as 0.57. Thus, $0.57 = 57\%$. Here is a general procedure for expressing a decimal number as a percent:

EXPRESSING A DECIMAL NUMBER AS A PERCENT
1. Move the decimal point two places to the right.
2. Attach a percent sign.

2 Express a decimal as a percent.

EXAMPLE 5 Expressing a Decimal as a Percent

Express 0.47 as a percent.

SOLUTION

> Move decimal point two places right.

$$0.47 \, \% \quad \longleftarrow \text{Attach a percent sign.}$$

Thus, $0.47 = 47\%$. ■

 CHECK POINT 5 Express 0.023 as a percent.

3 Express a percent as a decimal.

We reverse the procedure of Example 5 to express a percent as a decimal number.

EXPRESSING A PERCENT AS A DECIMAL NUMBER
1. Move the decimal point two places to the left.
2. Remove the percent sign.

EXAMPLE 6 Expressing Percents as Decimals

Express each percent as a decimal: **a.** 19%   **b.** 180%.

SOLUTION Use the two steps in the box.

a.

$$19\% = 19.\% = 0.19\%$$

The percent sign is removed.

The decimal point starts at the far right.

The decimal point is moved two places to the left.

Thus, 19% = 0.19.

b. 180% = 1.80% = 1.80 or 1.8.

 CHECK POINT 6 Express each percent as a decimal:

a. 67%   **b.** 250%.

4 Use the percent formula.

A Formula Involving Percent Percents are useful in comparing two numbers. To compare the number A to the number B using a percent P, the following formula is used:

A is P percent   of B.

$$A = P \cdot B.$$

In the formula

$$A = PB$$

B = the base number, P = the percent (in decimal form), and A = the number compared to B.

There are three basic types of percent problems that can be solved using the percent formula

$$A = PB. \quad \text{A is P percent of B.}$$

Question	Given	Percent Formula
What is P percent of B?	P and B	Solve for A.
A is P percent of what?	A and P	Solve for B.
A is what percent of B?	A and B	Solve for P.

Let's look at an example of each type of problem.

EXAMPLE 7 Using the Percent Formula: What Is P Percent of B?

What is 8% of 20?

SOLUTION We use the formula $A = PB$: A is P percent of B. We are interested in finding the quantity A in this formula.

What   is   8%   of   20?

$$A = 0.08 \cdot 20$$

Express 8% as 0.08.

$$A = 1.6$$

Multiply: $\begin{array}{r} .08 \\ \times\ 20 \\ \hline 1.60 \end{array}$

Thus, 1.6 is 8% of 20. The answer is 1.6.

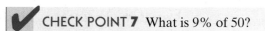 **CHECK POINT 7** What is 9% of 50?

EXAMPLE 8 Using the Percent Formula: A Is P Percent of What?

4 is 25% of what?

SOLUTION We use the formula $A = PB$: A is P percent of B. We are interested in finding the quantity B in this formula.

| 4 | is | 25% | of | what? |

$$4 = 0.25 \cdot B \qquad \text{Express 25\% as 0.25.}$$

$$\frac{4}{0.25} = \frac{0.25B}{0.25} \qquad \text{Divide both sides by 0.25.}$$

$$16 = B \qquad \text{Simplify: } 0.25\overline{)4.00}^{\,16.}$$

Thus, 4 is 25% of 16. The answer is 16. ∎

 CHECK POINT 8 9 is 60% of what?

EXAMPLE 9 Using the Percent Formula: A Is What Percent of B?

1.3 is what percent of 26?

SOLUTION We use the formula $A = PB$: A is P percent of B. We are interested in finding the quantity P in this formula.

| 1.3 | is | what percent | of | 26? |

$$1.3 = P \cdot 26$$

$$\frac{1.3}{26} = \frac{P \cdot 26}{26} \qquad \text{Divide both sides by 26.}$$

$$0.05 = P \qquad \text{Simplify: } 26\overline{)1.30}^{\,0.05}.$$

We change 0.05 to a percent by moving the decimal point two places to the right and adding a percent sign: 0.05 = 5%. Thus, 1.3 is 5% of 26. The answer is 5%. ∎

 CHECK POINT 9 18 is what percent of 50?

---

⑤ Solve applied problems involving percent change.

Applications Percents are used for comparing changes, such as increases or decreases in sales, population, prices, and production. If a quantity changes, its percent increase or percent decrease can be determined by asking the following question:

The change is what percent of the original amount?

The question is answered using the percent formula as follows:

EXAMPLE 10 Finding Percent Decrease

A jacket regularly sells for $135.00 The sale price is $60.75. Find the percent decrease in the jacket's price.

SOLUTION The percent decrease in price can be determined by asking the following question:

The price decrease is what percent of the original price ($135.00)?

The price decrease is the difference between the original price and the sale price ($60.75):

$$135.00 - 60.75 = 74.25.$$

Now we use the percent formula to find the percent decrease.

$$74.25 = P \cdot 135$$

$$\frac{74.25}{135} = \frac{P \cdot 135}{135} \qquad \text{Divide both sides by 135.}$$

$$0.55 = P \qquad \text{Simplify: } 135\overline{)74.25}^{\,0.55}.$$

We change 0.55 to a percent by moving the decimal point two places to the right and adding a percent sign: 0.55 = 55%. Thus, the percent decrease in the jacket's price is 55%. ∎

✔ **CHECK POINT 10** A television regularly sells for $940. The sale price is $611. Find the percent decrease in the television's price.

In our next example, we look at one of the many ways that percent can be used incorrectly.

EXAMPLE 11 Promises of a Politician

A politician states, "If you elect me to office, I promise to cut your taxes for each of my first three years in office by 10% each year, for a total reduction of 30%." Evaluate the accuracy of the politician's statement.

SOLUTION To make things simple, let's assume that a taxpayer paid $100 in taxes in the year previous to the politician's election. A 10% reduction during year 1 is 10% of $100.

10% of previous year's tax = 10% of $100 = 0.10 · $100 = $10

With a 10% reduction the first year, the taxpayer will pay only $100 − $10, or $90, in taxes during the politician's first year in office.

The table on the next page shows how we calculate the new, reduced tax for each of the first three years in office.

Year	Tax paid the year before	10% reduction	Taxes paid this year
1	$100	$0.10 \cdot \$100 = \$10$	$\$100 - \$10 = \$90$
2	$ 90	$0.10 \cdot \$90 = \$9$	$\$90 - \$9 = \$81$
3	$ 81	$0.10 \cdot \$81 = \$8.10$	$\$81 - \$8.10 = \$72.90$

The tax reduction is the amount originally paid, $100.00, minus the amount paid during the politician's third year in office, $72.90:

$$\$100.00 - \$72.90 = \$27.10.$$

Now we use the percent formula to determine the percent decrease in taxes over the three years.

The tax decrease   is   what percent   of   the original tax?

$$27.1 = P \cdot 100$$

$$\frac{27.1}{100} = \frac{P \cdot 100}{100} \qquad \text{Divide both sides by 100.}$$

$$0.271 = P \qquad \text{Simplify.}$$

Change 0.271 to a percent: $0.271 = 27.1\%$. The percent decrease is 27.1%. The taxes decline by 27.1%, not by 30%. The politician is ill-informed in saying that three consecutive 10% cuts add up to a total tax cut of 30%. In our calculation, which serves as a counterexample to the promise, the total tax cut is only 27.1%. ■

CHECK POINT 11 Suppose you paid $1200 in taxes. During year 1, taxes decrease by 20%. During year 2, taxes increase by 20%.

a. What do you pay in taxes for year 2?

b. How do your taxes for year 2 compare with what you originally paid, namely $1200? If the taxes are not the same, find the percent increase or decrease.

2.4 EXERCISE SET

 Student Solutions Manual   CD/Video   PH Math/Tutor Center   MathXL Tutorials on CD   MathXL®   MyMathLab   Interactmath.com

Practice Exercises

In Exercises 1–26, solve each formula for the specified variable. Do you recognize the formula? If so, what does it describe?

1. $d = rt$ for r

2. $d = rt$ for t

3. $I = Prt$ for P

4. $I = Prt$ for r

5. $C = 2\pi r$ for r

6. $C = \pi d$ for d

7. $E = mc^2$ for m

8. $V = \pi r^2 h$ for h

9. $y = mx + b$ for m

10. $y = mx + b$ for x

11. $T = D + pm$ for p

12. $P = C + MC$ for M

13. $A = \frac{1}{2}bh$ for b

14. $A = \frac{1}{2}bh$ for h

15. $M = \frac{n}{5}$ for n

16. $M = \frac{A}{740}$ for A

17. $\frac{c}{2} + 80 = 2F$ for c

18. $p = 15 + \frac{5d}{11}$ for d

19. $A = \frac{1}{2}(a + b)$ for a

20. $A = \frac{1}{2}(a + b)$ for b

21. $S = P + Prt$ for r

22. $S = P + Prt$ for t

23. $A = \frac{1}{2}h(a + b)$ for b

24. $A = \frac{1}{2}h(a + b)$ for a

25. $Ax + By = C$ for x

26. $Ax + By = C$ for y

In Exercises 27–34, express each decimal as a percent.

27. 0.89   **28.** 0.16   **29.** 0.002

30. 0.008   **31.** 4.78   **32.** 5.38

33. 100   **34.** 85

In Exercises 35–44, express each percent as a decimal.

35. 27%   **36.** 83%   **37.** 63.4%

38. 2.15%   **39.** 170%   **40.** 360%

41. 3%   **42.** 8%   **43.** $\frac{1}{2}$%

44. $\frac{1}{4}$%

Use the percent formula, $A = PB$: A is P percent of B, to solve Exercises 45–56.

45. What is 3% of 200?   **46.** What is 8% of 300?

47. What is 18% of 40?   **48.** What is 16% of 90?

49. 3 is 60% of what?   **50.** 8 is 40% of what?

51. 24% of what number is 40.8?

52. 32% of what number is 51.2?

53. 3 is what percent of 15?

54. 18 is what percent of 90?

55. What percent of 2.5 is 0.3?

56. What percent of 7.5 is 0.6?

57. If 5 is increased to 8, the increase is what percent of the original number?

58. If 5 is increased to 9, the increase is what percent of the original number?

59. If 4 is decreased to 1, the decrease is what percent of the original number?

60. If 8 is decreased to 6, the decrease is what percent of the original number?

Practice Plus

In Exercises 61–68, solve each equation for x.

61. $y = (a + b)x$

62. $y = (a - b)x$

63. $y = (a - b)x + 5$

64. $y = (a + b)x - 8$

65. $y = cx + dx$

66. $y = cx - dx$

67. $y = Ax - Bx - C$

68. $y = Ax + Bx + C$

Application Exercises

69. The average, or mean, A, of three exam grades, x, y, and z, is given by the formula
$$A = \frac{x + y + z}{3}.$$
   a. Solve the formula for z.
   b. Use the formula in part (a) to solve this problem. On your first two exams, your grades are 86% and 88%: $x = 86$ and $y = 88$. What must you get on the third exam to have an average of 90%?

70. The average, or mean, A, of four exam grades, x, y, z, and w, is given by the formula
$$A = \frac{x + y + z + w}{4}.$$
   a. Solve the formula for w.
   b. Use the formula in part (a) to solve this problem. On your first three exams, your grades are 76%, 78%, and 79%: $x = 76$, $y = 78$, and $z = 79$. What must you get on the fourth exam to have an average of 80%?

71. If you are traveling in your car at an average rate of r miles per hour for t hours, then the distance, d, in miles, that you travel is described by the formula $d = rt$: distance equals rate times time.
   a. Solve the formula for t.
   b. Use the formula in part (a) to find the time that you travel if you cover a distance of 100 miles at an average rate of 40 miles per hour.

72. The formula $F = \frac{9}{5}C + 32$ expresses the relationship between Celsius temperature, C, and Fahrenheit temperature, F.
   a. Solve the formula for C.
   b. Use the formula from part (a) to find the equivalent Celsius temperature for a Fahrenheit temperature of 59°.

A recent Time/CNN telephone poll included never-married single women between the ages of 18 and 49 and never-married single men between the ages of 18 and 49. The circle graphs show the results for one of the questions in the poll. Use this information to solve Exercises 73–74.

If You Couldn't Find the Perfect Mate, Would You Marry Someone Else?

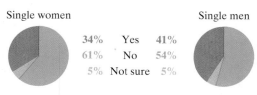

Single women   Single men

	34%	Yes	41%
	61%	No	54%
	5%	Not sure	5%

Source: *Time,* August 28, 2000

73. There were 1200 single women who participated in the poll. How many stated they would marry someone other than the perfect mate?

74. There were 1200 single men who participated in the poll. How many stated they would marry someone other than the perfect mate?

75. In 2002, the leading cause of death in the United States was heart disease, resulting in 710,760 deaths. If 30% of all deaths were caused by heart disease, find the total number of deaths in the United States in 2002.

(*Source*: Department of Health and Human Services)

76. In 2001, 4% of the Hispanic population in the United States was Cuban. If the Cuban-American population at that time was approximately 1.4 million, what was the approximate U.S. Hispanic population in 2001?

In 2002, there were approximately 12,000 hate crimes reported to the FBI in the United States. The circle graph shows the breakdown of this total number. Use this information to solve Exercises 77–78.

Motivation for 12,000 U.S. Hate-Crime Incidents

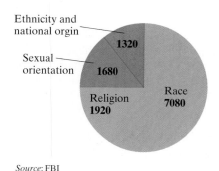

Ethnicity and national orgin — 1320
Sexual orientation — 1680
Religion 1920
Race 7080

Source: FBI

77. What percent of hate-crime incidents were motivated by race?

78. What percent of hate-crime incidents were motivated by sexual orientation?

79. A charity has raised $7500, with a goal of raising $60,000. What percent of the goal has been raised?

80. A charity has raised $225,000, with a goal of raising $500,000. What percent of the goal has been raised?

81. A restaurant bill came to $60. If 15% of this bill is left as a tip, how much was the tip?

82. If income tax is $3502 plus 28% of taxable income over $23,000, how much is the income tax on a taxable income of $35,000?

83. Suppose that the local sales tax rate is 6% and you buy a car for $16,800.
   a. How much tax is due?
   b. What is the car's total cost?

84. Suppose that the local sales tax rate is 7% and you buy a graphing calculator for $96.
   a. How much tax is due?
   b. What is the calculator's total cost?

85. An exercise machine with an original price of $860 is on sale at 12% off.
   a. What is the discount amount?
   b. What is the exercise machine's sale price?

86. A dictionary that normally sells for $16.50 is on sale at 40% off.
   a. What is the discount amount?
   b. What is the dictionary's sale price?

87. A sofa regularly sells for $840. The sale price is $714. Find the percent decrease in the sofa's price.

88. A fax machine regularly sells for $380. The sale price is $266. Find the percent decrease in the machine's price.

89. Suppose that you put $10,000 in a rather risky investment recommended by your financial advisor. During the first year, your investment decreases by 30% of its original value. During the second year, your investment increases by 40% of its first-year value. Your advisor tells you that there must have been a 10% overall increase of your original $10,000 investment. Is your financial advisor using percentages properly? If not, what is your actual percent gain or loss of your original $10,000 investment?

90. The price of a color printer is reduced by 30% of its original price. When it still does not sell, its price is reduced by 20% of the reduced price. The salesperson informs you that there has been a total reduction of 50%. Is the salesperson using percentages properly? If not, what is the actual percent reduction from the original price?

Writing in Mathematics

91. Explain what it means to solve a formula for a variable.

92. What is a percent?

93. Describe how to express a decimal number as a percent and give an example.

94. Describe how to express a percent as a decimal number and give an example.

95. What does the percent formula, $A = PB$, describe? Give an example of how the formula is used.

96. Describe one way in which you use percents in your daily life.

Critical Thinking Exercises

97. Which one of the following statements is true?

a. If $ax + b = 0$, then $x = \dfrac{b}{a}$.

b. If $A = lw$, then $w = \dfrac{l}{A}$.

c. If $A = \dfrac{1}{2}bh$, then $b = \dfrac{A}{2h}$.

d. Solving $x - y = -7$ for y gives $y = x + 7$.

98. In psychology, an intelligence quotient, Q, also called IQ, is measured by the formula

$$Q = \frac{100M}{C},$$

where M = mental age and C = chronological age. Solve the formula for C.

99. The height, h, in feet, of water in a fountain is described by the formula

$$h = -16t^2 + 64t$$

and the velocity, v, in feet per second, of water in the fountain is described by $v = -32t + 64$. Find the time when the water's velocity is 16 feet per second, and then find the water's height at that time.

Review Exercises

100. Solve and check: $5x + 20 = 8x - 16$. (Section 2.2, Example 7)

101. Solve and check: $5(2y - 3) - 1 = 4(6 + 2y)$. (Section 2.3, Example 3)

102. Simplify: $x - 0.3x$. (Section 1.4, Example 9)

✓ **MID-CHAPTER CHECK POINT**

CHAPTER 2

What You Know: We learned a step-by-step procedure for solving linear equations, including equations with fractions. We saw that some linear equations have no solution, whereas others have all real numbers as solutions. We used the addition and multiplication properties of equality to solve formulas for variables. Finally, we worked with the percent formula $A = PB$: A is P percent of B.

1. Solve: $\dfrac{x}{2} = 12 - \dfrac{x}{4}$.

2. Solve: $5x - 42 = -57$.

3. Solve for C: $H = \dfrac{EC}{825}$

4. What is 6% of 140?

5. Solve: $\dfrac{-x}{10} = -3$.

6. Solve: $1 - 3(y - 5) = 4(2 - 3y)$.

7. Solve for r: $S = 2\pi rh$

8. 12 is 30% of what?

9. Solve: $\dfrac{3y}{5} + \dfrac{y}{2} = \dfrac{5y}{4} - 3$.

10. Solve: $5z + 7 = 6(z - 2) - 4(2z - 3)$

11. Solve for x: $Ax - By = C$.

12. Solve: $6y + 7 + 3y = 3(3y - 1)$

13. The formula $D = 0.12x + 5.44$ models the number of children in the United States with physical disabilities, D, in millions, x years after 2000. According to this model, in which year will there be 6.4 million children in the United States with physical disabilities?

14. Solve: $10\left(\dfrac{1}{2}x + 3\right) = 10\left(\dfrac{3}{5}x - 1\right)$

15. 50 is what percent of 400?

16. Solve: $\dfrac{3(m + 2)}{4} = 2m + 3$.

17. If 40 is increased to 50, the increase is what percent of the original number?

18. Solve: $12w - 4 + 8w - 4 = 4(5w - 2)$.

Objectives

1 Translate English phrases into algebraic expressions.

2 Solve algebraic word problems using linear equations.

AN INTRODUCTION TO PROBLEM SOLVING

You started college with your best friend. This semester, you and your friend are taking two classes together. However, your friend often misses class and is not doing the necessary homework between classes to succeed. What can you say to your friend, who values your advice and who is in danger of flunking out of college if things continue on their present course?

Some problems have many plans for finding an answer. To solve your friend's problem, or any problem for that matter, we need to understand the problem fully, devise a plan for solving it, and then carry out the plan. However, problem solving in algebra is easier than solving the many problematic situations encountered in everyday life. Why? Algebra provides a step-by-step strategy for solving problems. As you become familiar with this strategy, you will learn to solve a wide variety of problems.

A Strategy for Solving Word Problems Using Equations Problem solving is an important part of algebra. The problems in this book are presented in English. We must translate from the ordinary language of English into the language of algebraic equations. To translate, however, we must understand the English prose and be familiar with the forms of algebraic language. Here are some general steps we will follow in solving word problems:

STRATEGY FOR SOLVING WORD PROBLEMS

Step 1 Read the problem carefully. Attempt to state the problem in your own words and state what the problem is looking for. Let x (or any variable) represent one of the unknown quantities in the problem.

Step 2 If necessary, write expressions for any other unknown quantities in the problem in terms of x.

Step 3 Write an equation in x that describes the conditions of the problem.

Step 4 Solve the equation and answer the problem's question.

Step 5 Check the solution *in the original wording* of the problem, not in the equation obtained from the words.

Take great care with step 1. Reading a word problem is not the same as reading a newspaper. Reading the problem involves slowly working your way through its parts, making notes on what is given, and perhaps rereading the problem a few times. Only at this point should you let x represent one of the quantities.

The most difficult step in this process is step 3 because it involves translating verbal conditions into an algebraic equation. Translations of some commonly used English phrases are listed in Table 2.1. We choose to use x to represent the variable, but we can use any letter.

1 Translate English phrases into algebraic expressions.

STUDY TIP

Cover the right column in Table 2.1 with a sheet of paper and attempt to formulate the algebraic expression in the column on your own. Then slide the paper down and check your answer. Work through the entire table in this manner.

Table 2.1 **Algebraic Translations of English Phrases**

English Phrase	Algebraic Expression
Addition	
The sum of a number and 7	$x + 7$
Five more than a number; a number plus 5	$x + 5$
A number increased by 6; 6 added to a number	$x + 6$
Subtraction	
A number minus 4	$x - 4$
A number decreased by 5	$x - 5$
A number subtracted from 8	$8 - x$
The difference between a number and 6	$x - 6$
The difference between 6 and a number	$6 - x$
Seven less than a number	$x - 7$
Seven minus a number	$7 - x$
Nine fewer than a number	$x - 9$
Multiplication	
Five times a number	$5x$
The product of 3 and a number	$3x$
Two-thirds of a number (used with fractions)	$\frac{2}{3}x$
Seventy-five percent of a number (used with decimals)	$0.75x$
Thirteen multiplied by a number	$13x$
A number multiplied by 13	$13x$
Twice a number	$2x$
Division	
A number divided by 3	$\dfrac{x}{3}$
The quotient of 7 and a number	$\dfrac{7}{x}$
The quotient of a number and 7	$\dfrac{x}{7}$
The reciprocal of a number	$\dfrac{1}{x}$
More than one operation	
The sum of twice a number and 7	$2x + 7$
Twice the sum of a number and 7	$2(x + 7)$
Three times the sum of 1 and twice a number	$3(1 + 2x)$
Nine subtracted from 8 times a number	$8x - 9$
Twenty-five percent of the sum of 3 times a number and 14	$0.25(3x + 14)$
Seven times a number, increased by 24	$7x + 24$
Seven times the sum of a number and 24	$7(x + 24)$

STUDY TIP

Here are three similar English phrases that have very different translations:

7 minus 10: $7 - 10$

7 less than 10: $10 - 7$

7 is less than 10: $7 < 10$.

Think carefully about what is expressed in English before you translate into the language of algebra.

EXAMPLE 1   Translating English Phrases into Algebraic Expressions

Write each English phrase as an algebraic expression. Let x represent the number.

a. Six subtracted from 5 times a number

b. The quotient of 9 and a number, decreased by 4 times the number

SOLUTION

a.

| Six subtracted from | 5 times a number |

$$5x \quad - \quad 6$$

The algebraic expression for "six subtracted from 5 times a number" is $5x - 6$.

b.

The quotient of 9 and a number,	decreased by	4 times the number

$$\frac{9}{x} \qquad - \qquad 4x$$

The algebraic expression for "the quotient of 9 and a number, decreased by 4 times the number" is $\frac{9}{x} - 4x$. ■

 CHECK POINT 1 Write each English phrase as an algebraic expression. Let x represent the number.

a. Four times a number, increased by 6

b. The quotient of a number decreased by 4 and 9.

---

2 Solve algebraic word problems using linear equations.

Applying the Strategy for Solving Word Problems Now that we've practiced writing algebraic expressions for English phrases, let's apply our five-step strategy for solving word problems.

EXAMPLE 2 Solving a Word Problem

Nine subtracted from eight times a number is 39. Find the number.

SOLUTION

Step 1.   **Let x represent one of the quantities.** Because we are asked to find a number, let
$$x = \text{the number.}$$

Step 2.   **Represent other quantities in terms of x.** There are no other unknown quantities to find, so we can skip this step.

Step 3.   **Write an equation in x that describes the conditions.**

Nine subtracted from	eight times a number	is	39.

$$8x \qquad -9 \qquad = \qquad 39$$

Step 4.   **Solve the equation and answer the question.**

$$8x - 9 = 39 \qquad \text{This is the equation for the problem's conditions.}$$
$$8x - 9 + 9 = 39 + 9 \qquad \text{Add 9 to both sides.}$$
$$8x = 48 \qquad \text{Simplify.}$$
$$\frac{8x}{8} = \frac{48}{8} \qquad \text{Divide both sides by 8.}$$
$$x = 6 \qquad \text{Simplify.}$$

The number is 6.

Step 5.   **Check the proposed solution in the original wording of the problem.** "Nine subtracted from eight times a number is 39." The proposed number is 6. Eight times 6 is $8 \cdot 6$, or 48. Nine subtracted from 48 is $48 - 9$, or 39. The proposed solution checks in the problem's wording, verifying that the number is 6. ■

 CHECK POINT 2 Four subtracted from six times a number is 68. Find the number.

EXAMPLE 3 Pet Population

Americans love their pets. The number of cats in the United States exceeds the number of dogs by 7.5 million. The number of cats and dogs combined is 114.7 million. Determine the number of dogs and cats in the United States.

SOLUTION

U.S. Pet Population

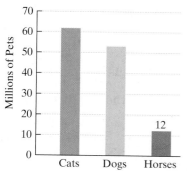

Americans spend more than $30 billion per year on their pets, from MRIs to spas to doggy diapers.

Source: American Veterinary Medical Association

Step 1. **Let *x* represent one of the quantities.** We know something about the number of cats: the cat population exceeds the dog population by 7.5 million. This means that there are 7.5 million more cats than dogs. We will let

$$x = \text{the number (in millions) of dogs in the United States.}$$

Step 2. **Represent other quantities in terms of *x*.** The other unknown quantity is the number of cats. Because there are 7.5 million more cats than dogs, let

$$x + 7.5 = \text{the number (in millions) of cats in the United States.}$$

Step 3. **Write an equation in *x* that describes the conditions.** The number of cats and dogs combined is 114.7 million.

The number (in millions) of dogs in the U.S.	plus	the number (in millions) of cats in the U.S.	equals	114.7 million.
x	$+$	$(x + 7.5)$	$=$	114.7

Step 4. **Solve the equation and answer the question.**

$$x + (x + 7.5) = 114.7 \qquad \text{This is the equation specified by the conditions of the problem.}$$

$$2x + 7.5 = 114.7 \qquad \text{Regroup and combine like terms on the left side.}$$

$$2x + 7.5 - 7.5 = 114.7 - 7.5 \qquad \text{Subtract 7.5 from both sides.}$$

$$2x = 107.2 \qquad \text{Simplify.}$$

$$\frac{2x}{2} = \frac{107.2}{2} \qquad \text{Divide both sides by 2.}$$

$$x = 53.6 \qquad \text{Simplify.}$$

Because *x* represents the number (in millions) of dogs, there are 53.6 million dogs in the United States. Because $x + 7.5$ represents the number (in millions) of cats, there are $53.6 + 7.5$, or 61.1 million cats in the United States.

Step 5. **Check the proposed solution in the original wording of the problem.** The problem states that the number of cats and dogs combined is 114.7 million. By adding 53.6 million, the dog population, and 61.1 million, the cat population, we do, indeed, obtain a sum of 114.7 million. ∎

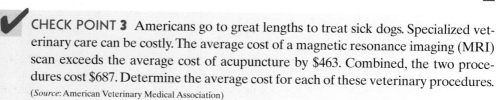

✔ CHECK POINT 3 Americans go to great lengths to treat sick dogs. Specialized veterinary care can be costly. The average cost of a magnetic resonance imaging (MRI) scan exceeds the average cost of acupuncture by $463. Combined, the two procedures cost $687. Determine the average cost for each of these veterinary procedures. (*Source*: American Veterinary Medical Association)

EXAMPLE 4 Consecutive Integers

Two pages that face each other in a book have 145 as the sum of their page numbers. What are the page numbers?

SOLUTION

Step 1. **Let x represent one of the quantities.** We will let

$$x = \text{the page number of the page on the left.}$$

Step 2. **Represent other quantities in terms of x.** The other unknown quantity is the page number of the facing page on the right. Page numbers on facing pages are consecutive integers. Thus,

$$x + 1 = \text{the page number of the page on the right.}$$

Step 3. **Write an equation in x that describes the conditions.** The two facing pages have 145 as the sum of their page numbers.

The page number on the left	plus	the page number on the right	equals	145.
x	$+$	$(x + 1)$	$=$	145

Step 4. **Solve the equation and answer the question.**

$x + (x + 1) = 145$	This is the equation for the problem's conditions.
$2x + 1 = 145$	Regroup and combine like terms.
$2x + 1 - 1 = 145 - 1$	Subtract 1 from both sides.
$2x = 144$	Simplify.
$\dfrac{2x}{2} = \dfrac{144}{2}$	Divide both sides by 2.
$x = 72$	Simplify.

Thus,

$$\text{the page number on the left} = x = 72$$

and

$$\text{the page number on the right} = x + 1 = 72 + 1 = 73.$$

The page numbers are 72 and 73.

Step 5. **Check the proposed solution in the original wording of the problem.** The problem states that the sum of the page numbers on the facing pages is 145. By adding 72, the page number on the left, and 73, the page number on the right, we do, indeed, obtain a sum of 145. ∎

 CHECK POINT 4 Two pages that face each other in a book have 193 as the sum of their page numbers. What are the page numbers?

Example 4 and Check Point 4 involved consecutive integers. By contrast, some word problems involve consecutive odd integers, such as 5, 7, and 9. Other word problems involve consecutive even integers, such as 6, 8, and 10. When working with consecutive even or consecutive odd integers, we must continually add 2 to move from one integer to the next successive integer in the list.

Table 2.2 should be helpful in solving consecutive integer problems.

Table 2.2   Consecutive Integers

English Phrase	Algebraic Expressions	Example
Two consecutive integers	$x, x + 1$	$13, 14$
Three consecutive integers	$x, x + 1, x + 2$	$-8, -7, -6$
Two consecutive even integers	$x, x + 2$	$40, 42$
Two consecutive odd integers	$x, x + 2$	$-37, -35$
Three consecutive even integers	$x, x + 2, x + 4$	$30, 32, 34$
Three consecutive odd integers	$x, x + 2, x + 4$	$9, 11, 13$

EXAMPLE 5   Renting a Car

Rent-a-Heap Agency charges $125 per week plus $0.20 per mile to rent a small car. How many miles can you travel for $335?

SOLUTION

Step 1.   **Let x represent one of the quantities.** Because we are asked to find the number of miles we can travel for $335, let

$$x = \text{the number of miles.}$$

Step 2.   **Represent other quantities in terms of x.** There are no other unknown quantities to find, so we can skip this step.

Step 3.   **Write an equation in x that describes the conditions.** Before writing the equation, let us consider a few specific values for the number of miles traveled. The rental charge is $125 plus $0.20 for each mile.

3 miles: The rental charge is $125 + $0.20(3).
30 miles: The rental charge is $125 + $0.20(30).
100 miles: The rental charge is $125 + $0.20(100).
x miles: The rental charge is $125 + 0.20x$.

The weekly charge of $125	plus	the charge of $0.20 per mile for x miles	equals	the total $335 rental charge.
125	+	0.20x	=	335

Step 4.   **Solve the equation and answer the question.**

$$125 + 0.20x = 335 \qquad \text{This is the equation specified by the conditions of the problem.}$$

$$125 + 0.20x - 125 = 335 - 125 \qquad \text{Subtract 125 from both sides.}$$

$$0.20x = 210 \qquad \text{Simplify.}$$

$$\frac{0.20x}{0.20} = \frac{210}{0.20} \qquad \text{Divide both sides by 0.20.}$$

$$x = 1050 \qquad \text{Simplify.}$$

You can travel 1050 miles for $335.

Step 5.   **Check the proposed solution in the original wording of the problem.** Traveling 1050 miles should result in a total rental charge of $335. The mileage charge of $0.20 per mile is

$$\$0.20(1050) = \$210.$$

Adding this to the $125 weekly charge gives a total rental charge of
$$\$125 + \$210 = \$335.$$

Because this results in the given rental charge of $335, this verifies that you can travel 1050 miles. ■

 CHECK POINT 5 A taxi charges $2.00 to turn on the meter plus $0.25 for each eighth of a mile. If you have $10.00, how many eighths of a mile can you go? How many miles is that?

We will be using the formula for the perimeter of a rectangle, $P = 2l + 2w$, in our next example. Twice the rectangle's length plus twice the rectangle's width is its perimeter.

EXAMPLE 6 Finding the Dimensions of a Soccer Field

A rectangular soccer field is twice as long as it is wide. If the perimeter of a soccer field is 300 yards, what are the field's dimensions?

SOLUTION

Step 1. **Let x represent one of the quantities.** We know something about the length; the field is twice as long as it is wide. We will let
$$x = \text{the width.}$$

Step 2. **Represent other quantities in terms of x.** Because the field is twice as long as it is wide, let
$$2x = \text{the length.}$$
Figure 2.7 illustrates the soccer field and its dimensions.

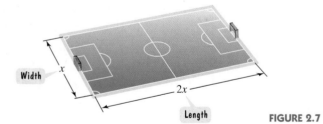

Width
x
$2x$
Length

FIGURE 2.7

Step 3. **Write an equation in x that describes the conditions.** Because the perimeter of a soccer field is 300 yards,

Twice the length	plus	twice the width	is	the perimeter.
$2 \cdot 2x$	$+$	$2 \cdot x$	$=$	$300.$

Step 4. **Solve the equation and answer the question.**

$$2\cdot 2x + 2 \cdot x = 300 \qquad \text{This is the equation for the problem's conditions.}$$
$$4x + 2x = 300 \qquad \text{Multiply.}$$
$$6x = 300 \qquad \text{Combine like terms.}$$
$$\frac{6x}{6} = \frac{300}{6} \qquad \text{Divide both sides by 6.}$$
$$x = 50 \qquad \text{Simplify.}$$

Thus,
$$\text{Width} = x = 50$$
$$\text{Length} = 2x = 2(50) = 100.$$
The dimensions of a soccer field are 50 yards by 100 yards.

Step 5. **Check the proposed solution in the original wording of the problem.** The perimeter of the soccer field using the dimensions that we found is 2(50 yards) + 2(100 yards) = 100 yards + 200 yards, or 300 yards. Because the problem's wording tells us that the perimeter is 300 yards, our dimensions are correct. ■

CHECK POINT 6 A rectangular swimming pool is three times as long as it is wide. If the perimeter of the pool is 320 feet, what are the pool's dimensions?

EXAMPLE 7 A Price Reduction

Your local computer store is having a sale. After a 30% price reduction, you purchase a new computer for $980. What was the computer's price before the reduction?

SOLUTION

Step 1. **Let x represent one of the quantities.** We will let x = the original price of the computer before the reduction.

Step 2. **Represent other quantities in terms of x.** There are no other unknown quantities to find, so we can skip this step.

Step 3. **Write an equation in x that describes the conditions.** The computer's original price minus the 30% reduction is the reduced price, $980.

Original price	minus	the reduction (30% of the original price)	is	the reduced price, $980.
x	$-$	$0.3x$	$=$	$980.$

Step 4. **Solve the equation and answer the question.**

$$x - 0.3x = 980 \qquad \text{This is the equation for the problem's conditions.}$$
$$0.7x = 980 \qquad \text{Combine like terms: } x - 0.3x = 1x - 0.3x = 0.7x.$$
$$\frac{0.7x}{0.7} = \frac{980}{0.7} \qquad \text{Divide both sides by 0.7.}$$
$$x = 1400 \qquad \text{Simplify: } 0.7\overline{)980.0}\,.$$

The computer's price before the reduction was $1400.

Step 5. **Check the proposed solution in the original wording of the problem.** The price before the reduction, $1400, minus the reduction in price should equal the reduced price given in the original wording, $980. The reduction in price is equal to 30% of the price before the reduction, $1400. To find the reduction, we multiply the decimal equivalent of 30%, 0.30 or 0.3, by the original price, $1400:

$$30\% \text{ of } \$1400 = (0.3)(\$1400) = \$420.$$

Now we can determine whether the calculation for the price before the reduction, $1400, minus the reduction, $420, is equal to the reduced price given in the problem, $980. We subtract:

$$\$1400 - \$420 = \$980.$$

This verifies that the price of the computer before the reduction was $1400. ■

CHECK POINT 7 After a 40% price reduction, an exercise machine sold for $564. What was the exercise machine's price before this reduction?

2.5 EXERCISE SET

Practice Exercises

In Exercises 1–14, let x represent the number. Write each English phrase as an algebraic expression.

1. The sum of a number and 7
2. A number increased by 15
3. A number subtracted from 25
4. 46 less than a number
5. 9 decreased by 4 times a number
6. 15 less than the product of 8 and a number
7. The quotient of 83 and a number
8. The quotient of a number and 83
9. The sum of twice a number and 40
10. Twice the sum of a number and 40
11. 93 subtracted from 9 times a number
12. The quotient of 13 and a number, decreased by 7 times the number
13. Eight times the sum of a number and 14
14. Nine times the difference of a number and 5

In Exercises 15–34, let x represent the number. Use the given conditions to write an equation. Solve the equation and find the number.

15. A number increased by 60 is equal to 410. Find the number.
16. The sum of a number and 43 is 107. Find the number.
17. A number decreased by 23 is equal to 214. Find the number.
18. The difference between a number and 17 is 96. Find the number.
19. The product of 7 and a number is 126. Find the number.
20. The product of 8 and a number is 272. Find the number.
21. The quotient of a number and 19 is 5. Find the number.
22. The quotient of a number and 14 is 8. Find the number.
23. The sum of four and twice a number is 56. Find the number.
24. The sum of five and three times a number is 59. Find the number.
25. Seven subtracted from five times a number is 178. Find the number.
26. Eight subtracted from six times a number is 298. Find the number.
27. A number increased by 5 is two times the number. Find the number.
28. A number increased by 12 is four times the number. Find the number.

29. Twice the sum of four and a number is 36. Find the number.
30. Three times the sum of five and a number is 48. Find the number.
31. Nine times a number is 30 more than three times that number. Find the number.
32. Five more than four times a number is that number increased by 35. Find the number.
33. If the quotient of three times a number and five is increased by four, the result is 34. Find the number.
34. If the quotient of three times a number and four is decreased by three, the result is nine. Find the number.

Application Exercises

In Exercises 35–62, use the five-step strategy to solve each problem.

35. Two of the most expensive movies ever made were *Titanic* and *Waterworld*. The cost to make *Titanic* exceeded the cost to make *Waterwold* by $25 million. The combined cost to make the two movies was $375 million. Find the cost of making each of these movies.

Paramount Pictures Corporation, Inc.

36. Each day, the number of births in the world exceeds the number of deaths by 229 thousand. The combined number of births and deaths is 521 thousand. Determine the number of births and the number of deaths per day.

Daily Growth of World Population

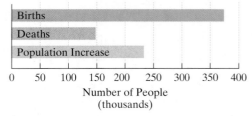

Number of People
(thousands)

Source: "Population Update" 2000

37. The circle graph shows the political ideology of U.S. college freshmen. The percentage of liberals exceeds twice that of conservatives by 4.4%. Using the displayed percents, it can be shown that liberals and conservatives combined account for 57.2% of college freshmen. Find the percentage of liberals and the percentage of conservatives.

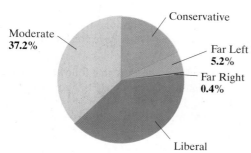

Political Ideology of U.S. College Freshmen

Moderate 37.2%
Conservative
Far Left 5.2%
Far Right 0.4%
Liberal

Source: The Chronicle of Higher Education

38. Commuters in one-third of the largest cities in the United States spend more than 40 hours per year, equivalent to one work week, sitting in traffic. The bar graph shows the number of hours in traffic per year for the average motorist in ten cities. The average motorist in Los Angeles spends 32 hours less than twice that of the average motorist in Miami stuck in traffic each year. In the two cities combined, 139 hours are spent by the average motorist per year in traffic. How many hours are wasted in traffic by the average motorist in Los Angeles and Miami?

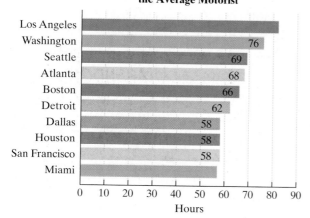

Hours in Traffic per Year for the Average Motorist

Los Angeles
Washington 76
Seattle 69
Atlanta 68
Boston 66
Detroit 62
Dallas 58
Houston 58
San Francisco 58
Miami

0 10 20 30 40 50 60 70 80 90
Hours

Source: Texas Transportation Institute

39. The sum of the page numbers on the facing pages of a book is 629. What are the page numbers?

40. The sum of the page numbers on the facing pages of a book is 525. What are the page numbers?

41. The highest-grossing North American concert tour was the Rolling Stones (1994), followed closely by Bruce Springsteen (2003). Combined, the two tours grossed $241 million. When expressed in millions, the earnings for the tours are consecutive integers. Find the gross, in millions, for the Rolling Stones tour and the Springsteen tour. *(Source: Rolling Stone)*

42. The first Super Bowl was played between the Green Bay Packers and the Kansas City Chiefs in 1967. Only once, in 1991, were the winning and losing scores in the Super Bowl consecutive integers. If the sum of the scores was 39, what were the scores?

43. Find two consecutive even integers whose sum is 66.

44. Find two consecutive odd integers whose sum is 72.

45. A car rental agency charges $200 per week plus $0.15 per mile to rent a car. How many miles can you travel in one week for $320?

46. A car rental agency charges $180 per week plus $0.25 per mile to rent a car. How many miles can you travel in one week for $395?

47. The average weight for female infants at birth is 7 pounds, with a monthly weight gain of 1.5 pounds. After how many months does a baby girl weigh 16 pounds?

48. The total revenue from Indian casinos in the United States has been increasing at approximately $2.2 billion per year. In 2002, Indian casinos reported a combined revenue of $12.7 billion. In which year will total revenue reach $28.1 billion? *(Source: National Indian Gaming Commission)*

49. A rectangular field is four times as long as it is wide. If the perimeter of the field is 500 yards, what are the field's dimensions?

50. A rectangular field is five times as long as it is wide. If the perimeter of the field is 288 yards, what are the field's dimensions?

51. An American football field is a rectangle with a perimeter of 1040 feet. The length is 200 feet more than the width. Find the width and length of the rectangular field.

52. A basketball court is a rectangle with a perimeter of 86 meters. The length is 13 meters more than the width. Find the width and length of the basketball court.

53. A bookcase is to be constructed as shown in the figure. The length is to be 3 times the height. If 60 feet of lumber is available for the entire unit, find the length and height of the bookcase.

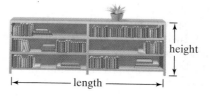

height

length

54. The height of the bookcase in the figure is 3 feet longer than the length of a shelf. If 18 feet of lumber is available for the entire unit, find the length and height of the unit.

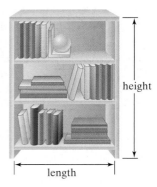

height

length

55. After a 20% reduction, you purchase a television for $320. What was the television's price before the reduction?

56. After a 30% reduction, you purchase a VCR for $98. What was the VCR's price before the reduction?

57. The average yearly earnings of pharmacists increased by 30% from 2001 to 2002. If salaries averaged $87,100 in 2002, what was the average salary in 2001?
(*Source*: Bureau of Labor Statistics)

58. The average yearly earnings of physical education teachers increased by 40% from 2001 to 2002. If salaries averaged $63,000 in 2002, what was the average salary in 2001?
(*Source*: Bureau of Labor Statistics)

59. Including 6% sales tax, a car sold for $15,370. Find the price of the car before the tax was added.

60. Including 8% sales tax, a bed-and-breakfast inn charges $172.80 per night. Find the inn's nightly cost before the tax is added.

61. An automobile repair shop charged a customer $448, listing $63 for parts and the remainder for labor. If the cost of labor is $35 per hour, how many hours of labor did it take to repair the car?

62. A repair bill on a sailboat came to $1603, including $532 for parts and the remainder for labor. If the cost of labor is $63 per hour, how many hours of labor did it take to repair the sailboat?

Writing in Mathematics

63. In your own words, describe a step-by-step approach for solving algebraic word problems.

64. Many students find solving linear equations much easier than solving algebraic word problems. Discuss some of the reasons why this is the case.

65. Did you have some difficulties solving some of the problems that were assigned in this exercise set? Discuss what you did if this happened to you. Did your course of action enhance your ability to solve algebraic word problems?

66. Write an original word problem that can be solved using a linear equation. Then solve the problem.

Critical Thinking Exercises

67. Which English statement given below is correctly translated into an algebraic equation?
  a. Ten pounds less than Bill's weight (x) equals 160 pounds: $10 - x = 160$.
  b. Four more than five times a number (x) is one less than six times that number: $5x + 4 = 1 - 6x$.
  c. Seven is three more than some number (x): $7 + 3 = x$.
  d. None of the above is correctly translated.

68. Explain how to use the three percents shown on the circle graph in Exercise 37 to determine the combined percentage of liberal and conservative college freshmen.

69. An HMO pamphlet contains the following recommended weight for women: "Give yourself 100 pounds for the first 5 feet plus 5 pounds for every inch over 5 feet tall." Using this description, which height corresponds to an ideal weight of 135 pounds?

70. The rate for a particular international telephone call is $0.55 for the first minute and $0.40 for each additional minute. Determine the length of a call that costs $6.95.

71. In a film, the actor Charles Coburn plays an elderly "uncle" character criticized for marrying a woman when he is 3 times her age. He wittily replies, "Ah, but in 20 years time I shall only be twice her age." How old is the "uncle" and the woman?

72. Answer the question in the following *Peanuts* cartoon strip. (*Note*: You may not use the answer given in the cartoon!)

PEANUTS reprinted by permission of United Features Syndicate, Inc.

Review Exercises

73. Solve and check: $\frac{4}{5}x = -16$. (Section 2.2, Example 3)

74. Solve and check: $6(y - 1) + 7 = 9y - y + 1$. (Section 2.3, Example 3)

75. Solve for w: $V = \frac{1}{3}lwh$. (Section 2.4, Example 4)

SOLVING LINEAR INEQUALITIES

Do you remember Rent-a-Heap, the car rental company that charged $125 per week plus $0.20 per mile to rent a small car? In Example 5 on page 143 we asked the question: How many miles can you travel for $335? We let x represent the number of miles and set up a linear equation as follows:

The weekly charge of $125	plus	the charge of $0.20 per mile for x miles	equals	the total $335 rental charge.
125	+	0.20x	=	335.

Because we are limited by how much money we can spend on everything from buying clothing to renting a car, it is also possible to ask: How many miles can you travel if you can spend *at most* $335? We again let x represent the number of miles. Spending *at most* $335 means that the amount spent on the weekly rental must be *less than or equal to* $335:

The weekly charge of $125	plus	the charge of $0.20 per mile for x miles	must be less than or equal to	$335.
125	+	0.20x	≤	335.

Using the commutative property of addition, we can express this inequality as

$$0.20x + 125 \le 335.$$

The form of this inequality is $ax + b \le c$, with $a = 0.20$, $b = 125$, and $c = 335$. Any inequality in this form is called a **linear inequality in one variable**. The symbol between $ax + b$ and c can be $\le$ (is less than or equal to), $<$ (is less than), $\ge$ (is greater than or equal to), or $>$ (is greater than). The greatest exponent on the variable in such an inequality is 1.

In this section, we will study how to solve linear inequalities such as $0.20x + 125 \le 335$. **Solving an inequality** is the process of finding the set of numbers that will make the inequality a true statement. These numbers are called the **solutions** of the inequality, and we say that they **satisfy** the inequality. The set of all solutions is called the **solution set** of the inequality. We begin by discussing how to graph and how to represent these solution sets.

Graphs of Inequalities There are infinitely many solutions to the inequality $x < 3$, namely, all real numbers that are less than 3. Although we cannot list all the solutions, we can make a drawing on a number line that represents these solutions. Such a drawing is called the **graph of the inequality**.

Graphs of solutions to linear inequalities are shown on a number line by shading all points representing numbers that are solutions. *Open dots* (∘) indicate endpoints that are *not solutions* and *closed dots* (·) indicate endpoints that *are solutions*.

EXAMPLE 1 Graphing Inequalities

Graph the solutions of each inequality: **a.** $x < 3$ **b.** $x \geq -1$ **c.** $-1 < x \leq 3$.

SOLUTION

a. The solutions of $x < 3$ are all real numbers that are less than 3. They are graphed on a number line by shading all points to the left of 3. The open dot at 3 indicates that 3 is not a solution, but numbers such as 2.9999 and 2.6 are. The arrow shows that the graph extends indefinitely to the left.

b. The solutions of $x \geq -1$ are all real numbers that are greater than or equal to -1. We shade all points to the right of -1 and the point for -1 itself. The closed dot at -1 shows that -1 is a solution of the given inequality. The arrow shows that the graph extends indefinitely to the right.

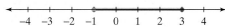

c. The inequality $-1 < x \leq 3$ is read "-1 is less than x *and* x is less than or equal to 3," or "x is greater than -1 *and* less than or equal to 3." The solutions of $-1 < x \leq 3$ are all real numbers between -1 and 3, not including -1 but including 3. The open dot at -1 indicates that -1 is not a solution. The closed dot at 3 shows that 3 is a solution. Shading indicates the other solutions.

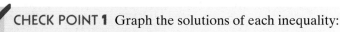

■

✔ **CHECK POINT 1** Graph the solutions of each inequality:

a. $x < 4$ **b.** $x \geq -2$ **c.** $-4 \leq x < 1$.

2 Use set-builder notation.

Solution Sets   The solutions of $x < 3$ are all real numbers that are less than 3. We can use the set concept introduced in Chapter 1 and state that the solution is the *set of all real numbers less than 3*. We use **set-builder notation** to write the solution set of $x < 3$ as

$$\{x \mid x < 3\}.$$

We read this as "the set of all x such that x is less than 3." Solutions of inequalities should be expressed in set-builder notation.

3 Understand properties used to solve linear inequalities.

Properties Used to Solve Linear Inequalities   Back to our question: How many miles can you drive your Rent-a-Heap car if you can spend at most $335 per week? We answer the question by solving

$$0.20x + 125 \leq 335$$

for x. The solution procedure is nearly identical to that for solving

$$0.20x + 125 = 335.$$

Our goal is to get x by itself on the left side. We do this by subtracting 125 from both sides to isolate $0.20x$:

$0.20x + 125 \leq 335$	This is the given inequality.
$0.20x + 125 - 125 \leq 335 - 125$	Subtract 125 from both sides.
$0.20x \leq 210.$	Simplify.

Finally, we isolate x from $0.20x$ by dividing both sides of the inequality by 0.20:

$$\frac{0.20x}{0.20} \leq \frac{210}{0.20} \qquad \textit{Divide both sides by 0.20.}$$

$$x \leq 1050. \qquad \textit{Simplify.}$$

With at most \$335 per week to spend, you can travel at most 1050 miles.

We started with the inequality $0.20x + 125 \leq 335$ and obtained the inequality $x \leq 1050$ in the final step. Both of these inequalities have the same solution set, namely $\{x \mid x \leq 1050\}$. Inequalities such as these, with the same solution set, are said to be **equivalent**.

We isolated x from $0.20x$ by dividing both sides of $0.20x \leq 210$ by 0.20, a positive number. Let's see what happens if we divide both sides of an inequality by a negative number. Consider the inequality $10 < 14$. Divide both 10 and 14 by -2:

$$\frac{10}{-2} = -5 \quad \text{and} \quad \frac{14}{-2} = -7.$$

Because -5 lies to the right of -7 on the number line, -5 is greater than -7:

$$-5 > -7.$$

Notice that the direction of the inequality symbol is reversed:

$$10 < 14$$

Dividing by -2 changes the direction of the inequality symbol.

$$-5 > -7$$

In general, **when we multiply or divide both sides of an inequality by a negative number, the direction of the inequality symbol is reversed**. When we reverse the direction of the inequality symbol, we say that we change the *sense* of the inequality.

We can isolate a variable in a linear inequality the same way we can isolate a variable in a linear equation. The following properties are used to create equivalent inequalities:

English Sentence	Inequality
x is at least 5.	$x \geq 5$
x is at most 5.	$x \leq 5$
x is between 5 and 7.	$5 < x < 7$
x is no more than 5.	$x \leq 5$
x is no less than 5.	$x \geq 5$

Properties of Inequalities

Property	The Property in Words	Example
The Addition Property of Inequality If $a < b$, then $a + c < b + c$. If $a < b$, then $a - c < b - c$.	If the same quantity is added to or subtracted from both sides of an inequality, the resulting inequality is equivalent to the original one.	$2x + 3 < 7$ Subtract 3: $2x + 3 - 3 < 7 - 3$ Simplify: $2x < 4$
The Positive Multiplication Property of Inequality If $a < b$ and c is positive, then $ac < bc$. If $a < b$ and c is positive, then $\dfrac{a}{c} < \dfrac{b}{c}$.	If we multiply or divide both sides of an inequality by the same positive quantity, the resulting inequality is equivalent to the original one.	$2x < 4$ Divide by 2: $\dfrac{2x}{2} < \dfrac{4}{2}$ Simplify: $x < 2$
The Negative Multiplication Property of Inequality If $a < b$ and c is negative, then $ac > bc$. If $a < b$ and c is negative, then $\dfrac{a}{c} > \dfrac{b}{c}$.	If we multiply or divide both sides of an inequality by the same negative quantity and reverse the direction of the inequality symbol, the resulting inequality is equivalent to the original one.	$-4x < 20$ Divide by -4 and reverse the sense of the inequality: $\dfrac{-4x}{-4} > \dfrac{20}{-4}$ Simplify: $x > -5$

4 Solve linear inequalities.

Solving Linear Inequalities Involving Only One Property of Inequality If you can solve a linear equation, it is likely that you can solve a linear inequality. Why? The procedure for solving linear inequalities is nearly the same as the procedure for solving linear equations, with one important exception: **When multiplying or dividing by a negative number, reverse the direction of the inequality symbol, changing the sense of the inequality.**

EXAMPLE 2 Solving a Linear Inequality

Solve and graph the solution set on a number line:
$$x + 3 < 8.$$

SOLUTION Our goal is to isolate x. We can do this by using the addition property, subtracting 3 from both sides.

$x + 3 < 8$	This is the given inequality.
$x + 3 - 3 < 8 - 3$	Subtract 3 from both sides.
$x < 5$	Simplify.

The solution set consists of all real numbers that are less than 5. We express this in set-builder notation as

$$\{x \mid x < 5\}. \qquad \text{This is read "the set of all x such that x is less than 5."}$$

The graph of the solution set is shown as follows:

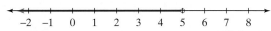

✔ **CHECK POINT 2** Solve and graph the solution set on a number line:
$$x + 6 < 9.$$

EXAMPLE 3 Solving a Linear Inequality

Solve and graph the solution set on a number line:
$$4x - 1 \geq 3x - 6.$$

SOLUTION Our goal is to isolate all terms involving x on one side and all numerical terms on the other side, exactly as we did when solving equations. Let's begin by using the addition property to isolate variable terms on the left.

$4x - 1 \geq 3x - 6$	This is the given inequality.
$4x - 3x - 1 \geq 3x - 3x - 6$	Subtract 3x from both sides.
$x - 1 \geq -6$	Simplify.

Now we isolate the numerical terms on the right. Use the addition property and add 1 to both sides.

$x - 1 + 1 \geq -6 + 1$	Add 1 to both sides.
$x \geq -5$	Simplify.

The solution set consists of all real numbers that are greater than or equal to -5. We express this in set-builder notation as

$$\{x \mid x \geq -5\}. \qquad \text{This is read "the set of all x such that x is greater than or equal to } -5."$$

The graph of the solution set is shown as follows:

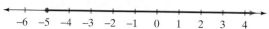

> ### CHECK POINT 3 Solve and graph the solution set on a number line:
>
> $$8x - 2 \geq 7x - 4.$$

We solved the inequalities in Examples 2 and 3 using the addition property of inequality. Now let's practice using the multiplication property of inequality. Do not forget to reverse the direction of the inequality symbol when multiplying or dividing both sides by a negative number.

EXAMPLE 4 Solving Linear Inequalities

Solve and graph the solution set on a number line: **a.** $\dfrac{1}{3}x < 5$ **b.** $-3x < 21$.

SOLUTION In each case, our goal is to isolate x. In the first inequality, this is accomplished by multiplying both sides by 3. In the second inequality, we can do this by dividing both sides by -3.

a. $\dfrac{1}{3}x < 5$ This is the given inequality.

$3 \cdot \dfrac{1}{3}x < 3 \cdot 5$ Isolate x by multiplying by 3 on both sides.
The symbol $<$ stays the same because we are multiplying both sides by a positive number.

$x < 15$ Simplify.

The solution set is $\{x \mid x < 15\}$. The graph of the solution set is shown as follows:

b. $-3x < 21$ This is the given inequality.

$\dfrac{-3x}{-3} > \dfrac{21}{-3}$ Isolate x by dividing by -3 on both sides.
The symbol $<$ must be reversed because we are dividing both sides by a negative number.

$x > -7$ Simplify.

The solution set is $\{x \mid x > -7\}$. The graph of the solution set is shown as follows:

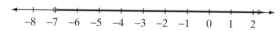

> ### CHECK POINT 4 Solve and graph the solution set on a number line:
>
> **a.** $\dfrac{1}{4}x < 2$ **b.** $-6x < 18$.

Inequalities Requiring Both the Addition and Multiplication Properties If an inequality does not contain fractions, it can be solved using the following procedure. Notice, again, how similar this procedure is to the procedure for solving an equation.

SOLVING A LINEAR INEQUALITY

1. Simplify the algebraic expression on each side.
2. Use the addition property of inequality to collect all the variable terms on one side and all the constant terms on the other side.
3. Use the multiplication property of inequality to isolate the variable and solve. Reverse the sense of the inequality when multiplying or dividing both sides by a negative number.
4. Express the solution set in set-builder notation and graph the solution set on a number line.

EXAMPLE 5 Solving a Linear Inequality

Solve and graph the solution set on a number line:

$$4y - 7 \geq 5.$$

SOLUTION

Step 1.   **Simplify each side.** Because each side is already simplified, we can skip this step.

Step 2.   **Collect variable terms on one side and constant terms on the other side.** The variable term, $4y$, is already on the left. We will collect constant terms on the right by adding 7 to both sides.

$$4y - 7 \geq 5 \qquad \text{This is the given inequality.}$$
$$4y - 7 + 7 \geq 5 + 7 \qquad \text{Add 7 to both sides.}$$
$$4y \geq 12 \qquad \text{Simplify.}$$

Step 3.   **Isolate the variable and solve.** We isolate the variable, y, by dividing both sides by 4. Because we are dividing by a positive number, we do not reverse the inequality symbol.

$$\frac{4y}{4} \geq \frac{12}{4} \qquad \text{Divide both sides by 4.}$$
$$y \geq 3 \qquad \text{Simplify.}$$

STUDY TIP

It is possible to perform a partial check for an inequality. Select one number from the solution set. Substitute that number into the original inequality and perform the resulting computations. You should obtain a true statement.

Step 4.   **Express the solution set in set-builder notation and graph the set on a number line.** The solution set consists of all real numbers that are greater than or equal to 3, expressed in set-builder notation as $\{y \mid y \geq 3\}$. The graph of the solution set is shown as follows:

CHECK POINT 5 Solve and graph the solution set on a number line:

$$5y - 3 \geq 17.$$

EXAMPLE 6 Solving a Linear Inequality

Solve and graph the solution set on a number line:

$$7x + 15 \geq 13x + 51.$$

STUDY TIP

You can solve

$$7x + 15 \geq 13x + 51$$

by isolating x on the right side. Subtract $7x$ from both sides:

$$7x + 15 - 7x$$
$$\geq 13x + 51 - 7x$$
$$15 \geq 6x + 51.$$

Now subtract 51 from both sides:

$$15 - 51 \geq 6x + 51 - 51$$
$$-36 \geq 6x.$$

Finally, divide both sides by 6:

$$\frac{-36}{6} \geq \frac{6x}{6}$$
$$-6 \geq x.$$

This last inequality means the same thing as

$$x \leq -6.$$

SOLUTION

Step 1. **Simplify each side.** Because each side is already simplified, we can skip this step.

Step 2. **Collect variable terms on one side and constant terms on the other side.** We will collect variable terms on the left and constant terms on the right.

$$7x + 15 \geq 13x + 51 \qquad \text{This is the given inequality.}$$
$$7x + 15 - 13x \geq 13x + 51 - 13x \qquad \text{Subtract 13x from both sides.}$$
$$-6x + 15 \geq 51 \qquad \text{Simplify.}$$
$$-6x + 15 - 15 \geq 51 - 15 \qquad \text{Subtract 15 from both sides.}$$
$$-6x \geq 36 \qquad \text{Simplify.}$$

Step 3. **Isolate the variable and solve.** We isolate the variable, x, by dividing both sides by -6. Because we are dividing by a negative number, we must reverse the inequality symbol.

$$\frac{-6x}{-6} \leq \frac{36}{-6} \qquad \text{Divide both sides by } -6 \text{ and reverse the sense of the inequality.}$$
$$x \leq -6 \qquad \text{Simplify.}$$

Step 4. **Express the solution set in set-builder notation and graph the set on a number line.** The solution set consists of all real numbers that are less than or equal to -6, expressed in set-builder notation as $\{x \mid x \leq -6\}$. The graph of the solution set is shown as follows:

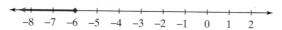

 CHECK POINT 6 Solve and graph the solution set: $6 - 3x \leq 5x - 2$.

EXAMPLE 7 Solving a Linear Inequality

Solve and graph the solution set on a number line:

$$2(x - 3) + 5x \leq 8(x - 1).$$

SOLUTION

Step 1. **Simplify each side.** We use the distributive property to remove parentheses. Then we combine like terms.

$$2(x - 3) + 5x \leq 8(x - 1) \qquad \text{This is the given inequality.}$$
$$2x - 6 + 5x \leq 8x - 8 \qquad \text{Use the distributive property.}$$
$$7x - 6 \leq 8x - 8 \qquad \text{Add like terms on the left.}$$

Step 2. **Collect variable terms on one side and constant terms on the other side.** We will collect variable terms on the left and constant terms on the right.

$$7x - 8x - 6 \leq 8x - 8x - 8 \qquad \text{Subtract 8x from both sides.}$$
$$-x - 6 \leq -8 \qquad \text{Simplify.}$$
$$-x - 6 + 6 \leq -8 + 6 \qquad \text{Add 6 to both sides.}$$
$$-x \leq -2 \qquad \text{Simplify.}$$

Step 3. **Isolate the variable and solve.** To isolate x in $-x \leq -2$, we must eliminate the negative sign in front of the x. Because $-x$ means $-1x$, we can do this by multiplying (or dividing) both sides of the inequality by -1. We are multiplying by a negative number. Thus, we must reverse the inequality symbol.

$$(-1)(-x) \geq (-1)(-2) \qquad \text{Multiply both sides of } -x \leq -2 \text{ by } -1 \text{ and reverse the sense of the inequality.}$$

$$x \geq 2 \qquad \text{Simplify.}$$

Step 4. **Express the solution set in set-builder notation and graph the set on a number line.** The solution set consists of all real numbers that are greater than or equal to 2, expressed in set-builder notation as $\{x \mid x \geq 2\}$. The graph of the solution set is shown as follows:

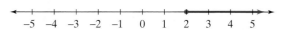

 CHECK POINT 7 Solve and graph the solution set:

$$2(x - 3) - 1 \leq 3(x + 2) - 14.$$

⑤ Identify inequalities with no solution or infinitely many solutions.

Inequalities with Unusual Solution Sets We have seen that some equations have no solution. This is also true for some inequalities. An example of such an inequality is

$$x > x + 1.$$

There is no number that is greater than itself plus 1. This inequality has no solution. Its solution set is written either as

$$\{ \ \} \qquad \text{or} \qquad \varnothing.$$

These symbols stand for the empty set, a set with no elements.

By contrast, some inequalities are true for all real numbers. An example of such an inequality is

$$x < x + 1.$$

Every real number is less than itself plus 1. The solution set is $\{x \mid x \text{ is a real number}\}$.

If you attempt to solve an inequality that has no solution, you will eliminate the variable and obtain a false statement, such as $0 > 1$. If you attempt to solve an inequality that is true for all real numbers, you will eliminate the variable and obtain a true statement, such as $0 < 1$.

EXAMPLE 8 Solving a Linear Inequality

Solve: $3(x + 1) > 3x + 5$.

SOLUTION

$$3(x + 1) > 3x + 5 \qquad \text{This is the given inequality.}$$

$$3x + 3 > 3x + 5 \qquad \text{Apply the distributive property.}$$

$$3x + 3 - 3x > 3x + 5 - 3x \qquad \text{Subtract 3x from both sides.}$$

Keep reading. $3 > 5$ is not the solution.

$$3 > 5 \qquad \text{Simplify.}$$

The original inequality is equivalent to the false statement $3 > 5$, which is false for every value of x. The inequality has no solution. The solution set is $\emptyset$, the empty set.

 CHECK POINT **8** Solve: $4(x + 2) > 4x + 15$.

EXAMPLE 9 Solving a Linear Inequality

Solve: $2(x + 5) \leq 5x - 3x + 14$.

SOLUTION

$$2(x + 5) \leq 5x - 3x + 14 \qquad \text{This is the given inequality.}$$
$$2x + 10 \leq 5x - 3x + 14 \qquad \text{Apply the distributive property.}$$
$$2x + 10 \leq 2x + 14 \qquad \text{Combine like terms.}$$
$$2x + 10 - 2x \leq 2x + 14 - 2x \qquad \text{Subtract 2x from both sides.}$$

| Keep reading. $10 \leq 14$ is not the solution. | $10 \leq 14$ | Simplify. |

The original inequality is equivalent to the true statement $10 \leq 14$, which is true for every value of x. The solution is the set of all real numbers, written

$$\{x \,|\, x \text{ is a real number}\}.$$

 CHECK POINT **9** Solve: $3(x + 1) \geq 2x + 1 + x$.

---

6 Solve problems using linear inequalities.

Applications As you know, different professors may use different grading systems to determine your final course grade. Some professors require a final examination; others do not. In our next example, a final exam is required *and* it also counts as two grades.

EXAMPLE 10 An Application: Final Course Grade

To earn an A in a course, you must have a final average of at least 90%. On the first four examinations, you have grades of 86%, 88%, 92%, and 84%. If the final examination counts as two grades, what must you get on the final to earn an A in the course?

SOLUTION We will use our five-step strategy for solving algebraic word problems.

Steps 1 and 2. **Represent unknown quantities in terms of x.** Let $x =$ your grade on the final examination.

Step 3. **Write an inequality in x that describes the conditions.** The average of the six grades is found by adding the grades and dividing the sum by 6.

$$\text{Average} = \frac{86 + 88 + 92 + 84 + x + x}{6}$$

Because the final counts as two grades, the x (your grade on the final examination) is added twice. This is also why the sum is divided by 6.

To get an A, your average must be at least 90. This means that your average must be greater than or equal to 90.

| Your average | must be greater than or equal to | 90. |

$$\frac{86 + 88 + 92 + 84 + x + x}{6} \geq 90$$

Step 4. **Solve the inequality and answer the problem's question.**

$$\frac{86 + 88 + 92 + 84 + x + x}{6} \geq 90 \qquad \text{This is the inequality for the given conditions.}$$

$$\frac{350 + 2x}{6} \geq 90 \qquad \text{Combine like terms in the numerator.}$$

$$6\left(\frac{350 + 2x}{6}\right) \geq 6(90) \qquad \text{Multiply both sides by 6, clearing the fraction.}$$

$$350 + 2x \geq 540 \qquad \text{Multiply.}$$

$$350 + 2x - 350 \geq 540 - 350 \qquad \text{Subtract 350 from both sides.}$$

$$2x \geq 190 \qquad \text{Simplify.}$$

$$\frac{2x}{2} \geq \frac{190}{2} \qquad \text{Divide both sides by 2.}$$

$$x \geq 95 \qquad \text{Simplify.}$$

You must get at least 95% on the final examination to earn an A in the course.

Step 5. **Check.** We can perform a partial check by computing the average with any grade that is at least 95. We will use 96. If you get 96% on the final examination, your average is

$$\frac{86 + 88 + 92 + 84 + 96 + 96}{6} = \frac{542}{6} = 90\frac{1}{3}.$$

Because $90\frac{1}{3} > 90$, you earn an A in the course.

CHECK POINT 10 To earn a B in a course, you must have a final average of at least 80%. On the first three examinations, you have grades of 82%, 74%, and 78%. If the final examination counts as two grades, what must you get on the final to earn a B in the course?

2.6 EXERCISE SET

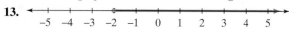
Student Solutions Manual   CD/Video   PH Math/Tutor Center   MathXL Tutorials on CD   MathXL®   MyMathLab   Interactmath.com

Practice Exercises

In Exercises 1–12, graph the solutions of each inequality on a number line.

1. $x > 5$ **2.** $x > -3$

3. $x < -2$ **4.** $x < 0$

5. $x \geq -4$ **6.** $x \geq -6$

7. $x \leq 4.5$ **8.** $x \leq 7.5$

9. $-2 < x \leq 6$ **10.** $-3 \leq x < 6$

11. $-1 < x < 3$ **12.** $-2 \leq x \leq 0$

Describe each graph in Exercises 13–18 using set-builder notation.

13.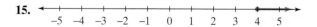

14.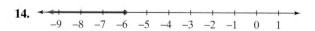

15.

16.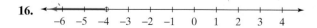

17.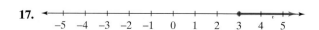

18.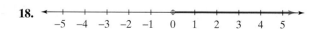

Use the addition property of inequality to solve each inequality in Exercises 19–36. Express the solution set in set-builder notation and graph the set on a number line.

19. $x - 3 > 4$

20. $x + 1 < 6$

21. $x + 4 \leq 10$

22. $x - 5 \geq 2$

23. $y - 2 < 0$

24. $y + 3 \geq 0$

25. $3x + 4 \leq 2x + 7$

26. $2x + 9 \leq x + 2$

27. $5x - 9 < 4x + 7$

28. $3x - 8 < 2x + 11$

29. $7x - 7 > 6x - 3$

30. $8x - 9 > 7x - 3$

31. $x - \dfrac{2}{3} > \dfrac{1}{2}$

32. $x - \dfrac{1}{3} \geq \dfrac{5}{6}$

33. $y + \dfrac{7}{8} \leq \dfrac{1}{2}$

34. $y + \dfrac{1}{3} \leq \dfrac{3}{4}$

35. $-15y + 13 > 13 - 16y$

36. $-12y + 17 > 20 - 13y$

Use the multiplication property of inequality to solve each inequality in Exercises 37–54. Express the solution set in set-builder notation and graph the set on a number line.

37. $\dfrac{1}{2}x < 4$

38. $\dfrac{1}{2}x > 3$

39. $\dfrac{x}{3} > -2$

40. $\dfrac{x}{4} < -1$

41. $4x < 20$

42. $6x < 18$

43. $3x \geq -21$

44. $7x \geq -56$

45. $-3x < 15$

46. $-7x > 21$

47. $-3x \geq 15$

48. $-7x \leq 21$

49. $-16x > -48$

50. $-20x > -140$

51. $-4y \leq \dfrac{1}{2}$

52. $-2y \leq \dfrac{1}{2}$

53. $-x < 4$

54. $-x > -3$

Use both the addition and multiplication properties of inequality to solve each inequality in Exercises 55–78. Express the solution set in set-builder notation and graph the set on a number line.

55. $2x - 3 > 7$

56. $3x + 2 \leq 14$

57. $3x + 3 < 18$

58. $8x - 4 > 12$

59. $3 - 7x \leq 17$

60. $5 - 3x \geq 20$

61. $-2x - 3 < 3$

62. $-3x + 14 < 5$

63. $5 - x \leq 1$

64. $3 - x \geq -3$

65. $2x - 5 > -x + 6$

66. $6x - 2 \geq 4x + 6$

67. $2y - 5 < 5y - 11$

68. $4y - 7 > 9y - 2$

69. $3(2y - 1) < 9$

70. $4(2y - 1) > 12$

71. $3(x + 1) - 5 < 2x + 1$

72. $4(x + 1) + 2 \geq 3x + 6$

73. $8x + 3 > 3(2x + 1) - x + 5$

74. $7 - 2(x - 4) < 5(1 - 2x)$

75. $\dfrac{x}{3} - 2 \geq 1$

76. $\dfrac{x}{4} - 3 \geq 1$

77. $1 - \dfrac{x}{2} > 4$

78. $1 - \dfrac{x}{2} < 5$

In Exercises 79–88, solve each inequality. Identify inequalities that have no solution, or inequalities that are true for all real numbers.

79. $4x - 4 < 4(x - 5)$

80. $3x - 5 < 3(x - 2)$

81. $x + 3 < x + 7$

82. $x + 4 < x + 10$

83. $7x \leq 7(x - 2)$

84. $3x + 1 \leq 3(x - 2)$

85. $2(x + 3) > 2x + 1$

86. $5(x + 4) > 5x + 10$

87. $5x - 4 \leq 4(x - 1)$

88. $6x - 3 \leq 3(x - 1)$

Practice Plus

In Exercises 89–92, use properties of inequality to rewrite each inequality so that x is isolated on one side.

89. $3x + a > b$ **90.** $-2x - a \leq b$

91. $y \leq mx + b$ and $m < 0$

92. $y > mx + b$ and $m > 0$

We know that $|x|$ represents the distance from 0 to x on a number line. In Exercises 93–96, use each sentence to describe all possible locations of x on a number line. Then rewrite the given sentence as an inequality involving $|x|$.

93. The distance from 0 to x on a number line is less than 2.

94. The distance from 0 to x on a number line is less than 3.

95. The distance from 0 to x on a number line is greater than 2.

96. The distance from 0 to x on a number line is greater than 3.

Application Exercises

The bar graph shows the percentage of wages full-time workers pay in income tax in eight selected countries. (The percents shown are averages for single earners without children.) Let x represent the percentage of wages workers pay in income tax. In Exercises 97–102, write the name of the country or countries described by the given inequality.

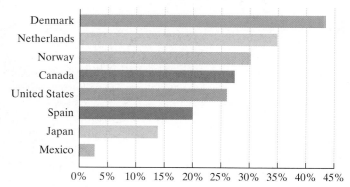

Percentage of Wages Full-Time Workers Pay in Income Tax

Source: The Washington Post

97. $x \geq 30\%$

98. $x > 30\%$

99. $x < 20\%$

100. $x \leq 20\%$

101. $25\% \leq x < 40\%$

102. $5\% < x \leq 25\%$

The line graph shows the declining consumption of cigarettes in the United States. The data shown by the graph can be modeled by

$$N = 550 - 9x,$$

where N is the number of cigarettes consumed, in billions, x years after 1988. Use this formula to solve Exercises 103–104.

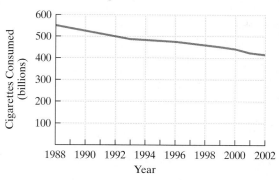

Consumption of Cigarettes in the U.S.

Source: Economic Research Service, USDA

103. Describe how many years after 1988 cigarette consumption will be less than 370 billion cigarettes each year. Which years are included in your description?

104. Describe how many years after 1988 cigarette consumption will be less than 325 billion cigarettes each year. Which years are included in your description?

105. On two examinations, you have grades of 86 and 88. There is an optional final examination, which counts as one grade. You decide to take the final in order to get a course grade of A, meaning a final average of at least 90.

   a. What must you get on the final to earn an A in the course?

   b. By taking the final, if you do poorly, you might risk the B that you have in the course based on the first two exam grades. If your final average is less than 80, you will lose your B in the course. Describe the grades on the final that will cause this to happen.

106. On three examinations, you have grades of 88, 78, and 86. There is still a final examination, which counts as one grade.

   a. In order to get an A, your average must be at least 90. If you get 100 on the final, compute your average and determine if an A in the course is possible.

   b. To earn a B in the course, you must have a final average of at least 80. What must you get on the final to earn a B in the course?

107. A car can be rented from Continental Rental for $80 per week plus 25 cents for each mile driven. How many miles can you travel if you can spend at most $400 for the week?

108. A car can be rented from Basic Rental for $60 per week plus 50 cents for each mile driven. How many miles can you travel if you can spend at most $600 for the week?

109. An elevator at a construction site has a maximum capacity of 3000 pounds. If the elevator operator weighs 245 pounds and each cement bag weighs 95 pounds, how many bags of cement can be safely lifted on the elevator in one trip?

110. An elevator at a construction site has a maximum capacity of 2800 pounds. If the elevator operator weighs 265 pounds and each cement bag weighs 65 pounds, how many bags of cement can be safely lifted on the elevator in one trip?

Writing in Mathematics

111. When graphing the solutions of an inequality, what is the difference between an open dot and a closed dot?

112. When solving an inequality, when is it necessary to change the direction of the inequality symbol? Give an example.

113. Describe ways in which solving a linear inequality is similar to solving a linear equation.

114. Describe ways in which solving a linear inequality is different from solving a linear equation.

115. Using current trends, future costs of Medicare can be modeled by $C = 18x + 250$, where x represents the number of years after 2000 and C represents the cost of Medicare, in billions of dollars. Use the formula to write a word problem that can be solved using a linear inequality. Then solve the problem.

Critical Thinking Exercises

116. Which one of the following statements is true?
   a. The inequality $x - 3 > 0$ is equivalent to $x < 3$.
   b. The statement "x is at most 5" is written $x < 5$.
   c. The inequality $-4x < -20$ is equivalent to $x > -5$.
   d. The statement "the sum of x and 6% of x is at least 80" is written $x + 0.06x \geq 80$.

117. A car can be rented from Basic Rental for $260 per week with no extra charge for mileage. Continental charges $80 per week plus 25 cents for each mile driven to rent the same car. How many miles should be driven in a week to make the rental cost for Basic Rental a better deal than Continental's?

118. Membership in a fitness club costs $500 yearly plus $1 per hour spent working out. A competing club charges $440 yearly plus $1.75 per hour for use of their equipment. How many hours must a person work out yearly to make membership in the first club cheaper than membership in the second club?

Technology Exercises

Solve each inequality in Exercises 119–120. Use a calculator to help with the arithmetic.

119. $1.45 - 7.23x > -1.442$

120. $126.8 - 9.4y \leq 4.8y + 34.5$

Review Exercises

121. 8 is 40% of what number? (Section 2.4, Example 8)

122. The length of a rectangle exceeds the width by 5 inches. The perimeter is 34 inches. What are the rectangle's dimensions? (Section 2.5, Example 6)

123. Solve and check: $5x + 16 = 3(x + 8)$. (Section 2.3, Example 2)

GROUP PROJECT

CHAPTER

2

One of the best ways to learn how to *solve* a word problem in algebra is to *design* word problems of your own. Creating a word problem makes you very aware of precisely how much information is needed to solve the problem. You must also focus on the best way to present information to a reader and on how much information to give. As you write your problem, you gain skills that will help you solve problems created by others.

   The group should design five different word problems that can be solved using an algebraic equation. All of the problems should be on different topics. For example, the group should not have more than one problem on finding a number. The group should turn in both the problems and their algebraic solutions.

CHAPTER 2 SUMMARY

Definitions and Concepts	Examples
Section 2.1 The Addition Property of Equality	
A linear equation in one variable can be written in the form $ax + b = c$, where a is not zero.	$3x + 7 = 9$ is a linear equation.
Equivalent equations have the same solution.	$2x - 4 = 6, 2x = 10,$ and $x = 5$ are equivalent equations.
The Addition Property of Equality Adding the same number (or algebraic expression) to or subtracting the same number (or algebraic expression) from both sides of an equation does not change its solution.	• $\quad x - 3 = 8$ $x - 3 + 3 = 8 + 3$ $x = 11$ • $\quad x + 4 = 10$ $x + 4 - 4 = 10 - 4$ $x = 6$
Section 2.2 The Multiplication Property of Equality	
The Multiplication Property of Equality Multiplying both sides or dividing both sides of an equation by the same nonzero real number (or algebraic expression) does not change the solution.	• $\quad \dfrac{x}{-5} = 6$ $-5\left(\dfrac{x}{-5}\right) = -5(6)$ $x = -30$ • $\quad -50 = -5y$ $\dfrac{-50}{-5} = \dfrac{-5y}{-5}$ $10 = y$
Equations and Coefficients of -1 If $-x = c$, multiply both sides by -1 to solve for x. The solution is the additive inverse of c.	$-x = -12$ $(-1)(-x) = (-1)(-12)$ $x = 12$
Using the Addition and Multiplication Properties If an equation does not contain fractions, • Use the addition property to isolate the variable term. • Use the multiplication property to isolate the variable.	$-2x - 5 = 11$ $-2x - 5 + 5 = 11 + 5$ $-2x = 16$ $\dfrac{-2x}{-2} = \dfrac{16}{-2}$ $x = -8$
Section 2.3 Solving Linear Equations	
Solving a Linear Equation **1.** Simplify each side.	Solve: $\quad 7 - 4(x - 1) = x + 1.$ $7 - 4x + 4 = x + 1$ $-4x + 11 = x + 1$
2. Collect all the variable terms on one side and all the constant terms on the other side.	$-4x - x + 11 = x - x + 1$ $-5x + 11 = 1$ $-5x + 11 - 11 = 1 - 11$ $-5x = -10$

Definitions and Concepts	Examples

Section 2.3 Solving Linear Equations (continued)

3. Isolate the variable and solve. (If the variable is eliminated and a false statement results, the inconsistent equation has no solution. If a true statement results, all real numbers are solutions of the identity.)	$$\frac{-5x}{-5} = \frac{-10}{-5}$$ $$x = 2$$
4. Check the proposed solution in the original equation.	$7 - 4(x - 1) = x + 1$ $7 - 4(2 - 1) \stackrel{?}{=} 2 + 1$ $7 - 4(1) \stackrel{?}{=} 2 + 1$ $7 - 4 \stackrel{?}{=} 2 + 1$ $3 = 3$, true The solution is 2.
Equations Containing Fractions Multiply both sides (all terms) by the least common denominator. This clears the equation of fractions.	$$\frac{x}{5} + \frac{1}{2} = \frac{x}{2} - 1$$ $$10\left(\frac{x}{5} + \frac{1}{2}\right) = 10\left(\frac{x}{2} - 1\right)$$ $$10 \cdot \frac{x}{5} + 10 \cdot \frac{1}{2} = 10 \cdot \frac{x}{2} - 10 \cdot 1$$ $$2x + 5 = 5x - 10$$ $$-3x = -15$$ $$x = 5$$ The solution is 5.

Section 2.4 Formulas and Percents

To solve a formula for one of its variables, use the steps for solving a linear equation and isolate the specified variable on one side of the equation.	Solve for l: $$w = \frac{P - 2l}{2}$$ $$2w = 2\left(\frac{P - 2l}{2}\right)$$ $$2w = P - 2l$$ $$2w - P = P - P - 2l$$ $$2w - P = -2l$$ $$\frac{2w - P}{-2} = \frac{-2l}{-2}$$ $$\frac{2w - P}{-2} = l$$
The word *percent* means *per hundred*. The symbol % denotes percent.	$47\% = \dfrac{47}{100}$ $\quad$ $3\% = \dfrac{3}{100}$
To express a decimal as a percent, move the decimal point two places to the right and add a percent sign.	$0.37 = 37\%$ $0.006 = 0.6\%$
To express a percent as a decimal, move the decimal point two places to the left and remove the percent sign.	$250\% = 250\% = 2.5$ $4\% = 04.\% = 0.04$

Definitions and Concepts	Examples

Section 2.4 Formulas and Percents (continued)

A Formula Involving Percent

A is P percent of B.

$$A = P \cdot B$$

In the formula $A = PB$, P is expressed as a decimal.

• What is 5% of 20?

$$A = 0.05 \cdot 20$$
$$A = 1$$

Thus, 1 is 5% of 20.

• 6 is 30% of what?

$$6 = 0.3 \cdot B$$
$$\frac{6}{0.3} = B$$
$$20 = B$$

Thus, 6 is 30% of 20.

• 33 is what percent of 75?

$$33 = P \cdot 75$$
$$\frac{33}{75} = P$$
$$P = 0.44 = 44\%$$

Thus, 33 is 44% of 75.

Section 2.5 An Introduction to Problem Solving

Strategy for Solving Word Problems

The length of a rectangle exceeds the width by 3 inches. The perimeter is 26 inches. What are the rectangle's dimensions?

Step 1 Let x represent one of the quantities.

Let $x =$ the width.

Step 2 Represent other quantities in terms of x.

$x + 3 =$ the length

Step 3 Write an equation that describes the conditions.

Twice length plus twice width is perimeter.

$$2(x + 3) + 2x = 26$$

Step 4 Solve the equation and answer the question.

$$2(x + 3) + 2x = 26$$
$$2x + 6 + 2x = 26$$
$$4x + 6 = 26$$
$$4x = 20$$
$$x = 5$$

The width (x) is 5 inches and the length $(x + 3)$ is $5 + 3$, or 8 inches.

Step 5 Check the proposed solution in the original wording of the problem.

$$\text{Perimeter} = 2(5 \text{ in.}) + 2(8 \text{ in.})$$
$$= 10 \text{ in.} + 16 \text{ in.} = 26 \text{ in.}$$

This checks with the given perimeter.

Definitions and Concepts	Examples

Section 2.6 Solving Linear Inequalities

A linear inequality in one variable can be written in one of these forms:

$$ax + b < c \qquad ax + b \le c$$
$$ax + b > c \qquad ax + b \ge c$$

where a is not 0.

$3x + 6 > 12$ is a linear inequality.

Set-Builder Notation and Graphs

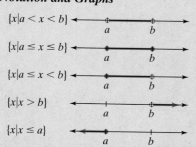

$\{x|a < x < b\}$

$\{x|a \le x \le b\}$

$\{x|a \le x < b\}$

$\{x|x > b\}$

$\{x|x \le a\}$

• Graph the solution of $x < 4$.

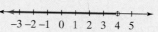

$-3\ -2\ -1\ \ 0\ \ 1\ \ 2\ \ 3\ \ 4\ \ 5$

• Graph the solution of $-2 < x \le 1$.

$-4\ -3\ -2\ -1\ \ 0\ \ 1\ \ 2\ \ 3\ \ 4$

The Addition Property of Inequality

Adding the same number to or subtracting the same number from both sides of an inequality does not change the solutions.

$$x + 3 < 8$$
$$x + 3 - 3 < 8 - 3$$
$$x < 5$$

The Positive Multiplication Property of Inequality

Multiplying or dividing both sides of an inequality by the same positive number does not change the solutions.

$$\frac{x}{6} \ge 5$$
$$6 \cdot \frac{x}{6} \ge 6 \cdot 5$$
$$x \ge 30$$

The Negative Multiplication Property of Inequality

Multiplying or dividing both sides of an inequality by the same negative number and reversing the direction of the inequality sign does not change the solutions.

$$-3x \le 12$$
$$\frac{-3x}{-3} \ge \frac{12}{-3}$$
$$x \ge -4$$

Solving Linear Inequalities

Use the procedure for solving linear equations. When multiplying or dividing by a negative number, reverse the direction of the inequality symbol. Express the solution set in set-builder notation and graph the set on a number line. If the variable is eliminated and a false statement results, the inequality has no solution. The solution set is ∅, the empty set. If a true statement results, the solution is the set of all real numbers: $\{x|x$ is a real number$\}$.

Solve:

$$x + 4 \ge 6x - 16$$
$$x + 4 - 6x \ge 6x - 16 - 6x$$
$$-5x + 4 \ge -16$$
$$-5x + 4 - 4 \ge -16 - 4$$
$$-5x \ge -20$$
$$\frac{-5x}{-5} \le \frac{-20}{-5}$$
$$x \le 4$$
$$\{x|x \le 4\}$$

$-3\ -2\ -1\ \ 0\ \ 1\ \ 2\ \ 3\ \ 4\ \ 5$

CHAPTER 2 REVIEW EXERCISES

2.1 *Solve each equation in Exercises 1–5 using the addition property of equality. Be sure to check proposed solutions.*

1. $x - 10 = 22$

2. $-14 = y + 8$

3. $7z - 3 = 6z + 9$

4. $4(x + 3) = 3x - 10$

5. $6x - 3x - 9 + 1 = -5x + 7x - 3$

2.2 *Solve each equation in Exercises 6–13 using the multiplication property of equality. Be sure to check proposed solutions.*

6. $\dfrac{x}{8} = 10$

7. $\dfrac{y}{-8} = 7$

8. $7z = 77$

9. $-36 = -9y$

10. $\dfrac{3}{5}x = -9$

11. $30 = -\dfrac{5}{2}y$

12. $-x = 25$

13. $\dfrac{-x}{10} = -1$

Solve each equation in Exercises 14–18 using both the addition and multiplication properties of equality. Check proposed solutions.

14. $4x + 9 = 33$

15. $-3y - 2 = 13$

16. $5z + 20 = 3z$

17. $5x - 3 = x + 5$

18. $3 - 2x = 9 - 8x$

19. The formula $F = 1.2x + 21.6$ models the average family income, F, in thousands of dollars, for Puerto Ricans x years after 1990. How many years after 1990 is the average family income expected to reach $40.8 thousand? In which year is this expected to occur?

2.3 *Solve and check each equation in Exercises 20–28.*

20. $5x + 9 - 7x + 6 = x + 18$

21. $3(x + 4) = 5x - 12$

22. $1 - 2(6 - y) = 3y + 2$

23. $2(x - 4) + 3(x + 5) = 2x - 2$

24. $-2(y - 4) - (3y - 2) = -2 - (6y - 2)$

25. $\dfrac{2x}{3} = \dfrac{x}{6} + 1$

26. $\dfrac{x}{2} - \dfrac{1}{10} = \dfrac{x}{5} + \dfrac{1}{2}$

27. $3(8x - 1) = 6(5 + 4x)$

28. $4(2x - 3) + 4 = 8x - 8$

29. The optimum heart rate that a person should achieve during exercise for the exercise to be most beneficial is modeled by $r = 0.6(220 - a)$, where a represents a person's age and r represents that person's optimum heart rate, in beats per minute. If the optimum heart rate is 120 beats per minute, how old is that person?

2.4 *In Exercises 30–34, solve each formula for the specified variable.*

30. $I = Pr$ for r

31. $V = \dfrac{1}{3}Bh$ for h

32. $P = 2l + 2w$ for w

33. $A = \dfrac{B + C}{2}$ for B

34. $T = D + pm$ for m

In Exercises 35–36, express each decimal as a percent.

35. 0.72

36. 0.0035

In Exercises 37–39, express each percent as a decimal.

37. 65%

38. 150%

39. 3%

40. What is 8% of 120?

41. 90 is 45% of what?

42. 36 is what percent of 75?

43. If 6 is increased to 12, the increase is what percent of the original number?

44. If 5 is decreased to 3, the decrease is what percent of the original number?

45. A college that had 40 students for each lecture course increased the number to 45 students. What is the percent increase in the number of students in a lecture course?

46. Consider the following statement:

   My portfolio fell 10% last year, but then it rose 10% this year, so at least I recouped my losses.

   Is this statement true? In particular, suppose you invested $10,000 in the stock market last year. How much money would be left in your portfolio with a 10% fall and then a 10% rise? If there is a loss, what is the percent decrease, to the nearest tenth of a percent, in your portfolio?

47. The radius is one of two bones that connect the elbow and the wrist. The formula $r = \dfrac{h}{7}$ models the length of a woman's radius, r, in inches, and her height, h, in inches.

   a. Solve the formula for h.

   b. Use the formula in part (a) to find a woman's height if her radius is 9 inches long.

48. Every day, the average U.S. household uses 91 gallons of water flushing toilets. The circle graph on the next page shows that this represents 26% of the total number of gallons

of water used per day. How many gallons of water does the average U.S. household use per day?

Where U.S. Households Use Water

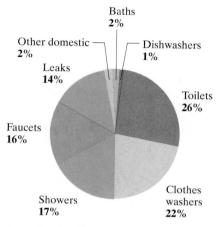

Source: American Water Works Association

2.5 *In Exercises 49–56, use the five-step strategy to solve each problem.*

49. Six times a number, decreased by 20, is four times the number. Find the number.

50. On average, the number of unhealthy air days per year in Los Angeles exceeds three times that of New York City by 48 days. If Los Angeles and New York combined have 268 unhealthy air days per year, determine the number of unhealthy days for the two cities.
(*Source*: Environmental Protection Agency)

51. Two pages that face each other in a book have 93 as the sum of their page numbers. What are the page numbers?

52. The two female artists in the United States with the most platinum albums are Barbra Streisand followed by Madonna. (A platinum album represents one album sold per 266 people.) The number of platinum albums by these two singers form consecutive odd integers. Combined, they have 96 platinum albums. Determine the number of platinum albums by Streisand and the number of platinum albums by Madonna.

53. In 2003, the average weekly salary for workers in the United States was $612. If this amount is increasing by $15 yearly, in how many years after 2003 will the average salary reach $747? In which year will that be?

54. A bank's total monthly charge for a checking account is $6 plus $0.05 per check. If your total monthly charge is $6.90, how many checks did you write during that month?

55. A rectangular field is three times as long as it is wide. If the perimeter of the field is 400 yards, what are the field's dimensions?

56. After a 25% reduction, you purchase a table for $180. What was the table's price before the reduction?

2.6 *In Exercises 57–58, graph the solution of each inequality on a number line.*

57. $x < -1$

58. $-2 < x \leq 4$

Describe each graph in Exercises 59–60 using set-builder notation.

59.

$$\xleftarrow{\hspace{0.5cm}} \overset{-3 \;\; -2 \;\; -1 \;\;\; 0 \;\;\; 1 \;\;\; 2 \;\;\; 3 \;\;\; 4 \;\;\; 5 \;\;\; 6 \;\;\; 7}{\rule{6cm}{0.4pt}} \xrightarrow{\hspace{0.5cm}}$$

60.

$$\xleftarrow{\hspace{0.5cm}} \overset{-5 \;\; -4 \;\; -3 \;\; -2 \;\; -1 \;\;\; 0 \;\;\; 1 \;\;\; 2 \;\;\; 3 \;\;\; 4 \;\;\; 5}{\rule{6cm}{0.4pt}} \xrightarrow{\hspace{0.5cm}}$$

Solve each inequality in Exercises 61–68. Express the solution set in set-builder notation and graph the set on a number line. If the inequality has no solution or is true for all real numbers, so state. It is not necessary to graph solution sets for these inequalities.

61. $2x - 5 < 3$

62. $\dfrac{x}{2} > -4$

63. $3 - 5x \leq 18$

64. $4x + 6 < 5x$

65. $6x - 10 \geq 2(x + 3)$

66. $4x + 3(2x - 7) \leq x - 3$

67. $2(2x + 4) > 4(x + 2) - 6$

68. $-2(x - 4) \leq 3x + 1 - 5x$

69. To pass a course, a student must have an average on three examinations of at least 60. If a student scores 42 and 74 on the first two tests, what must be earned on the third test to pass the course?

70. A long distance telephone service charges 10¢ for the first minute and 5¢ for each minute thereafter. The cost, C, in cents, for a call lasting x minutes is modeled by the formula

$$C = 10 + 5(x - 1).$$

How many minutes can you talk on the phone if you do not want the cost to exceed $5, or 500¢?

CHAPTER 2 TEST

Remember to use your Chapter Test Prep Video CD to see the worked-out solutions to the test questions you want to review.

In Exercises 1–6, solve each equation.

1. $4x - 5 = 13$
2. $12x + 4 = 7x - 21$
3. $8 - 5(x - 2) = x + 26$
4. $3(2y - 4) = 9 - 3(y + 1)$
5. $\frac{3}{4}x = -15$
6. $\frac{x}{10} + \frac{1}{3} = \frac{x}{5} + \frac{1}{2}$
7. The formula $N = 2.4x + 180$ models U.S. population, N, in millions x years after 1960. How many years after 1960 is the U.S. population expected to reach 324 million? In which year is this expected to occur?

In Exercises 8–9, solve each formula for the specified variable.

8. $V = \pi r^2 h$ for h
9. $l = \dfrac{P - 2w}{2}$ for w
10. What is 6% of 140?
11. 120 is 80% of what?
12. 12 is what percent of 240?

In Exercises 13–17, solve each problem.

13. The product of 5 and a number, decreased by 9, is 306. What is the number?
14. In New York City, a fitness trainer earns $22,870 more per year than a preschool teacher. The yearly average salaries for fitness trainers and preschool teachers combined are $79,030. Determine the average yearly salary of a fitness trainer and a preschool teacher in New York City. (*Source: Time*, April 14, 2003)

15. A long-distance telephone plan has a monthly fee of $15.00 and a rate of $0.05 per minute. How many minutes can you chat long distance in a month for a total cost, including the $15.00, of $45.00?
16. A rectangular field is twice as long as it is wide. If the perimeter of the field is 450 yards, what are the field's dimensions?
17. After a 20% reduction, you purchase a new Stephen King novel for $28. What was the book's price before the reduction?

In Exercises 18–19, graph the solution of each inequality on a number line.

18. $x > -2$
19. $-4 \leq x < 1$
20. Use set-builder notation to describe the following graph.

$$\xleftarrow{\quad}\overset{-5\ -4\ -3\ -2\ -1\ \ 0\ \ 1\ \ 2\ \ 3\ \ 4\ \ 5}{\longleftarrow\!\!\!\!\!+\!\!+\!\!+\!\!+\!\!+\!\!\bullet\!\!+\!\!+\!\!+\!\!+\!\!+\!\!\longrightarrow}$$

Solve each inequality in Exercises 21–23. Express the solution set in set-builder notation and graph the set on a number line.

21. $\dfrac{x}{2} < -3$
22. $6 - 9x \geq 33$
23. $4x - 2 > 2(x + 6)$
24. A student has grades on three examinations of 76, 80, and 72. What must the student earn on a fourth examination to have an average of at least 80?
25. The length of a rectangle is 20 inches. For what widths is the perimeter greater than 56 inches?

CUMULATIVE REVIEW EXERCISES (CHAPTERS 1–2)

In Exercises 1–3, perform the indicated operation or operations.

1. $-8 - (12 - 16)$
2. $(-3)(-2) + (-2)(4)$
3. $(8 - 10)^3(7 - 11)^2$
4. Simplify: $2 - 5[x + 3(x + 7)]$.
5. List all the rational numbers in this set:
$$\left\{-4, -\frac{1}{3}, 0, \sqrt{2}, \sqrt{4}, \frac{\pi}{2}, 1063\right\}.$$

6. Plot $(-2, -1)$ in a rectangular coordinate system. In which quadrant does the point lie?
7. Insert either $<$ or $>$ in the box to make a true statement:
$$-10{,}000 \ \square \ -2.$$
8. Use the distributive property to rewrite without parentheses:
$$6(4x - 1 - 5y).$$

The graph shows the unemployment rate in the United States from 1990 through mid-2003. Use the information in the graph to solve Exercises 9–10.

U.S. Unemployment Rate

Source: Bureau of Labor Statistics

9. For the period shown, in which year was the unemployment rate at a minimum? What percentage of the work force was unemployed in that year?

10. For the period shown, during which year did the unemployment rate reach a maximum? Estimate the percentage of the work force unemployed, to the nearest tenth of a percent, at that time.

In Exercises 11–12, solve each equation.

11. $5 - 6(x + 2) = x - 14$

12. $\dfrac{x}{5} - 2 = \dfrac{x}{3}$

13. Solve for A: $V = \dfrac{1}{3}Ah$.

14. 48 is 30% of what?

15. The length of a rectangular parking lot is 10 yards less than twice its width. If the perimeter of the lot is 400 yards, what are its dimensions?

16. A gas station owner makes a profit of 40 cents per gallon of gasoline sold. How many gallons of gasoline must be sold in a year to make a profit of $30,000 from gasoline sales?

17. Graph the solution set of $-2 < x \le 3$ on a number line.

Solve each inequality in Exercises 18–19. Express the solution set in set-builder notation and graph the set on a number line.

18. $3 - 3x > 12$

19. $5 - 2(3 - x) \le 2(2x + 5) + 1$

20. You take a summer job selling medical supplies. You are paid $600 per month plus 4% of the sales price of all the supplies you sell. If you want to earn more than $2500 per month, what value of medical supplies must you sell?

On a summer job, you are delivering three large cylindrical tanks filled with water to a remote country house. Your truck weighs 12 tons and you are worried about the extra weight of the tanks on these poorly paved, winding roads. Up ahead you see a warning sign at the entrance to a wobbly-looking small bridge: Weight Not to Exceed 20 Tons. Knowing that making an error would not be a good idea, you pull over to compute the weight of your truck. Your boss told you every cubic inch of water that you are carrying weighs about 0.036 pound. You need to compute the volume of the tanks and then determine the combined weight of the tanks and your truck. What else do you need to know to decide whether to cross the bridge or go home without pay?

This problem appears as the group project on page 201.

Problem Solving

The ability to solve problems using thought and reasoning is indispensable to every area of our lives. We are all problem solvers. Our work in solving problems in school, on the job, and even in our personal lives involves understanding the problem, devising a plan for solving it, and then carrying out the plan. The aim of this chapter is to help you refine your problem-solving skills. By the time you complete the chapter, you will be a better, more confident problem solver.

SECTION 3.1

Objectives

1 Solve simple interest problems.

2 Solve mixture problems.

3 Solve motion problems.

FURTHER PROBLEM SOLVING

George Polya (1887–1985), author of *How to Solve It*

Understand the problem. Devise a plan for solving the problem. Carry out the plan and solve the problem. Look back and check the answer. These are the guidelines for problem solving offered by the charismatic teacher and mathematician George Polya in *How to Solve It*. First published in 1945, the book has sold more than one million copies and is available in 17 languages. Polya's book demonstrates how to think clearly in any field.

In this section, you will learn to solve problems involving simple interest, mixtures, and motion. To solve these problems, we will use the five-step strategy that was introduced in Chapter 2. We will also use three problem-solving tips from *How to Solve It*:

- Organize the given information in a table.
- Make a sketch to illustrate the problem.
- Relate the problem to similar problems that you have seen before.

1 Solve simple interest problems.

Problems Involving Simple Interest **Interest** is the dollar amount that we get paid for lending money or pay for borrowing money. When we deposit money in a savings institution, the institution pays us interest for its use. When we borrow money, interest is the price we pay for the privilege of using the money until we repay it.

The amount of money that we deposit or borrow is called the **principal**. For example, if you deposit $2000 in a savings account, then $2000 is the principal. The amount of interest depends on the principal, the interest **rate**, which is given as a percent and varies from bank to bank, and the length of time for which the money is deposited. In this section, the rate is assumed to be annual (per year).

Simple interest involves interest calculated only on the principal. The following formula is used to find simple interest:

CALCULATING SIMPLE INTEREST

$$\text{Interest} = \text{principal} \times \text{rate} \times \text{time}$$

$$I = Prt$$

The rate, r, is expressed as a decimal when calculating simple interest.

EXAMPLE 1 Calculating Simple Interest for a Year

You deposit $2000 in a savings account at Hometown Bank, which has a rate of 6%. Find the interest at the end of the first year.

SOLUTION To find the interest at the end of the first year, we use the simple interest formula.

$$I \ = \ Prt \ = \ (2000)\,(0.06)\,(1) \ = \ 120$$

| Principal, or amount deposited, is $2000. | Rate is 6% = 0.06. | Time is 1 year. |

At the end of the first year, the interest is $120. You can withdraw the $120 in interest, and you still have $2000 in the savings account. ∎

 CHECK POINT 1 You deposit $3000 in a savings account at Yourtown Bank, which has a rate of 5%. Find the interest at the end of the first year.

Our next problem again involves simple interest earned for one year. Because the time is one year, we substitute 1 for t in the simple interest formula, $I = Prt$. The formula becomes $I = Pr \cdot 1$, or simply $I = Pr$. Simple interest earned in one year is the product of the principal and the rate.

EXAMPLE 2 Solving a Simple Interest Problem

You inherited $16,000 with the stipulation that for the first year the money must be invested in two stocks paying 6% and 8% annual interest, respectively. How much did you invest at each rate if the total interest earned for the year was $1180?

SOLUTION

Step 1. **Let x represent one of the quantities.**

Let x = the amount invested at 6%.

Step 2. **Represent other quantities in terms of x.** The other quantity that we seek is the amount invested at 8%. Because the total amount invested is $16,000, and we already used up x,

16,000 − x = the amount invested at 8%.

Step 3. **Write an equation in x that describes the conditions.** We can use a table to organize the information given in the problem.

	Principal	×	Rate	=	Interest
6% Investment	x		0.06		$0.06x$
8% Investment	16,000 − x		0.08		$0.08(16,000 − x)$

The interest for the two investments combined is $1180.

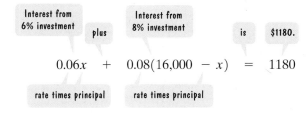

$$0.06x \ + \ 0.08(16,000 − x) \ = \ 1180$$

Step 4. **Solve the equation and answer the question.**

$$0.06x + 0.08(16,000 - x) = 1180$$

This is the equation that describes the problem's conditions.

$$0.06x + 1280 - 0.08x = 1180$$

Use the distributive property.

$$-0.02x + 1280 = 1180$$

Combine like terms.

$$-0.02x + 1280 - 1280 = 1180 - 1280$$

Subtract 1280 from both sides.

$$-0.02x = -100$$

Simplify.

$$\frac{-0.02x}{-0.02} = \frac{-100}{-0.02}$$

Divide both sides by −0.02.

$$x = 5000$$

Simplify.

Because x represents the amount invested at 6%, $5000 was invested at 6%. Because $16,000 - x$ represents the amount invested at 8%, $16,000 − $5000, or $11,000, was invested at 8%.

Step 5. **Check the proposed solution in the original wording of the problem.** The problem states that the total interest was $1180. The interest earned on $5000 at 6% is ($5000)(0.06), or $300. The interest earned on $11,000 at 8% is ($11,000)(0.08), or $880. The total interest is $300 + $880, or $1180, exactly the amount given in the problem. ■

 CHECK POINT 2 Suppose that you invested $25,000, part at 9% simple interest and the remainder at 12%. If the total yearly interest from these investments was $2550, find the amount invested at each rate.

---

2 Solve mixture problems.

Problems Involving Mixtures Chemists and pharmacists often have to change the concentration of solutions and other mixtures. In these situations, the amount of a particular ingredient in the solution or mixture is expressed as a percentage of the total solution.

EXAMPLE 3 Finding the Amount of Acid in a Solution

A 40-milliliter solution of acid in water contains 35% acid. How much acid is in the solution?

SOLUTION

| Amount of acid in the solution | is | 35% | of | total number of milliliters in the solution. |

$$= (0.35) \cdot (40)$$

$$= 14$$

There are 14 milliliters of acid in the solution. ■

 CHECK POINT 3 A 60-milliliter solution of acid in water contains 45% acid. How much acid is in the solution?

We can express our computation in Example 3 as follows:

$$(40)\,(0.35) \quad = \quad 14.$$

total solution amount: 40 milliliters	rate of acid's concentration: 35%	amount of acid: 14 milliliters

The total amount times the rate gives the amount. Can you see that this is the same idea we used to compute simple interest?

total money amount	rate of interest	amount of interest

Simple Interest Problems: principal · rate = interest

Percents are expressed as decimals in these equations.

Mixture Problems: solution · concentration = ingredient

total solution amount	percent of ingredient, or rate of concentration	amount of ingredient

In our simple interest problem, we were mixing two investments. In the mixture problem that follows, we will be mixing two liquids. The equations in these problems are obtained from similar conditions:

Simple Interest Problems				**Mixture Problems**		
Interest from investment 1	+	Interest from investment 2	= amount of interest from mixed investments.	Ingredient amount in solution 1	+ Ingredient amount in solution 2	= amount of ingredient in mixture.

We will continue using our five-step strategy to solve mixture problems. Organizing the given information in a table helps obtain an equation, which is usually the most difficult step in our strategy.

STUDY TIP

"Relate the problem to similar problems that you have seen before."
George Polya, *How to Solve It*

Being aware of the similarities between interest and mixture problems should make you a better problem solver in a variety of situations that involve mixtures. Mixing investments or liquids, combining two blends of coffee with different prices into a single blend, mixing coins or stamps of different values: These problems have nearly identical solutions. Only the nouns and the amounts change from situation to situation.

EXAMPLE 4   Solving a Solution Mixture Problem

A chemist needs to mix an 18% acid solution with a 45% acid solution to obtain 12 liters of a 36% acid solution. How many liters of each of the acid solutions must be used?

SOLUTION

Step 1.   **Let x represent one of the quantities.**

Let x = the number of liters of the 18% acid
solution to be used in the mixture.

Step 2. **Represent other quantities in terms of *x*.** The other quantity that we seek is the number of liters of the 45% acid solution to be used in the mixture. Because the total to be obtained is 12 liters, and we already have used up *x* liters,

$$12 - x = \text{the number of liters of the 45\% acid}$$
$$\text{solution to be used in the mixture.}$$

Step 3. **Write an equation in *x* that describes the conditions.** The situation is illustrated in Figure 3.1.

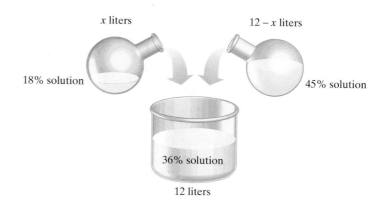

x liters   12 − *x* liters

18% solution   45% solution

36% solution

12 liters

FIGURE 3.1   Obtaining a 12-liter 36% acid mixture

The sum of the amounts of acid in the 18% solution and the 45% solution must equal the amount of acid in the 36%, 12-liter, mixture. We form a table that shows the amount of acid in each of the three solutions.

	Number of Liters	×	Percent of Acid	=	Amount of Acid
18% Acid Solution	x		18% = 0.18		$0.18x$
45% Acid Solution	$12 - x$		45% = 0.45		$0.45(12 - x)$
36% Acid Mixture	12		36% = 0.36		$0.36(12)$

Amount of acid in the 18% solution   **plus**   amount of acid in the 45% solution   **equals**   amount of acid in the 36% mixture.

$$0.18x \ + \ 0.45(12 - x) \ = \ 0.36(12)$$

Step 4. **Solve the equation and answer the question.**

$$0.18x + 0.45(12 - x) = 0.36(12)$$ This is the equation that describes the problem's conditions.

$$0.18x + 5.4 - 0.45x = 4.32$$ Use the distributive property.

$$-0.27x + 5.4 = 4.32$$ Combine like terms.

$$-0.27x = -1.08$$ Subtract 5.4 from both sides.

$$\frac{-0.27x}{-0.27} = \frac{-1.08}{-0.27}$$ Divide both sides by − 0.27.

$$x = 4$$ Simplify.

Because x represents the number of liters of the 18% solution, 4 liters of the 18% solution should be used in the mixture. Because $12 - x$ represents the number of liters of the 45% solution, $12 - 4$, or 8 liters of the 45% solution should be used in the mixture. Thus, the chemist should mix 4 liters of the 18% acid solution with 8 liters of the 45% acid solution.

Step 5. **Check the proposed solution in the original wording of the problem.** The problem states that the 12-liter mixture should be 36% acid. The amount of acid in this mixture is 0.36(12), or 4.32 liters of acid. The amount of acid in 4 liters of the 18% solution is 0.18(4), or 0.72 liter. The amount of acid in 8 liters of the 45% solution is 0.45(8), or 3.6 liters. The amount of acid in the two solutions used in the mixture is 0.72 liter + 3.6 liters, or 4.32 liters, exactly as it should be. ■

✓ CHECK POINT **4** A chemist needs to mix a 10% acid solution with a 60% acid solution to obtain 50 milliliters of a 30% acid solution. How many milliliters of each of the acid solutions must be used?

3 Solve motion problems.

Problems Involving Motion Suppose that you ride your bike at an average speed of 12 miles per hour. What distance do you cover in 2 hours? Your distance is the product of your speed and the time that you travel:

$$\frac{12 \text{ miles}}{\text{hour}} \times 2 \text{ hours} = 24 \text{ miles}.$$

Your distance is 24 miles. Notice how the hour units cancel. The distance is expressed in miles.

In general, the distance covered by any moving body is the product of its average speed, or rate, and its time in motion:

A FORMULA FOR MOTION

$$d = rt$$

Distance equals rate times time.

EXAMPLE 5 Using the Motion Formula

Your nonstop flight from San Francisco to New York, a distance of about 2700 miles, takes 5 hours. What is the plane's average rate?

SOLUTION Use the formula $d = rt$. The distance, d, is 2700 miles. The time, t, is 5 hours. The problem asks us to find r, the plane's average rate.

$$d = rt \qquad \text{Use the motion formula.}$$
$$2700 = r \cdot 5 \qquad d = 2700 \text{ and } t = 5.$$
$$\frac{2700}{5} = \frac{r \cdot 5}{5} \qquad \text{Divide both sides of the equation by 5.}$$
$$540 = r \qquad \text{Simplify.}$$

The plane's average rate, or speed, is 540 miles per hour. ■

✓ CHECK POINT **5** Riding your bike a distance of 27 miles takes 3 hours. What is your average rate?

The five-step strategy can be used to solve motion problems involving two rates. For example, a problem might involve two planes traveling at different speeds.

Because there are two speeds, there are also two distances. The equations in these problems are obtained from the relationship between these distances. Depending on the problem's conditions, these distances are often added or set equal to each other.

EXAMPLE 6 Solving a Problem Involving Motion

Two planes leave the same airport at the same time, flying in opposite directions. The rate of the faster plane is 300 miles per hour. The rate of the slower plane is 200 miles per hour. After how many hours will the planes be 1000 miles apart?

200 miles per hour

300 miles per hour

SOLUTION

Step 1. **Let x represent one of the quantities.** Let x = the time when the planes are 1000 miles apart.

Step 2. **Represent other quantities in terms of x.** Because there are no other unknown quantities that we are asked to find, we can skip this step.

Step 3. **Write an equation in x that describes the conditions.** We obtain this equation by finding a relationship between the distances of the faster and slower planes. Here is a table that summarizes the problem's information and includes these distances:

	Rate	×	Time	=	Distance
Faster Plane	300		x		$300x$
Slower Plane	200		x		$200x$

We are interested in finding x, the time when the planes are 1000 miles apart. At this time, the distance traveled by the faster plane plus the distance traveled by the slower plane equals 1000 miles.

Distance traveled by the faster plane	plus	distance traveled by the slower plane	equals	1000 miles.

$$300x \quad + \quad 200x \quad = \quad 1000$$

Step 4. **Solve the equation and answer the question.**

$300x + 200x = 1000$ This is the equation that describes the problem's conditions.

$500x = 1000$ Add like terms.

$\dfrac{500x}{500} = \dfrac{1000}{500}$ Divide both sides by 500.

$x = 2$ Simplify.

Because x represents the time when the planes are 1000 miles apart, we can conclude that they will be 1000 miles apart after 2 hours.

Step 5.   **Check the proposed solution in the original wording of the problem.** The problem states that the planes are to be 1000 miles apart. After 2 hours, the distance covered by the plane averaging 300 miles per hour is $300 \cdot 2$, or 600 miles. The distance covered by the plane averaging 200 miles per hour is $200 \cdot 2$, or 400 miles. The distance between the planes is 600 miles + 400 miles, or 1000 miles, as specified by the problem's conditions. ∎

 CHECK POINT 6 Two cars leave from the same place, traveling in opposite directions. The rate of the faster car is 55 miles per hour. The rate of the slower car is 50 miles per hour. In how many hours will the cars be 420 miles apart?

STUDY TIP

In the section opener, we listed three problem-solving tips from *How to Solve It* that we used throughout this section. Here are some additional tips from Polya's book that you may find helpful in many problem-solving situations:

• Look for a pattern.

• Make a systematic list.

• Make an educated guess at the solution. Check the guess against the problem's conditions and work backward to determine the solution.

• Try expressing the problem more simply and solve a similar simpler problem.

• Use trial and error.

• Look for a "catch" if the answer seems too obvious. Perhaps the problem involves some sort of trick question deliberately intended to lead the problem solver in the wrong direction.

• Use the given information to eliminate possibilities.

• Use common sense.

3.1 EXERCISE SET

 Student Solutions Manual   CD/Video   PH Math/Tutor Center   MathXL Tutorials on CD   Math XL MathXL®   MyMathLab MyMathLab   Interactmath.com

Practice and Application Exercises

Exercises 1–10 involve simple interest.

1. You deposit $4000 in a savings account, which has a rate of 4%. Find the interest at the end of the first year.

2. You deposit $9200 in a savings account, which has a rate of 4%. Find the interest at the end of the first year.

3. You invest $20,000 in two accounts paying 7% and 8% annual interest, respectively. How much should be invested at each rate if the total interest earned for the year is to be $1520? Begin by filling in the missing entry in the following table. Then use the fact that the interest for the two investments combined must be $1520.

	Principal $\times$	Rate	= Interest
7% Investment	x	0.07	$0.07x$
8% Investment	$20{,}000 - x$	0.08	

4. You invest $20,000 in two accounts paying 7% and 9% annual interest, respectively. If the total interest earned for the year is $1550, how much was invested at each rate? Begin by filling in the missing entry in the table at the top of the next page. Then use the fact that the interest for the two investments combined was $1550.

	Principal	×	Rate	=	Interest
7% Investment	x		0.07		$0.07x$
9% Investment	$20,000 - x$		0.09		

5. A bank loaned out $120,000, part of it at the rate of 8% annual mortgage interest and the rest at the rate of 18% annual credit card interest. The interest received on both loans totaled $10,000. How much was loaned at each rate? Organize your work in the following table.

	Principal	×	Rate	=	Interest
8% Loan	x		0.08		
18% Loan			0.18		

6. A bank loaned out $250,000, part of it at the rate of 8% annual mortgage interest and the rest at the rate of 18% annual credit card interest. The interest received on both loans totaled $23,000. How much was loaned at each rate? Organize your work in the following table.

	Principal	×	Rate	=	Interest
8% Loan	x		0.08		
18% Loan			0.18		

7. You invest $6000 in two accounts paying 6% and 9% annual interest, respectively. At the end of the year, the accounts earn the same interest. How much was invested at each rate?

8. You invest $7200 in two accounts paying 8% and 10% annual interest, respectively. At the end of the year, the accounts earn the same interest. How much was invested at each rate?

9. Your grandmother needs your help. She has $50,000 to invest. Part of this money is to be invested in noninsured bonds paying 15% annual interest. The rest of this money is to be invested in a government-insured certificate of deposit paying 7% annual interest. She told you that she requires a total of $6000 per year in extra income from these investments. How much money should be placed in each investment?

10. Things did not go quite as planned. You invested part of $8000 in a stock that paid 12% annual interest. However, the rest of the money suffered a 5% loss. If the total annual income from both investments was $620, how much was invested at each rate?

Exercises 11–18 involve mixtures.

11. A 20 milliliter solution of acid in water contains 30% acid. How much acid is in the solution?

12. A 70 milliliter solution of acid in water contains 60% acid. How much acid is in the solution?

13. A lab technician needs to mix a 5% fungicide solution with a 10% fungicide solution to obtain a 50-liter mixture consisting of 8% fungicide. How many liters of each of the fungicide –solutions must be used? Begin by filling in the missing entries in the following table. Then use the fact that the amount of fungicide in the 5% solution plus the amount of fungicide in the 10% solution must equal the amount of fungicide in the 8% mixture.

	Number of Liters	×	Percent of Fungicide	=	Amount of Fungicide
5% Fungicide Solution	x		0.05		$0.05x$
10% Fungicide Solution	$50 - x$		0.10		
8% Fungicide Mixture	50		0.08		

14. A candy company needs to mix a 20% fat-content chocolate with a 15% fat-content chocolate to obtain 50 kilograms of a 16% fat-content chocolate. How many kilograms of each kind of chocolate must be used? Begin by filling in the missing entries in the following table. Then use the fact that the amount of fat in the 20% fat-content chocolate plus the amount of fat in the 15% fat-content chocolate must equal the amount of fat in the 16% mixture.

	Number of Kilograms	×	Percent of Fat	=	Amount of Fat
20% Fat Chocolate	x		0.20		$0.20x$
15% Fat Chocolate	$50 - x$		0.15		
16% Fat Mixture	50		0.16		

15. How many ounces of a 15% alcohol solution must be mixed with 4 ounces of a 20% alcohol solution to make a 17% alcohol solution?

16. How many ounces of a 50% alcohol solution must be mixed with 80 ounces of a 20% alcohol solution to make a 40% alcohol solution?

17. At the north campus of a performing arts school, 10% of the students are music majors. At the south campus, 90% of the students are music majors. The campuses are merged into one east campus. If 42% of the 1000 students at the east campus are music majors, how many students did each of the north and south campuses have before the merger?

18. At the north campus of a small liberal arts college, 10% of the students are women. At the south campus, 50% of the students are women. The campuses are merged into one east campus. If 40% of the 1200 students at the east campus are women, how many students did each of the north and south campuses have before the merger?

Exercises 19–26 involve motion.

19. Two cyclists, one averaging 10 miles per hour and the other 12 miles per hour, start from the same town at the same time. If they travel in opposite directions, after how long will they be 66 miles apart? Organize your work in the following table.

	Rate	×	Time	=	Distance
Slower Cyclist	10		x		
Faster Cyclist	12		x		

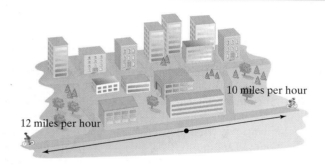

12 miles per hour

10 miles per hour

20. Two cars start from the same place at the same time. They travel in opposite directions. One averages 45 miles per hour and the other averages 30 miles per hour. After how long will they be 225 miles apart? Organize your work in the following table.

	Rate	×	Time	=	Distance
Faster Car	45		x		
Slower Car	30		x		

21. Two trucks leave a warehouse at the same time, traveling in opposite directions. The rate of the faster truck exceeds that of the slower truck by 5 miles per hour. After 5 hours, they are 600 miles apart. What are the rates of the trucks?

	Rate	×	Time	=	Distance
Faster Truck	$x + 5$		5		
Slower Truck	x		5		

22. Two buses leave a station at the same time, traveling in opposite directions. The rate of the faster bus exceeds that of the slower bus by 15 miles per hour. After 3 hours, they are 345 miles apart. What are the rates of the buses?

	Rate	×	Time	=	Distance
Faster Bus	$x + 15$		3		
Slower Bus	x		3		

23. New York City and Washington, D.C., are about 240 miles apart. A car leaves New York City traveling toward Washington, D.C., at 55 miles per hour. At the same time, a bus leaves Washington D.C., bound for New York City at 45 miles per hour. How long will it take before they meet?

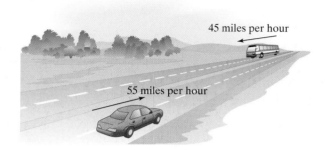

45 miles per hour

55 miles per hour

24. Two cities are 315 miles apart. A car leaves one of the cities traveling toward the second city at 50 miles per hour. At the same time, a bus leaves the second city bound for the first city at 55 miles per hour. How long will it take for them to meet?

25. A bus traveled at an average rate of 50 miles per hour and then reduced its average rate to 40 miles per hour for the rest of the trip. If the 220-mile trip took 5 hours, determine how long the bus traveled at each rate.

26. A car traveled at an average rate of 40 miles per hour and then reduced its average rate to 30 miles per hour for the rest of the trip. If the 180-mile trip took 5 hours, determine how long the car traveled at each rate.

Practice and Application Plus

In Exercises 27–36, you will use what you learned solving interest and mixture problems to solve new kinds of mixture problems. Although tables may help you organize the information, use of tables is not necessary to obtain the equations.

*Exercises 27–30 are called **value-mixture problems**. In these problems, two ingredients that have different values are combined into a single blend. The value of each ingredient in the blend is determined as follows:*

> amount of ingredient × cost per unit of the ingredient
> = value of the ingredient.

Equations are obtained from the following condition:

> value of ingredient 1 + value of ingredient 2 = value of blend.

Use this information to solve Exercises 27–30.

27. The manager of a Starbucks store plans to mix A-grade coffee that costs $9.50 per pound with B-grade coffee that costs $7.00 per pound to create a 20-pound blend that will sell for $8.50 per pound. How many pounds of each grade coffee are required?

28. The manager of a store selling tea plans to mix a more expensive tea that costs $5 per pound with a less expensive tea that costs $3 per pound to create a 100-pound blend that will sell for $4.50 per pound. How many pounds of each type of tea are required?

29. How many pounds of cashews selling for $7.40 per pound must be mixed with 6 pounds of peanuts selling for $2.60 per pound to create a mixture that sells for $5.80 per pound?

30. How many pounds of cashews selling for $8.96 per pound must be mixed with 12 pounds of chocolates selling for $4.48 per pound to create a mixture that sells for $7.28 per pound?

Exercises 31–36 involve mixtures of items with different monetary values, such as coins. If the collection contains coins, the total monetary value of each coin denomination is determined as follows:

number of coins × value of one coin, in cents
= total denomination value, in cents.

Equations are obtained from the following condition:

$$\boxed{\text{total value of denomination 1}} + \boxed{\text{total value of denomination 2}} = \boxed{\text{total value of all the coins.}}$$

Use this information to solve Exercises 31–36.

31. A coin purse contains a mixture of 32 coins in dimes and quarters. The coins have a total value of $5, or 500¢. Determine the number of dimes and the number of quarters in the purse.

32. A coin purse contains a mixture of 46 coins in dimes and quarters. The coins have a total value of $7, or 700¢. Determine the number of dimes and the number of quarters in the purse.

33. A bank teller cashed a check for $200 using $5 bills and $10 bills. The teller handed 25 more $5 bills than $10 bills to the customer. Determine how many bills of each denomination were handed to the customer.

34. A collection of stamps consists of 25¢ stamps and 40¢ stamps. The number of 40¢ stamps is 5 more than 3 times the number of 25¢ stamps. The total value of the stamps is $10.70. Determine how many stamps of each kind are in the collection.

35. Tickets to a community theater cost $25 for adults and $5 for children. At a recent performance, a total of 305 tickets were sold for $6025. Determine how many tickets of each kind were sold.

36. The manager of a men's clothing store sold out the entire stock of $60 jackets and $50 jackets for a total of $3080. If the manager sold 22 more $50 jackets than $60 jackets, how many jackets of each type were sold?

Writing in Mathematics

37. What is interest?

38. How is simple interest calculated?

39. Describe how to find the amount of an ingredient in a mixture if the percentage of the ingredient in the mixture is known.

40. Describe what the formula for motion, $d = rt$, means.

41. Is it necessary to use tables and charts to solve the problems discussed in this section? What advantage is there in using charts to solve interest, mixture, and motion problems?

42. Describe two similarities between interest and mixture problems.

43. Must the concentration of a mixture always be greater than the concentration of one of the ingredients and less than the concentration of the other ingredient? Explain your answer.

Critical Thinking Exercises

44. You have $70,000 to invest. Part of the money is to be placed in a certificate of deposit paying 8% per year. The rest is to be placed in corporate bonds paying 12% per year. If you wish to obtain an overall return of 9% per year, how much should you place in each investment?

45. How many milliliters of pure water should a pharmacist add to 50 milliliters of a 15% minoxidil solution (used to treat male pattern baldness) to make a 10% minoxidil solution?

46. Two thieves rob a jewelry store and make their getaway at 10:00 A.M., traveling at 85 miles per hour. At noon, the police leave the store, in hot pursuit of the robbers. If the police travel at 105 miles per hour, how long does it take for the cops to catch the robbers? At what time of the day does this occur? How far from the jewelry store is the arrest made?

Review Exercises

47. Simplify: $3^2 - 5[4 - 3(2 - 6)]$. (Section 1.8, Example 7)

48. List the integers in this set: $\left\{-100, -\frac{3}{4}, 0, \sqrt{3}, \sqrt{16}\right\}$. (Section 1.2, Example 5)

49. Solve: $3(x + 3) = 5(2x + 6)$. (Section 2.3, Example 3)

SECTION 3.2

Objectives

1 Find ratios.

2 Solve proportions.

3 Solve problems using proportions.

RATIO AND PROPORTION

The possibility of seeing a blue whale, the largest mammal ever to grace the earth, increases the excitement of gazing out over the ocean's swell of waves. Blue whales were hunted to near extinction in the last half of the nineteenth and the first half of the twentieth centuries. Using a method for estimating wildlife populations that we discuss in this section, by the mid-1960s it was determined that the world population of blue whales was less than 1000. This led the International Whaling Commission to ban the killing of blue whales to prevent their extinction. A dramatic increase in blue whale sightings indicates an ongoing increase in their population and the success of the killing ban.

1 Find ratios.

Ratios   A **ratio** compares quantities by division. For example, this year's entering class at a medical school contains 60 women and 30 men. The ratio of women to men is $\frac{60}{30}$. We can express this ratio as a fraction reduced to lowest terms:

$$\frac{60}{30} = \frac{\cancel{30} \cdot 2}{\cancel{30} \cdot 1} = \frac{2}{1}.$$

This ratio can be expressed as 2 : 1, or 2 to 1.

EXAMPLE 1   Finding Ratios

The bar graph in Figure 3.2 shows the distribution, by branch and gender, of the 1.37 million, or 1370 thousand, active-duty personnel in the U.S. military. Use the numbers shown to find each of the following ratios:

a. the ratio of men to women in the Army.

b. the ratio of women to total personnel in the Army.

SOLUTION   Figure 3.2 indicates that in the Army there are 400 thousand men and 70 thousand women, for a total of 470 thousand active-duty personnel.

a. The ratio of men to women in the Army is determined as follows:

$$\frac{\text{number of Army men}}{\text{number of Army women}} = \frac{400}{70} = \frac{\cancel{10} \cdot 40}{\cancel{10} \cdot 7} = \frac{40}{7}.$$

> Numbers are expressed in thousands.

The ratio of men to women in the Army is 40 : 7, or 40 to 7.

Active-Duty U.S. Military Personnel

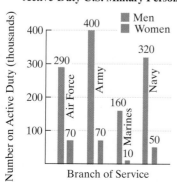

■ Men
■ Women

Branch of Service

FIGURE 3.2

Source: U.S. Defense Department

b. The ratio of women to total personnel in the Army is determined as follows:

$$\frac{\text{number of Army women}}{\text{total Army personnel}} = \frac{70}{470} = \frac{10 \cdot 7}{10 \cdot 47} = \frac{7}{47}.$$

men + women = 400 + 70 = 470

The ratio of women to total Army personnel is $7:47$, or 7 to 47. ■

 CHECK POINT 1 Use the numbers shown in Figure 3.2 to find each of the following ratios:

  a. the ratio of women to men in the Marines.

  b. the ratio of men to all personnel in the Marines.

ENRICHMENT ESSAY

Ratios in Baseball

A baseball player's batting average is the ratio of the number of hits to the number of times at bat. For example, a player with 60 hits and 200 times at bat has a batting average of

$$\frac{60}{200} = \frac{20 \cdot 3}{20 \cdot 10} = \frac{3}{10}$$

or 0.300. Joe DiMaggio was the American League batting champion in 1939 with a 0.381 batting average. The leading batter of all time was left-handed hitter Ty Cobb. Cobb's batting average over 24 years (4189 hits and 11,434 times at bat) was 0.366.

Joe DiMaggio's (1914–1999) grace rivaled that of the great sailing ships, earning him the nickname the "Yankee Clipper."

---

2 Solve proportions.

Proportions A **proportion** is a statement that says that two ratios are equal. If the ratios are $\frac{a}{b}$ and $\frac{c}{d}$, then the proportion is

$$\frac{a}{b} = \frac{c}{d}.$$

We can clear this equation of fractions by multiplying both sides by bd, the least common denominator:

$$\frac{a}{b} = \frac{c}{d} \qquad \text{This is the given proportion.}$$

$$bd \cdot \frac{a}{b} = bd \cdot \frac{c}{d} \qquad \text{Multiply both sides by } bd (b \neq 0 \text{ and } d \neq 0). \text{ Then simplify.}$$

$$ad = bc \qquad \text{On the left, } \frac{bd}{1} \cdot \frac{a}{b} = da = ad. \text{ On the right, } \frac{bd}{1} \cdot \frac{c}{d} = bc.$$

We see that the following principle is true for any proportion:

THE CROSS-PRODUCTS PRINCIPLE FOR PROPORTIONS

If $\frac{a}{b} = \frac{c}{d}$, then $ad = bc$. ($b \neq 0$ and $d \neq 0$.)

The cross products ad and bc are equal.

$$bc$$
$$\frac{a}{b} = \frac{c}{d}$$
$$ad$$

The cross-products principle: $ad = bc$

For example, since $\frac{2}{3} = \frac{6}{9}$, we see that $2 \cdot 9 = 3 \cdot 6$, or $18 = 18$.

If three of the numbers in a proportion are known, the value of the missing quantity can be found by using the cross-products principle. This idea is illustrated in Example 2.

EXAMPLE 2 Solving Proportions

Solve each proportion for x:

a. $\dfrac{63}{x} = \dfrac{7}{5}$ **b.** $\dfrac{-8}{3} = \dfrac{32}{x}$.

SOLUTION

a.
$$\dfrac{63}{x} = \dfrac{7}{5}$$ This is the given proportion.

$63 \cdot 5 = 7x$ Apply the cross-products principle.

$315 = 7x$ Simplify.

$\dfrac{315}{7} = \dfrac{7x}{7}$ Divide both sides by 7.

$45 = x$ Simplify.

Cross products: $\dfrac{63}{x} = \dfrac{7}{5}$, $7x$, $63 \cdot 5$

This means that the ratio of 63 to 45 is the same as the ratio of 7 to 5. This can be checked by simplifying $\frac{63}{45}$, dividing the numerator and denominator by 9 as follows:

$$\dfrac{63}{45} = \dfrac{9 \cdot 7}{9 \cdot 5} = \dfrac{7}{5}.$$

The solution to the proportion is 45.

b.
$$\dfrac{-8}{3} = \dfrac{32}{x}$$ This is the given proportion.

$-8x = 3 \cdot 32$ Apply the cross-products principle.

$-8x = 96$ Simplify.

$\dfrac{-8x}{-8} = \dfrac{96}{-8}$ Divide both sides by -8.

$x = -12$ Simplify.

This means that the ratio of 32 to -12 is the same as the ratio of -8 to 3.

$$\dfrac{32}{-12} = \dfrac{4 \cdot 8}{-4 \cdot 3} = \dfrac{8}{-3} = \dfrac{-8}{3}$$

The solution to the proportion is -12. ■

✔ **CHECK POINT 2** Solve each proportion for x:

a. $\dfrac{10}{x} = \dfrac{2}{3}$ **b.** $\dfrac{-7}{9} = \dfrac{21}{x}$.

3 Solve problems using proportions.

Applications of Proportions We now turn to practical application problems that can be solved using proportions. Here is a procedure for solving these problems:

SOLVING APPLIED PROBLEMS USING PROPORTIONS

1. Read the problem and represent the unknown quantity by x (or any letter).

2. Set up a proportion by listing the given ratio on one side and the ratio with the unknown quantity on the other side. Each respective quantity should occupy the same corresponding position on each side of the proportion.

3. Drop units and apply the cross-products principle.

4. Solve for x and answer the question.

EXAMPLE 3 Applying Proportions: Calculating Taxes

The property tax on a house with an assessed value of $65,000 is $825. Determine the property tax on a house with an assessed value of $180,000, assuming the same tax rate.

SOLUTION

STUDY TIP

Here are three other correct proportions you can use in step 2:

- $\dfrac{\$65{,}000 \text{ value}}{\$825 \text{ tax}} = \dfrac{\$180{,}000 \text{ value}}{\$x \text{ tax}}$

- $\dfrac{\$65{,}000 \text{ value}}{\$180{,}000 \text{ value}} = \dfrac{\$825 \text{ tax}}{\$x \text{ tax}}$

- $\dfrac{\$180{,}000 \text{ value}}{\$65{,}000 \text{ value}} = \dfrac{\$x \text{ tax}}{\$825 \text{ tax}}.$

Each proportion gives the same cross product obtained in step 3.

Step 1. **Represent the unknown by x.** Let $x =$ the tax on the $180,000 house.

Step 2. **Set up a proportion.** We will set up a proportion comparing taxes to assessed value.

Tax on $65,000 house / Assessed value ($65,000) **equals** Tax on $180,000 house / Assessed value ($180,000)

$$\text{Given ratio}\left\{\dfrac{\$825}{\$65{,}000}\right. = \dfrac{\$x}{\$180{,}000} \begin{array}{l}\longleftarrow \text{Unknown} \\ \longleftarrow \text{Given quantity}\end{array}$$

Step 3. **Drop the units and apply the cross-products principle.** We drop the dollar signs and begin to solve for x.

$$\dfrac{825}{65{,}000} = \dfrac{x}{180{,}000} \qquad \text{This is the proportion for the problem's conditions.}$$

$$65{,}000x = (825)(180{,}000) \qquad \text{Apply the cross-products principle.}$$
$$65{,}000x = 148{,}500{,}000 \qquad \text{Multiply.}$$

Step 4. **Solve for x and answer the question.**

$$\dfrac{65{,}000x}{65{,}000} = \dfrac{148{,}500{,}000}{65{,}000} \qquad \text{Divide both sides by 65,000.}$$

$$x \approx 2284.62 \qquad \text{Round the value of } x \text{ to the nearest cent.}$$

The property tax on the $180,000 house is approximately $2284.62. ∎

 CHECK POINT 3 The property tax on a house with an assessed value of $45,000 is $600. Determine the property tax on a house with an assessed value of $112,500, assuming the same tax rate.

Sampling in Nature The method that was used to estimate the blue whale population described in the section opener is called the **capture-recapture method**. Because it is impossible to count each individual animal within a population, wildlife biologists randomly catch and tag a given number of animals. Sometime later they recapture a second sample of animals and count the number of recaptured tagged animals. The total size of the wildlife population is then estimated using the following proportion:

Initially unknown (x) ⟶ $\dfrac{\text{Original number of tagged animals}}{\text{Total number of animals in the population}} = \dfrac{\text{Number of recaptured tagged animals}}{\text{Number of animals in second sample}}$ } Known ratio .

Although this is called the capture-recapture method, it is not necessary to recapture animals to observe whether or not they are tagged. This could be done from a distance, with binoculars for instance.

EXAMPLE 4   Applying Proportions: Estimating Wildlife Population

Wildlife biologists catch, tag, and then release 135 deer back into a wildlife refuge. Two weeks later they observe a sample of 140 deer, 30 of which are tagged. Assuming the ratio of tagged deer in the sample holds for all deer in the refuge, approximately how many deer are in the refuge?

SOLUTION

Step 1.   Represent the unknown by x. Let x = the total number of deer in the refuge.

Step 2.   Set up a proportion.

Unknown ⟶ $\dfrac{\text{Original number of tagged deer}}{\text{Total number of deer}}$ equals $\dfrac{\text{Number of tagged deer in the observed sample}}{\text{Total number of deer in the observed sample}}$ } Known ratio

$$\frac{135}{x} = \frac{30}{140}$$

Steps 3 and 4.   Apply the cross-products principle, solve, and answer the question.

$$\frac{135}{x} = \frac{30}{140}$$ This is the proportion for the problem's conditions.

$$(135)(140) = 30x$$ Apply the cross-products principle.

$$18,900 = 30x$$ Multiply.

$$\frac{18,900}{30} = \frac{30x}{30}$$ Divide both sides by 30.

$$630 = x$$ Simplify.

There are approximately 630 deer in the refuge.   ■

✔ **CHECK POINT 4** Wildlife biologists catch, tag, and then release 120 deer back into a wildlife refuge. Two weeks later they observe a sample of 150 deer, 25 of which are tagged. Assuming the ratio of tagged deer in the sample holds for all deer in the refuge, approximately how many deer are in the refuge?

3.2 EXERCISE SET

Student Solutions Manual   CD/Video   PH Math/Tutor Center   MathXL Tutorials on CD   MathXL®   MyMathLab   Interactmath.com

Practice Exercises

In Exercises 1–6, express each ratio as a fraction reduced to lowest terms.

1. 24 to 48

2. 14 to 49

3. 48 to 20

4. 24 to 15

5. 27:36

6. 25:40

In a class, there are 20 men and 10 women. Find each ratio in Exercises 7–10. First express the ratio as a fraction reduced to lowest terms. Then rewrite the ratio using the reduced fraction and a colon.

7. Find the ratio of the number of men to the number of women.

8. Find the ratio of the number of women to the number of men.

9. Find the ratio of the number of women to the number of students in the class.

10. Find the ratio of the number of men to the number of students in the class.

Solve each proportion in Exercises 11–22.

11. $\dfrac{20}{x} = \dfrac{5}{3}$

12. $\dfrac{27}{x} = \dfrac{9}{4}$

13. $\dfrac{x}{3} = \dfrac{5}{2}$

14. $\dfrac{x}{6} = \dfrac{3}{2}$

15. $\dfrac{x}{12} = -\dfrac{3}{4}$

16. $\dfrac{x}{64} = -\dfrac{9}{16}$

17. $\dfrac{x-2}{12} = \dfrac{8}{3}$

18. $\dfrac{x-4}{10} = \dfrac{3}{5}$

19. $\dfrac{x}{7} = \dfrac{x+14}{5}$

20. $\dfrac{x}{5} = \dfrac{x-3}{2}$

21. $\dfrac{x+9}{5} = \dfrac{x-10}{11}$

22. $\dfrac{x+7}{-4} = \dfrac{x-12}{6}$

Practice Plus

23. The following proportion compares boxes of CDs and their prices:

$$\frac{3 \text{ boxes}}{\text{price of 3 boxes}} = \frac{8 \text{ boxes}}{\text{price of 8 boxes}}.$$

Write three other correct proportions.

24. Given that $\frac{a}{b} = \frac{c}{d}$, write three other correct proportions involving $a, b, c,$ and d.

In Exercises 25–30, solve each proportion for x.

25. $\dfrac{x}{a} = \dfrac{b}{c}$

26. $\dfrac{a}{x} = \dfrac{b}{c}$

27. $\dfrac{a+b}{c} = \dfrac{x}{d}$

28. $\dfrac{a-b}{c} = \dfrac{x}{d}$

29. $\dfrac{x+a}{a} = \dfrac{b+c}{c}$

30. $\dfrac{ax-b}{b} = \dfrac{c-d}{d}$

Application Exercises

The bar graph indicates countries where ten or more languages have become extinct. Use the graph to find each of the ratios in Exercises 31–32. First express the ratio as a fraction reduced to lowest terms. Then rewrite the ratio using the reduced fraction and a colon.

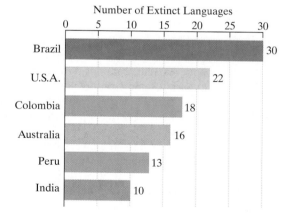

Countries Where Ten or More Languages Have Become Extinct

Source: Grimes

31. The number of extinct languages in Brazil to that of the United States

32. The number of extinct languages in Australia and India combined to that of Colombia

The points in the graph show costs for private and public four-year colleges projected through the year 2017. According to these projections, your daughter's college education at a private four-year school could cost about $250,000. Use the graph to estimate each of the ratios in Exercises 33–34. First express the estimated ratio as a fraction reduced to lowest terms. Then rewrite the ratio using the reduced fraction and a colon.

Cost of a Four-Year College

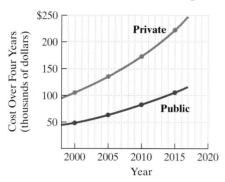

Source: U.S. Department of Education

33. The projected cost of a public four-year college to that of a private four-year college in 2010

34. The projected cost of a public four-year college to that of a private four-year college in 2015

Use a proportion to solve each problem in Exercises 35–44.

35. The tax on a property with an assessed value of $65,000 is $720. Find the tax on a property with an assessed value of $162,500.

36. The maintenance bill for a shopping center containing 180,000 square feet is $45,000. What is the bill for a store in the center that is 4800 square feet?

37. St. Paul Island in Alaska has 12 fur seal rookeries (breeding places). In 1961, to estimate the fur seal pup population in the Gorbath rookery, 4963 fur seal pups were tagged in early August. In late August, a sample of 900 pups was observed and 218 of these were found to have been previously tagged. Estimate the total number of fur seal pups in this rookery.

38. To estimate the number of bass in a lake, wildlife biologists tagged 50 bass and released them in the lake. Later they netted 108 bass and found that 27 of them were tagged. Approximately how many bass are in the lake?

39. The ratio of monthly child support to a father's yearly income is $1:40$. How much should a father earning $38,000 annually pay in monthly child support?

40. The amount of garbage is proportional to the population. Dallas, Texas, has a population of 1.2 million and creates 38.4 million pounds of garbage each week. Find the amount of weekly garbage produced by New York City with a population of 8 million.

41. Height is proportional to foot length. A person whose foot length is 10 inches is 67 inches tall. In 1951, photos of large footprints were published. Some believed that these footprints were made by the "Abominable Snowman." Each footprint was 23 inches long. If indeed they belonged to the Abominable Snowman, how tall is the critter?

Roger Patterson comparing his foot with a plaster cast of a footprint of the purported "Bigfoot" that Mr. Patterson said he sighted in a California forest in 1967

42. A person's hair length is proportional to the number of years it has been growing. After 2 years, a person's hair grows 8 inches. The longest moustache on record was grown by Kalyan Sain of India. Sain grew his moustache for 17 years. How long was it?

43. The formulas

$$C_{\text{Gas}} = 12,000 + 700x$$
$$C_{\text{Solar}} = 30,000 + 150x$$

model the total cost, in dollars, for gas and solar heating systems x years after installation. What is the ratio of the total cost for gas heating to the total cost for solar heating 5 years after installation? What is this ratio 40 years after installation?

44. The formulas

$$C_{\text{Electric}} = 5000 + 1100x$$
$$C_{\text{Solar}} = 30,000 + 150x$$

model the total cost, in dollars, for electric and solar heating systems x years after installation. What is the ratio of the total cost for electric heating to the total cost for solar heating 5 years after installation? What is this ratio 40 years after installation?

Writing in Mathematics

45. What is a ratio? Give an example with your description.

46. What is a proportion? Give an example with your description.

47. Explain the difference between a ratio and a proportion.

48. Explain how to solve a proportion. Illustrate your explanation with an example.

49. Explain the meaning of this statement: A company's monthly sales is proportional to its advertising budget.

50. Use the graph for Exercises 33–34 to describe what is projected to happen to the ratio of the cost of a public four-year college to that of a private four-year college over time. How is this projection shown by the points in the graph?

Critical Thinking Exercises

51. Which one of the following is true?

    a. A statement that two proportions are equal is called a ratio.

    b. The ratio of 1.25 to 2 is equal to the ratio $5:8$.

    c. The ratio of heights of a 12-inch plant to a 4-foot plant is $3:1$.

    d. If 30 people out of 70 are men, then the ratio of women to men is $40:100$.

52. The front sprocket on a bicycle has 60 teeth and the rear sprocket has 20 teeth. For mountain biking, an owner needs a 5-to-1 front-to-rear ratio. If only one of the sprockets is to be replaced, describe the two ways in which this can be done.

53. Use the information in the essay on page 183 to solve this exercise. A baseball player has 10 hits out of 20 times at bat. How many consecutive pitches must be hit to raise the player's batting average to 0.600?

Technology Exercises

Use a calculator to solve Exercises 54–55. Round your answer to two decimal places.

54. Solve: $\dfrac{7.32}{x} = \dfrac{-19.03}{28}$.

55. On a map, 2 inches represent 13.47 miles. How many miles does a person plan to travel if the distance on the map is 9.85 inches?

Review Exercises

56. Solve and graph the solution set: $-6x + 2 \le 2(5 - x)$. (Section 2.6, Example 7)

57. 112 is 40% of what? (Section 2.4, Example 8)

58. Simplify: $9 - 2[4(x - 3) + 7]$. (Section 1.8, Example 11)

✔ MID-CHAPTER CHECK POINT

CHAPTER 3

What You Know: We learned to solve problems involving simple interest, mixtures, and motion, using tables to organize the information. We used the cross-products principle to solve proportions and used proportions to solve a variety of practical problems.

1. If you can type 600 pages of manuscript in 21 days, how many days will it take to type 230 pages working at the same rate?

2. How many liters of a 30% alcohol solution must be mixed with a 70% alcohol solution to obtain 12 liters of a 60% alcohol solution?

3. Solve: $\dfrac{4x}{9} = \dfrac{5}{-15}$.

4. You invested $25,000 in two accounts paying 8% and 9% annual interest, respectively. At the end of the year, the total interest from these investments was $2135. How much was invested at each rate?

5. In a class, there are 25 men and 35 women. Find the ratio of the number of men to the number of students in the class. First express the ratio as a fraction reduced to lowest terms. Then rewrite the ratio using a second method.

6. Two planes leave an airport at the same time and fly in opposite directions. The rate of the faster plane exceeds that of the slower plane by 100 miles per hour. After 5 hours, they are 4250 miles apart. What are the rates of the planes?

7. To estimate the number of elk in a wildlife refuge, park rangers catch, tag, and then release 40 elk. One month later they observe a sample of 80 elk, 4 of which are tagged. Assuming the ratio of tagged elk in the sample holds for all elk in the refuge, approximately how many elk are in the refuge?

8. How many quarts of $\frac{1}{2}$%-fat milk should be added to 4 quarts of 2%-fat milk to obtain 1%-fat milk?

9. Solve: $\dfrac{2}{x + 1} = \dfrac{3}{2x + 3}$.

10. You invested $4000. On part of this investment, you earned 4% interest. On the remainder of the investment, you lost 3%. Combining earnings and losses, the annual income from these investments was $55. How much was invested at each rate?

11. To reach your vacation destination, you travel part of the distance by plane averaging 450 miles per hour and the remainder of the distance by car averaging 50 miles per hour. You spend 2 hours less traveling by car than by plane. If the entire trip is 1900 miles, how long do you travel by plane and how long do you travel by car?

12. The distance that a spring stretches is proportional to the weight attached to the spring. When a 60-pound weight is attached, it stretches the spring 4 inches. What distance will an 80-pound weight stretch the spring?

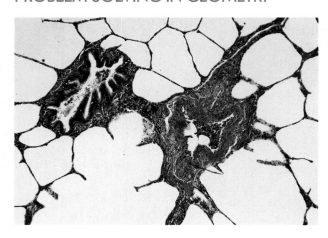

SECTION 3.3

Objectives

1 Solve problems using formulas for perimeter and area.

2 Solve problems using formulas for a circle's area and circumference.

3 Solve problems using formulas for volume.

4 Solve problems involving the angles of a triangle.

5 Solve problems involving complementary and supplementary angles.

1 Solve problems using formulas for perimeter and area.

PROBLEM SOLVING IN GEOMETRY

A portion of the human lung magnified 160 times

Geometry is about the space you live in and the shapes that surround you. You're even made of it. The human lung consists of nearly 300 spherical air sacs, geometrically designed to provide the greatest surface area within the limited volume of our bodies. Viewed in this way, geometry becomes an intimate experience.

For thousands of years, people have studied geometry in some form to obtain a better understanding of the world in which they live. A study of the shape of your world will provide you with many practical applications that will help to increase your problem-solving skills.

Geometric Formulas for Perimeter and Area Solving geometry problems often requires using basic geometric formulas. Formulas for perimeter and area are summarized in Table 3.1. Remember that perimeter is measured in linear units, such as feet or meters, and area is measured in square units, such as square feet, ft^2, or square meters, m^2.

Table 3.1   **Common Formulas for Perimeter and Area**

Square	Rectangle	Triangle	Trapezoid
$A = s^2$	$A = lw$	$A = \frac{1}{2}bh$	$A = \frac{1}{2}h(a + b)$
$P = 4s$	$P = 2l + 2w$		

FIGURE 3.3 Finding the height of a triangular sail

Area 30 ft^2
$h = ?$
$b = 12$ ft

EXAMPLE 1   Using the Formula for the Area of a Triangle

A sailboat has a triangular sail with an area of 30 square feet and a base that is 12 feet long. (See Figure 3.3.) Find the height of the sail.

SOLUTION   We begin with the formula for the area of a triangle given in Table 3.1.

$$A = \frac{1}{2}bh \qquad \text{The area of a triangle is } \tfrac{1}{2} \text{ the product of its base and height.}$$

$$30 = \frac{1}{2}(12)h \qquad \text{Substitute 30 for A and 12 for b.}$$

$$30 = 6h \qquad \text{Simplify.}$$

$$\frac{30}{6} = \frac{6b}{6} \qquad \text{Divide both sides by 6.}$$

$$5 = h \qquad \text{Simplify.}$$

The height of the sail is 5 feet.

Check

The area is $A = \frac{1}{2}bh = \frac{1}{2}(12 \text{ feet})(5 \text{ feet}) = 30$ square feet. ■

 CHECK POINT 1 A sailboat has a triangular sail with an area of 24 square feet and a base that is 4 feet long. Find the height of the sail.

---

2 Solve problems using formulas for a circle's area and circumference.

The point at which a pebble hits a flat surface of water becomes the center of a number of circular ripples.

Geometric Formulas for Circumference and Area of a Circle It's a good idea to know your way around a circle. Clocks, angles, maps, and compasses are based on circles. Circles occur everywhere in nature: in ripples on water, patterns on a butterfly's wings, and cross sections of trees. Some consider the circle to be the most pleasing of all shapes.

A **circle** is a set of points in the plane equally distant from a given point, its center. Figure 3.4 shows two circles. A **radius** (plural: radii), *r*, is a line segment from the center to any point on the circle. For a given circle, all radii have the same length. A **diameter**, *d*, is a line segment through the center whose endpoints both lie on the circle. For a given circle, all diameters have the same length. In any circle, **the length of a diameter is twice the length of a radius.**

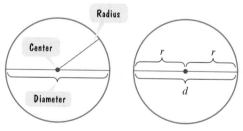

FIGURE 3.4

The words *radius* and *diameter* refer to both the line segments in Figure 3.4 as well as their linear measures. The distance around a circle (its perimeter) is called its **circumference**. Formulas for the area and circumference of a circle are given in terms of π and appear in Table 3.2. We have seen that π is an irrational number and is approximately equal to 3.14.

Table 3.2	Formulas for Circles	
Circle	**Area**	**Circumference**
	$A = \pi r^2$	$C = 2\pi r$

When computing a circle's area or circumference by hand, round π to 3.14. When using a calculator, use the $\boxed{\pi}$ key, which gives the value of π rounded to approximately 11 decimal places. In either case, calculations involving π give approximate answers. These answers can vary slightly depending on how π is rounded. The symbol $\approx$ (is approximately equal to) will be written in these calculations.

EXAMPLE 2   Finding the Area and Circumference of a Circle

Find the area and circumference of a circle with a diameter measuring 20 inches.

SOLUTION   The radius is half the diameter, so $r = \frac{20}{2} = 10$ inches.

$$A = \pi r^2 \qquad\qquad C = 2\pi r \qquad\qquad \text{\small Use the formulas for area and circumference of a circle.}$$
$$A = \pi(10)^2 \qquad C = 2\pi(10) \qquad \text{\small Substitute 10 for } r.$$
$$A = 100\pi \qquad\quad C = 20\pi$$

The area of the circle is 100π square inches and the circumference is 20π inches. Using the fact that $\pi \approx 3.14$, the area is approximately $100(3.14)$, or 314 square inches. The circumference is approximately $20(3.14)$, or 62.8 inches. ■

 CHECK POINT 2   The diameter of a circular landing pad for helicopters is 40 feet. Find the area and circumference of the landing pad. Express answers in terms of π. Then round answers to the nearest square foot and foot, respectively.

EXAMPLE 3   Problem Solving Using the Formula for a Circle's Area

Which one of the following is the better buy: a large pizza with a 16-inch diameter for $15.00 or a medium pizza with an 8-inch diameter for $7.50?

SOLUTION   The better buy is the pizza with the lower price per square inch. The radius of the large pizza is $\frac{1}{2} \cdot 16$ inches, or 8 inches, and the radius of the medium pizza is $\frac{1}{2} \cdot 8$ inches, or 4 inches. The area of the surface of each circular pizza is determined using the formula for the area of a circle.

$$\text{Large pizza:} \quad A = \pi r^2 = \pi(8 \text{ in.})^2 = 64\pi \text{ in.}^2 \approx 201 \text{ in.}^2$$
$$\text{Medium pizza:} \quad A = \pi r^2 = \pi(4 \text{ in.})^2 = 16\pi \text{ in.}^2 \approx 50 \text{ in.}^2$$

For each pizza, the price per square inch is found by dividing the price by the area:

$$\text{Price per square inch for large pizza} = \frac{\$15.00}{64\pi \text{ in.}^2} \approx \frac{\$15.00}{201 \text{ in.}^2} \approx \frac{\$0.07}{\text{in.}^2}$$

$$\text{Price per square inch for medium pizza} = \frac{\$7.50}{16\pi \text{ in.}^2} \approx \frac{\$7.50}{50 \text{ in.}^2} = \frac{\$0.15}{\text{in.}^2}.$$

The large pizza costs approximately $0.07 per square inch and the medium pizza costs approximately $0.15 per square inch. Thus, the large pizza is the better buy. ■

USING TECHNOLOGY

You can use your calculator to obtain the price per square inch for each pizza in Example 3. The price per square inch for the large pizza, $\frac{15}{64\pi}$, is approximated by one of the following keystrokes:

Many Scientific Calculators

$15 \boxed{\div} \boxed{(} 64 \boxed{\times} \boxed{\pi} \boxed{)} \boxed{=}$

Many Graphing Calculators

$15 \boxed{\div} \boxed{(} 64 \boxed{\pi} \boxed{)} \boxed{\text{ENTER}}$

In Example 3, did you at first think that the price per square inch would be the same for the large and the medium pizzas? After all, the radius of the large pizza is twice that of the medium pizza, and the cost of the large is twice that of the medium. However, the large pizza's area, 64π square inches, is *four times the area* of the medium pizza, 16π square inches. Doubling the radius of a circle increases its area by four times the original amount.

✓ **CHECK POINT 3** Which one of the following is the better buy: a large pizza with an 18-inch diameter for $20.00 or a medium pizza with a 14-inch diameter for $14.00?

3 Solve problems using formulas for volume.

Geometric Formulas for Volume   A shoe box and a basketball are examples of three-dimensional figures. **Volume** refers to the amount of space occupied by such a figure. To measure this space, we begin by selecting a cubic unit. One such cubic unit, 1 cubic centimeter (cm^3), is shown in Figure 3.5.

The edges of a cube all have the same length. Other cubic units used to measure volume include 1 cubic inch ($in.^3$) and 1 cubic foot (ft^3). The volume of a solid is the number of cubic units that can be contained in the solid.

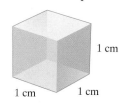

FIGURE 3.5

Formulas for volumes of three-dimensional figures are given in Table 3.3.

Table 3.3	**Common Formulas for Volume**

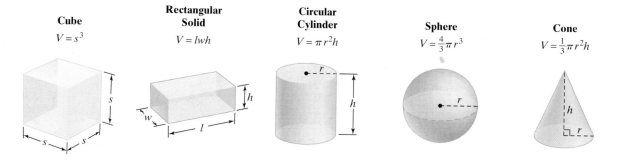

Cube	Rectangular Solid	Circular Cylinder	Sphere	Cone
$V = s^3$	$V = lwh$	$V = \pi r^2 h$	$V = \frac{4}{3}\pi r^3$	$V = \frac{1}{3}\pi r^2 h$

EXAMPLE 4   Using the Formula for the Volume of a Cylinder

A cylinder with a radius of 2 inches and a height of 6 inches has its radius doubled. (See Figure 3.6.) How many times greater is the volume of the larger cylinder than the volume of the smaller cylinder?

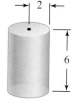

Radius: 2 inches
Height: 6 inches

Radius: 4 inches
Height: 6 inches

FIGURE 3.6   Doubling a cylinder's radius

SOLUTION   We begin with the formula for the volume of a cylinder given in Table 3.3. Find the volume of the smaller cylinder and the volume of the larger cylinder. To compare the volumes, divide the volume of the larger cylinder by the volume of the smaller cylinder.

$$V = \pi r^2 h$$

Use the formula for the volume of a cylinder.

Radius is doubled.

$$V_{Smaller} = \pi(2)^2(6) \quad V_{Larger} = \pi(4)^2(6) \qquad \text{Substitute the given values.}$$
$$V_{Smaller} = \pi(4)(6) \quad V_{Larger} = \pi(16)(6)$$
$$V_{Smaller} = 24\pi \qquad V_{Larger} = 96\pi$$

The volume of the smaller cylinder is 24π cubic inches. The volume of the larger cylinder is 96π cubic inches. We use division to compare the volumes:

$$\frac{V_{Larger}}{V_{Smaller}} = \frac{96\pi}{24\pi} = \frac{4}{1}.$$

Thus, the volume of the larger cylinder is 4 times the volume of the smaller cylinder. ∎

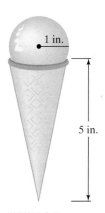

1 in.

5 in.

FIGURE 3.7

✔ **CHECK POINT 4** A cylinder with a radius of 3 inches and a height of 5 inches has its height doubled. How many times greater is the volume of the larger cylinder than the volume of the smaller cylinder?

EXAMPLE 5 Applying Volume Formulas

An ice cream cone is 5 inches deep and has a radius of 1 inch. A spherical scoop of ice cream also has a radius of 1 inch. (See Figure 3.7.) If the ice cream melts into the cone, will it overflow?

SOLUTION The ice cream will overflow if the volume of the ice cream, a sphere, is greater than the volume of the cone. Find the volume of each.

$$V_{\text{cone}} = \frac{1}{3}\pi r^2 h = \frac{1}{3}\pi(1 \text{ in.})^2 \cdot 5 \text{ in.} = \frac{5\pi}{3} \text{ in.}^3 \approx 5 \text{ in.}^3$$

$$V_{\text{sphere}} = \frac{4}{3}\pi r^3 = \frac{4}{3}\pi(1 \text{ in.})^3 = \frac{4\pi}{3} \text{ in.}^3 \approx 4 \text{ in.}^3$$

The volume of the spherical scoop of ice cream is less than the volume of the cone, so there will be no overflow. ■

✔ **CHECK POINT 5** A basketball has a radius of 4.5 inches. If the ball is filled with 350 cubic inches of air, is this enough air to fill it completely?

4 Solve problems involving the angles of a triangle.

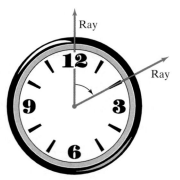

FIGURE 3.8 Clock with rays rotating to form an angle

The Angles of a Triangle The hour hand of a clock moves from 12 to 2. The hour hand suggests a **ray**, a part of a line that has only one endpoint and extends forever in the opposite direction. An *angle* is formed as the ray in Figure 3.8 rotates from 12 to 2.

An **angle**, symbolized ∡, is made up of two rays that have a common endpoint. Figure 3.9 shows an angle. The common endpoint, *B* in the figure, is called the **vertex** of the angle. The two rays that form the angle are called its **sides**. The four ways of naming the angle are shown to the right of Figure 3.9.

FIGURE 3.9 An angle: two rays with a common endpoint

One way to measure angles is in **degrees**, symbolized by a small, raised circle °. Think of the hour hand of a clock. From 12 noon to 12 midnight, the hour hand moves around in a complete circle. By definition, the ray has rotated through 360 degrees, or 360°. Using 360° as the amount of rotation of a ray back onto itself, a degree, 1°, is $\frac{1}{360}$ of a complete rotation.

Our next problem is based on the relationship among the three angles of any triangle.

THE ANGLES OF A TRIANGLE

The sum of the measures of the three angles of any triangle is 180°.

EXAMPLE 6 Angles of a Triangle

In a triangle, the measure of the first angle is twice the measure of the second angle. The measure of the third angle is 20° less than the second angle. What is the measure of each angle?

SOLUTION

Step 1. **Let x represent one of the quantities.** Let

$$x = \text{the measure of the second angle.}$$

Step 2. **Represent other quantities in terms of x.** The measure of the first angle is twice the measure of the second angle. Thus, let

$$2x = \text{the measure of the first angle.}$$

The measure of the third angle is 20° less than the second angle. Thus, let

$$x - 20 = \text{the measure of the third angle.}$$

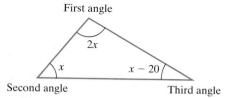

Step 3. **Write an equation in x that describes the conditions.** Because we are working with a triangle, the sum of the measures of its three angles is 180°.

Measure of first angle	plus	measure of second angle	plus	measure of third angle	equals	180°.
$2x$	$+$	x	$+$	$(x - 20)$	$=$	180

Step 4. **Solve the equation and answer the question.**

$2x + x + (x - 20) = 180$ This is the equation that describes the sum of the measures of the angles.

$4x - 20 = 180$ Regroup and combine like terms.

$4x - 20 + 20 = 180 + 20$ Add 20 to both sides.

$4x = 200$ Simplify.

$\dfrac{4x}{4} = \dfrac{200}{4}$ Divide both sides by 4.

$x = 50$ Simplify.

Measure of first angle $= 2x = 2 \cdot 50 = 100$

Measure of second angle $= x = 50$

Measure of third angle $= x - 20 = 50 - 20 = 30$

The angles measure 100°, 50°, and 30°.

Step 5. **Check the proposed solution in the original wording of the problem.** The problem tells us that we are working with a triangle's angles. Thus, the sum of the measures should be 180°. Adding the three measures, we obtain $100° + 50° + 30°$, giving the required sum of 180°. ∎

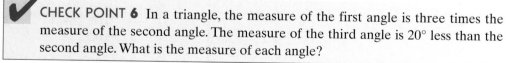

CHECK POINT 6 In a triangle, the measure of the first angle is three times the measure of the second angle. The measure of the third angle is 20° less than the second angle. What is the measure of each angle?

5 Solve problems involving complementary and supplementary angles.

Complementary and Supplementary Angles Two angles with measures having a sum of 90° are called **complementary angles**. For example, angles measuring 70° and 20° are complementary angles because $70° + 20° = 90°$. For angles such as those measuring 70° and 20°, each angle is a **complement** of the other: The 70° angle is the complement of the 20° angle and the 20° angle is the complement of the 70° angle. The measure of the complement can be found by subtracting the angle's measure from 90°. For example, we can find the complement of a 25° angle by subtracting 25° from 90°: $90° - 25° = 65°$. Thus, an angle measuring 65° is the complement of one measuring 25°.

Two angles with measures having a sum of 180° are called **supplementary angles**. For example, angles measuring 110° and 70° are supplementary angles because $110° + 70° = 180°$. For angles such as those measuring 110° and 70°, each angle is a **supplement** of the other: The 110° angle is the supplement of the 70° angle and the 70° angle is the supplement of the 110° angle. The measure of the supplement can be found by subtracting the angle's measure from 180°. For example, we can find the supplement of a 25° angle by subtracting 25° from 180°: $180° - 25° = 155°$. Thus, an angle measuring 155° is the supplement of one measuring 25°.

ALGEBRAIC EXPRESSIONS FOR COMPLEMENTS AND SUPPLEMENTS

Measure of an angle: x

Measure of the angle's complement: $90 - x$

Measure of the angle's supplement: $180 - x$

EXAMPLE 7 Angle Measures and Complements

The measure of an angle is 40° less than four times the measure of its complement. What is the angle's measure?

SOLUTION

Step 1.   **Let x represent one of the quantities.** Let

$$x = \text{the measure of the angle.}$$

Step 2.   **Represent other unknown quantities in terms of x.** Because this problem involves an angle and its complement, let

$$90 - x = \text{the measure of the complement.}$$

Step 3.   **Write an equation in x that describes the conditions.**

The angle's measure	is	40° less than	four times the measure of the complement.
x	$=$	$4(90 - x)$	$- 40$

Step 4.   **Solve the equation and answer the question.**

$x = 4(90 - x) - 40$ This is the equation that describes the problem's conditions.

$x = 360 - 4x - 40$ Use the distributive property.

$x = 320 - 4x$ Simplify: $360 - 40 = 320$.

$x + 4x = 320 - 4x + 4x$ Add 4x to both sides.

$$5x = 320 \qquad \text{Simplify.}$$

$$\frac{5x}{5} = \frac{320}{5} \qquad \text{Divide both sides by 5.}$$

$$x = 64 \qquad \text{Simplify.}$$

The angle measures 64°.

Step 5. **Check the proposed solution in the original wording of the problem.** The measure of the complement is 90° − 64° = 26°. Four times the measure of the complement is 4 · 26°, or 104°. The angle's measure, 64°, is 40° less than 104°: 104° − 40° = 64°. As specified by the problem's wording, the angle's measure is 40° less than four times the measure of its complement. ■

✔ CHECK POINT **7** The measure of an angle is twice the measure of its complement. What is the angle's measure?

3.3 EXERCISE SET

Student Solutions Manual   CD/Video   PH Math/Tutor Center   MathXL Tutorials on CD   MathXL®   MyMathLab   Interactmath.com

Practice Exercises

Use the formulas for perimeter and area in Table 3.1 on page 190 to solve Exercises 1–12.
In Exercises 1–2, find the perimeter and area of each rectangle.

1.

3 m
6 m

2.

3 ft
4 ft

In Exercises 3–4, find the area of each triangle.

3.
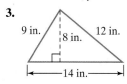
9 in.   8 in.   12 in.
14 in.

4.
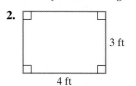
36 m   33 m   36 m
30 m

In Exercises 5–6, find the area of each trapezoid.

5.
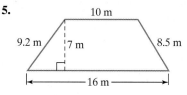
10 m
9.2 m   7 m   8.5 m
16 m

6.
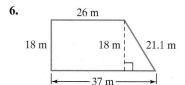
26 m
18 m   18 m   21.1 m
37 m

7. A rectangular swimming pool has a width of 25 feet and an area of 1250 square feet. What is the pool's length?

8. A rectangular swimming pool has a width of 35 feet and an area of 2450 square feet. What is the pool's length?

9. A triangle has a base of 5 feet and an area of 20 square feet. Find the triangle's height.

10. A triangle has a base of 6 feet and an area of 30 square feet. Find the triangle's height.

11. A rectangle has a width of 44 centimeters and a perimeter of 188 centimeters. What is the rectangle's length?

12. A rectangle has a width of 46 centimeters and a perimeter of 208 centimeters. What is the rectangle's length?

Use the formulas for the area and circumference of a circle in Table 3.2 on page 191 to solve Exercises 13–18.
In Exercises 13–16, find the area and circumference of each circle. Express answers in terms of π. Then round to the nearest whole number.

13.
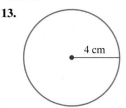
4 cm

14.
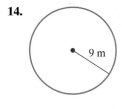
9 m

15.

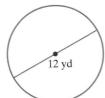

12 yd

16.

40 ft

24.

24 in.

17. The circumference of a circle is 14π inches. Find the circle's radius and diameter.

18. The circumference of a circle is 16π inches. Find the circle's radius and diameter.

Use the formulas for volume in Table 3.3 on page 193 to solve Exercises 19–30.
In Exercises 19–26, find the volume of each figure. Where applicable, express answers in terms of π. Then round to the nearest whole number.

19.

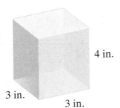

4 in.

3 in.

3 in.

20.

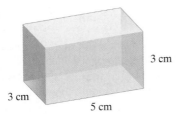

3 cm

3 cm

5 cm

21.

5 cm

6 cm

22.

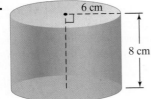

6 cm

8 cm

23.

18 cm

25.

9 m

4 m

26.

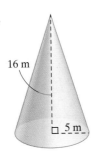

16 m

5 m

27. Solve the formula for the volume of a circular cylinder for h.

28. Solve the formula for the volume of a cone for h.

29. A cylinder with radius 3 inches and height 4 inches has its radius tripled. How many times greater is the volume of the larger cylinder than the smaller cylinder?

30. A cylinder with radius 2 inches and height 3 inches has its radius quadrupled. How many times greater is the volume of the larger cylinder than the smaller cylinder?

Use the relationship among the three angles of any triangle to solve Exercises 31–36.

31. Two angles of a triangle have the same measure and the third angle is $30°$ greater than the measure of the other two. Find the measure of each angle.

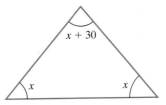

$x + 30$

x x

32. One angle of a triangle is three times as large as another. The measure of the third angle is $40°$ more than that of the smallest angle. Find the measure of each angle.

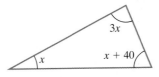

$3x$

x $x + 40$

Find the measure of each angle whose degree measure is represented in terms of x in the triangles in Exercises 33–34.

33.

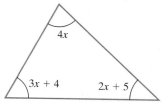

4x

3x + 4   2x + 5

34.

x

4x   5x

35. One angle of a triangle is twice as large as another. The measure of the third angle is 20° more than that of the smallest angle. Find the measure of each angle.

36. One angle of a triangle is three times as large as another. The measure of the third angle is 30° greater than that of the smallest angle. Find the measure of each angle.

In Exercises 37–40, find the measure of the complement of each angle.

37. 58°   **38.** 41°   **39.** 88°   **40.** 2°

In Exercises 41–44, find the measure of the supplement of each angle.

41. 132°   **42.** 93°   **43.** 90°   **44.** 179.5°

In Exercises 45–50, use the five-step problem-solving strategy to find the measure of the angle described.

45. The angle's measure is 60° more than that of its complement.

46. The angle's measure is 78° less than that of its complement.

47. The angle's measure is three times that of its supplement.

48. The angle's measure is 16° more than triple that of its supplement.

49. The measure of the angle's supplement is 10° more than three times that of its complement.

50. The measure of the angle's supplement is 52° more than twice that of its complement.

Practice Plus

In Exercises 51–53, find the area of each figure.

51.

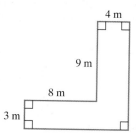

4 m

9 m

8 m

3 m

52.

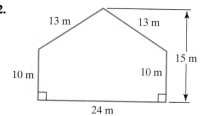

13 m   13 m

10 m   15 m   10 m

24 m

53.

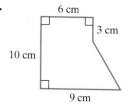

6 cm

3 cm

10 cm

9 cm

54. Find the area of the shaded region in terms of π.

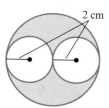

2 cm

In Exercises 55–56, find the volume of the darkly shaded region. In Exercise 55, use the fact that the volume of a pyramid is $\frac{1}{3}$ the volume of a rectangular solid having the same base and the same height. In Exercise 56, express the answer in terms of π.

55.

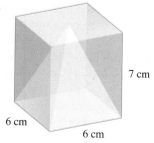

7 cm

6 cm   6 cm

56.

2 in.

10 in.

6 in.

Application Exercises

Use the formulas for perimeter and area in Table 3.1 on page 190 to solve Exercises 57–58.

57. Taxpayers with an office in their home may deduct a percentage of their home-related expenses. This percentage is based on the ratio of the office's area to the area of the home. A taxpayer with an office in a 2200-square-foot home maintains a 20-foot by 16-foot office. If the yearly electricity bills for the home come to $4800, how much of this is deductible?

58. The lot in the figure shown, except for the house, shed, and driveway, is lawn. One bag of lawn fertilizer costs $25.00 and covers 4000 square feet.

   a. Determine the minimum number of bags of fertilizer needed for the lawn.

   b. Find the total cost of the fertilizer.

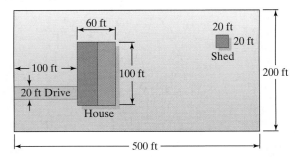

Use the formulas for the area and the circumference of a circle in Table 3.2 on page 191 to solve Exercises 59–64. Round all circumference and area calculations to the nearest whole number.

59. Which one of the following is a better buy: a large pizza with a 14-inch diameter for $12.00 or a medium pizza with a 7-inch diameter for $5.00?

60. Which one of the following is a better buy: a large pizza with a 16-inch diameter for $12.00 or two small pizzas, each with a 10-inch diameter, for $12.00?

61. If asphalt pavement costs $0.80 per square foot, find the cost to pave the circular road in the figure shown. Round to the nearest dollar.

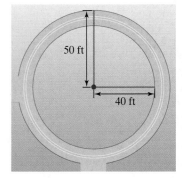

62. Hardwood flooring costs $10.00 per square foot. How much will it cost (to the nearest dollar) to cover the dance floor shown in the figure at the top of the next column with hardwood flooring?

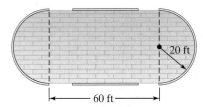

63. A glass window is to be placed in a house. The window consists of a rectangle, 6 feet high by 3 feet wide, with a semicircle at the top. Approximately how many feet of stripping, to the nearest tenth of a foot, will be needed to frame the window?

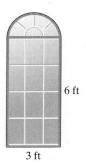

64. How many plants spaced every 6 inches are needed to surround a circular garden with a 30-foot radius?

Use the formulas for volume in Table 3.3 on page 193 to solve Exercises 65–69. When necessary, round all volume calculations to the nearest whole number.

65. A water reservoir is shaped like a rectangular solid with a base that is 50 yards by 30 yards, and a vertical height of 20 yards. At the start of a three-month period of no rain, the reservoir was completely full. At the end of this period, the height of the water was down to 6 yards. How much water was used in the three-month period?

66. A building contractor is to dig a foundation 4 yards long, 3 yards wide, and 2 yards deep for a toll booth's foundation. The contractor pays $10 per load for trucks to remove the dirt. Each truck holds 6 cubic yards. What is the cost to the contractor to have all the dirt hauled away?

67. Two cylindrical cans of soup sell for the same price. One can has a diameter of 6 inches and a height of 5 inches. The other has a diameter of 5 inches and a height of 6 inches. Which can contains more soup and, therefore, is the better buy?

68. The tunnel under the English Channel that connects England and France is one of the world's longest tunnels. The Chunnel, as it is known, consists of three separate tunnels built side by side. Each is a half-cylinder that is 50,000 meters long and 4 meters high. How many cubic meters of dirt had to be removed to build the Chunnel?

69. You are about to sue your contractor who promised to install a water tank that holds 500 gallons of water. You know that 500 gallons is the capacity of a tank that holds 67 cubic feet. The cylindrical tank has a radius of 3 feet and a height of 2 feet 4 inches. Does the evidence indicate you can win the case against the contractor if it goes to court?

Writing in Mathematics

70. Using words only, describe how to find the area of a triangle.

71. Describe the difference between the following problems: How much fencing is needed to enclose a garden? How much fertilizer is needed for the garden?

72. Describe how volume is measured. Explain why linear or square units cannot be used.

73. What is an angle?

74. If the measures of two angles of a triangle are known, explain how to find the measure of the third angle.

75. Can a triangle contain two 90° angles? Explain your answer.

76. What are complementary angles? Describe how to find the measure of an angle's complement.

77. What are supplementary angles? Describe how to find the measure of an angle's supplement?

78. Describe an application of a geometric formula involving area or volume.

79. Write and solve an original problem involving the measures of the three angles of a triangle.

Critical Thinking Exercises

80. Which one of the following is true?

a. It is not possible to have a circle whose circumference is numerically equal to its area.

b. When the measure of a given angle is added to three times the measure of its complement, the sum equals the sum of the measures of the complement and supplement of the angle.

c. The complement of an angle that measures less than 90° is an angle that measures more than 90°.

d. Two complementary angles cannot be equal in measure.

81. Suppose you know the cost for building a rectangular deck measuring 8 feet by 10 feet. If you decide to increase the dimensions to 12 feet by 15 feet, by how many times will the cost increase?

82. A rectangular swimming pool measures 14 feet by 30 feet. The pool is surrounded on all four sides by a path that is 3 feet wide. If the cost to resurface the path is $2 per square foot, what is the total cost of resurfacing the path?

83. What happens to the volume of a sphere if its radius is doubled?

84. A scale model of a car is constructed so that its length, width, and height are each $\frac{1}{10}$ the length, width, and height of the actual car. By how many times does the volume of the car exceed its scale model?

85. Find the measure of the angle of inclination, denoted by x in the figure, for the road leading to the bridge.

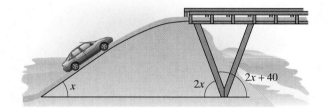

Review Exercises

86. Solve for s: $P = 2s + b$. (Section 2.4, Example 3)

87. Solve for x: $\dfrac{x}{2} + 7 = 13 - \dfrac{x}{4}$ (Section 2.3, Example 4)

88. Simplify: $[3(12 \div 2^2 - 3)^2]^2$. (Section 1.8, Example 8)

GROUP PROJECT

CHAPTER 3

a. As a group, write a problem that is based on the situation that opened this chapter. Group members will need to assign realistic values, in inches, to the radius and height of each cylindrical tank. You can decide if each tank is the same size, or if they differ in size. Use these assigned numbers, the formula for the volume of a cylinder, and the information given to the left of the photograph on page 170 to determine if it is safe to cross the bridge.

b. We have seen that geometry has many practical applications. Now, group members need to take this one step further. Describe as many situations as possible in which a knowledge of geometry can prevent a disaster (or at least a serious mishap) from happening. Be as creative as possible. It is not necessary to assign specific numbers in each situation or to solve the problem, as you were asked to do in part (a).

CHAPTER 3 SUMMARY

Definitions and Concepts	Examples

Section 3.1 Further Problem Solving

Simple interest involves interest calculated only on the principal. Information in these problems can be organized in a table.

	Principal × Rate = Interest
Investment 1	?
Investment 2	?

Represent the principal invested at each rate in terms of x. Set the sum of the interests from the final column equal to the given amount for the total interest from the two investments.

You invested $16,000 in two investments paying 6% and 8% annual interest. At the end of the year total interest was $1180. How much was invested at each rate?

$$x = \text{amount invested at } 6\%$$

$$16{,}000 - x = \text{amount invested at } 8\%$$

	Principal	× Rate	= Interest
6% Investment	x	0.06	$0.06x$
8% Investment	$16{,}000 - x$	0.08	$0.08(16{,}000 - x)$

$$0.06x + 0.08(16{,}000 - x) = 1180$$
(See Example 2 in Section 3.1.)

$$x = 5000$$

$$16{,}000 - x = 16{,}000 - 5000 = 11{,}000$$

Thus, $5000 was invested at 6% and $11,000 at 8%.

Mixture problems can be solved by organizing information in a table. The amount of the substance in solution 1 plus the amount of the substance in solution 2 equals the amount of the substance in the mixture.

How many liters of a 50% acid solution must be mixed with a 20% acid solution to obtain 12 liters of a 30% solution?

$$x = \text{number of liters of } 50\% \text{ solution}$$

$$12 - x = \text{number of liters of } 20\% \text{ solution}$$

	Number of Liters	× Percent of Acid	= Amount of Acid
50% Solution	x	0.5	$0.5x$
20% Solution	$12 - x$	0.2	$0.2(12 - x)$
30% Mixture	12	0.3	$0.3(12)$

$$0.5x + 0.2(12 - x) = 0.3(12)$$

Solving: $x = 4$.
Use 4 liters of the 50% solution and $12 - x = 12 - 4 = 8$ liters of the 20% solution.

Definitions and Concepts	Examples

Section 3.1 Further Problem Solving (continued)

Motion problems can be solved by organizing information in a table.

Rate × Time = Distance
Moving object 1
Moving object 2

The algebraic expressions for distance in the final column are used to set up an equation based on the problem's conditions.

Two runners averaging 5 miles per hour and 6 miles per hour, respectively, start in the same place and run in opposite directions. After how many hours will they be 22 miles apart?

$$x = \text{time when runners are 22 miles apart}$$

	Rate	×	Time	=	Distance
Runner 1	5		x		$5x$
Runner 2	6		x		$6x$

$$5x + 6x = 22$$
$$11x = 22$$
$$x = 2$$

The runners will be 22 miles apart after 2 hours.

Section 3.2 Ratio and Proportion

A ratio compares quantities by division. The ratio of a to b can be expressed as

$$\frac{a}{b} \quad \text{or} \quad a : b.$$

20 men, 30 women

Ratio of women to men

$$= \frac{30}{20} = \frac{3}{2} = 3 : 2 \ (3 \text{ to } 2)$$

Ratio of men to all people

$$= \frac{20}{50} = \frac{2}{5} = 2 : 5 \ (2 \text{ to } 5)$$

A proportion is a statement in the form $\frac{a}{b} = \frac{c}{d}$. The cross-products principal states that if $\frac{a}{b} = \frac{c}{d}$, then $ad = bc$ ($b \neq 0$ and $d \neq 0$).

Solve: $\frac{-3}{4} = \frac{x}{12}.$

$$4x = -3(12)$$
$$4x = -36$$
$$x = -9$$

The solution is -9.

Solving Applied Problems Using Proportions

1. Read the problem and represent the unknown quantity by x (or any letter).

2. Set up a proportion by listing the given ratio on one side and the ratio with the unknown quantity on the other side.

3. Drop units and apply the cross-products principle.

4. Solve for x and answer the question.

30 elk are tagged and released. Sometime later, a sample of 80 elk are observed and 10 are tagged. How many elk are there?

$$x = \text{number of elk}$$

$$\begin{matrix} \text{Tagged} \\ \text{Total} \end{matrix} \quad \frac{30}{x} = \frac{10}{80}$$

$$10x = 30 \cdot 80$$
$$10x = 2400$$
$$x = 240$$

There are 240 elk.

Definitions and Concepts	Examples

Section 3.3 Problem Solving in Geometry

Solving geometry problems often requires using basic geometric formulas. Formulas for perimeter, area, circumference, and volume are given in Tables 3.1 (page 190), 3.2 (page 191), and 3.3 (page 193) in Section 3.3.	A sailboat's triangular sail has an area of 24 ft^2 and a base of 8 ft. Find its height.$$A = \frac{1}{2}bh$$$$24 = \frac{1}{2}(8)h$$$$24 = 4h$$$$6 = h$$The sail's height is 6 ft.
The sum of the measures of the three angles of any triangle is 180°.	In a triangle, the first angle measures 3 times the second and the third measures 40° less than the second. Find each angle's measure.$$\text{Second angle} = x$$$$\text{First angle} = 3x$$$$\text{Third angle} = x - 40$$Sum of measures is 180°.$$x + 3x + (x - 40) = 180$$$$5x - 40 = 180$$$$5x = 220$$$$x = 44$$The angles measure $x = 44$, $3x = 3 \cdot 44 = 132$, and $x - 40 = 44 - 40 = 4$.The angles measure 44°, 132°, and 4°.
Two complementary angles have measures whose sum is 90°. Two supplementary angles have measures whose sum is 180°. If an angle measures x, its complement measures $90 - x$, and its supplement measures $180 - x$.	An angle measures five times its complement. Find the angle's measure.$$x = \text{angle's measure}$$$$90 - x = \text{measure of complement}$$$$x = 5(90 - x)$$$$x = 450 - 5x$$$$6x = 450$$$$x = 75$$The angle measures 75°.

CHAPTER 3   REVIEW EXERCISES

3.1

1. You invested $10,000 in two stocks paying 8% and 10% annual interest, respectively. At the end of the year, the total interest from these investments was $940. How much was invested at each rate?

2. You invested money in two stocks that paid 10% and 12% annual interest. The amount invested at 12% exceeded the amount invested at 10% by $6000. At the end of the year, the total interest from these investments was $2480. How much was invested at each rate?

3. A chemist needs to mix a 75% saltwater solution with a 50% saltwater solution to obtain 10 gallons of a 60% saltwater solution. How many gallons of each of the solutions must be used?

4. Is it possible to mix an alloy that is 30% copper with one that is 50% copper to obtain 50 ounces of an alloy that is 70% copper? If it is possible, how many ounces of each kind of alloy must be used?

5. A university is merging two departments into one. At the north campus, 5% of the students in the department are men. At the south campus, 80% of the students in the department are men. After the merger, 50% of the 150 students in the department are men. How many students were in the department at the north and south campuses before the merger?

6. Two trains start from the same place at the same time. They travel in opposite directions. One averages 60 miles per hour and the other averages 80 miles per hour. After how long will they be 420 miles apart?

7. Two buses leave a station at the same time, traveling in opposite directions. The rate of the faster bus exceeds that of the slower bus by 10 miles per hour. After 3 hours, they are 210 miles apart. What are the rates of the buses?

8. A wallet contains $5 bills and $10 bills. There are 15 bills in the wallet with a total value of $120. Determine the number of $5 bills and the number of $10 bills in the wallet.

In Exercises 9–11, determine if there is **exactly enough information** *or* **not enough information** *to solve each problem. It is not necessary actually to solve any of these problems.*

9. You invested $10,000 in stocks paying 6% annual interest and the remainder of your savings in bonds paying 8% annual interest. At the end of the year, the total interest from these investments was $730. How much was invested at 8%?

10. A chemist needs a 15% benzene solution. The stockroom has only 12% and 20% solutions. How much of each of these solutions must be used?

11. Downtown Books is having a sale on hardcover and softcover books. All hardcover books are priced at $5 per book and all softcover books at $2 per book. The store sold 30 more softcover than hardcover books and had total sales of $570. How many books of each type did the store sell?

3.2

12. In a class, there are 15 men and 10 women. Find the ratio of the number of women to the number of students in the class. First express the ratio as a fraction reduced to lowest terms. Then rewrite the ratio using a second method.

The circle graph indicates the average number of hours Americans sleep each night. Use the graph to find each of the ratios in Exercises 13–14. First express the ratio as a fraction reduced to lowest terms. Then rewrite the ratio using a second method.

**Average Number of Hours
Americans Sleep Each Night**

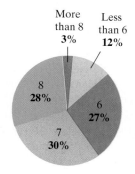

Source: American Medical Association

13. The percentage of Americans who sleep 7 hours per night to the percentage of Americans who sleep 8 hours per night.

14. The percentage of Americans who sleep less than 6 hours per night to the percentage of Americans who sleep more than 8 hours per night.

Solve each proportion in Exercises 15–18.

15. $\dfrac{3}{x} = \dfrac{15}{25}$

16. $\dfrac{-7}{5} = \dfrac{91}{x}$

17. $\dfrac{x+2}{3} = \dfrac{4}{5}$

18. $\dfrac{5}{x+7} = \dfrac{3}{x+3}$

Use a proportion to solve each problem in Exercises 19–21.

19. If a school board determines that there should be 3 teachers for every 50 students, how many teachers are needed for an enrollment of 5400 students?

20. To determine the number of trout in a lake, a conservationist catches 112 trout, tags them, and returns them to the lake. Later, 82 trout are caught, and 32 of them are found to be tagged. How many trout are in the lake?

21. The owners of Skaters Now determine that the monthly sales of their skates is proportional to their advertising budget. When $60,000 is spent on advertising, the company sells 12,000 skates in a month. What will be the monthly sales if they spend $96,000 on advertising?

3.3 *Use a formula for area to find the area of each figure in Exercises 22–24.*

22.

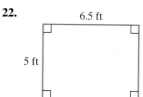

23.

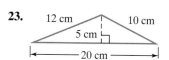

24.

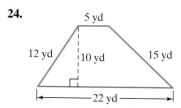

25. Find the circumference and the area of a circle with a diameter of 20 meters. Round answers to the nearest whole number.

26. A sailboat has a triangular sail with an area of 42 square feet and a base that measures 14 feet. Find the height of the sail.

27. A rectangular kitchen floor measures 12 feet by 15 feet. A stove on the floor has a rectangular base measuring 3 feet by 4 feet, and a refrigerator covers a rectangular area of the floor measuring 3 feet by 4 feet. How many square feet of tile will be needed to cover the kitchen floor not counting the area used by the stove and the refrigerator?

28. A yard that is to be covered with mats of grass is shaped like a trapezoid. The bases are 80 feet and 100 feet, and the height is 60 feet. What is the cost of putting the grass mats on the yard if the landscaper charges $0.35 per square foot?

29. Which one of the following is a better buy: a medium pizza with a 14-inch diameter for $6.00 or two small pizzas, each with an 8-inch diameter, for $6.00?

Use a formula for volume to find the volume of each figure in Exercises 30–32. Where applicable, express answers in terms of π. Then round to the nearest whole number.

30.

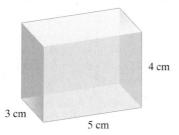

4 cm

3 cm

5 cm

31.

8 yd

4 yd

32.

6 m

33. A train is being loaded with freight containers. Each box is 8 meters long, 4 meters wide, and 3 meters high. If there are 50 freight containers, how much space is needed?

34. A cylindrical fish tank has a diameter of 6 feet and a height of 3 feet. How many tropical fish can be put in the tank if each fish needs 5 cubic feet of water?

35. Find the measure of each angle of the triangle shown in the figure.

C

3*x*

2*x*   *x*

A   *B*

36. In a triangle, the measure of the first angle is 15° more than twice the measure of the second angle. The measure of the third angle exceeds that of the second angle by 25°. What is the measure of each angle?

37. Find the measure of the complement of a 57° angle.

38. Find the measure of the supplement of a 75° angle.

39. How many degrees are there in an angle that measures 25° more than the measure of its complement?

40. The measure of the supplement of an angle is 45° less than four times the measure of the angle. Find the measure of the angle and its supplement.

CHAPTER 3   TEST

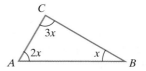

Remember to use your Chapter Test Prep Video CD to see the worked-out solutions to the test questions you want to review.

1. You invested $6000 in two stocks paying 9% and 6% annual interest, respectively. At the end of the year, the total interest from these investments was $480. How much was invested at each rate?

2. A chemist needs to mix a 50% acid solution with an 80% acid solution to obtain 100 milliliters of a 68% acid solution. How many milliliters of each of the solutions must be used?

3. Two cars that are 400 miles apart are traveling directly toward each other on the same road. One is averaging 45 miles per hour and the other averages 35 miles per hour. After how long will they meet?

4. In a class, there are 20 men and 15 women. Find the ratio of the number of men to the number of students in the class.

First express the ratio as a fraction reduced to lowest terms. Then rewrite the ratio using a second method.

5. Solve the proportion: $\dfrac{-5}{8} = \dfrac{x}{12}$.

6. Park rangers catch, tag, and release 200 tule elk back into a wildlife refuge. Two weeks later they observe a sample of 150 elk, of which 5 are tagged. Assuming that the ratio of tagged elk in the sample holds for all elk in the refuge, how many elk are there in the park?

7. The pressure of water on an object below the surface is proportional to its distance below the surface. If a submarine experiences a pressure of 25 pounds per square inch 60 feet below the surface, how much pressure will it experience 330 feet below the surface?

In Exercises 8–9, find the area of each figure.

8.

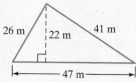

9.

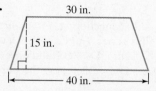

In Exercises 10–11, find the volume of each figure. Where applicable, express answers in terms of π. Then round to the nearest whole number.

10.

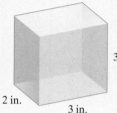

11.

12. What will it cost to cover a rectangular floor measuring 40 feet by 50 feet with square tiles that measure 2 feet on each side if a package of 10 tiles costs $13 per package?

13. A sailboat has a triangular sail with an area of 56 square feet and a base that measures 8 feet. Find the height of the sail.

14. In a triangle, the measure of the first angle is three times that of the second angle. The measure of the third angle is $30°$ less than the measure of the second angle. What is the measure of each angle?

15. How many degrees are there in an angle that measures $16°$ more than the measure of its complement?

CUMULATIVE REVIEW EXERCISES (CHAPTERS 1–3)

1. Perform the indicated operations:
$$\frac{-9(3-6)}{(-12)(3)+(-3-5)(8-4)}.$$

2. Simplify:
$$8 - 3[2(x-1)+5].$$

3. List all the integers in this set:
$$\left\{-3, -\frac{1}{2}, 0, \sqrt{3}, \sqrt{9}\right\}.$$

4. Plot $(4, -2)$ in a rectangular coordinate system. In which quadrant does the point lie?

5. Evaluate $x^2 - 10x$ for $x = -3$.

6. Insert either $<$ or $>$ in the shaded area to make a true statement:
$$-2000 \ \blacksquare \ -3.$$

7. On February 8, the temperature in Manhattan at 10 P.M. was $-4°F$. By 3 A.M. the next day, the temperature had fallen $11°$, but by noon the temperature increased by $21°$. What was the temperature at noon?

In Exercises 8–9, solve each equation.

8. $10(2x - 1) = 8(2x + 1) + 14$

9. $\dfrac{x}{5} + \dfrac{2x}{3} = x + \dfrac{1}{15}$

10. Solve for m: $A = \dfrac{m + n}{2}$.

11. The amount of money owed by doctors graduating from medical school is on the rise. The formula
$$D = 4x + 30$$
models the average debt, D, in thousands of dollars, of indebted medical-school graduates x years after 1985. Use the formula to determine in which year this debt will reach $150 thousand.

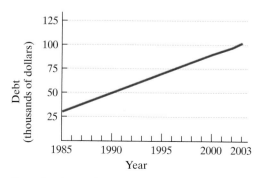

Average Debt of Indebted Medical School Graduates

Source: American Medical Association

12. 144 is 60% of what?

13. The length of a rectangular football field is 14 meters more than twice the width. If the perimeter is 346 meters, find the field's dimensions.

14. After a 10% weight loss, a person weighed 180 pounds. What was the weight before the loss?

In Exercises 15–16, solve each inequality. Express the solution set in set-builder notation and graph the set on a number line.

15. $5x - 5 \leq -5$

16. $-5x + 9 > -2x + 6$

17. Two runners start at the same point and run in opposite directions. One runs at 6 miles per hour and the other runs at 8 miles per hour. In how many hours will they be 21 miles apart?

18. Solve the proportion: $\dfrac{-5}{20} = \dfrac{x}{21}$.

19. At the time they took office, Ronald Reagan and James Buchanan were among the oldest U.S. presidents. Reagan was 4 years older than Buchanan. The sum of their ages was 134. Determine Reagan's age and Buchanan's age at the time each man took office.

20. John Tesh, while he was still coanchoring *Entertainment Tonight*, reported that the PBS series *The Civil War* had an audience of 13% versus the usual 4% PBS audience, "an increase of more than 300%." Did Tesh report the percent increase correctly?

You are in Yosemite National Park in California, surrounded by evergreen forests, alpine meadows, and sheer walls of granite. The beauty of soaring cliffs, plunging waterfalls, gigantic trees, rugged canyons, mountains, and valleys is overwhelming. This is so different from where you live and attend college, a region in which grasslands predominate.

This discussion is developed algebraically in Example 4 in Section 4.6.

Linear Equations and Inequalities in Two Variables

Most things in life depend on many variables. Temperature and precipitation are two variables that affect whether regions are forests, grasslands, or deserts. In this chapter, you will learn graphing and modeling methods for situations involving two variables. With these methods, you will be able to create formulas that model the data of your world.

Objectives

1 Determine whether an ordered pair is a solution of an equation.

2 Find solutions of an equation in two variables.

3 Use point plotting to graph linear equations.

4 Use point plotting to graph other kinds of equations.

5 Use graphs of linear equations to solve problems.

GRAPHING EQUATIONS IN TWO VARIABLES

A picture, as they say, is worth a thousand words. Have you seen pictures of gas-guzzling cars from the 1950s, with their huge fins and overstated designs? The worst year for automobile fuel efficiency was 1958, when U.S. cars averaged a dismal 12.4 miles per gallon.

There is a formula that models fuel efficiency of U.S. cars over time. The formula is

$$y = 0.0075x^2 - 0.2672x + 14.8.$$

The variable x represents the number of years after 1940. The variable y represents the average number of miles per gallon for U.S. automobiles. Looking at the formula does not make it obvious that 1958, 18 years after 1940, was the worst year for fuel efficiency. However, if we could somehow make a picture of the formula, such as the one shown in Figure 4.1, the lowest point on the picture would reveal approximately 1958 as the year in which gas-guzzling cars averaged less than 13 miles per gallon. The shape of the graph also shows decreasing fuel efficiency from 1940 through 1958 and increasing fuel efficiency after 1958. In this chapter, we will be making pictures of equations. We can use these pictures to visualize the behavior of the variables in the equation.

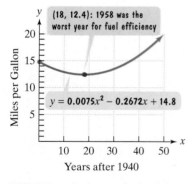

FIGURE 4.1 A picture of a formula

Solutions of Equations The rectangular coordinate system allows us to visualize relationships between two variables by connecting any equation in two variables with a geometric figure. Consider, for example, the following equation in two variables:

$$x + y = 10.$$

> The sum of two numbers, x and y, is 10.

Many pairs of numbers fit the description in the voice balloon, such as $x = 1$ and $y = 9$, or $x = 3$ and $y = 7$. The phrase "$x = 1$ and $y = 9$" is abbreviated using the ordered pair $(1, 9)$. Similarly, the phrase "$x = 3$ and $y = 7$" is abbreviated using the ordered pair $(3, 7)$.

A **solution of an equation in two variables**, x and y, is an ordered pair of real numbers with the following property: When the x-coordinate is substituted for x and the y-coordinate is substituted for y in the equation, we obtain a true statement. For example, $(1, 9)$ is a solution of the equation $x + y = 10$. When 1 is substituted for x and 9 is substituted for y, we obtain the true statement $1 + 9 = 10$, or $10 = 10$. Because there

1 Determine whether an ordered pair is a solution of an equation.

are infinitely many pairs of numbers that have a sum of 10, the equation $x + y = 10$ has infinitely many solutions. Each ordered-pair solution is said to **satisfy** the equation. Thus, $(1, 9)$ satisfies the equation $x + y = 10$.

EXAMPLE 1 Deciding Whether an Ordered Pair Satisfies an Equation

Determine whether each ordered pair is a solution of the equation

$$x - 4y = 14:$$

a. $(2, -3)$ **b.** $(12, 1)$.

SOLUTION

a. To determine whether $(2, -3)$ is a solution of the equation, we substitute 2 for x and -3 for y.

$x - 4y = 14$	This is the given equation.
$2 - 4(-3) \stackrel{?}{=} 14$	Substitute 2 for x and -3 for y.
$2 - (-12) \stackrel{?}{=} 14$	Multiply: $4(-3) = -12$.
$14 = 14$	Subtract: $2 - (-12) = 2 + 12 = 14$.

This statement is true.

Because we obtain a true statement, we conclude that $(2, -3)$ is a solution of the equation $x - 4y = 14$. Thus, $(2, -3)$ satisfies the equation.

b. To determine whether $(12, 1)$ is a solution of the equation, we substitute 12 for x and 1 for y.

$x - 4y = 14$	This is the given equation.
$12 - 4(1) \stackrel{?}{=} 14$	Substitute 12 for x and 1 for y.
$12 - 4 \stackrel{?}{=} 14$	Multiply: $4(1) = 4$.
$8 = 14$	Subtract: $12 - 4 = 8$.

This statement is false.

Because we obtain a false statement, we conclude that $(12, 1)$ is not a solution of $x - 4y = 14$. The ordered pair $(12, 1)$ does not satisfy the equation. ■

✓ **CHECK POINT 1** Determine whether each ordered pair is a solution of the equation $x - 3y = 9$:

a. $(3, -2)$ **b.** $(-2, 3)$.

In this chapter, we will use x and y to represent the variables of an equation in two variables. However, any two letters can be used. Solutions are still ordered pairs. The first number in an ordered pair usually replaces the variable that occurs first alphabetically. The second number in an ordered pair usually replaces the variable that occurs last alphabetically.

How do we find ordered pairs that are solutions of an equation in two variables, x and y?

2 Find solutions of an equation in two variables.

- Select a value for one of the variables.
- Substitute that value into the equation and find the corresponding value of the other variable.
- Use the values of the two variables to form an ordered pair (x, y). This pair is a solution of the equation.

EXAMPLE 2 Finding Solutions of an Equation

Find five solutions of
$$y = 2x - 1.$$

Select integers for x, starting with -2 and ending with 2.

SOLUTION We organize the process of finding solutions in the following table of values.

Start with these values of x.	Substitute x into $y = 2x - 1$ and compute y.	Use values for x and y to form an ordered-pair solution.

Any numbers can be selected for x. There is nothing special about integers from -2 to 2, inclusive. We chose these values to include two negative numbers, 0, and two positive numbers. We also wanted to keep the resulting computations for y relatively simple.

x	$y = 2x - 1$	(x, y)
-2	$y = 2(-2) - 1 = -4 - 1 = -5$	$(-2, -5)$
-1	$y = 2(-1) - 1 = -2 - 1 = -3$	$(-1, -3)$
0	$y = 2 \cdot 0 - 1 = 0 - 1 = -1$	$(0, -1)$
1	$y = 2 \cdot 1 - 1 = 2 - 1 = 1$	$(1, 1)$
2	$y = 2 \cdot 2 - 1 = 4 - 1 = 3$	$(2, 3)$

Look at the ordered pairs in the last column. Five solutions of $y = 2x - 1$ are $(-2, -5)$, $(-1, -3)$, $(0, -1)$, $(1, 1)$, and $(2, 3)$. ■

 CHECK POINT 2 Find five solutions of $y = 3x + 2$. Select integers for x, starting with -2 and ending with 2.

---

3 Use point plotting to graph linear equations.

Graphing Linear Equations in the Form $y = mx + b$ In Example 2, we found five solutions of $y = 2x - 1$. We can generate as many ordered-pair solutions as desired to $y = 2x - 1$ by substituting numbers for x and then finding the corresponding values for y. The **graph of the equation** is the set of all points whose coordinates satisfy the equation.

One method for graphing an equation such as $y = 2x - 1$ is the **point-plotting method**.

> **THE POINT-PLOTTING METHOD FOR GRAPHING AN EQUATION IN TWO VARIABLES**
>
> 1. Find several ordered pairs that are solutions of the equation.
> 2. Plot these ordered pairs as points in the rectangular coordinate system.
> 3. Connect the points with a smooth curve or line.

EXAMPLE 3 Graphing an Equation Using the Point-Plotting Method

Graph the equation: $y = 3x$.

SOLUTION

Step 1. **Find several ordered pairs that are solutions of the equation.** Because there are infinitely many solutions, we cannot list them all. To find some solutions of the equation, we select integers for x, starting with -2 and ending with 2.

Start with these values of x.	Substitute x into $y = 3x$ and compute y.	These are some solutions of $y = 3x$.

x	$y = 3x$	(x, y)
-2	$y = 3(-2) = -6$	$(-2, -6)$
-1	$y = 3(-1) = -3$	$(-1, -3)$
0	$y = 3 \cdot 0 = 0$	$(0, 0)$
1	$y = 3 \cdot 1 = 3$	$(1, 3)$
2	$y = 3 \cdot 2 = 6$	$(2, 6)$

Step 2. **Plot these ordered pairs as points in the rectangular coordinate system.** The five ordered pairs in the table of values are plotted in Figure 4.2(a).

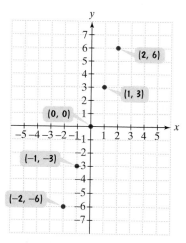

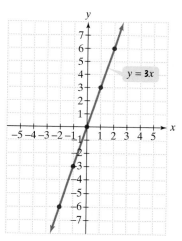

FIGURE 4.2 (a) Some solutions of $y = 3x$ plotted as points

FIGURE 4.2 (b) The graph of $y = 3x$

Step 3. **Connect the points with a smooth curve or line.** The points lie along a straight line. The graph of $y = 3x$ is shown in Figure 4.2(b). The arrows on both ends of the line indicate that it extends indefinitely in both directions. ∎

 CHECK POINT 3 Graph the equation: $y = 2x$.

Equations like $y = 3x$ and $y = 2x$ are called **linear equations in two variables** because the graph of each equation is a line. Any equation that can be written in the form $y = mx + b$, where m and b are constants, is a linear equation in two variables. Here are examples of linear equations in two variables:

$$y = 3x \qquad\qquad y = 3x - 2$$

$$\text{or} \quad y = 3x + 0 \qquad \text{or} \quad y = 3x + (-2)$$

This is in the form of $y = mx + b$ with $m = 3$ and $b = 0$.	This is in the form of $y = mx + b$ with $m = 3$ and $b = -2$.

Can you guess how the graph of the linear equation $y = 3x - 2$ compares with the graph of $y = 3x$? In Example 3, we graphed $y = 3x$. Now, let's graph the equation $y = 3x - 2$.

EXAMPLE 4 Graphing a Linear Equation in Two Variables

Graph the equation: $y = 3x - 2$.

SOLUTION

Step 1. **Find several ordered pairs that are solutions of the equation.** To find some solutions, we select integers for x, starting with -2 and ending with 2.

Start with x.   Compute y.   Form the ordered pair (x, y).

x	$y = 3x - 2$	(x, y)
-2	$y = 3(-2) - 2 = -6 - 2 = -8$	$(-2, -8)$
-1	$y = 3(-1) - 2 = -3 - 2 = -5$	$(-1, -5)$
0	$y = 3 \cdot 0 - 2 = 0 - 2 = -2$	$(0, -2)$
1	$y = 3 \cdot 1 - 2 = 3 - 2 = 1$	$(1, 1)$
2	$y = 3 \cdot 2 - 2 = 6 - 2 = 4$	$(2, 4)$

Step 2. **Plot these ordered pairs as points in the rectangular coordinate system.** The five ordered pairs in the table of values are plotted in Figure 4.3(a).

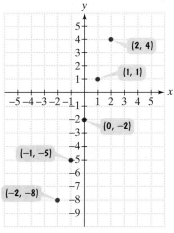

FIGURE 4.3 (a) Some solutions of $y = 3x - 2$ plotted as points

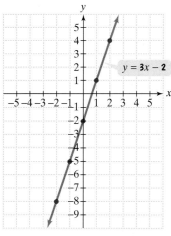

FIGURE 4.3 (b) The graph of $y = 3x - 2$

Step 3. **Connect the points with a smooth curve or line.** The points lie along a straight line. The graph of $y = 3x - 2$ is shown in Figure 4.3(b). ∎

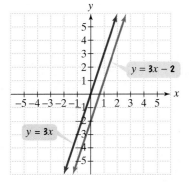

FIGURE 4.4

Now we are ready to compare the graphs of $y = 3x - 2$ and $y = 3x$. The graphs of both linear equations are shown in the same rectangular coordinate system in Figure 4.4. Can you see that the blue graph of $y = 3x - 2$ is parallel to the red graph of $y = 3x$ and shifted 2 units down? Instead of crossing the y-axis at $(0, 0)$, the graph now crosses the y-axis at $(0, -2)$.

COMPARING GRAPHS OF LINEAR EQUATIONS If the value of m does not change,

- The graph of $y = mx + b$ is the graph of $y = mx$ shifted b units up when b is a positive number.
- The graph of $y = mx + b$ is the graph of $y = mx$ shifted b units down when b is a negative number.

 CHECK POINT 4 Graph the equation: $y = 2x - 2$.

EXAMPLE 5 Graphing a Linear Equation in Two Variables

Graph the equation: $y = \dfrac{2}{3}x + 1$.

SOLUTION

Step 1. **Find several ordered pairs that are solutions of the equation.** Notice that m, the coefficient of x, is $\frac{2}{3}$. When m is a fraction, we will select values of x that are multiples of the denominator. In this way, we can avoid values of y that are fractions. Because the denominator of $\frac{2}{3}$ is 3, we select multiples of 3 for x. Let's use $-6, -3, 0, 3,$ and 6.

Start with multiples of 3 for x.   Compute y.   Form the ordered pair (x, y).

x	$y = \dfrac{2}{3}x + 1$	(x, y)
-6	$y = \dfrac{2}{3}(-6) + 1 = -4 + 1 = -3$	$(-6, -3)$
-3	$y = \dfrac{2}{3}(-3) + 1 = -2 + 1 = -1$	$(-3, -1)$
0	$y = \dfrac{2}{3} \cdot 0 + 1 = 0 + 1 = 1$	$(0, 1)$
3	$y = \dfrac{2}{3} \cdot 3 + 1 = 2 + 1 = 3$	$(3, 3)$
6	$y = \dfrac{2}{3} \cdot 6 + 1 = 4 + 1 = 5$	$(6, 5)$

Step 2. **Plot these ordered pairs as points in the rectangular coordinate system.** The five ordered pairs in the table of values are plotted in Figure 4.5.

Step 3. **Connect the points with a smooth curve or line.** The points lie along a straight line. The graph of $y = \frac{2}{3}x + 1$ is shown in Figure 4.5.

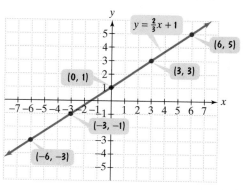

FIGURE 4.5 The graph of $y = \frac{2}{3}x + 1$

✔ CHECK POINT 5 Graph the equation: $y = \frac{1}{2}x + 2$.

4 Use point plotting to graph other kinds of equations.

Graphing Nonlinear Equations in Two Variables Look at the picture of this gymnast. He has created a perfect balance in which the two halves of his body are mirror images of each other. Is it possible for graphs to have mirrorlike qualities? Yes. Although our next graph is not a straight line, we can obtain its cuplike U-shape using the point-plotting method for graphing an equation in two variables.

EXAMPLE 6 Graphing a Nonlinear Equation in Two Variables

Graph the equation: $y = x^2 - 4$.

SOLUTION The given equation involves two variables, x and y. However, because the variable x is squared, it is not a linear equation in two variables.

$$y = x^2 - 4$$

This is not in the form $y = mx + b$ because x is squared.

Although the graph is not a line, it is still a picture of all the ordered-pair solutions of $y = x^2 - 4$. Thus, we can use the point-plotting method to obtain the graph.

Step 1. **Find several ordered pairs that are solutions of the equation.** To find some solutions, we select integers for x, starting with -3 and ending with 3.

Start with x. Compute y. Form the ordered pair (x, y).

x	$y = x^2 - 4$	(x,y)
-3	$y = (-3)^2 - 4 = 9 - 4 = 5$	$(-3, 5)$
-2	$y = (-2)^2 - 4 = 4 - 4 = 0$	$(-2, 0)$
-1	$y = (-1)^2 - 4 = 1 - 4 = -3$	$(-1, -3)$
0	$y = 0^2 - 4 = 0 - 4 = -4$	$(0, -4)$
1	$y = 1^2 - 4 = 1 - 4 = -3$	$(1, -3)$
2	$y = 2^2 - 4 = 4 - 4 = 0$	$(2, 0)$
3	$y = 3^2 - 4 = 9 - 4 = 5$	$(3, 5)$

STUDY TIP

If the graph of an equation is not a straight line, use more solutions than when graphing lines. These extra solutions are needed to get a better general idea of the graph's shape.

Step 2. **Plot these ordered pairs as points in the rectangular coordinate system.** The seven ordered pairs in the table of values are plotted in Figure 4.6(a).

Step 3. **Connect the points with a smooth curve.** The seven points are joined with a smooth curve in Figure 4.6(b). The graph of $y = x^2 - 4$ is a curve where the part of the graph to the right of the y-axis is a reflection of the part to the left of it, and vice versa. The arrows on both ends of the curve indicate that it extends indefinitely in both directions.

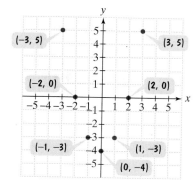

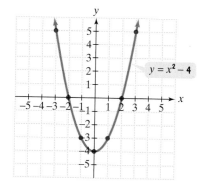

FIGURE 4.6 (a) Some solutions of $y = x^2 - 4$ plotted as points

FIGURE 4.6 (b) The graph of $y = x^2 - 4$

 CHECK POINT 6 Graph the equation: $y = x^2 - 1$. Select integers for x, starting with -3 and ending with 3.

---

5 Use graphs of linear equations to solve problems.

Applications Part of the beauty of the rectangular coordinate system is that it allows us to "see" mathematical formulas and visualize the solution to a problem. This idea is demonstrated in Example 7.

EXAMPLE 7 An Application Using Graphs of Linear Equations

The toll to a bridge costs $2.50. Commuters who use the bridge frequently have the option of purchasing a monthly coupon book for $21.00. With the coupon book, the toll is reduced to $1.00. The monthly cost, y, of using the bridge x times can be described by the following formulas:

Without the coupon book:

$$y = 2.50x$$ The monthly cost, y, is $2.50 times the number of times, x, that the bridge is used.

With the coupon book:

$$y = 21 + 1 \cdot x$$ The monthly cost, y, is $21 for the book plus $1 times the number of times, x, that the bridge is used.

$$y = 21 + x$$

a. Let $x = 0, 2, 4, 10, 12, 14,$ and 16. Make a table of values for each linear equation showing seven solutions for the equation.

b. Graph the equations in the same rectangular coordinate system.

c. What are the coordinates of the intersection point for the two graphs? Interpret the coordinates in practical terms.

SOLUTION

a. Tables of values showing seven solutions for each equation follow.

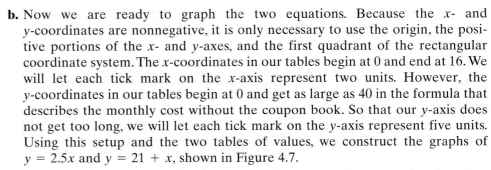

Without the Coupon Book				With the Coupon Book		
x	$y = 2.5x$	(x, y)		x	$y = 21 + x$	(x, y)
0	$y = 2.5(0) = 0$	$(0, 0)$		0	$y = 21 + 0 = 21$	$(0, 21)$
2	$y = 2.5(2) = 5$	$(2, 5)$		2	$y = 21 + 2 = 23$	$(2, 23)$
4	$y = 2.5(4) = 10$	$(4, 10)$		4	$y = 21 + 4 = 25$	$(4, 25)$
10	$y = 2.5(10) = 25$	$(10, 25)$		10	$y = 21 + 10 = 31$	$(10, 31)$
12	$y = 2.5(12) = 30$	$(12, 30)$		12	$y = 21 + 12 = 33$	$(12, 33)$
14	$y = 2.5(14) = 35$	$(14, 35)$		14	$y = 21 + 14 = 35$	$(14, 35)$
16	$y = 2.5(16) = 40$	$(16, 40)$		16	$y = 21 + 16 = 37$	$(16, 37)$

b. Now we are ready to graph the two equations. Because the x- and y-coordinates are nonnegative, it is only necessary to use the origin, the positive portions of the x- and y-axes, and the first quadrant of the rectangular coordinate system. The x-coordinates in our tables begin at 0 and end at 16. We will let each tick mark on the x-axis represent two units. However, the y-coordinates in our tables begin at 0 and get as large as 40 in the formula that describes the monthly cost without the coupon book. So that our y-axis does not get too long, we will let each tick mark on the y-axis represent five units. Using this setup and the two tables of values, we construct the graphs of $y = 2.5x$ and $y = 21 + x$, shown in Figure 4.7.

c. The graphs intersect at $(14, 35)$. This means that if the bridge is used 14 times in a month, the total monthly cost without the coupon book is the same as the total monthly cost with the coupon book, namely $35.

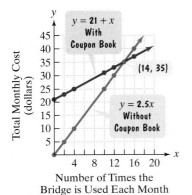

FIGURE 4.7 Options for a toll

In Figure 4.7, look at the two graphs to the right of the intersection point $(14, 35)$. The red graph of $y = 21 + x$ lies below the blue graph of $y = 2.5x$. This means that if the bridge is used more than 14 times in a month ($x > 14$), the (red) monthly cost, y, with the coupon book is cheaper than the (blue) monthly cost, y, without the coupon book.

 CHECK POINT 7 The toll to a bridge costs $2.00. If you use the bridge x times in a month, the monthly cost, y, is $y = 2x$. With a $10 coupon book, the toll is reduced to $1.00. The monthly cost, y, of using the bridge x times in a month with the coupon book is $y = 10 + x$.

a. Let $x = 0, 2, 4, 6, 8, 10$, and 12. Make tables of values showing seven solutions of $y = 2x$ and seven solutions of $y = 10 + x$.

b. Graph the equations in the same rectangular coordinate system.

c. What are the coordinates of the intersection point for the two graphs? Interpret the coordinates in practical terms.

Mathematical Blossom

Graph of an equation in a three-dimensional Cartesian plane

This picture is a graph in a three-dimensional rectangular coordinate system. For many, the picture is more interesting than its equation:

$$z = (|x| - |y|)^2 + \frac{2|xy|}{\sqrt{x^2 + y^2}}.$$

Sometimes it's pleasant to simply get the "feel" of an equation by seeing its picture. Turning equations into visual images hints at the beauty that lies within mathematics.

USING TECHNOLOGY

Graphing calculators or graphing software packages for computers are referred to as **graphing utilities** or graphers. A graphing utility is a powerful tool that quickly generates the graph of an equation in two variables. Figure 4.8 shows two such graphs for the equations in Examples 3 and 4.

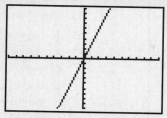

FIGURE 4.8(a) The graph of $y = 3x$

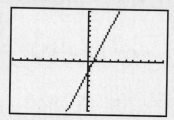

FIGURE 4.8(b) The graph of $y = 3x - 2$

What differences do you notice between these graphs and the graphs that we drew by hand? They do seem a bit "jittery." Arrows do not appear on both ends of the graphs. Furthermore, numbers are not given along the axes. For both graphs in Figure 4.8, the x-axis extends from -10 to 10 and the y-axis also extends from -10 to 10. The distance represented by each consecutive tick mark is one unit. We say that the **viewing rectangle**, or the **viewing window**, is $[-10, 10, 1]$ by $[-10, 10, 1]$.

$[-10,$ $10,$ $1]$ by $[-10,$ $10,$ $1].$

| The minimum x-value along the x-axis is -10. | The maximum x-value along the x-axis is 10. | The scale on the x-axis is 1 unit per tick mark. | The minimum y-value along the y-axis is -10. | The maximum y-value along the y-axis is 10. | The scale on the y-axis is 1 unit per tick mark. |

To graph an equation in x and y using a graphing utility, enter the equation and specify the size of the viewing rectangle. The size of the viewing rectangle sets minimum and maximum values for both the x- and y-axes. Enter these values, as well as the values between consecutive tick marks, on the respective axes. The $[-10, 10, 1]$ by $[-10, 10, 1]$ viewing rectangle used in Figure 4.8 is called the **standard viewing rectangle**.

On most graphing utilities, the display screen is two-thirds as high as it is wide. By using a square setting, you can equally space the x and y tick marks. (This does not occur in the standard viewing rectangle.) Graphing utilities can also *zoom in* and *zoom out*. When you zoom in, you see a smaller portion of the graph, but you do so in greater detail. When you zoom out, you see a larger portion of the graph. Thus, zooming out may help you to develop a better understanding of the overall character of the graph. With practice, you will become more comfortable with graphing equations in two variables using your graphing utility. You will also develop a better sense of the size of the viewing rectangle that will reveal needed information about a particular graph.

4.1 EXERCISE SET

Student Solutions Manual   CD/Video   PH Math/Tutor Center   MathXL Tutorials on CD   MathXL®   MyMathLab   Interactmath.com

Practice Exercises

In Exercises 1–12, determine whether each ordered pair is a solution of the given equation.

1. $y = 3x$ $(2, 3), (3, 2), (-4, -12)$

2. $y = 4x$ $(3, 12), (12, 3), (-5, -20)$

3. $y = -4x$ $(-5, -20), (0, 0), (9, -36)$

4. $y = -3x$ $(-5, 15), (0, 0), (7, -21)$

5. $y = 2x + 6$ $(0, 6), (-3, 0), (2, -2)$

6. $y = 8 - 4x$ $(8, 0), (16, -2), (3, -4)$

7. $3x + 5y = 15$ $(-5, 6), (0, 5), (10, -3)$

8. $2x - 5y = 0$ $(-2, 0), (-10, 6), (5, 0)$

9. $x + 3y = 0$ $(0, 0), \left(1, \frac{1}{3}\right), \left(2, -\frac{2}{3}\right)$

10. $x + 5y = 0$ $(0, 0), \left(1, \frac{1}{5}\right), \left(2, -\frac{2}{5}\right)$

11. $x - 4 = 0$ $(4, 7), (3, 4), (0, -4)$

12. $y + 2 = 0$ $(0, 2), (2, 0), (0, -2)$

In Exercises 13–20, find five solutions of each equation. Select integers for x, starting with −2 and ending with 2. Organize your work in a table of values.

13. $y = 12x$

14. $y = 14x$

15. $y = -10x$

16. $y = -20x$

17. $y = 8x - 5$

18. $y = 6x - 4$

19. $y = -3x + 7$

20. $y = -5x + 9$

In Exercises 21–44, graph each linear equation in two variables. Find at least five solutions in your table of values for each equation.

21. $y = x$

22. $y = x + 1$

23. $y = x - 1$

24. $y = x - 2$

25. $y = 2x + 1$

26. $y = 2x - 1$

27. $y = -x + 2$

28. $y = -x + 3$

29. $y = -3x - 1$

30. $y = -3x - 2$

31. $y = \frac{1}{2}x$

32. $y = -\frac{1}{2}x$

33. $y = -\frac{1}{4}x$

34. $y = \frac{1}{4}x$

35. $y = \frac{1}{3}x + 1$

36. $y = \frac{1}{3}x - 1$

37. $y = -\frac{3}{2}x + 1$

38. $y = -\frac{3}{2}x + 2$

39. $y = -\frac{5}{2}x - 1$

40. $y = -\frac{5}{2}x + 1$

41. $y = x + \frac{1}{2}$

42. $y = x - \frac{1}{2}$

43. $y = 4$, or $y = 0x + 4$

44. $y = 3$, or $y = 0x + 3$

Graph each equation in Exercises 45–50. Find seven solutions in your table of values for each equation by using integers for x, starting with −3 and ending with 3.

45. $y = x^2$

46. $y = x^2 - 2$

47. $y = x^2 + 1$

48. $y = x^2 + 2$

49. $y = 4 - x^2$

50. $y = 9 - x^2$

Practice Plus

In Exercises 51–54, write each sentence as a linear equation in two variables. Then graph the equation.

51. The *y*-variable is 3 more than the *x*-variable.

52. The *y*-variable exceeds the *x*-variable by 4.

53. The *y*-variable exceeds twice the *x*-variable by 5.

54. The *y*-variable is 2 less than 3 times the *x*-variable.

55. At the beginning of a semester, a student purchased eight pens and six pads for a total cost of $14.50.

    a. If *x* represents the cost of one pen and *y* represents the cost of one pad, write an equation in two variables that reflects the given conditions.

    b. If pads cost $0.75 each, find the cost of one pen.

56. A nursery offers a package of three small orange trees and four small grapefruit trees for $22.

    a. If *x* represents the cost of one orange tree and *y* represents the cost of one grapefruit tree, write an equation in two variables that reflects the given conditions.

    b. If a grapefruit tree costs $2.50, find the cost of an orange tree.

Application Exercises

The graph shows the percentage of divorced Americans, by race.

Divorced Americans, by Race

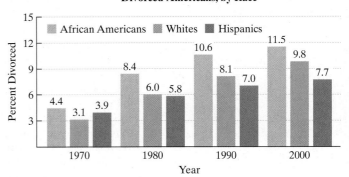

Source: U.S. Census Bureau

The data can be modeled by linear equations in two variables:

African Americans → $y = 0.24x + 5.2$

Whites → $y = 0.22x + 3.4$

Hispanics → $y = 0.13x + 4.2$

In each model, x is the number of years after 1970 and y is the percentage of divorced Americans in the racial group. Use this information to solve Exercises 57–58.

57. a. Find four solutions of the linear equation that models the data for African Americans. Use 0, 10, 20, and 30 for *x*. Organize your work in a table of values. How well does the linear equation model the data shown in the bar graph?

b. Repeat part (a) for the linear equation that models the data for whites.

c. Repeat part (a) for the linear equation that models the data for Hispanics.

58. a. Use the tables of values from Exercise 57 to graph any one of the three linear equations.

b. If you have not yet done so, extend your graph to include $x = 40$. Use the line to predict the percentage of divorced Americans, rounded to the nearest tenth of a percent, in this group in 2010.

59. A rental company charges $40.00 a day plus $0.35 per mile to rent a moving truck. The total cost, y, for a day's rental if x miles are driven is described by $y = 40 + 0.35x$. A second company charges $36.00 a day plus $0.45 per mile, so the daily cost, y, if x miles are driven is described by $y = 36 + 0.45x$. The graphs of the two equations are shown in the same rectangular coordinate system.

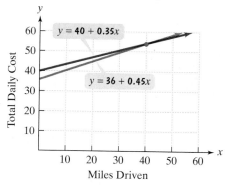

a. What is the x-coordinate of the intersection point of the graphs? Describe what this x-coordinate means in practical terms.

b. What is a reasonable estimate for the y-coordinate of the intersection point?

c. Substitute the x-coordinate of the intersection point into each of the equations and find the corresponding value for y. Describe what this value represents in practical terms. How close is this value to your estimate from part (b)?

60. The linear equation in two variables $y = 166x + 1781$ models the cost, y, in tuition and fees per year, of a four-year public college x years after 1990.

a. Find five solutions of $y = 166x + 1781$. Use 0, 5, 10, 15, and 20 for x. Organize your work in a table of values.

b. Use the solutions in part (a) to graph $y = 166x + 1781$. What does the shape of the graph indicate about the cost of a four-year public college?

61. The linear equation in two variables $y = 50x + 30,000$ models the total weekly cost, y, in dollars, for a business that manufactures x racing bicycles each week. The equation indicates that the business has weekly fixed costs of $30,000 plus a cost of $50 to manufacture each bicycle.

a. Find five solutions of $y = 50x + 30,000$. Use 0, 10, 20, 30, and 40 for x. Organize your work in a table of values.

b. Use the solutions in part (a) to graph $y = 50x + 30,000$.

Writing in Mathematics

62. How do you determine whether an ordered pair is a solution of an equation in two variables, x and y?

63. Explain how to find ordered pairs that are solutions of an equation in two variables, x and y.

64. What is the graph of an equation?

65. Explain how to graph an equation in two variables in the rectangular coordinate system.

Critical Thinking Exercises

66. Which one of the following is true?

a. The graph of $y = 3x + 1$ is parallel to the graph of $y = 2x$, but shifted up 1 unit.

b. The graph of any equation in the form $y = mx + b$ passes through the point $(0, b)$.

c. The ordered pair $(3, 4)$ satisfies the equation

$$2y - 3x = -6.$$

d. If $(2, 5)$ satisfies an equation, then $(5, 2)$ also satisfies the equation.

Graph each equation in Exercises 67–68. Find seven solutions in your table of values by using integers for x, starting with −3 and ending with 3.

67. $y = |x|$

68. $y = |x| + 1$

69. Although the level of air pollution varies from day to day and from hour to hour, during the summer the level of air pollution depends on the time of day. The equation in two variables

$$y = 0.1x^2 - 0.4x + 0.6$$

describes the level of air pollution, in parts per million (ppm), where x corresponds to the number of hours after 9 A.M.

a. Find six solutions of the equation. Select integers for x, starting with 0 and ending with 5.

b. Researchers have determined that a level of 0.3 ppm or more of pollutants in the air can be hazardous to your health. Based on the six solutions in part (a), at what times between 9 A.M. and 2 P.M. should runners exercise to avoid unsafe air?

Technology Exercises

Use a graphing utility to graph each equation in Exercises 70–73 in a standard viewing rectangle. Then use the TRACE *feature to trace along the line and find the coordinates of two points.*

70. $y = 2x - 1$

71. $y = -3x + 2$

72. $y = \frac{1}{2}x$ **73.** $y = \frac{3}{4}x - 2$

74. Use a graphing utility to verify any five of your hand-drawn graphs in Exercises 21–44. Use an appropriate viewing rectangle and the $\boxed{\text{ZOOM SQUARE}}$ feature to make the graph look like the one you drew by hand.

75. The linear equation $y = 2.4x + 180$ models U.S. population, y, in millions, x years after 1960. Use a graphing utility to graph the equation in a $[0, 90, 10]$ by $[0, 500, 100]$ viewing rectangle.

What does the shape of the graph indicate about changing U.S. population over time?

Review Exercises

76. Solve: $3x + 5 = 4(2x - 3) + 7$. (Section 2.3, Example 3)

77. Simplify: $3(1 - 2 \cdot 5) - (-28)$. (Section 1.8, Example 7)

78. Solve for h: $V = \frac{1}{3}Ah$. (Section 2.4, Example 4)

SECTION

4.2

Objectives

1 Use a graph to identify intercepts.

2 Graph a linear equation in two variables using intercepts.

3 Graph horizontal or vertical lines.

GRAPHING LINEAR EQUATIONS USING INTERCEPTS

Despite the nutritional disaster floating within the cans, the international sale of carbonated soft drinks continues to flourish. A can of Coca-Cola is sold every six seconds throughout the world.

The carbonated soft drinks market in the United States appears to be losing its fizz. As consumption of carbonated soft drinks declines, there is an emerging market for alternative beverages, including juices, teas, bottled waters, and sports drinks. The line graphs in Figure 4.9 show the projected U.S. beverage market share for carbonated and noncarbonated alternative beverages through 2010.

The decreasing market share for carbonated beverages can be modeled by

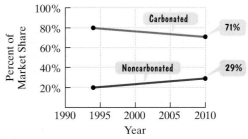

FIGURE 4.9

Source: Beverage Digest

$$y = -0.57x + 80.$$

The variable x represents the number of years after 1994. The variable y represents the market share of carbonated beverages, measured as a percentage of all sales.

There is another way that we can write the equation $y = -0.57x + 80$. We will collect the x- and y-terms on the left side. This is done by adding $0.57x$ to both sides:

$$0.57x + y = 80.$$

The form of this equation is $Ax + By = C$.

$$0.57x \ + \ y \ = \ 80$$

A, the coefficient of x, is **0.57**. B, the coefficient of y, is **1**. C, the constant on the right, is **80**.

All equations of the form $Ax + By = C$ are straight lines when graphed as long as A and B are not both zero. To graph linear equations of this form, we will use two important points: the *intercepts*.

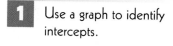
Use a graph to identify intercepts.

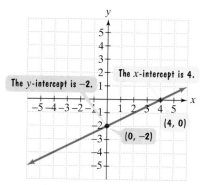

FIGURE 4.10 The graph of $2x - 4y = 8$

Intercepts An *x-intercept* of a graph is the x-coordinate of a point where the graph intersects the x-axis. For example, look at the graph of $2x - 4y = 8$ in Figure 4.10. The graph crosses the x-axis at $(4, 0)$. Thus, the x-intercept is 4. **The y-coordinate corresponding to an x-intercept is always zero.**

A *y-intercept* of a graph is the y-coordinate of a point where the graph intersects the y-axis. The graph of $2x - 4y = 8$ in Figure 4.10 shows that the graph crosses the y-axis at $(0, -2)$. Thus, the y-intercept is -2. **The x-coordinate corresponding to a y-intercept is always zero.**

STUDY TIP

Mathematicians tend to use two ways to describe intercepts. Did you notice that we are using single numbers? If a graph's x-intercept is a, it passes through the point $(a, 0)$. If a graph's y-intercept is b, it passes through the point $(0, b)$.

Some books state that the x-intercept is the *point* $(a, 0)$ and the x-intercept is *at a* on the x-axis. Similarly, the y-intercept is the *point* $(0, b)$ and the y-intercept is *at b* on the y-axis. In these descriptions, the intercepts are the actual points where a graph crosses the axes.

Although we'll describe intercepts as single numbers, we'll immediately state the point on the x- or y-axis that the graph passes through. Here's the important thing to keep in mind:

x-intercept: The corresponding y is 0.

y-intercept: The corresponding x is 0.

EXAMPLE 1 Identifying Intercepts

Identify the x- and y-intercepts.

a.

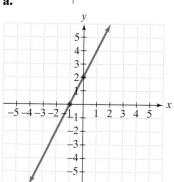

b.

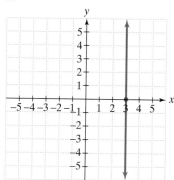

c.

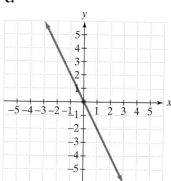

SOLUTION

a. The graph crosses the x-axis at $(-1, 0)$. Thus, the x-intercept is -1. The graph crosses the y-axis at $(0, 2)$. Thus, the y-intercept is 2.

b. The graph crosses the x-axis at $(3, 0)$, so the x-intercept is 3. This vertical line does not cross the y-axis. Thus, there is no y-intercept.

c. This graph crosses the x- and y-axes at the same point, the origin. Because the graph crosses both axes at $(0, 0)$, the x-intercept is 0 and the y-intercept is 0. ■

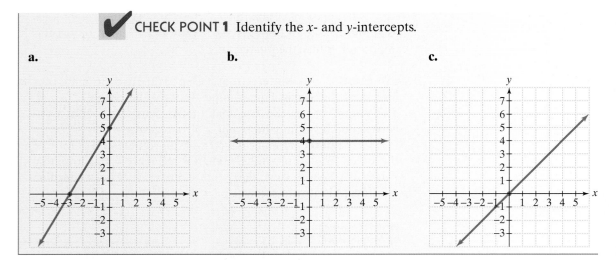

✓ **CHECK POINT 1** Identify the *x*- and *y*-intercepts.

a. **b.** **c.**

---

2 Graph a linear equation in two variables using intercepts.

Graphing Using Intercepts An equation of the form $Ax + By = C$, where A, B, and C are integers, is called the **standard form** of the equation of a line. The equation can be graphed by finding the *x*- and *y*-intercepts, plotting the intercepts, and drawing a straight line through these points. How do we find the intercepts of a line, given its equation? Because the *y*-coordinate of the *x*-intercept is 0, to find the *x*-intercept,

- Substitute 0 for *y* in the equation.
- Solve for *x*.

EXAMPLE 2 Finding the *x*-Intercept

Find the *x*-intercept of the graph of $3x - 4y = 24$.

SOLUTION To find the *x*-intercept, let $y = 0$ and solve for *x*.

$$3x - 4y = 24 \qquad \text{This is the given equation.}$$
$$3x - 4 \cdot 0 = 24 \qquad \text{Let } y = 0.$$
$$3x = 24 \qquad \text{Simplify: } 4 \cdot 0 = 0 \text{ and } 3x - 0 = 3x.$$
$$x = 8 \qquad \text{Divide both sides by 3.}$$

The *x*-intercept is 8. The graph of $3x - 4y = 24$ passes through the point $(8, 0)$. ∎

✓ **CHECK POINT 2** Find the *x*-intercept of the graph of $4x - 3y = 12$.

Because the *x*-coordinate of the *y*-intercept is 0, to find the *y*-intercept,

- Substitute 0 for *x* in the equation.
- Solve for *y*.

EXAMPLE 3 Finding the *y*-Intercept

Find the *y*-intercept of the graph of $3x - 4y = 24$.

SOLUTION To find the *y*-intercept, let $x = 0$ and solve for *y*.

$$3x - 4y = 24 \qquad \text{This is the given equation.}$$
$$3 \cdot 0 - 4y = 24 \qquad \text{Let } x = 0.$$
$$-4y = 24 \qquad \text{Simplify: } 3 \cdot 0 = 0 \text{ and } 0 - 4y = -4y.$$
$$y = -6 \qquad \text{Divide both sides by } -4.$$

The y-intercept is -6. The graph of $3x - 4y = 24$ passes through the point $(0, -6)$. ■

✔ **CHECK POINT 3** Find the y-intercept of the graph of $4x - 3y = 12$.

When graphing using intercepts, it is a good idea to use a third point, a check-point, before drawing the line. A checkpoint can be obtained by selecting a value for either variable, other than 0, and finding the corresponding value for the other variable. The checkpoint should lie on the same line as the x- and y-intercepts. If it does not, recheck your work and find the error.

> **USING INTERCEPTS TO GRAPH $Ax + By = C$**
>
> 1. Find the x-intercept. Let $y = 0$ and solve for x.
> 2. Find the y-intercept. Let $x = 0$ and solve for y.
> 3. Find a checkpoint, a third ordered-pair solution.
> 4. Graph the equation by drawing a line through the three points.

EXAMPLE 4 Using Intercepts to Graph a Linear Equation

Graph: $3x + 2y = 6$.

SOLUTION

Step 1. Find the x-intercept. Let $y = 0$ and solve for x.

$$3x + 2 \cdot 0 = 6$$
$$3x = 6$$
$$x = 2$$

The x-intercept is 2, so the line passes through $(2, 0)$.

Step 2. Find the y-intercept. Let $x = 0$ and solve for y.

$$3 \cdot 0 + 2y = 6$$
$$2y = 6$$
$$y = 3$$

The y-intercept is 3, so the line passes through $(0, 3)$.

Step 3. Find a checkpoint, a third ordered-pair solution. For our checkpoint, we will let $x = 1$ and find the corresponding value for y.

$3x + 2y = 6$	This is the given equation.
$3 \cdot 1 + 2y = 6$	Substitute 1 for x.
$3 + 2y = 6$	Simplify.
$2y = 3$	Subtract 3 from both sides.
$y = \dfrac{3}{2}$	Divide both sides by 2.

The checkpoint is the ordered pair $\left(1, \dfrac{3}{2}\right)$, or $(1, 1.5)$.

Step 4. Graph the equation by drawing a line through the three points. The three points in Figure 4.11 lie along the same line. Drawing a line through the three points results in the graph of $3x + 2y = 6$. ■

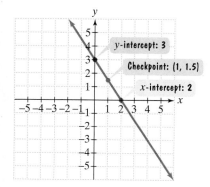

FIGURE 4.11 The graph of $3x + 2y = 6$

USING TECHNOLOGY

You can use a graphing utility to graph equations of the form $Ax + By = C$. Begin by solving the equation for y. For example, to graph $3x + 2y = 6$, solve the equation for y.

$$3x + 2y = 6 \qquad \text{This is the equation to be graphed.}$$
$$3x - 3x + 2y = -3x + 6 \qquad \text{Subtract 3x from both sides.}$$
$$2y = -3x + 6 \qquad \text{Simplify.}$$
$$\frac{2y}{2} = \frac{-3x + 6}{2} \qquad \text{Divide both sides by 2.}$$
$$y = -\frac{3}{2}x + 3 \qquad \text{Simplify.}$$

This is the equation to enter into your graphing utility. The graph of $y = -\frac{3}{2}x + 3$ or, equivalently, $3x + 2y = 6$, is shown below in a $[-6, 6, 1]$ by $[-6, 6, 1]$ viewing rectangle.

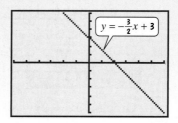

$$y = -\frac{3}{2}x + 3$$

✔ **CHECK POINT 4** Graph: $2x + 3y = 6$.

EXAMPLE 5 Using Intercepts to Graph a Linear Equation

Graph: $2x - y = 4$.

SOLUTION

Step 1. Find the x-intercept. Let $y = 0$ and solve for x.

$$2x - 0 = 4$$
$$2x = 4$$
$$x = 2$$

The x-intercept is 2, so the line passes through $(2, 0)$.

Step 2. Find the y-intercept. Let $x = 0$ and solve for y.

$$2 \cdot 0 - y = 4$$
$$-y = 4$$
$$y = -4$$

The y-intercept is -4, so the line passes through $(0, -4)$.

Step 3. Find a checkpoint, a third ordered-pair solution. For our checkpoint, we will let $x = 1$ and find the corresponding value for y.

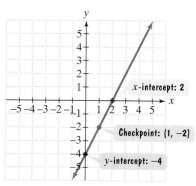

FIGURE 4.12   The graph of $2x - y = 4$

$$2x - y = 4 \qquad \text{This is the given equation.}$$
$$2 \cdot 1 - y = 4 \qquad \text{Substitute 1 for } x.$$
$$2 - y = 4 \qquad \text{Simplify.}$$
$$-y = 2 \qquad \text{Subtract 2 from both sides.}$$
$$y = -2 \qquad \text{Multiply (or divide) both sides by } -1.$$

The checkpoint is $(1, -2)$.

Step 4.   Graph the equation by drawing a line through the three points. The three points in Figure 4.12 lie along the same line. Drawing a line through the three points results in the graph of $2x - y = 4$. ∎

✔ **CHECK POINT 5** Graph: $x - 2y = 4$.

We have seen that not all lines have two different intercepts. Some lines pass through the origin. Thus, they have an x-intercept of 0 and a y-intercept of 0. Is it possible to recognize these lines by their equations? Yes. **The graph of the linear equation $Ax + By = 0$ passes through the origin**. Notice that the constant on the right side of this equation is 0.

An equation of the form $Ax + By = 0$ can be graphed by using the origin as one point on the line. Find two other points by finding two other solutions of the equation. Select values for either variable, other than 0, and find the corresponding values for the other variable.

EXAMPLE 6   Graphing a Linear Equation of the Form $Ax + By = 0$

Graph: $x + 2y = 0$.

SOLUTION   Because the constant on the right is 0, the graph passes through the origin. The x- and y-intercepts are both 0. Remember that we are using two points and a checkpoint to determine a line. Thus, we still want to find two other points. Let $y = -1$ to find a second ordered-pair solution. Let $y = 1$ to find a third ordered-pair (checkpoint) solution.

$x + 2y = 0$	$x + 2y = 0$
Let $y = -1$.	Let $y = 1$.
$x + 2(-1) = 0$	$x + 2 \cdot 1 = 0$
$x + (-2) = 0$	$x + 2 = 0$
$x = 2$	$x = -2$

The solutions are $(2, -1)$ and $(-2, 1)$. Plot these two points, as well as the origin—that is, $(0, 0)$. The three points in Figure 4.13 lie along the same line. Drawing a line through the three points results in the graph of $x + 2y = 0$. ∎

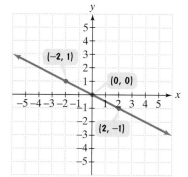

FIGURE 4.13   The graph of $x + 2y = 0$

✔ **CHECK POINT 6** Graph: $x + 3y = 0$.

3 Graph horizontal or vertical lines.

Equations of Horizontal and Vertical Lines Some things change very little. For example, from 1985 to the present, the number of Americans participating in downhill skiing has remained relatively constant, indicated by the graph shown in Figure 4.14. Shown in the figure is a horizontal line that passes through or near most of the data points.

We can use the horizontal line in Figure 4.14 to write an equation that reasonably models the data. The y-intercept of the line is 15, so the graph passes through $(0, 15)$. Furthermore, all points on the line have a value of y that is always 15. Thus, an equation that models the number of participants in downhill skiing for the period shown is

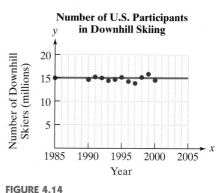

FIGURE 4.14

Source: National Ski Areas Association

$$y = 15.$$

The popularity of downhill skiing has remained relatively constant in the United States at approximately 15 million participants each year.

The equation $y = 15$ can be expressed as $0x + 1y = 15$. We know that the graph of any equation of the form $Ax + By = C$ is a line as long as A and B are not both zero. The graph of $y = 15$ suggests that when A is zero, the line is horizontal.

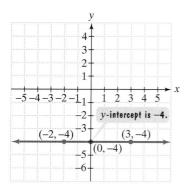

FIGURE 4.15 The graph of $y = -4$

EXAMPLE 7 Graphing a Horizontal Line

Graph the linear equation: $y = -4$.

SOLUTION All ordered pairs that are solutions of $y = -4$ have a value of y that is always -4. Any value can be used for x. Let's select three of the possible values for x: $-2, 0,$ and 3. Using these values of x, three ordered pairs that are solutions of $y = -4$ are $(-2, -4), (0, -4),$ and $(3, -4)$. Plot each of these points. Drawing a line that passes through the three points gives the horizontal line shown in Figure 4.15. ∎

✔ **CHECK POINT 7** Graph the linear equation: $y = 3$.

Next, let's see what we can discover about the graph of an equation of the form $Ax + By = C$ when B is zero.

EXAMPLE 8 Graphing a Vertical Line

Graph the linear equation: $x = 5$.

SOLUTION All ordered pairs that are solutions of $x = 5$ have a value of x that is always 5. Any value can be used for y. Let's select three of the possible values for y: $-2, 0,$ and 3. Using these values of y, three ordered pairs that are solutions of $x = 5$ are $(5, -2), (5, 0),$ and $(5, 3)$. Drawing a line that passes through the three points gives the vertical line shown in Figure 4.16. ∎

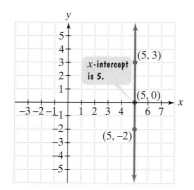

FIGURE 4.16 The graph $x = 5$

STUDY TIP

Do not confuse two-dimensional graphing and one-dimensional graphing of $x = 5$. The graph of $x = 5$ in a two-dimensional rectangular coordinate system is the vertical line in Figure 4.16. By contrast, the graph of $x = 5$ on a one-dimensional number line representing values of x is a single point at 5:

✔ **CHECK POINT 8** Graph the linear equation: $x = -2$.

HORIZONTAL AND VERTICAL LINES The graph of a linear equation in one variable is a horizontal or vertical line.

The graph of $y = b$ is a horizontal line. The y-intercept is b.

The graph of $x = a$ is a vertical line. The x-intercept is a.

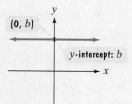

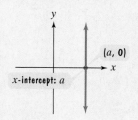

4.2 EXERCISE SET

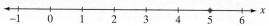

Student Solutions Manual CD/Video PH Math/Tutor Center MathXL Tutorials on CD MathXL® MyMathLab Interactmath.com

Practice Exercises

In Exercises 1–8, use the graph to identify the
 a. *x-intercept, or state that there is no x-intercept;*   **b.** *y-intercept, or state that there is no y-intercept.*

1.
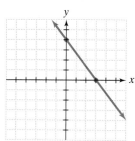

2.
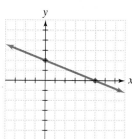

5.

6.
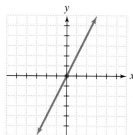

3.
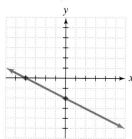

4.
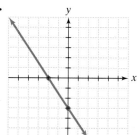

7.
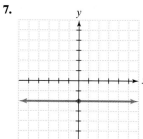

8.
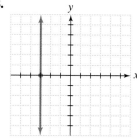

In Exercises 9–18, find the x-intercept and the y-intercept of the graph of each equation. Do not graph the equation.

9. $2x + 5y = 20$

10. $2x + 6y = 30$

11. $2x - 3y = 15$

12. $4x - 5y = 10$

13. $-x + 3y = -8$

14. $-x + 3y = -10$

15. $7x - 9y = 0$

16. $8x - 11y = 0$

17. $2x = 3y - 11$

18. $2x = 4y - 13$

In Exercises 19–40, use intercepts and a checkpoint to graph each equation.

19. $x + y = 5$

20. $x + y = 6$

21. $x + 3y = 6$

22. $2x + y = 4$

23. $6x - 9y = 18$

24. $6x - 2y = 12$

25. $-x + 4y = 6$

26. $-x + 3y = 10$

27. $2x - y = 7$

28. $2x - y = 5$

29. $3x = 5y + 15$

30. $2x = 3y + 6$

31. $25y = 100 - 50x$

32. $10y = 60 - 40x$

33. $2x - 8y = 12$

34. $3x - 6y = 15$

35. $x + 2y = 0$

36. $2x + y = 0$

37. $y - 3x = 0$

38. $y - 4x = 0$

39. $2x - 3y = -11$

40. $3x - 2y = -7$

In Exercises 41–46, write an equation for each graph.

41.

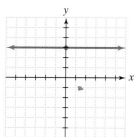

42.

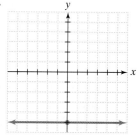

43.

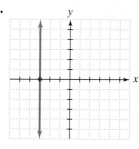

44.

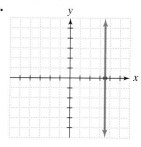

45.

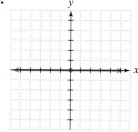

46.

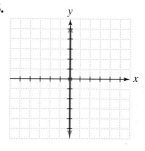

In Exercises 47–62, graph each equation.

47. $y = 4$

48. $y = 2$

49. $y = -2$

50. $y = -3$

51. $x = 2$

52. $x = 4$

53. $x + 1 = 0$

54. $x + 5 = 0$

55. $y - 3.5 = 0$

56. $y - 2.5 = 0$

57. $x = 0$

58. $y = 0$

59. $3y = 9$

60. $5y = 20$

61. $12 - 3x = 0$

62. $12 - 4x = 0$

Practice Plus

In Exercises 63–68, match each equation with one of the graphs shown in Exercises 1–8.

63. $3x + 2y = -6$

64. $x + 2y = -4$

65. $y = -2$

66. $x = -3$

67. $4x + 3y = 12$

68. $2x + 5y = 10$

In Exercises 69–70,

a. *Write a linear equation in standard form satisfying the given condition. Assume that all measurements shown in each figure are in feet.*

b. *Graph the equation in part (a). Because x and y must be non-negative (why?), limit your final graph to quadrant I and its boundaries.*

69. The perimeter of the larger rectangle is 58 feet.

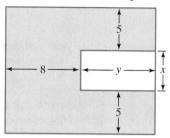

70. The perimeter of the shaded trapezoid is 84 feet.

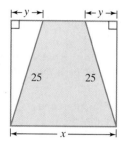

Application Exercises

The flight of a vulture is observed for 30 seconds. The graph shows the vulture's height, in meters, during this period of time. Use the graph to solve Exercises 71–75.

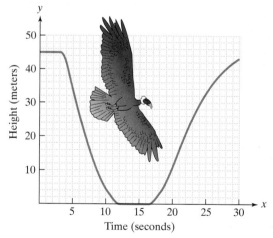

Time (seconds)

71. During which period of time is the vulture's height decreasing?

72. During which period of time is the vulture's height increasing?

73. What is the *y*-intercept? What does this mean about the vulture's height at the beginning of the observation?

74. During the first three seconds of observation, the vulture's flight is graphed as a horizontal line. Write the equation of the line. What does this mean about the vulture's flight pattern during this time?

75. Use integers to write five *x*-intercepts of the graph. What is the vulture doing during these times?

Too late for that flu shot now! It's only 8 A.M. and you're feeling lousy. Fascinated by the way that algebra models the world (your author is projecting a bit here), you decide to construct a graph showing your body temperature from 8 A.M. through 3 P.M. You decide to let x represent the number of hours after 8 A.M. and y represent your temperature at time x. The graph is shown. Use it to solve Exercises 76–80.

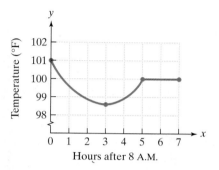

Hours after 8 A.M.

76. What is the *y*-intercept? What does this mean about your temperature at 8 A.M.?

77. During which period of time is your temperature decreasing?

78. Estimate your minimum temperature during the time period shown. How many hours after 8 A.M. does this occur? At what time does this occur?

79. During which period of time is your temperature increasing?

80. From five hours after 8 A.M. until seven hours after 8 A.M., your temperature is graphed as a portion of a horizontal line. Write the equation of the line. What does this mean about your temperature over this period of time?

81. In the section opener, we saw that the linear equation

$$0.57x + y = 80$$

models the percentage of market share, *y*, for carbonated beverages of soft drinks sold in the United States *x* years after 1994.

a. Find the *y*-intercept of the equation's graph. Describe what this means in terms of the variables in the equation.

b. How well does your description in part (a) model the appropriate portion of the data shown in Figure 4.9 on page 222?

c. Find the *x*-intercept of the equation's graph. Round to the nearest whole number. Describe what this means in terms of the variables in the equation. Does this seem reasonable or has model breakdown occurred?

82. As shown in the bar graph, the percentage of people in the United States satisfied with their lives remains relatively constant for all age groups. If x represents a person's age and y represents the percentage of people satisfied with their lives at that age, write an equation that reasonably models the data.

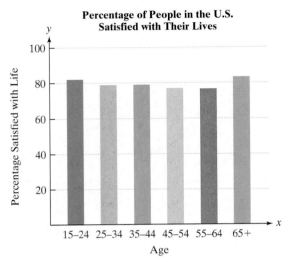

Percentage of People in the U.S. Satisfied with Their Lives

Source: Culture Shift in Advanced Industrial Society, Princeton University Press

Writing in Mathematics

83. What is an x-intercept of a graph?

84. What is a y-intercept of a graph?

85. If you are given an equation of the form $Ax + By = C$, explain how to find the x-intercept.

86. If you are given an equation of the form $Ax + By = C$, explain how to find the y-intercept.

87. Explain how to graph $Ax + By = C$ if C is not equal to zero.

88. Explain how to graph a linear equation of the form $Ax + By = 0$.

89. How many points are needed to graph a line? How many should actually be used? Explain.

90. Describe the graph of $y = 200$.

91. Describe the graph of $x = -100$.

92. We saw that the number of skiers in the United States has remained constant over time. Exercise 82 showed that the percentage of people satisfied with their lives remains constant for all age groups. Give another example of a real-world phenomenon that has remained relatively constant. Try writing an equation that models this phenomenon.

Critical Thinking Exercises

93. Write the equation of the line passing through the point $(5, 6)$ and parallel to the line whose equation is $y = -1$.

In Exercises 94–95, find the coefficients that must be placed in each shaded area so that the equation's graph will be a line with the specified intercepts.

94. $\boxed{}\ x + \boxed{}\ y = 10$; x-intercept $= 5$; y-intercept $= 2$

95. $\boxed{}\ x + \boxed{}\ y = 12$; x-intercept $= -2$; y-intercept $= 4$

Technology Exercises

96. Use a graphing utility to verify any five of your hand-drawn graphs in Exercises 19–40. Solve the equation for y before entering it.

In Exercises 97–100, use a graphing utility to graph each equation. You will need to solve the equation for y before entering it. Use the equation displayed on the screen to identify the x-intercept and the y-intercept.

97. $2x + y = 4$

98. $3x - y = 9$

99. $2x + 3y = 30$

100. $4x - 2y = -40$

Review Exercises

101. Find the absolute value: $|-13.4|$. (Section 1.2, Example 8)

102. Simplify: $7x - (3x - 5)$. (Section 1.7, Example 7)

103. Graph: $-2 \le x < 4$. (Section 2.6, Example 1)

SECTION

4.3

Objectives

1 Compute a line's slope.

2 Use slope to show that lines are parallel.

3 Calculate rate of change in applied situations.

Number of People in the U.S. Living Alone

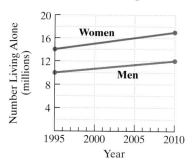

FIGURE 4.17

Source: Forrester Research

SLOPE

A best guess at the look of our nation in the next decades indicates that the number of men and women living alone will increase each year. Figure 4.17 shows that by 2010, approximately 12 million men and 17 million women will be living alone.

By looking at Figure 4.17, can you tell that the green graph representing women is steeper than the blue graph representing men? This indicates a greater yearly increase in the millions of women living alone than in the millions of men living alone. In this section, we will study the idea of a line's steepness and see what that has to do with how its variables are changing.

The Slope of a Line Mathematicians have developed a useful measure of the steepness of a line, called the **slope** of the line. Slope compares the vertical change (the **rise**) to the horizontal change (the **run**) when moving from one fixed point to another along the line. To calculate the slope of a line, we use a ratio that compares the change in *y* (the rise) to the change in *x* (the run).

DEFINITION OF SLOPE The **slope** of the line through the distinct points (x_1, y_1) and (x_2, y_2) is

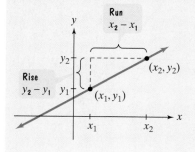

$$\frac{\text{Change in } y}{\text{Change in } x} = \frac{\text{Rise}}{\text{Run}}$$

$$= \frac{y_2 - y_1}{x_2 - x_1},$$

where $x_2 - x_1 \neq 0$.

It is common notation to let the letter *m* represent the slope of a line. The letter *m* is used because it is the first letter of the French verb *monter*, meaning to rise, or to ascend.

1 Compute a line's slope.

EXAMPLE 1 Using the Definition of Slope

Find the slope of the line passing through each pair of points:

a. $(-3, -1)$ and $(-2, 4)$ **b.** $(-3, 4)$ and $(2, -2)$.

SOLUTION

a. Let $(x_1, y_1) = (-3, -1)$ and $(x_2, y_2) = (-2, 4)$. We obtain the slope as follows:

$$m = \frac{\text{Change in } y}{\text{Change in } x} = \frac{y_2 - y_1}{x_2 - x_1} = \frac{4 - (-1)}{-2 - (-3)} = \frac{5}{1} = 5.$$

The situation is illustrated in Figure 4.18(a). The slope of the line is 5, indicating that there is a vertical change, a rise, of 5 units for each horizontal change, a run, of 1 unit. The slope is positive and the line rises from left to right.

STUDY TIP

When computing slope, it makes no difference which point you call (x_1, y_1) and which point you call (x_2, y_2). If we let $(x_1, y_1) = (-2, 4)$ and $(x_2, y_2) = (-3, -1)$, the slope is still 5:

$$m = \frac{\text{Change in } y}{\text{Change in } x} = \frac{y_2 - y_1}{x_2 - x_1} = \frac{-1 - 4}{-3 - (-2)} = \frac{-5}{-1} = 5.$$

However, you should not subtract in one order in the numerator $(y_2 - y_1)$ and then in a different order in the denominator $(x_1 - x_2)$. The slope is *not* -5:

$$\frac{-1 - 4}{-2 - (-3)} = \frac{-5}{1} = -5. \qquad \text{Incorrect}$$

b. We can let $(x_1, y_1) = (-3, 4)$ and $(x_2, y_2) = (2, -2)$. The slope of the line shown in Figure 4.18(b) is computed as follows:

$$m = \frac{\text{Change in } y}{\text{Change in } x} = \frac{y_2 - y_1}{x_2 - x_1} = \frac{-2 - 4}{2 - (-3)} = \frac{-6}{5} = -\frac{6}{5}.$$

The slope of the line is $-\frac{6}{5}$. For every vertical change of -6 units (6 units down), there is a corresponding horizontal change of 5 units. The slope is negative and the line falls from left to right.

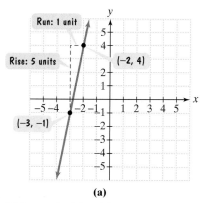

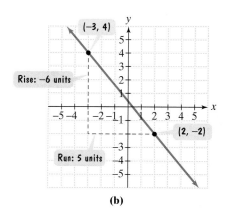

(a)   (b)

FIGURE 4.18 Visualizing slope

✔ **CHECK POINT 1** Find the slope of the line passing through each pair of points:

a. $(-3, 4)$ and $(-4, -2)$ **b.** $(4, -2)$ and $(-1, 5)$.

> ### EXAMPLE 2   Using the Definition of Slope for Horizontal and Vertical Lines
>
> Find the slope of the line passing through each pair of points:
>
>   **a.** $(5, 4)$ and $(3, 4)$ **b.** $(2, 5)$ and $(2, 1)$.
>
> **SOLUTION**
>
>   **a.** Let $(x_1, y_1) = (5, 4)$ and $(x_2, y_2) = (3, 4)$. We obtain the slope as follows:
>
> $$m = \frac{\text{Change in } y}{\text{Change in } x} = \frac{y_2 - y_1}{x_2 - x_1} = \frac{4 - 4}{3 - 5} = \frac{0}{-2} = 0.$$
>
> The situation is illustrated in Figure 4.19(a). Can you see that the line is horizontal? Because any two points on a horizontal line have the same y-coordinate, these lines neither rise nor fall from left to right. The change in y, $y_2 - y_1$, is always zero. Thus, **the slope of any horizontal line is zero**.
>
>   **b.** We can let $(x_1, y_1) = (2, 5)$ and $(x_2, y_2) = (2, 1)$. Figure 4.19(b) shows that these points are on a vertical line. We attempt to compute the slope as follows:
>
> $$m = \frac{\text{Change in } y}{\text{Change in } x} = \frac{1 - 5}{2 - 2} = \frac{-4}{0} \quad \text{Division by zero is undefined.}$$
>
> Because division by zero is undefined, the slope of the vertical line in Figure 4.19(b) is undefined. In general, **the slope of any vertical line is undefined**.
>
> 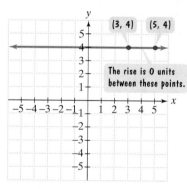   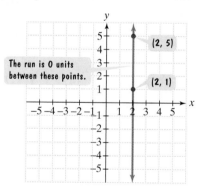
>
> (a) Horizontal lines have no vertical change.   (b) Vertical lines have no horizontal change.
>
> **FIGURE 4.19**   Visualizing slope

Table 4.1 summarizes four possibilities for the slope of a line.

Table 4.1   Possibilities for a Line's Slope

Positive Slope	Negative Slope	Zero Slope	Undefined Slope
$m > 0$	$m < 0$	$m = 0$	m is undefined.
Line rises from left to right.	Line falls from left to right.	Line is horizontal.	Line is vertical.

✔ **CHECK POINT 2** Find the slope of the line passing through each pair of points or state that the slope is undefined:

a. $(6, 5)$ and $(2, 5)$ **b.** $(1, 6)$ and $(1, 4)$.

2 Use slope to show that lines are parallel.

Slope and Parallel Lines Two nonintersecting lines that lie in the same plane are **parallel**. If two lines do not intersect, the ratio of the vertical change to the horizontal change is the same for each line. Because two parallel lines have the same "steepness," they must have the same slope.

> **SLOPE AND PARALLEL LINES**
>
> **1.** If two nonvertical lines are parallel, then they have the same slope.
> **2.** If two distinct nonvertical lines have the same slope, then they are parallel.
> **3.** Two distinct vertical lines, each with undefined slope, are parallel.

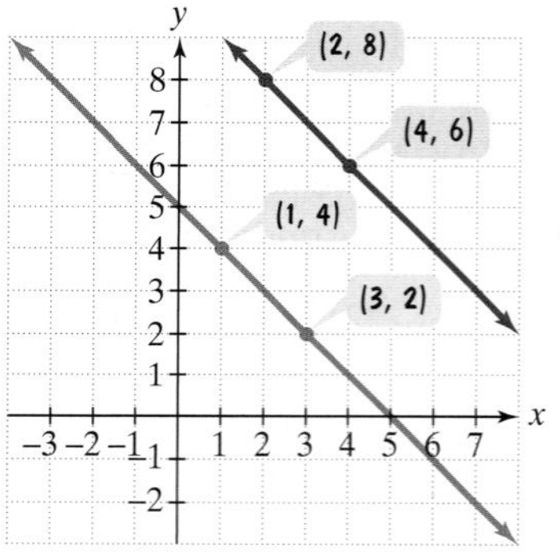

FIGURE 4.20 Using slope to show that lines are parallel

EXAMPLE 3 Using Slope to Show That Lines Are Parallel

Show that the line passing through $(1, 4)$ and $(3, 2)$ is parallel to the line passing through $(2, 8)$ and $(4, 6)$.

SOLUTION The situation is illustrated in Figure 4.20. The lines certainly look like they are parallel. Let's use equal slopes to confirm this fact. For each line, we compute the ratio of the difference in y-coordinates to the difference in x-coordinates. Be sure to subtract the coordinates in the same order.

Slope of the line through $(1, 4)$ and $(3, 2)$ is

$$\frac{4 - 2}{1 - 3} = \frac{2}{-2} = -1.$$

Slope of the line through $(2, 8)$ and $(4, 6)$ is

$$\frac{8 - 6}{2 - 4} = \frac{2}{-2} = -1.$$

Because the slopes are equal, the lines are parallel. ■

✔ **CHECK POINT 3** Show that the line passing through $(4, 2)$ and $(6, 6)$ is parallel to the line passing through $(0, -2)$ and $(1, 0)$.

3 Calculate rate of change in applied situations.

Slope as Rate of Change Slope is defined as the ratio of a change in y to a corresponding change in x. It tells how fast y is changing with respect to x. Thus, the slope of a line represents its rate of change.

Our next example shows how slope can be interpreted as a rate of change in an applied situation. When calculating slope in applied problems, keep track of the units in the numerator and the denominator.

EXAMPLE 4 Slope as a Rate of Change

The line graphs for the number of women and men living alone are shown again in Figure 4.21 at the top of the next page. Find the slope of the line segment for the women. Describe what this slope represents.

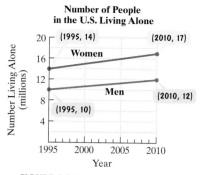

Number of People in the U.S. Living Alone

FIGURE 4.21
Source: Forrester Research

SOLUTION We let x represent a year and y the number of women living alone in that year. The two points shown on the line segment for women have the following coordinates:

$$(1995, 14) \quad \text{and} \quad (2010, 17).$$

In 1995, 14 million U.S. women lived alone.

In 2010, 17 million U.S. women are projected to live alone.

Now we compute the slope:

$$m = \frac{\text{Change in } y}{\text{Change in } x} = \frac{17 - 14}{2010 - 1995}$$

The unit in the numerator is *million women*.

$$= \frac{3}{15} = \frac{1}{5} = \frac{0.2 \text{ million people}}{\text{year}}.$$

The unit in the denominator is *year*.

The slope indicates that the number of U.S. women living alone is projected to increase by 0.2 million each year. The rate of change is 0.2 million women per year. ∎

CHECK POINT 4 Use the graph in Example 4 to find the slope of the line segment for the men. Express the slope correct to two decimal places and describe what it represents.

In Check Point 4, did you find that the slope of the line segment for the men is different from that of the women? The rate of change for men living alone is not equal to the rate of change for women living alone. Because of these different slopes, if you extend the line segments in Figure 4.21, the resulting lines will intersect. They are not parallel.

Railroads and Highways

The steepest part of Mt. Washington Cog Railway in New Hampshire has a 37% grade. This is equivalent to a slope of $\frac{37}{100}$. For every horizontal change of 100 feet, the railroad ascends 37 feet vertically. Engineers denote slope by grade, expressing slope as a percentage.

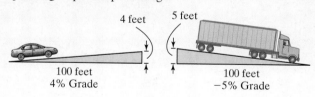

100 feet
4% Grade

100 feet
−5% Grade

Railroad grades are usually less than 2%, although in the mountains they may go as high as 4%. The grade of the Mt. Washington Cog Railway is phenomenal, making it necessary for locomotives to *push* single cars up its steepest part.

A Mount Washington Cog Railway locomotive pushing a single car up the steepest part of the railroad. The locomotive is about 120 years old.

4.3 EXERCISE SET

Student Solutions Manual   CD/Video   PH Math/Tutor Center   MathXL Tutorials on CD   MathXL®   MyMathLab   Interactmath.com

Practice Exercises

In Exercises 1–10, find the slope of the line passing through each pair of points or state that the slope is undefined. Then indicate whether the line through the points rises, falls, is horizontal, or is vertical.

1. $(4, 7)$ and $(8, 10)$

2. $(2, 1)$ and $(3, 4)$

3. $(-2, 1)$ and $(2, 2)$

4. $(-1, 3)$ and $(2, 4)$

5. $(4, -2)$ and $(3, -2)$

6. $(4, -1)$ and $(3, -1)$

7. $(-2, 4)$ and $(-1, -1)$

8. $(6, -4)$ and $(4, -2)$

9. $(5, 3)$ and $(5, -2)$

10. $(3, -4)$ and $(3, 5)$

In Exercises 11–22, find the slope of each line, or state that the slope is undefined.

11.

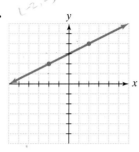

12.

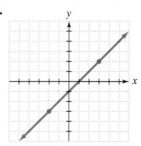

13.

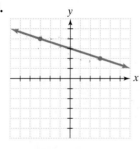

14.

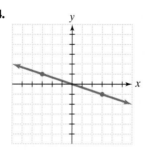

15.

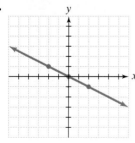

16.

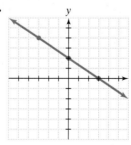

17.

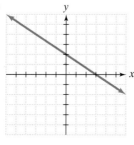

18.

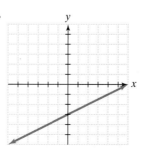

19. **20.**

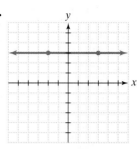

21. **22.**

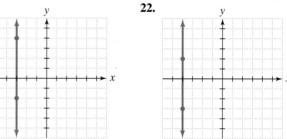

In Exercises 23–26, determine whether the distinct lines through each pair of points are parallel.

23. $(-2, 0)$ and $(0, 6)$; $(1, 8)$ and $(0, 5)$

24. $(2, 4)$ and $(6, 1)$; $(-3, 1)$ and $(1, -2)$

25. $(0, 3)$ and $(1, 5)$; $(-1, 7)$ and $(1, 10)$

26. $(-7, 6)$ and $(0, 4)$; $(-9, -3)$ and $(1, 5)$

Practice Plus

27. On the same set of axes, draw lines passing through the origin with slopes -1, $-\frac{1}{2}$, 0, $\frac{1}{3}$, and 2.

28. On the same set of axes, draw lines with y-intercept 4 and slopes -1, $-\frac{1}{2}$, 0, $\frac{1}{3}$, and 2.

Use slopes to solve Exercises 29–30.

29. Show that the points whose coordinates are $(-3, -3)$, $(2, -5)$, $(5, -1)$, and $(0, 1)$ are the vertices of a four-sided figure whose opposite sides are parallel. (Such a figure is called a *parallelogram*.)

30. Show that the points whose coordinates are $(-3, 6)$, $(2, -3)$, $(11, 2)$, and $(6, 11)$ are the vertices of a four-sided figure whose opposite sides are parallel.

31. The line passing through $(5, y)$ and $(1, 0)$ is parallel to the line joining $(2, 3)$ and $(-2, 1)$. Find y.

32. The line passing through $(1, y)$ and $(7, 12)$ is parallel to the line joining $(-3, 4)$ and $(-5, -2)$. Find y.

In Exercises 33–34, use slopes to determine whether the given points lie on a line. (Such points are said to be collinear.)

33. (1, 2), (3, 12), (6, 27)

34. (1, 3), (3, 11), (6, 22)

Application Exercises

The graph shows online shopping per U.S. online household through 2004. Use the information provided by the graph to solve Exercises 35–36.

Online Spending: Yearly Spending per Online Household

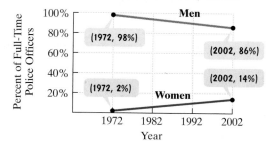

Source: Forrester Research

35. Find the slope of the line passing through (1999, 1000) and (2001, 1500). What does this represent in terms of the increase in online shopping per year?

36. Find the slope of the line passing through (2001, 1500) and (2004, 3900). What does this represent in terms of the increase in online shopping per year?

The graphs approximate the percentage of full-time police officers in the United States, by gender, y, for year x, where 1972 ≤ x ≤ 2002. Use this information to solve Exercises 37–38.

Police Officers in the U.S., by Gender

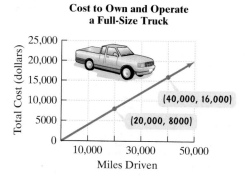

Source: National Center for Women and Policing

37. Find the slope of the line segment representing men. Then express the slope as a rate of change with the proper units attached.

38. Find the slope of the line segment representing women. Then express the slope as a rate of change with the proper units attached.

The graphs approximate the mean, or average, income for well-to-do U.S. families (the highest 20%) and the lowest-income U.S. families (the lowest 20%), y, for year x, where 1980 ≤ x ≤ 2002. Use this information to solve Exercises 39–40.

Mean Income of U.S. Families in 2002 Dollars

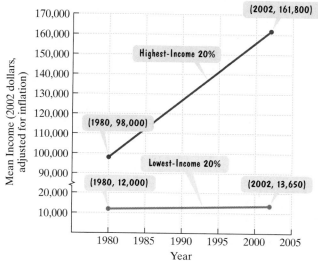

Source: U.S. Census Bureau

39. Find the slope of the line segment representing the highest-income 20%. Then express the slope as a rate of change with the proper units attached.

40. Find the slope of the line segment representing the lowest-income 20%. Then express the slope as a rate of change with the proper units attached.

In Exercises 41–42, find the slope of each line. Then express the slope as a rate of change with the proper units attached.

41. The graph shows the cost to own and operate a full-size truck, *y*, in dollars, in terms of the miles the truck is driven, *x*.

Cost to Own and Operate a Full-Size Truck

(40,000, 16,000)

(20,000, 8000)

Source: Federal Highway Administration

42. The graph shows the cost to own and operate a compact car, y, in dollars, in terms of the miles the car is driven, x.

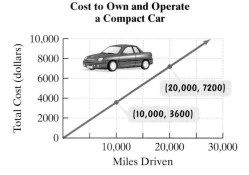

**Cost to Own and Operate
a Compact Car**

(20,000, 7200)

(10,000, 3600)

Source: Federal Highway Administration

The pitch of a roof refers to the absolute value of its slope. In Exercises 43–44, find the pitch of each roof shown.

43.

6 feet ←18 feet→

44.

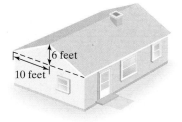

6 feet
10 feet

The grade of a road or ramp refers to its slope expressed as a percent. Use this information to solve Exercises 45–46.

45. Construction laws are very specific when it comes to access ramps for the disabled. Every vertical rise of 1 foot requires a horizontal run of 12 feet. What is the grade of such a ramp? Round to the nearest tenth of a percent.

1 foot

12 feet

46. A college campus goes beyond the standards described in Exercise 45. All wheelchair ramps on campus are designed so that every vertical rise of 1 foot is accompanied by a horizontal run of 14 feet. What is the grade of such a ramp? Round to the nearest tenth of a percent.

Writing in Mathematics

47. What is the slope of a line?

48. Describe how to calculate the slope of a line passing through two points.

49. What does it mean if the slope of a line is zero?

50. What does it mean if the slope of a line is undefined?

51. If two lines are parallel, describe the relationship between their slopes.

52. Look back at the graph for Exercises 35–36. Do you think that the line through the points corresponding to the year 2001 and the year 2004 will model online spending per online household in the year 2040? Explain your answer.

Critical Thinking Exercises

53. Which one of the following is true?

   a. Slope is run divided by rise.

   b. The line through $(2, 2)$ and the origin has slope 1.

   c. A line with slope 3 can be parallel to a line with slope -3.

   d. The line through $(3, 1)$ and $(3, -5)$ has zero slope.

In Exercises 54–55, use the figure shown to make the indicated list.

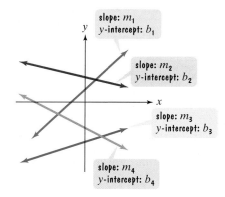

slope: m_1
y-intercept: b_1

slope: m_2
y-intercept: b_2

slope: m_3
y-intercept: b_3

slope: m_4
y-intercept: b_4

54. List the slopes m_1, m_2, m_3, and m_4 in order of decreasing size.

55. List the y-intercepts b_1, b_2, b_3, and b_4 in order of decreasing size.

Technology Exercises

Use a graphing utility to graph each equation in Exercises 56–59. Then use the $\boxed{\text{TRACE}}$ *feature to trace along the line and find the coordinates of two points. Use these points to compute the line's slope.*

56. $y = 2x + 4$ **57.** $y = -3x + 6$

58. $y = -\frac{1}{2}x - 5$

59. $y = \frac{3}{4}x - 2$

60. In Exercises 56–59, compare the slope that you found with the line's equation. What relationship do you observe between the line's slope and one of the constants in the equation?

Review Exercises

61. A 36-inch board is cut into two pieces. One piece is twice as long as the other. How long are the pieces? (Section 2.5, Example 3)

62. Simplify: $-10 + 16 \div 2(-4)$. (Section 1.8, Example 4)

63. Solve and graph the solution set on a number line: $2x - 3 \le 5$. (Section 2.6, Example 5)

SECTION 4.4

THE SLOPE-INTERCEPT FORM OF THE EQUATION OF A LINE

Objectives

1 Find a line's slope and y-intercept from its equation.

2 Graph lines in slope-intercept form.

3 Use slope and y-intercept to graph $Ax + By = C$.

4 Use slope and y-intercept to model data.

Children in Kampala, Uganda, discuss HIV/AIDS issues.

AIDS, famine, malaria, and civil wars have all taken their toll on Africa. In 2003. Secretary of State Colin Powell declared, "HIV is now more destructive than any army, any conflict, any weapon of mass destruction."

In this section, you will study the form of an equation that will enable you to develop a model for the growth of HIV/AIDS in Africa. Using this model, you will understand the need to fight AIDS on a continent ravaged by the virus.

1 Find a line's slope and y-intercept from its equation.

The Slope-Intercept Form of the Equation of a Line Let's begin with an example that shows how easy it is to find a line's slope and y-intercept from its equation.

Figure 4.22 shows the graph of $y = 2x + 4$. Verify that the x-intercept is -2 by setting y equal to 0 and solving for x. Similarly, verify that the y-intercept is 4 by setting x equal to 0 and solving for y.

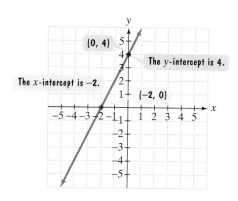

FIGURE 4.22 The graph of $y = 2x + 4$

Now that we have two points on the line, $(-2, 0)$ and $(0, 4)$, we can calculate the slope of the graph of $y = 2x + 4$.

$$\text{Slope} = \frac{\text{Change in } y}{\text{Change in } x}$$

$$= \frac{4 - 0}{0 - (-2)} = \frac{4}{2} = 2$$

We see that the slope of the line is 2, the same as the coefficient of x in the equation $y = 2x + 4$. The y-intercept is 4, the same as the constant in the equation $y = 2x + 4$.

$$y = 2x + 4$$

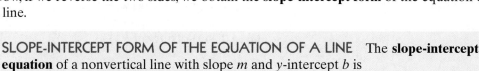

The slope is 2.   The y-intercept is 4.

It is not merely a coincidence that the x-coefficient is the line's slope and the constant term is the y-intercept. Let's find the equation of any nonvertical line with slope m and y-intercept b. Because the y-intercept is b, the point $(0, b)$ lies on the line. Now, let (x, y) represent any other point on the line, shown in Figure 4.23. Keep in mind that the point (x, y) is arbitrary and is not in one fixed position. By contrast, the point $(0, b)$ is fixed.

Regardless of where the point (x, y) is located, the steepness of the line in Figure 4.23 remains the same. Thus, the ratio for slope stays a constant m. This means that for all points along the line

$$m = \frac{\text{Change in } y}{\text{Change in } x} = \frac{y - b}{x - 0} = \frac{y - b}{x}.$$

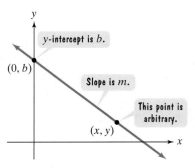

FIGURE 4.23 A line with slope m and y-intercept b

We can clear the fraction by multiplying both sides by x, the least common denominator.

$$m = \frac{y - b}{x} \qquad \text{This is the slope of the line in Figure 4.23.}$$

$$mx = \frac{y - b}{x} \cdot x \qquad \text{Multiply both sides by } x.$$

$$mx = y - b \qquad \text{Simplify: } \frac{y - b}{\cancel{x}} \cdot \cancel{x} = y - b.$$

$$mx + b = y - b + b \qquad \text{Add } b \text{ to both sides and solve for } y.$$

$$mx + b = y \qquad \text{Simplify.}$$

Now, if we reverse the two sides, we obtain the **slope-intercept form** of the equation of a line.

> **SLOPE-INTERCEPT FORM OF THE EQUATION OF A LINE**   The **slope-intercept equation** of a nonvertical line with slope m and y-intercept b is
>
> $$y = mx + b.$$

Thus, if a line's equation is written with y isolated on one side, the x-coefficient is the line's slope and the constant term is the y-intercept.

EXAMPLE 1   Finding a Line's Slope and y-Intercept from Its Equation

Find the slope and the y-intercept of the line with the given equation:

a. $y = 2x - 4$ **b.** $y = \frac{1}{2}x + 2$ **c.** $5x + y = 4$.

SOLUTION

a. We write $y = 2x - 4$ as $y = 2x + (-4)$. The slope is the x-coefficient and the y-intercept is the constant term.

$$y = 2x + (-4)$$

The slope is 2. The y-intercept is −4.

b. The equation $y = \frac{1}{2}x + 2$ is in the form $y = mx + b$. We can find the slope, m, by identifying the coefficient of x. We can find the y-intercept, b, by identifying the constant term.

$$y = \frac{1}{2}x + 2$$

The slope is $\frac{1}{2}$. The y-intercept is 2.

c. The equation $5x + y = 4$ is not in the form $y = mx + b$. We can obtain this form by isolating y on one side. We isolate y on the left side by subtracting $5x$ from both sides.

$$5x + y = 4 \qquad \text{This is the given equation.}$$
$$5x - 5x + y = -5x + 4 \qquad \text{Subtract 5x from both sides.}$$
$$y = -5x + 4 \qquad \text{Simplify.}$$

Now, the equation is in the form $y = mx + b$. The slope is the coefficient of x and the y-intercept is the constant term.

$$y = -5x + 4$$

The slope is −5. The y-intercept is 4.

CHECK POINT 1 Find the slope and the y-intercept of the line with the given equation:

a. $y = 5x - 3$ **b.** $y = \frac{2}{3}x + 4$

c. $7x + y = 6$.

2 Graph lines in slope-intercept form.

Graphing $y = mx + b$ **by Using the Slope and y-Intercept** If a line's equation is written with y isolated on one side, we can use the y-intercept and the slope to obtain its graph.

GRAPHING $y = mx + b$ BY USING THE SLOPE AND y-INTERCEPT

1. Plot the point containing the y-intercept on the y-axis. This is the point $(0, b)$.
2. Obtain a second point using the slope, m. Write m as a fraction, and use rise over run, starting at the point containing the y-intercept, to plot this point.
3. Use a straightedge to draw a line through the two points. Draw arrowheads at the ends of the line to show that the line continues indefinitely in both directions.

EXAMPLE 2 Graphing by Using the Slope and y-Intercept

Graph the line whose equation is $y = 4x - 3$.

SOLUTION We write $y = 4x - 3$ in the form $y = mx + b$.

$$y = 4x + (-3)$$

The slope is 4.   The y-intercept is –3.

Now that we have identified the slope and the y-intercept, we use the three steps in the box to graph the equation.

Step 1. **Plot the point containing the y-intercept on the y-axis.** The y-intercept is −3. We plot the point $(0, -3)$, shown in Figure 4.24(a).

Step 2. **Obtain a second point using the slope, m. Write m as a fraction, and use rise over run, starting at the point containing the y-intercept, to plot this point.** The slope, 4, written as a fraction is $\frac{4}{1}$.

$$m = \frac{4}{1} = \frac{\text{Rise}}{\text{Run}}$$

We plot the second point on the line by starting at $(0, -3)$, the first point. Based on the slope, we move 4 units *up* (the rise) and 1 unit to the *right* (the run). This puts us at a second point on the line, $(1, 1)$, shown in Figure 4.24(b).

Step 3. **Use a straightedge to draw a line through the two points.** The graph of $y = 4x - 3$ is shown in Figure 4.24(c).

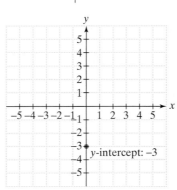

(a) The y-intercept is –3, so $(0, -3)$ is a point on the line.

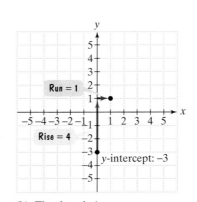

(b) The slope is 4.

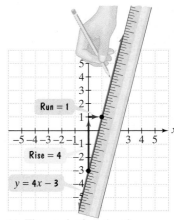

(c) The graph of $y = 4x - 3$

FIGURE 4.24 Graphing $y = 4x - 3$ using the y-intercept and slope

✔ **CHECK POINT 2** Graph the line whose equation is $y = 3x - 2$.

EXAMPLE 3 Graphing by Using the Slope and y-Intercept

Graph the line whose equation is $y = \frac{2}{3}x + 2$.

SOLUTION The equation of the line is in the form $y = mx + b$. We can find the slope, m, by identifying the coefficient of x. We can find the y-intercept, b, by identifying the constant term.

$$y = \frac{2}{3}x + 2$$

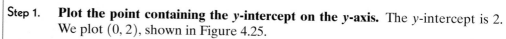

The slope is $\frac{2}{3}$.   The y-intercept is **2**.

Now that we have identified the slope and the y-intercept, we use the three-step procedure to graph the equation.

Step 1. **Plot the point containing the y-intercept on the y-axis.** The y-intercept is 2. We plot $(0, 2)$, shown in Figure 4.25.

Step 2. **Obtain a second point using the slope, m. Write m as a fraction, and use rise over run, starting at the point containing the y-intercept, to plot this point.** The slope, $\frac{2}{3}$, is already written as a fraction.

$$m = \frac{2}{3} = \frac{\text{Rise}}{\text{Run}}$$

We plot the second point on the line by starting at $(0, 2)$, the first point. Based on the slope, we move 2 units *up* (the rise) and 3 units to the *right* (the run). This puts us at a second point on the line, $(3, 4)$, shown in Figure 4.25.

Step 3. **Use a straightedge to draw a line through the two points.** The graph of $y = \frac{2}{3}x + 2$ is shown in Figure 4.25. ■

✔ CHECK POINT **3** Graph the line whose equation is $y = \frac{3}{5}x + 1$.

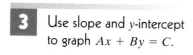

FIGURE 4.25 The graph of $y = \frac{2}{3}x + 2$

---

3 Use slope and y-intercept to graph $Ax + By = C$.

Graphing $Ax + By = C$ by Using the Slope and y-Intercept Earlier in this chapter, we considered linear equations of the form $Ax + By = C$. We used x- and y-intercepts, as well as checkpoints, to graph these equations. It is also possible to obtain the graphs by using the slope and y-intercept. To do this, begin by solving $Ax + By = C$ for y. This will put the equation in slope-intercept form. Then use the three-step procedure to graph the equation. This is illustrated in Example 4.

EXAMPLE 4 Graphing by Using the Slope and y-Intercept

Graph the linear equation $2x + 5y = 0$ by using the slope and y-intercept.

SOLUTION We put the equation in slope-intercept form by solving for y.

$2x + 5y = 0$	This is the given equation.
$2x - 2x + 5y = -2x + 0$	Subtract 2x from both sides.
$5y = -2x + 0$	Simplify.
$\dfrac{5y}{5} = \dfrac{-2x + 0}{5}$	Divide both sides by 5.
$y = \dfrac{-2x}{5} + \dfrac{0}{5}$	Divide each term in the numerator by 5.
$y = -\dfrac{2}{5}x + 0$	Simplify.

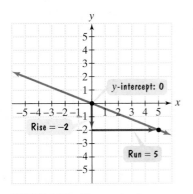

FIGURE 4.26 The graph of $2x + 5y = 0$, or $y = -\frac{2}{5}x + 0$

Now that the equation is in slope-intercept form, we can use the slope and y-intercept to obtain its graph. Examine the slope-intercept form:

$$y = -\frac{2}{5}x + 0.$$

slope: $-\frac{2}{5}$ y-intercept: 0

Note that the slope is $-\frac{2}{5}$ and the y-intercept is 0. Use the y-intercept to plot $(0, 0)$ on the y-axis. Then locate a second point by using the slope.

$$m = -\frac{2}{5} = \frac{-2}{5} = \frac{\text{Rise}}{\text{Run}}$$

Because the rise is -2 and the run is 5, move *down* 2 units and to the *right* 5 units, starting at the point $(0, 0)$. This puts us at a second point on the line, $(5, -2)$. The graph of $2x + 5y = 0$ is the line drawn through these points, shown in Figure 4.26. ∎

DISCOVER FOR YOURSELF

Obtain a second point in Example 4 by writing the slope as follows:

$$m = \frac{2}{-5} = \frac{\text{Rise}}{\text{Run}}$$

$-\frac{2}{5}$ can be expressed as $\frac{-2}{5}$ or $\frac{2}{-5}$.

Obtain a second point in Figure 4.26 by moving *up* 2 units and to the *left* 5 units, starting at $(0, 0)$. What do you observe once you graph the line?

 CHECK POINT 4 Graph the linear equation $3x + 4y = 0$ by using the slope and y-intercept.

4 Use slope and y-intercept to model data.

Modeling with the Slope-Intercept Form of the Equation of a Line If an equation in slope-intercept form models some physical situation, then the slope and y-intercept have physical interpretations. For the equation $y = mx + b$, the y-intercept, b, tells us what is happening to y when x is 0. If x represents time, the y-intercept describes the value of y at the beginning, or when time equals 0. The slope represents the rate of change in y per unit change in x.

Let's see how we can use these ideas to develop a model for data. In the previous section, we looked at line graphs for the number of U.S. men and women living alone, repeated in Figure 4.27. Let's develop a model for the data for women living alone. We let x represent the number of years after 1995. At the beginning of our data, or 0 years after 1995, 14 million women lived alone. Thus, $b = 14$. In Example 4 in the previous section, we found that $m = 0.2$ (rate of change is 0.2 million women per year). An equation of the form $y = mx + b$ that models the data is

$$y = 0.2x + 14,$$

where y is the number, in millions, of U.S. women living alone x years after 1995.

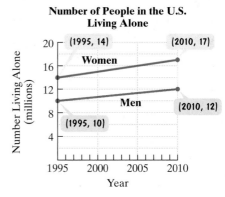

FIGURE 4.27

4.4 EXERCISE SET

Student Solutions Manual   CD/Video   PH Math/Tutor Center   MathXL Tutorials on CD   MathXL®   MyMathLab   Interactmath.com

Practice Exercises

In Exercises 1–12, find the slope and the y-intercept of the line with the given equation.

1. $y = 3x + 2$

2. $y = 9x + 4$

3. $y = 3x - 5$

4. $y = 4x - 2$

5. $y = -\frac{1}{2}x + 5$

6. $y = -\frac{3}{4}x + 6$

7. $y = 7x$

8. $y = 10x$

9. $y = 10$

10. $y = 7$

11. $y = 4 - x$

12. $y = 5 - x$

In Exercises 13–26, begin by solving the linear equation for y. This will put the equation in slope-intercept form. Then find the slope and the y-intercept of the line with this equation.

13. $-5x + y = 7$

14. $-9x + y = 5$

15. $x + y = 6$

16. $x + y = 8$

17. $6x + y = 0$

18. $8x + y = 0$

19. $3y = 6x$

20. $3y = -9x$

21. $2x + 7y = 0$

22. $2x + 9y = 0$

23. $3x + 2y = 3$

24. $4x + 3y = 4$

25. $3x - 4y = 12$

26. $5x - 2y = 10$

In Exercises 27–38, graph each linear equation using the slope and y-intercept.

27. $y = 2x + 4$

28. $y = 3x + 1$

29. $y = -3x + 5$

30. $y = -2x + 4$

31. $y = \frac{1}{2}x + 1$

32. $y = \frac{1}{3}x + 2$

33. $y = \frac{2}{3}x - 5$

34. $y = \frac{3}{4}x - 4$

35. $y = -\frac{3}{4}x + 2$

36. $y = -\frac{2}{3}x + 4$

37. $y = -\frac{5}{3}x$

38. $y = -\frac{4}{3}x$

In Exercises 39–46,

a. *Put the equation in slope-intercept form by solving for y.*

b. *Identify the slope and the y-intercept.*

c. *Use the slope and y-intercept to graph the equation.*

39. $3x + y = 0$

40. $2x + y = 0$

41. $3y = 4x$

42. $4y = 5x$

43. $2x + y = 3$

44. $3x + y = 4$

45. $7x + 2y = 14$

46. $5x + 3y = 15$

In Exercises 47–52, graph both linear equations in the same rectangular coordinate system. If the lines are parallel, explain why.

47. $y = 3x + 1$
$y = 3x - 3$

48. $y = 2x + 4$
$y = 2x - 3$

49. $y = -3x + 2$
$y = 3x + 2$

50. $y = -2x + 1$
$y = 2x + 1$

51. $x - 2y = 2$
$2x - 4y = 3$

52. $x - 3y = 9$
$3x - 9y = 18$

Practice Plus

In Exercises 53–58, write an equation in the form $y = mx + b$ of the line that is described.

53. The y-intercept is 5 and the line is parallel to the line whose equation is $3x + y = 6$.

54. The y-intercept is -4 and the line is parallel to the line whose equation is $2x + y = 8$.

55. The line has the same y-intercept as the line whose equation is $16y = 8x + 32$ and is parallel to the line whose equation is $3x + 3y = 9$.

56. The line has the same y-intercept as the line whose equation is $2y = 6x + 8$ and is parallel to the line whose equation is $4x + 4y = 20$.

57. The line rises from left to right. It passes through the origin and a second point with equal x- and y-coordinates.

58. The line falls from left to right. It passes through the origin and a second point with opposite x- and y-coordinates.

Application Exercises

59. The formula $y = -0.4x + 38$ models the percentage of U.S. men, y, smoking cigarettes x years after 1980.

a. Use the formula to find the percentage of men who smoked in 1980, 1981, 1982, 1983, 1990, and 2000.

b. What is the slope of this model? What does it represent in this situation?

c. What is the y-intercept of this model? What does it represent in this situation?

60. A salesperson receives a fixed salary plus a percentage of all sales. The linear equation $y = 0.05x + 500$ describes the weekly salary, y, in dollars, in terms of weekly sales, x, also in dollars.

a. Use the formula to find the weekly salary for sales of $0, $1, $2, $3, $4, $5, $100, and $1000.

b. What is the slope of this equation? What does it represent in this situation?

c. What is the y-intercept of this equation? What does it represent in this situation?

61. In 2002, approximately 29 million people, including 3 million children, were living with HIV/AIDS in sub-Saharan Africa. The region accounted for 70% of the world's new HIV infections. The bar graph shows the number of people, in millions, living with the virus from 1997 through 2002. The graph in the rectangular coordinate system approximates the data.

Millions of People Living with HIV/AIDS in Sub-Saharan Africa

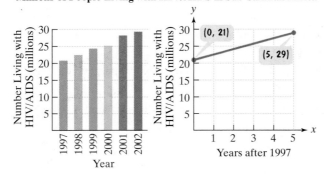

Source: U.N. AIDS

a. According to the rectangular coordinate graph, what is the y-intercept? Describe what this represents in this situation.

b. Use the coordinates of the two points shown to compute the slope. What does this mean in terms of rate of change?

c. Use the y-intercept from part (a) and the slope from part (b) to write an equation that models the number of people, in millions, living with HIV in sub-Saharan Africa, y, x years after 1997. Write your model in the form $y = mx + b$.

d. If worldwide efforts to fight AIDS in sub-Saharan Africa are not successful, use your model from part (c) to predict the number of people, in millions, who will be living with the virus in 2006.

62. The graph shows that the percentage of married women in the U.S. labor force who have children under 6 has been increasing steadily since 1960.

Source: James M. Henslin, *Essentials of Sociology*, Allyn and Bacon, 2002

a. According to the graph, what is the y-intercept? Describe what this represents in this situation.

b. Use the coordinates of the two points shown to compute the slope. What does this mean in terms of rate of change?

c. Use the y-intercept from part (a) and the slope from part (b) to write an equation that models the percentage of married women in the labor force with preschool children, y, x years after 1960. Write your model in the form $y = mx + b$.

d. Use your model from part (c) to predict the percentage of married women in the labor force with preschool children in 2006.

Writing in Mathematics

63. Describe how to find the slope and the y-intercept of a line whose equation is given.

64. Describe how to graph a line using the slope and y-intercept. Provide an original example with your description.

65. A formula in the form $y = mx + b$ models the cost, y, of a four-year college x years after 2003. Would you expect m to be positive, negative, or zero? Explain your answer.

Critical Thinking Exercises

66. Which one of the following is true?

    a. The equation $y = mx + b$ shows that no line can have a y-intercept that is numerically equal to its slope.

    b. Every line in the rectangular coordinate system has an equation that can be expressed in slope-intercept form.

    c. The line $3x + 2y = 5$ has slope $-\frac{3}{2}$.

    d. The line $2y = 3x + 7$ has a y-intercept of 7.

67. The relationship between Celsius temperature, C, and Fahrenheit temperature, F, can be described by a linear equation in the form $F = mC + b$. The graph of this equation contains the point $(0, 32)$: Water freezes at $0°C$ or at $32°F$. The line also contains the point $(100, 212)$: Water boils at $100°C$ or at $212°F$. Write the linear equation expressing Fahrenheit temperature in terms of Celsius temperature.

68. The graph in the next column indicates that lower fertility rates (the number of births per woman) are related to the percentage of the population using contraceptives. A line that best fits the data is shown. Estimate the y-intercept and the slope of this line. Then write the line's slope-intercept equation. Use the equation to find the number of births per woman if 90% of married women of child-bearing age used contraceptives.

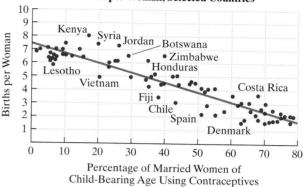

Contraceptive Prevalence and Average Number of Births per Woman, Selected Countries

Source: Population Reference Bureau

Review Exercises

69. Solve: $\dfrac{x}{2} + 7 = 13 - \dfrac{x}{4}$. (Section 2.3, Example 4)

70. Simplify: $3(12 \div 2^2 - 3)^2$. (Section 1.8, Example 6)

71. 14 is 25% of what? (Section 2.4, Example 8)

✔ MID-CHAPTER CHECK POINT

CHAPTER

4

What You Know: We learned to graph equations in two variables using point plotting, as well as a variety of other techniques. We used intercepts and a checkpoint to graph linear equations in the form $Ax + By = C$. We saw that the graph of a linear equation in one variable is a horizontal or a vertical line: $y = b$ graphs as a horizontal line and $x = a$ graphs as a vertical line. We determined a line's steepness, or rate of change, by computing its slope and we saw that lines with the same slope are parallel. Finally, we learned to graph linear equations in slope-intercept form, $y = mx + b$, using the slope, m, and the y-intercept, b.

In Exercises 1–3, use each graph to determine

    a. *the x-intercept, or state that there is no x-intercept.*

    b. *the y-intercept, or state that there is no y-intercept.*

    c. *the line's slope, or state that the slope is undefined.*

1.

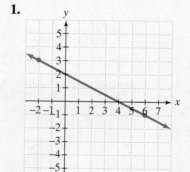

2.

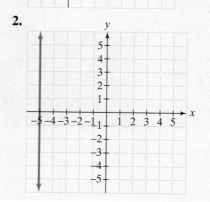

3.

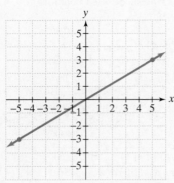

In Exercises 4–15, graph each equation in a rectangular coordinate system.

4. $y = -2x$

5. $y = -2$

6. $x + y = -2$

7. $y = \frac{1}{3}x - 2$

8. $x = 3.5$

9. $4x - 2y = 8$

10. $y = 3x + 2$

11. $3x + y = 0$

12. $y = x^2 - 4$

13. $y = x - 4$

14. $5y = -3x$

15. $5y = 20$

16. Find the slope and the y-intercept of the line whose equation is $5x - 2y = 10$.

17. Determine whether the line through $(2, -4)$ and $(7, 0)$ is parallel to a second line through $(-4, 2)$ and $(1, 6)$.

18. The graph shows the percentage of U.S. colleges that offered distance learning by computer for selected years from 1995 through 2002.

Percentage of U.S. Colleges Offering Distance Learning by Computer

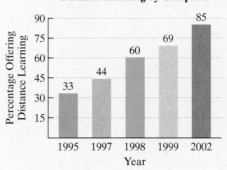

Source: International Data Corporation

The data for the years 1995 through 2002 can be modeled by the linear equation

$$y = 7.8x + 33$$

where x is the number of years after 1995 and y is the percentage of U.S. Colleges offering distance learning.

a. Find the y-intercept of this line.

b. Describe what this y-intercept represents.

c. Find the slope of this line.

d. Describe the meaning of this slope as a rate of change.

SECTION

4.5

THE POINT-SLOPE FORM OF THE EQUATION OF A LINE

Objectives

1 Use the point-slope form to write equations of a line.

2 Write linear equations that model data and make predictions.

Surprised by the number of people smoking cigarettes in movies and television shows made in the 1940s and 1950s? At that time, there was little awareness of the relationship between tobacco use and numerous diseases. Cigarette smoking was seen as a healthy way to relax and help digest a hearty meal. Then, in 1964, a linear equation changed everything. To understand the mathematics behind this turning point in public health, we explore another form of a line's equation.

Point-Slope Form   We can use the slope of a line to obtain another useful form of the line's equation. Consider a nonvertical line that has slope m and contains the point (x_1, y_1). Now, let (x, y) represent any other point on the line, shown in Figure 4.28. Keep in mind that the point (x, y) is arbitrary and is not in one fixed position. By contrast, the point (x_1, y_1) is fixed.

Regardless of where the point (x, y) is located, the steepness of the line in Figure 4.28 remains the same. Thus, the ratio for slope stays a constant m. This means that for all points along the line,

$$m = \frac{\text{Change in } y}{\text{Change in } x} = \frac{y - y_1}{x - x_1}.$$

We can clear the fraction by multiplying both sides by $x - x_1$, the least common denominator.

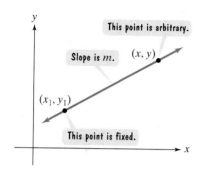

FIGURE 4.28   A line passing through (x_1, y_1) with slope m

$$m = \frac{y - y_1}{x - x_1} \qquad \text{This is the slope of the line in Figure 4.28.}$$

$$m(x - x_1) = \frac{y - y_1}{x - x_1} \cdot (x - x_1) \qquad \text{Multiply both sides by } x - x_1.$$

$$m(x - x_1) = y - y_1 \qquad \text{Simplify: } \frac{y - y_1}{x - x_1} \cdot x - x_1 = y - y_1$$

Now, if we reverse the two sides, we obtain the **point-slope form** of the equation of a line.

> **POINT-SLOPE FORM OF THE EQUATION OF A LINE**   The **point-slope equation** of a nonvertical line with slope m that passes through the point (x_1, y_1) is
>
> $$y - y_1 = m(x - x_1).$$

For example, the point-slope equation of the line passing through $(1, 4)$ with slope 2 $(m = 2)$ is

$$y - 4 = 2(x - 1).$$

① Use the point-slope form to write equations of a line.

Using the Point-Slope Form to Write a Line's Equation   If we know the slope of a line and a point not containing the y-intercept through which the line passes, the point-slope form is the equation that we should use. Once we have obtained this equation, it is customary to solve for y and write the equation in slope-intercept form. Examples 1 and 2 illustrate these ideas.

EXAMPLE 1   Writing the Point-Slope Form and the Slope-Intercept Form

Write the point-slope form and the slope-intercept form of the equation of the line with slope 4 that passes through the point $(-1, 3)$.

SOLUTION   We begin with the point-slope equation of a line with $m = 4$, $x_1 = -1$, and $y_1 = 3$.

$$y - y_1 = m(x - x_1) \qquad \text{This is the point-slope form of the equation.}$$

$$y - 3 = 4[x - (-1)] \qquad \text{Substitute the given values.}$$

$$y - 3 = 4(x + 1) \qquad \text{We now have the point-slope form of the equation of the given line.}$$

Now we solve this equation for y and write an equivalent equation in slope-intercept form ($y = mx + b$).

$$y - 3 = 4(x + 1)$$ This is the point-slope equation.

$$y - 3 = 4x + 4$$ Use the distributive property.

$$y = 4x + 7$$ Add 3 to both sides.

The slope-intercept form of the line's equation is $y = 4x + 7$. ■

✔ **CHECK POINT 1** Write the point-slope form and the slope-intercept form of the equation of the line with slope 6 that passes through the point $(2, -5)$.

EXAMPLE 2 Writing the Point-Slope Form and the Slope-Intercept Form

A line passes through the points $(4, -3)$ and $(-2, 6)$. (See Figure 4.29.) Find the equation of the line

a. in point-slope form. **b.** in slope-intercept form.

SOLUTION

a. To use the point-slope form, we need to find the slope. The slope is the change in the y-coordinates divided by the corresponding change in the x-coordinates.

$$m = \frac{6 - (-3)}{-2 - 4} = \frac{9}{-6} = -\frac{3}{2}$$ This is the definition of slope using $(4, -3)$ and $(-2, 6)$.

We can take either point on the line to be (x_1, y_1). Let's use $(x_1, y_1) = (4, -3)$. Now, we are ready to write the point-slope equation.

$$y - y_1 = m(x - x_1)$$ This is the point-slope form of the equation.

$$y - (-3) = -\frac{3}{2}(x - 4)$$ Substitute: $(x_1, y_1) = (4, -3)$ and $m = -\frac{3}{2}$.

$$y + 3 = -\frac{3}{2}(x - 4)$$ Simplify.

This equation is the point-slope form of the equation of the line shown in Figure 4.29.

b. Now, we solve this equation for y and write an equivalent equation in slope-intercept form ($y = mx + b$).

$$y + 3 = -\frac{3}{2}(x - 4)$$ This is the point-slope equation.

$$y + 3 = -\frac{3}{2}x + 6$$ Use the distributive property.

$$y = -\frac{3}{2}x + 3$$ Subtract 3 from both sides.

This equation is the slope-intercept form of the equation of the line shown in Figure 4.29. ■

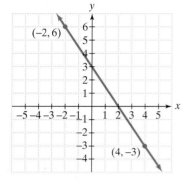

FIGURE 4.29

DISCOVER FOR YOURSELF

If you are given two points on a line, you can use either point for (x_1, y_1) when you write its point-slope equation. Rework Example 2 using $(-2, 6)$ for (x_1, y_1). Once you solve for y, you should obtain the same slope-intercept equation as the one shown in the last line of the solution to Example 2.

> ✔ **CHECK POINT 2** A line passes through the points $(-2, -1)$ and $(-1, -6)$. Find the equation of the line
> **a.** in point-slope form. **b.** in slope-intercept form.

In Examples 1 and 2, we eventually write a line's equation in slope-intercept form? But where do we start our work?

Starting with $y = mx + b$	Starting with $y - y_1 = m(x - x_1)$
Begin with the slope-intercept form if you know	Begin with the point-slope form if you know
1. The slope of the line and the y-intercept.	1. The slope of the line and a point on the line not containing the y-intercept or
	2. Two points on the line, neither of which contains the y-intercept.

② Write linear equations that model data and make predictions.

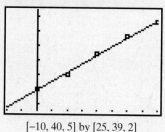
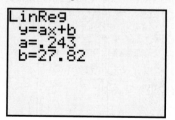
Applications Linear equations are useful for modeling data that fall on or near a line. For example, the bar graph in Figure 4.30(a) gives the median age of the U.S. population in the indicated year. (The median age is the age in the middle when all the ages of the U.S. population are arranged from youngest to oldest.) The data are displayed as a set of five points in a rectangular coordinate system in Figure 4.30(b).

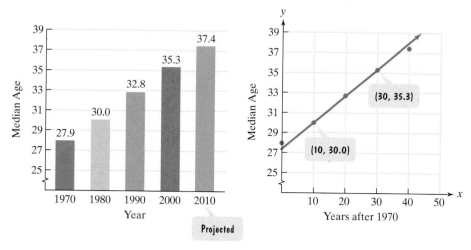

The Graying of America: Median Age of the U.S. Population

FIGURE 4.30(a)

Source: U.S. Census Bureau

FIGURE 4.30(b)

A set of points representing data is called a **scatter plot**. Also shown on the scatter plot in Figure 4.30(b) is a line that passes through or near the five points. By writing the equation of this line, we can obtain a model of the data and make predictions about the median age of the U.S. population in the future.

EXAMPLE 3 Modeling the Graying of America

Write the slope-intercept equation of the line shown in Figure 4.30(b). Use the equation to predict the median age of the U.S. population in 2020.

SOLUTION The line in Figure 4.30(b) passes through $(10, 30.0)$ and $(30, 35.3)$. We start by finding its slope.

$$m = \frac{\text{Change in } y}{\text{Change in } x} = \frac{35.3 - 30.0}{30 - 10} = \frac{5.3}{20} = 0.265$$

The slope indicates that each year the median age of the U.S. population is increasing by 0.265 years.

Now, we write the line's slope-intercept equation.

$y - y_1 = m(x - x_1)$	Begin with the point-slope form.
$y - 30.0 = 0.265(x - 10)$	Either ordered pair can be (x_1, y_1). Let $(x_1, y_1) = (10, 30.0)$. From above, $m = 0.265$.
$y - 30.0 = 0.265x - 2.65$	Apply the distributive property.
$y = 0.265x + 27.35$	Add 30 to both sides and solve for y.

A linear equation that models the median age of the U.S. population, y, x years after 1970 is

$$y = 0.265x + 27.35.$$

Now, let's use this equation to predict the median age in 2020. Because 2020 is 50 years after 1970, substitute 50 for x and compute y.

$$y = 0.265(50) + 27.35 = 40.6$$

Our model predicts that the median age of the U.S. population in 2020 will be 40.6. ■

CHECK POINT 3 Use the data points $(10, 30.0)$ and $(20, 32.8)$ from Figure 4.30(b) on page 253 to write a slope-intercept equation that models the median age of the U.S. population x years after 1970. Use this model to predict the median age in 2020.

ENRICHMENT ESSAY

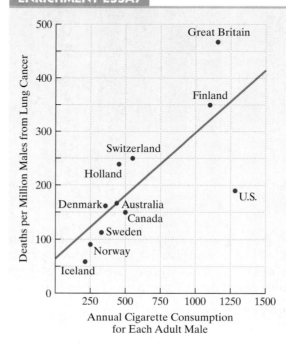

Annual Cigarette Consumption for Each Adult Male

Source: Smoking and Health, Washington, D.C., 1964

Cigarettes and Lung Cancer

This scatter plot shows a relationship between cigarette consumption among males and deaths due to lung cancer per million males. The data are from 11 countries and date back to a 1964 report by the U.S. Surgeon General. The scatter plot can be modeled by a line whose slope indicates an increasing death rate from lung cancer with increased cigarette consumption. At that time, the tobacco industry argued that in spite of this regression line, tobacco use is not the cause of cancer. Recent data do, indeed, show a causal effect between tobacco use and numerous diseases.

4.5 EXERCISE SET

Student Solutions Manual   CD/Video   PH Math/Tutor Center   MathXL Tutorials on CD   MathXL®   MyMathLab   Interactmath.com

Practice Exercises

Write the point-slope form of the line satisfying each of the conditions in Exercises 1–28. Then use the point-slope form of the equation to write the slope-intercept form of the equation.

1. Slope = 3, passing through $(2, 5)$

2. Slope = 6, passing through $(3, 1)$

3. Slope = 5, passing through $(-2, 6)$

4. Slope = 7, passing through $(-4, 9)$

5. Slope = -8, passing through $(-3, -2)$

6. Slope = -4, passing through $(-5, -2)$

7. Slope = -12, passing through $(-8, 0)$

8. Slope = -11, passing through $(0, -3)$

9. Slope = -1, passing through $\left(-\frac{1}{2}, -2\right)$

10. Slope = -1, passing through $\left(-4, -\frac{1}{4}\right)$

11. Slope = $\frac{1}{2}$, passing through the origin

12. Slope = $\frac{1}{3}$, passing through the origin

13. Slope = $-\frac{2}{3}$, passing through $(6, -2)$

14. Slope = $-\frac{3}{5}$, passing through $(10, -4)$

15. Passing through $(1, 2)$ and $(5, 10)$

16. Passing through $(3, 5)$ and $(8, 15)$

17. Passing through $(-3, 0)$ and $(0, 3)$

18. Passing through $(-2, 0)$ and $(0, 2)$

19. Passing through $(-3, -1)$ and $(2, 4)$

20. Passing through $(-2, -4)$ and $(1, -1)$

21. Passing through $(-4, -1)$ and $(3, 4)$

22. Passing through $(-6, 1)$ and $(2, -5)$

23. Passing through $(-3, -1)$ and $(4, -1)$

24. Passing through $(-2, -5)$ and $(6, -5)$

25. Passing through $(2, 4)$ with x-intercept = -2

26. Passing through $(1, -3)$ with x-intercept = -1

27. x-intercept = $-\frac{1}{2}$ and y-intercept = 4

28. x-intercept = 4 and y-intercept = -2

Practice Plus

In Exercises 29–36, write an equation in slope-intercept form of the line satisfying the given conditions.

29. The line passes through $(-3, 2)$ and is parallel to the line whose equation is $y = 4x + 1$.

30. The line passes through $(5, -3)$ and is parallel to the line whose equation is $y = 2x + 1$.

31. The line passes through $(-1, -5)$ and is parallel to the line whose equation is $3x + y = 6$.

32. The line passes through $(-4, -7)$ and is parallel to the line whose equation is $6x + y = 8$.

33. The line passes through $(2, 4)$ and has the same y-intercept as the line whose equation is $x - 4y = 8$.

34. The line passes through $(2, 6)$ and has the same y-intercept as the line whose equation is $x - 3y = 18$.

35. The line has an x-intercept at -4 and is parallel to the line containing $(3, 1)$ and $(2, 6)$.

36. The line has an x-intercept at -6 and is parallel to the line containing $(4, -3)$ and $(2, 2)$.

Application Exercises

37. We seem to be fed up with being lectured at about our waistlines. The points in the graph show the average weight of American adults from 1990 through 2002. Also shown is a line that passes through or near the points.

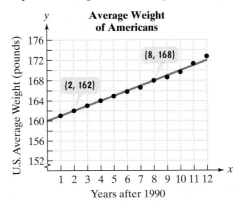

Source: Diabetes Care

(Be sure to refer to the graph on page 255.)

a. Use the two points whose coordinates are shown by the voice balloons to find the point-slope equation of the line that models average weight of Americans, y, in pounds, x years after 1990.

b. Write the equation in part (a) in slope-intercept form.

c. Use the slope-intercept equation to predict the average weight of Americans in 2010.

38. Films may not be getting any better, but in this era of moviegoing, the number of screens available for new films and the classics has exploded. The points in the graph show the number of screens in the United States from 1995 through 2002. Also shown is a line that passes through or near the points.

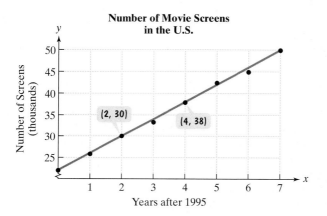

Number of Movie Screens in the U.S.

(2, 30) (4, 38)

Years after 1995

Source: Motion Picture Association of America

a. Use the two points whose coordinates are shown by the voice balloons to find the point-slope equation of the line that models the number of screens, y, in thousands, x years after 1995.

b. Write the equation in part (a) in slope-intercept form.

c. Use the slope-intercept equation to predict the number of screens, in thousands, in 2010.

39. Is there a relationship between education and prejudice? With increased education, does a person's level of prejudice tend to decrease? The scatter plot at the top of the next column shows ten data points, each representing the number of years of school completed and the score on a test measuring prejudice for each subject. Higher scores on this 1-to-10 test indicate greater prejudice. Also shown is the regression line, the line that best fits the data. Use two points on this line to write both its point-slope and slope-intercept equations. Then use the slope-intercept equation to predict the score on the prejudice test for a person with seven years of education.

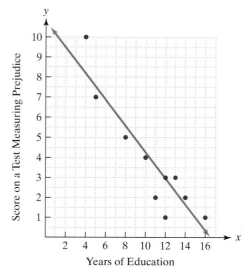

Years of Education

40. A business discovers a linear relationship between the number of shirts it can sell and the price per shirt. In particular, 20,000 shirts per week can be sold at $19 each. Raising the price to $55 causes the sales to fall to 2000 shirts per week. Write the point-slope and slope-intercept equations of the *demand line* through the ordered pairs

(20,000 shirts, $19) and (2000 shirts, $55).

Then determine the number of shirts that can be sold per week at $50 each.

41. The scatter plot shows the average number of minutes each that 16 people exercise per week and the average number of headaches per month each person experiences.

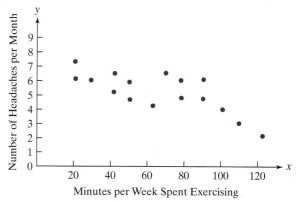

Minutes per Week Spent Exercising

a. Draw a line that fits the data so that the spread of the data points around the line is as small as possible.

b. Use the coordinates of two points along your line to write its point-slope and slope-intercept equations.

c. Use the equation in part (b) to predict the number of headaches per month for a person exercising 130 minutes per week.

d. What general observation can you make about the number of headaches per month as the number of minutes per week spent exercising increases?

Writing in Mathematics

42. Describe how to write the equation of a line if its slope and a point along the line are known.

43. Describe how to write the equation of a line if two points along the line are known.

44. Take a second look at the scatter plot in Exercise 39. Although there is a relationship between education and prejudice, we cannot necessarily conclude that increased education causes a person's level of prejudice to decrease. Offer two or more possible explanations for the data in the scatter plot.

Critical Thinking Exercises

45. Which one of the following is true?
   a. If a line has undefined slope, then it has no equation.
   b. The line whose equation is $y - 3 = 7(x + 2)$ passes through $(-3, 2)$.
   c. The point-slope form cannot be applied to the line through the points $(2, -5)$ and $(2, 6)$.
   d. The slope of the line whose equation is $3x + y = 7$ is 3.

46. Excited about the success of celebrity stamps, post office officials were rumored to have put forth a plan to institute two new types of thermometers. On these new scales, $°E$ represents degrees Elvis and $°M$ represents degrees Madonna. If it is known that $40°E = 25°M$, $280°E = 125°M$, and degrees Elvis is linearly related to degrees Madonna, write an equation expressing E in terms of M.

Technology Exercises

47. Use a graphing utility to graph $y = 1.75x - 2$. Select the best viewing rectangle possible by experimenting with the range settings to show that the line's slope is $\frac{7}{4}$.

48. Use a graphing utility to graph the slope-intercept equation that you wrote in Exercise 40. Then select an appropriate

range setting and use the $\boxed{\text{TRACE}}$ feature to graphically show the number of shirts that can be sold at $50 each.

49. a. Use the statistical menu of a graphing utility to enter the ten data points shown in the scatter plot in Exercise 39.

   b. Use the $\boxed{\text{DRAW}}$ menu and the scatter plot capability to draw a scatter plot of the data points like the one shown in Exercise 39.

   c. Select the linear regression option. Use your utility to obtain values for a and b for the equation of the regression line, $y = ax + b$. You may also be given a *correlation coefficient*, r. Values of r close to 1 indicate that the points can be described by a linear relationship and the regression line has a positive slope. Values of r close to -1 indicate that the points can be described by a linear relationship and the regression line has a negative slope. Values of r close to 0 indicate no linear relationship between the variables.

   d. Use the appropriate sequence (consult your manual) to graph the regression equation on top of the points in the scatter plot.

Review Exercises

50. How many sheets of paper, weighing 2 grams each, can be put in an envelope weighing 4 grams if the total weight must not exceed 29 grams? (Section 2.6, Example 8, and Section 2.5, Example 5)

51. List all the natural numbers in this set:
$$\left\{ -2, 0, \frac{1}{2}, 1, \sqrt{3}, \sqrt{4} \right\}.$$
(Section 1.2, Example 5)

52. Use intercepts to graph $3x - 5y = 15$. (Section 4.2, Example 4)

SECTION 4.6

LINEAR INEQUALITIES IN TWO VARIABLES

Objectives

1 Determine whether an ordered pair is a solution of an inequality.

2 Graph a linear inequality in two variables.

3 Solve applied problems involving linear inequalities in two variables.

We opened this chapter with the observation that temperature and precipitation are two variables that affect whether regions are forests, grasslands, or deserts. In this section, you will see how linear inequalities in two variables describe some of the most magnificent places in our nation's landscape.

1 Determine whether an ordered pair is a solution of an inequality.

Linear Inequalities in Two Variables and Their Solutions We have seen that equations in the form $Ax + By = C$ are straight lines when graphed. If we change the = sign to $>$, $<$, $\geq$, or $\leq$, we obtain a **linear inequality in two variables**. Some examples of linear inequalities in two variables are $x + y > 2$, $3x - 5y \leq 15$, and $2x - y < 4$.

A **solution of an inequality in two variables**, x and y, is an ordered pair of real numbers with the following property: When the x-coordinate is substituted for x and the y-coordinate is substituted for y in the inequality, we obtain a true statement. For example, $(3, 2)$ is a solution of the inequality $x + y > 1$. When 3 is substituted for x and 2 is substituted for y, we obtain the true statement $3 + 2 > 1$, or $5 > 1$. Because there are infinitely many pairs of numbers that have a sum greater than 1, the inequality $x + y > 1$ has infinitely many solutions. Each ordered-pair solution is said to **satisfy** the inequality. Thus, $(3, 2)$ satisfies the inequality $x + y > 1$.

EXAMPLE 1 Deciding Whether an Ordered Pair Satisfies an Inequality

Determine whether each ordered pair is a solution of the inequality

$$2x - 3y \geq 6:$$

a. $(0, 0)$ **b.** $(3, -1)$.

SOLUTION

a. To determine whether $(0, 0)$ is a solution of the inequality, we substitute 0 for x and 0 for y.

$2x - 3y \geq 6$	This is the given inequality.
$2 \cdot 0 - 3 \cdot 0 \overset{?}{\geq} 6$	Substitute 0 for x and 0 for y.
$0 - 0 \overset{?}{\geq} 6$	Multiply: $2 \cdot 0 = 0$ and $3 \cdot 0 = 0$.
This statement is false. $\quad 0 \geq 6$	Subtract: $0 - 0 = 0$.

Because we obtain a false statement, we conclude that $(0, 0)$ is not a solution of $2x - 3y \geq 6$. The ordered pair $(0, 0)$ does not satisfy the inequality.

b. To determine whether $(3, -1)$ is a solution of the inequality, we substitute 3 for x and -1 for y.

$2x - 3y \geq 6$	This is the given inequality.
$2 \cdot 3 - 3(-1) \overset{?}{\geq} 6$	Substitute 3 for x and -1 for y.
$6 - (-3) \overset{?}{\geq} 6$	Multiply: $2 \cdot 3 = 6$ and $3(-1) = -3$.
This statement is true. $\quad 9 \geq 6$	Subtract: $6 - (-3) = 6 + 3 = 9$.

Because we obtain a true statement, we conclude that $(3, -1)$ is a solution of $2x - 3y \geq 6$. The ordered pair $(3, -1)$ satisfies the inequality. ■

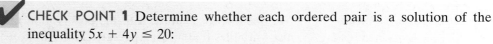 **CHECK POINT 1** Determine whether each ordered pair is a solution of the inequality $5x + 4y \leq 20$:

a. $(0, 0)$ **b.** $(6, 2)$.

② Graph a linear inequality in two variables.

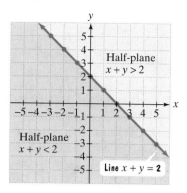

FIGURE 4.31

The Graph of a Linear Inequality in Two Variables We know that the graph of an equation in two variables is the set of all points whose coordinates satisfy the equation. Similarly, the **graph of an inequality in two variables** is the set of all points whose coordinates satisfy the inequality.

Let's use Figure 4.31 to get an idea of what the graph of a linear inequality in two variables looks like. Part of the figure shows the graph of the linear equation $x + y = 2$. The line divides the points in the rectangular coordinate system into three sets. First, there is the set of points along the line, satisfying $x + y = 2$. Next, there is the set of points in the green region above the line. Points in the green region satisfy the linear inequality $x + y > 2$. Finally, there is the set of points in the purple region below the line. Points in the purple region satisfy the linear inequality $x + y < 2$.

A **half-plane** is the set of all the points on one side of a line. In Figure 4.31, the green region is a half-plane. The purple region is also a half-plane. A half-plane is the graph of a linear inequality that involves > or <. The graph of an inequality that involves ≥ or ≤ is a half-plane and a line. A solid line is used to show that the line is part of the graph. A dashed line is used to show that a line is not part of a graph.

GRAPHING A LINEAR INEQUALITY IN TWO VARIABLES

1. Replace the inequality symbol with an equal sign and graph the corresponding linear equation. Draw a solid line if the original inequality contains a ≤ or ≥ symbol. Draw a dashed line if the original inequality contains a < or > symbol.

2. Choose a test point in one of the half-planes that is not on the line. Substitute the coordinates of the test point into the inequality.

3. If a true statement results, shade the half-plane containing this test point. If a false statement results, shade the half-plane not containing this test point.

EXAMPLE 2 Graphing a Linear Inequality in Two Variables

Graph: $3x - 5y < 15$.

SOLUTION

Step 1. **Replace the inequality symbol with = and graph the linear equation.** We need to graph $3x - 5y = 15$. We can use intercepts to graph this line.

We set $y = 0$ to find the x-intercept:	We set $x = 0$ to find the y-intercept:
$3x - 5y = 15$	$3x - 5y = 15$
$3x - 5 \cdot 0 = 15$	$3 \cdot 0 - 5y = 15$
$3x = 15$	$-5y = 15$
$x = 5.$	$y = -3.$

The x-intercept is 5, so the line passes through $(5, 0)$. The y-intercept is -3, so the line passes through $(0, -3)$. The graph of the equation is indicated by a dashed line because the inequality $3x - 5y < 15$ contains a < symbol, rather than ≤. The graph of the line is shown in Figure 4.32.

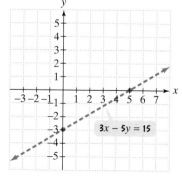

FIGURE 4.32 Preparing to graph $3x - 5y < 15$

Step 2. **Choose a test point in one of the half-planes that is not on the line. Substitute its coordinates into the inequality.** The line $3x - 5y = 15$ divides the plane into three parts—the line itself and two half-planes. The points in one half-plane

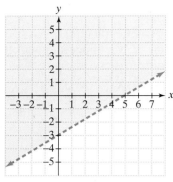

FIGURE 4.33 The graph of $3x - 5y < 15$

satisfy $3x - 5y > 15$. The points in the other half-plane satisfy $3x - 5y < 15$. We need to find which half-plane is the solution. To do so, we test a point from either half-plane. The origin, $(0, 0)$, is the easiest point to test.

$$3x - 5y < 15 \qquad \text{This is the given inequality.}$$
$$3 \cdot 0 - 5 \cdot 0 \overset{?}{<} 15 \qquad \text{Test } (0, 0) \text{ by substituting 0 for } x \text{ and 0 for } y.$$
$$0 - 0 \overset{?}{<} 15 \qquad \text{Multiply.}$$
$$0 < 15 \qquad \text{This statement is true.}$$

Step 3. **If a true statement results, shade the half-plane containing the test point.** Because 0 is less than 15, the test point $(0, 0)$ is part of the solution set. All the points on the same side of the line $3x - 5y = 15$ as the point $(0, 0)$ are members of the solution set. The solution set is the half-plane that contains the point $(0, 0)$, indicated by shading this half-plane. The graph is shown using green shading and a dashed blue line in Figure 4.33. ∎

✔ **CHECK POINT 2** Graph: $2x - 4y < 8$.

When graphing a linear inequality, test a point that lies in one of the half-planes and *not on the line dividing the half-planes*. The test point $(0, 0)$ is convenient because it is easy to calculate when 0 is substituted for each variable. However, if $(0, 0)$ lies on the dividing line and not in a half-plane, a different test point must be selected.

EXAMPLE 3 Graphing a Linear Inequality

Graph: $y \leq \dfrac{2}{3} x$.

SOLUTION

Step 1. **Replace the inequality symbol with = and graph the linear equation.** We need to graph $y = \frac{2}{3} x$. We can use the slope and the y-intercept to graph this line.

$$y = \frac{2}{3} x + 0$$

slope $= \frac{2}{3} = \frac{\text{rise}}{\text{run}}$ y-intercept is 0.

The y-intercept is 0, so the line passes through $(0, 0)$. Using the y-intercept and the slope, the line is shown in Figure 4.34 as a solid line. This is because the inequality $y \leq \frac{2}{3} x$ contains a $\leq$ symbol, in which equality is included.

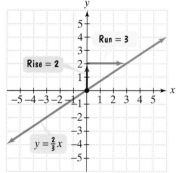

FIGURE 4.34 Preparing to graph $y \leq \dfrac{2}{3} x$

Step 2. **Choose a test point in one of the half-planes that is not on the line. Substitute its coordinates into the inequality.** We cannot use $(0, 0)$ as a test point because it lies on the line and not in a half-plane. Let's use $(1, 1)$, which lies in the half-plane above the line.

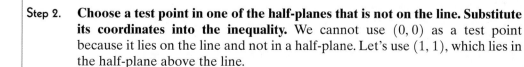

$$y \leq \frac{2}{3} x \qquad \text{This is the given inequality.}$$
$$1 \overset{?}{\leq} \frac{2}{3} \cdot 1 \qquad \text{Test } (1, 1) \text{ by substituting 1 for } x \text{ and 1 for } y.$$
$$1 \leq \frac{2}{3} \qquad \text{This statement is false.}$$

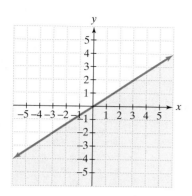

FIGURE 4.35 The graph of $y \leq \frac{2}{3}x$

Step 3. **If a false statement results, shade the half-plane not containing the test point.** Because 1 is not less than or equal to $\frac{2}{3}$, the test point $(1, 1)$ is not part of the solution set. Thus, the half-plane below the solid line $y = \frac{2}{3}x$ is part of the solution set. The solution set is the line and the half-plane that does not contain the point $(1, 1)$, indicated by shading this half-plane. The graph is shown using green shading and a blue line in Figure 4.35. ∎

✔ **CHECK POINT 3** Graph: $y \geq \frac{1}{2}x$.

STUDY TIP

You can graph inequalities in the form $y > mx + b$ or $y < mx + b$ without using test points. The inequality symbol indicates which half-plane to shade.
- If $y > mx + b$, shade above the line $y = mx + b$.
- If $y < mx + b$, shade below the line $y = mx + b$.

Continue using test points to graph inequalities in the form $Ax + By > C$ or $Ax + By < C$. The graph of $Ax + By > C$ can lie above or below the line $Ax + By = C$, depending on the value of B. The same comment applies to the graph of $Ax + By < C$.

In Section 4.2, we learned that $y = b$ graphs as a horizontal line, where b is the y-intercept. Similarly, the graph of $x = a$ is a vertical line, where a is the x-intercept. Half-planes can be separated by horizontal or vertical lines. For example, Figure 4.36 shows the graph of $y \leq 2$. Because $(0,0)$ satisfies this inequality ($0 \leq 2$ is true), the graph consists of the half-plane below the line $y = 2$ and the line. Similarly, Figure 4.37 shows the graph of $x < 4$.

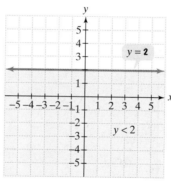

FIGURE 4.36 The graph of $y \leq 2$

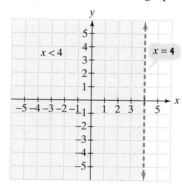

FIGURE 4.37 The graph of $x < 4$

It is not necessary to use test points when graphing inequalities involving vertical or horizontal lines.

For the Vertical Line $x = a$:
- If $x > a$, shade to the right of $x = a$.
- If $x < a$, shade to the left of $x = a$.

For the Horizontal Line $y = b$:
- If $y > b$, shade above $y = b$.
- If $y < b$, shade below $y = b$.

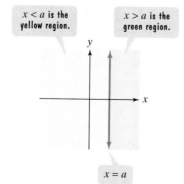

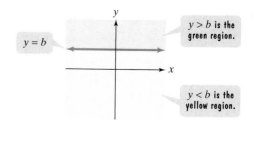

3 Solve applied problems involving linear inequalities in two variables.

Applications Temperature and precipitation affect whether or not trees and forests can grow. At certain levels of precipitation and temperature, only grasslands and deserts will exist. Figure 4.38 shows three kinds of regions—deserts, grasslands, and forests—that result from various ranges of temperature and precipitation. Notice that the horizontal axis is labeled T, for temperature, rather than x. The vertical axis is labeled P, for precipitation, rather than y. We can use inequalities in two variables, T and P, to describe the regions in the figure.

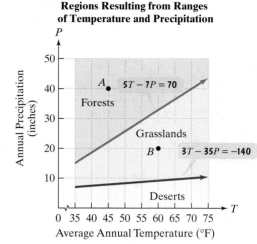

Regions Resulting from Ranges of Temperature and Precipitation

FIGURE 4.38

Source: A. Miller and J. Thompson, *Elements of Meteorology*

EXAMPLE 4 Forests and Inequalities

a. Use Figure 4.38 to find the coordinates of point A. What does this mean in terms of the kind of region that occurs?

b. For average annual temperatures that exceed 35°F, the inequality

$$5T - 7P < 70$$

models where forests occur. Show that the coordinates of point A satisfy this inequality.

SOLUTION

a. Point A has coordinates $(45, 40)$. This means that if a region has an average annual temperature of 45°F and an average annual precipitation of 40 inches, then a forest occurs.

b. We can show that $(45, 40)$ satisfies the inequality for forests by substituting 45 for T and 40 for P.

$$5T - 7P < 70 \qquad \text{This is the given inequality.}$$
$$5 \cdot 45 - 7 \cdot 40 \overset{?}{<} 70 \qquad \text{Substitute 45 for } T \text{ and 40 for } P.$$
$$225 - 280 \overset{?}{<} 70 \qquad \text{Multiply: } 5 \cdot 45 = 225 \text{ and } 7 \cdot 40 = 280.$$

$$\underset{\text{This statement is true.}}{\quad} -55 < 70 \qquad \text{Subtract: } 225 - 280 = -55.$$

The coordinates $(45, 40)$ make the inequality true. Thus, $(45, 40)$ satisfies the inequality. ∎

CHECK POINT 4

a. Use Figure 4.38 to find the coordinates of point B. What does this mean in terms of the kind of region that occurs?

b. For average annual temperatures that exceed 35°F, the inequalities $5T - 7P \geq 70$ and $3T - 35P \leq -140$ model where grasslands occur. Show that the coordinates of point B satisfy both of these inequalities.

4.6 EXERCISE SET

Student Solutions Manual CD/Video PH Math/Tutor Center MathXL Tutorials on CD MathXL® MyMathLab Interactmath.com

Practice Exercises

In Exercises 1–8, determine whether each ordered pair is a solution of the given inequality.

1. $x + y > 4$: $(2, 2)$, $(3, 2)$, $(-3, 8)$
2. $2x - y < 3$: $(0, 0)$, $(3, 0)$, $(-4, -15)$
3. $2x + y \geq 5$: $(4, 0)$, $(1, 3)$, $(0, 0)$
4. $3x - 5y \geq -12$: $(2, -3)$, $(2, 8)$, $(0, 0)$
5. $y \geq -2x + 4$: $(4, 0)$, $(1, 3)$, $(-2, -4)$
6. $y \leq -x + 5$: $(5, 0)$, $(0, 5)$, $(8, -4)$
7. $y > -2x + 1$: $(2, 3)$, $(0, 0)$, $(0, 5)$
8. $x < -y - 2$: $(-1, -1)$, $(0, 0)$, $(4, -5)$

In Exercises 9–36, graph each inequality.

9. $x + y \geq 3$
10. $x + y \geq 4$
11. $x - y < 5$
12. $x - y < 2$
13. $x + 2y > 4$
14. $2x + y > 6$
15. $3x - y \leq 6$
16. $x - 3y \leq -6$
17. $3x - 2y \leq 8$
18. $2x - 3y \geq 8$
19. $4x + 3y > 15$
20. $5x + 10y > 15$
21. $5x - y < -7$
22. $x - 5y < -7$
23. $y \leq \frac{1}{3}x$
24. $y \leq \frac{1}{4}x$
25. $y > 2x$
26. $y > 4x$
27. $y > 3x + 2$
28. $y > 2x - 1$
29. $y < \frac{3}{4}x - 3$
30. $y < \frac{2}{3}x - 1$
31. $x \leq 1$
32. $x \leq -3$
33. $y > 1$
34. $y > -3$
35. $x \geq 0$
36. $y \leq 0$

Practice Plus

In Exercises 37–44, write each sentence as a linear inequality in two variables. Then graph the inequality.

37. The sum of the x-variable and the y-variable is at least 2.

38. The difference between the x-variable and the y-variable is at least 3.

39. The difference between 5 times the x-variable and 2 times the y-variable is at most 10.

40. The sum of 4 times the x-variable and 2 times the y-variable is at most 8.

41. The y-variable is no less than $\frac{1}{2}$ of the x-variable.

42. The y-variable is no less than $\frac{1}{4}$ of the x-variable.

43. The y-variable is no more than -1.

44. The y-variable is no more than -2.

Application Exercises

45. Bottled water and medical supplies are to be shipped to victims of a hurricane by plane. Each plane can carry no more than 80,000 pounds. The bottled water weighs 20 pounds per container and each medical kit weighs 10 pounds. Let x represent the number of bottles of water to be shipped. Let y represent the number of medical kits. The plane's weight limitations can be described by the following inequality:

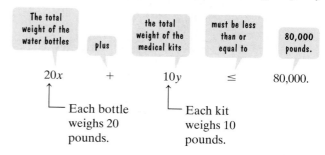

a. Graph the inequality. Because x and y must be positive, limit the graph to quadrant I only.

b. Select an ordered pair satisfying the inequality. What are its coordinates and what do they represent in this situation?

46. Bottled water and medical supplies are to be shipped to victims of a hurricane by plane. Each plane can carry a total volume that does not exceed 6000 cubic feet. Each water bottle is 2 cubic feet and each medical kit has a volume of 1 cubic foot. Let x represent the number of bottles of water to be shipped. Let y represent the number of medical kits. The plane's volume limitations can be described by the following inequality:

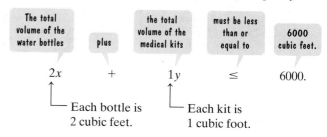

a. Graph the inequality. Because x and y must be positive, limit the graph to quadrant I only.

b. Select an ordered pair satisfying the inequality. What are its coordinates and what do they represent in this situation?

47. Many elevators have a capacity of 2000 pounds.

   a. If a child averages 50 pounds and an adult 150 pounds, write an inequality that describes when x children and y adults will cause the elevator to be overloaded.

   b. Graph the inequality. Because x and y must be positive, limit the graph to quadrant I only.

   c. Select an ordered pair satisfying the inequality. What are its coordinates and what do they represent in this situation?

48. A patient is not allowed to have more than 330 milligrams of cholesterol per day from a diet of eggs and meat. Each egg provides 165 milligrams of cholesterol. Each ounce of meat provides 110 milligrams of cholesterol.

   a. Write an inequality that describes the patient's dietary restrictions for x eggs and y ounces of meat.

   b. Graph the inequality. Because x and y must be positive, limit the graph to quadrant I only.

   c. Select an ordered pair satisfying the inequality. What are its coordinates and what do they represent in this situation?

The graph of a linear inequality in two variables is a region in the rectangular coordinate system. Regions in coordinate systems have numerous applications. For example, the regions in the following two graphs indicate whether a young person is overweight, borderline overweight, or normal weight.

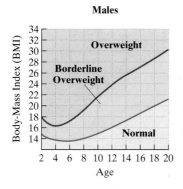

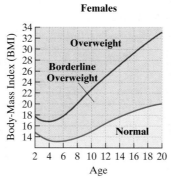

Source: Centers for Disease Control and Prevention

The horizontal axis shows a person's age. The vertical axis shows that person's body-mass index (BMI), computed using the following formula:

$$\text{BMI} = \frac{703W}{H^2}.$$

The variable W represents weight, in pounds. The variable H represents height, in inches. Use this information to solve Exercises 49–50.

49. A man is 20 years old, 72 inches (6 feet) tall, and weighs 200 pounds.

   a. Compute the man's BMI. Round to the nearest tenth.

   b. Use the man's age and his BMI to locate this information as a point in the coordinate system for males. Is this person overweight, borderline overweight, or normal weight?

50. A girl is 10 years old, 50 inches (4 feet, 2 inches) tall, and weighs 100 pounds.

   a. Compute the girl's BMI. Round to the nearest tenth.

   b. Use the girl's age and her BMI to locate this information as a point in the coordinate system for females. Is this person overweight, borderline overweight, or normal weight?

Writing in Mathematics

51. What is a linear inequality in two variables? Provide an example with your description.

52. How do you determine whether an ordered pair is a solution of an inequality in two variables, x and y?

53. What is a half-plane?

54. What does a solid line mean in the graph of an inequality?

55. What does a dashed line mean in the graph of an inequality?

56. Explain how to graph $2x - 3y < 6$.

57. Compare the graphs of $3x - 2y > 6$ and $3x - 2y \le 6$. Discuss similarities and differences between the graphs.

Critical Thinking Exercises

58. Which one of the following is true?

   a. The ordered pair $(0, -3)$ satisfies $y > 2x - 3$.

   b. The graph of $x < y + 1$ is the half-plane below the line $x = y + 1$.

   c. In graphing $y \ge 4x$, a dashed line is used.

   d. The graph of $x < 4$ is the half-plane to the left of the vertical line described by $x = 4$.

In Exercises 59–60, write an inequality that represents each graph.

59.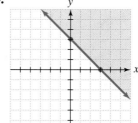

60.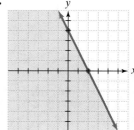

Technology Exercises

Graphing utilities can be used to shade regions in the rectangular coordinate system, thereby graphing an inequality in two variables. Read the section of the user's manual for your graphing utility that describes how to shade a region. Then use your graphing utility to graph the inequalities in Exercises 61–64.

61. $y \le 4x + 4$

62. $y \ge x - 2$

63. $y \ge \dfrac{1}{2}x + 4$

64. $y \le -\dfrac{1}{2}x + 4$

Review Exercises

65. Solve for h: $V = lwh$. (Section 2.4, Example 1)

66. Find the quotient: $\dfrac{2}{3} \div \left(-\dfrac{5}{4}\right)$. (Section 1.7, Example 5)

67. Evaluate $x^2 - 4$ for $x = -3$. (Section 1.8, Example 10)

GROUP PROJECT

CHAPTER 4

In Example 3 on pages 253–254, we used the data in Figure 4.30 to develop a linear equation that modeled the graying of America. For this group exercise, you might find it helpful to pattern your work after Figure 4.30 and the solution to Example 3. Group members should begin by consulting an almanac, newspaper, magazine, or the Internet to find data that lie approximately on or near a straight line. Working by hand or using a graphing utility, construct a scatter plot for the data. If working by hand, draw a line that approximately fits the data and then write its equation. If using a graphing utility, obtain the equation of the regression line. Then use the equation of the line to make a prediction about what might happen in the future. Are there circumstances that might affect the accuracy of this prediction? List some of these circumstances.

CHAPTER 4 SUMMARY

Definitions and Concepts	Examples

Section 4.1 Graphing Equations in Two Variables

An ordered pair is a solution of an equation in two variables if replacing the variables by the coordinates of the ordered pair results in a true statement.

Is $(-1, 4)$ a solution of $2x + 5y = 18$?

$$2(-1) + 5 \cdot 4 \overset{?}{=} 18$$
$$-2 + 20 \overset{?}{=} 18$$
$$18 = 18, \text{ true}$$

Thus, $(-1, 4)$ is a solution.

One method for graphing an equation in two variables is point plotting. Find several ordered-pair solutions, plot them as points, and connect the points with a smooth curve or line.

Graph: $y = 2x + 1$.

x	$y = 2x + 1$	(x, y)
-2	$y = 2(-2) + 1 = -3$	$(-2, -3)$
-1	$y = 2(-1) + 1 = -1$	$(-1, -1)$
0	$y = 2 \cdot 0 + 1 = 1$	$(0, 1)$
1	$y = 2 \cdot 1 + 1 = 3$	$(1, 3)$
2	$y = 2 \cdot 2 + 1 = 5$	$(2, 5)$

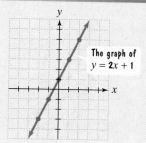

The graph of $y = 2x + 1$

Definitions and Concepts	Examples

Section 4.2 Graphing Linear Equations Using Intercepts

If a graph intersects the x-axis at $(a, 0)$, then a is an x-intercept.
If a graph intersects the y-axis at $(0, b)$, then b is a y-intercept.

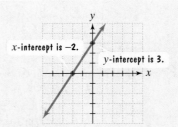

An equation of the form $Ax + By = C$, where A, B, and C are integers, is called the standard form of the equation of a line. The graph of $Ax + By = C$ is a line that can be obtained using intercepts. To find the x-intercept, let $y = 0$ and solve for x. To find the y-intercept, let $x = 0$ and solve for y. Find a checkpoint, a third ordered-pair solution. Graph the equation by drawing a line through the three points.

Graph using intercepts: $4x + 3y = 12$.

x-intercept: $4x = 12$
          $x = 3$
y-intercept: $3y = 12$
          $y = 4$
Checkpoint: Let $x = 2$.
      $8 + 3y = 12$
       $3y = 4$
        $y = \frac{4}{3}$

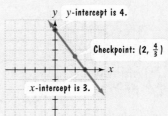

The graph of $Ax + By = 0$ is a line that passes through the origin. Find two other points by finding two other solutions of the equation. Graph the equation by drawing a line through the origin and these two points.

Graph: $x + 2y = 0$.

$x = 2$: $2 + 2y = 0$
       $2y = -2$
        $y = -1$
$y = 1$: $x + 2(1) = 0$
        $x = -2$

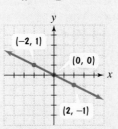

Horizontal and Vertical Lines
The graph of $y = b$ is a horizontal line. The y-intercept is b.
The graph of $x = a$ is a vertical line. The x-intercept is a.

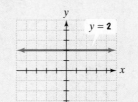
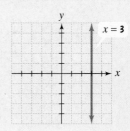

Definitions and Concepts	Examples

Section 4.3 Slope

The slope, m, of the line through the points (x_1, y_1) and (x_2, y_2) is

$$m = \frac{y_2 - y_1}{x_2 - x_1}, \quad x_2 - x_1 \neq 0.$$

If the slope is positive, the line rises from left to right. If the slope is negative, the line falls from left to right.

The slope of a horizontal line is 0. The slope of a vertical line is undefined.

If two distinct nonvertical lines have the same slope, then they are parallel.

Find the slope of the line passing through the points shown.

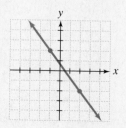

Let $(x_1, y_1) = (-1, 2)$ and $(x_2, y_2) = (2, -2)$.

$$m = \frac{y_2 - y_1}{x_2 - x_1} = \frac{-2 - 2}{2 - (-1)} = \frac{-4}{3} = -\frac{4}{3}$$

Section 4.4 The Slope-Intercept Form of the Equation of a Line

The slope-intercept equation of a nonvertical line with slope m and y-intercept b is

$$y = mx + b.$$

Find the slope and the y-intercept of the line with the given equation.

- $y = -2x + 5$

  Slope −2. y-intercept is 5.

- $2x + 3y = 9$ (Solve for y)

  $3y = -2x + 9$ Subtract $2x$.

  $y = -\frac{2}{3}x + 3$ Divide by 3.

  Slope is $-\frac{2}{3}$. y-intercept is 3.

Graphing $y = mx + b$ Using the Slope and y-Intercept

1. Plot the point containing the y-intercept on the y-axis. This is the point $(0, b)$.

2. Use the slope, m, to obtain a second point. Write m as a fraction, and use rise over run, starting at the point containing the y-intercept, to plot this point.

3. Graph the equation by drawing a line through the two points.

Graph: $y = -\frac{3}{4}x + 1$.

Slope is $-\frac{3}{4}$. y-intercept is 1.

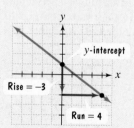

Definitions and Concepts	Examples

Section 4.5 The Point-Slope Form of the Equation of a Line

The point-slope equation of a nonvertical line with slope m that passes through the point (x_1, y_1) is $$y - y_1 = m(x - x_1).$$	Slope $= -3$, passing through $(-1, 5)$ $m = -3 \qquad x_1 = -1 \qquad y_1 = 5$ The line's point-slope equation is $$y - 5 = -3[x - (-1)].$$ Simplify: $$y - 5 = -3(x + 1).$$
To write the point-slope form of the line passing through two points, begin by using the points to compute the slope, m. Use either given point as (x_1, y_1) and write the point-slope equation: $$y - y_1 = m(x - x_1).$$ Solving this equation for y gives the slope-intercept form of the line's equation.	Write an equation in point-slope form and slope-intercept form of the line passing through $(-1, -3)$ and $(4, 2)$. $$m = \frac{2 - (-3)}{4 - (-1)} = \frac{2 + 3}{4 + 1} = \frac{5}{5} = 1$$ Using $(4, 2)$ as (x_1, y_1), the point-slope equation is $$y - 2 = 1(x - 4).$$ Solve for y to obtain the slope-intercept form. $$y = x - 2 \qquad \text{Add 2 to both sides.}$$

Section 4.6 Linear Inequalities in Two Variables

A linear inequality in two variables can be written in one of the following forms: $\quad Ax + By < C \qquad Ax + By \le C$ $\quad Ax + By > C \qquad Ax + By \ge C.$ An ordered pair is a solution if replacing the variables by the coordinates of the ordered pair results in a true statement.	Is $(2, -6)$ a solution of $3x - 4y \le 7$? $$3 \cdot 2 - 4(-6) \overset{?}{\le} 7$$ $$6 - (-24) \overset{?}{\le} 7$$ $$6 + 24 \overset{?}{\le} 7$$ $$30 \le 7, \text{false}$$ Thus, $(2, -6)$ is not a solution.
Graphing a Linear Inequality in Two Variables 1. Replace the inequality symbol with an equal sign and graph the boundary line. Use a solid line for $\le$ or $\ge$ and a dashed line for $<$ or $>$. 2. Choose a test point not on the line and substitute its coordinates into the inequality. 3. If a true statement results, shade the half-plane containing the test point. If a false statement results, shade the half-plane not containing the test point.	Graph: $x - 2y \le 4$. 1. Graph $x - 2y = 4$. Use a solid line because the inequality symbol is $\le$. 2. Test $(0, 0)$. $$0 - 2 \cdot 0 \overset{?}{\le} 4$$ $$0 \le 4, \text{true}$$ 3. The inequality is true. Shade the half-plane containing $(0, 0)$. 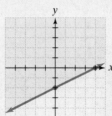

CHAPTER 4 REVIEW EXERCISES

4.1 *In Exercises 1–2, determine whether each ordered pair is a solution of the given equation.*

1. $y = 3x + 6$ $(-3, 3)$, $(0, 6)$, $(1, 9)$

2. $3x - y = 12$ $(0, 4)$, $(4, 0)$, $(-1, 15)$

In Exercises 3–4,

    a. *Find five solutions of each equation. Organize your work in a table of values.*

    b. *Use the five solutions in the table to graph each equation.*

3. $y = 2x - 3$ **4.** $y = \frac{1}{2}x + 1$

5. Graph the equation: $y = x^2 - 3$. Select integers for x, starting with -3 and ending with 3.

6. The linear equation in two variables $y = 5x - 41$ models the percentage of U.S. adults, y, with x years of education who are doing volunteer work.

    a. Find four solutions of the equation. Use 10, 12, 14, and 16 for x. Organize your work in a table of values.

    b. How well does the given equation model the data shown in the following table? Explain your answer.

Years of Education	10	12	14	16
Percentage Doing Volunteer Work	8.3%	18.8%	28.1%	38.4%

Source: U.S. Bureau of Labor

4.2 *In Exercises 7–9, use the graph to identify the*

    a. *x-intercept, or state that there is no x-intercept.*

    b. *y-intercept, or state that there is no y-intercept.*

7.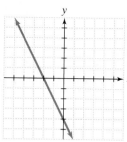

8.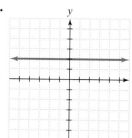

9.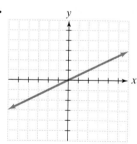

In Exercises 10–13, use intercepts to graph each equation.

10. $2x + y = 4$ **11.** $3x - 2y = 12$

12. $3x = 6 - 2y$ **13.** $3x - y = 0$

In Exercises 14–17, graph each equation.

14. $x = 3$ **15.** $y = -5$ **16.** $y + 3 = 5$ **17.** $2x = -8$

18. The graph shows the Fahrenheit temperature, y, x hours after noon.

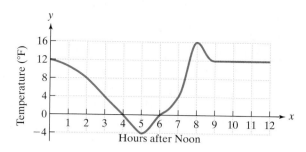

    a. At what time did the minimum temperature occur? What is the minimum temperature?

    b. At what time did the maximum temperature occur? What is the maximum temperature?

    c. What are the x-intercepts? In terms of time and temperature, interpret the meaning of these intercepts.

    d. What is the y-intercept? What does this mean in terms of time and temperature?

    e. From 9 P.M. until midnight, the graph is shown as a horizontal line. What does this mean about the temperature over this period of time?

4.3 *In Exercises 19–22, calculate the slope of the line passing through the given points. If the slope is undefined, so state. Then indicate whether the line rises, falls, is horizontal, or is vertical.*

19. $(3, 2)$ and $(5, 1)$

20. $(-1, 2)$ and $(-3, -4)$

21. $(-3, 4)$ and $(6, 4)$

22. $(5, 3)$ and $(5, -3)$

In Exercises 23–26, find the slope of each line, or state that the slope is undefined.

23.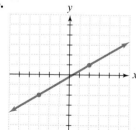

24.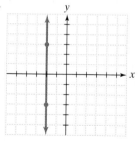

25.

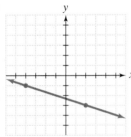

26.

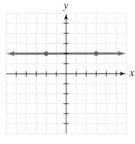

In Exercises 27–28, determine whether the distinct lines through each pair of points are parallel.

27. $(-1, -3)$ and $(2, -8)$ **28.** $(5, 4)$ and $(9, 7)$
    $(8, -7)$ and $(9, 10)$ $(-6, 0)$ and $(-2, 3)$

29. The graph shows new AIDS diagnoses among the general U.S. population, y, for year x, where $1999 \le x \le 2003$.

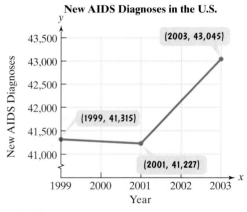

New AIDS Diagnoses in the U.S.

Source: Centers for Disease Control

a. Find the slope of the line passing through $(1999, 41,315)$ and $(2001, 41,227)$. Then express the slope as a rate of change with the proper units attached.

b. Find the slope of the line passing through $(2001, 41,227)$ and $(2003, 43,045)$. Then express the slope as a rate of change.

c. Draw a line passing through $(1999, 41,315)$ and $(2003, 43,045)$ and find its slope. Is the slope the average of the slopes of the lines that you found in parts (a) and (b)? Explain your answer.

4.4 *In Exercises 30–33, find the slope and the y-intercept of the line with the given equation.*

30. $y = 5x - 7$ **31.** $y = 6 - 4x$

32. $y = 3$ **33.** $2x + 3y = 6$

In Exercises 34–36, graph each linear equation using the slope and y-intercept.

34. $y = 2x - 4$ **35.** $y = \frac{1}{2}x - 1$

36. $y = -\frac{2}{3}x + 5$

In Exercises 37–38, write each equation in slope-intercept form. Then use the slope and y-intercept to graph the equation.

37. $y - 2x = 0$

38. $\frac{1}{3}x + y = 2$

39. Graph $y = -\frac{1}{2}x + 4$ and $y = -\frac{1}{2}x - 1$ in the same rectangular coordinate system. Are the lines parallel? If so, explain why.

40. The graph shows the average age of U.S. whites, African Americans, and Americans of Hispanic origin from 1990 through 2000.

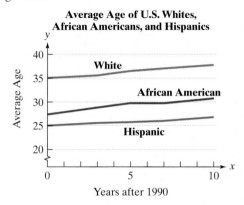

Average Age of U.S. Whites, African Americans, and Hispanics

Source: U.S. Census Bureau

a. What is the smallest y-intercept? Describe what this represents in this situation.

b. The average age of the group with the greatest y-intercept was approximately 38 in 2000. Use the points $(0, 35)$ and $(10, 38)$ to compute the slope for this group. What does this mean about their average age for the period shown?

c. Use the slope from part (b) and the y-intercept for this group shown by the graph to write an equation that models the group's average age, y, x years after 1990. Write your model in $y = mx + b$ form.

d. Use your model from part (c) to predict the average age for the group in 2010.

4.5 *Write the point-slope form of the line satisfying the conditions in Exercises 41–42. Then use the point-slope form of the equation to write the slope-intercept form.*

41. Slope = 6, passing through $(-4, 7)$

42. Passing through $(3, 4)$ and $(2, 1)$

43. You can click a mouse and bet the house. The points in the graph show the dizzying growth of online gambling. With more than 1800 sites, the industry has become the Web's biggest moneymaker.

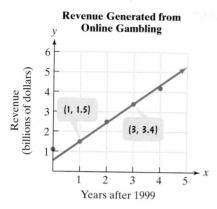

Revenue Generated from Online Gambling

(1, 1.5)

(3, 3.4)

Years after 1999

Source: Newsweek

a. Use the two points whose coordinates are shown by the voice balloons to find the point-slope equation of the line that models the revenue from online gambling, y, in billions of dollars, x years after 1999.

b. Write the equation in part (a) in slope-intercept form.

c. In 2003, nearly \$3.5 billion was lost on Internet bets, triggering a sharp backlash that threatened to shut down Internet wagering. If this crackdown on the industry is not successful, use your slope-intercept model to predict the billions of dollars in revenue from online gambling in 2009.

4.6

44. Determine whether each ordered pair is a solution of $3x - 4y > 7$:

$$(0, 0), (3, -6), (-2, -5), (-3, 4).$$

In Exercises 45–52, graph each inequality.

45. $x - 2y > 6$

46. $4x - 6y \leq 12$

47. $y > 3x + 2$

48. $y \leq \frac{1}{3}x - 1$

49. $y < -\frac{1}{2}x$

50. $x < 4$

51. $y \geq -2$

52. $x + 2y \leq 0$

CHAPTER 4 TEST

Remember to use your Chapter Test Prep Video CD to see the worked-out solutions to the test questions you want to review.

1. Determine whether each ordered pair is a solution of $4x - 2y = 10$:

$$(0, -5), \quad (-2, 1), \quad (4, 3).$$

2. Find five solutions of $y = 3x + 1$. Organize your work in a table of values. Then use the five solutions in the table to graph the equation.

3. Graph: $y = x^2 - 1$. Select integers for x, starting with -3 and ending with 3.

4. Use the graph to identify the

a. x-intercept, or state that there is no x-intercept.

b. y-intercept, or state that there is no y-intercept.

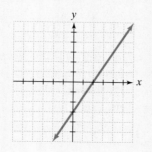

5. Use intercepts to graph $4x - 2y = -8$.

6. Graph $y = 4$ in a rectangular coordinate system.

In Exercises 7–8, calculate the slope of the line passing through the given points. If the slope is undefined, so state. Then indicate whether the line rises, falls, is horizontal, or is vertical.

7. $(-3, 4)$ and $(-5, -2)$

8. $(6, -1)$ and $(6, 3)$

9. Find the slope of the line in the figure shown or state that the slope is undefined.

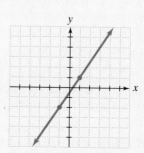

10. Determine whether the line through $(2, 4)$ and $(6, 1)$ is parallel to a second line through $(-3, 1)$ and $(1, -2)$.

In Exercises 11–12, find the slope and the y-intercept of the line with the given equation.

11. $y = -x + 10$

12. $2x + y = 6$

In Exercises 13–14, graph each linear equation using the slope and y-intercept.

13. $y = \frac{2}{3}x - 1$

14. $y = -2x + 3$

In Exercises 15–16, use the given conditions to write an equation for each line in point-slope form and slope-intercept form.

15. Slope $= -2$, passing through $(-1, 4)$

16. Passing through $(2, 1)$ and $(-1, -8)$

In Exercises 17–19, graph each inequality.

17. $3x - 2y < 6$

18. $y \geq 2x - 2$

19. $x > -1$

20. The graph shows spending per pupil in public schools. Find the slope of the line passing through $(1970, 2100)$ and $(2000, 5280)$. Describe what the slope represents in this situation.

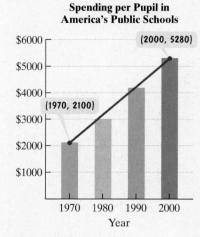

Spending per Pupil in America's Public Schools

Source: National Education Association

CUMULATIVE REVIEW EXERCISES (CHAPTERS 1–4)

1. Perform the indicated operations:

$$\frac{10 - (-6)}{3^2 - (4 - 3)}.$$

2. Simplify: $6 - 2[3(x - 1) + 4]$.

3. List all the irrational numbers in this set: $\left\{-3, 0, 1, \sqrt{4}, \sqrt{5}, \frac{11}{2}\right\}$.

In Exercises 4–5, solve each equation.

4. $6(2x - 1) - 6 = 11x + 7$

5. $x - \frac{3}{4} = \frac{1}{2}$

6. Solve for x: $y = mx + b$.

7. 120 is 15% of what?

8. The formula $y = 4.5x - 46.7$ models the stopping distance, y, in feet, for a car traveling x miles per hour. If the stopping distance is 133.3 feet, how fast was the car traveling?

In Exercises 9–10, solve each inequality. Express the solution set in set-builder notation and graph the set on a number line.

9. $2 - 6x \geq 2(5 - x)$

10. $6(2 - x) > 12$

11. A plumber charged a customer $228, listing $18 for parts and the remainder for labor. If the cost of the labor is $35 per hour, how many hours did the plumber work?

12. A chemist needs to mix a 40% acid solution with a 70% acid solution to obtain 12 liters of a 50% acid solution. How many liters of each of the acid solutions must be used?

13. Two people leave by car from the same point of departure and travel in opposite directions. One car averages 40 miles per hour and the other averages 60 miles per hour. After how many hours will the cars be 350 miles apart?

14. In a triangle, the measure of the second angle is 20° greater than the measure of the first angle. The measure of the third angle is twice that of the first. What is the measure of each angle?

15. Solve the proportion: $\frac{45}{2} = \frac{360}{x}$.

In Exercises 16–20, graph each equation or inequality in the rectangular coordinate system.

16. $2x - y = 4$

17. $y = x^2 - 5$

18. $y = -4x + 3$

19. $3x - 2y < -6$

20. $y \geq -1$

Mary Katherine Campbell,
Miss America 1922

Ericka Dunlap,
Miss America 2004

You are not a great fan of beauty pageants. However, as you were channel surfing, you tuned into the Miss America festivities. Is it your imagination, or does the icon of American beauty have that lean and hungry look?

This question is addressed in the Enrichment Essay "Missing America" in Section 5.1.

Systems of Linear Equations and Inequalities

Television, movies, and magazines place great emphasis on physical beauty. Our culture emphasizes physical appearance to such an extent that it is a central factor in the perception and judgment of others. The modern emphasis on thinness as the ideal body shape has been suggested as a major cause of eating disorders among adolescent women. In this chapter, you will learn how systems of linear equations in two variables reveal the hidden patterns of your world, including a relationship between our changing cultural values of physical attractiveness and undernutrition.

SECTION **5.1**

Objectives

1 Decide whether an ordered pair is a solution of a linear system.

2 Solve systems of linear equations by graphing.

3 Use graphing to identify systems with no solution or infinitely many solutions.

SOLVING SYSTEMS OF LINEAR EQUATIONS BY GRAPHING

Key West residents Brian Goss (left), George Wallace, and Michael Mooney (right) hold on to each other as they battle 90 mph winds along Houseboat Row in Key West, FL., on Friday, Sept. 25, 1998. The three had sought shelter behind a Key West hotel as Hurricane Georges descended on the Florida Keys, but were forced to seek other shelter when the storm conditions became too rough. Hundreds of people were killed by the storm when it swept through the Caribbean.

Problems ranging from scheduling airline flights to controlling traffic flow to routing phone calls over the nation's communication network often require solutions in a matter of moments. The solution to these real-world problems can involve solving thousands of equations having thousands of variables. AT&T's domestic long-distance network involves 800,000 variables! Meteorologists describing atmospheric conditions surrounding a hurricane must solve problems involving thousands of equations rapidly and efficiently. The difference between a two-hour warning and a two-day warning is a life-and-death issue for people in the path of one of nature's most destructive forces.

Although we will not be solving 800,000 equations with 800,000 variables, we will turn our attention to two equations with two variables, such as

$$2x - 3y = -4$$
$$2x + y = 4.$$

The methods that we consider for solving such problems provide the foundation for solving far more complex systems with many variables.

1 Decide whether an ordered pair is a solution of a linear system.

Systems of Linear Equations and Their Solutions We have seen that all equations in the form $Ax + By = C$ are straight lines when graphed. Two such equations, such as those listed above, are called a **system of linear equations** or a **linear system**. A **solution to a system of two linear equations in two variables** is an ordered pair that satisfies all equations in the system. For example, $(3, 4)$ satisfies the system

$$x + y = 7 \quad \text{(3 + 4 is, indeed, 7.)}$$
$$x - y = -1. \quad \text{(3 - 4 is, indeed, -1.)}$$

Thus, $(3, 4)$ satisfies both equations and is a solution of the system. The solution can be described by saying that $x = 3$ and $y = 4$. The solution can also be described using the ordered pair $(3, 4)$.

A system of linear equations can have exactly one solution, no solution, or infinitely many solutions. We begin with systems having exactly one solution.

EXAMPLE 1 Determining Whether Ordered Pairs Are Solutions of a System

Consider the system:

$$x + 2y = 2$$
$$x - 2y = 6.$$

Determine if each ordered pair is a solution of the system:

a. $(4, -1)$ **b.** $(-4, 3)$.

SOLUTION

a. We begin by determining whether $(4, -1)$ is a solution. Because 4 is the x-coordinate and -1 is the y-coordinate of $(4, -1)$, we replace x with 4 and y with -1.

$$x + 2y = 2 \qquad\qquad x - 2y = 6$$
$$4 + 2(-1) \stackrel{?}{=} 2 \qquad\qquad 4 - 2(-1) \stackrel{?}{=} 6$$
$$4 + (-2) \stackrel{?}{=} 2 \qquad\qquad 4 - (-2) \stackrel{?}{=} 6$$
$$2 = 2, \quad \text{true} \qquad\qquad 4 + 2 \stackrel{?}{=} 6$$
$$\qquad\qquad\qquad\qquad\qquad 6 = 6, \quad \text{true}$$

The pair $(4, -1)$ satisfies both equations: It makes each equation true. Thus, the ordered pair is a solution of the system.

b. To determine whether $(-4, 3)$ is a solution, we replace x with -4 and y with 3.

$$x + 2y = 2 \qquad\qquad x - 2y = 6$$
$$-4 + 2 \cdot 3 \stackrel{?}{=} 2 \qquad\qquad -4 - 2 \cdot 3 \stackrel{?}{=} 6$$
$$-4 + 6 \stackrel{?}{=} 2 \qquad\qquad -4 - 6 \stackrel{?}{=} 6$$
$$2 = 2, \quad \text{true} \qquad\qquad -10 = 6, \quad \text{false}$$

The pair $(-4, 3)$ fails to satisfy *both* equations: It does not make both equations true. Thus, the ordered pair is not a solution of the system. ∎

 CHECK POINT 1 Consider the system:

$$2x - 3y = -4$$
$$2x + y = 4.$$

Determine if each ordered pair is a solution of the system:

a. $(1, 2)$ **b.** $(7, 6)$.

2 Solve systems of linear equations by graphing.

Solving Linear Systems by Graphing The solution of a system of linear equations can be found by graphing both of the equations in the same rectangular coordinate system. For a system with one solution, **the coordinates of the point of intersection give the system's solution**. For example, the system in Example 1,

$$x + 2y = 2$$
$$x - 2y = 6$$

is graphed in Figure 5.1. The solution of the system, $(4, -1)$, corresponds to the point of intersection of the lines.

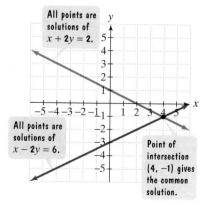

FIGURE 5.1 Visualizing a system's solution

SOLVING SYSTEMS OF TWO LINEAR EQUATIONS IN TWO VARIABLES, x AND y, BY GRAPHING

1. Graph the first equation.
2. Graph the second equation on the same axes.
3. If the lines representing the two graphs intersect at a point, determine the coordinates of this point of intersection. The ordered pair is the solution of the system.
4. Check the solution in both equations.

EXAMPLE 2   Solving a Linear System by Graphing

Solve by graphing:

$$2x + 3y = 6$$
$$2x + y = -2.$$

SOLUTION

Step 1.   **Graph the first equation.** We use intercepts to graph $2x + 3y = 6$.

x-intercept (Set $y = 0$.)	y-intercept (Set $x = 0$.)
$2x + 3 \cdot 0 = 6$	$2 \cdot 0 + 3y = 6$
$2x = 6$	$3y = 6$
$x = 3$	$y = 2$

The x-intercept is 3, so the line passes through $(3, 0)$. The y-intercept is 2, so the line passes through $(0, 2)$. The graph of $2x + 3y = 6$ is shown as the red line in Figure 5.2.

Step 2.   **Graph the second equation on the same axes.** We use intercepts to graph $2x + y = -2$.

x-intercept (Set $y = 0$.)	y-intercept (Set $x = 0$.)
$2x + 0 = -2$	$2 \cdot 0 + y = -2$
$2x = -2$	$y = -2$
$x = -1$	

The x-intercept is -1, so the line passes through $(-1, 0)$. The y-intercept is -2, so the line passes through $(0, -2)$. The graph of $2x + y = -2$ is shown as the blue line in Figure 5.2.

Step 3.   **Determine the coordinates of the intersection point. This ordered pair is the system's solution.** Using Figure 5.2, it appears that the lines intersect at $(-3, 4)$. The "apparent" solution of the system is $(-3, 4)$.

Step 4.   **Check the solution in both equations.**

Check $(-3, 4)$ in $2x + 3y = 6$:	Check $(-3, 4)$ in $2x + y = -2$:
$2(-3) + 3 \cdot 4 \overset{?}{=} 6$	$2(-3) + 4 \overset{?}{=} -2$
$-6 + 12 \overset{?}{=} 6$	$-6 + 4 \overset{?}{=} -2$
$6 = 6$, true	$-2 = -2$, true

Because both equations are satisfied, $(-3, 4)$ is the solution of the system. ∎

✔ **CHECK POINT 2** Solve by graphing:

$$2x + y = 6$$
$$2x - y = -2.$$

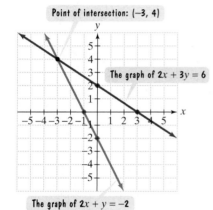

Point of intersection: $(-3, 4)$

The graph of $2x + 3y = 6$

The graph of $2x + y = -2$

FIGURE 5.2

Must two lines intersect at exactly one point? Sketch two lines that have less than one intersection point. Now sketch two lines that have more than one intersection point. What does this say about each of these systems?

STUDY TIP

When solving linear systems by graphing, neatly drawn graphs are essential for determining points of intersection.

- Use rectangular coordinate graph paper.
- Use a ruler or straightedge.
- Use a pencil with a sharp point.

EXAMPLE 3 Solving a Linear System by Graphing

Solve by graphing:

$$y = -3x + 2$$
$$y = 5x - 6.$$

SOLUTION Each equation is in the form $y = mx + b$. Thus, we use the y-intercept, b, and the slope, m, to graph each line.

Step 1. Graph the first equation.

$$y = -3x + 2$$

The slope is –3. The y-intercept is 2.

The y-intercept is 2, so the line passes through $(0, 2)$. The slope is $-\frac{3}{1}$. Start at the y-intercept and move 3 units down (the rise) and 1 unit to the right (the run). The graph of $y = -3x + 2$ is shown as the red line in Figure 5.3.

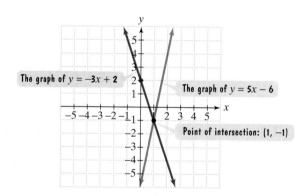

The graph of $y = -3x + 2$

The graph of $y = 5x - 6$

Point of intersection: (1, –1)

FIGURE 5.3

Step 2. Graph the second equation on the same axes.

$$y = 5x - 6$$

The slope is 5. The y-intercept is –6.

The y-intercept is -6, so the line passes through $(0, -6)$. The slope is $\frac{5}{1}$. Start at the y-intercept and move 5 units up (the rise) and 1 unit to the right (the run). The graph of $y = 5x - 6$ is shown as the blue line in Figure 5.3.

Step 3. Determine the coordinates of the intersection point. This ordered pair is the system's solution. Using Figure 5.3, it appears that the lines intersect at $(1, -1)$. The "apparent" solution of the system is $(1, -1)$.

Step 4. **Check the solution in both equations.**

Check $(1, -1)$ in　　　　**Check $(1, -1)$ in**
$y = -3x + 2$:　　　　　$y = 5x - 6$:

$$-1 \stackrel{?}{=} -3 \cdot 1 + 2 \qquad\qquad -1 \stackrel{?}{=} 5 \cdot 1 - 6$$

$$-1 \stackrel{?}{=} -3 + 2 \qquad\qquad -1 \stackrel{?}{=} 5 - 6$$

$$-1 = -1, \;\; \text{true} \qquad\qquad -1 = -1, \;\; \text{true}$$

Because both equations are satisfied, $(1, -1)$ is the solution.

CHECK POINT 3 Solve by graphing:

$$y = -x + 6$$
$$y = 3x - 6.$$

3 Use graphing to identify systems with no solution or infinitely many solutions.

Linear Systems Having No Solution or Infinitely Many Solutions We have seen that a system of linear equations in two variables represents a pair of lines. The lines either intersect, are parallel, or are identical. Thus, there are three possibilities for the number of solutions to a system of two linear equations.

THE NUMBER OF SOLUTIONS TO A SYSTEM OF TWO LINEAR EQUATIONS
The number of solutions to a system of two linear equations in two variables is given by one of the following. (See Figure 5.4.)

Number of Solutions	What This Means Graphically
Exactly one ordered-pair solution	The two lines intersect at one point.
No solution	The two lines are parallel.
Infinitely many solutions	The two lines are identical.

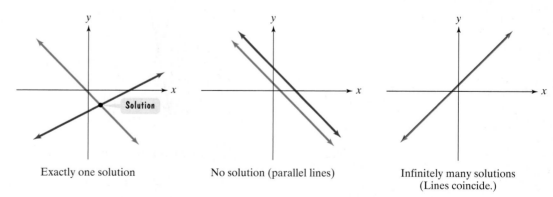

Exactly one solution　　　　No solution (parallel lines)　　　　Infinitely many solutions (Lines coincide.)

FIGURE 5.4 Possible graphs for a system of two linear equations in two variables

A linear system with no solution is called an **inconsistent system**. If you attempt to solve such a system by graphing, you will obtain two parallel lines.

EXAMPLE 4 A System with No Solution

Solve by graphing:

$$y = 2x - 1$$
$$y = 2x + 3.$$

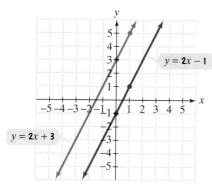

FIGURE 5.5 The graph of an inconsistent system

SOLUTION Compare the slopes and y-intercepts in the two equations.

> The lines have the same slope, 2.

$$y = 2x - 1 \qquad y = 2x + 3$$

> The lines have different y-intercepts, −1 and 3.

Figure 5.5 shows the graphs of the two equations. Because both equations have the same slope, 2, but different y-intercepts, the lines are parallel. The system is inconsistent and has no solution. ■

 CHECK POINT 4 Solve by graphing:

$$y = 3x - 2$$
$$y = 3x + 1.$$

EXAMPLE 5 A System with Infinitely Many Solutions

Solve by graphing:

$$2x + \ y = 3$$
$$4x + 2y = 6.$$

SOLUTION We use intercepts to graph each equation.

• $2x + y = 3$

x-intercept	**y-intercept**
$2x + 0 = 3$	$2 \cdot 0 + y = 3$
$2x = 3$	$y = 3$
$x = \dfrac{3}{2}$	

> Graph $\left(\frac{3}{2}, 0\right)$ and $(0, 3)$.

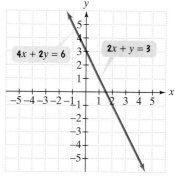

FIGURE 5.6 The graph of a system with infinitely many solutions

• $4x + 2y = 6$

x-intercept	**y-intercept**
$4x + 2 \cdot 0 = 6$	$4 \cdot 0 + 2y = 6$
$4x = 6$	$2y = 6$
$x = \dfrac{6}{4} = \dfrac{3}{2}$	$y = 3$

> Graph $\left(\frac{3}{2}, 0\right)$ and $(0, 3)$.

Both lines have the same x-intercept, $\frac{3}{2}$, or 1.5, and the same y-intercept, 3. Thus, the graphs of the two equations in the system are the same line, shown in Figure 5.6. The two equations have the same solutions. Any ordered pair that is a solution of one equation is a solution of the other, and, consequently, a solution of the system. The system has an infinite number of solutions, namely all points that are solutions of either line. ■

Take a second look at the two equations, $2x + y = 3$ and $4x + 2y = 6$, in Example 5. If you multiply both sides of the first equation, $2x + y = 3$, by 2, you will obtain the second equation, $4x + 2y = 6$.

$$2x + y = 3 \qquad \text{This is the first equation in the system.}$$

$$2(2x + y) = 2 \cdot 3 \qquad \text{Multiply both sides by 2.}$$

$$2 \cdot 2x + 2y = 2 \cdot 3 \qquad \text{Use the distributive property.}$$

This is the second equation in the system. $4x + 2y = 6 \qquad$ Simplify.

Because $2x + y = 3$ and $4x + 2y = 6$ are different forms of the same equation, these equations are called *dependent equations*. In general, the equations in a linear system with infinitely many solutions are called **dependent equations**.

✔ CHECK POINT 5 Solve by graphing:

$$x + y = 3$$
$$2x + 2y = 6.$$

ENRICHMENT ESSAY

Missing America

Here she is, Miss America, the icon of American beauty. Always thin, she is becoming more so. The scatter plot in the figure shows Miss America's body-mass index, a ratio comparing weight divided by the square of height. Two lines are also shown: a line that passes near the data points and a horizontal line representing the World Health Organization's cutoff point for undernutrition. The intersection point indicates that in approximately 1978, Miss America reached this cutoff. There she goes: If the trend continues, Miss America's body-mass index could reach zero in about 320 years.

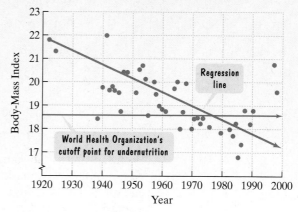

Body-Mass Index of Miss America

Source: Johns Hopkins School of Public Health

5.1 EXERCISE SET

Student Solutions Manual   CD/Video   PH Math/Tutor Center   MathXL Tutorials on CD   MathXL®   MyMathLab   Interactmath.com

Practice Exercises

In Exercises 1–10, determine whether the given ordered pair is a solution of the system.

1. $(2, -3)$
$2x + 3y = -5$
$7x - 3y = 23$

2. $(-2, -5)$
$6x - 2y = -2$
$3x + y = -11$

3. $\left(\dfrac{2}{3}, \dfrac{1}{9}\right)$
$x + 3y = 1$
$4x + 3y = 3$

4. $\left(\dfrac{7}{25}, -\dfrac{1}{25}\right)$
$4x + 3y = 1$
$3x - 4y = 1$

5. $(-5, 9)$
$5x + 3y = 2$
$x + 4y = 14$

6. $(10, 7)$
$6x - 5y = 25$
$4x + 15y = 13$

7. $(1400, 450)$
$x - 2y = 500$
$0.03x + 0.02y = 51$

8. $(200, 700)$
$-4x + y = -100$
$0.05x - 0.06y = -32$

9. $(8, 5)$
$5x - 4y = 20$
$3y = 2x + 1$

10. $(5, -2)$
$4x - 3y = 26$
$x = 15 - 5y$

In Exercises 11–42, solve each system by graphing. If there is no solution or an infinite number of solutions, so state.

11. $x + y = 6$
$x - y = 2$

12. $x + y = 2$
$x - y = 4$

13. $x + y = 1$
$y - x = 3$

14. $x + y = 4$
$y - x = 4$

15. $2x - 3y = 6$
$4x + 3y = 12$

16. $x + 2y = 2$
$x - y = 2$

17. $4x + y = 4$
$3x - y = 3$

18. $5x - y = 10$
$2x + y = 4$

19. $y = x + 5$
$y = -x + 3$

20. $y = x + 1$
$y = 3x - 1$

21. $y = 2x$
$y = -x + 6$

22. $y = 2x + 1$
$y = -2x - 3$

23. $y = -2x + 3$
$y = -x + 1$

24. $y = 3x - 4$
$y = -2x + 1$

25. $y = 2x - 1$
$y = 2x + 1$

26. $y = 3x - 1$
$y = 3x + 2$

27. $x + y = 4$
$x = -2$

28. $x + y = 6$
$y = -3$

29. $x - 2y = 4$
$2x - 4y = 8$

30. $2x + 3y = 6$
$4x + 6y = 12$

31. $y = 2x - 1$
$x - 2y = -4$

32. $y = -2x - 4$
$4x - 2y = 8$

33. $x + y = 5$
$2x + 2y = 12$

34. $x - y = 2$
$3x - 3y = -6$

35. $x - y = 0$
$y = x$

36. $2x - y = 0$
$y = 2x$

37. $x = 2$
$y = 4$

38. $x = 3$
$y = 5$

39. $x = 2$
$x = -1$

40. $x = 3$
$x = -2$

41. $y = 0$
$y = 4$

42. $y = 0$
$y = 5$

Practice Plus

Solve Exercises 43–50 without graphing. Find the slope and the y-intercept for the graph of each equation in the given system. Use this information (and not the equations' graphs) to determine if the system has no solution, one solution, or an infinite number of solutions.

43. $y = \dfrac{1}{2}x - 3$
$y = \dfrac{1}{2}x - 5$

44. $y = \dfrac{3}{4}x - 2$
$y = \dfrac{3}{4}x + 1$

45. $y = -\dfrac{1}{2}x + 4$
$3x - y = -4$

46. $y = -\dfrac{1}{4}x + 3$
$4x - y = -3$

47. $3x - y = 6$
$x = \dfrac{y}{3} + 2$

48. $2x - y = 4$
$x = \dfrac{y}{2} + 2$

49. $3x + y = 0$
$y = -3x + 1$

50. $2x + y = 0$
$y = -2x + 1$

Application Exercises

51. The graph shows the number of births in Massachusetts for women under 30 years old and women 30 years or older.

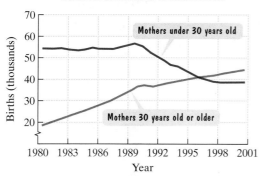

Number of Births in Massachusetts

Source: Massachusetts Department of Public Health

a. Estimate the coordinates of the point of intersection. What does this mean in terms of the number of babies born to older mothers?

b. Describe what is happening to the number of births in Massachusetts to the right of the intersection point.

52. The figure shows scatter plots for the men's and women's winning times, in seconds, in the Olympic 100-meter freestyle swimming event. Also shown are lines that best fit the data, one for the men and one for the women.

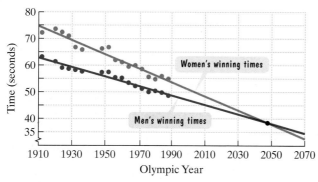

Winning Times in the Olympic 100-Meter Freestyle Swimming Event

a. Estimate the coordinates of the point of intersection. What does this mean in terms of the women's time and the men's time?

b. Make a prediction about the swimming event in the Olympic years to the right of the intersection point.

Writing in Mathematics

53. What is a system of linear equations? Provide an example with your description.

54. What is a solution of a system of linear equations?

55. Explain how to determine if an ordered pair is a solution of a system of linear equations.

56. Explain how to solve a system of linear equations by graphing.

57. What is an inconsistent system? What happens if you attempt to solve such a system by graphing?

58. Explain how a linear system can have infinitely many solutions.

59. What are dependent equations? Provide an example with your description.

60. The following system models the winning times, y, in seconds, in the Olympic 500-meter speed skating event x years after 1970:

$$y = -0.19x + 43.7 \quad \text{Women}$$
$$y = -0.16x + 39.9 \quad \text{Men}$$

Use the slope of each model to explain why the system has a solution. What does this solution represent?

Critical Thinking Exercises

61. Which one of the following statements is true?

a. If a linear system has graphs with equal slopes, the system must be inconsistent.

b. If a linear system has graphs with equal y-intercepts, the system must have infinitely many solutions.

c. If a linear system has two distinct points that are solutions, then the graphs of the system's equations have equal slopes and equal y-intercepts.

d. It is possible for a linear system with one solution to have graphs with equal slopes.

62. Write a system of linear equations whose solution is $(5, 1)$. How many different systems are possible? Explain.

63. Write a system of equations with one solution, a system of equations with no solution, and a system of equations with infinitely many solutions. Explain how you were able to think of these systems.

64. Graph $y = x^2$ and $y = x + 2$ on the same axes. Find two ordered pairs that satisfy the system. Check that your answers satisfy both equations in the system.

Technology Exercises

65. Verify your solutions to any five exercises from Exercises 11 through 36 by using a graphing utility to graph the two equations in the system in the same viewing rectangle. After entering the two equations, one as y_1 and the other as y_2, and graphing them, use the TRACE and ZOOM features to find the coordinates of the intersection point. (It may first be necessary to solve the equation for y before entering it.) Many graphing utilities have a special INTERSECTION feature that displays the coordinates of the intersection point once the equations are graphed. Consult your manual.

Read Exercise 65. Then use a graphing utility to solve the systems in Exercises 66–73.

66. $y = 2x + 2$
$y = -2x + 6$

67. $y = -x + 5$
$y = x - 7$

68. $x + 2y = 2$
$x - y = 2$

69. $2x - 3y = 6$
$4x + 3y = 12$

70. $3x - y = 5$
$-5x + 2y = -10$

71. $2x - 3y = 7$
$3x + 5y = 1$

72. $y = \dfrac{1}{3}x + \dfrac{2}{3}$
$y = \dfrac{5}{7}x - 2$

73. $y = -\dfrac{1}{2}x + 2$
$y = \dfrac{3}{4}x + 7$

Review Exercises

In Exercises 74–76, perform the indicated operation.

74. $-3 + (-9)$ (Section 1.7, Table 1.6)
75. $-3 - (-9)$ (Section 1.7, Table 1.6)
76. $-3(-9)$ (Section 1.7, Table 1.6)

SECTION

5.2

Objectives

1 Solve linear systems by the substitution method.

2 Use the substitution method to identify systems with no solution or infinitely many solutions.

3 Solve problems using the substitution method.

SOLVING SYSTEMS OF LINEAR EQUATIONS BY THE SUBSTITUTION METHOD

Other than outrage, what is going on at the gas pumps? Is surging demand creating the increasing oil prices? Like all things in a free market economy, the price of a commodity is based on supply and demand. In this section, we use a second method for solving linear systems, the *substitution method*, to understand this economic phenomenon.

1 Solve linear systems by the substitution method.

Eliminating a Variable Using the Substitution Method Finding the solution of a linear system by graphing equations may not be easy to do. For example, a solution of $\left(-\frac{2}{3}, \frac{157}{29}\right)$ would be difficult to "see" as an intersection point on a graph.

Let's consider a method that does not depend on finding a system's solution visually: the substitution method. This method involves converting the system to one equation in one variable by an appropriate substitution.

EXAMPLE 1 Solving a System by Substitution

Solve by the substitution method:

$$y = -x - 1$$
$$4x - 3y = 24.$$

SOLUTION

Step 1. **Solve either of the equations for one variable in terms of the other.** This step has already been done for us. The first equation, $y = -x - 1$, is solved for y in terms of x.

Step 2. **Substitute the expression from step 1 into the other equation.** We substitute the expression $-x - 1$ for y into the other equation:

$$y = \boxed{-x - 1} \qquad 4x - 3\boxed{y} = 24. \qquad \text{Substitute } -x - 1 \text{ for } y.$$

This gives us an equation in one variable, namely

$$4x - 3(-x - 1) = 24.$$

The variable y has been eliminated.

Step 3. **Solve the resulting equation containing one variable.**

$4x - 3(-x - 1) = 24$	This is the equation containing one variable.
$4x + 3x + 3 = 24$	Apply the distributive property.
$7x + 3 = 24$	Combine like terms.
$7x = 21$	Subtract 3 from both sides.
$x = 3$	Divide both sides by 7.

Step 4. **Back-substitute the obtained value into the equation from step 1.** We now know that the x-coordinate of the solution is 3. To find the y-coordinate, we back-substitute the x-value into the equation from step 1,

$$y = -x - 1. \qquad \text{This is the equation from step 1.}$$

$$\text{Substitute 3 for } x.$$

$$y = -3 - 1$$

$$y = -4 \qquad \text{Simplify.}$$

With $x = 3$ and $y = -4$, the proposed solution is $(3, -4)$.

Step 5. **Check the proposed solution in both of the system's given equations.** Replace x with 3 and y with -4.

$y = -x - 1$	$4x - 3y = 24$
$-4 \overset{?}{=} -3 - 1$	$4(3) - 3(-4) \overset{?}{=} 24$
$-4 = -4, \quad \text{true}$	$12 + 12 \overset{?}{=} 24$
	$24 = 24, \quad \text{true}$

The pair $(3, -4)$ satisfies both equations. The system's solution is $(3, -4)$. ∎

 CHECK POINT 1 Solve by the substitution method:

$$y = 5x - 13$$
$$2x + 3y = 12.$$

Before considering additional examples, let's summarize the steps used in the substitution method.

USING TECHNOLOGY

A graphing utility can be used to solve the system in Example 1. Graph each equation and use the intersection feature. The utility displays the solution $(3, -4)$ as $x = 3$, $y = -4$.

```
Intersection
X=3        Y=-4
```
[−10, 10, 1] by [−10, 10, 1]

STUDY TIP

In step 1, you can choose which variable to isolate in which equation. If possible, solve for a variable whose coefficient is 1 or −1 to avoid working with fractions.

SOLVING LINEAR SYSTEMS BY SUBSTITUTION

1. Solve either of the equations for one variable in terms of the other. (If one of the equations is already in this form, you can skip this step.)
2. Substitute the expression found in step 1 into the *other* equation. This will result in an equation in one variable.
3. Solve the equation containing one variable.
4. Back-substitute the value found in step 3 into the equation from step 1. Simplify and find the value of the remaining variable.
5. Check the proposed solution in both of the system's given equations.

EXAMPLE 2 Solving a System by Substitution

Solve by the substitution method:

$$5x - 4y = 9$$
$$x - 2y = -3.$$

SOLUTION

Step 1. **Solve either of the equations for one variable in terms of the other.** We begin by isolating one of the variables in either of the equations. By solving for x in the second equation, which has a coefficient of 1, we can avoid fractions.

$$x - 2y = -3 \qquad \text{This is the second equation in the given system.}$$
$$x = 2y - 3 \qquad \text{Solve for } x \text{ by adding } 2y \text{ to both sides.}$$

Step 2. **Substitute the expression from step 1 into the other equation.** We substitute $2y - 3$ for x in the first equation.

$$x = \boxed{2y - 3} \quad 5\boxed{x} - 4y = 9$$

This gives us an equation in one variable, namely

$$5(2y - 3) - 4y = 9.$$

The variable x has been eliminated.

Step 3. **Solve the resulting equation containing one variable.**

$$5(2y - 3) - 4y = 9 \qquad \text{This is the equation containing one variable.}$$
$$10y - 15 - 4y = 9 \qquad \text{Apply the distributive property.}$$
$$6y - 15 = 9 \qquad \text{Combine like terms.}$$
$$6y = 24 \qquad \text{Add 15 to both sides.}$$
$$y = 4 \qquad \text{Divide both sides by 6.}$$

Step 4. **Back-substitute the obtained value into the equation from step 1.** Now that we have the y-coordinate of the solution, we back-substitute 4 for y in the equation $x = 2y - 3$.

$$x = 2y - 3 \qquad \text{Use the equation obtained in step 1.}$$
$$x = 2(4) - 3 \qquad \text{Substitute 4 for } y.$$
$$x = 8 - 3 \qquad \text{Multiply.}$$
$$x = 5 \qquad \text{Subtract.}$$

STUDY TIP

Get into the habit of checking ordered-pair solutions in *both* equations of the system.

With $x = 5$ and $y = 4$, the proposed solution is $(5, 4)$.

Step 5. **Check.** Take a moment to show that $(5, 4)$ satisfies both given equations. The solution is $(5, 4)$. ■

 CHECK POINT 2 Solve by the substitution method:

$$3x + 2y = -1$$
$$x - y = 3.$$

2 Use the substitution method to identify systems with no solution or infinitely many solutions.

A graphing utility was used to graph the equations in Example 3. The lines are parallel and have no point of intersection. This verifies that the system is inconsistent.

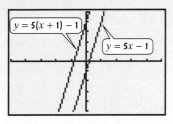

$[-5, 5, 1]$ by $[-5, 5, 1]$

The Substitution Method with Linear Systems Having No Solution or Infinitely Many Solutions Recall that a linear system with no solution is called an **inconsistent system**. If you attempt to solve such a system by substitution, you will eliminate both variables. A false statement such as $0 = 17$ will be the result.

EXAMPLE 3 Using the Substitution Method on an Inconsistent System

Solve the system:

$$y + 1 = 5(x + 1)$$
$$y = 5x - 1.$$

SOLUTION The variable y is isolated in the second equation. We use the substitution method and substitute the expression for y in the first equation.

$$\boxed{y} + 1 = 5(x + 1) \qquad y = \boxed{5x - 1} \qquad \text{Substitute } 5x - 1 \text{ for } y.$$

$$(5x - 1) + 1 = 5(x + 1) \qquad \text{This substitution gives an equation in one variable.}$$

$$5x = 5x + 5 \qquad \text{Simplify on the left side. Use the distributive property on the right side.}$$

There are no values of x and y for which $0 = 5$. $0 = 5,$ false   Subtract $5x$ from both sides.

The false statement $0 = 5$ indicates that the system is inconsistent and has no solution. ■

✔ **CHECK POINT 3** Solve the system:

$$3x + y = -5$$
$$y = -3x + 3.$$

Do you remember that the equations in a linear system with infinitely many solutions are called **dependent**? If you attempt to solve such a system by substitution, you will eliminate both variables. However, a true statement such as $5 = 5$ will be the result.

EXAMPLE 4 Using the Substitution Method on a System with Infinitely Many Solutions

Solve the system:

$$y = 3 - 2x$$
$$4x + 2y = 6.$$

SOLUTION The variable y is isolated in the first equation. We use the substitution method and substitute the expression for y in the second equation.

$$y = \boxed{3 - 2x} \qquad 4x + 2\boxed{y} = 6 \qquad \text{Substitute } 3 - 2x \text{ for } y.$$

$$4x + 2(3 - 2x) = 6 \qquad \text{This substitution gives an equation in one variable.}$$

$$4x + 6 - 4x = 6 \qquad \text{Apply the distributive property:}$$

$$2(3 - 2x) = 2 \cdot 3 - 2 \cdot 2x = 6 - 4x.$$

$$6 = 6, \quad \text{true} \qquad \text{Simplify: } 4x - 4x = 0.$$

This true statement indicates that the system contains dependent equations and has infinitely many solutions. ■

STUDY TIP

Although the system in Example 4 has infinitely many solutions, this does not mean that any ordered pair of numbers you can form will be a solution. The ordered pair (x, y) must satisfy either of the system's equations, $y = 3 - 2x$ or $4x + 2y = 6$, and there are infinitely many such ordered pairs. Because the graphs are coinciding lines, the ordered pairs that are solutions to one of the equations are also solutions to the other equation.

 CHECK POINT 4 Solve the system:

$$y = 3x - 4$$
$$9x - 3y = 12.$$

3 Solve problems using the substitution method.

Applications An important application of systems of equations arises in connection with supply and demand. As the price of a product increases, the demand for that product decreases. However, at higher prices suppliers are willing to produce greater quantities of the product.

EXAMPLE 5 Supply and Demand Models

A chain of video stores specializes in cult films. The weekly demand and supply models for *The Rocky Horror Picture Show* are given by

$$N = -13p + 760 \quad \text{Demand model}$$
$$N = 2p + 430 \quad \text{Supply model}$$

in which p is the price of the video and N is the number of copies of the video sold or supplied each week to the chain of stores.

    a. How many copies of the video can be sold and supplied at $18 per copy?

    b. Find the price at which supply and demand are equal. At this price, how many copies of *Rocky Horror* can be supplied and sold each week?

SOLUTION

    a. To find how many copies of the video can be sold and supplied at $18 per copy, we substitute 18 for p in the demand and supply models.

Demand Model	**Supply Model**
$N = -13p + 760$	$N = 2p + 430$
Substitute 18 for p.	Substitute 18 for p.
$N = -13 \cdot 18 + 760 = 526$	$N = 2 \cdot 18 + 430 = 466$

At $18 per video, the chain can sell 526 copies of *Rocky Horror* in a week. The manufacturer is willing to supply 466 copies per week. This will result in a shortage of copies of the video. Under these conditions, the retail chain is likely to raise the price of the video.

b. We can find the price at which supply and demand are equal by solving the demand-supply linear system. We will use substitution, substituting $-13p + 760$ for N in the second equation.

$$N = \boxed{-13p + 760} \quad \boxed{N} = 2p + 430 \qquad \text{Substitute } -13p + 760 \text{ for } N.$$

$$-13p + 760 = 2p + 430 \qquad \text{The resulting equation contains only one variable.}$$

$$-15p + 760 = 430 \qquad \text{Subtract } 2p \text{ from both sides.}$$

$$-15p = -330 \qquad \text{Subtract } 760 \text{ from both sides.}$$

$$p = 22 \qquad \text{Divide both sides by } -15.$$

The price at which supply and demand are equal is $22 per video. To find the value of N, the number of videos supplied and sold weekly at this price, we back-substitute 22 for p into either the demand or the supply model. We'll use both models to make sure we get the same number in each case.

Demand Model

$$N = -13p + 760$$

Substitute **22** for p.

$$N = -13 \cdot 22 + 760 = 474$$

Supply Model

$$N = 2p + 430$$

Substitute **22** for p.

$$N = 2 \cdot 22 + 430 = 474$$

At a price of $22 per video, 474 units of the video can be supplied and sold weekly. The intersection point, $(22, 474)$, is shown in Figure 5.7.

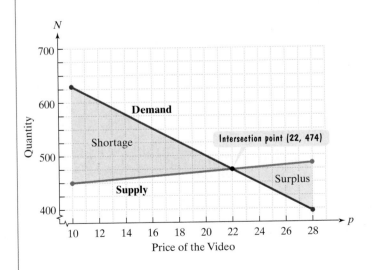

FIGURE 5.7 Priced at $22 per video, 474 copies of the video can be supplied and sold weekly.

CHECK POINT 5 The demand for a product is modeled by $N = -20p + 1000$ and the supply for the product by $N = 5p + 250$. In these models, p is the price of the product and N is the number of units of the product supplied or sold weekly. At what price will supply equal demand? At that price, how many units of the product will be supplied and sold each week?

5.2 EXERCISE SET

Student Solutions Manual   CD/Video   PH Math/Tutor Center   MathXL Tutorials on CD   Math XL   MathXL®   MyMathLab   MyMathLab   Interactmath.com

Practice Exercises

In Exercises 1–32, solve each system by the substitution method. If there is no solution or an infinite number of solutions, so state.

1. $x + y = 4$
$y = 3x$

2. $x + y = 6$
$y = 2x$

3. $x + 3y = 8$
$y = 2x - 9$

4. $2x - 3y = -13$
$y = 2x + 7$

5. $x + 3y = 5$
$4x + 5y = 13$

6. $x + 2y = 5$
$2x - y = -15$

7. $2x - y = -5$
$x + 5y = 14$

8. $2x + 3y = 11$
$x - 4y = 0$

9. $2x - y = 3$
$5x - 2y = 10$

10. $-x + 3y = 10$
$2x + 8y = -6$

11. $-3x + y = -1$
$x - 2y = 4$

12. $-4x + y = -11$
$2x - 3y = 5$

13. $x = 9 - 2y$
$x + 2y = 13$

14. $6x + 2y = 7$
$y = 2 - 3x$

15. $y = 3x - 5$
$21x - 35 = 7y$

16. $9x - 3y = 12$
$y = 3x - 4$

17. $5x + 2y = 0$
$x - 3y = 0$

18. $4x + 3y = 0$
$2x - y = 0$

19. $2x - y = 6$
$3x + 2y = 5$

20. $2x - y = 4$
$3x - 5y = 2$

21. $2(x - 1) - y = -3$
$y = 2x + 3$

22. $x + y - 1 = 2(y - x)$
$y = 3x - 1$

23. $x = 2y + 9$
$x = 7y + 10$

24. $x = 5y - 3$
$x = 8y + 4$

25. $4x - y = 100$
$0.05x - 0.06y = -32$

26. $x + 6y = 8000$
$0.3x - 0.6y = 0$

27. $y = \frac{1}{3}x + \frac{2}{3}$
$y = \frac{5}{7}x - 2$

28. $y = -\frac{1}{2}x + 2$
$y = \frac{3}{4}x + 7$

29. $\frac{x}{6} - \frac{y}{2} = \frac{1}{3}$
$x + 2y = -3$

30. $\frac{x}{4} - \frac{y}{4} = -1$
$x + 4y = -9$

31. $2x - 3y = 8 - 2x$
$3x + 4y = x + 3y + 14$

32. $3x - 4y = x - y + 4$
$2x + 6y = 5y - 4$

Practice Plus

In Exercises 33–38, write a system of equations describing the given conditions. Then solve the system by the substitution method and find the two numbers.

33. The sum of two numbers is 81. One number is 41 more than the other. Find the numbers.

34. The sum of two numbers is 62. One number is 12 more than the other. Find the numbers.

35. The difference between two numbers is 5. Four times the larger number is 6 times the smaller number. Find the numbers.

36. The difference between two numbers is 25. Two times the larger number is 12 times the smaller number. Find the numbers.

37. The difference between two numbers is 1. The sum of the larger number and twice the smaller number is 7. Find the numbers.

38. The difference between two numbers is 5. The sum of the larger number and twice the smaller number is 14. Find the numbers.

In Exercises 39–40, multiply each equation in the system by an appropriate number so that the coefficients are integers. Then solve the system by the substitution method.

39. $0.7x - 0.1y = 0.6$
$0.8x - 0.3y = -0.8$

40. $1.25x - 0.01y = 4.5$
$0.5x - 0.02y = 1$

Application Exercises

41. At a price of p dollars per ticket, the number of tickets to a rock concert that can be sold is given by the demand model $N = -25p + 7500$. At a price of p dollars per ticket, the number of tickets that the concert's promoters are willing to make available is given by the supply model $N = 5p + 6000$.

　a. How many tickets can be sold and supplied for $40 per ticket?

　b. Find the ticket price at which supply and demand are equal. At this price, how many tickets will be supplied and sold?

42. The weekly demand and supply models for a particular brand of scientific calculator for a chain of stores are given by the demand model $N = -53p + 1600$ and the supply model $N = 75p + 320$. In these models, p is the price of the calculator and N is the number of calculators sold or supplied each week to the stores.

　a. How many calculators can be sold and supplied at $12 per calculator?

　b. Find the price at which supply and demand are equal. At this price, how many calculators of this type can be supplied and sold each week?

A business breaks even when the cost for running the business is equal to the money taken in by the business. In Exercises 43–44, determine how many units must be sold so that the business breaks even, experiencing neither loss nor profit.

43. A gasoline station has weekly costs and revenue (the money taken in by the station) that depend on the number of gallons of gasoline purchased and sold. If x gallons are purchased and sold, weekly costs are given by $y = 1.2x + 1080$ and weekly revenue by $y = 1.6x$. How many gallons of gasoline must be sold weekly for the station to break even?

44. An artist has monthly costs and revenue (the money taken in by the artist) that depend on the number of ceramic pieces produced and sold. If x ceramic pieces are produced and sold, monthly costs are given by $y = 4x + 2000$ and monthly revenue by $y = 9x$. How many ceramic pieces must be sold monthly for the artist to break even?

45. Infant mortality for African Americans is decreasing at a faster rate than it is for whites, shown by the graphs below. Infant mortality for African Americans can be modeled by $M = -0.41x + 22$ and for whites by $M = -0.18x + 10$. In both models, x is the number of years after 1980 and M is infant mortality, measured in deaths per 1000 live births. Use these models to project when infant mortality for African Americans and whites will be the same. What will be the infant mortality for both groups at that time?

Infant Mortality

Source: National Center for Health Statistics

46. The equation $x + 10y = 2120$ models deaths from gunfire in the United States, y, in deaths per hundred thousand Americans, in year x. The equation $7x + 8y = 14,065$ models deaths from car accidents in the United States, y, in deaths per hundred thousand Americans, in year x. Solve the linear system formed by the two models. Then describe what the solution means in terms of the variables in the given models.

Writing in Mathematics

47. Describe a problem that might arise when solving a system of equations using graphing. Assume that both equations in the system have been graphed correctly and the system has exactly one solution.

48. Explain how to solve a system of equations using the substitution method. Use $y = 3 - 3x$ and $3x + 4y = 6$ to illustrate your explanation.

49. When using the substitution method, how can you tell if a system of linear equations has no solution?

50. When using the substitution method, how can you tell if a system of linear equations has infinitely many solutions?

51. The law of supply and demand states that, in a free market economy, a commodity tends to be sold at its equilibrium price. At this price, the amount that the seller will supply is the same amount that the consumer will buy. Explain how systems of equations can be used to determine the equilibrium price.

Critical Thinking Exercises

52. Which one of the following is true?

a. Solving an inconsistent system by substitution results in a true statement.

b. The line passing through the intersection of the graphs of $x + y = 4$ and $x - y = 0$ with slope $= 3$ has an equation given by $y - 2 = 3(x - 2)$.

c. Unlike the graphing method, where solutions cannot be seen, the substitution method provides a way to visualize solutions as intersection points.

d. To solve the system

$$2x - y = 5$$
$$3x + 4y = 7$$

by substitution, replace y in the second equation with $5 - 2x$.

53. If $x = 3 - y - z, 2x + y - z = -6$, and $3x - y + z = 11$, find the values for $x, y,$ and z.

54. Find the value of m that makes

$$y = mx + 3$$
$$5x - 2y = 7$$

an inconsistent system.

Review Exercises

55. Graph: $4x + 6y = 12$. (Section 4.2, Example 4)

56. Solve: $4(x + 1) = 25 + 3(x - 3)$. (Section 2.3, Example 3)

57. List all the integers in this set: $\left\{-73, -\dfrac{2}{3}, 0, \dfrac{3}{1}, \dfrac{3}{2}, \dfrac{\pi}{1}\right\}$. (Section 1.2, Example 5)

SOLVING SYSTEMS OF LINEAR EQUATIONS BY THE ADDITION METHOD

<table>
<tr><td>
SECTION

5.3

Objectives

1 Solve linear systems by the addition method.

2 Use the addition method to identify systems with no solution or infinitely many solutions.

3 Determine the most efficient method for solving a linear system.
</td></tr>
</table>

SOLVING SYSTEMS OF LINEAR EQUATIONS BY THE ADDITION METHOD

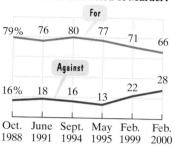

Are You in Favor of the Death Penalty for a Person Convicted of Murder?

For
79%   76   80   77   71   66

Against
16%   18   16   13   22   28

Oct.   June   Sept.   May   Feb.   Feb.
1988   1991   1994   1995   1999   2000

The graphs shown above are based on 543 adults polled nationally by *Newsweek*. If these trends continue, when will the percentage of Americans in favor of the death penalty be the same as the percentage of those who oppose it? The question can be answered by modeling the data with a system of linear equations and solving the system. However, the substitution method is not always the easiest way to solve linear systems. In this section we consider a third method for solving these systems.

1 Solve linear systems by the addition method.

Eliminating a Variable Using the Addition Method   The substitution method is most useful if one of the given equations has an isolated variable. A third, and frequently the easiest, method for solving a linear system is the addition method. Like the substitution method, the addition method involves eliminating a variable and ultimately solving an equation containing only one variable. However, this time we eliminate a variable by adding the equations.

For example, consider the following system of equations:

$$3x - 4y = 11$$
$$-3x + 2y = -7.$$

When we add these two equations, the x-terms are eliminated. This occurs because the coefficients of the x-terms, 3 and -3, are opposites (additive inverses) of each other.

$$
\begin{array}{r}
3x - 4y = 11 \\
-3x + 2y = -7 \\
\hline
\text{Add:} \quad -2y = 4
\end{array}
$$

> The sum is an equation in one variable.

$$y = -2 \qquad \text{Solve for } y \text{ by dividing both sides by } -2.$$

Now we can back-substitute -2 for y into one of the original equations to find x. It does not matter which equation you use; you will obtain the same value for x in either case. If we use either equation, we can show that $x = 1$ and the solution $(1, -2)$ satisfies both equations in the system.

When we use the addition method, we want to obtain two equations whose sum is an equation containing only one variable. The key step is to **obtain, for one of the variables, coefficients that differ only in sign**. To do this, we may need to multiply one or both equations by some nonzero number so that the coefficients of one of the variables, x or y, become opposites. Then when the two equations are added, this variable is eliminated.

EXAMPLE 1 Solving a System by the Addition Method

Solve by the addition method:

$$x + y = 4$$
$$x - y = 6.$$

SOLUTION The coefficients of y in the two equations, 1 and -1, differ only in sign. Therefore, by adding the two left sides and the two right sides, we can eliminate the y-terms.

$$
\begin{aligned}
x + y &= 4 \\
\underline{x - y} &= \underline{6} \\
\text{Add:}\quad 2x + 0y &= 10 \\
2x &= 10 \qquad \text{Simplify.}
\end{aligned}
$$

Now y is eliminated and we can solve $2x = 10$ for x.

$$
\begin{aligned}
2x &= 10 \\
x &= 5 \qquad \text{Divide both sides by 2 and solve for } x.
\end{aligned}
$$

We back-substitute 5 for x into one of the original equations to find y. We will use both equations to show that we obtain the same value for y in either case.

Use the first equation:	**Use the second equation:**
$x + y = 4$	$x - y = 6$
$5 + y = 4$	$5 - y = 6$ Replace x with 5.
$y = -1.$	$-y = 1$ Solve for y.
	$y = -1.$

Thus, $x = 5$ and $y = -1$. The proposed solution, $(5, -1)$, can be shown to satisfy both equations in the system. Consequently, the solution is $(5, -1)$. ∎

 CHECK POINT 1 Solve by the addition method:

$$x + y = 5$$
$$x - y = 9.$$

EXAMPLE 2 Solving a System by the Addition Method

Solve by the addition method:

$$3x - y = 11$$
$$2x + 5y = 13.$$

SOLUTION We must rewrite one or both equations in equivalent forms so that the coefficients of the same variable (either x or y) differ only in sign. Consider the terms in y in each equation, that is, $-1y$ and $5y$. To eliminate y, we can multiply each term of the first equation by 5 and then add the equations.

$$
\begin{array}{lcl}
3x - y = 11 & \xrightarrow{\text{Multiply by 5.}} & 15x - 5y = 55 \\
2x + 5y = 13 & \xrightarrow{\text{No change}} & \underline{2x + 5y = 13} \\
& \text{Add:} & 17x + 0y = 68 \qquad \text{Simplify.} \\
& & 17x = 68 \qquad \text{Divide both sides by} \\
& & x = 4 \qquad \text{17 and solve for } x.
\end{array}
$$

Thus, $x = 4$. To find y, we back-substitute 4 for x into either one of the given equations. We'll use the second equation.

$$2x + 5y = 13 \qquad \text{This is the second equation in the given system.}$$
$$2 \cdot 4 + 5y = 13 \qquad \text{Substitute 4 for } x.$$
$$8 + 5y = 13 \qquad \text{Multiply: } 2 \cdot 4 = 8.$$
$$5y = 5 \qquad \text{Subtract 8 from both sides.}$$
$$y = 1 \qquad \text{Divide both sides by 5.}$$

The solution is $(4, 1)$. Check to see that it satisfies both of the original equations in the system. ■

 CHECK POINT 2 Solve by the addition method:

$$4x - y = 22$$
$$3x + 4y = 26.$$

Before considering additional examples, let's summarize the steps for solving linear systems by the addition method.

SOLVING LINEAR SYSTEMS BY ADDITION

1. If necessary, rewrite both equations in the form $Ax + By = C$.

2. If necessary, multiply either equation or both equations by appropriate nonzero numbers so that the sum of the x-coefficients or the sum of the y-coefficients is 0.

3. Add the equations in step 2. The sum is an equation in one variable.

4. Solve the equation in one variable.

5. Back-substitute the value obtained in step 4 into either of the given equations and solve for the other variable.

6. Check the solution in both of the original equations.

EXAMPLE 3 Solving a System by the Addition Method

Solve by the addition method:

$$3x + 2y = 48$$
$$9x - 8y = -24.$$

SOLUTION

Step 1. **Rewrite both equations in the form $Ax + By = C$.** Both equations are already in this form. Variable terms appear on the left and constants appear on the right.

Step 2. **If necessary, multiply either equation or both equations by appropriate numbers so that the sum of the x-coefficients or the sum of the y-coefficients is 0.** We can eliminate x or y. Let's eliminate x. Consider the terms in x in each equation, that is, $3x$ and $9x$. To eliminate x, we can multiply each term of the first equation by -3 and then add the equations.

$$
\begin{array}{ll}
3x + 2y = 48 & \xrightarrow{\text{Multiply by } -3.} \quad -9x - 6y = -144 \\
9x - 8y = -24 & \xrightarrow{\text{No change}} \quad \underline{9x - 8y = -24} \\
\end{array}
$$

Step 3. **Add the equations.** $\qquad\qquad$ Add: $\quad -14y = -168$

Step 4. **Solve the equation in one variable.** We solve $-14y = -168$ by dividing both sides by -14.

$$\frac{-14y}{-14} = \frac{-168}{-14} \qquad \text{Divide both sides by } -14.$$

$$y = 12 \qquad \text{Simplify.}$$

Step 5. **Back-substitute and find the value for the other variable.** We can back-substitute 12 for y into either one of the given equations. We'll use the first one.

$$3x + 2y = 48 \qquad \text{This is the first equation in the given system.}$$

$$3x + 2(12) = 48 \qquad \text{Substitute 12 for y.}$$

$$3x + 24 = 48 \qquad \text{Multiply.}$$

$$3x = 24 \qquad \text{Subtract 24 from both sides.}$$

$$x = 8 \qquad \text{Divide both sides by 3.}$$

The solution is $(8, 12)$.

Step 6. **Check.** Take a few minutes to show that $(8, 12)$ satisfies both of the original equations in the system. ∎

 CHECK POINT 3 Solve by the addition method:

$$4x + 5y = 3$$
$$2x - 3y = 7.$$

Some linear systems have solutions that are not integers. If the value of one variable turns out to be a "messy" fraction, back-substitution might lead to cumbersome arithmetic. If this happens, you can return to the original system and use addition to find the value of the other variable.

EXAMPLE 4 Solving a System by the Addition Method

Solve by the addition method:

$$2x = 7y - 17$$
$$5y = 17 - 3x.$$

SOLUTION

Step 1. **Rewrite both equations in the form $Ax + By = C$.** We first arrange the system so that variable terms appear on the left and constants appear on the right. We obtain

$$2x - 7y = -17 \qquad \text{Subtract 7y from both sides of the first equation.}$$

$$3x + 5y = 17. \qquad \text{Add 3x to both sides of the second equation.}$$

Step 2. **If necessary, multiply either equation or both equations by appropriate numbers so that the sum of the x-coefficients or the sum of the y-coefficients is 0.** We can eliminate x or y. Let's eliminate x by multiplying the first equation by 3 and the second equation by -2.

$$
\begin{array}{l}
2x - 7y = -17 \xrightarrow{\text{Multiply by 3.}} 6x - 21y = -51 \\
3x + 5y = 17 \xrightarrow{\text{Multiply by } -2} \underline{-6x - 10y = -34} \\
\phantom{3x + 5y = 17 \xrightarrow{aaaa}} \text{Add:} -31y = -85
\end{array}
$$

Step 3. **Add the equations.**

Step 4. **Solve the equation in one variable.** We solve $-31y = -85$ by dividing both sides by -31.

$$\frac{-31y}{-31} = \frac{-85}{-31} \qquad \text{Divide both sides by } -31.$$

$$y = \frac{85}{31} \qquad \text{Simplify.}$$

Step 5. **Back-substitute and find the value for the other variable.** Back-substitution of $\frac{85}{31}$ for y into either of the given equations results in cumbersome arithmetic. Instead, let's use the addition method on the given system in the form $Ax + By = C$ to find the value for x. Thus, we eliminate y by multiplying the first equation by 5 and the second equation by 7.

$$
\begin{array}{l}
2x - 7y = -17 \quad \xrightarrow{\text{Multiply by 5.}} \quad 10x - 35y = -85 \\
3x + 5y = 17 \quad \xrightarrow{\text{Multiply by 7.}} \quad \underline{21x + 35y = 119} \\
\phantom{3x + 5y = 17 \quad \text{Multiply by 7.}} \text{Add: } 31x = 34
\end{array}
$$

$$x = \frac{34}{31} \qquad \text{Divide both sides by 31.}$$

The solution is $\left(\dfrac{34}{31}, \dfrac{85}{31}\right)$.

Step 6. **Check.** For this system, a calculator is helpful in showing that $\left(\frac{34}{31}, \frac{85}{31}\right)$ satisfies both of the original equations in the system. ∎

✔ CHECK POINT 4 Solve by the addition method:

$$2x = 9 + 3y$$
$$4y = 8 - 3x.$$

---

2 Use the addition method to identify systems with no solution or infinitely many solutions.

The Addition Method with Linear Systems Having No Solution or Infinitely Many Solutions As with the substitution method, if the addition method results in a false statement, the linear system is inconsistent and has no solution.

EXAMPLE 5 Using the Addition Method on an Inconsistent System

Solve the system:

$$4x + 6y = 12$$
$$6x + 9y = 12.$$

SOLUTION We can eliminate x or y. Let's eliminate x by multiplying the first equation by 3 and the second equation by -2.

$$
\begin{array}{l}
4x + 6y = 12 \quad \xrightarrow{\text{Multiply by 3.}} \quad 12x + 18y = 36 \\
6x + 9y = 12 \quad \xrightarrow{\text{Multiply by } -2.} \quad \underline{-12x - 18y = -24} \\
\phantom{6x + 9y = 12 \quad \text{Multiply by } -2.} \text{Add: } 0 = 12
\end{array}
$$

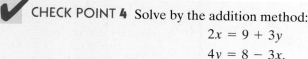

There are no values of x and y for which $0 = 12$.

The false statement $0 = 12$ indicates that the system is inconsistent and has no solution. The graphs of the system's equations are shown in Figure 5.8. The lines are parallel and have no point of intersection. ∎

✔ CHECK POINT 5 Solve the system:

$$x + 2y = 4$$
$$3x + 6y = 13.$$

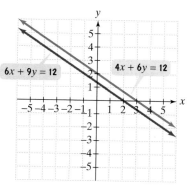

FIGURE 5.8 Visualizing the inconsistent system in Example 5

If you use the addition method, how can you tell if a system has infinitely many solutions? As with the substitution method, you will eliminate both variables and obtain a true statement.

EXAMPLE 6 Using the Addition Method on a System with Infinitely Many Solutions

Solve by the addition method:

$$2x - y = 3$$
$$-4x + 2y = -6.$$

SOLUTION We can eliminate y by multiplying the first equation by 2.

$$
\begin{array}{llll}
2x - y = 3 & \xrightarrow{\text{Multiply by 2.}} & 4x - 2y = 6 \\
-4x + 2y = -6 & \xrightarrow{\text{No change}} & -4x + 2y = -6 \\
& \text{Add:} & 0 = 0
\end{array}
$$

The true statement $0 = 0$ indicates that the system contains dependent equations and has infinitely many solutions. Any ordered pair that satisfies the first equation also satisfies the second equation. ∎

 CHECK POINT 6 Solve by the addition method:

$$x - 5y = 7$$
$$3x - 15y = 21.$$

---

3 Determine the most efficient method for solving a linear system.

Comparing the Three Solution Methods The following chart compares the graphing, substitution, and addition methods for solving systems of linear equations. With increased practice, it becomes easier for you to select the best method for solving a particular linear system.

Comparing Solution Methods

Method	Advantages	Disadvantages
Graphing	You can see the solutions.	If the solutions do not involve integers or are too large to be seen on the graph, it's impossible to tell exactly what the solutions are.
Substitution	Gives exact solutions. Easy to use if a variable is on one side by itself.	Solutions cannot be seen. Can introduce extensive work with fractions when no variable has a coefficient of 1 or −1.
Addition	Gives exact solutions. Easy to use even if no variable has a coefficient of 1 or −1.	Solutions cannot be seen.

5.3 EXERCISE SET

Student Solutions Manual CD/Video PH Math/Tutor Center MathXL Tutorials on CD MathXL® MyMathLab Interactmath.com

Practice Exercises

In Exercises 1–44, solve each system by the addition method. If there is no solution or an infinite number of solutions, so state.

1. $x + y = -3$
$x - y = 11$

2. $x + y = 6$
$x - y = -2$

3. $2x + 3y = 6$
$2x - 3y = 6$

4. $3x + 2y = 14$
$3x - 2y = 10$

5. $x + 2y = 7$
$-x + 3y = 18$

6. $2x + y = -2$
$-2x - 3y = -6$

7. $5x - y = 14$
$-5x + 2y = -13$

8. $7x - 4y = 13$
$-7x + 6y = -11$

9. $3x + y = 7$
$2x - 5y = -1$

10. $3x - y = 11$
$2x + 5y = 13$

11. $x + 3y = 4$
$4x + 5y = 2$

12. $x + 2y = -1$
$4x - 5y = 22$

13. $-3x + 7y = 14$
$2x - y = -13$

14. $2x - 5y = -1$
$3x + y = 7$

15. $3x - 14y = 6$
$5x + 7y = 10$

16. $5x - 4y = 19$
$3x + 2y = 7$

17. $3x - 4y = 11$
$2x + 3y = -4$

18. $2x + 3y = -16$
$5x - 10y = 30$

19. $3x + 2y = -1$
$-2x + 7y = 9$

20. $5x + 3y = 27$
$7x - 2y = 13$

21. $3x = 2y + 7$
$5x = 2y + 13$

22. $9x = 25 + y$
$2y = 4 - 9x$

23. $2x = 3y - 4$
$-6x + 12y = 6$

24. $5x = 4y - 8$
$3x + 7y = 14$

25. $2x - y = 3$
$4x + 4y = -1$

26. $3x - y = 22$
$4x + 5y = -21$

27. $4x = 5 + 2y$
$2x + 3y = 4$

28. $3x = 4y + 1$
$4x + 3y = 1$

29. $3x - y = 1$
$3x - y = 2$

30. $4x - 9y = -2$
$-4x + 9y = -2$

31. $x + 3y = 2$
$3x + 9y = 6$

32. $4x - 2y = 2$
$2x - y = 1$

33. $7x - 3y = 4$
$-14x + 6y = -7$

34. $2x + 4y = 5$
$3x + 6y = 6$

35. $5x + y = 2$
$3x + y = 1$

36. $2x - 5y = -1$
$2x - y = 1$

37. $x = 5 - 3y$
$2x + 6y = 10$

38. $4x = 36 + 8y$
$3x - 6y = 27$

39. $4(3x - y) = 0$
$3(x + 3) = 10y$

40. $2(2x + 3y) = 0$
$7x = 3(2y + 3) + 2$

41. $x + y = 11$
$\dfrac{x}{5} + \dfrac{y}{7} = 1$

42. $x - y = -3$
$\dfrac{x}{9} - \dfrac{y}{7} = -1$

43. $\dfrac{4}{5}x - y = -1$
$\dfrac{2}{5}x + y = 1$

44. $\dfrac{x}{3} + y = 3$
$\dfrac{x}{2} - \dfrac{y}{4} = 1$

In Exercises 45–56, solve each system by the method of your choice. If there is no solution or an infinite number of solutions, so state. Explain why you selected one method over the other two.

45. $3x - 2y = 8$
$x = -2y$

46. $2x - y = 10$
$y = 3x$

47. $3x + 2y = -3$
$2x - 5y = 17$

48. $2x - 7y = 17$
$4x - 5y = 25$

49. $3x - 2y = 6$
$y = 3$

50. $2x + 3y = 7$
$x = 2$

51. $y = 2x + 1$
$y = 2x - 3$

52. $y = 2x + 4$
$y = 2x - 1$

53. $2(x + 2y) = 6$
$3(x + 2y - 3) = 0$

54. $2(x + y) = 4x + 1$
$3(x - y) = x + y - 3$

55. $3y = 2x$
$2x + 9y = 24$

56. $4y = -5x$
$5x + 8y = 20$

Practice Plus

In Exercises 57–64, solve each system or state that the system is inconsistent or dependent.

57. $\dfrac{3x}{5} + \dfrac{4y}{5} = 1$
$\dfrac{x}{4} - \dfrac{3y}{8} = -1$

58. $\dfrac{x}{3} - \dfrac{y}{2} = \dfrac{2}{3}$
$\dfrac{2x}{3} + y = \dfrac{4}{3}$

59. $5(x + 1) = 7(y + 1) - 7$
$6(x + 1) + 5 = 5(y + 1)$

60. $6x = 5(x + y + 3) - x$
$3(x - y) + 4y = 5(y + 1)$

61. $0.4x + y = 2.2$
$0.5x - 1.2y = 0.3$

62. $1.25x - 1.5y = 2$
$3.5x - 1.75y = 10.5$

63. $\dfrac{x}{2} = \dfrac{y + 8}{3}$
$\dfrac{x + 2}{2} = \dfrac{y + 11}{3}$

64. $\dfrac{x}{2} = \dfrac{y + 8}{4}$
$\dfrac{x + 3}{2} = \dfrac{y + 5}{4}$

Application Exercises

We opened this section with data from 1988 through 2000 showing the percentage of Americans for and against the death penalty for a person convicted of murder. The data can be modeled by the following system of equations:

$13x + 12y = 992$ → The percent, y, in favor of the death penalty x years after 1988

$-x + y = 16.$ → The percent, y, against the death penalty x years after 1988

Use this system to solve Exercises 65–66.

65. Use the addition method to determine in which year the percentage of Americans in favor of the death penalty will be the same as the percentage of Americans who oppose it. For that year, what percent will be for the death penalty and what percent will be against it?

66. Use the substitution method to solve Exercise 65.

Writing in Mathematics

67. Explain how to solve a system of equations using the addition method. Use $3x + 5y = -2$ and $2x + 3y = 0$ to illustrate your explanation.

68. When using the addition method, how can you tell if a system of linear equations has no solution?

69. When using the addition method, how can you tell if a system of linear equations has infinitely many solutions?

70. Take a second look at the data about the death penalty shown on page 291. Do you think that these trends will continue? Explain your answer.

71. The formula $3239x + 96y = 134{,}014$ models the number of daily evening newspapers, y, x years after 1980. The formula $-665x + 36y = 13{,}800$ models the number of daily morning newspapers, y, x years after 1980. What is the most efficient method for solving this system? Explain why. What does the solution mean in terms of the variables in the formulas? (It is not necessary to actually solve the system.)

Critical Thinking Exercises

72. Which one of the following statements is true?
   a. If x can be eliminated by the addition method, y cannot be eliminated by using the original equations of the system.
   b. If $Ax + 2y = 2$ and $2x + By = 10$ have graphs that intersect at $(2, -2)$, then $A = -3$ and $B = 3$.
   c. The equations $y = x - 1$ and $x = y + 1$ are dependent.
   d. If the two equations in a linear system are $5x - 3y = 7$ and $4x + 9y = 11$, multiplying the first equation by 4, the second by 5, and then adding equations will eliminate x.

73. Solve by expressing x and y in terms of a and b:
$$x - y = a$$
$$y = 2x + b.$$

74. The point of intersection of the graphs of the equations $Ax - 3y = 16$ and $3x + By = 7$ is $(5, -2)$. Find A and B.

Technology Exercises

75. Some graphing utilities can give the solution to a system of linear equations. (Consult your manual for details.) This capability is usually accessed with the $\boxed{\text{SIMULT}}$ (simultaneous equations) feature. First, you must enter 2, for two equations in two variables. With each equation in $Ax + By = C$ form, you must then enter the coefficients for x and y and the constant term, one equation at a time. After entering all six numbers, press $\boxed{\text{SOLVE}}$. The solution will be displayed on the screen. (The x-value may be displayed as $x_1 =$ and the y-value as $x_2 =$.) Use this capability to verify the solution to any five of the exercises you solved in the practice exercises of this exercise set. Describe what happens when you use your graphing utility on a system with no solution or infinitely many solutions.

If your graphing utility has the feature described in Exercise 75, use it to solve each system in Exercises 76–78.

76. $\dfrac{1}{4}x - \dfrac{1}{4}y = -1$
$-3x + 7y = 8$

77. $x = 5y$
$2x - 3y = 7$

78. $0.6x + 0.08y = 4$
$3x + 2y = 4$

Review Exercises

79. For which number is 5 times the number equal to the number increased by 40? (Section 2.5, Example 2)

80. In which quadrant is $\left(-\frac{3}{2}, 15\right)$ located? (Section 1.3, Example 1)

81. Solve: $29{,}700 + 150x = 5000 + 1100x$. (Section 2.2, Example 7)

✔ **MID-CHAPTER CHECK POINT**

CHAPTER **5**

What You Know: We learned how to solve systems of linear equations by graphing, by the substitution method, and by the addition method. We saw that some systems, called inconsistent systems, have no solution, whereas other systems, called dependent systems, have infinitely many solutions.

In Exercises 1–3, solve each system by graphing.

1. $3x + 2y = 6$
$2x - y = 4$

2. $y = 2x - 1$
$y = 3x - 2$

3. $y = 2x - 1$
$6x - 3y = 12$

In Exercises 4–15, solve each system by the method of your choice.

4. $5x - 3y = 1$
$y = 3x - 7$

5. $6x + 5y = 7$
$3x - 7y = 13$

6. $x = \dfrac{y}{3} - 1$
$6x + y = 21$

7. $3x - 4y = 6$
$5x - 6y = 8$

8. $3x - 2y = 32$
$\dfrac{x}{5} + 3y = -1$

9. $x - y = 3$
$2x = 4 + 2y$

10. $x = 2(y - 5)$
$4x + 40 = y - 7$

11. $y = 3x - 2$
$y = 2x - 9$

12. $2x - 3y = 4$
$3x + 4y = 0$

13. $y - 2x = 7$
$4x = 2y - 14$

14. $4(x + 3) = 3y + 7$
$2(y - 5) = x + 5$

15. $\dfrac{x}{2} - \dfrac{y}{5} = 1$
$y - \dfrac{x}{3} = 8$

SECTION **5.4**

Objective

1 Solve problems using linear systems.

PROBLEM SOLVING USING SYSTEMS OF EQUATIONS

Enjoy chatting long distance on the phone? Telecommunication companies want your business. You can go online and get a list of their plans. Is there a monthly fee? What is the rate per minute? Does the plan involve a monthly minimum? In this section, you will learn to use systems of equations to select a plan that will save you the most money.

1 Solve problems using linear systems.

A Strategy for Solving Word Problems Using Systems of Equations When we solved problems in Chapters 2 and 3, we let x represent a quantity that was unknown. Problems in this section involve two unknown quantities. We will let x and

y represent these quantities. We then translate from the verbal conditions of the problem to a *system* of linear equations.

EXAMPLE 1 The World's Longest Snakes

The royal python and the anaconda are the world's longest snakes. The maximum length for each of these snakes is implied by the following description:

> Three royal pythons and two anacondas measure 161 feet. The royal python's length increased by triple the anaconda's length is 119 feet. Find the maximum length for each of these snakes.

SOLUTION

Step 1. **Use variables to represent unknown quantities.** Let *x* represent the royal python's length, in feet. Let *y* represent the anaconda's length, in feet.

Step 2. **Write a system of equations describing the problem's conditions.**

Three royal pythons	plus	two anacondas	measure	161 feet.
$3x$	$+$	$2y$	$=$	161

The royal python's length	increased by	triple the anaconda's length	is	119 feet.
x	$+$	$3y$	$=$	119

Step 3. **Solve the system and answer the problem's question.** The system

$$3x + 2y = 161$$
$$x + 3y = 119$$

can be solved by substitution or addition. Substitution works well because *x* in the second equation has a coefficient of 1. We can solve for *x* by subtracting $3y$ from both sides, thereby avoiding fractions. Addition also works well; if we multiply the second equation by -3, adding equations will eliminate *x*. We will use addition.

$$
\begin{array}{ll}
3x + 2y = 161 & \xrightarrow{\text{No change}} \\
x + 3y = 119 & \xrightarrow{\text{Multiply by } -3.}
\end{array}
\qquad
\begin{array}{l}
3x + 2y = 161 \\
\underline{-3x - 9y = -357} \\
 -7y = -196 \\
y = \dfrac{-196}{-7} = 28
\end{array}
$$

Add:

Because *y* represents the anaconda's length, we see that the anaconda is 28 feet long. Now we can find *x*, the royal python's length. We do so by back-substituting 28 for *y* in either of the system's equations.

$$
\begin{array}{ll}
x + 3y = 119 & \text{We'll use the second equation.} \\
x + 3 \cdot 28 = 119 & \text{Back-substitute 28 for y.} \\
x + 84 = 119 & \text{Multiply: } 3 \cdot 28 = 84. \\
x = 35 & \text{Subtract 84 from both sides.}
\end{array}
$$

Because $x = 35$ and $y = 28$, the royal python is 35 feet long and the anaconda is 28 feet long.

Step 4. **Check the proposed answers in the original wording of the problem.** Three royal pythons and two anacondas should measure 161 feet:

$$3(35 \text{ ft}) + 2(28 \text{ ft}) = 105 \text{ ft} + 56 \text{ ft} = 161 \text{ ft}.$$

The royal python's length increased by triple the anaconda's length should be 119 feet:

$$35 \text{ ft} + 3(28 \text{ ft}) = 35 \text{ ft} + 84 \text{ ft} = 119 \text{ ft}.$$

This verifies that the royal python's and the anaconda's length are 35 feet and 28 feet, respectively. ∎

✔ **CHECK POINT 1** The bustard and the condor are the world's heaviest flying bird and the world's heaviest bird of prey, respectively. The maximum weight for each of these birds is implied by the following description:

Two bustards and three condors weigh 173 pounds. The bustard's weight increased by double the condor's weight is 100 pounds. What is the maximum weight for each of these birds?

EXAMPLE 2 Cholesterol and Heart Disease

The verdict is in. After years of research, the nation's health experts agree that high cholesterol in the blood is a major contributor to heart disease. Thus, cholesterol intake should be limited to 300 milligrams or less each day. Fast foods provide a cholesterol carnival. All together, two McDonald's Quarter Pounders and three Burger King Whoppers with cheese contain 520 milligrams of cholesterol. Three Quarter Pounders and one Whopper with cheese exceed the suggested daily cholesterol intake by 53 milligrams. Determine the cholesterol content in each item.

SOLUTION

Step 1. **Use variables to represent the unknown quantities.** Let x represent the cholesterol content, in milligrams, of a Quarter Pounder. Let y represent the cholesterol content, in milligrams, of a Whopper with cheese.

Step 2. **Write a system of equations describing the problem's conditions.**

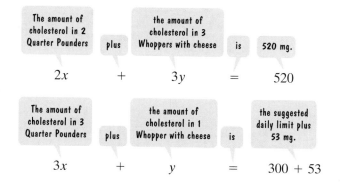

The amount of cholesterol in 2 Quarter Pounders	plus	the amount of cholesterol in 3 Whoppers with cheese	is	520 mg.
$2x$	$+$	$3y$	$=$	520

The amount of cholesterol in 3 Quarter Pounders	plus	the amount of cholesterol in 1 Whopper with cheese	is	the suggested daily limit plus 53 mg.
$3x$	$+$	y	$=$	$300 + 53$

Step 3. **Solve the system and answer the problem's question.** The system

$$2x + 3y = 520$$
$$3x + \ y = 353$$

About 15 million hamburgers are eaten every day in the United States.

can be solved by substitution or addition. We will use addition; if we multiply the second equation by -3, adding equations will eliminate y.

$$
\begin{array}{rcl}
2x + 3y = 520 & \xrightarrow{\text{No change}} & 2x + 3y = 520 \\
3x + y = 353 & \xrightarrow{\text{Multiply by }-3.} & -9x - 3y = -1059 \\
\hline
& \text{Add:} & -7x = -539
\end{array}
$$

$$x = \frac{-539}{-7} = 77$$

Because x represents the cholesterol content of a Quarter Pounder, we see that a Quarter Pounder contains 77 milligrams of cholesterol. Now we can find y, the cholesterol content of a Whopper with cheese. We do so by back-substituting 77 for x in either of the system's equations.

$$
\begin{array}{ll}
3x + y = 353 & \textit{We'll use the second equation.} \\
3(77) + y = 353 & \textit{Back-substitute 77 for x.} \\
231 + y = 353 & \textit{Multiply.} \\
y = 122 & \textit{Subtract 231 from both sides.}
\end{array}
$$

Because $x = 77$ and $y = 122$, a Quarter Pounder contains 77 milligrams of cholesterol and a Whopper with cheese contains 122 milligrams of cholesterol.

Step 4. **Check the proposed answers in the original wording of the problem.** Two Quarter Pounders and three Whoppers with cheese contain

$$2(77 \text{ mg}) + 3(122 \text{ mg}) = 520 \text{ mg},$$

which checks with the given conditions. Furthermore, three Quarter Pounders and one Whopper with cheese contain

$$3(77 \text{ mg}) + 1(122 \text{ mg}) = 353 \text{ mg},$$

which does exceed the daily limit of 300 milligrams by 53 milligrams. ∎

✔ **CHECK POINT 2** How do the Quarter Pounder and Whopper with cheese measure up in the calorie department? Actually, not too well. Two Quarter Pounders and three Whoppers with cheese provide 2607 calories. Even combining one of each provides enough calories to bring tears to Jenny Craig's eyes—9 calories in excess of what is allowed on a 1000 calorie-a-day diet. Find the caloric content of each item.

EXAMPLE 3 Fencing a Waterfront Lot

You just purchased a rectangular waterfront lot along a river's edge. The perimeter of the lot is 1000 feet. To create a sense of privacy, you decide to fence along three sides, excluding the side that fronts the river (see Figure 5.9). An expensive fencing along the

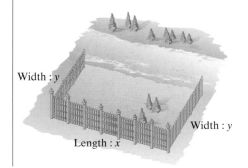

Width : y

Width : y

Length : x

FIGURE 5.9

lot's front length costs $25 per foot. An inexpensive fencing along the two side widths costs only $5 per foot. The total cost of the fencing along three sides comes to $9500.

a. What are the lot's dimensions?

b. You are considering using the expensive fencing on all three sides of the lot. However, you are limited by a $16,000 budget. Can this be done within your budget constraints?

SOLUTION

a. We begin by finding the lot's dimensions.

Step 1.   **Use variables to represent unknown quantities.** Let x represent the lot's length, in feet. Let y represent the lot's width, in feet. (If you prefer, you can use the variable l for length and w for width.)

Step 2.   **Write a system of equations describing the problem's conditions.**

- The lot's perimeter is 1000 feet.

Twice the length	plus	twice the width	is	the perimeter.
$2x$	$+$	$2y$	$=$	1000

- The cost of fencing three sides of the lot is $9500.

Fencing along the front length	plus	fencing along the two side widths	costs	$9500.

Cost per foot	$\cdot$	number of feet	$+$	cost per foot	$\cdot$	number of feet	is	$9500.
25	$\cdot$	x	$+$	5	$\cdot$	$2y$	$=$	9500

Step 3.   **Solve the system and answer the problem's question.** The system

$$2x + 2y = 1000$$
$$25x + 10y = 9500$$

can be solved most easily by addition. If we multiply the first equation by -5, adding equations will eliminate y.

$$\begin{array}{ll} 2x + 2y = 1000 & \xrightarrow{\text{Multiply by } -5.} & -10x - 10y = -5000 \\ 25x + 10y = 9500 & \xrightarrow{\text{No change}} & \underline{25x + 10y = 9500} \\ & \text{Add:} & 15x \qquad\quad = 4500 \\ & & x = \dfrac{4500}{15} = 300 \end{array}$$

Because x represents length, we see that the lot is 300 feet long. Now we can find y, the lot's width. We do so by back-substituting 300 for x in either of the system's equations.

$2x + 2y = 1000$	We'll use the first equation.
$2 \cdot 300 + 2y = 1000$	Back-substitute 300 for x.
$600 + 2y = 1000$	Multiply: 2 · 300 = 600.
$2y = 400$	Subtract 600 from both sides.
$y = 200$	Divide both sides by 2.

Because $x = 300$ and $y = 200$, the lot is 300 feet long and 200 feet wide. Its dimensions are 300 feet by 200 feet.

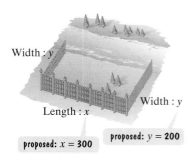

Width : y

Length : x

Width : y

proposed: $x = 300$

proposed: $y = 200$

FIGURE 5.9 (repeated)
Showing proposed values for x and y

Step 4. **Check the proposed answers in the original wording of the problem.** The perimeter should be 1000 feet:

$$2(300 \text{ ft}) + 2(200 \text{ ft}) = 600 \text{ ft} + 400 \text{ ft} = 1000 \text{ ft.}$$

The cost of fencing three sides of the lot should be $9500:

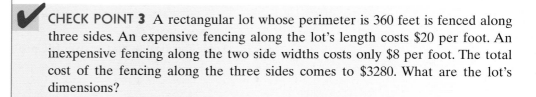

Fencing cost: front length

Fencing cost: two side widths

$$300 \cdot \$25 + 400 \cdot \$5 = \$7500 + \$2000 = \$9500$$

b. Can you use the expensive fencing along all three sides of the lot and stay within a $16,000 budget? The fencing is to be placed along the length, 300 feet, and the two side widths, 2 · 200 feet, or 400 feet. Thus, you will fence 700 feet. The expensive fencing costs $25 per foot. The cost is

$$700 \text{ feet} \cdot \frac{\$25}{\text{foot}} = 700 \cdot \$25 = \$17,500.$$

Limited by a $16,000 budget, you cannot use the expensive fencing on three sides of the lot. ∎

✔ **CHECK POINT 3** A rectangular lot whose perimeter is 360 feet is fenced along three sides. An expensive fencing along the lot's length costs $20 per foot. An inexpensive fencing along the two side widths costs only $8 per foot. The total cost of the fencing along the three sides comes to $3280. What are the lot's dimensions?

EXAMPLE 4 Solar and Electric Heating Systems

The costs for two different kinds of heating systems for a three-bedroom home are given in the following table.

System	Cost to Install	Operating Cost/Year
Solar	$29,700	$150
Electric	$5000	$1100

After how many years will total costs for solar heating and electric heating be the same? What will be the cost at that time?

SOLUTION

Step 1. **Use variables to represent unknown quantities.** Let x represent the number of years the heating system is used. Let y represent the total cost for the heating system.

System	Cost to Install	Operating Cost/Year
Solar	$29,700	$150
Electric	$5000	$1100

Costs for heating systems, repeated

Step 2. **Write a system of equations describing the problem's conditions.**

Total cost for the solar system **equals** installation cost **plus** yearly operating cost **times** the number of years the system is used.

$$y = 29{,}700 + 150 \cdot x$$

Total cost for the electric system **equals** installation cost **plus** yearly operating cost **times** the number of years the system is used.

$$y = 5000 + 1100 \cdot x$$

Step 3. **Solve the system and answer the problem's question.** We want to know after how many years the total costs for the two systems will be the same. We must solve the system

$$y = 29{,}700 + 150x$$
$$y = 5000 + 1100x.$$

Substitution works well because y is isolated in each equation.

$$y = \boxed{29{,}700 + 150x} \quad \boxed{y} = 5000 + 1100x \qquad \text{Substitute } 29{,}700 + 150x \text{ for } y.$$

$$29{,}700 + 150x = 5000 + 1100x \qquad \text{This substitution gives an equation in one variable.}$$

$$29{,}700 = 5000 + 950x \qquad \text{Subtract } 150x \text{ from both sides.}$$

$$24{,}700 = 950x \qquad \text{Subtract } 5000 \text{ from both sides.}$$

$$26 = x \qquad \text{Divide both sides by 950: } \dfrac{24{,}700}{950} = 26.$$

Because x represents the number of years the heating system is used, we see that after 26 years, the total costs for the two systems will be the same. Now we can find y, the total cost. Back-substitute 26 for x in either of the system's equations. We will use the second equation, $y = 5000 + 1100x$.

$$y = 5000 + 1100 \cdot 26 = 5000 + 28{,}600 = 33{,}600$$

Because $x = 26$ and $y = 33{,}600$, after 26 years, the total costs for the two systems will be the same. The cost for each system at that time will be $33,600.

Step 4. **Check the proposed answers in the original wording of the problem.** Let's verify that after 26 years, the two systems will cost the same amount. The installation cost for the solar system is $29,700 and the yearly operating cost is $150. Thus, the total cost after 26 years is

$$\$29{,}700 + \$150(26) = \$29{,}700 + \$3900 = \$33{,}600.$$

The installation cost for the electric system is $5000 and the yearly operating cost is $1100. Thus, the total cost after 26 years is

$$\$5000 + \$1100(26) = \$5000 + \$28{,}600 = \$33{,}600.$$

This verifies that after 26 years the two systems will cost the same amount, $33,600.

The graphs in Figure 5.10 give us a way of visualizing the solution to Example 4. The total cost of solar heating over 40 years is represented by the blue line. The total cost of electric heating over 40 years is represented by the red line. The lines intersect at (26, 33,600): After 26 years, the cost for each system is the same, $33,600.

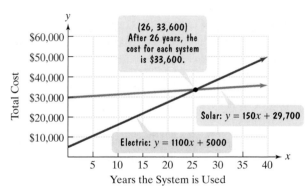

FIGURE 5.10 Visualizing total costs of solar and electric heating

Can you see that to the right of the intersection point, (26, 33,600), the blue graph representing solar costs lies below the red graph representing electric costs? Thus, after 26 years, or when $x > 26$, the cost for solar heating is less than the cost for electric heating.

 CHECK POINT 4 Costs for two different kinds of heating systems for a three-bedroom house are given in the following table.

System	Cost to Install	Operating Cost/Year
Electric	$5000	$1100
Gas	$12,000	$700

After how long will total costs for electric heating and gas heating be the same? What will be the cost at that time?

5.4 EXERCISE SET

Student Solutions Manual   CD/Video   PH Math/Tutor Center   MathXL Tutorials on CD   MathXL®   MyMathLab   Interactmath.com

Practice Exercises

In Exercises 1–4, let x represent one number and let y represent the other number. Use the given conditions to write a system of equations. Solve the system and find the numbers.

1. The sum of two numbers is 17. If one number is subtracted from the other, their difference is −3. Find the numbers.

2. The sum of two numbers is 5. If one number is subtracted from the other, their difference is 13. Find the numbers.

3. Three times a first number decreased by a second number is −1. The first number increased by twice the second number is 23. Find the numbers.

4. The sum of three times a first number and twice a second number is 43. If the second number is subtracted from twice the first number, the result is −4. Find the numbers.

Application Exercises

5. The graph makes Super Bowl Sunday look like a day of snack food bingeing in the United States. Combined, we wolf down 10.4 million pounds of potato chips and tortilla chips.

The difference between consumption of potato chips and tortilla chips is 1.2 million pounds. How many millions of pounds of potato chips and tortilla chips are consumed on Super Bowl Sunday?

Millions of Pounds of Snack Food Consumed on Super Bowl Sunday

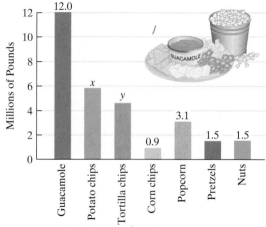

Source: Association of American Snack Foods

6. The graph shows the places with the greatest number of documented fatal shark attacks from 1580 through 2002. Combined, 112 fatal attacks were recorded in the continental United States and South Africa. The difference between the number of attacks in the two countries was 26. How many documented fatal shark attacks were there in the continental United States and in South Africa from 1580 through 2002?

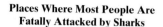

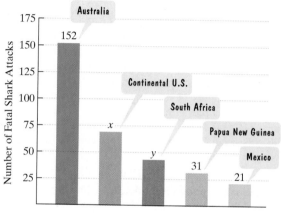

Places Where Most People Are Fatally Attacked by Sharks

Source: International Shark Attack File

The graph shows the calories in some favorite fast foods. Use the information in Exercises 7–8 to find the exact caloric content of the specified foods.

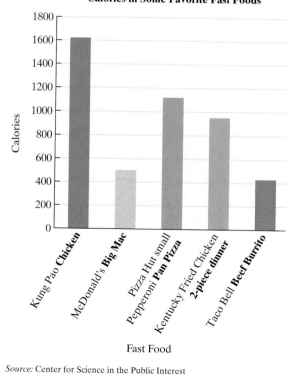

Calories in Some Favorite Fast Foods

Source: Center for Science in the Public Interest

7. One pan pizza and two beef burritos provide 1980 calories. Two pan pizzas and one beef burrito provide 2670 calories. Find the caloric content of each item.

8. One Kung Pao chicken and two Big Macs provide 2620 calories. Two Kung Pao chickens and one Big Mac provide 3740 calories. Find the caloric content of each item.

9. Cholesterol intake should be limited to 300 mg or less each day. One serving of scrambled eggs from McDonalds and one Double Beef Whopper from Burger King exceed this intake by 241 mg. Two servings of scrambled eggs and three Double Beef Whoppers provide 1257 mg of cholesterol. Determine the cholesterol content in each item.

10. Two medium eggs and three cups of ice cream contain 701 milligrams of cholesterol. One medium egg and one cup of ice cream exceed the suggested daily cholesterol intake of 300 milligrams by 25 milligrams. Determine the cholesterol content in each item.

11. In a discount clothing store, all sweaters are sold at one fixed price and all shirts are sold at another fixed price. If one sweater and three shirts cost $42, while three sweaters and two shirts cost $56, find the price of one sweater and the price of one shirt.

12. A restaurant purchased eight tablecloths and five napkins for $106. A week later, a tablecloth and six napkins were bought for $24. Find the cost of one tablecloth and the cost of one napkin, assuming the same prices for both purchases.

13. The perimeter of a badminton court is 128 feet. After a game of badminton, a player's coach estimates that the athlete has run a total of 444 feet, which is equivalent to six times the court's length plus nine times its width. What are the dimensions of a standard badminton court?

14. The perimeter of a tennis court is 228 feet. After a round of tennis, a player's coach estimates that the athlete has run a total of 690 feet, which is equivalent to 7 times the court's length plus four times its width. What are the dimensions of a standard tennis court?

15. A rectangular lot whose perimeter is 320 feet is fenced along three sides. An expensive fencing along the lot's length costs $16 per foot. An inexpensive fencing along the two side widths costs only $5 per foot. The total cost of the fencing along the three sides comes to $2140. What are the lot's dimensions?

16. A rectangular lot whose perimeter is 1600 feet is fenced along three sides. An expensive fencing along the lot's length costs $20 per foot. An inexpensive fencing along the two side widths costs only $5 per foot. The total cost of the fencing along the three sides comes to $13,000. What are the lot's dimensions?

17. You are choosing between two long-distance telephone plans. Plan A has a monthly fee of $20 with a charge of $0.05 per minute for all long-distance calls. Plan B has a monthly fee of $5 with a charge of $0.10 per minute for all long-distance calls.

a. For how many minutes of long-distance calls will the costs for the two plans be the same? What will be the cost for each plan?

b. If you make approximately 10 long-distance calls per month, each averaging 20 minutes, which plan should you select? Explain your answer.

18. You are choosing between two long-distance telephone plans. Plan A has a monthly fee of $15 with a charge of $0.08 per minute for all long-distance calls. Plan B has a monthly fee of $3 with a charge of $0.12 per minute for all long-distance calls.

a. For how many minutes of long-distance calls will the costs for the two plans be the same? What will be the cost for each plan?

b. If you make approximately 15 long-distance calls per month, each averaging 30 minutes, which plan should you select? Explain your answer.

19. You are choosing between two plans at a discount warehouse. Plan A offers an annual membership fee of $100 and you pay 80% of the manufacturer's recommended list price. Plan B offers an annual membership fee of $40 and you pay 90% of the manufacturer's recommended list price. How many dollars of merchandise would you have to purchase in a year to pay the same amount under both plans? What will be the cost for each plan?

20. You are choosing between two plans at a discount warehouse. Plan A offers an annual membership fee of $300 and you pay 70% of the manufacturer's recommended list price. Plan B offers an annual membership fee of $40 and you pay 90% of the manufacturer's recommended list price. How many dollars of merchandise would you have to purchase in a year to pay the same amount under both plans? What will be the cost for each plan?

The graphs show average weekly earnings of full-time wage and salary workers 25 and older, by educational attainment. Exercises 21–22 involve the information in these graphs.

Average Weekly Earnings by Educational Attainment

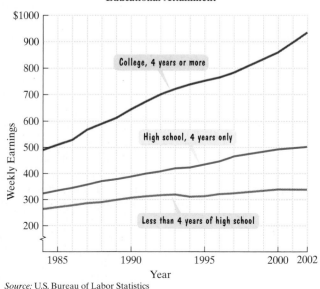

Source: U.S. Bureau of Labor Statistics

21. In 1985, college graduates averaged $508 in weekly earnings. This amount has increased by approximately $25 in weekly earnings per year. By contrast, in 1985, high school graduates averaged $345 in weekly earnings. This amount has only increased by approximately $9 in weekly earnings per year. How many years after 1985 will college graduates be earning twice the amount per week that high school graduates earn? In which year will this occur? What will be the weekly earnings for each group at that time?

22. In 1985, college graduates averaged $508 in weekly earnings. This amount has increased by approximately $25 in weekly earnings per year. By contrast, in 1985, people with less than four years of high school averaged $270 in weekly earnings. This amount has only increased by approximately $4 in weekly earnings per year. How many years after 1985 will college graduates be earning three times the amount per week that people with less than four years of high school earn? (Round to the nearest whole number.) In which year will this occur? What will be the weekly earnings for each group at that time?

23. Nutritional information for macaroni and broccoli is given in the table. How many servings of each would it take to get exactly 14 grams of protein and 48 grams of carbohydrates?

	Macaroni	**Broccoli**
Protein (grams/serving)	3	2
Carbohydrates (grams/serving)	16	4

24. The calorie-nutrient information for an apple and an avocado is given in the table. How many of each should be eaten to get exactly 1000 calories and 100 grams of carbohydrates?

	One Apple	**One Avocado**
Calories	100	350
Carbohydrates (grams)	24	14

In Exercises 25–26, an isosceles triangle with angles B and C having the same measure is shown. Find the measure of each angle whose degree measure is represented with variables.

25.

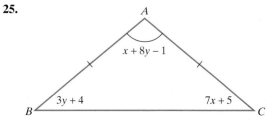

26.

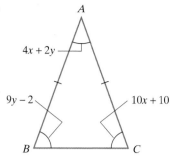

27. When a crew rows with the current, it travels 16 miles in 2 hours. Against the current, the crew rows 8 miles in 2 hours. Let x = the crew's rowing rate in still water and let y = the rate of the current. The following chart summarizes this information.

	Rate	× Time	= Distance
Rowing with current	$x + y$	2	16
Rowing against current	$x - y$	2	8

Find the rate of rowing in still water and the rate of the current.

28. When an airplane flies with the wind, it travels 800 miles in 4 hours. Against the wind, it takes 5 hours to cover the same distance. Let x = the plane's rate in still air and let y = the rate of the wind. The following chart summarizes this information.

	Rate	× Time	= Distance
Flying with the wind	$x + y$	4	800
Flying against the wind	$x - y$	5	800

Find the plane's rate in still air and the rate of the wind.

Writing in Mathematics

29. Describe the conditions in a problem that enable it to be solved using a system of linear equations.

30. Write a word problem that can be solved by translating to a system of linear equations. Then solve the problem.

31. Exercises 17–20 involve using systems of linear equations to compare costs of long-distance telephone plans and plans at a discount warehouse. Describe another situation that involves choosing between two options that can be modeled and solved with a linear system.

Critical Thinking Exercises

32. A set of identical twins can only be recognized by the characteristic that one always tells the truth and the other always lies. One twin tells you of a lucky number pair: "When I multiply my first lucky number by 3 and my second lucky number by 6, the addition of the resulting numbers produces a sum of 12. When I add my first lucky number and twice my second lucky number, the sum is 5." Which twin is talking?

33. Tourist: "How many birds and lions do you have in your zoo?" Zookeeper: "There are 30 heads and 100 feet." Tourist: "I can't tell from that." Zookeeper: "Oh, yes, you can!" Can you? Find the number of each.

34. Find the measure of each angle whose degree measure is represented with a variable.

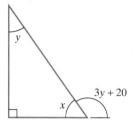

35. One apartment is directly above a second apartment. The resident living downstairs calls his neighbor living above him and states, "If one of you is willing to come downstairs, we'll have the same number of people in both apartments." The upstairs resident responds, "We're all too tired to move. Why don't one of you come up here? Then we'll have twice as many people up here as you've got down there." How many people are in each apartment?

36. In Lewis Carroll's *Through the Looking Glass*, the following dialogue takes place:

> **Tweedledum (to Tweedledee):** The sum of your weight and twice mine is 361 pounds.

> **Tweedledee (to Tweedledum):** Contrawise, the sum of your weight and twice mine is 362 pounds.

Find the weight of each of the two characters.

Technology Exercise

37. Select any two problems that you solved from Exercises 5–24. Use a graphing utility to graph the system of equations that you wrote for that problem. Then use the $\boxed{\text{TRACE}}$ or $\boxed{\text{INTERSECTION}}$ feature to show the point on the graphs that corresponds to the problem's solution.

Review Exercises

38. Graph: $2x - y < 4$. (Section 4.6, Example 2)

39. Graph: $y \geq x + 1$. (Section 4.6, Example 3)

40. Graph: $x \geq 2$. (Section 4.6, Figure 4.37)

SECTION

5.5

Objectives

1 Graph the solutions of systems of linear inequalities.

2 Solve applied problems involving systems of inequalities.

SYSTEMS OF LINEAR INEQUALITIES

Had a good workout lately? If so, could you tell if you were overdoing it or not pushing yourself hard enough? In this section, we will use systems of inequalities in two variables to help you establish a target zone for your workouts.

---

1 Graph the solutions of systems of linear inequalities.

Systems of Linear Inequalities In Section 4.6, we graphed the solutions of a linear inequality in two variables, such as $2x - y < 4$. Just as two linear equations make up a system of linear equations, two (or more) linear inequalities make up a **system of linear inequalities**. Here is an example of a system of linear inequalities:

$$2x - y < 4$$
$$x + y \geq -1.$$

A **solution of a system of linear inequalities** in two variables is an ordered pair that satisfies each inequality in the system. The set of all such ordered pairs is the **solution set** of the system. Thus, to graph a system of inequalities in two variables, begin by graphing each individual inequality in the same rectangular coordinate system. Then find the region, if there is one, that is common to every graph in the system. This region of overlap gives a picture of the system's solution set.

EXAMPLE 1 Graphing a System of Linear Inequalities

Graph the solutions of the system:

$$2x - y < 4$$
$$x + y \geq -1.$$

SOLUTION We begin by graphing $2x - y < 4$. Because the inequality contains a $<$ symbol, rather than $\leq$, we graph $2x - y = 4$ as a dashed line, shown in blue in Figure 5.11. [If $x = 0$, then $y = -4$, and if $y = 0$, then $x = 2$. The x-intercept is 2 and the y-intercept is -4, so the line passes through $(2, 0)$ and $(0, -4)$.] Because $(0, 0)$ makes the inequality $2x - y < 4$ true, we shade the half-plane containing $(0, 0)$, shown in yellow in Figure 5.11.

Now we graph $x + y \geq -1$ in the same rectangular coordinate system. Because the inequality contains a $\geq$ symbol, in which equality is included, we graph $x + y = -1$ as a solid line, shown in red in Figure 5.12. [If $x = 0$, then $y = -1$, and if $y = 0$, then $x = -1$. The x-intercept and y-intercept are both -1, so the red line passes through $(-1, 0)$ and $(0, -1)$.] Because $(0, 0)$ makes the inequality true, we shade the half-plane containing $(0, 0)$. This is shown in Figure 5.12 using green vertical shading.

The solution set of the system is shown graphically by the intersection (the overlap) of the two half-planes. This is shown in Figure 5.12 as the region in which the yellow shading and the green vertical shading overlap. The solutions of the system are shown again in Figure 5.13.

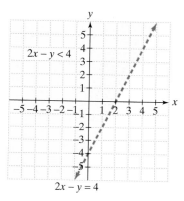

FIGURE 5.11 The graph of $2x - y < 4$

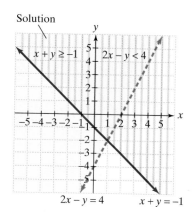

FIGURE 5.12 Adding the graph of $x + y \geq -1$

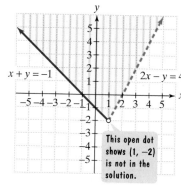

FIGURE 5.13 The graph of $2x - y < 4$ and $x + y \geq -1$ ■

✔ **CHECK POINT 1** Graph the solutions of the system:

$$x + 2y > 4$$
$$2x - 3y \leq -6.$$

EXAMPLE 2 Graphing a System of Linear Inequalities

Graph the solutions of the system:

$$y \geq x + 1$$
$$x \geq 2.$$

SOLUTION We begin by graphing $y \geq x + 1$. Because the inequality contains a $\geq$ symbol, in which equality is included, we graph $y = x + 1$ as a solid line, shown in blue in Figure 5.14.

STUDY TIP

The two inequalities in this system can be graphed without using $(0, 0)$ as a test point. See the study tip on page 261 and the paragraph following Figure 4.37 on the same page.

$$y = 1x + 1 \qquad \text{The form of this equation is } y = mx + b.$$

Slope is $\frac{1}{1}$. y-intercept is 1.

Using the y-intercept, 1, and the slope, 1, the line is shown in Figure 5.14. Because $(0, 0)$ makes the inequality $y \geq x + 1$ false, we shade the half-plane that does not contain $(0, 0)$. This half-plane is shown in yellow in Figure 5.14.

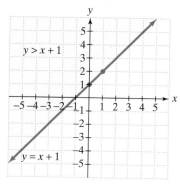

FIGURE 5.14 The graph of $y \geq x + 1$

Now we graph $x \geq 2$ in the same rectangular coordinate system. Because the inequality contains a $\geq$ symbol, in which equality is included, we graph $x = 2$ as a solid line. This vertical line has an x-intercept of 2 and is shown in red in Figure 5.15. Because $(0, 0)$ makes the inequality $x \geq 2$ false ($0 \geq 2$ is false), we shade the half-plane that does not contain $(0, 0)$. This is shown in Figure 5.15 using green vertical shading.

The solution set of the system is shown graphically by the intersection (the overlap) of the two half-planes. This is shown in Figure 5.15 as the region in which the yellow shading and the green vertical shading overlap. The solutions of the system are shown again in Figure 5.16.

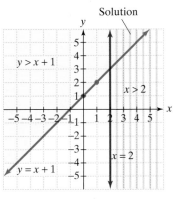

FIGURE 5.15 Adding the graph of $x \geq 2$

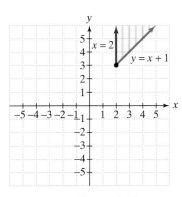

FIGURE 5.16 The graph of $y \geq x + 1$ and $x \geq 2$

✔ **CHECK POINT 2** Graph the solutions of the system:

$$y \geq x + 2$$
$$x \geq 1.$$

2 Solve applied problems involving systems of inequalities.

Applications Now we are ready to use a system of inequalities to establish a target zone for your workouts.

EXAMPLE 3 Inequalities and Aerobic Exercise

For people between ages 10 and 70, inclusive, the target zone for aerobic exercise is given by the following system of inequalities in which a represents one's age and p is one's pulse rate:

$$2a + 3p \geq 450$$
$$a + p \leq 190.$$

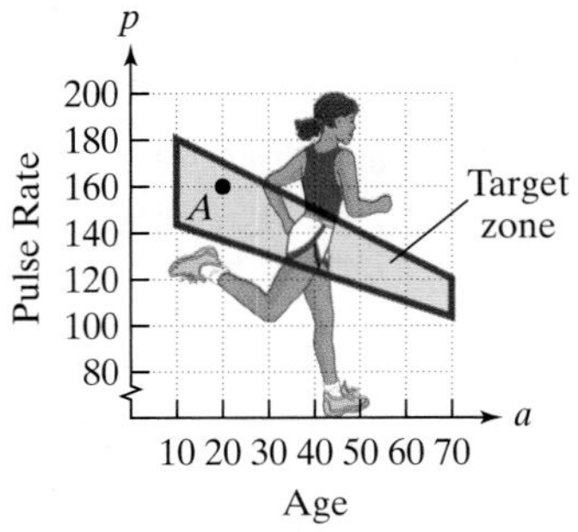

The graph of this target zone is shown in Figure 5.17. Find your age. The line segments on the top and bottom of the shaded region indicate upper and lower limits for your pulse rate, in beats per minute, when engaging in aerobic exercise.

a. What are the coordinates of point A and what does this mean in terms of age and pulse rate?

b. Show that the coordinates of point A satisfy each inequality in the system.

SOLUTION

a. Point A has coordinates $(20, 160)$. This means that a pulse rate of 160 beats per minute is within the target zone for a 20-year-old person engaged in aerobic exercise.

FIGURE 5.17

b. We can show that $(20, 160)$ satisfies each inequality by substituting 20 for a and 160 for p.

$$2a + 3p \geq 450 \qquad\qquad a + p \leq 190$$
$$2(20) + 3(160) \overset{?}{\geq} 450 \qquad\qquad 20 + 160 \overset{?}{\geq} 190$$
$$40 + 480 \overset{?}{\geq} 450 \qquad\qquad 180 \leq 190, \quad \text{true}$$
$$520 \geq 450, \quad \text{true}$$

The pair $(20, 160)$ makes each inequality true, so it satisfies each inequality in the system. Equivalently, $(20, 160)$ is a solution of the system. ∎

CHECK POINT 3 Identify a point other than A in the target zone in Figure 5.17.

a. What are the coordinates of this point and what does this mean in terms of age and pulse rate?

b. Show that the coordinates of the point satisfy each inequality in the system in Example 3.

5.5 EXERCISE SET

Student Solutions Manual   CD/Video   PH Math/Tutor Center   MathXL Tutorials on CD   MathXL®   MyMathLab   Interactmath.com

Practice Exercises

In Exercises 1–36, graph the solutions of each system of linear inequalities.

1. $x + y \leq 4$
$x - y \leq 2$

2. $x + y \geq 4$
$x - y \leq 2$

3. $2x - 4y \leq 8$
$x + y \geq -1$

4. $4x + 3y \leq 12$
$x - 2y \leq 4$

5. $x + 3y \leq 6$
$x - 2y \leq 4$

6. $2x + y \leq 4$
$2x - y \leq 6$

7. $x - 2y > 4$
$2x + y \geq 6$

8. $3x + y < 6$
$x + 2y \geq 2$

9. $x + y > 1$
$x + y < 4$

10. $x - y > 1$
$x - y < 3$

11. $y \geq 2x + 1$
$y \leq 4$

12. $y \geq \frac{1}{2}x + 2$
$y \leq 2$

13. $y > x - 1$
$x > 5$

14. $y > x - 2$
$x > 3$

15. $y \geq 2x - 3$
$y \leq 2x + 1$

16. $y \geq 3x - 2$
$y \leq 3x + 1$

17. $y > 2x - 3$
$y \leq -x + 6$

18. $y < -2x + 4$
$y \leq x - 4$

19. $x + 2y \leq 4$
$y \geq x - 3$

20. $x + y \leq 4$
$y \geq 2x - 4$

21. $x \leq 3$
$y \geq -2$

22. $x \leq 4$
$y \leq -3$

23. $x \geq 3$
$y < 2$

24. $x \geq -2$
$y < -1$

25. $x \geq 0$
$y \leq 0$

26. $x \leq 0$
$y \geq 0$

27. $x \geq 0$
$y > 0$

28. $x \leq 0$
$y < 0$

29. $x + y \leq 5$
$x \geq 0$
$y \geq 0$

30. $2x + y \leq 4$
$x \geq 0$
$y \geq 0$

31. $4x - 3y > 12$
$x \geq 0$
$y \leq 0$

32. $2x - 6y > 12$
$x \leq 0$
$y \leq 0$

33. $0 \leq x \leq 3$
$0 \leq y \leq 3$

34. $0 \leq x \leq 5$
$0 \leq y \leq 5$

35. $x - y \leq 4$
$x + 2y \leq 4$

36. $x - y \leq 3$
$2x + y \leq 4$

Practice Plus

In Exercises 37–44, graph the solutions of each system of linear inequalities. If the system has no solutions, state this and explain why.

37. $x + y \geq 1$
$x - y \geq 1$
$x \geq 4$

38. $x - y \leq 3$
$x + y \leq 3$
$x \geq -2$

39. $x + 2y < 6$
$y > 2x - 2$
$y \geq 2$

40. $2x - 3y < 6$
$2x - 3y > -6$
$-3 \leq x \leq 2$

41. $y \leq -3x + 3$
$y \geq -x - 1$
$y < x + 7$

42. $y \geq -3x + 5$
$y \geq -x + 3$
$y \geq \frac{1}{2}x$
$x \geq 0$
$y \geq 0$

43. $y \geq 2x + 2$
$y < 2x - 3$
$x \geq 2$

44. $y \geq -3x + 2$
$y < -3x$
$x \geq 1$

Application Exercises

45. Use Figure 5.17 on page 312 to solve this exercise.

a. Find a pulse rate that lies within the target zone for a person your age engaged in aerobic exercise.

b. Express your answer in part (a) as an ordered pair. Show that the coordinates of this ordered pair satisfy each inequality in the system.

46. Suppose a patient is not allowed to have more than 330 milligrams of cholesterol per day from a diet of eggs and meat. Each egg provides 165 milligrams of cholesterol and each ounce of meat provides 110 milligrams of cholesterol. Thus, $165x + 110y \leq 330$, where x is the number of eggs and y the number of ounces of meat. Furthermore, the patient must have at least 165 milligrams of cholesterol from the diet. Graph the system of inequalities in the first quadrant. Give the coordinates of any two points in the solution set. Describe what each set of coordinates means in terms of the variables in the problem.

The shaded region in the figure shows recommended weight and height combinations based on information from the Department of Agriculture. Use this region to solve Exercises 47–50.

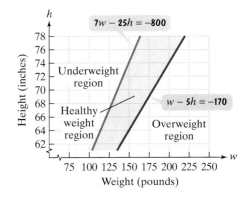

47. Is a person who is 70 inches tall weighing 175 pounds within the healthy weight region?

48. Is a person who is 64 inches tall weighing 105 pounds within the healthy weight region?

49. Estimate the recommended weight range for a person who is 6 feet tall.

50. Write a system of linear inequalities that describes the healthy weight region.

Writing in Mathematics

51. What does the graph of a system of linear inequalities represent?

52. Look at the shaded region showing recommended weight and height combinations in the figure for Exercises 47–50. Describe why a system of inequalities, rather than an equation, is better suited to give the recommended combinations.

Critical Thinking Exercises

53. Write a system of inequalities that has no solutions.

54. A person plans to invest no more than $15,000, placing the money in two investments. One investment is high risk, high yield; the other is low risk, low yield. At least $2000 is to be placed in the high-risk investment. Furthermore, the amount invested at low risk should be at least three times the amount invested at high risk. Write and graph a system of inequalities that describes all possibilities for placing the money in the high- and low-risk investments.

55. Promoters of a rock concert must sell at least 25,000 tickets priced at $35 and $50 per ticket. Furthermore, the promoters must take in at least $1,025,000 in ticket sales. Write and graph a system of inequalities that describes all possibilities for selling the $35 tickets and the $50 tickets.

Review Exercises

56. Find the slope of the line containing the points $(-6, 1)$ and $(2, -1)$. (Section 4.3, Example 1)

57. Add: $\frac{1}{5} + \left(-\frac{3}{4}\right)$. (Section 1.5, Example 4)

58. Graph: $y = x^2$. (Section 4.1, Example 6)

GROUP PROJECT

Group members should go online and obtain a list of telecommunication companies that provide residential long-distance service. The list should contain the monthly fee, the monthly minimum, and the rate per minute for each service provider.

a. For each provider in the list, write an equation that describes the total monthly cost, y, of x minutes of long-distance phone calls.

b. Compare two of the plans. After how many minutes of long-distance calls will the costs for the two plans be the same? Solve a linear system to obtain your answer. What will be the cost for each plan?

c. Repeat part (b) for another two of the plans.

d. Each person should estimate the number of minutes he or she spends talking long distance each month. Group members should assist that person in selecting the plan that will save the most amount of money. Be sure to factor in the monthly minimum, if any, when choosing a plan. Whenever possible, use the equations that you wrote in part (a).

Caution: We've left something out! Your comparisons do not take into account the in-state rates of the plan. Furthermore, if you make international calls, international rates should also be one of your criteria in choosing a plan. Problem solving in real life can get fairly complicated.

CHAPTER 5 SUMMARY

Definitions and Concepts	Examples

Section 5.1 Solving Systems of Linear Equations by Graphing

A system of linear equations in two variables, x and y, consists of two equations of the form $Ax + By = C$. A solution is an ordered pair of numbers that satisfies both equations.

Determine whether $(3, -1)$ is a solution of

$$2x + 5y = 1$$
$$4x + y = 11.$$

Replace x with 3 and y with -1 in both equations.

$$2x + 5y = 1 \qquad\qquad 4x + y = 11$$
$$2\cdot 3 + 5(-1) \stackrel{?}{=} 1 \qquad 4\cdot 3 + (-1) \stackrel{?}{=} 11$$
$$6 + (-5) \stackrel{?}{=} 1 \qquad 12 + (-1) \stackrel{?}{=} 11$$
$$1 = 1, \quad \text{true} \qquad 11 = 11, \quad \text{true}$$

Thus, $(3, -1)$ is a solution of the system.

Using the graphing method, a solution of a linear system is a point common to the graphs of both equations in the system. If the graphs are parallel lines, the system has no solution and is called inconsistent. If the graphs are the same line, the system has infinitely many solutions. The equations are called dependent.

Solve by graphing: $2x + y = 4$
$\qquad\qquad\qquad\; x + y = 2.$

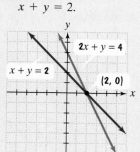

The solution is $(2, 0)$.

Definitions and Concepts	Examples

Section 5.2 Solving Systems of Linear Equations by the Substitution Method

To solve a linear system by the substitution method,

1. Solve one equation for one variable in terms of the other.
2. Substitute the expression for that variable into the other equation. This will result in an equation in one variable.
3. Solve the equation in one variable.
4. Back-substitute the value of the variable found in step 3 in the equation from step 1. Simply and find the value of the remaining variable.
5. Check the proposed solution in both of the system's given equations.

If both variables are eliminated and a false statement results, the system has no solution. If both variables are eliminated and a true statement results, the system has infinitely many solutions.

Solve by the substitution method:
$$y = 2x + 3$$
$$7x - 5y = -18.$$
Substitute $2x + 3$ for y in the second equation.
$$7x - 5(2x + 3) = -18$$
$$7x - 10x - 15 = -18$$
$$-3x - 15 = -18$$
$$-3x = -3$$
$$x = 1$$
Find y. Substitute 1 for x in $y = 2x + 3$.
$$y = 2 \cdot 1 + 3 = 2 + 3 = 5$$
The solution, $(1, 5)$, checks.

Section 5.3 Solving Systems of Linear Equations by the Addition Method

To solve a linear system by the addition method,

1. Write equations in $Ax + By = C$ form.
2. Multiply one or both equations by nonzero numbers so that coefficients of a variable are opposites.
3. Add equations.
4. Solve the resulting equation for a variable.
5. Back-substitute the value of the variable into either original equation and find the value of the remaining variable.
6. Check the proposed solution in both the original equations.

If both variables are eliminated and a false statement results, the system has no solution. If both variables are eliminated and a true statement results, the system has infinitely many solutions.

Solve by the addition method:
$$3x + y = -11$$
$$6x - 2y = -2.$$
Eliminate y. Multiply both sides of the first equation by 2.
$$6x + 2y = -22$$
$$\underline{6x - 2y = \quad -2}$$
Add: $12x \qquad = -24$
$$x = -2$$
Find y. Back-substitute -2 for x. We'll use the first equation.
$$3(-2) + y = -11$$
$$-6 + y = -11$$
$$y = -5$$
The solution, $(-2, -5)$, checks.

Section 5.4 Problem Solving Using Systems of Equations

A Problem-Solving Strategy

1. Use variables, usually x and y, to represent unknown quantities.
2. Write a system of equations describing the problem's conditions.
3. Solve the system and answer the problem's question.
4. Check proposed answers in the problem's wording.

A rectangular garden has a perimeter of 50 yards. Fencing along the length costs $3 per yard and along the width $2 per yard. The total cost of the fencing is $140. Find the length and width of the rectangle. Let x = length and y = width.

Perimeter	is	50 yards.

$$2x + 2y = 50$$

Cost to fence two lengths	plus	cost to fence two widths	is	$140.

$$3 \cdot 2x + 2 \cdot 2y = 140$$

Solve
$$2x + 2y = 50$$
$$6x + 4y = 140$$
by addition. Multiply the first equation by -2 and add. Solving gives $x = 20$ and $y = 5$, so the length is 20 yards and the width is 5 yards.

Definitions and Concepts	Examples

Section 5.5 Systems of Linear Inequalities

Two (or more) linear inequalities make up a system of linear inequalities. A solution is an ordered pair satisfying each of the inequalities in the system. To graph a system of inequalities, graph each inequality in the system in the same rectangular coordinate system. The overlapping region represents the solutions of the system.

Graph the solutions of the system:

$$y \leq -2x$$
$$x - y \geq 3.$$

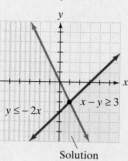

Solution

CHAPTER 5 REVIEW EXERCISES

5.1 *In Exercises 1–2, determine whether the given ordered pair is a solution of the system.*

1. $(1, -5)$
$4x - y = 9$
$2x + 3y = -13$

2. $(-5, 2)$
$2x + 3y = -4$
$x - 4y = -10$

3. Does the graphing-utility screen show the solution for the following system? Explain.

$$x + y = 2$$
$$2x + y = -5$$

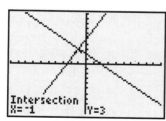

In Exercises 4–14, solve each system by graphing. If there is no solution or an infinite number of solutions, so state.

4. $x + y = 2$
$x - y = 6$

5. $2x - 3y = 12$
$-2x + y = -8$

6. $3x + 2y = 6$
$3x - 2y = 6$

7. $y = \dfrac{1}{2}x$
$y = 2x - 3$

8. $x + 2y = 2$
$y = x - 5$

9. $x + 2y = 8$
$3x + 6y = 12$

10. $2x - 4y = 8$
$x - 2y = 4$

11. $y = 3x - 1$
$y = 3x + 2$

12. $x - y = 4$
$x = -2$

13. $x = 2$
$y = 5$

14. $x = 2$
$x = 5$

5.2 *In Exercises 15–23, solve each system by the substitution method. If there is no solution or an infinite number of solutions, so state.*

15. $2x - 3y = 7$
$y = 3x - 7$

16. $2x - y = 6$
$x = 13 - 2y$

17. $2x - 5y = 1$
$3x + y = -7$

18. $3x + 4y = -13$
$5y - x = -21$

19. $y = 39 - 3x$
$y = 2x - 61$

20. $4x + y = 5$
$12x + 3y = 15$

21. $4x - 2y = 10$
$y = 2x + 3$

22. $x - 4 = 0$
$9x - 2y = 0$

23. $8y = 4x$
$7x + 2y = -8$

24. The weekly demand and supply models for the video *Titanic* at a chain of stores that sells videos are given by the demand model $N = -60p + 1000$ and the supply model $N = 4p + 200$, in which p is the price of the video and N is the number of videos sold or supplied each week to the chain of stores. Find the price at which supply and demand are equal. At this price, how many copies of *Titanic* can be supplied and sold each week?

5.3 *In Exercises 25–35, solve each system by the addition method. If there is no solution or an infinite number of solutions, so state.*

25. $x + y = 6$
$2x + y = 8$

26. $3x - 4y = 1$
$12x - y = -11$

27. $3x - 7y = 13$
$6x + 5y = 7$

28. $8x - 4y = 16$
$4x + 5y = 22$

29. $5x - 2y = 8$
$3x - 5y = 1$

30. $2x + 7y = 0$
$7x + 2y = 0$

31. $x + 3y = -4$
$3x + 2y = 3$

32. $2x + y = 5$
$2x + y = 7$

33. $3x - 4y = -1$
$-6x + 8y = 2$

34. $2x = 8y + 24$
$3x + 5y = 2$

35. $5x - 7y = 2$
$3x = 4y$

In Exercises 36–41, solve each system by the method of your choice. If there is no solution or an infinite number of solutions, so state.

36. $3x + 4y = -8$
$2x + 3y = -5$

37. $6x + 8y = 39$
$y = 2x - 2$

38. $x + 2y = 7$
$2x + y = 8$

39. $y = 2x - 3$
$y = -2x - 1$

40. $3x - 6y = 7$
$3x = 6y$

41. $y - 7 = 0$
$7x - 3y = 0$

5.4

42. The bar graph shows the five countries with the longest healthy life expectancy at birth. Combined, people in Japan and Switzerland can expect to spend 146.4 years in good health. The difference between healthy life expectancy between these two countries is 0.8 years. Find the healthy life expectancy at birth in Japan and Switzerland.

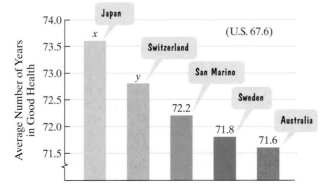

Countries with the Longest Healthy Life Expectancy

Source: World Health Organization

43. The gorilla and orangutan are the heaviest of the world's apes. Two gorillas and three orangutans weigh 1465 pounds. A gorilla's weight increased by twice an orangutan's weight is 815 pounds. Find the weight for each of these primates.

44. Health experts agree that cholesterol intake should be limited to 300 milligrams or less each day. Three ounces of shrimp and 2 ounces of scallops contain 156 milligrams of cholesterol. Five ounces of shrimp and 3 ounces of scallops contain 45 milligrams of cholesterol less than the suggested maximum daily intake. Determine the cholesterol content in an ounce of each item.

45. The perimeter of a table tennis top is 28 feet. The difference between 4 times the length and 3 times the width is 21 feet. Find the dimensions.

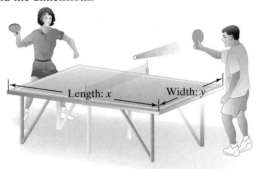

46. A rectangular garden has a perimeter of 24 yards. Fencing across the length costs $3 per yard and along the width $2 per yard. The total cost of the fencing is $62. Find the length and width of the rectangle.

47. A travel agent offers two package vacation plans. The first plan costs $360 and includes 3 days at a hotel and a rental car for 2 days. The second plan costs $500 and includes 4 days at a hotel and a rental car for 3 days. The daily charge for the room is the same under each plan, as is the daily charge for the car. Find the cost per day for the room and for the car.

48. You are choosing between two long-distance telephone plans. One plan has a monthly fee of $15 with a charge of $0.05 per minute for all long-distance calls. The other plan has a monthly fee of $10 with a charge of $0.075 per minute for all long-distance calls. For how many minutes of long-distance calls will the costs for the two plans be the same? What will be the cost for each plan?

5.5 *In Exercises 49–55, graph the solutions of each system of linear inequalities.*

49. $3x - y \leq 6$
$x + y \geq 2$

50. $x + y < 4$
$x - y < 4$

51. $y < 2x - 2$
$x \geq 3$

52. $4x + 6y \leq 24$
$y > 2$

53. $x \leq 3$
$y \geq -2$

54. $y \geq \frac{1}{2}x - 2$
$y \leq \frac{1}{2}x + 1$

55. $x \leq 0$
$y \geq 0$

CHAPTER 5 TEST

Remember to use your Chapter Test Prep Video CD to see the worked-out solutions to the test questions you want to review.

In Exercises 1–2, determine whether the given ordered pair is a solution of the system.

1. $(5, -5)$
$$2x + y = 5$$
$$x + 3y = -10$$

2. $(-3, 2)$
$$x + 5y = 7$$
$$3x - 4y = 1$$

In Exercises 3–4, solve each system by graphing. If there is no solution or an infinite number of solutions, so state.

3. $x + y = 6$
$\quad 4x - y = 4$

4. $2x + y = 8$
$\quad y = 3x - 2$

In Exercises 5–7, solve each system by the substitution method. If there is no solution or an infinite number of solutions, so state.

5. $x = y + 4$
$\quad 3x + 7y = -18$

6. $2x - y = 7$
$\quad 3x + 2y = 0$

7. $2x - 4y = 3$
$\quad x = 2y + 4$

In Exercises 8–10, solve each system by the addition method. If there is no solution or an infinite number of solutions, so state.

8. $2x + y = 2$
$\quad 4x - y = -8$

9. $2x + 3y = 1$
$\quad 3x + 2y = -6$

10. $3x - 2y = 2$
$\quad -9x + 6y = -6$

11. According to the U.S. Census Bureau, the two most popular female first names in the United States are Mary and Patricia. Combined, these two names account for 3.7% of all female first names. The difference between the percentage of women named Mary and the percentage of women named Patricia is 1.5%. What percentage of all first names in the United States are Mary and what percentage are Patricia?

12. A rectangular garden has a perimeter of 34 yards. Fencing across the length costs $2 per yard and along the width $1 per yard. The total cost of the fencing is $58. Find the length and width of the rectangle.

13. You are choosing between two long-distance telephone plans. One plan has a monthly fee of $15 with a charge of $0.05 per minute. The other plan has a monthly fee of $5 with a charge of $0.07 per minute. For how many minutes of long-distance calls will the costs for the two plans be the same? What will be the cost for each plan?

In Exercises 14–15, graph the solutions of each system of linear inequalities.

14. $x - 3y > 6$
$\quad 2x + 4y \leq 8$

15. $y \geq 2x - 4$
$\quad x < 2$

CUMULATIVE REVIEW EXERCISES (CHAPTERS 1–5)

1. Perform the indicated operations:
$$-14 - [18 - (6 - 10)].$$

2. Simplify: $6(3x - 2) - (x - 1)$.

In Exercises 3–4, solve each equation.

3. $17(x + 3) = 13 + 4(x - 10)$

4. $\dfrac{x}{4} - 1 = \dfrac{x}{5}$

5. Solve for t: $A = P + Prt$.

6. Solve and graph the solution set on a number line:
$2x - 5 < 5x - 11$.

In Exercises 7–9, graph each equation in the rectangular coordinate system.

7. $x - 3y = 6$

8. $y = 4 - x^2$

9. $y = -\dfrac{3}{5}x + 2$

In Exercises 10–11, solve each linear system.

10. $3x - 4y = 8$
$\quad 4x + 5y = -10$

11. $2x - 3y = 9$
$\quad y = 4x - 8$

12. Find the slope of the line passing through $(5, -6)$ and $(6, -5)$.

13. Write the point-slope form and the slope-intercept form of the equation of the line passing through $(-1, 6)$ with slope $= -4$.

14. The area of a triangle is 80 square feet. Find the height if the base is 16 feet.

15. If 10 pens and 15 pads cost $26, and 5 of the same pens and 10 of the same pads cost $16, find the cost of a pen and a pad.

16. List all the integers in this set:
$$\left\{-93, -\frac{7}{3}, 0, \sqrt{3}, \frac{7}{1}, \sqrt{100}\right\}.$$

The graphs show the percentage of U.S. households with one computer and multiple computers. Use the information provided by the graphs to solve Exercises 17–20.

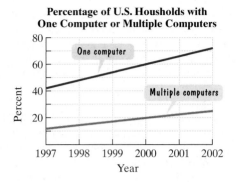

Percentage of U.S. Housholds with One Computer or Multiple Computers

One computer

Multiple computers

Source: Forrester Research. Inc.

17. What percentage of U.S. households had multiple computers in 2000?

18. Which graph has the greater slope? What does this mean in terms of the variables in this situation?

19. In 1997, 42% of U.S. households had one computer. This is increasing by approximately 6% per year. If this trend continues, in how many years after 1997 will 90% of U.S. households have one computer? In which year will that be?

20. The formula $y = \frac{8}{3}x + 12$ models the percentage of U.S. households, y, with multiple computers x years after 1997. Use the formula to find in which year 52% of U.S. households will have multiple computers.

One of the joys of your life is your dog, your very special buddy. Lately, however, you've noticed that your companion is slowing down a bit. He is now 8 years old and you wonder how this translates into human years. You remember something about every year of a dog's life being equal to seven years for a human. Is there a more accurate description?

This question is addressed in Exercises 103–106 in Exercise Set 6.1.

Exponents and Polynomials

There is a formula that models the age in human years, y, of a dog that is x years old:

$$y = -0.001618x^4 + 0.077326x^3$$
$$-1.2367x^2 + 11.460x + 2.914.$$

The algebraic expression on the right side of the formula contains variables to powers that are whole numbers and is an example of a polynomial. Much of what we do in algebra involves operations with polynomials. In this chapter, we study these operations, as well as the many applications of polynomials.

Objectives

1 Understand the vocabulary used to describe polynomials.

2 Add polynomials.

3 Subtract polynomials.

4 Use mathematical models that contain polynomials.

ADDING AND SUBTRACTING POLYNOMIALS

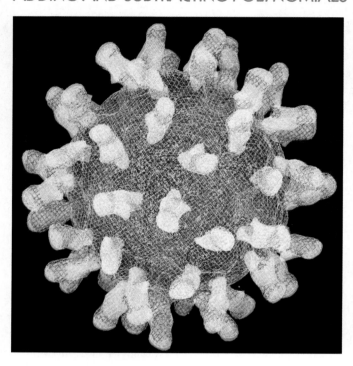

This computer-simulated model of the common cold virus was developed by researchers at Purdue University. Their discovery of how the virus infects human cells could lead to more effective treatment for the illness.

Runny nose? Sneezing? You are probably familiar with the unpleasant onset of a cold. We "catch cold" when the cold virus enters our bodies, where it multiplies. Fortunately, at a certain point the virus begins to die. The algebraic expression $-0.75x^4 + 3x^3 + 5$ describes the billions of viral particles in our bodies after x days of invasion. The expression enables mathematicians to determine the day on which there is a maximum number of viral particles and, consequently, the day we feel sickest.

The algebraic expression $-0.75x^4 + 3x^3 + 5$ is an example of a *polynomial*. A **polynomial** is a single term or the sum of two or more terms containing variables in the numerator with whole-number exponents. This particular polynomial contains three terms. Equations containing polynomials are used in such diverse areas as science, business, medicine, psychology, and sociology. In this section, we present basic ideas about polynomials. We then use our knowledge of combining like terms to find sums and differences of polynomials.

Describing Polynomials Consider the polynomial

$$7x^3 - 9x^2 + 13x - 6.$$

We can express this polynomial as

$$7x^3 + (-9x^2) + 13x + (-6).$$

The polynomial contains four terms. It is customary to write the terms in the order of descending powers of the variables. This is the **standard form** of a polynomial.

We begin this chapter by limiting our discussion to polynomials containing only one variable. Each term of such a polynomial in x is of the form ax^n. The **degree** of ax^n is n. For example, the degree of the term $7x^3$ is 3.

1 Understand the vocabulary used to describe polynomials.

STUDY TIP

We can express 0 in many ways, including $0x$, $0x^2$, and $0x^3$. It is impossible to assign a unique exponent to the variable. This is why 0 has no defined degree.

THE DEGREE OF ax^n If $a \neq 0$, the degree of ax^n is n. The degree of a nonzero constant term is 0. The constant 0 has no defined degree.

Here is an example of a polynomial and the degree of each of its four terms:

$$6x^4 - 3x^3 - 2x - 5.$$

| degree 4 | degree 3 | degree 1 | degree of nonzero constant: 0 |

Notice that the exponent on x for the term $2x$, meaning $2x^1$, is understood to be 1. For this reason, the degree of $2x$ is 1.

A polynomial which when simplified has exactly one term is called a **monomial**. A **binomial** is a polynomial that has two terms, each with a different exponent. A **trinomial** is a polynomial with three terms, each with a different exponent. Polynomials with four or more terms have no special names.

The degree of a polynomial is the highest degree of all the terms of the polynomial. For example, $4x^2 + 3x$ is a binomial of degree 2 because the degree of the first term is 2, and the degree of the other term is less than 2. Also, $7x^5 - 2x^2 + 4$ is a trinomial of degree 5 because the degree of the first term is 5, and the degrees of the other terms are less than 5.

Up to now, we have used x to represent the variable in a polynomial. However, any letter can be used. For example,

- $7x^5 - 3x^3 + 8$ is a polynomial (in x) of degree 5. Because there are three terms, the polynomial is a trinomial.

- $6y^3 + 4y^2 - y + 3$ is a polynomial (in y) of degree 3. Because there are four terms, the polynomial has no special name.

- $z^7 + \sqrt{2}$ is a polynomial (in z) of degree 7. Because there are two terms, the polynomial is a binomial.

2 Add polynomials.

Adding Polynomials Recall that *like terms* are terms containing exactly the same variables to the same powers. Polynomials are added by combining like terms. For example, we can add the monomials $-9x^3$ and $13x^3$ as follows:

$$-9x^3 + 13x^3 = (-9 + 13)x^3 = 4x^3.$$

| These like terms both contain x to the third power. | Add coefficients and keep the same variable factor, x^3. |

EXAMPLE 1 Adding Polynomials

Add: $(-9x^3 + 7x^2 - 5x + 3) + (13x^3 + 2x^2 - 8x - 6)$.

SOLUTION   The like terms are $-9x^3$ and $13x^3$, containing the same variable to the same power (x^3), as well as $7x^2$ and $2x^2$ (both contain x^2), $-5x$ and $-8x$ (both contain x), and the constant terms 3 and -6. We begin by grouping these pairs of like terms.

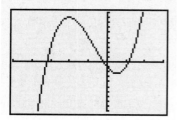
$$(-9x^3 + 7x^2 - 5x + 3) + (13x^3 + 2x^2 - 8x - 6)$$

$$= (-9x^3 + 13x^3) + (7x^2 + 2x^2) + (-5x - 8x) + (3 - 6) \quad \text{Group like terms.}$$

$$= 4x^3 + 9x^2 + (-13x) + (-3) \qquad\qquad \text{Combine like terms.}$$

$$= 4x^3 + 9x^2 - 13x - 3 \qquad\qquad\qquad ■$$

✔ **CHECK POINT 1** Add: $(-11x^3 + 7x^2 - 11x - 5) + (16x^3 - 3x^2 + 3x - 15)$.

Polynomials can be added by arranging like terms in columns. Then combine like terms, column by column.

EXAMPLE 2 Adding Polynomials Vertically

Add: $(-9x^3 + 7x^2 - 5x + 3) + (13x^3 + 2x^2 - 8x - 6)$.

SOLUTION

$$
\begin{array}{llll}
-9x^3 & 7x^2 & -5x & 3 \\
\underline{13x^3} & \underline{2x^2} & \underline{-8x} & \underline{-6} \\
4x^3 & 9x^2 & -13x & -3
\end{array}
$$

We consider each term separately and write like terms in columns.

Add, column by column.

Now add the four sums together:

$$4x^3 + 9x^2 + (-13x) + (-3) = 4x^3 + 9x^2 - 13x - 3.$$

This is the same answer that we found in Example 1. ■

✔ **CHECK POINT 2** Add the polynomials in Check Point 1 using a vertical format. Begin by arranging like terms in columns.

3 Subtract polynomials.

Subtracting Polynomials We subtract real numbers by adding the opposite, or additive inverse, of the number being subtracted. For example,

$$8 - 3 = 8 + (-3) = 5.$$

Subtraction of polynomials also involves opposites. If the sum of two polynomials is 0, the polynomials are **opposites**, or **additive inverses**, of each other. Here is an example:

$$(4x^2 - 6x - 7) + (-4x^2 + 6x + 7) = 0.$$

The opposite of $4x^2 - 6x - 7$ is $-4x^2 + 6x + 7$, and vice-versa.

Observe that the opposite of $4x^2 - 6x - 7$ can be obtained by changing the sign of each of its coefficients:

Polynomial

$4x^2 - 6x - 7$

Change 4 to −4, change −6 to 6, and change −7 to 7.

Opposite

$-4x^2 + 6x + 7$

In general, **the opposite of a polynomial is that polynomial with the sign of every coefficient changed**. Just as we did with real numbers, we subtract one polynomial from another by adding the opposite of the polynomial being subtracted.

> **SUBTRACTING POLYNOMIALS** To subtract two polynomials, add the first polynomial and the opposite of the polynomial being subtracted.

EXAMPLE 3 Subtracting Polynomials

Subtract: $(7x^2 + 3x - 4) - (4x^2 - 6x - 7)$.

SOLUTION

$$(7x^2 + 3x - 4) - (4x^2 - 6x - 7)$$

Change the sign of each coefficient.

$$= (7x^2 + 3x - 4) + (-4x^2 + 6x + 7)$$ Add the opposite of the polynomial being subtracted.

$$= (7x^2 - 4x^2) + (3x + 6x) + (-4 + 7)$$ Group like terms.

$$= 3x^2 + 9x + 3$$ Combine like terms. ■

CHECK POINT 3 Subtract: $(9x^2 + 7x - 2) - (2x^2 - 4x - 6)$.

STUDY TIP

Be careful of the order in Example 4. For example, subtracting 2 from 5 means $5 - 2$. In general, subtracting B from A means $A - B$. The order of the resulting algebraic expression is not the same as the order in English.

EXAMPLE 4 Subtracting Polynomials

Subtract $2x^3 - 6x^2 - 3x + 9$ from $7x^3 - 8x^2 + 9x - 6$.

SOLUTION

$$(7x^3 - 8x^2 + 9x - 6) - (2x^3 - 6x^2 - 3x + 9)$$

Change the sign of each coefficient.

$$= (7x^3 - 8x^2 + 9x - 6) + (-2x^3 + 6x^2 + 3x - 9)$$ Add the opposite of the polynomial being subtracted.

$$= (7x^3 - 2x^3) + (-8x^2 + 6x^2)$$

$$+ (9x + 3x) + (-6 - 9)$$ Group like terms.

$$= 5x^3 + (-2x^2) + 12x + (-15)$$ Combine like terms.

$$= 5x^3 - 2x^2 + 12x - 15$$ ■

CHECK POINT 4 Subtract $3x^3 - 8x^2 - 5x + 6$ from $10x^3 - 5x^2 + 7x - 2$.

Subtraction can also be performed in columns.

EXAMPLE 5 Subtracting Polynomials Vertically

Use the method of subtracting by columns to find

$$(12y^3 - 9y^2 - 11y - 3) - (4y^3 - 5y + 8).$$

SOLUTION Arrange like terms in columns.

$$
\begin{array}{r}
12y^3 - 9y^2 - 11y - 3 \\
-(4y^3 \qquad - 5y + 8)
\end{array}
$$

Add the opposite of the polynomial being subtracted.

Leave space for the missing term.

$$
\begin{array}{r}
12y^3 - 9y^2 - 11y - 3 \\
+ -4y^3 \qquad + 5y - 8 \\
\hline
8y^3 - 9y^2 - 6y - 11
\end{array}
$$

Change the sign of each coefficient of $4y^3 - 5y + 8$.

Combine like terms. ∎

✔ **CHECK POINT 5** Use the method of subtracting by columns to find

$$(8y^3 - 10y^2 - 14y - 2) - (5y^3 - 3y + 6).$$

4 Use mathematical models that contain polynomials.

Applications Polynomials often appear in formulas that model real-world situations.

EXAMPLE 6 An Application: Death Rate

The formula

$$y = 0.036x^2 - 2.8x + 58.14$$

models the number of deaths per year per thousand people, y, for people who are x years old, $40 \le x \le 60$. Approximately how many people per thousand who are 50 years old die each year?

SOLUTION Because we are interested in people who are 50 years old, substitute 50 for x in the formula's polynomial.

$y = 0.036x^2 - 2.8x + 58.14$ This is the given formula.

$y = 0.036(50)^2 - 2.8(50) + 58.14$ Substitute 50 for x.

$y = 0.036(2500) - 2.8(50) + 58.14$ Evaluate the exponential expression: $50^2 = 50 \cdot 50 = 2500$.

$y = 90 - 140 + 58.14$ Perform the multiplications.

$y = 8.14 \approx 8$ Simplify.

Approximately 8 people per thousand who are 50 years old die each year. ∎

We can use point plotting or a graphing utility to graph formulas that contain polynomials. These graphs contain only rounded curves with no sharp corners. For example, the graph of $y = 0.036x^2 - 2.8x + 58.14$, the formula from Example 6 that models the number of deaths per thousand, is shown in Figure 6.1. Our work in Example 6 can be visualized as a point on the curve.

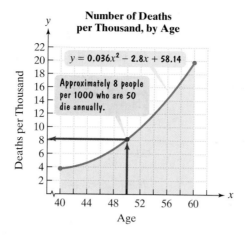

FIGURE 6.1 The graph of a formula containing a polynomial

CHECK POINT 6 Use the formula $y = 0.036x^2 - 2.8x + 58.14$ to answer this question: Approximately how many people per thousand who are 40 years old die annually? Identify your solution as a point on the curve in Figure 6.1.

6.1 EXERCISE SET

Student Solutions Manual CD/Video PH Math/Tutor Center MathXL Tutorials on CD   MathXL®   MyMathLab   Interactmath.com

Practice Exercises

In Exercises 1–16, identify each polynomial as a monomial, a binomial, or a trinomial. Give the degree of the polynomial.

1. $3x + 7$

2. $5x - 2$

3. $x^3 - 2x$

4. $x^5 - 7x$

5. $8x^2$

6. $10x^2$

7. 5

8. 9

9. $x^2 - 3x + 4$

10. $x^2 - 9x + 2$

11. $7y^2 - 9y^4 + 5$

12. $3y^2 - 14y^5 + 6$

13. $15x - 7x^3$

14. $9x - 5x^3$

15. $-9y^{23}$

16. $-11y^{26}$

In Exercises 17–38, add the polynomials.

17. $(9x + 8) + (-17x + 5)$

18. $(8x - 5) + (-13x + 9)$

19. $(4x^2 + 6x - 7) + (8x^2 + 9x - 2)$

20. $(11x^2 + 7x - 4) + (27x^2 + 10x - 20)$

21. $(7x^2 - 11x) + (3x^2 - x)$

22. $(-3x^2 + x) + (4x^2 + 8x)$

23. $(4x^2 - 6x + 12) + (x^2 + 3x + 1)$

24. $(-7x^2 + 8x + 3) + (2x^2 + x + 8)$

25. $(4y^3 + 7y - 5) + (10y^2 - 6y + 3)$

26. $(2y^3 + 3y + 10) + (3y^2 + 5y - 22)$

27. $(2x^2 - 6x + 7) + (3x^3 - 3x)$

28. $(4x^3 + 5x + 13) + (-4x^2 + 22)$

29. $(4y^2 + 8y + 11) + (-2y^3 + 5y + 2)$

30. $(7y^3 + 5y - 1) + (2y^2 - 6y + 3)$

31. $(-2y^6 + 3y^4 - y^2) + (-y^6 + 5y^4 + 2y^2)$

32. $(7r^4 + 5r^2 + 2r) + (-18r^4 - 5r^2 - r)$

33. $\left(9x^3 - x^2 - x - \dfrac{1}{3}\right) + \left(x^3 + x^2 + x + \dfrac{4}{3}\right)$

34. $\left(12x^3 - x^2 - x + \dfrac{4}{3}\right) + \left(x^3 + x^2 + x - \dfrac{1}{3}\right)$

35. $\left(\dfrac{1}{5}x^4 + \dfrac{1}{3}x^3 + \dfrac{3}{8}x^2 + 6\right) +$
$\left(-\dfrac{3}{5}x^4 + \dfrac{2}{3}x^3 - \dfrac{1}{2}x^2 - 6\right)$

36. $\left(\dfrac{2}{5}x^4 + \dfrac{2}{3}x^3 + \dfrac{5}{8}x^2 + 7\right) +$
$\left(-\dfrac{4}{5}x^4 + \dfrac{1}{3}x^3 - \dfrac{1}{4}x^2 - 7\right)$

37. $(0.03x^5 - 0.1x^3 + x + 0.03) +$
$(-0.02x^5 + x^4 - 0.7x + 0.3)$

38. $(0.06x^5 - 0.2x^3 + x + 0.05) +$
$(-0.04x^5 + 2x^4 - 0.8x + 0.5)$

In Exercises 39–54, use a vertical format to add the polynomials.

39. $\begin{aligned}5y^3 - 7y^2 \\ 6y^3 + 4y^2\end{aligned}$

40. $\begin{aligned}13x^4 - \;\; x^2 \\ 7x^4 + 2x^2\end{aligned}$

41.
$$3x^2 - 7x + 4$$
$$\underline{-5x^2 + 6x - 3}$$

42.
$$7x^2 - 5x - 6$$
$$\underline{-9x^2 + 4x + 6}$$

43.
$$\tfrac{1}{4}x^4 - \tfrac{2}{3}x^3 - 5$$
$$\underline{-\tfrac{1}{2}x^4 + \tfrac{1}{5}x^3 + 4.7}$$

44.
$$\tfrac{1}{3}x^9 - \tfrac{1}{5}x^5 - 2.7$$
$$\underline{-\tfrac{3}{4}x^9 + \tfrac{2}{3}x^5 + 1}$$

45.
$$y^3 + 5y^2 - 7y - 3$$
$$\underline{-2y^3 + 3y^2 + 4y - 11}$$

46.
$$y^3 + \ y^2 - 7y + \ 9$$
$$\underline{-y^3 - 6y^2 - 8y + 11}$$

47.
$$4x^3 - 6x^2 + 5x - 7$$
$$\underline{-9x^3 \qquad\ -\ 4x + 3}$$

48.
$$-4y^3 + 6y^2 - 8y + 11$$
$$\underline{\ 2y^3 \qquad\ +\ 9y - \ 3}$$

49.
$$7x^4 - 3x^3 + x^2$$
$$\underline{\qquad\ x^3 - x^2 + 4x - 2}$$

50.
$$7y^5 - 3y^3 + y^2$$
$$\underline{\qquad\ 2y^3 - y^2 - 4y - 3}$$

51.
$$7x^2 - \ 9x + 3$$
$$4x^2 + 11x - 2$$
$$\underline{-3x^2 + \ 5x - 6}$$

52.
$$7y^2 - 11y - 6$$
$$8y^2 + \ 3y + 4$$
$$\underline{-9y^2 - \ 5y + 2}$$

53.
$$1.2x^3 - \ 3x^2 + 9.1$$
$$7.8x^3 - 3.1x^2 + \ 8$$
$$\underline{\qquad\ 1.2x^2 - \ 6}$$

54.
$$7.9x^3 - 6.8x^2 + 3.3$$
$$6.1x^3 - 2.2x^2 + 7$$
$$\underline{\qquad\ 4.3x^2 - 5}$$

In Exercises 55–74, subtract the polynomials.

55. $(x - 8) - (3x + 2)$

56. $(x - 2) - (7x + 9)$

57. $(x^2 - 5x - 3) - (6x^2 + 4x + 9)$

58. $(3x^2 - 8x - 2) - (11x^2 + 5x + 4)$

59. $(x^2 - 5x) - (6x^2 - 4x)$

60. $(3x^2 - 2x) - (5x^2 - 6x)$

61. $(x^2 - 8x - 9) - (5x^2 - 4x - 3)$

62. $(x^2 - 5x + 3) - (x^2 - 6x - 8)$

63. $(y - 8) - (3y - 2)$

64. $(y - 2) - (7y - 9)$

65. $(6y^3 + 2y^2 - y - 11) - (y^2 - 8y + 9)$

66. $(5y^3 + y^2 - 3y - 8) - (y^2 - 8y + 11)$

67. $(7n^3 - n^7 - 8) - (6n^3 - n^7 - 10)$

68. $(2n^2 - n^7 - 6) - (2n^3 - n^7 - 8)$

69. $(y^6 - y^3) - (y^2 - y)$

70. $(y^5 - y^3) - (y^4 - y^2)$

71. $(7x^4 + 4x^2 + 5x) - (-19x^4 - 5x^2 - x)$

72. $(-3x^6 + 3x^4 - x^2) - (-x^6 + 2x^4 + 2x^2)$

73. $\left(\dfrac{3}{7}x^3 - \dfrac{1}{5}x - \dfrac{1}{3}\right) - \left(-\dfrac{2}{7}x^3 + \dfrac{1}{4}x - \dfrac{1}{3}\right)$

74. $\left(\dfrac{3}{8}x^2 - \dfrac{1}{3}x - \dfrac{1}{4}\right) - \left(-\dfrac{1}{8}x^2 + \dfrac{1}{2}x - \dfrac{1}{4}\right)$

In Exercises 75–88, use a vertical format to subtract the polynomials.

75.
$$7x + 1$$
$$\underline{-(3x - 5)}$$

76.
$$4x + 2$$
$$\underline{-(3x - 5)}$$

77.
$$7x^2 - 3$$
$$\underline{-(-3x^2 + 4)}$$

78.
$$9y^2 - 6$$
$$\underline{-(-5y^2 + 2)}$$

79.
$$7y^2 - 5y + 2$$
$$\underline{-(11y^2 + 2y - 3)}$$

80.
$$3x^5 - 5x^3 + 6$$
$$\underline{-(7x^5 + 4x^3 - 2)}$$

81.
$$7x^3 + 5x^2 - 3$$
$$\underline{-(-2x^3 - 6x^2 + 5)}$$

82.
$$3y^4 - 4y^2 + \ 7$$
$$\underline{-(-5y^4 - 6y^2 - 13)}$$

83.
$$5y^3 + 6y^2 - 3y + 10$$
$$\underline{-(6y^3 - 2y^2 - 4y - \ 4)}$$

84.
$$4y^3 + 5y^2 + 7y + 11$$
$$\underline{-(-5y^3 + 6y^2 - 9y - \ 3)}$$

85.
$$7x^4 - 3x^3 + 2x^2$$
$$\underline{-(\qquad\ -\ x^3 - x^2 + x - 2)}$$

86.
$$5y^6 - 3y^3 + 2y^2$$
$$\underline{-(\qquad\ -\ y^3 - y^2 - y - 1)}$$

87.
$$0.07x^3 - 0.01x^2 + 0.02x$$
$$\underline{-(0.02x^3 - 0.03x^2 - \qquad x)}$$

88.
$$0.04x^3 - 0.03x^2 + 0.05x$$
$$\underline{-(0.02x^3 - 0.06x^2 - \qquad x)}$$

Practice Plus

In Exercises 89–92, perform the indicated operations.

89. $[(4x^2 + 7x - 5) - (2x^2 - 10x + 3)] - (x^2 + 5x - 8)$

90. $[(10x^3 - 5x^2 + 4x + 3) - (-3x^3 - 4x^2 + x)] -$
$\quad (7x^3 - 5x + 4)$

91. $[(4y^2 - 3y + 8) - (5y^2 + 7y - 4)] -$
$\quad [(8y^2 + 5y - 7) + (-10y^2 + 4y + 3)]$

92. $[(7y^2 - 4y + 2) - (12y^2 + 3y - 5)] -$
$\quad [(5y^2 - 2y - 8) + (-7y^2 + 10y - 13)]$

93. Subtract $x^3 - 2x^2 + 2$ from the sum of $4x^3 + x^2$ and $-x^3 + 7x - 3$.

94. Subtract $-3x^3 - 7x + 5$ from the sum of $2x^2 + 4x - 7$ and $-5x^3 - 2x - 3$.

95. Subtract $-y^2 + 7y^3$ from the difference between $-5 + y^2 + 4y^3$ and $-8 - y + 7y^3$. Express the answer in standard form.

96. Subtract $-2y^2 + 8y^3$ from the difference between $-6 + y^2 + 5y^3$ and $-12 - y + 13y^3$. Express the answer in standard form.

Application Exercises

97. The common cold is caused by a rhinovirus. The polynomial

$$-0.75x^4 + 3x^3 + 5$$

describes the billions of viral particles in our bodies after x days of invasion. Find the number of viral particles, in billions, after 0 days (the time of the cold's onset when we are still feeling well), 1 day, 2 days, 3 days, and 4 days. After how many days is the number of viral particles at a maximum and, consequently, the day we feel the sickest? By when should we feel completely better?

98. The polynomial $-0.02A^2 + 2A + 22$ is used by coaches to get athletes fired up so that they can perform well. The polynomial represents the performance level related to various levels of enthusiasm, from $A = 1$ (almost no enthusiasm) to $A = 100$ (maximum level of enthusiasm). Evaluate the polynomial when $A = 20$, $A = 50$, and $A = 80$. Describe what happens to performance as we get more and more fired up.

The graph shows cigarette consumption per U.S. adult from 1900 through 2001. The data from 1940 through 2001 can be modeled by the formula

$$y = -2.3x^2 + 135.3x + 2191,$$

where x represents years after 1940 and y represents cigarette consumption per U.S. adult. Use the formula to solve Exercises 99–100.

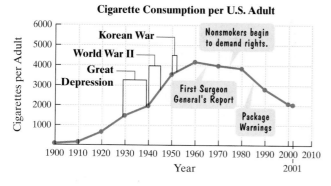

Cigarette Consumption per U.S. Adult

Source: U.S. Department of Health and Human Services

99. What was cigarette consumption per adult in 2000? How well does the formula model the actual data shown by the graph?

100. What was cigarette consumption per adult in 1980? How well does the formula model the actual data shown by the graph?

The wage gap is used to compare the status of women's earnings relative to men's. The wage gap is expressed as a percent and is calculated by dividing the median, or middlemost, annual earnings for women by the median annual earnings for men. The line graph shows the wage gap from 1960 through 2002. The data can be modeled by the formula

$$y = 0.012x^2 - 0.16x + 60,$$

where x represents years after 1960 and y represents median women's earnings as a percentage of median men's earnings. Use this information to solve Exercises 101–102.

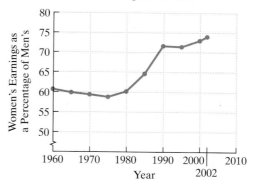

Median Women's Earnings as a Percentage of Median Men's Earnings in the U.S.

Source: U.S. Women's Bureau

101. a. Use the graph to estimate, to the nearest percent, women's earnings as a percentage of men's in 2000.

b. Use the mathematical model to find women's earnings as a percentage of men's in 2000.

c. In 2000, median annual earnings for U.S. women and men were $27,355 and $37,339, respectively. What were women's earnings as a percentage of men's? Use a calculator and round to the nearest tenth of a percent. How well do your answers in parts (a) and (b) model the actual data?

102. a. Use the graph to estimate, to the nearest percent, women's earnings as a percentage of men's in 1970.

b. Use the mathematical model to find women's earnings as a percentage of men's in 1970.

c. In 1970, median annual earnings for U.S. women and men were $5440 and $9184, respectively. What were women's earnings as a percentage of men's? Use a calculator and round to the nearest tenth of a percent. How well do your answers in parts (a) and (b) model the actual data?

The formula

$$y = -0.001618x^4 + 0.077326x^3 - 1.2367x^2 + 11.460x + 2.914$$

models the age in human years, y, of a dog that is x years old, where x > 1. The coefficients make it difficult to use this formula when computing by hand. However, a graph of the formula makes approximations possible. Use this information to solve Exercises 103–106.

Dog's Age in Human Years

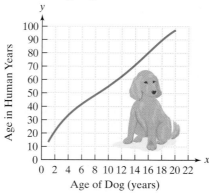

Source: U.C. Davis

103. a. If your dog is 6 years old, use the graph to estimate the equivalent age in human years.

 b. Use the given mathematical model to verify your estimate in part (a). Use a calculator and round to the nearest tenth of a human year.

104. a. If your dog is 16 years old, use the graph to estimate the equivalent age in human years.

 b. Use the given mathematical model to verify your estimate in part (a). Use a calculator and round to the nearest tenth of a human year.

105. If you are 25, use the graph to find the equivalent age for dogs.

106. If you are 45, use the graph to find the equivalent age for dogs.

Writing in Mathematics

107. What is a polynomial?

108. What is a monomial? Give an example with your explanation.

109. What is a binomial? Give an example with your explanation.

110. What is a trinomial? Give an example with your explanation.

111. What is the degree of a polynomial? Provide an example with your explanation.

112. Explain how to add polynomials.

113. Explain how to subtract polynomials.

114. A friend who is blind is having difficulty visualizing the relationship between age and deaths per thousand. Describe this relationship for your friend as age increases from 40 through 60. Use Figure 6.1 on page 327.

115. For Exercise 98, explain why performance levels do what they do as we get more and more fired up. If possible, describe an example of a time when you were too enthused and thus did poorly at something when you were hoping to do well.

Critical Thinking Exercises

116. Which one of the following is true?

 a. In the polynomial $3x^2 - 5x + 13$, the coefficient of x is 5.

 b. The degree of $3x^2 - 7x + 9x^3 + 5$ is 2.

 c. $\dfrac{1}{5x^2} + \dfrac{1}{3x}$ is a binomial.

 d. $(2x^2 - 8x + 6) - (x^2 - 3x + 5) = x^2 - 5x + 1$ for any value of x.

117. What polynomial must be subtracted from $5x^2 - 2x + 1$ so that the difference is $8x^2 - x + 3$?

118. The number of people who catch a cold t weeks after January 1 is $5t - 3t^2 + t^3$. The number of people who recover t weeks after January 1 is $t - t^2 + \frac{1}{3}t^3$. Write a polynomial in standard form for the number of people who are still ill with a cold t weeks after January 1.

119. Explain why it is not possible to add two polynomials of degree 3 and get a polynomial of degree 4.

Review Exercises

120. Simplify: $(-10)(-7) \div (1 - 8)$. (Section 1.8, Example 8)

121. Subtract: $-4.6 - (-10.2)$. (Section 1.6, Example 2)

122. Solve: $3(x - 2) = 9(x + 2)$. (Section 2.3, Example 3)

MULTIPLYING POLYNOMIALS

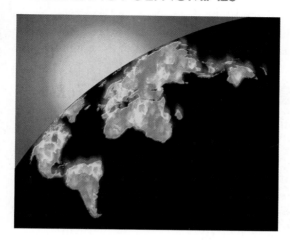

Recent advances in our understanding of climate have changed global warming from a subject for a disaster movie (the Statue of Liberty up to its chin in water) to a serious scientific and policy issue. Global warming appears to be related to the burning of fossil fuels, which adds carbon dioxide to the atmosphere. In the next few decades, we will see whether our use of fossil fuels will add enough carbon dioxide to the atmosphere to change it (and our climate) in significant ways. In this section's essay, you will see how a polynomial models trends in global warming through 2040. In the section itself, you will learn to multiply these algebraic expressions that play a significant role in modeling your world.

Before studying how polynomials are multiplied, we must develop some rules for working with exponents.

1 Use the product rule for exponents.

The Product Rule for Exponents We have seen that exponents are used to indicate repeated multiplication. For example, 2^4, where 2 is the base and 4 is the exponent, indicates that 2 occurs as a factor four times:

$$2^4 = 2 \cdot 2 \cdot 2 \cdot 2.$$

Now consider the multiplication of two exponential expressions, such as $2^4 \cdot 2^3$. We are multiplying 4 factors of 2 and 3 factors of 2. We have a total of 7 factors of 2:

4 factors of 2   3 factors of 2

$$2^4 \cdot 2^3 = (2 \cdot 2 \cdot 2 \cdot 2) \cdot (2 \cdot 2 \cdot 2)$$

Total: 7 factors of 2

Thus, $$2^4 \cdot 2^3 = 2^7.$$

Caution: $2^4 \cdot 2^3$ is not equal to $2^{4 \cdot 3}$, or 2^{12}, as might be expected.

We can quickly find the exponent, 7, of the product by adding 4 and 3, the original exponents:

$$2^4 \cdot 2^3 = 2^{4+3} = 2^7.$$

This suggests the following rule:

THE PRODUCT RULE

$$b^m \cdot b^n = b^{m+n}$$

When multiplying exponential expressions with the same base, add the exponents. Use this sum as the exponent of the common base.

EXAMPLE 1　Using the Product Rule

Multiply each expression using the product rule:

a. $2^2 \cdot 2^3$　　　**b.** $x^7 \cdot x^9$　　　**c.** $y \cdot y^5$　　　**d.** $y^3 \cdot y^2 \cdot y^5$.

STUDY TIP

The product rule does not apply to exponential expressions with different bases:

- $x^7 \cdot y^9$, or $x^7 y^9$, cannot be simplified.

SOLUTION

a. $2^2 \cdot 2^3 = 2^{2+3} = 2^5$ or 32
b. $x^7 \cdot x^9 = x^{7+9} = x^{16}$
c. $y \cdot y^5 = y^1 \cdot y^5 = y^{1+5} = y^6$
d. $y^3 \cdot y^2 \cdot y^5 = y^{3+2+5} = y^{10}$ ■

 CHECK POINT 1 Multiply each expression using the product rule:

a. $2^2 \cdot 2^4$　　　**b.** $x^6 \cdot x^4$　　　**c.** $y \cdot y^7$　　　**d.** $y^4 \cdot y^3 \cdot y^2$.

2　Use the power rule for exponents.

The Power Rule for Exponents　The next property of exponents applies when an exponential expression is raised to a power. Here is an example:

$$(3^2)^4.$$

> The exponential expression 3^2 is raised to the fourth power.

There are 4 factors 3^2. Thus,

$$(3^2)^4 = 3^2 \cdot 3^2 \cdot 3^2 \cdot 3^2 = 3^{2+2+2+2} = 3^8.$$

> Add exponents when multiplying with the same base.

We can obtain the answer, 3^8, by multiplying the exponents:

$$(3^2)^4 = 3^{2 \cdot 4} = 3^8.$$

This suggests the following rule:

THE POWER RULE (POWERS TO POWERS)

$$(b^m)^n = b^{mn}$$

When an exponential expression is raised to a power, multiply the exponents. Place the product of the exponents on the base and remove the parentheses.

EXAMPLE 2　Using the Power Rule

Simplify each expression using the power rule:

a. $(2^3)^5$　　　**b.** $(x^6)^4$　　　**c.** $[(-3)^7]^5$.

3 Use the products-to-powers rule.

SOLUTION

a. $(2^3)^5 = 2^{3 \cdot 5} = 2^{15}$

b. $(x^6)^4 = x^{6 \cdot 4} = x^{24}$

c. $[(-3)^7]^5 = (-3)^{7 \cdot 5} = (-3)^{35}$ ■

✔ **CHECK POINT 2** Simplify each expression using the power rule:

a. $(3^4)^5$ **b.** $(x^9)^{10}$ **c.** $[(-5)^7]^3$.

The Products-to-Powers Rule for Exponents The next property of exponents applies when we are raising a product to a power. Here is an example:

$$(2x)^4.$$

> The product $2x$ is raised to the fourth power.

There are four factors of $2x$. Thus,

$$(2x)^4 = 2x \cdot 2x \cdot 2x \cdot 2x = 2 \cdot 2 \cdot 2 \cdot 2 \cdot x \cdot x \cdot x \cdot x = 2^4 x^4.$$

We can obtain the answer, $2^4 x^4$, by raising each factor within the parentheses to the fourth power:

$$(2x)^4 = 2^4 x^4.$$

This suggests the following rule:

PRODUCTS TO POWERS

$$(ab)^n = a^n b^n$$

When a product is raised to a power, raise each factor to the power.

EXAMPLE 3 Using the Products-to-Powers Rule

Simplify each expression using the products-to-powers rule:

a. $(5x)^3$ **b.** $(-2y^4)^5$.

SOLUTION

a. $(5x)^3 = 5^3 x^3$ Raise each factor to the third power.

$\qquad = 125x^3$ $5^3 = 5 \cdot 5 \cdot 5 = 125$

b. $(-2y^4)^5 = (-2)^5 (y^4)^5$ Raise each factor to the fifth power.

$\qquad = (-2)^5 y^{4 \cdot 5}$ To raise an exponential expression to a power, multiply exponents: $(b^m)^n = b^{mn}$.

$\qquad = -32y^{20}$ $(-2)^5 = (-2)(-2)(-2)(-2)(-2) = -32$ ■

✔ **CHECK POINT 3** Simplify each expression using the products-to-powers rule:

a. $(2x)^4$ **b.** $(-4y^2)^3$.

STUDY TIP

Try to avoid the following common errors that can occur when simplifying exponential expressions.

Correct	Incorrect	Description of Error
$b^3 \cdot b^4 = b^{3+4} = b^7$	$b^3 \cdot b^4 = b^{12}$	Exponents should be added, not multiplied.
$3^2 \cdot 3^4 = 3^{2+4} = 3^6$	$3^2 \cdot 3^4 = 9^{2+4} = 9^6$	The common base should be retained, not multiplied.
$(x^5)^3 = x^{5 \cdot 3} = x^{15}$	$(x^5)^3 = x^{5+3} = x^8$	Exponents should be multiplied, not added, when raising a power to a power.
$(4x)^3 = 4^3 x^3 = 64x^3$	$(4x)^3 = 4x^3$	Both factors should be cubed.

4 Multiply monomials.

Multiplying Monomials Now that we have developed three properties of exponents, we are ready to turn to polynomial multiplication. We begin with the product of two monomials, such as $-8x^6$ and $5x^3$. This product is obtained by multiplying the coefficients, -8 and 5, and then multiplying the variables using the product rule for exponents.

$$(-8x^6)(5x^3) = -8 \cdot 5 \cdot x^6 \cdot x^3 = -8 \cdot 5x^{6+3} = -40x^9$$

Multiply coefficients and add exponents.

MULTIPLYING MONOMIALS To multiply monomials with the same variable base, multiply the coefficients and then multiply the variables. Use the product rule for exponents to multiply the variables: Keep the variable and add the exponents.

EXAMPLE 4 Multiplying Monomials

Multiply: **a.** $(2x)(4x^2)$ **b.** $(-10x^6)(6x^{10})$.

SOLUTION

a. $(2x)(4x^2) = (2 \cdot 4)(x \cdot x^2)$ Multiply the coefficients and multiply the variables.

$$= 8x^{1+2}$$ Add exponents: $b^m \cdot b^n = b^{m+n}$.

$$= 8x^3$$ Simplify.

b. $(-10x^6)(6x^{10}) = (-10 \cdot 6)(x^6 \cdot x^{10})$ Multiply the coefficients and multiply the variables.

$$= -60x^{6+10}$$ Add exponents: $b^m \cdot b^n = b^{m+n}$.

$$= -60x^{16}$$ Simplify.

✔ **CHECK POINT 4** Multiply: **a.** $(7x^2)(10x)$ **b.** $(-5x^4)(4x^5)$.

5   Multiply a monomial and a polynomial.

Multiplying a Monomial and a Polynomial That Is Not a Monomial   We use the distributive property to multiply a monomial and a polynomial that is not a monomial. For example,

$$3x^2(2x^3 + 5x) = 3x^2 \cdot 2x^3 + 3x^2 \cdot 5x = 3 \cdot 2x^{2+3} + 3 \cdot 5x^{2+1} = 6x^5 + 15x^3.$$

Monomial   Binomial   Multiply coefficients and add exponents.

MULTIPLYING A MONOMIAL AND A POLYNOMIAL THAT IS NOT A MONOMIAL   To multiply a monomial and a polynomial, use the distributive property to multiply each term of the polynomial by the monomial.

EXAMPLE 5   Multiplying a Monomial and a Polynomial

Multiply: **a.** $2x(x + 4)$ **b.** $3x^2(4x^3 - 5x + 2)$.

SOLUTION

a. $2x(x + 4) = 2x \cdot x + 2x \cdot 4$ Use the distributive property.

$ = 2 \cdot 1x^{1+1} + 2 \cdot 4x$ To multiply the monomials, multiply coefficients and add exponents.

$ = 2x^2 + 8x$ Simplify.

b. $3x^2(4x^3 - 5x + 2)$

$ = 3x^2 \cdot 4x^3 - 3x^2 \cdot 5x + 3x^2 \cdot 2$ Use the distributive property.

$ = 3 \cdot 4x^{2+3} - 3 \cdot 5x^{2+1} + 3 \cdot 2x^2$ To multiply the monomials, multiply coefficients and add exponents.

$ = 12x^5 - 15x^3 + 6x^2$ Simplify.

Rectangles often make it possible to visualize polynomial multiplication. For example, Figure 6.2 shows a rectangle with length $2x$ and width $x + 4$. The area of the large rectangle is

$$2x(x + 4).$$

The sum of the areas of the two smaller rectangles is

$$2x^2 + 8x.$$

Conclusion:

$$2x(x + 4) = 2x^2 + 8x.$$

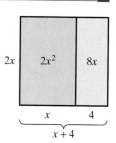

FIGURE 6.2

 CHECK POINT 5 Multiply:

   a. $3x(x + 5)$ **b.** $6x^2(5x^3 - 2x + 3)$.

6 Multiply polynomials when neither is a monomial.

Multiplying Polynomials When Neither Is a Monomial How do we multiply two polynomials if neither is a monomial? For example, consider

$$(2x + 3)(x^2 + 4x + 5).$$

Binomial   Trinomial

One way to perform this multiplication is to distribute $2x$ throughout the trinomial

$$2x(x^2 + 4x + 5)$$

and 3 throughout the trinomial

$$3(x^2 + 4x + 5).$$

Then combine the like terms that result. In general, the product of two polynomials is the polynomial obtained by multiplying each term of one polynomial by each term of the other polynomial and then combining like terms.

> **MULTIPLYING POLYNOMIALS WHEN NEITHER IS A MONOMIAL** Multiply each term of one polynomial by each term of the other polynomial. Then combine like terms.

EXAMPLE 6 Multiplying Binomials

Multiply: **a.** $(x + 3)(x + 2)$ **b.** $(3x + 7)(2x - 4)$.

SOLUTION We begin by multiplying each term of the second binomial by each term of the first binomial.

a. $(x + 3)(x + 2)$

$= x(x + 2) + 3(x + 2)$ Multiply the second binomial by each term of the first binomial.

$= x \cdot x + x \cdot 2 + 3 \cdot x + 3 \cdot 2$ Use the distributive property.

$= x^2 + 2x + 3x + 6$ Multiply. Note that $x \cdot x = x^1 \cdot x^1 = x^{1+1} = x^2$.

$= x^2 + 5x + 6$ Combine like terms.

b. $(3x + 7)(2x - 4)$

$= 3x(2x - 4) + 7(2x - 4)$ Multiply the second binomial by each term of the first binomial.

$= 3x \cdot 2x - 3x \cdot 4 + 7 \cdot 2x - 7 \cdot 4$ Use the distributive property.

$= 6x^2 - 12x + 14x - 28$ Multiply.

$= 6x^2 + 2x - 28$ Combine like terms. ■

CHECK POINT 6 Multiply:

a. $(x + 4)(x + 5)$ **b.** $(5x + 3)(2x - 7)$.

STUDY TIP

You can visualize the polynomial multiplication in Example 6(a) using the rectangle with dimensions $x + 3$ and $x + 2$.

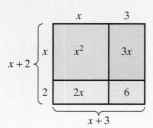

Area of large rectangle

$= (x + 3)(x + 2)$

Sum of areas of smaller rectangles

$= x^2 + 3x + 2x + 6$

$= x^2 + 5x + 6$

Conclusion:

$(x + 3)(x + 2) = x^2 + 5x + 6.$

USING TECHNOLOGY

A graphing utility can be used to see if a polynomial operation has been performed correctly. For example, to check

$$(x + 3)(x + 2) = x^2 + 5x + 6,$$

graph the left and right sides on the same screen, using

$$y_1 = (x + 3)(x + 2)$$

and

$$y_2 = x^2 + 5x + 6.$$

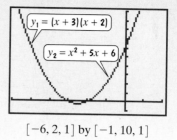

$y_1 = (x + 3)(x + 2)$

$y_2 = x^2 + 5x + 6$

$[-6, 2, 1]$ by $[-1, 10, 1]$

As shown in the figure, both graphs are the same, verifying that the binomial multiplication was performed correctly.

EXAMPLE 7   Multiplying a Binomial and a Trinomial

Multiply: $(2x + 3)(x^2 + 4x + 5)$.

SOLUTION

$(2x + 3)(x^2 + 4x + 5)$

$= 2x(x^2 + 4x + 5) + 3(x^2 + 4x + 5)$ Multiply the trinomial by each term of the binomial.

$= 2x \cdot x^2 + 2x \cdot 4x + 2x \cdot 5 + 3x^2 + 3 \cdot 4x + 3 \cdot 5$ Use the distributive property.

$= 2x^3 + 8x^2 + 10x + 3x^2 + 12x + 15$ Multiply monomials: Multiply coefficients and add exponents.

$= 2x^3 + 11x^2 + 22x + 15$ Combine like terms:
$8x^2 + 3x^2 = 11x^2$ and
$10x + 12x = 22x.$ ■

✔ **CHECK POINT 7** Multiply: $(5x + 2)(x^2 - 4x + 3)$.

Another method for solving Example 7 is to use a vertical format similar to that used for multiplying whole numbers.

$$
\begin{array}{r}
x^2 + 4x + 5 \\
2x + 3 \\
\hline
3x^2 + 12x + 15 \\
2x^3 + 8x^2 + 10x \\
\hline
2x^3 + 11x^2 + 22x + 15
\end{array}
$$

Write like terms in the same column.

$3(x^2 + 4x + 5)$

$2x(x^2 + 4x + 5)$

Combine like terms.

EXAMPLE 8   Multiplying Polynomials Using a Vertical Format

Multiply: $(2x^2 - 3x)(5x^3 - 4x^2 + 7x)$.

SOLUTION   To use the vertical format, it is most convenient to write the polynomial with the greatest number of terms in the top row.

$$
\begin{array}{r}
5x^3 - 4x^2 + 7x \\
2x^2 - 3x
\end{array}
$$

We now multiply each term in the top polynomial by the last term in the bottom polynomial.

$$5x^3 - 4x^2 + 7x$$
$$2x^2 - 3x$$
$$\overline{-15x^4 + 12x^3 - 21x^2} \quad \text{—} \quad -3x(5x^3 - 4x^2 + 7x)$$

Then we multiply each term in the top polynomial by $2x^2$, the first term in the bottom polynomial. Like terms are placed in columns because the final step involves combining them.

$$5x^3 - 4x^2 + 7x$$
$$2x^2 - 3x$$

Write like terms in the same column.

$$\overline{-15x^4 + 12x^3 - 21x^2} \quad -3x(5x^3 - 4x^2 + 7x)$$
$$10x^5 - 8x^4 + 14x^3 \quad\quad 2x^2(5x^3 - 4x^2 + 7x)$$
$$\overline{10x^5 - 23x^4 + 26x^3 - 21x^2}$$

Combine like terms, which are lined up in columns.

✔ **CHECK POINT 8** Multiply using a vertical format: $(3x^2 - 2x)(2x^3 - 5x^2 + 4x)$.

ENRICHMENT ESSAY

Is It Hot in Here Or Is It Just Me?

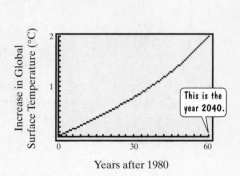

This is the year 2040.

Increase in Global Surface Temperature (°C)

Years after 1980

In the 1980s, a rising trend in global surface temperature was observed and the term "global warming" was coined. Scientists are more convinced than ever that burning coal, oil, and gas results in a buildup of gases and particles that trap heat and raise the planet's temperature. The average increase in global surface temperature, y, in degrees Celsius, x years after 1980 can be modeled by the polynomial formula

$$y = \frac{21}{5{,}000{,}000}x^3 - \frac{127}{1{,}000{,}000}x^2 + \frac{1293}{50{,}000}x.$$

The graph of this formula is shown above in a $[0, 60, 3]$ by $[0, 2, 0.1]$ viewing rectangle. The graph illustrates that the model predicts global warming will increase through the year 2040. Furthermore, the increasing steepness of the curve shows that global warming will increase at greater rates near the middle of the twenty-first century.

6.2 EXERCISE SET

Student Solutions Manual   CD/Video   PH Math/Tutor Center   MathXL Tutorials on CD   MathXL®   MyMathLab   Interactmath.com

Practice Exercises

In Exercises 1–8, multiply each expression using the product rule.

1. $x^{15} \cdot x^3$ **2.** $x^{12} \cdot x^4$

3. $y \cdot y^{11}$ **4.** $y \cdot y^{19}$

5. $x^2 \cdot x^6 \cdot x^3$ **6.** $x^4 \cdot x^3 \cdot x^5$

7. $7^9 \cdot 7^{10}$ **8.** $8^7 \cdot 8^{10}$

In Exercises 9–14, simplify each expression using the power rule.

9. $(6^9)^{10}$ **10.** $(6^7)^{10}$

11. $(x^{15})^3$ **12.** $(x^{12})^4$

13. $[(-20)^3]^3$ **14.** $[(-50)^4]^4$

In Exercises 15–24, simplify each expression using the products-to-powers rule.

15. $(2x)^3$ **16.** $(4x)^3$

17. $(-5x)^2$ **18.** $(-6x)^2$

19. $(4x^3)^2$ **20.** $(6x^3)^2$

21. $(-2y^6)^4$ **22.** $(-2y^5)^4$

23. $(-2x^7)^5$ **24.** $(-2x^{11})^5$

In Exercises 25–34, multiply the monomials.

25. $(7x)(2x)$ **26.** $(8x)(3x)$

27. $(6x)(4x^2)$ **28.** $(10x)(3x^2)$

29. $(-5y^4)(3y^3)$ **30.** $(-6y^4)(2y^3)$

31. $\left(-\dfrac{1}{2}a^3\right)\left(-\dfrac{1}{4}a^2\right)$ **32.** $\left(-\dfrac{1}{3}a^4\right)\left(-\dfrac{1}{2}a^2\right)$

33. $(2x^2)(-3x)(8x^4)$ **34.** $(3x^3)(-2x)(5x^6)$

In Exercises 35–54, find each product of the monomial and the polynomial.

35. $4x(x + 3)$ **36.** $6x(x + 5)$

37. $x(x - 3)$ **38.** $x(x - 7)$

39. $2x(x - 6)$ **40.** $3x(x - 5)$

41. $-4y(3y + 5)$

42. $-5y(6y + 7)$

43. $4x^2(x + 2)$ **44.** $5x^2(x + 6)$

45. $2y^2(y^2 + 3y)$ **46.** $4y^2(y^2 + 2y)$

47. $2y^2(3y^2 - 4y + 7)$

48. $4y^2(5y^2 - 6y + 3)$

49. $(3x^3 + 4x^2)(2x)$

50. $(4x^3 + 5x^2)(2x)$

51. $(x^2 + 5x - 3)(-2x)$

52. $(x^3 - 2x + 2)(-4x)$

53. $-3x^2(-4x^2 + x - 5)$

54. $-6x^2(3x^2 - 2x - 7)$

In Exercises 55–78, find each product. In each case, neither factor is a monomial.

55. $(x + 3)(x + 5)$

56. $(x + 4)(x + 6)$

57. $(2x + 1)(x + 4)$

58. $(2x + 5)(x + 3)$

59. $(x + 3)(x - 5)$

60. $(x + 4)(x - 6)$

61. $(x - 11)(x + 9)$

62. $(x - 12)(x + 8)$

63. $(2x - 5)(x + 4)$

64. $(3x - 4)(x + 5)$

65. $\left(\dfrac{1}{4}x + 4\right)\left(\dfrac{3}{4}x - 1\right)$

66. $\left(\dfrac{1}{5}x + 5\right)\left(\dfrac{3}{5}x - 1\right)$

67. $(x + 1)(x^2 + 2x + 3)$

68. $(x + 2)(x^2 + x + 5)$

69. $(y - 3)(y^2 - 3y + 4)$

70. $(y - 2)(y^2 - 4y + 3)$

71. $(2a - 3)(a^2 - 3a + 5)$

72. $(2a - 1)(a^2 - 4a + 3)$

73. $(x + 1)(x^3 + 2x^2 + 3x + 4)$

74. $(x + 1)(x^3 + 4x^2 + 7x + 3)$

75. $\left(x - \dfrac{1}{2}\right)(4x^3 - 2x^2 + 5x - 6)$

76. $\left(x - \dfrac{1}{3}\right)(3x^3 - 6x^2 + 5x - 9)$

77. $(x^2 + 2x + 1)(x^2 - x + 2)$

78. $(x^2 + 3x + 1)(x^2 - 2x - 1)$

In Exercises 79–92, use a vertical format to find each product.

79. $x^2 - 5x + 3$
$\underline{ x + 8}$

80. $x^2 - 7x + 9$
$\underline{ x + 4}$

81. $x^2 - 3x + 9$
$\underline{ 2x - 3}$

82. $y^2 - 5y + 3$
$\underline{\qquad 4y - 5}$

83. $2x^3 + x^2 + 2x + 3$
$\underline{\qquad\qquad x + 4}$

84. $3y^3 + 2y^2 + y + 4$
$\underline{\qquad\qquad y + 3}$

85. $4z^3 - 2z^2 + 5z - 4$
$\underline{\qquad\qquad 3z - 2}$

86. $5z^3 - 3z^2 + 4z - 3$
$\underline{\qquad\qquad 2z - 4}$

87. $7x^3 - 5x^2 + 6x$
$\underline{\qquad\qquad 3x^2 - 4x}$

88. $9y^3 - 7y^2 + 5y$
$\underline{\qquad\qquad 3y^2 + 5y}$

89. $2y^5 - 3y^3 + y^2 - 2y + 3$
$\underline{\qquad\qquad\qquad 2y - 1}$

90. $n^4 - n^3 + n^2 - n + 1$
$\underline{\qquad\qquad\qquad 2n + 3}$

91. $x^2 + 7x - 3$
$x^2 - x - 1$

92. $x^2 + 6x - 4$
$x^2 - x - 2$

Practice Plus

In Exercises 93–100, perform the indicated operations.

93. $(x + 4)(x - 5) - (x + 3)(x - 6)$

94. $(x + 5)(x - 6) - (x + 2)(x - 9)$

95. $4x^2(5x^3 + 3x - 2) - 5x^3(x^2 - 6)$

96. $3x^2(6x^3 + 2x - 3) - 4x^3(x^2 - 5)$

97. $(y + 1)(y^2 - y + 1) + (y - 1)(y^2 + y + 1)$

98. $(y + 1)(y^2 - y + 1) - (y - 1)(y^2 + y + 1)$

99. $(y + 6)^2 - (y - 2)^2$

100. $(y + 5)^2 - (y - 4)^2$

Application Exercises

101. Find a trinomial for the area of the rectangular rug shown below whose sides are $x + 5$ feet and $2x - 3$ feet.

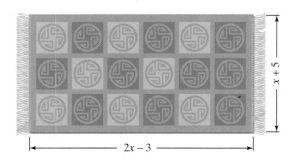

102. The base of a triangular sail is $4x$ feet and its height is $3x + 10$ feet. Write a binomial in terms of x for the area of the sail.

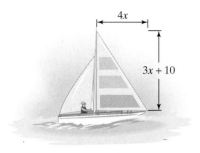

In Exercises 103–104,

a. *Express the area of the large rectangle as the product of two binomials.*

b. *Find the sum of the areas of the four smaller rectangles.*

c. *Use polynomial multiplication to show that your expressions for area in parts (a) and (b) are equal.*

103.

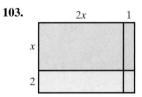

104.

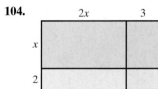

Writing in Mathematics

105. Explain the product rule for exponents. Use $2^3 \cdot 2^5$ in your explanation.

106. Explain the power rule for exponents. Use $(3^2)^4$ in your explanation.

107. Explain how to simplify an expression that involves a product raised to a power. Provide an example with your explanation.

108. Explain how to multiply monomials. Give an example.

109. Explain how to multiply a monomial and a polynomial that is not a monomial. Give an example.

110. Explain how to multiply polynomials when neither is a monomial. Give an example.

111. Explain the difference between performing these two operations:

$$2x^2 + 3x^2 \quad \text{and} \quad (2x^2)(3x^2).$$

112. Discuss situations in which a vertical format, rather than a horizontal format, is useful for multiplying polynomials.

113. Describe one change that might alter the prediction about global warming given by the model and graph on page 338.

Critical Thinking Exercises

114. Which one of the following is true?

a. $4x^3 \cdot 3x^4 = 12x^{12}$

b. $5x^2 \cdot 4x^6 = 9x^8$

c. $(y - 1)(y^2 + y + 1) = y^3 - 1$

d. Some polynomial multiplications can only be performed by using a vertical format.

115. Find a polynomial in descending powers of x representing the area of the shaded region.

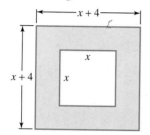

116. Find each of the products in parts (a)–(c).

a. $(x - 1)(x + 1)$

b. $(x - 1)(x^2 + x + 1)$

c. $(x - 1)(x^3 + x^2 + x + 1)$

d. Using the pattern found in parts (a)–(c), find $(x - 1)(x^4 + x^3 + x^2 + x + 1)$ without actually multiplying.

117. Find the missing factor.

$$(\underline{})\left(-\frac{1}{4}xy^3\right) = 2x^5y^3$$

Review Exercises

118. Solve: $4x - 7 > 9x - 2$. (Section 2.6, Example 6)

119. Graph $3x - 2y = 6$ using intercepts. (Section 4.2, Example 4)

120. Find the slope of the line passing through the points $(-2, 8)$ and $(1, 6)$. (Section 4.3, Example 1)

Objectives

1 Use FOIL in polynomial multiplication.

2 Multiply the sum and difference of two terms.

3 Find the square of a binomial sum.

4 Find the square of a binomial difference.

SPECIAL PRODUCTS

Let's cut to the chase. Are there fast methods for finding products of polynomials? Yes. In this section, we use the distributive property to develop patterns that will let you multiply certain binomials quite rapidly.

1 Use FOIL in polynomial multiplication.

The Product of Two Binomials: FOIL   Frequently, we need to find the product of two binomials. One way to perform this multiplication is to distribute each term in the first binomial through the second binomial. For example, we can find the product of the binomials $3x + 2$ and $4x + 5$ as follows:

$$(3x + 2)(4x + 5) = 3x(4x + 5) + 2(4x + 5)$$

Distribute $3x$ over $4x + 5$.   Distribute 2 over $4x + 5$.

$$= 3x(4x) + 3x(5) + 2(4x) + 2(5)$$

$$= 12x^2 + 15x + 8x + 10.$$

We can also find the product of $3x + 2$ and $4x + 5$ using a method called FOIL, which is based on our work shown above. Any two binomials can be quickly multiplied by using the FOIL method, in which **F** represents the product of the **first** terms in each binomial, **O** represents the product of the **outside** terms, **I** represents the product of the **inside** terms, and **L** represents the product of the **last**, or second, terms in each binomial. For example, we can use the FOIL method to find the product of the binomials $3x + 2$ and $4x + 5$ as follows:

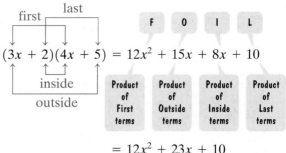

$$(3x + 2)(4x + 5) = 12x^2 + 15x + 8x + 10$$

$$= 12x^2 + 23x + 10 \qquad \text{\textit{Combine like terms.}}$$

In general, here's how to use the FOIL method to find the product of $ax + b$ and $cx + d$:

USING THE FOIL METHOD TO MULTIPLY BINOMIALS

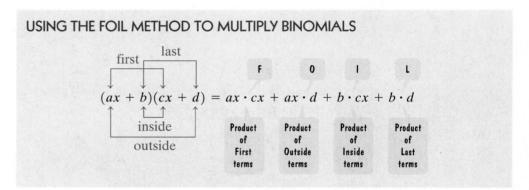

$$(ax + b)(cx + d) = ax \cdot cx + ax \cdot d + b \cdot cx + b \cdot d$$

EXAMPLE 1   Using the FOIL Method

Multiply: $(x + 3)(x + 4)$.

SOLUTION

F: First terms $= x \cdot x = x^2$ $(x + 3)(x + 4)$

O: Outside terms $= x \cdot 4 = 4x$ $(x + 3)(x + 4)$

I: Inside terms $= 3 \cdot x = 3x$ $(x + 3)(x + 4)$

L: Last terms $= 3 \cdot 4 = 12$ $(x + 3)(x + 4)$

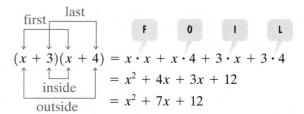

$$(x + 3)(x + 4) = x \cdot x + x \cdot 4 + 3 \cdot x + 3 \cdot 4$$

$$= x^2 + 4x + 3x + 12$$

$$= x^2 + 7x + 12 \qquad \text{\textit{Combine like terms.}}$$

CHECK POINT **1** Multiply: $(x + 5)(x + 6)$.

EXAMPLE 2 Using the FOIL Method

Multiply: $(3x + 4)(5x - 3)$.

SOLUTION

$$(3x + 4)(5x - 3) = 3x \cdot 5x + 3x(-3) + 4 \cdot 5x + 4(-3)$$
$$= 15x^2 - 9x + 20x - 12$$
$$= 15x^2 + 11x - 12 \qquad \text{Combine like terms.} \blacksquare$$

 CHECK POINT **2** Multiply: $(7x + 5)(4x - 3)$.

EXAMPLE 3 Using the FOIL Method

Multiply: $(2 - 5x)(3 - 4x)$.

SOLUTION

$$(2 - 5x)(3 - 4x) = 2 \cdot 3 + 2(-4x) + (-5x)(3) + (-5x)(-4x)$$
$$= 6 - 8x - 15x + 20x^2$$
$$= 6 - 23x + 20x^2 \qquad \text{Combine like terms.}$$

The product can also be expressed in standard form as $20x^2 - 23x + 6$. $\blacksquare$

CHECK POINT **3** Multiply: $(4 - 2x)(5 - 3x)$.

2 Multiply the sum and difference of two terms.

Multiplying the Sum and Difference of Two Terms We can use the FOIL method to multiply $A + B$ and $A - B$ as follows:

$$(A + B)(A - B) = A^2 - AB + AB - B^2 = A^2 - B^2.$$

Notice that the outside and inside products have a sum of 0 and the terms cancel. The FOIL multiplication provides us with a quick rule for multiplying the sum and difference of two terms, referred to as a special-product formula.

THE PRODUCT OF THE SUM AND DIFFERENCE OF TWO TERMS

$$(A + B)(A - B) = A^2 - B^2$$

| The product of the sum and the difference of the same two terms | is | the square of the first term minus the square of the second term. |

EXAMPLE 4 Finding the Product of the Sum and Difference of Two Terms

Find each product by using the preceding rule:

    a. $(4y + 3)(4y - 3)$ **b.** $(3x - 7)(3x + 7)$ **c.** $(5a^4 + 6)(5a^4 - 6)$.

SOLUTION Use the special-product formula shown.

$$(A + B)(A - B) = A^2 - B^2$$

First term squared $-$ Second term squared $=$ Product

    a. $(4y + 3)(4y - 3) = (4y)^2 - 3^2 = 16y^2 - 9$
    b. $(3x - 7)(3x + 7) = (3x)^2 - 7^2 = 9x^2 - 49$
    c. $(5a^4 + 6)(5a^4 - 6) = (5a^4)^2 - 6^2 = 25a^8 - 36$ ■

✔ **CHECK POINT 4** Find each product:

    a. $(7y + 8)(7y - 8)$ **b.** $(4x - 5)(4x + 5)$
    c. $(2a^3 + 3)(2a^3 - 3)$.

---

3   Find the square of a binomial sum.

The Square of a Binomial Let's now find $(A + B)^2$, the square of a binomial sum. To do so, we begin with the FOIL method and look for a general rule.

                    F   O   I   L

$$(A + B)^2 = (A + B)(A + B) = A \cdot A + A \cdot B + A \cdot B + B \cdot B$$
$$= A^2 + 2AB + B^2$$

This result implies the following rule, which is another example of a special-product formula:

STUDY TIP

Caution! The square of a sum is *not* the sum of the squares.

$$(A + B)^2 \neq A^2 + B^2$$

The middle term $2AB$ is missing.

$$(x + 3)^2 \neq x^2 + 9$$

Incorrect!

Show that $(x + 3)^2$ and $x^2 + 9$ are not equal by substituting 5 for x in each expression and simplifying.

THE SQUARE OF A BINOMIAL SUM

$$(A + B)^2 = A^2 + 2AB + B^2$$

The square of a binomial sum   is   first term squared   plus   2 times the product of the terms   plus   last term squared.

EXAMPLE 5 Finding the Square of a Binomial Sum

Square each binomial using the preceding rule:

    a. $(x + 3)^2$ **b.** $(3x + 7)^2$.

SOLUTION   Use the special-product formula shown.

$$(A + B)^2 = A^2 + 2AB + B^2$$

	(First Term)^2	+	2 · Product of the Terms	+	(Last Term)^2	= Product
a. $(x + 3)^2 =$	x^2	+	$2 \cdot x \cdot 3$	+	3^2	$= x^2 + 6x + 9$
b. $(3x + 7)^2 =$	$(3x)^2$	+	$2(3x)(7)$	+	7^2	$= 9x^2 + 42x + 49$

■

 CHECK POINT 5 Square each binomial:

   a. $(x + 10)^2$ **b.** $(5x + 4)^2$.

---

4 Find the square of a binomial difference.

Using the FOIL method on $(A - B)^2$, the square of a binomial difference, we obtain the following rule:

THE SQUARE OF A BINOMIAL DIFFERENCE

$$(A - B)^2 = A^2 - 2AB + B^2$$

| The square of a binomial difference | is | first term squared | minus | 2 times the product of the terms | plus | last term squared. |

EXAMPLE 6   Finding the Square of a Binomial Difference

Square each binomial using the preceding rule:

   a. $(x - 4)^2$ **b.** $(5y - 6)^2$.

SOLUTION   Use the special-product formula shown.

$$(A - B)^2 = A^2 - 2AB + B^2$$

	(First Term)^2	−	2 · Product of the Terms	+	(Last Term)^2	= Product
a. $(x - 4)^2 =$	x^2	−	$2 \cdot x \cdot 4$	+	4^2	$= x^2 - 8x + 16$
b. $(5y - 6)^2 =$	$(5y)^2$	−	$2(5y)(6)$	+	6^2	$= 25y^2 - 60y + 36$

■

 CHECK POINT 6 Square each binomial:

   a. $(x - 9)^2$ **b.** $(7x - 3)^2$.

Figure 6.3 makes it possible to visualize the square of a binomial sum. The area of the large square is

$$(A + B)(A + B) \quad \text{or} \quad (A + B)^2.$$

The sum of the areas of the four smaller rectangles that make up the large square is

$$A^2 + AB + AB + B^2$$

or

$$A^2 + 2AB + B^2.$$

Conclusion:

$$(A + B)^2 = A^2 + 2AB + B^2.$$

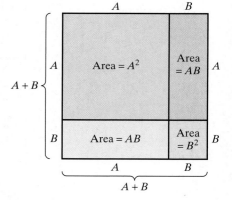

FIGURE 6.3

The following box summarizes the FOIL method and the three special products. The special products occur so frequently in algebra that it is convenient to memorize the form or pattern of these formulas.

FOIL and Special Products
Let A, B, C, and D be real numbers, variables, or algebraic expressions.

FOIL	*Example*
F O I L $(A + B)(C + D) = AC + AD + BC + BD$	F O I L $(2x + 3)(4x + 5) = (2x)(4x) + (2x)(5) + (3)(4x) + (3)(5)$ $= 8x^2 + 10x + 12x + 15$ $= 8x^2 + 22x + 15$
Sum and Difference of Two Terms $(A + B)(A - B) = A^2 - B^2$	*Example* $(2x + 3)(2x - 3) = (2x)^2 - 3^2$ $= 4x^2 - 9$
Square of a Binomial $(A + B)^2 = A^2 + 2AB + B^2$ $(A - B)^2 = A^2 - 2AB + B^2$	*Example* $(2x + 3)^2 = (2x)^2 + 2(2x)(3) + 3^2$ $= 4x^2 + 12x + 9$ $(2x - 3)^2 = (2x)^2 - 2(2x)(3) + 3^2$ $= 4x^2 - 12x + 9$

6.3 EXERCISE SET

Student Solutions Manual   CD/Video   PH Math/Tutor Center   MathXL Tutorials on CD   MathXL®   MyMathLab   Interactmath.com

Practice Exercises

In Exercises 1–24, use the FOIL method to find each product. Express the product in descending powers of the variable.

1. $(x + 4)(x + 6)$

2. $(x + 8)(x + 2)$

3. $(y - 7)(y + 3)$

4. $(y - 3)(y + 4)$

5. $(2x - 3)(x + 5)$

6. $(3x - 5)(x + 7)$

7. $(4y + 3)(y - 1)$

8. $(5y + 4)(y - 2)$

9. $(2x - 3)(5x + 3)$

10. $(2x - 5)(7x + 2)$

11. $(3y - 7)(4y - 5)$

12. $(4y - 5)(7y - 4)$

13. $(7 + 3x)(1 - 5x)$

14. $(2 + 5x)(1 - 4x)$

15. $(5 - 3y)(6 - 2y)$

16. $(7 - 2y)(10 - 3y)$

17. $(5x^2 - 4)(3x^2 - 7)$

18. $(7x^2 - 2)(3x^2 - 5)$

19. $(6x - 5)(2 - x)$

20. $(4x - 3)(2 - x)$

21. $(x + 5)(x^2 + 3)$

22. $(x + 4)(x^2 + 5)$

23. $(8x^3 + 3)(x^2 + 5)$

24. $(7x^3 + 5)(x^2 + 2)$

In Exercises 25–44, multiply using the rule for finding the product of the sum and difference of two terms.

25. $(x + 3)(x - 3)$

26. $(y + 5)(y - 5)$

27. $(3x + 2)(3x - 2)$

28. $(2x + 5)(2x - 5)$

29. $(3r - 4)(3r + 4)$

30. $(5z - 2)(5z + 2)$

31. $(3 + r)(3 - r)$

32. $(4 + s)(4 - s)$

33. $(5 - 7x)(5 + 7x)$

34. $(4 - 3y)(4 + 3y)$

35. $\left(2x + \dfrac{1}{2} \right)\left(2x - \dfrac{1}{2} \right)$

36. $\left(3y + \dfrac{1}{3} \right)\left(3y - \dfrac{1}{3} \right)$

37. $(y^2 + 1)(y^2 - 1)$

38. $(y^2 + 2)(y^2 - 2)$

39. $(r^3 + 2)(r^3 - 2)$

40. $(m^3 + 4)(m^3 - 4)$

41. $(1 - y^4)(1 + y^4)$

42. $(2 - s^5)(2 + s^5)$

43. $(x^{10} + 5)(x^{10} - 5)$

44. $(x^{12} + 3)(x^{12} - 3)$

In Exercises 45–62, multiply using the rule for the square of a binomial.

45. $(x + 2)^2$

46. $(x + 5)^2$

47. $(2x + 5)^2$

48. $(5x + 2)^2$

49. $(x - 3)^2$

50. $(x - 6)^2$

51. $(3y - 4)^2$

52. $(4y - 3)^2$

53. $(4x^2 - 1)^2$

54. $(5x^2 - 3)^2$

55. $(7 - 2x)^2$

56. $(9 - 5x)^2$

57. $\left(2x + \dfrac{1}{2} \right)^2$

58. $\left(3x + \dfrac{1}{3} \right)^2$

59. $\left(4y - \dfrac{1}{4} \right)^2$

60. $\left(2y - \dfrac{1}{2} \right)^2$

61. $(x^8 + 3)^2$

62. $(x^8 + 5)^2$

In Exercises 63–82, multiply using the method of your choice.

63. $(x - 1)(x^2 + x + 1)$

64. $(x + 1)(x^2 - x + 1)$

65. $(x - 1)^2$

66. $(x + 1)^2$

67. $(3y + 7)(3y - 7)$

68. $(4y + 9)(4y - 9)$

69. $3x^2(4x^2 + x + 9)$

70. $5x^2(7x^2 + x + 6)$

71. $(7y + 3)(10y - 4)$

72. $(8y + 3)(10y - 5)$

73. $(x^2 + 1)^2$

74. $(x^2 + 2)^2$

75. $(x^2 + 1)(x^2 + 2)$

76. $(x^2 + 2)(x^2 + 3)$

77. $(x^2 + 4)(x^2 - 4)$

78. $(x^2 + 5)(x^2 - 5)$

79. $(2 - 3x^5)^2$

80. $(2 - 3x^6)^2$

81. $\left(\frac{1}{4}x^2 + 12\right)\left(\frac{3}{4}x^2 - 8\right)$

82. $\left(\frac{1}{4}x^2 + 16\right)\left(\frac{3}{4}x^2 - 4\right)$

In Exercises 83–88, find the area of each shaded region. Write the answer as a polynomial in descending powers of x.

83.

$x + 1$

$x + 1$

84.

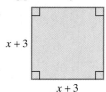

$x + 3$

$x + 3$

85.

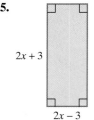

$2x + 3$

$2x - 3$

86.

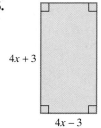

$4x + 3$

$4x - 3$

87.

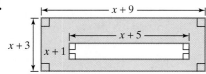

$x + 9$

$x + 5$

$x + 3$ $x + 1$

88.

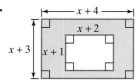

$x + 4$

$x + 2$

$x + 3$ $x + 1$

Practice Plus

In Exercises 89–96, multiply by the method of your choice.

89. $[(2x + 3)(2x - 3)]^2$

90. $[(3x + 2)(3x - 2)]^2$

91. $(4x^2 + 1)[(2x + 1)(2x - 1)]$

92. $(9x^2 + 1)[(3x + 1)(3x - 1)]$

93. $(x + 2)^3$

94. $(x + 4)^3$

95. $[(x + 3) - y][(x + 3) + y]$

96. $[(x + 5) - y][(x + 5) + y]$

Application Exercises

The square garden shown in the figure measures x yards on each side. The garden is to be expanded so that one side is increased by 2 yards and an adjacent side is increased by 1 yard. The graph shows the area of the expanded garden, y, in terms of the length of one of its original square sides, x. Use this information to solve Exercises 97–100.

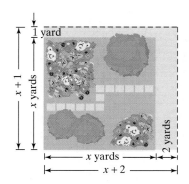

1 yard

$x + 1$ x yards

x yards

$x + 2$

2 yards

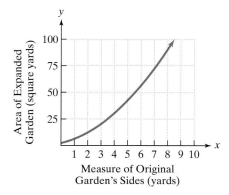

97. Write a product of two binomials that expresses the area of the larger garden.

98. Write a polynomial in descending powers of x that expresses the area of the larger garden.

99. If the original garden measures 6 yards on a side, use your expression from Exercise 97 to find the area of the larger garden. Then identify your solution as a point on the graph shown.

100. If the original garden measures 8 yards on a side, use your polynomial from Exercise 98 to find the area of the larger garden. Then identify your solution as a point on the graph shown.

The square painting in the figure measures x inches on each side. The painting is uniformly surrounded by a frame that measures 1 inch wide. Use this information to solve Exercises 101–102.

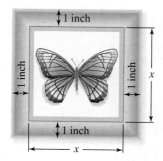

101. Write a polynomial in descending powers of x that expresses the area of the square that includes the painting and the frame.

102. Write an algebraic expression that describes the area of the frame. (*Hint:* The area of the frame is the area of the square that includes the painting and the frame minus the area of the painting.)

Writing in Mathematics

103. Explain how to multiply two binomials using the FOIL method. Give an example with your explanation.

104. Explain how to find the product of the sum and difference of two terms. Give an example with your explanation.

105. Explain how to square a binomial sum. Give an example with your explanation.

106. Explain how to square a binomial difference. Give an example with your explanation.

107. Explain why the graph for Exercises 97–100 is shown only in quadrant I.

Critical Thinking Exercises

108. Which one of the following is true?

  a. $(3 + 4)^2 = 3^2 + 4^2$

  b. $(2y + 7)^2 = 4y^2 + 28y + 49$

  c. $(3x^2 + 2)(3x^2 - 2) = 9x^2 - 4$

  d. $(x - 5)^2 = x^2 - 5x + 25$

109. What two binomials must be multiplied using the FOIL method to give a product of $x^2 - 8x - 20$?

110. Express the volume of the box as a polynomial in standard form.

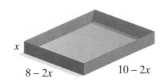

111. Express the area of the plane figure shown as a polynomial in standard form.

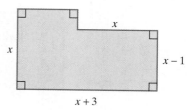

Technology Exercises

In Exercises 112–115, use a graphing utility to graph each side of the equation in the same viewing rectangle. (Call the left side y_1 and the right side y_2.) If the graphs coincide, verify that the multiplication has been performed correctly. If the graphs do not appear to coincide, this indicates that the multiplication is incorrect. In these exercises, correct the right side of the equation. Then graph the left side and the corrected right side to verify that the graphs coincide.

112. $(x + 1)^2 = x^2 + 1$; Use a $[-5, 5, 1]$ by $[0, 20, 1]$ viewing rectangle.

113. $(x + 2)^2 = x^2 + 2x + 4$; Use a $[-6, 5, 1]$ by $[0, 20, 1]$ viewing rectangle.

114. $(x + 1)(x - 1) = x^2 - 1$; Use a $[-6, 5, 1]$ by $[-2, 18, 1]$ viewing rectangle.

115. $(x - 2)(x + 2) + 4 = x^2$; Use a $[-6, 5, 1]$ by $[-2, 18, 1]$ viewing rectangle.

Review Exercises

In Exercises 116–117, solve each system by the method of your choice.

116. $2x + 3y = 1$

  $y = 3x - 7$

  (Section 5.2, Example 1)

117. $3x + 4y = 7$

  $2x + 7y = 9$

  (Section 5.3, Example 3)

118. Graph: $y \le \dfrac{1}{3}x$.

  (Section 4.6, Example 3)

SECTION 6.4

Objectives

1 Evaluate polynomials in several variables.

2 Understand the vocabulary of polynomials in two variables.

3 Add and subtract polynomials in several variables.

4 Multiply polynomials in several variables.

POLYNOMIALS IN SEVERAL VARIABLES

The next time you visit a lumberyard and go rummaging through piles of wood, think *polynomials*, although polynomials a bit different from those we have encountered so far. The construction industry uses a polynomial in two variables to determine the number of board feet that can be manufactured from a tree with a diameter of x inches and a length of y feet. This polynomial is

$$\frac{1}{4}x^2y - 2xy + 4y.$$

We call a polynomial containing two or more variables a **polynomial in several variables**. These polynomials can be evaluated, added, subtracted, and multiplied just like polynomials that contain only one variable.

1 Evaluate polynomials in several variables.

Evaluating a Polynomial in Several Variables Two steps can be used to evaluate a polynomial in several variables.

EVALUATING A POLYNOMIAL IN SEVERAL VARIABLES

1. Substitute the given value for each variable.
2. Perform the resulting computation using the order of operations.

EXAMPLE 1 Evaluating a Polynomial in Two Variables

Evaluate $2x^3y + xy^2 + 7x - 3$ for $x = -2$ and $y = 3$.

SOLUTION We begin by substituting -2 for x and 3 for y in the polynomial.

$2x^3y + xy^2 + 7x - 3$ — This is the given polynomial.

$= 2(-2)^3 \cdot 3 + (-2) \cdot 3^2 + 7(-2) - 3$ — Replace x with -2 and y with 3.

$= 2(-8) \cdot 3 + (-2) \cdot 9 + 7(-2) - 3$ — Evaluate exponential expressions: $(-2)^3 = (-2)(-2)(-2) = -8$ and $3^2 = 3 \cdot 3 = 9$.

$= -48 + (-18) + (-14) - 3$ — Perform the indicated multiplications.

$= -83$ — Add from left to right. ■

CHECK POINT 1 Evaluate $3x^3y + xy^2 + 5y + 6$ for $x = -1$ and $y = 5$.

2 Understand the vocabulary of polynomials in two variables.

Describing Polynomials in Two Variables In this section, we will limit our discussion of polynomials in several variables to two variables.

In general, a **polynomial in two variables**, x and y, contains the sum of one or more monomials in the form ax^ny^m. The constant, a, is the **coefficient**. The exponents, n and m, represent whole numbers. The **degree** of the monomial ax^ny^m is $n + m$. We'll use the polynomial from the construction industry to illustrate these ideas.

The coefficients are $\frac{1}{4}$, -2, and **4**.

$$\frac{1}{4}x^2y \quad - 2xy \quad + 4y$$

| Degree of monomial: $2 + 1 = 3$ | Degree of monomial: $1 + 1 = 2$ | Degree of monomial: 1 |

The **degree of a polynomial in two variables** is the highest degree of all its terms. For the preceding polynomial, the degree is 3.

EXAMPLE 2 Using the Vocabulary of Polynomials

Determine the coefficient of each term, the degree of each term, and the degree of the polynomial:

$$7x^2y^3 - 17x^4y^2 + xy - 6y^2 + 9.$$

SOLUTION

Think of xy as $1x^1y^1$.

Term	Coefficient	Degree (Sum of Exponents on the Variables)
$7x^2y^3$	7	$2 + 3 = 5$
$-17x^4y^2$	-17	$4 + 2 = 6$
xy	1	$1 + 1 = 2$
$-6y^2$	-6	2
9	9	0

The degree of the polynomial is the highest degree of all its terms, which is 6. ■

 CHECK POINT 2 Determine the coefficient of each term, the degree of each term, and the degree of the polynomial:

$$8x^4y^5 - 7x^3y^2 - x^2y - 5x + 11.$$

---

3 Add and subtract polynomials in several variables.

Adding and Subtracting Polynomials in Several Variables Polynomials in several variables are added by combining like terms. For example, we can add the monomials $-7xy^2$ and $13xy^2$ as follows:

$$-7xy^2 + 13xy^2 = (-7 + 13)xy^2 = 6xy^2.$$

These like terms both contain the variable factors x and y^2.

Add coefficients and keep the same variable factors, xy^2.

EXAMPLE 3 Adding Polynomials in Two Variables

Add: $(6xy^2 - 5xy + 7) + (9xy^2 + 2xy - 6)$.

SOLUTION

$(6xy^2 - 5xy + 7) + (9xy^2 + 2xy - 6)$

$= (6xy^2 + 9xy^2) + (-5xy + 2xy) + (7 - 6)$ Group like terms.

$= 15xy^2 - 3xy + 1$ Combine like terms by adding coefficients and keeping the same variable factors.

✔ **CHECK POINT 3** Add: $(-8x^2y - 3xy + 6) + (10x^2y + 5xy - 10)$.

We subtract polynomials in two variables just as we did when subtracting polynomials in one variable. Add the first polynomial and the opposite of the polynomial being subtracted.

EXAMPLE 4 Subtracting Polynomials in Two Variables

Subtract:

$$(5x^3 - 9x^2y + 3xy^2 - 4) - (3x^3 - 6x^2y - 2xy^2 + 3).$$

SOLUTION

$(5x^3 - 9x^2y + 3xy^2 - 4) - (3x^3 - 6x^2y - 2xy^2 + 3)$

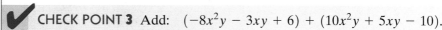
Change the sign of each coefficient.

$= (5x^3 - 9x^2y + 3xy^2 - 4) + (-3x^3 + 6x^2y + 2xy^2 - 3)$ Add the opposite of the polynomial being subtracted.

$= (5x^3 - 3x^3) + (-9x^2y + 6x^2y) + (3xy^2 + 2xy^2) + (-4 - 3)$ Group like terms.

$= 2x^3 - 3x^2y + 5xy^2 - 7$ Combine like terms by adding coefficients and keeping the same variable factors.

✔ **CHECK POINT 4** Subtract: $(7x^3 - 10x^2y + 2xy^2 - 5) - (4x^3 - 12x^2y - 3xy^2 + 5)$.

4 Multiply polynomials in several variables.

Multiplying Polynomials in Several Variables The product of monomials forms the basis of polynomial multiplication. As with monomials in one variable, multiplication can be done mentally by multiplying coefficients and adding exponents on variables with the same base.

EXAMPLE 5 Multiplying Monomials

Multiply: $(7x^2y)(5x^3y^2)$.

SOLUTION

$$(7x^2y)(5x^3y^2)$$
$$= (7 \cdot 5)(x^2 \cdot x^3)(y \cdot y^2)$$ This regrouping can be worked mentally.
$$= 35x^{2+3}y^{1+2}$$ Multiply coefficients and add exponents on variables with same base. ∎
$$= 35x^5y^3$$ Simplify.

 CHECK POINT **5** Multiply: $(6xy^3)(10x^4y^2)$.

How do we multiply a monomial and a polynomial that is not a monomial? As we did with polynomials in one variable, multiply each term of the polynomial by the monomial.

EXAMPLE 6 Multiplying a Monomial and a Polynomial

Multiply: $3x^2y(4x^3y^2 - 6x^2y + 2)$.

SOLUTION

$$3x^2y(4x^3y^2 - 6x^2y + 2)$$
$$= 3x^2y \cdot 4x^3y^2 - 3x^2y \cdot 6x^2y + 3x^2y \cdot 2$$ Use the distributive property.
$$= 12x^{2+3}y^{1+2} - 18x^{2+2}y^{1+1} + 6x^2y$$ Multiply coefficients and add exponents on variables with the same base.
$$= 12x^5y^3 - 18x^4y^2 + 6x^2y$$ Simplify. ∎

 CHECK POINT **6** Multiply: $6xy^2(10x^4y^5 - 2x^2y + 3)$.

FOIL and the special-products formulas can be used to multiply polynomials in several variables.

EXAMPLE 7 Multiplying Polynomials in Two Variables

Multiply: **a.** $(x + 4y)(3x - 5y)$ **b.** $(5x + 3y)^2$.

SOLUTION We will perform the multiplication in part (a) using the FOIL method. We will multiply in part (b) using the formula for the square of a binomial, $(A + B)^2$.

a. $(x + 4y)(3x - 5y)$ Multiply these binomials using the FOIL method.

F O I L

$$= (x)(3x) + (x)(-5y) + (4y)(3x) + (4y)(-5y)$$
$$= 3x^2 - 5xy + 12xy - 20y^2$$
$$= 3x^2 + 7xy - 20y^2$$ Combine like terms.

$$(A + B)^2 = A^2 + 2 \cdot A \cdot B + B^2$$

b. $(5x + 3y)^2 = (5x)^2 + 2(5x)(3y) + (3y)^2$
$$= 25x^2 + 30xy + 9y^2$$ ∎

 CHECK POINT **7** Multiply:

a. $(7x - 6y)(3x - y)$ **b.** $(2x + 4y)^2$.

EXAMPLE 8 Multiplying Polynomials in Two Variables

Multiply: **a.** $(4x^2y + 3y)(4x^2y - 3y)$ **b.** $(x + y)(x^2 - xy + y^2)$.

SOLUTION We perform the multiplication in part (a) using the formula for the product of the sum and difference of two terms. We perform the multiplication in part (b) by multiplying each term of the trinomial, $x^2 - xy + y^2$, by x and y, respectively, and then adding like terms.

$$(A + B) \cdot (A - B) \quad = \quad A^2 \quad - \quad B^2$$

a. $(4x^2y + 3y)(4x^2y - 3y) = (4x^2y)^2 - (3y)^2$
$$= 16x^4y^2 - 9y^2$$

b. $(x + y)(x^2 - xy + y^2)$
$$= x(x^2 - xy + y^2) + y(x^2 - xy + y^2)$$ Multiply the trinomial by each term of the binomial.

$$= x \cdot x^2 - x \cdot xy + x \cdot y^2 + y \cdot x^2 - y \cdot xy + y \cdot y^2$$ Use the distributive property.

$$= x^3 - x^2y + xy^2 + x^2y - xy^2 + y^3$$ Add exponents on variables with the same base.

$$= x^3 + y^3$$ Combine like terms: $-x^2y + x^2y = 0$ and $xy^2 - xy^2 = 0$. ■

✔ **CHECK POINT 8** Multiply:

a. $(6xy^2 + 5x)(6xy^2 - 5x)$
b. $(x - y)(x^2 + xy + y^2)$.

6.4 EXERCISE SET

 Student Solutions Manual   CD/Video   PH Math/Tutor Center   MathXL Tutorials on CD   Math XL MathXL®   MyMathLab MyMathLab   Interactmath.com

Practice Exercises

In Exercises 1–6, evaluate each polynomial for $x = 2$ and $y = -3$.

1. $x^2 + 2xy + y^2$ **2.** $x^2 + 3xy + y^2$

3. $xy^3 - xy + 1$ **4.** $x^3y - xy + 2$

5. $2x^2y - 5y + 3$ **6.** $3x^2y - 4y + 5$

In Exercises 7–8, determine the coefficient of each term, the degree of each term, and the degree of the polynomial.

7. $x^3y^2 - 5x^2y^7 + 6y^2 - 3$ **8.** $12x^4y - 5x^3y^7 - x^2 + 4$

In Exercises 9–20, add or subtract as indicated.

9. $(5x^2y - 3xy) + (2x^2y - xy)$

10. $(-2x^2y + xy) + (4x^2y + 7xy)$

11. $(4x^2y + 8xy + 11) + (-2x^2y + 5xy + 2)$

12. $(7x^2y + 5xy + 13) + (-3x^2y + 6xy + 4)$

13. $(7x^4y^2 - 5x^2y^2 + 3xy) + (-18x^4y^2 - 6x^2y^2 - xy)$

14. $(6x^4y^2 - 10x^2y^2 + 7xy) + (-12x^4y^2 - 3x^2y^2 - xy)$

15. $(x^3 + 7xy - 5y^2) - (6x^3 - xy + 4y^2)$

16. $(x^4 - 7xy - 5y^3) - (6x^4 - 3xy + 4y^3)$

17. $(3x^4y^2 + 5x^3y - 3y) - (2x^4y^2 - 3x^3y - 4y + 6x)$

18. $(5x^4y^2 + 6x^3y - 7y) - (3x^4y^2 - 5x^3y - 6y + 8x)$

19. $(x^3 - y^3) - (-4x^3 - x^2y + xy^2 + 3y^3)$

20. $(x^3 - y^3) - (-6x^3 + x^2y - xy^2 + 2y^3)$

21. Add: $\quad 5x^2y^2 - 4xy^2 + 6y^2$
$\qquad \underline{-8x^2y^2 + 5xy^2 - \quad y^2}$

22. Add: $\quad 7a^2b^2 - 5ab^2 + 6b^2$
$\qquad \underline{-10a^2b^2 + 6ab^2 + 6b^2}$

23. Subtract: $\quad 3a^2b^4 - 5ab^2 + 7ab$
$\qquad \underline{-(-5a^2b^4 - 8ab^2 - \quad ab)}$

24. Subtract: $\quad 13x^2y^4 - 17xy^2 + xy$
$\qquad \underline{-(-7x^2y^4 - \quad 8xy^2 - xy)}$

25. Subtract $11x - 5y$ from the sum of $7x + 13y$ and $-26x + 19y$.

26. Subtract $23x - 5y$ from the sum of $6x + 15y$ and $x - 19y$.

In Exercises 27–76, find each product.

27. $(5x^2y)(8xy)$

28. $(10x^2y)(5xy)$

29. $(-8x^3y^4)(3x^2y^5)$

30. $(7x^4y^5)(-10x^7y^{11})$

31. $9xy(5x + 2y)$

32. $7xy(8x + 3y)$

33. $5xy^2(10x^2 - 3y)$

34. $6x^2y(5x^2 - 9y)$

35. $4ab^2(7a^2b^3 + 2ab)$

36. $2ab^2(20a^2b^3 + 11ab)$

37. $-b(a^2 - ab + b^2)$

38. $-b(a^3 - ab + b^3)$

39. $(x + 5y)(7x + 3y)$

40. $(x + 9y)(6x + 7y)$

41. $(x - 3y)(2x + 7y)$

42. $(3x - y)(2x + 5y)$

43. $(3xy - 1)(5xy + 2)$

44. $(7xy + 1)(2xy - 3)$

45. $(2x + 3y)^2$

46. $(2x + 5y)^2$

47. $(xy - 3)^2$

48. $(xy - 5)^2$

49. $(x^2 + y^2)^2$

50. $(2x^2 + y^2)^2$

51. $(x^2 - 2y^2)^2$

52. $(x^2 - y^2)^2$

53. $(3x + y)(3x - y)$

54. $(x + 5y)(x - 5y)$

55. $(ab + 1)(ab - 1)$

56. $(ab + 2)(ab - 2)$

57. $(x + y^2)(x - y^2)$

58. $(x^2 + y)(x^2 - y)$

59. $(3a^2b + a)(3a^2b - a)$

60. $(5a^2b + a)(5a^2b - a)$

61. $(3xy^2 - 4y)(3xy^2 + 4y)$

62. $(7xy^2 - 10y)(7xy^2 + 10y)$

63. $(a + b)(a^2 - b^2)$

64. $(a - b)(a^2 + b^2)$

65. $(x + y)(x^2 + 3xy + y^2)$

66. $(x + y)(x^2 + 5xy + y^2)$

67. $(x - y)(x^2 - 3xy + y^2)$

68. $(x - y)(x^2 - 4xy + y^2)$

69. $(xy + ab)(xy - ab)$

70. $(xy + ab^2)(xy - ab^2)$

71. $(x^2 + 1)(x^4y + x^2 + 1)$

72. $(x^2 + 1)(xy^4 + y^2 + 1)$

73. $(x^2y^2 - 3)^2$

74. $(x^2y^2 - 5)^2$

75. $(x + y + 1)(x + y - 1)$

76. $(x + y + 1)(x - y + 1)$

In Exercises 77–80 write a polynomial in two variables that describes the total area of each shaded region. Express each polynomial as the sum or difference of terms.

77.
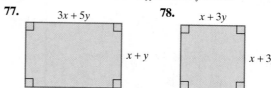
$3x + 5y$
$x + y$

78.
$x + 3y$
$x + 3y$

79.
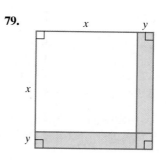
x
y
x
y

80.

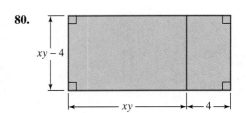

Practice Plus

In Exercises 81–86, find each product. As we said in the Section 6.3 opener, cut to the chase in each part of the polynomial multiplication: Use only the special-product formula for the sum and difference of two terms or the formulas for the square of a binomial.

81. $[(x^3y^3 + 1)(x^3y^3 - 1)]^2$

82. $[(1 - a^3b^3)(1 + a^3b^3)]^2$

83. $(xy - 3)^2(xy + 3)^2$ (Do not begin by squaring a binomial.)

84. $(ab - 4)^2(ab + 4)^2$ (Do not begin by squaring a binomial.)

85. $[x + y + z][x - (y + z)]$

86. $(a - b - c)(a + b + c)$

Application Exercises

87. The number of board feet, N, that can be manufactured from a tree with a diameter of x inches and a length of y feet is modeled by the formula

$$N = \frac{1}{4}x^2y - 2xy + 4y.$$

A building contractor estimates that 3000 board feet of lumber is needed for a job. The lumber company has just milled a fresh load of timber from 20 trees that averaged 10 inches in diameter and 16 feet in length. Is this enough to complete the job? If not, how many additional board feet of lumber is needed?

88. The storage shed shown in the figure has a volume given by the polynomial

$$2x^2y + \frac{1}{2}\pi x^2y.$$

a. A small business is considering having a shed installed like the one shown in the figure. The shed's height, $2x$, is 26 feet and its length, y, is 27 feet. Using $x = 13$ and $y = 27$, find the volume of the storage shed.

b. The business requires at least 18,000 cubic feet of storage space. Should they construct the storage shed described in part (a)?

An object that is falling or vertically projected into the air has its height, in feet, above the ground given by

$$s = -16t^2 + v_0t + s_0,$$

where s is the height, in feet, v_0 is the original velocity of the object, in feet per second, t is the time the object is in motion, in seconds, and s_0 is the height, in feet, from which the object is dropped or projected. The figure shows that a ball is thrown straight up from a rooftop at an original velocity of 80 feet per second from a height of 96 feet. The ball misses the rooftop on its way down and eventually strikes the ground. Use the formula and this information to solve Exercises 89–91.

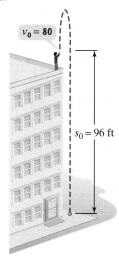

89. How high above the ground will the ball be 2 seconds after being thrown?

90. How high above the ground will the ball be 4 seconds after being thrown?

91. How high above the ground will the ball be 6 seconds after being thrown? Describe what this means in practical terms.

The graph visually displays the information about the thrown ball described in Exercises 89–91. The horizontal axis represents the ball's time in motion, in seconds. The vertical axis represents the ball's height above the ground, in feet. Use the graph to solve Exercises 92–97.

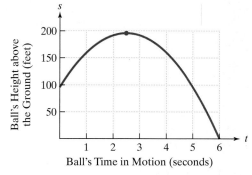

92. During which time period is the ball rising?

93. During which time period is the ball falling?

94. Identify your answer from Exercise 90 as a point on the graph.

95. Identify your answer from Exercise 89 as a point on the graph.

96. After how many seconds does the ball strike the ground?

97. After how many seconds does the ball reach its maximum height above the ground? What is a reasonable estimate of this maximum height?

Writing in Mathematics

98. What is a polynomial in two variables? Provide an example with your description.

99. Explain how to find the degree of a polynomial in two variables.

100. Suppose that you take up sky diving. Explain how to use the formula for Exercises 89–91 to determine your height above the ground at every instant of your fall.

Critical Thinking Exercises

101. Which one of the following is true?

    a. The degree of $5x^{24} - 3x^{16}y^9 - 7xy^2 + 6$ is 24.

    b. In the polynomial $4x^2y + x^3y^2 + 3x^2y^3 + 7y$, the term x^3y^2 has degree 5 and no numerical coefficient.

    c. $(2x + 3 - 5y)(2x + 3 + 5y) = 4x^2 + 12x + 9 - 25y^2$

    d. $(6x^2y - 7xy - 4) - (6x^2y + 7xy - 4) = 0$

In Exercises 102–103, find a polynomial in two variables that describes the area of the shaded region of each figure. Write the polynomial as the sum or difference of terms.

102.

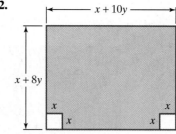

103.

104. Use the formulas for the volume of a rectangular solid and a cylinder to derive the polynomial in Exercise 88 that describes the volume of the storage building.

Review Exercises

105. Solve for W: $R = \dfrac{L + 3W}{2}$. (Section 2.4, Example 4)

106. Subtract: $-6.4 - (-10.2)$. (Section 1.6, Example 2)

107. Solve the proportion: $\dfrac{63}{x} = \dfrac{3}{5}$. (Section 3.2, Example 2)

MID-CHAPTER CHECK POINT

CHAPTER 6

What You Know: We learned to add, subtract, and multiply polynomials. We used a number of fast methods for finding products of polynomials, including the FOIL method for multiplying binomials, a special-product formula for the product of the sum and difference of two terms $[(A + B)(A - B) = A^2 - B^2]$, and special-product formulas for squaring binomials $[(A + B)^2 = A^2 + 2AB + B^2; (A - B)^2 = A^2 - 2AB + B^2]$. Finally, we applied all of these operations to polynomials in several variables.

In Exercises 1–21, perform the indicated operations.

1. $(11x^2y^3)(-5x^2y^3)$

2. $11x^2y^3 - 5x^2y^3$

3. $(3x + 5)(4x - 7)$

4. $(3x + 5) - (4x - 7)$

5. $(2x - 5)(x^2 - 3x + 1)$

6. $(2x - 5) + (x^2 - 3x + 1)$

7. $(8x - 3)^2$

8. $(-10x^4)(-7x^5)$

9. $(x^2 + 2)(x^2 - 2)$

10. $(x^2 + 2)^2$

11. $(9a - 10b)(2a + b)$

12. $7x^2(10x^3 - 2x + 3)$

13. $(3a^2b^3 - ab + 4b^2) - (-2a^2b^3 - 3ab + 5b^2)$

14. $2(3y - 5)(3y + 5)$

15. $(-9x^3 + 5x^2 - 2x + 7) + (11x^3 - 6x^2 + 3x - 7)$

16. $10x^2 - 8xy - 3(y^2 - xy)$

17. $(-2x^5 + x^4 - 3x + 10) - (2x^5 - 6x^4 + 7x - 13)$

18. $(x + 3y)(x^2 - 3xy + 9y^2)$

19. $(5x^4 + 4)(2x^3 - 1)$

20. $(y - 6z)^2$

21. $(2x + 3)(2x - 3) - (5x + 4)(5x - 4)$

SECTION

6.5

Objectives

1 Use the quotient rule for exponents.

2 Use the zero-exponent rule.

3 Use the quotients-to-powers rule.

4 Divide monomials.

5 Check polynomial division.

6 Divide a polynomial by a monomial.

DIVIDING POLYNOMIALS

To play the part of Charlie Chaplin, actor Robert Downey Jr. (1965–) learned to pantomime, speak two British dialects, and play left-handed tennis. His problems with substance abuse have fueled the debate over whether the illness should be handled by our criminal justice system or by health care professionals.

As you learn more mathematics, you will discover new ways to describe your world. Almost anything that you can think of involving variables can be modeled by a formula. For example, a polynomial models the annual number of drug convictions in the United States and another polynomial models drug arrests. By dividing the respective polynomials, we obtain an algebraic expression that describes the conviction rate for drug arrests.

In the next two sections, you will learn how to divide polynomials. Before turning to polynomial division, we must develop some additional rules for working with exponents.

1 Use the quotient rule for exponents.

The Quotient Rule for Exponents Consider the quotient of two exponential expressions, such as the quotient of 2^7 and 2^3. We are dividing 7 factors of 2 by 3 factors of 2. We are left with 4 factors of 2:

7 factors of 2

$$\frac{2^7}{2^3} = \frac{2 \cdot 2 \cdot 2 \cdot 2 \cdot 2 \cdot 2 \cdot 2}{2 \cdot 2 \cdot 2} = \frac{\cancel{2} \cdot \cancel{2} \cdot \cancel{2} \cdot 2 \cdot 2 \cdot 2 \cdot 2}{\cancel{2} \cdot \cancel{2} \cdot \cancel{2}} = 2 \cdot 2 \cdot 2 \cdot 2$$

3 factors of 2   Divide out pairs of factors: $\frac{2}{2} = 1$.   4 factors of 2

Thus,

$$\frac{2^7}{2^3} = 2^4.$$

We can quickly find the exponent, 4, on the quotient by subtracting the original exponents:

$$\frac{2^7}{2^3} = 2^{7-3}.$$

This suggests the following rule:

THE QUOTIENT RULE

$$\frac{b^m}{b^n} = b^{m-n}, \quad b \neq 0$$

When dividing exponential expressions with the same nonzero base, subtract the exponent in the denominator from the exponent in the numerator. Use this difference as the exponent of the common base.

EXAMPLE 1 Using the Quotient Rule

Divide each expression using the quotient rule:

a. $\dfrac{2^8}{2^4}$ **b.** $\dfrac{x^{13}}{x^3}$ **c.** $\dfrac{y^{15}}{y}$.

SOLUTION

a. $\dfrac{2^8}{2^4} = 2^{8-4} = 2^4$ or 16

b. $\dfrac{x^{13}}{x^3} = x^{13-3} = x^{10}$

c. $\dfrac{y^{15}}{y} = \dfrac{y^{15}}{y^1} = y^{15-1} = y^{14}$

■

✔ **CHECK POINT 1** Divide each expression using the quotient rule:

a. $\dfrac{5^{12}}{5^4}$ **b.** $\dfrac{x^9}{x^2}$ **c.** $\dfrac{y^{20}}{y}$.

---

2 Use the zero-exponent rule.

Zero as an Exponent A nonzero base can be raised to the 0 power. The quotient rule can be used to help determine what zero as an exponent should mean. Consider the quotient of b^4 and b^4, where b is not zero. We can determine this quotient in two ways.

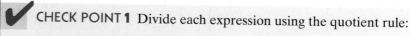

$$\frac{b^4}{b^4} = 1 \qquad\qquad \frac{b^4}{b^4} = b^{4-4} = b^0$$

Any nonzero expression divided by itself is 1.

Use the quotient rule and subtract exponents.

This means that b^0 must equal 1.

> **THE ZERO-EXPONENT RULE** If b is any real number other than 0,
> $$b^0 = 1.$$

EXAMPLE 2 Using the Zero-Exponent Rule

Use the zero-exponent rule to simplify each expression:

a. 7^0 **b.** $(-5)^0$ **c.** -5^0 **d.** $10x^0$ **e.** $(10x)^0$.

SOLUTION

a. $7^0 = 1$ Any nonzero number raised to the 0 power is 1.

b. $(-5)^0 = 1$ Any nonzero number raised to the 0 power is 1.

c. $-5^0 = -1$ $\qquad\qquad$ $-5^0 = -(5^0) = -1$

> Only 5 is raised to the 0 power.

d. $10x^0 = 10 \cdot 1 = 10$

> Only x is raised to the 0 power.

e. $(10x)^0 = 1$

> The entire expression, 10x, is raised to the 0 power.

✔ **CHECK POINT 2** Use the zero-exponent rule to simplify each expression:

a. 14^0 $\qquad$ **b.** $(-10)^0$ $\qquad$ **c.** -10^0 $\qquad$ **d.** $20x^0$ $\qquad$ **e.** $(20x)^0$.

3 Use the quotients-to-powers rule.

The Quotients-to-Powers Rule for Exponents We have seen that when a product is raised to a power, we raise every factor in the product to the power:

$$(ab)^n = a^n b^n.$$

There is a similar property for raising a quotient to a power.

> **QUOTIENTS TO POWERS** If a and b are real numbers and b is nonzero, then
> $$\left(\frac{a}{b}\right)^n = \frac{a^n}{b^n}.$$
> When a quotient is raised to a power, raise the numerator to the power and divide by the denominator raised to the power.

EXAMPLE 3 Using the Quotients-to-Powers Rule

Simplify each expression using the quotients-to-powers rule:

a. $\left(\dfrac{x}{4}\right)^2$ $\qquad$ **b.** $\left(\dfrac{x^2}{5}\right)^3$ $\qquad$ **c.** $\left(\dfrac{2a^3}{b^4}\right)^5$.

SOLUTION

a. $\left(\dfrac{x}{4}\right)^2 = \dfrac{x^2}{4^2} = \dfrac{x^2}{16}$ $\qquad\qquad$ Square the numerator and the denominator.

b. $\left(\dfrac{x^2}{5}\right)^3 = \dfrac{(x^2)^3}{5^3} = \dfrac{x^{2\cdot3}}{5\cdot5\cdot5} = \dfrac{x^6}{125}$ $\qquad\qquad$ Cube the numerator and the denominator.

c. $\left(\dfrac{2a^3}{b^4}\right)^5 = \dfrac{(2a^3)^5}{(b^4)^5}$ $\qquad\qquad$ Raise the numerator and the denominator to the fifth power.

$\qquad\quad = \dfrac{2^5(a^3)^5}{(b^4)^5}$ $\qquad\qquad$ Raise each factor in the numerator to the fifth power.

$\qquad\quad = \dfrac{2^5 a^{3\cdot5}}{b^{4\cdot5}}$ $\qquad\qquad$ To raise exponential expressions to powers, multiply exponents: $(b^m)^n = b^{mn}$.

$\qquad\quad = \dfrac{32a^{15}}{b^{20}}$ $\qquad\qquad$ Simplify.

✔ **CHECK POINT 3** Simplify each expression using the quotients-to-powers rule:

a. $\left(\dfrac{x}{5}\right)^2$ **b.** $\left(\dfrac{x^4}{2}\right)^3$ **c.** $\left(\dfrac{2a^{10}}{b^3}\right)^4$.

STUDY TIP

Try to avoid the following common errors that can occur when simplifying exponential expressions.

Correct	Incorrect	Description of Error
$\dfrac{2^{20}}{2^4} = 2^{20-4} = 2^{16}$	$\dfrac{2^{20}}{2^4} = 2^5$	Exponents should be subtracted, not divided.
$-8^0 = -1$	$-8^0 = 1$	Only 8 is raised to the 0 power.
$\left(\dfrac{x}{5}\right)^2 = \dfrac{x^2}{5^2} = \dfrac{x^2}{25}$	$\left(\dfrac{x}{5}\right)^2 = \dfrac{x^2}{5}$	The numerator and denominator must both be squared.

---

4 Divide monomials.

Dividing Monomials Now that we have developed three additional properties of exponents, we are ready to turn to polynomial division. We begin with the quotient of two monomials, such as $16x^{14}$ and $8x^2$. This quotient is obtained by dividing the coefficients, 16 and 8, and then dividing the variables using the quotient rule for exponents.

$$\frac{16x^{14}}{8x^2} = \frac{16}{8}x^{14-2} = 2x^{12}$$

Divide coefficients and subtract exponents.

> **DIVIDING MONOMIALS** To divide monomials, divide the coefficients and then divide the variables. Use the quotient rule for exponents to divide the variables: Keep the variable and subtract the exponents.

EXAMPLE 4 Dividing Monomials

Divide: **a.** $\dfrac{-12x^8}{4x^2}$ **b.** $\dfrac{2x^3}{8x^3}$ **c.** $\dfrac{15x^5y^4}{3x^2y}$.

SOLUTION

a. $\dfrac{-12x^8}{4x^2} = \dfrac{-12}{4}x^{8-2} = -3x^6$

b. $\dfrac{2x^3}{8x^3} = \dfrac{2}{8}x^{3-3} = \dfrac{1}{4}x^0 = \dfrac{1}{4}\cdot 1 = \dfrac{1}{4}$

c. $\dfrac{15x^5y^4}{3x^2y} = \dfrac{15}{3}x^{5-2}y^{4-1} = 5x^3y^3$

■

STUDY TIP

Look at the solution to Example 4(b). Rather than subtracting exponents for division that results in a 0 exponent, you might prefer to divide out x^3.

$$\frac{2x^3}{8x^3} = \frac{2}{8} = \frac{1}{4}$$

✔ **CHECK POINT 4** Divide:

a. $\dfrac{-20x^{12}}{10x^4}$ **b.** $\dfrac{3x^4}{15x^4}$ **c.** $\dfrac{9x^6y^5}{3xy^2}$.

5 Check polynomial division.

Checking Division of Polynomial Problems   The answer to a division problem can be checked. For example, consider the following problem:

Dividend: the polynomial you are dividing into

$$\frac{15x^5y^4}{3x^2y} = 5x^3y^3.$$

Quotient: the answer to your division problem

Divisor: the polynomial you are dividing by

The quotient is correct if the product of the divisor and the quotient is the dividend. Is the quotient shown in the preceding equation correct?

$$(3x^2y)(5x^3y^3) = 3 \cdot 5x^{2+3}y^{1+3} = 15x^5y^4.$$

Divisor   Quotient   This is the dividend.

Because the product of the divisor and the quotient is the dividend, the answer to the division problem is correct.

> **CHECKING DIVISION OF POLYNOMIALS**   To check a quotient in a division problem, multiply the divisor and the quotient. If this product is the dividend, the quotient is correct.

6 Divide a polynomial by a monomial.

Dividing a Polynomial That Is Not a Monomial by a Monomial   To divide a polynomial by a monomial, we divide each term of the polynomial by the monomial. For example,

polynomial dividend

$$\frac{10x^8 + 15x^6}{5x^3} = \frac{10x^8}{5x^3} + \frac{15x^6}{5x^3} = \frac{10}{5}x^{8-3} + \frac{15}{5}x^{6-3} = 2x^5 + 3x^3.$$

monomial divisor   Divide the first term by $5x^3$.   Divide the second term by $5x^3$.

STUDY TIP

Try to avoid this common error:

Incorrect:

$$\frac{x^4 - \overset{1}{\cancel{x}}}{\underset{1}{\cancel{x}}} = \frac{x^4 - 1}{1} = x^4 - 1$$

Correct:

$$\frac{x^4 - x}{x} = \frac{x^4}{x} - \frac{x}{x}$$
$$= x^{4-1} - x^{1-1}$$
Don't leave out the 1.
$$= x^3 - x^0$$
$$= x^3 - 1$$

Is the quotient correct? Multiply the divisor and the quotient.

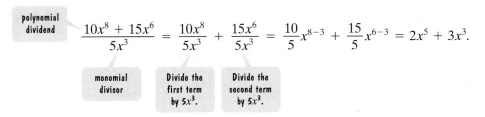

$$5x^3(2x^5 + 3x^3) = 5x^3 \cdot 2x^5 + 5x^3 \cdot 3x^3$$
$$= 5 \cdot 2x^{3+5} + 5 \cdot 3x^{3+3} = 10x^8 + 15x^6$$

Because this product gives the dividend, the quotient is correct.

> **DIVIDING A POLYNOMIAL THAT IS NOT A MONOMIAL BY A MONOMIAL**   To divide a polynomial by a monomial, divide each term of the polynomial by the monomial.

EXAMPLE 5 Dividing a Polynomial by a Monomial

Find the quotient: $(-12x^8 + 4x^6 - 8x^3) \div 4x^2$.

SOLUTION

$$\frac{-12x^8 + 4x^6 - 8x^3}{4x^2}$$
Rewrite the division in a vertical format.

$$= \frac{-12x^8}{4x^2} + \frac{4x^6}{4x^2} - \frac{8x^3}{4x^2}$$
Divide each term of the polynomial by the monomial.

$$= \frac{-12}{4}x^{8-2} + \frac{4}{4}x^{6-2} - \frac{8}{4}x^{3-2}$$
Divide coefficients and subtract exponents.

$$= -3x^6 + x^4 - 2x$$
Simplify.

To check the answer, multiply the divisor and the quotient.

$$4x^2(-3x^6 + x^4 - 2x) = 4x^2(-3x^6) + 4x^2 \cdot x^4 - 4x^2(2x)$$

Divisor   Quotient

$$= 4(-3)x^{2+6} + 4x^{2+4} - 4 \cdot 2x^{2+1}$$

$$= -12x^8 + 4x^6 - 8x^3$$

This is the dividend.

Because the product of the divisor and the quotient is the dividend, the answer—that is, the quotient—is correct. ∎

 CHECK POINT 5 Find the quotient: $(-15x^9 + 6x^5 - 9x^3) \div 3x^2$.

EXAMPLE 6 Dividing a Polynomial by a Monomial

Divide: $\dfrac{16x^5 - 9x^4 + 8x^3}{2x^3}$.

SOLUTION

$$\frac{16x^5 - 9x^4 + 8x^3}{2x^3}$$
This is the given polynomial division.

$$= \frac{16x^5}{2x^3} - \frac{9x^4}{2x^3} + \frac{8x^3}{2x^3}$$
Divide each term by $2x^3$.

$$= \frac{16}{2}x^{5-3} - \frac{9}{2}x^{4-3} + \frac{8}{2}x^{3-3}$$
Divide coefficients and subtract exponents. Did you immediately write the last term as 4?

$$= 8x^2 - \frac{9}{2}x + 4x^0$$
Simplify.

$$= 8x^2 - \frac{9}{2}x + 4$$
$x^0 = 1$, so $4x^0 = 4 \cdot 1 = 4$.

Check the answer by showing that the product of the divisor and the quotient is the dividend. ∎

 **CHECK POINT 6** Divide: $\dfrac{25x^9 - 7x^4 + 10x^3}{5x^3}$.

EXAMPLE 7 Dividing Polynomials in Two Variables

Divide: $(15x^5y^4 - 3x^3y^2 + 9x^2y) \div 3x^2y$.

SOLUTION

$$\frac{15x^5y^4 - 3x^3y^2 + 9x^2y}{3x^2y}$$
Rewrite the division in a vertical format.

$$= \frac{15x^5y^4}{3x^2y} - \frac{3x^3y^2}{3x^2y} + \frac{9x^2y}{3x^2y}$$
Divide each term of the polynomial by the monomial.

$$= \frac{15}{3}x^{5-2}y^{4-1} - \frac{3}{3}x^{3-2}y^{2-1} + \frac{9}{3}x^{2-2}y^{1-1}$$
Divide coefficients and subtract exponents.

$$= 5x^3y^3 - xy + 3$$
Simplify.

Check the answer by showing that the product of the divisor and the quotient is the dividend. ■

✔ **CHECK POINT 7** Divide: $(18x^7y^6 - 6x^2y^3 + 60xy^2) \div 6xy^2$.

6.5 EXERCISE SET

Student Solutions Manual CD/Video PH Math/Tutor Center MathXL Tutorials on CD MathXL® MyMathLab Interactmath.com

Practice Exercises

In Exercises 1–10, divide each expression using the quotient rule. Express any numerical answers in exponential form.

1. $\dfrac{3^{20}}{3^5}$ **2.** $\dfrac{3^{30}}{3^{10}}$ **3.** $\dfrac{x^6}{x^2}$

4. $\dfrac{x^8}{x^4}$ **5.** $\dfrac{y^{13}}{y^5}$ **6.** $\dfrac{y^{19}}{y^6}$

7. $\dfrac{5^6 \cdot 2^8}{5^3 \cdot 2^4}$ **8.** $\dfrac{3^6 \cdot 2^8}{3^3 \cdot 2^4}$ **9.** $\dfrac{x^{100}y^{50}}{x^{25}y^{10}}$

10. $\dfrac{x^{200}y^{40}}{x^{25}y^{10}}$

In Exercises 11–24, use the zero-exponent rule to simplify each expression.

11. 2^0 **12.** 4^0 **13.** $(-2)^0$

14. $(-4)^0$ **15.** -2^0 **16.** -4^0

17. $100y^0$ **18.** $200y^0$ **19.** $(100y)^0$

20. $(200y)^0$ **21.** $-5^0 + (-5)^0$ **22.** $-6^0 + (-6)^0$

23. $-\pi^0 - (-\pi)^0$ **24.** $-\sqrt{3^0} - (-\sqrt{3})^0$

In Exercises 25–36, simplify each expression using the quotients-to-powers rule. If possible, evaluate exponential expressions.

25. $\left(\dfrac{x}{3}\right)^2$ **26.** $\left(\dfrac{x}{5}\right)^2$

27. $\left(\dfrac{x^2}{4}\right)^3$ **28.** $\left(\dfrac{x^2}{3}\right)^3$

29. $\left(\dfrac{2x^3}{5}\right)^2$ **30.** $\left(\dfrac{3x^4}{7}\right)^2$

31. $\left(\dfrac{-4}{3a^3}\right)^3$ **32.** $\left(\dfrac{-5}{2a^3}\right)^3$

33. $\left(\dfrac{-2a^7}{b^4}\right)^5$ **34.** $\left(\dfrac{-2a^8}{b^3}\right)^5$

35. $\left(\dfrac{x^2y^3}{2z}\right)^4$ **36.** $\left(\dfrac{x^3y^2}{2z}\right)^4$

In Exercises 37–52, divide the monomials. Check each answer by showing that the product of the divisor and the quotient is the dividend.

37. $\dfrac{30x^{10}}{10x^5}$ **38.** $\dfrac{45x^{12}}{15x^4}$

39. $\dfrac{-8x^{22}}{4x^2}$ **40.** $\dfrac{-15x^{40}}{3x^4}$

41. $\dfrac{-9y^8}{18y^5}$ **42.** $\dfrac{-15y^{13}}{45y^9}$

43. $\dfrac{7y^{17}}{5y^5}$ **44.** $\dfrac{9y^{19}}{7y^{11}}$

45. $\dfrac{30x^7y^5}{5x^2y}$ **46.** $\dfrac{40x^9y^5}{2x^2y}$

47. $\dfrac{-18x^{14}y^2}{36x^2y^2}$ **48.** $\dfrac{-15x^{16}y^2}{45x^2y^2}$

49. $\dfrac{9x^{20}y^{20}}{7x^{20}y^{20}}$ **50.** $\dfrac{7x^{30}y^{30}}{15x^{30}y^{30}}$

51. $\dfrac{-5x^{10}y^{12}z^6}{50x^2y^3z^2}$

52. $\dfrac{-8x^{12}y^{10}z^4}{40x^2y^3z^2}$

In Exercises 53–78, divide the polynomial by the monomial. Check each answer by showing that the product of the divisor and the quotient is the dividend.

53. $\dfrac{10x^4 + 2x^3}{2}$

54. $\dfrac{20x^4 + 5x^3}{5}$

55. $\dfrac{14x^4 - 7x^3}{7x}$

56. $\dfrac{24x^4 - 8x^3}{8x}$

57. $\dfrac{y^7 - 9y^2 + y}{y}$

58. $\dfrac{y^8 - 11y^3 + y}{y}$

59. $\dfrac{24x^3 - 15x^2}{-3x}$

60. $\dfrac{10x^3 - 20x^2}{-5x}$

61. $\dfrac{18x^5 + 6x^4 + 9x^3}{3x^2}$

62. $\dfrac{18x^5 + 24x^4 + 12x^3}{6x^2}$

63. $\dfrac{12x^4 - 8x^3 + 40x^2}{4x}$

64. $\dfrac{49x^4 - 14x^3 + 70x^2}{-7x}$

65. $(4x^2 - 6x) \div x$

66. $(16y^2 - 8y) \div y$

67. $\dfrac{30z^3 + 10z^2}{-5z}$

68. $\dfrac{12y^4 - 42y^2}{-4y}$

69. $\dfrac{8x^3 + 6x^2 - 2x}{2x}$

70. $\dfrac{9x^3 + 12x^2 - 3x}{3x}$

71. $\dfrac{25x^7 - 15x^5 - 5x^4}{5x^3}$

72. $\dfrac{49x^7 - 28x^5 - 7x^4}{7x^3}$

73. $\dfrac{18x^7 - 9x^6 + 20x^5 - 10x^4}{-2x^4}$

74. $\dfrac{25x^8 - 50x^7 + 3x^6 - 40x^5}{-5x^5}$

75. $\dfrac{12x^2y^2 + 6x^2y - 15xy^2}{3xy}$

76. $\dfrac{18a^3b^2 - 9a^2b - 27ab^2}{9ab}$

77. $\dfrac{20x^7y^4 - 15x^3y^2 - 10x^2y}{-5x^2y}$

78. $\dfrac{8x^6y^3 - 12x^8y^2 - 4x^{14}y^6}{-4x^6y^2}$

Practice Plus

In Exercises 79–82, simplify each expression.

79. $\dfrac{2x^3(4x + 2) - 3x^2(2x - 4)}{2x^2}$

80. $\dfrac{6x^3(3x - 1) + 5x^2(6x - 3)}{3x^2}$

81. $\left(\dfrac{18x^2y^4}{9xy^2}\right) - \left(\dfrac{15x^5y^6}{5x^4y^4}\right)$

82. $\left(\dfrac{9x^3 + 6x^2}{3x}\right) - \left(\dfrac{12x^2y^2 - 4xy^2}{2xy^2}\right)$

83. Divide the sum of $(y + 5)^2$ and $(y + 5)(y - 5)$ by $2y$.

84. Divide the sum of $(y + 4)^2$ and $(y + 4)(y - 4)$ by $2y$.

In Exercises 85–86, the variable n in each exponent represents a natural number. Divide the polynomial by the monomial. Then use polynomial multiplication to check the quotient.

85. $\dfrac{12x^{15n} - 24x^{12n} + 8x^{3n}}{4x^{3n}}$

86. $\dfrac{35x^{10n} - 15x^{8n} + 25x^{2n}}{5x^{2n}}$

Application Exercises

87. The polynomial
$$28t^4 - 711t^3 + 5963t^2 - 1695t + 27{,}424$$
models the annual number of drug arrests in the United States t years after 1984. The polynomial
$$6t^4 - 207t^3 + 2128t^2 - 6622t + 15{,}220$$
models the annual number of drug convictions in the United States t years after 1984.

a. Write an algebraic expression that describes the conviction rate for drug arrests in the United States t years after 1984.

b. Can the polynomial division for the model in part (a) be performed using the methods that you learned in this section? Explain your answer.

88. The polynomial
$$0.0067t^2 + 2.56t + 250.39$$
models the U.S. population, in millions, t years after 1990. The polynomial
$$-35t^2 + 1160t + 10{,}890$$
models revenue, in millions of dollars, from video sales and rentals in the United States t years after 1990.

a. Write an algebraic expression that describes the average amount, in dollars, that each person in the United States spent on video sales and rentals t years after 1990.

b. Can the polynomial division for the model in part (a) be performed using the methods that you learned in this section? Explain your answer.

Writing in Mathematics

89. Explain the quotient rule for exponents. Use $\dfrac{3^6}{3^2}$ in your explanation.

90. Explain how to find any nonzero number to the 0 power.

91. Explain the difference between $(-7)^0$ and -7^0.

92. Explain how to simplify an expression that involves a quotient raised to a power. Provide an example with your explanation.

93. Explain how to divide monomials. Give an example.

94. Explain how to divide a polynomial that is not a monomial by a monomial. Give an example.

95. Are the expressions
$$\frac{12x^2 + 6x}{3x} \quad \text{and} \quad 4x + 2$$
equal for every value of x? Explain.

Critical Thinking Exercises

96. Which one of the following is true?

 a. $x^{10} \div x^2 = x^5$ for all nonzero real numbers x.

 b. $\dfrac{12x^3 - 6x}{2x} = 6x^2 - 6x$

 c. $\dfrac{x^2 + x}{x} = x$

 d. If a polynomial in x of degree 6 is divided by a monomial in x of degree 2, the degree of the quotient is 4.

97. What polynomial, when divided by $3x^2$, yields the trinomial $6x^6 - 9x^4 + 12x^2$ as a quotient?

In Exercises 98–99, find the missing coefficients and exponents designated by question marks.

98. $\dfrac{?x^8 - ?x^6}{3x^?} = 3x^5 - 4x^3$

99. $\dfrac{3x^{14} - 6x^{12} - ?x^7}{?x^?} = -x^7 + 2x^5 + 3$

Review Exercises

100. Find the absolute value: $|-20.3|$. (Section 1.2, Example 8)

101. Express $\frac{7}{8}$ as a decimal. (Section 1.2, Example 4)

102. Graph: $y = \dfrac{1}{3}x + 2$. (Section 4.4, Example 3)

SECTION

6.6

Objective

1 Divide polynomials by binomials.

DIVIDING POLYNOMIALS BY BINOMIALS

For those of you who are dog lovers, you might still be thinking of the polynomial formula that models the age in human years, y, of a dog that is x years old, namely
$$y = -0.001618x^4 + 0.077326x^3 - 1.2367x^2 + 11.460x + 2.914.$$

Suppose that you are in your twenties, say 25. What is Fido's equivalent age? To answer this question, we must substitute 25 for y and solve the resulting polynomial equation for x:
$$25 = -0.001618x^4 + 0.077326x^3 - 1.2367x^2 + 11.460x + 2.914.$$

Don't panic! We won't be solving an equation as complicated as this—yet. You will learn to solve polynomial equations in which the highest power of the variable is 4 in more advanced algebra courses. Part of the method for solving such an equation involves the division of a polynomial by a binomial, such as

$$x + 3 \overline{)x^2 + 10x + 21}.$$

Divisor has two terms and is a binomial.

The polynomial dividend has three terms and is a trinomial.

In this section, you will learn how to perform such divisions.

1 Divide polynomials by binomials.

The Steps in Dividing a Polynomial by a Binomial Dividing a polynomial by a binomial may remind you of long division.

Divide 3983 by 26 without the use of a calculator. Describe the process of the division using the four steps—*divide, multiply, subtract,* and *bring down.* What do you observe about this process? When does it come to an end?

When a divisor is a binomial, the four steps used to divide whole numbers—**divide, multiply, subtract, bring down the next term**—form the repetitive procedure for dividing a polynomial by a binomial.

EXAMPLE 1 Dividing a Polynomial by a Binomial

Divide $x^2 + 10x + 21$ by $x + 3$.

SOLUTION The following steps illustrate how polynomial division is very similar to numerical division.

$$x + 3\overline{)x^2 + 10x + 21}$$

Arrange the terms of the dividend $(x^2 + 10x + 21)$ and the divisor $(x + 3)$ in descending powers of x.

$$\begin{array}{r} x \\ x + 3\overline{)x^2 + 10x + 21} \end{array}$$

Divide x^2 (the first term in the dividend) by x (the first term in the divisor): $\dfrac{x^2}{x} = x$. Align like terms.

$x(x + 3) = x^2 + 3x.$

$$\begin{array}{r} x \\ x + 3\overline{)x^2 + 10x + 21} \\ x^2 + 3x \end{array}$$

Multiply each term in the divisor $(x + 3)$ by x, aligning terms of the product under like terms in the dividend.

$$\begin{array}{r} x \\ x + 3\overline{)x^2 + 10x + 21} \\ \ominus x^2 \ominus 3x \\ \hline 7x \end{array}$$

Change signs of the polynomial being subtracted.

Subtract $x^2 + 3x$ from $x^2 + 10x$ by changing the sign of each term in the lower expression and adding.

$$\begin{array}{r} x \\ x + 3\overline{)x^2 + 10x + 21} \\ x^2 + 3x \\ \hline 7x + 21 \end{array}$$

Bring down 21 from the original dividend and add algebraically to form a new dividend.

$$\begin{array}{r} x + 7 \\ x + 3\overline{)x^2 + 10x + 21} \\ x^2 + 3x \\ \hline 7x + 21 \end{array}$$

Find the second term of the quotient. Divide the first term of $7x + 21$ by x, the first term of the divisor: $\dfrac{7x}{x} = 7$.

$$7(x + 3) = 7x + 21$$

$$
\begin{array}{r}
x + 7 \\
x + 3 \overline{)x^2 + 10x + 21} \\
x^2 + 3x \\
\hline
7x + 21 \\
7x + 21 \\
\hline
0
\end{array}
$$

Remainder

Multiply the divisor $(x + 3)$ by 7, aligning under like terms in the new dividend. Then subtract to obtain the remainder of 0.

The quotient is $x + 7$ and the remainder is 0. We will not list a remainder of 0 in the answer. Thus,

$$\frac{x^2 + 10x + 21}{x + 3} = x + 7.$$

When dividing polynomials by binomials, the answer can be checked. Find the product of the divisor and the quotient and add the remainder. If the result is the dividend, the answer to the division problem is correct. For example, let's check our work in Example 1.

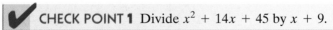

$$(x^2 + 10x + 21) \div (x + 3) = x + 7$$

Dividend   Divisor   Quotient we wish to check

Multiply the divisor and the quotient and add the remainder, 0:

$$(x + 3)(x + 7) + 0 = x^2 + 7x + 3x + 21 + 0 = x^2 + 10x + 21$$

Divisor   Quotient   Remainder   This is the dividend.

Because we obtained the dividend, the quotient is correct.

CHECK POINT 1 Divide $x^2 + 14x + 45$ by $x + 9$.

Before considering additional examples, let's summarize the general procedure for dividing a polynomial by a binomial.

DIVIDING A POLYNOMIAL BY A BINOMIAL

1. **Arrange the terms** of both the dividend and the divisor in descending powers of any variable.
2. **Divide** the first term in the dividend by the first term in the divisor. The result is the first term of the quotient.
3. **Multiply** every term in the divisor by the first term in the quotient. Write the resulting product beneath the dividend with like terms lined up.
4. **Subtract** the product from the dividend.
5. **Bring down** the next term in the original dividend and write it next to the remainder to form a new dividend.
6. Use this new expression as the dividend and repeat this process until the remainder can no longer be divided. This will occur when the degree of the remainder (the highest exponent on a variable in the remainder) is less than the degree of the divisor.

In our next division, we will obtain a nonzero remainder.

EXAMPLE 2 Dividing a Polynomial by a Binomial

Divide: $\dfrac{7x - 9 - 4x^2 + 4x^3}{2x - 1}$.

SOLUTION We begin by writing the dividend in descending powers of x.

$$7x - 9 - 4x^2 + 4x^3 = 4x^3 - 4x^2 + 7x - 9$$

> Think of 9 as $9x^0$. The powers descend from 3 to 0.

$$2x - 1 \overline{)4x^3 - 4x^2 + 7x - 9}$$

This is the problem with the dividend in descending powers of x.

$$\begin{array}{r} 2x^2 \\ 2x - 1 \overline{)4x^3 - 4x^2 + 7x - 9} \end{array}$$

Divide: $\dfrac{4x^3}{2x} = 2x^2$.

$2x^2(2x - 1) = 4x^3 - 2x^2$

$$\begin{array}{r} 2x^2 \\ 2x - 1 \overline{)4x^3 - 4x^2 + 7x - 9} \\ 4x^3 - 2x^2 \end{array}$$

Multiply: $2x^2(2x - 1) = 4x^3 - 2x^2$.

$$\begin{array}{r} 2x^2 \\ 2x - 1 \overline{)4x^3 - 4x^2 + 7x - 9} \\ {}^{\ominus}4x^3 {}^{\oplus} 2x^2 \\ \hline - 2x^2 \end{array}$$

Subtract: $4x^3 - 4x^2 - (4x^3 - 2x^2)$
$= 4x^3 - 4x^2 - 4x^3 + 2x^2$
$= -2x^2$.

> Change signs of the polynomial being subtracted.

$$\begin{array}{r} 2x^2 \\ 2x - 1 \overline{)4x^3 - 4x^2 + 7x - 9} \\ 4x^3 - 2x^2 \\ \hline - 2x^2 + 7x \end{array}$$

Bring down $7x$. The new dividend is $-2x^2 + 7x$.

$$\begin{array}{r} 2x^2 - x \\ 2x - 1 \overline{)4x^3 - 4x^2 + 7x - 9} \\ 4x^3 - 2x^2 \\ \hline - 2x^2 + 7x \end{array}$$

Divide: $\dfrac{-2x^2}{2x} = -x$.

$-x(2x - 1) = -2x^2 + x$

$$\begin{array}{r} 2x^2 - x \\ 2x - 1 \overline{)4x^3 - 4x^2 + 7x - 9} \\ 4x^3 - 2x^2 \\ \hline - 2x^2 + 7x \\ - 2x^2 + x \end{array}$$

Multiply: $-x(2x - 1) = -2x^2 + x$.

$$\begin{array}{r} 2x^2 - x \\ 2x - 1 \overline{)4x^3 - 4x^2 + 7x - 9} \\ 4x^3 - 2x^2 \\ \hline - 2x^2 + 7x \\ {}^{\oplus} 2x^2 {}^{\ominus} x \\ \hline 6x \end{array}$$

Subtract:
$-2x^2 + 7x - (-2x^2 + x)$
$= -2x^2 + 7x + 2x^2 - x$
$= 6x$.

$$
\begin{array}{r}
2x^2 - x \\
2x - 1 \overline{)\, 4x^3 - 4x^2 + 7x - 9} \\
\underline{4x^3 - 2x^2} \\
-2x^2 + 7x \\
\underline{-2x^2 + x} \\
6x - 9
\end{array}
$$

Bring down -9. The new dividend is $6x - 9$.

$$
\begin{array}{r}
2x^2 - x + 3 \\
2x - 1 \overline{)\, 4x^3 - 4x^2 + 7x - 9} \\
\underline{4x^3 - 2x^2} \\
-2x^2 + 7x \\
\underline{-2x^2 + x} \\
6x - 9
\end{array}
$$

Divide: $\dfrac{6x}{2x} = 3.$

$3(2x - 1) = 6x - 3$

$$
\begin{array}{r}
2x^2 - x + 3 \\
2x - 1 \overline{)\, 4x^3 - 4x^2 + 7x - 9} \\
\underline{4x^3 - 2x^2} \\
-2x^2 + 7x \\
\underline{-2x^2 + x} \\
6x - 9 \\
6x - 3
\end{array}
$$

Multiply: $3(2x - 1) = 6x - 3.$

$$
\begin{array}{r}
2x^2 - x + 3 \\
2x - 1 \overline{)\, 4x^3 - 4x^2 + 7x - 9} \\
\underline{4x^3 - 2x^2} \\
-2x^2 + 7x \\
\underline{-2x^2 + x} \\
6x - 9 \\
\underline{\ominus 6x \oplus 3} \\
-6
\end{array}
$$

Subtract:
$6x - 9 - (6x - 3)$
$= 6x - 9 - 6x + 3$
$= -6$

Remainder

The quotient is $2x^2 - x + 3$ and the remainder is -6. When there is a nonzero remainder, as in this example, list the quotient, plus the remainder above the divisor. Thus,

$$
\frac{7x - 9 - 4x^2 + 4x^3}{2x - 1} = 2x^2 - x + 3 + \frac{-6}{2x - 1}
$$

Remainder above divisor

Quotient

or

$$
\frac{7x - 9 - 4x^2 + 4x^3}{2x - 1} = 2x^2 - x + 3 - \frac{6}{2x - 1}.
$$

Check this result by showing that the product of the divisor and the quotient,

$$
(2x - 1)(2x^2 - x + 3),
$$

plus the remainder, -6, is the dividend, $7x - 9 - 4x^2 + 4x^3$. ∎

✔ CHECK POINT **2** Divide: $\dfrac{6x + 8x^2 - 12}{2x + 3}$.

If a power of a variable is missing in a dividend, add that power of the variable with a coefficient of 0 and then divide. In this way, like terms will be aligned as you carry out the division.

EXAMPLE 3 Dividing a Polynomial with Missing Terms

Divide: $\dfrac{8x^3 - 1}{2x - 1}$.

SOLUTION   We write the dividend, $8x^3 - 1$, as

$$8x^3 + 0x^2 + 0x - 1.$$

> Use a coefficient of 0 with missing terms.

By doing this, we will keep all like terms aligned.

$4x^2(2x - 1) = 8x^3 - 4x^2$

$$
\begin{array}{r}
4x^2 \\
2x - 1\overline{\smash{)}8x^3 + 0x^2 + 0x - 1} \\
\underline{8x^3 \ominus 4x^2} \\
4x^2 + 0x
\end{array}
$$

Divide $\left(\dfrac{8x^3}{2x} = 4x^2\right)$, multiply, subtract, and bring down the next term.
The new dividend is $4x^2 + 0x$.

$2x(2x - 1) = 4x^2 - 2x$

$$
\begin{array}{r}
4x^2 + 2x \\
2x - 1\overline{\smash{)}8x^3 + 0x^2 + 0x - 1} \\
\underline{8x^3 - 4x^2} \\
4x^2 + 0x \\
\underline{\ominus 4x^2 \oplus 2x} \\
2x - 1
\end{array}
$$

Divide $\left(\dfrac{4x^2}{2x} = 2x\right)$, multiply $[2x(2x - 1) = 4x^2 - 2x]$, subtract, and bring down the next term.
The new dividend is $2x - 1$.

$1(2x - 1) = 2x - 1$

$$
\begin{array}{r}
4x^2 + 2x + 1 \\
2x - 1\overline{\smash{)}8x^3 + 0x^2 + 0x - 1} \\
\underline{8x^3 - 4x^2} \\
4x^2 + 0x \\
\underline{4x^2 - 2x} \\
2x - 1 \\
\underline{\ominus 2x \oplus 1} \\
0
\end{array}
$$

Divide $\left(\dfrac{2x}{2x} = 1\right)$, multiply $[1(2x - 1) = 2x - 1]$, and subtract. The remainder is 0.

Thus,

$$\frac{8x^3 - 1}{2x - 1} = 4x^2 + 2x + 1.$$

Check this result by showing that the product of the divisor and the quotient,

$$(2x - 1)(4x^2 + 2x + 1)$$

plus the remainder, 0, is the dividend, $8x^3 - 1$. ∎

USING TECHNOLOGY

The graphs of $y_1 = \dfrac{8x^3 - 1}{2x - 1}$ and $y_2 = 4x^2 + 2x + 1$ are shown below.

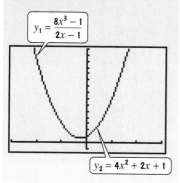

$y_1 = \dfrac{8x^3 - 1}{2x - 1}$

$y_2 = 4x^2 + 2x + 1$

$[-3, 3, 1]$ by $[-1, 15, 1]$

The graphs coincide. Thus,

$$\frac{8x^3 - 1}{2x - 1} = 4x^2 + 2x + 1.$$

✔ **CHECK POINT 3** Divide: $\dfrac{x^3 - 1}{x - 1}$.

6.6 EXERCISE SET

    Student Solutions Manual CD/Video PH Math/Tutor Center MathXL Tutorials on CD  MathXL® MyMathLab   Interactmath.com

Practice Exercises

In Exercises 1–36, divide as indicated. Check each answer by showing that the product of the divisor and the quotient, plus the remainder, is the dividend.

1. $\dfrac{x^2 + 6x + 8}{x + 2}$

2. $\dfrac{x^2 + 7x + 10}{x + 5}$

3. $\dfrac{2x^2 + x - 10}{x - 2}$

4. $\dfrac{2x^2 + 13x + 15}{x + 5}$

5. $\dfrac{x^2 - 5x + 6}{x - 3}$

6. $\dfrac{x^2 - 2x - 24}{x + 4}$

7. $\dfrac{2y^2 + 5y + 2}{y + 2}$

8. $\dfrac{2y^2 - 13y + 21}{y - 3}$

9. $\dfrac{x^2 - 3x + 4}{x + 2}$

10. $\dfrac{x^2 - 7x + 5}{x + 3}$

11. $\dfrac{5y + 10 + y^2}{y + 2}$

12. $\dfrac{-8y + y^2 - 9}{y - 3}$

13. $\dfrac{x^3 - 6x^2 + 7x - 2}{x - 1}$

14. $\dfrac{x^3 + 3x^2 + 5x + 3}{x + 1}$

15. $\dfrac{12y^2 - 20y + 3}{2y - 3}$

16. $\dfrac{4y^2 - 8y - 5}{2y + 1}$

17. $\dfrac{4a^2 + 4a - 3}{2a - 1}$

18. $\dfrac{2b^2 - 9b - 5}{2b + 1}$

19. $\dfrac{3y - y^2 + 2y^3 + 2}{2y + 1}$

20. $\dfrac{9y + 18 - 11y^2 + 12y^3}{4y + 3}$

21. $\dfrac{6x^2 - 5x - 30}{2x - 5}$

22. $\dfrac{4y^2 + 8y + 3}{2y - 1}$

23. $\dfrac{x^3 + 4x - 3}{x - 2}$

24. $\dfrac{x^3 + 2x^2 - 3}{x - 2}$

25. $\dfrac{4y^3 + 8y^2 + 5y + 9}{2y + 3}$

26. $\dfrac{2y^3 - y^2 + 3y + 2}{2y + 1}$

27. $\dfrac{6y^3 - 5y^2 + 5}{3y + 2}$

28. $\dfrac{4y^3 + 3y + 5}{2y - 3}$

29. $\dfrac{27x^3 - 1}{3x - 1}$

30. $\dfrac{8x^3 + 27}{2x + 3}$

31. $\dfrac{81 - 12y^3 + 54y^2 + y^4 - 108y}{y - 3}$

32. $\dfrac{8y^3 + y^4 + 16 + 32y + 24y^2}{y + 2}$

33. $\dfrac{4y^2 + 6y}{2y - 1}$

34. $\dfrac{10x^2 - 3x}{x + 3}$

35. $\dfrac{y^4 - 2y^2 + 5}{y - 1}$

36. $\dfrac{y^4 - 6y^2 + 3}{y - 1}$

Practice Plus

In Exercises 37–42, divide as indicated.

37. $\dfrac{4x^3 - 3x^2 + x + 1}{x^2 + 2}$

38. $\dfrac{3x^3 + 4x^2 + x + 7}{x^2 + 1}$

39. $\dfrac{x^3 - a^3}{x - a}$

40. $\dfrac{x^4 - a^4}{x - a}$

41. $\dfrac{6x^4 - 5x^3 - 8x^2 + 16x - 8}{3x^2 + 2x - 4}$

42. $\dfrac{2x^4 + 5x^3 - 11x^2 - 20x + 12}{x^2 + x - 6}$

43. Divide the difference between $4x^3 + x^2 - 2x + 7$ and $3x^3 - 2x^2 - 7x + 4$ by $x + 1$.

44. Divide the difference between $4x^3 + 2x^2 - x - 1$ and $2x^3 - x^2 + 2x - 5$ by $x + 2$.

Application Exercises

45. Write a simplified polynomial that represents the length of the rectangle.

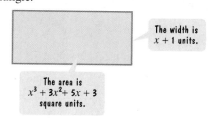

The width is $x + 1$ units.

The area is $x^3 + 3x^2 + 5x + 3$ square units.

46. Write a simplified polynomial that represents the measure of the base of the parallelogram.

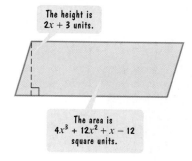

The height is $2x + 3$ units.

The area is $4x^3 + 12x^2 + x - 12$ square units.

You just signed a contract for a new job. The salary for the first year is $30,000 and there is to be a percent increase in your salary each year. The algebraic expression

$$\frac{30{,}000x^n - 30{,}000}{x - 1}$$

describes your total salary over n years, where x is the sum of 1 and the yearly percent increase, expressed as a decimal. Use this information to solve Exercises 47–48.

47. a. Use the given expression and write a quotient of polynomials that describes your total salary over three years.

  b. Simplify the expression in part (a) by performing the division.

  c. Suppose you are to receive an increase of 5% per year. Thus, x is the sum of 1 and 0.05, or 1.05. Substitute 1.05 for x in the expression in part (a) as well as in the simplified form of the expression in part (b). Evaluate each expression. What is your total salary over the three-year period?

48. a. Use the given expression and write a quotient of polynomials that describes your total salary over four years.

  b. Simplify the expression in part (a) by performing the division.

  c. Suppose you are to receive an increase of 8% per year. Thus, x is the sum of 1 and 0.08, or 1.08. Substitute 1.08 for x in the expression in part (a) as well as in the simplified

form of the expression in part (b). Evaluate each expression. What is your total salary over the four-year period?

Writing in Mathematics

49. In your own words, explain how to divide a polynomial by a binomial. Use $\dfrac{x^2 + 4}{x + 2}$ in your explanation.

50. When dividing a polynomial by a binomial, explain when to stop dividing.

51. After dividing a polynomial by a binomial, explain how to check the answer.

52. When dividing a binomial into a polynomial with missing terms, explain the advantage of writing the missing terms with zero coefficients.

Critical Thinking Exercises

53. Which one of the following is true?
  a. If $4x^2 + 25x - 3$ is divided by $4x + 1$, the remainder is 9.
  b. If polynomial division results in a remainder of zero, then the product of the divisor and the quotient is the dividend.
  c. The degree of a polynomial is the power of the term that appears in the first position.
  d. When a polynomial is divided by a binomial, the division process stops when the last term of the dividend is brought down.

54. When a certain polynomial is divided by $2x + 4$, the quotient is

$$x - 3 + \frac{17}{2x + 4}.$$

What is the polynomial?

55. Find the number k such that when $16x^2 - 2x + k$ is divided by $2x - 1$, the remainder is 0.

56. Describe the pattern that you observe in the following quotients and remainders.

$$\frac{x^3 - 1}{x + 1} = x^2 - x + 1 - \frac{2}{x + 1}$$

$$\frac{x^5 - 1}{x + 1} = x^4 - x^3 + x^2 - x + 1 - \frac{2}{x + 1}$$

Use this pattern to find $\dfrac{x^7 - 1}{x + 1}$. Verify your result by dividing.

Technology Exercises

In Exercises 57–61, use a graphing utility to determine whether the divisions have been performed correctly. Graph each side of the given equation in the same viewing rectangle. The graphs should coincide. If they do not, correct the expression on the right side by using polynomial division. Then use your graphing utility to show that the division has been performed correctly.

57. $\dfrac{x^2 - 4}{x - 2} = x + 2$

58. $\dfrac{x^2 - 25}{x - 5} = x - 5$

59. $\dfrac{2x^2 + 13x + 15}{x - 5} = 2x + 3$

60. $\dfrac{6x^2 + 16x + 8}{3x + 2} = 2x - 4$

61. $\dfrac{x^3 + 3x^2 + 5x + 3}{x + 1} = x^2 - 2x + 3$

Review Exercises

62. Graph the solutions of the system:

$$2x - y \geq 4$$
$$x + y \leq -1.$$

(Section 5.5, Example 1)

63. What is 6% of 20? (Section 2.4, Example 7)

64. Solve: $\dfrac{x}{3} + \dfrac{2}{5} = \dfrac{x}{5} - \dfrac{2}{5}$ (Section 2.3, Example 4)

SECTION 6.7

NEGATIVE EXPONENTS AND SCIENTIFIC NOTATION

Objectives

1 Use the negative exponent rule.

2 Simplify exponential expressions.

3 Convert from scientific notation to decimal notation.

4 Convert from decimal notation to scientific notation.

5 Compute with scientific notation.

6 Solve applied problems with scientific notation.

You are listening to a discussion of the country's 6.8 trillion dollar deficit. It seems that this is a real problem, but then you realize that you don't really know what this number means. How can you look at this deficit in the proper perspective? If the national debt were evenly divided among all citizens of the country, how much would each citizen have to pay?

In the new millennium, literacy with numbers, called **numeracy**, will be a necessary skill for functioning in a meaningful way personally, professionally, and as a citizen. In this section, you will learn to use exponents to provide a way of putting large and small numbers into perspective.

Negative Integers as Exponents A nonzero base can be raised to a negative power. The quotient rule can be used to help determine what a negative integer as an exponent should mean. Consider the quotient of b^3 and b^5, where b is not zero. We can determine this quotient in two ways.

1 Use the negative exponent rule.

$$\dfrac{b^3}{b^5} = \dfrac{1 \cdot \cancel{b} \cdot \cancel{b} \cdot \cancel{b}}{\cancel{b} \cdot \cancel{b} \cdot \cancel{b} \cdot b \cdot b} = \dfrac{1}{b^2} \qquad \dfrac{b^3}{b^5} = b^{3-5} = b^{-2}$$

After dividing out pairs of factors, we have two factors of b in the denominator.

Use the quotient rule and subtract exponents.

Notice that $\dfrac{b^3}{b^5}$ equals both b^{-2} and $\dfrac{1}{b^2}$. This means that b^{-2} must equal $\dfrac{1}{b^2}$. This example is a special case of the **negative exponent rule**.

THE NEGATIVE EXPONENT RULE If b is any real number other than 0 and n is a natural number, then

$$b^{-n} = \frac{1}{b^n}.$$

EXAMPLE 1 Using the Negative Exponent Rule

Use the negative exponent rule to write each expression with a positive exponent. Then simplify the expression.

a. 7^{-2} **b.** 4^{-3} **c.** $(-2)^{-4}$ **d.** -2^{-4} **e.** 5^{-1}

SOLUTION

a. $7^{-2} = \dfrac{1}{7^2} = \dfrac{1}{7 \cdot 7} = \dfrac{1}{49}$

b. $4^{-3} = \dfrac{1}{4^3} = \dfrac{1}{4 \cdot 4 \cdot 4} = \dfrac{1}{64}$

c. $(-2)^{-4} = \dfrac{1}{(-2)^4} = \dfrac{1}{(-2)(-2)(-2)(-2)} = \dfrac{1}{16}$

d. $-2^{-4} = -\dfrac{1}{2^4} = -\dfrac{1}{2 \cdot 2 \cdot 2 \cdot 2} = -\dfrac{1}{16}$

> The negative is not inside parentheses and is not taken to the -4 power.

e. $5^{-1} = \dfrac{1}{5^1} = \dfrac{1}{5}$

✔ **CHECK POINT 1** Use the negative exponent rule to write each expression with a positive exponent. Then simplify the expression.

a. 6^{-2} **b.** 5^{-3}

c. $(-3)^{-4}$ **d.** -3^{-4} **e.** 8^{-1}

Negative exponents can also appear in denominators. For example,

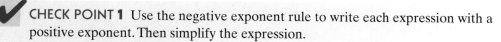

$$\frac{1}{2^{-10}} = \frac{1}{\dfrac{1}{2^{10}}} = 1 \div \frac{1}{2^{10}} = 1 \cdot \frac{2^{10}}{1} = 2^{10}.$$

In general, if a negative exponent appears in a denominator, an expression can be written with a positive exponent using

$$\frac{1}{b^{-n}} = b^n.$$

For example,

$$\frac{1}{2^{-3}} = 2^3 = 8 \qquad \text{and} \qquad \frac{1}{(-6)^{-2}} = (-6)^2 = 36.$$

> Change only the sign of the exponent and not the sign of the base, −6.

NEGATIVE EXPONENTS IN NUMERATORS AND DENOMINATORS If b is any real number other than 0 and n is a natural number, then

$$b^{-n} = \frac{1}{b^n} \qquad \text{and} \qquad \frac{1}{b^{-n}} = b^n.$$

When a negative number appears as an exponent, switch the position of the base (from numerator to denominator or from denominator to numerator) and make the exponent positive. The sign of the base does not change.

EXAMPLE 2 Using Negative Exponents

Write each expression with positive exponents only. Then simplify, if possible.

a. $\dfrac{4^{-3}}{5^{-2}}$ **b.** $\left(\dfrac{3}{4}\right)^{-2}$ **c.** $\dfrac{1}{4x^{-3}}$ **d.** $\dfrac{x^{-5}}{y^{-1}}$

SOLUTION

a. $\dfrac{4^{-3}}{5^{-2}} = \dfrac{5^2}{4^3} = \dfrac{5 \cdot 5}{4 \cdot 4 \cdot 4} = \dfrac{25}{64}$

> Switch the position of the bases and make the exponents positive.

b. $\left(\dfrac{3}{4}\right)^{-2} = \dfrac{3^{-2}}{4^{-2}} = \dfrac{4^2}{3^2} = \dfrac{4 \cdot 4}{3 \cdot 3} = \dfrac{16}{9}$

> Switch the position of the bases and make the exponents positive.

c. $\dfrac{1}{4x^{-3}} = \dfrac{x^3}{4}$

> Switch the position of the base and make the exponent positive. Note that only x is raised to the −3 power.

d. $\dfrac{x^{-5}}{y^{-1}} = \dfrac{y^1}{x^5} = \dfrac{y}{x^5}$ ∎

✔ **CHECK POINT 2** Write each expression with positive exponents only. Then simplify, if possible.

a. $\dfrac{2^{-3}}{7^{-2}}$ **b.** $\left(\dfrac{4}{5}\right)^{-2}$ **c.** $\dfrac{1}{7y^{-2}}$ **d.** $\dfrac{x^{-1}}{y^{-8}}$

2 Simplify exponential expressions.

Simplifying Exponential Expressions Properties of exponents are used to simplify exponential expressions. An exponential expression is **simplified** when

- No parentheses appear.
- No powers are raised to powers.
- Each base occurs only once.
- No negative or zero exponents appear.

SIMPLIFYING EXPONENTIAL EXPRESSIONS

1. If necessary, remove parentheses using

$$(ab)^n = a^n b^n \quad \text{or} \quad \left(\frac{a}{b}\right)^n = \frac{a^n}{b^n}.$$

   Example

   $$(xy)^3 = x^3 y^3$$

2. If necessary, simplify powers to powers using

$$(b^m)^n = b^{mn}.$$

   $$(x^4)^3 = x^{4\cdot 3} = x^{12}$$

3. If necessary, be sure that each base appears only once, using

$$b^m \cdot b^n = b^{m+n} \quad \text{or} \quad \frac{b^m}{b^n} = b^{m-n}.$$

   $$x^4 \cdot x^3 = x^{4+3} = x^7$$

4. If necessary, rewrite exponential expressions with zero powers as 1 ($b^0 = 1$). Furthermore, write the answer with positive exponents using

$$b^{-n} = \frac{1}{b^n} \quad \text{or} \quad \frac{1}{b^{-n}} = b^n.$$

   $$\frac{x^5}{x^8} = x^{5-8} = x^{-3} = \frac{1}{x^3}$$

The following examples show how to simplify exponential expressions. In each example, assume that any variable in a denominator is not equal to zero.

STUDY TIP

There is often more than one way to simplify an exponential expression. For example, you may prefer to simplify Example 3 as follows:

$$x^{-9} \cdot x^4 = \frac{x^4}{x^9} = x^{4-9} = x^{-5} = \frac{1}{x^5}.$$

EXAMPLE 3 Simplifying an Exponential Expression

Simplify: $x^{-9} \cdot x^4$.

SOLUTION

$$x^{-9} \cdot x^4 = x^{-9+4} \qquad b^m \cdot b^n = b^{m+n}$$
$$= x^{-5} \qquad \text{The base, } x, \text{ now appears only once.}$$
$$= \frac{1}{x^5} \qquad b^{-n} = \frac{1}{b^n}$$

✓ **CHECK POINT 3** Simplify: $x^{-12} \cdot x^2$.

EXAMPLE 4 Simplifying Exponential Expressions

Simplify:

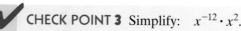

   a. $\dfrac{x^4}{x^9}$ **b.** $\dfrac{25x^6}{5x^8}$ **c.** $\dfrac{10y^7}{-2y^{10}}$.

SOLUTION

a. $\dfrac{x^4}{x^9} = x^{4-9} = x^{-5} = \dfrac{1}{x^5}$

b. $\dfrac{25x^6}{5x^8} = \dfrac{25}{5} \cdot \dfrac{x^6}{x^8} = 5x^{6-8} = 5x^{-2} = \dfrac{5}{x^2}$

c. $\dfrac{10y^7}{-2y^{10}} = \dfrac{10}{-2} \cdot \dfrac{y^7}{y^{10}} = -5y^{7-10} = -5y^{-3} = -\dfrac{5}{y^3}$ ■

CHECK POINT 4 Simplify:

a. $\dfrac{x^2}{x^{10}}$ **b.** $\dfrac{75x^3}{5x^9}$ **c.** $\dfrac{50y^8}{-25y^{14}}.$

EXAMPLE 5 Simplifying an Exponential Expression

Simplify: $\dfrac{(5x^3)^2}{x^{10}}.$

SOLUTION

$\dfrac{(5x^3)^2}{x^{10}} = \dfrac{5^2(x^3)^2}{x^{10}}$ Raise each factor in the product to the second power. Parentheses are removed using $(ab)^n = a^n b^n$.

$= \dfrac{5^2 x^{3 \cdot 2}}{x^{10}}$ Multiply powers to powers using $(b^m)^n = b^{mn}$.

$= \dfrac{25x^6}{x^{10}}$ Simplify.

$= 25x^{6-10}$ When dividing with the same base, subtract exponents: $\dfrac{b^m}{b^n} = b^{m-n}$.

$= 25x^{-4}$ Simplify. The base, x, now appears only once.

$= \dfrac{25}{x^4}$ Rewrite with a positive exponent using $b^{-n} = \dfrac{1}{b^n}$.   ■

CHECK POINT 5 Simplify: $\dfrac{(6x^4)^2}{x^{11}}.$

EXAMPLE 6 Simplifying an Exponential Expression

Simplify: $\left(\dfrac{x^5}{x^2}\right)^{-3}.$

SOLUTION
Method 1.   Remove parentheses first by raising the numerator and denominator to the −3 power.

$$\left(\frac{x^5}{x^2}\right)^{-3} = \frac{(x^5)^{-3}}{(x^2)^{-3}}$$

Use $\left(\frac{a}{b}\right)^n = \frac{a^n}{b^n}$ and raise the numerator and denominator to the -3 power.

$$= \frac{x^{5(-3)}}{x^{2(-3)}}$$

Multiply powers to powers using $(b^m)^n = b^{mn}$.

$$= \frac{x^{-15}}{x^{-6}}$$

Simplify.

$$= x^{-15-(-6)}$$

When dividing with the same base, subtract the exponent in the denominator from the exponent in the numerator: $\frac{b^m}{b^n} = b^{m-n}$.

$$= x^{-9}$$

Subtract: $-15 - (-6) = -15 + 6 = -9$. The base, x, now appears only once.

$$= \frac{1}{x^9}$$

Rewrite with a positive exponent using $b^{-n} = \frac{1}{b^n}$.

Method 2. First perform the division within the parentheses.

$$\left(\frac{x^5}{x^2}\right)^{-3} = (x^{5-2})^{-3}$$

Within parentheses, divide by subtracting exponents: $\frac{b^m}{b^n} = b^{m-n}$.

$$= (x^3)^{-3}$$

Simplify. The base, x, now appears only once.

$$= x^{3(-3)}$$

Multiply powers to powers: $(b^m)^n = b^{mn}$.

$$= x^{-9}$$

Simplify.

$$= \frac{1}{x^9}$$

Rewrite with a positive exponent using $b^{-n} = \frac{1}{b^n}$.

Which method do you prefer? ■

✔ CHECK POINT **6** Simplify: $\left(\dfrac{x^8}{x^4}\right)^{-5}$.

Scientific Notation By the end of 2003, the national debt of the United States was about $6.8 trillion. This is the amount of money the government has had to borrow over the years, mostly by selling bonds, because it has spent more than it has collected in taxes. A stack of $1 bills equaling the national debt would rise to twice the distance from Earth to the moon. Because a trillion is 10^{12}, the national debt can be expressed as

$$6.8 \times 10^{12}.$$

The number 6.8×10^{12} is written in a form called *scientific notation*.

SCIENTIFIC NOTATION A positive number is written in **scientific notation** when it is expressed in the form

$$a \times 10^n,$$

where *a* is a number greater than or equal to 1 and less than 10 ($1 \leq a < 10$) and *n* is an integer.

It is customary to use the multiplication symbol, $\times$, rather than a dot, when writing a number in scientific notation.

Here are two examples of numbers in scientific notation:

- Each day, 2.6×10^7 pounds of dust from the atmosphere settle on Earth.
- The length of the AIDS virus is 1.1×10^{-4} millimeter.

3 Convert from scientific notation to decimal notation.

We can use n, the exponent on the 10 in $a \times 10^n$, to change a number in scientific notation to decimal notation. If n is **positive**, move the decimal point in a to the **right** n places. If n is **negative**, move the decimal point in a to the **left** $|n|$ places.

EXAMPLE 7 Converting from Scientific to Decimal Notation

Write each number in decimal notation:

   a. 2.6×10^7 **b.** 1.1×10^{-4}

SOLUTION In each case, we use the exponent on the 10 to move the decimal point. In part (a), the exponent is positive, so we move the decimal point to the right. In part (b), the exponent is negative, so we move the decimal point to the left.

   a. $2.6 \times 10^7 = 26,000,000$

      $n = 7$

         Move the decimal point 7 places to the right.

   b. $1.1 \times 10^{-4} = 0.00011$

      $n = -4$

         Move the decimal point $|-4|$ places, or 4 places, to the left.

 CHECK POINT 7 Write each number in decimal notation:

   a. 7.4×10^9 **b.** 3.017×10^{-6}.

---

4 Convert from decimal notation to scientific notation.

To convert a positive number from decimal notation to scientific notation, we reverse the procedure of Example 7.

CONVERTING FROM DECIMAL TO SCIENTIFIC NOTATION Write the number in the form $a \times 10^n$.

- Determine a, the numerical factor. Move the decimal point in the given number to obtain a number greater than or equal to 1 and less than 10.
- Determine n, the exponent on 10^n. The absolute value of n is the number of places the decimal point was moved. The exponent n is positive if the given number is greater than 10 and negative if the given number is between 0 and 1.

Section 6.7 • Negative Exponents and Scientific Notation • 381

USING TECHNOLOGY

You can use your calculator's $\boxed{\text{EE}}$ (enter exponent) or $\boxed{\text{EXP}}$ key to convert from decimal to scientific notation. Here is how it's done for 0.000023:

Many Scientific Calculators

Keystrokes	Display
.000023 $\boxed{\text{EE}}$ $\boxed{=}$	2.3 – 05

Many Graphing Calculators

Use the mode setting for scientific notation.

Keystrokes	Display
.000023 $\boxed{\text{ENTER}}$	2.3E−5

EXAMPLE 8 Converting from Decimal Notation to Scientific Notation

Write each number in scientific notation:

a. 4,600,000 **b.** 0.000023.

SOLUTION

a. $4{,}600{,}000 = 4.6 \times 10^6$

This number is greater than 10, so n is positive in $a \times 10^n$.	Move the decimal point in 4,600,000 to get $1 \le a < 10$.	The decimal point moved 6 places from 4,600,000 to 4.6.

b. $0.000023 = 2.3 \times 10^{-5}$

This number is less than 1, so n is negative in $a \times 10^n$.	Move the decimal point in 0.000023 to get $1 \le a < 10$.	The decimal point moved 5 places from 0.000023 to 2.3.

 CHECK POINT 8 Write each number in scientific notation:

a. 7,410,000,000 **b.** 0.000000092.

Computations with Scientific Notation Properties of exponents are used to perform computations with numbers that are expressed in scientific notation.

5 Compute with scientific notation.

COMPUTATIONS WITH NUMBERS IN SCIENTIFIC NOTATION

Multiplication

$$(a \times 10^n) \times (b \times 10^m) = (a \times b) \times 10^{n+m}$$

Add the exponents on 10 and multiply the other parts of the numbers separately.

Division

$$\frac{a \times 10^n}{b \times 10^m} = \left(\frac{a}{b}\right) \times 10^{n-m}$$

Subtract the exponents on 10 and divide the other parts of the numbers separately.

Exponentiation

$$(a \times 10^n)^m = a^n \times 10^{nm}$$

Multiply exponents on 10 and raise the other part of the number to the power.

After the computation is completed, the answer may require an adjustment before it is back in scientific notation.

EXAMPLE 9 Computations with Scientific Notation

Perform the indicated computations, writing the answers in scientific notation:

a. $(4 \times 10^5)(2 \times 10^9)$ **b.** $\dfrac{1.2 \times 10^6}{4.8 \times 10^{-3}}$ **c.** $(5 \times 10^{-4})^3$.

SOLUTION

a. $(4 \times 10^5)(2 \times 10^9) = (4 \times 2) \times (10^5 \times 10^9)$ — Regroup.

$= 8 \times 10^{5+9}$ — Add the exponents on 10 and multiply the other parts.

$= 8 \times 10^{14}$ — Simplify.

b. $\dfrac{1.2 \times 10^6}{4.8 \times 10^{-3}} = \left(\dfrac{1.2}{4.8}\right) \times \left(\dfrac{10^6}{10^{-3}}\right)$ — Regroup.

$= 0.25 \times 10^{6-(-3)}$ — Subtract the exponents on 10 and divide the other parts.

$= 0.25 \times 10^9$ — Simplify. Because 0.25 is not between 1 and 10, it must be written in scientific notation.

$= 2.5 \times 10^{-1} \times 10^9$ — $0.25 = 2.5 \times 10^{-1}$

$= 2.5 \times 10^{-1+9}$ — Add the exponents on 10.

$= 2.5 \times 10^8$ — Simplify.

c. $(5 \times 10^{-4})^3 = 5^3 \times (10^{-4})^3$ — $(ab)^n = a^n b^n$. Cube each factor in parentheses.

$= 125 \times 10^{-4 \cdot 3}$ — Multiply the exponents and cube the other part of the number.

$= 125 \times 10^{-12}$ — Simplify. 125 must be written in scientific notation.

$= 1.25 \times 10^2 \times 10^{-12}$ — $125 = 1.25 \times 10^2$

$= 1.25 \times 10^{2+(-12)}$ — Add the exponents on 10.

$= 1.25 \times 10^{-10}$ — Simplify. ∎

✔ **CHECK POINT 9** Perform the indicated computation, writing the answers in scientific notation:

a. $(3 \times 10^8)(2 \times 10^2)$ **b.** $\dfrac{8.4 \times 10^7}{4 \times 10^{-4}}$

c. $(4 \times 10^{-2})^3$.

6 Solve applied problems with scientific notation.

Applications: Putting Numbers in Perspective Due to tax cuts and spending increases, the United States began accumulating large deficits in the 1980s. To finance the deficit, the government had borrowed $6.8 trillion as of the end of 2003. The graph in Figure 6.4 shows the national debt increasing over time.

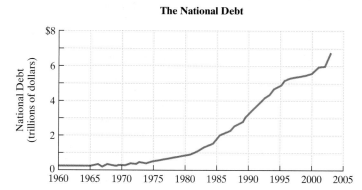

The National Debt

FIGURE 6.4

Source: Office of Management and Budget

Example 10 shows how we can use scientific notation to comprehend the meaning of a number such as 6.8 trillion.

EXAMPLE 10 The National Debt

As of November 2003, the national debt was \$6.8 trillion, or 6.8×10^{12} dollars. At that time, the U.S. population was approximately 290,000,000 (290 million), or 2.9×10^8. If the national debt was evenly divided among every individual in the United States, how much would each citizen have to pay?

SOLUTION The amount each citizen must pay is the total debt, 6.8×10^{12} dollars, divided by the number of citizens, 2.9×10^8.

$$\frac{6.8 \times 10^{12}}{2.9 \times 10^8} = \left(\frac{6.8}{2.9}\right) \times \left(\frac{10^{12}}{10^8}\right)$$

$$\approx 2.3 \times 10^{12-8}$$

$$= 2.3 \times 10^4$$

$$= 23{,}000$$

Every U.S. citizen would have to pay about \$23,000 to the federal government to pay off the national debt. ∎

 CHECK POINT 10 Approximately 2×10^4 people run in the New York City Marathon each year. Each runner runs a distance of 26 miles. Write the total distance covered by all the runners (assuming that each person completes the marathon) in scientific notation.

ENRICHMENT ESSAY

Earthquakes and Exponents

The earthquake that ripped through northern California on October 17, 1989, measured 7.1 on the Richter scale, killed more than 60 people, and injured more than 2400. Shown here is San Francisco's Marina district, where shock waves tossed houses off their foundations and into the street.

The Richter scale is misleading because it is not actually a 1 to 8, but rather a 1 to 10 million scale. Each level indicates a tenfold increase in magnitude from the previous level, making a 7.0 earthquake a million times greater than a 1.0 quake. The following is a translation of the Richter scale:

Richter number (R)	Magnitude (10^{R-1})
1	$10^{1-1} = 10^0 = 1$
2	$10^{2-1} = 10^1 = 10$
3	$10^{3-1} = 10^2 = 100$
4	$10^{4-1} = 10^3 = 1000$
5	$10^{5-1} = 10^4 = 10{,}000$
6	$10^{6-1} = 10^5 = 100{,}000$
7	$10^{7-1} = 10^6 = 1{,}000{,}000$
8	$10^{8-1} = 10^7 = 10{,}000{,}000$

6.7 EXERCISE SET

 Student Solutions Manual   CD/Video   PH Math/Tutor Center   MathXL Tutorials on CD   Math XL MathXL®   MyMathLab MyMathLab   Interactmath.com

Practice Exercises

In Exercises 1–28, write each expression with positive exponents only. Then simplify, if possible.

1. 8^{-2}

2. 9^{-2}

3. 5^{-3}

4. 4^{-3}

5. $(-6)^{-2}$

6. $(-7)^{-2}$

7. -6^{-2}

8. -7^{-2}

9. 4^{-1}

10. 6^{-1}

11. $2^{-1} + 3^{-1}$

12. $3^{-1} - 6^{-1}$

13. $\dfrac{1}{3^{-2}}$

14. $\dfrac{1}{4^{-3}}$

15. $\dfrac{1}{(-3)^{-2}}$

16. $\dfrac{1}{(-2)^{-2}}$

17. $\dfrac{2^{-3}}{8^{-2}}$

18. $\dfrac{4^{-3}}{2^{-2}}$

19. $\left(\dfrac{1}{4}\right)^{-2}$

20. $\left(\dfrac{1}{5}\right)^{-2}$

21. $\left(\dfrac{3}{5}\right)^{-3}$

22. $\left(\dfrac{3}{4}\right)^{-3}$

23. $\dfrac{1}{6x^{-5}}$

24. $\dfrac{1}{8x^{-6}}$

25. $\dfrac{x^{-8}}{y^{-1}}$

26. $\dfrac{x^{-12}}{y^{-1}}$

27. $\dfrac{3}{(-5)^{-3}}$

28. $\dfrac{4}{(-3)^{-3}}$

In Exercises 29–78, simplify each exponential expression. Assume that variables represent nonzero real numbers.

29. $x^{-8} \cdot x^3$

30. $x^{-11} \cdot x^5$

31. $(4x^{-5})(2x^2)$

32. $(5x^{-7})(3x^3)$

33. $\dfrac{x^3}{x^9}$

34. $\dfrac{x^5}{x^{12}}$

35. $\dfrac{y}{y^{100}}$

36. $\dfrac{y}{y^{50}}$

37. $\dfrac{30z^5}{10z^{10}}$

38. $\dfrac{45z^4}{15z^{12}}$

39. $\dfrac{-8x^3}{2x^7}$

40. $\dfrac{-15x^4}{3x^9}$

41. $\dfrac{-9a^5}{27a^8}$

42. $\dfrac{-15a^8}{45a^{13}}$

43. $\dfrac{7w^5}{5w^{13}}$

44. $\dfrac{7w^8}{9w^{14}}$

45. $\dfrac{x^3}{(x^4)^2}$

46. $\dfrac{x^5}{(x^3)^2}$

47. $\dfrac{y^{-3}}{(y^4)^2}$

48. $\dfrac{y^{-5}}{(y^3)^2}$

49. $\dfrac{(4x^3)^2}{x^8}$

50. $\dfrac{(5x^3)^2}{x^7}$

51. $\dfrac{(6y^4)^3}{y^{-5}}$

52. $\dfrac{(4y^5)^3}{y^{-4}}$

53. $\left(\dfrac{x^4}{x^2}\right)^{-3}$

54. $\left(\dfrac{x^6}{x^2}\right)^{-3}$

55. $\left(\dfrac{4x^5}{2x^2}\right)^{-4}$

56. $\left(\dfrac{6x^7}{2x^2}\right)^{-4}$

57. $(3x^{-1})^{-2}$

58. $(4x^{-1})^{-2}$

59. $(-2y^{-1})^{-3}$

60. $(-3y^{-1})^{-3}$

61. $\dfrac{2x^5 \cdot 3x^7}{15x^6}$

62. $\dfrac{3x^3 \cdot 5x^{14}}{20x^{14}}$

63. $(x^3)^5 \cdot x^{-7}$

64. $(x^4)^3 \cdot x^{-5}$

65. $(2y^3)^4 y^{-6}$

66. $(3y^4)^3 y^{-7}$

67. $\dfrac{(y^3)^4}{(y^2)^7}$

68. $\dfrac{(y^2)^5}{(y^3)^4}$

69. $(y^{10})^{-5}$

70. $(y^{20})^{-5}$

71. $(a^4 b^5)^{-3}$

72. $(a^5 b^3)^{-4}$

73. $(a^{-2} b^6)^{-4}$

74. $(a^{-7} b^2)^{-5}$

75. $\left(\dfrac{x^2}{2}\right)^{-2}$

76. $\left(\dfrac{x^2}{2}\right)^{-3}$

77. $\left(\dfrac{x^2}{y^3}\right)^{-3}$

78. $\left(\dfrac{x^3}{y^2}\right)^{-4}$

In Exercises 79–90, write each number in decimal notation without the use of exponents.

79. 8.7×10^2

80. 2.75×10^3

81. 9.23×10^5

82. 7.24×10^4

83. 3.4×10^0

84. 9.115×10^0

85. 7.9×10^{-1}

86. 8.6×10^{-1}

87. 2.15×10^{-2}

88. 3.14×10^{-2}

89. 7.86×10^{-4}

90. 4.63×10^{-5}

In Exercises 91–106, write each number in scientific notation.

91. 32,400

92. 327,000

93. 220,000,000

94. 370,000,000,000

95. 713

96. 623

97. 6751

98. 9832

99. 0.0027

100. 0.00083

101. 0.0000202

102. 0.00000103

103. 0.005

104. 0.006

105. 3.14159

106. 2.71828

In Exercises 107–126, perform the indicated computations. Write the answers in scientific notation.

107. $(2 \times 10^3)(3 \times 10^2)$

108. $(3 \times 10^4)(3 \times 10^2)$

109. $(2 \times 10^5)(8 \times 10^3)$

110. $(4 \times 10^3)(5 \times 10^4)$

111. $\dfrac{12 \times 10^6}{4 \times 10^2}$

112. $\dfrac{20 \times 10^{20}}{10 \times 10^{10}}$

113. $\dfrac{15 \times 10^4}{5 \times 10^{-2}}$

114. $\dfrac{18 \times 10^2}{9 \times 10^{-3}}$

115. $\dfrac{15 \times 10^{-4}}{5 \times 10^2}$

116. $\dfrac{18 \times 10^{-2}}{9 \times 10^3}$

117. $\dfrac{180 \times 10^6}{2 \times 10^3}$

118. $\dfrac{180 \times 10^8}{2 \times 10^4}$

119. $\dfrac{3 \times 10^4}{12 \times 10^{-3}}$

120. $\dfrac{5 \times 10^2}{20 \times 10^{-3}}$

121. $(5 \times 10^2)^3$

122. $(4 \times 10^3)^2$

123. $(3 \times 10^{-2})^4$

124. $(2 \times 10^{-3})^5$

125. $(4 \times 10^6)^{-1}$

126. $(5 \times 10^4)^{-1}$

Practice Plus

In Exercises 127–134, simplify each exponential expression. Assume that variables represent nonzero real numbers.

127. $\dfrac{(x^{-2}y)^{-3}}{(x^2y^{-1})^3}$

128. $\dfrac{(xy^{-2})^{-2}}{(x^{-2}y)^{-3}}$

129. $(2x^{-3}yz^{-6})(2x)^{-5}$

130. $(3x^{-4}yz^{-7})(3x)^{-3}$

131. $\left(\dfrac{x^3y^4z^5}{x^{-3}y^{-4}z^{-5}}\right)^{-2}$

132. $\left(\dfrac{x^4y^5z^6}{x^{-4}y^{-5}z^{-6}}\right)^{-4}$

133. $\dfrac{(2^{-1}x^{-2}y^{-1})^{-2}(2x^{-4}y^3)^{-2}(16x^{-3}y^3)^0}{(2x^{-3}y^{-5})^2}$

134. $\dfrac{(2^{-1}x^{-3}y^{-1})^{-2}(2x^{-6}y^4)^{-2}(9x^3y^{-3})^0}{(2x^{-4}y^{-6})^2}$

In Exercises 135–138, perform the indicated computations. Express answers in scientific notation.

135. $(5 \times 10^3)(1.2 \times 10^{-4}) \div (2.4 \times 10^2)$

136. $(2 \times 10^2)(2.6 \times 10^{-3}) \div (4 \times 10^3)$

137. $\dfrac{(1.6 \times 10^4)(7.2 \times 10^{-3})}{(3.6 \times 10^8)(4 \times 10^{-3})}$

138. $\dfrac{(1.2 \times 10^6)(8.7 \times 10^{-2})}{(2.9 \times 10^6)(3 \times 10^{-3})}$

Application Exercises

In Exercises 139–142, rewrite the number in each statement in scientific notation.

139. King Mongkut of Siam (the king in the musical *The King and I*) had 9200 wives.

140. The top-selling music album of all time is "Their Greatest Hits" by the Eagles, selling 28,000,000 copies.

141. The volume of a bacterium is 0.00000000000000025 cubic meter.

142. Home computers can perform a multiplication in 0.00000000036 second.

In Exercises 143–146, use 10^6 for one million and 10^9 for one billion to rewrite the number in each statement in scientific notation.

143. In 2003, Americans consumed 600 million Big Macs at McDonald's.

144. In 2002, the United States imported 550 million barrels of crude oil from Saudi Arabia.

In Exercises 145–146, which appear on the next page, use the numbers displayed in the following graph.

Defense Spending: U.S. and 20 Next Top-Spending Nations

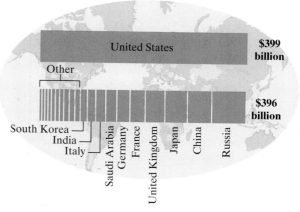

Source: U.S. Defense Department; Center for Defense Information

145. In 2003, the United States spent $399 billion on defense.

146. In 2003, the United States spent more on defense than the 20 next top-spending nations combined. These nations spent $396 billion on defense.

147. In 2002, the United States government spent approximately $2 trillion, or 2×10^{12} dollars. Use the graph to write the amount that it spent on Social Security in scientific notation.

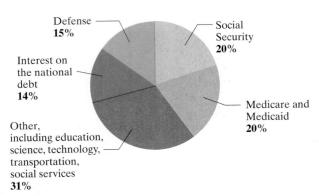

2002 Federal Budget

Defense 15%
Social Security 20%
Interest on the national debt 14%
Other, including education, science, technology, transportation, social services 31%
Medicare and Medicaid 20%

Source: U.S. Office of Management and Budget

148. Americans say they lead active lives, but for the 205 million of us ages 18 and older, walking is often as strenuous as it gets. Use the graph to write the number of Americans whose lifestyle is not very active. Express the answer in scientific notation.

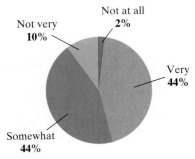

How Active Is Your Lifestyle?

Not at all 2%
Not very 10%
Very 44%
Somewhat 44%

Source: Discovery Health Media

149. If the population of the United States is 2.9×10^8 and each person spends about $120 per year on ice cream, express the total annual spending on ice cream in scientific notation.

150. A human brain contains 3×10^{10} neurons and a gorilla brain contains 7.5×10^9 neurons. How many times as many neurons are in the brain of a human as in the brain of a gorilla?

Use the motion formula $d = rt$, distance equals rate times time, and the fact that light travels at the rate of 1.86×10^5 miles per second, to solve Exercises 151–152.

151. If the moon is approximately 2.325×10^5 miles from Earth, how many seconds does it take moonlight to reach Earth?

152. If the sun is approximately 9.14×10^7 miles from Earth, how many seconds, to the nearest tenth of a second, does it take sunlight to reach Earth?

Writing in Mathematics

153. Explain the negative exponent rule and give an example.

154. How do you know if an exponential expression is simplified?

155. How do you know if a number is written in scientific notation?

156. Explain how to convert from scientific to decimal notation and give an example.

157. Explain how to convert from decimal to scientific notation and give an example.

158. Describe one advantage of expressing a number in scientific notation over decimal notation.

Critical Thinking Exercises

159. Which one of the following is true?
   a. $4^{-2} < 4^{-3}$
   b. $5^{-2} > 2^{-5}$
   c. $(-2)^4 = 2^{-4}$
   d. $5^2 \cdot 5^{-2} > 2^5 \cdot 2^{-5}$

160. Which one of the following is true?
   a. $534.7 = 5.347 \times 10^3$
   b. $\dfrac{8 \times 10^{30}}{4 \times 10^{-5}} = 2 \times 10^{25}$
   c. $(7 \times 10^5) + (2 \times 10^{-3}) = 9 \times 10^2$
   d. $(4 \times 10^3) + (3 \times 10^2) = 4.3 \times 10^3$

161. Give an example of a number where there is no advantage to using scientific notation instead of decimal notation. Explain why this is the case.

162. The mad Dr. Frankenstein has gathered enough bits and pieces (so to speak) for $2^{-1} + 2^{-2}$ of his creature-to-be. Write a fraction that represents the amount of his creature that must still be obtained.

Technology Exercises

163. Use a calculator in a fraction mode to check any five of your answers in Exercises 1–22.

164. Use a calculator to check any three of your answers in Exercises 79–90.

165. Use a calculator to check any three of your answers in Exercises 91–106.

166. Use a calculator with an EE or EXP key to check any four of your computations in Exercises 107–126. Display the result of the computation in scientific notation.

Review Exercises

167. Park rangers catch, tag, and then release 25 deer back into a state park. Two weeks later, they select a sample of 36 deer, 4 of which are tagged. Assuming the ratio of tagged deer in the sample holds for all deer in the park, approximately how many deer are in the park? (Section 3.2, Example 4)

168. Simplify: $24 \div 8 \cdot 3 + 28 \div (-7)$. (Section 1.8, Example 8)

169. List the whole numbers in this set:

$$\left\{ -4, -\frac{1}{5}, 0, \pi, \sqrt{16}, \sqrt{17} \right\}.$$

(Section 1.2, Example 5)

GROUP PROJECT

CHAPTER 6

Putting Numbers into Perspective. A large number can be put into perspective by comparing it with another number. For example, we put the $6.8 trillion national debt into perspective by comparing it to the number of U.S. citizens. The total distance covered by all the runners in the New York City Marathon (Check Point 10 on page 383) can be put into perspective by comparing this distance with, say, the distance from New York to San Francisco.

For this project, each group member should consult an almanac, a newspaper, or the World Wide Web to find a number greater than one million. Explain to other members of the group the context in which the large number is used. Express the number in scientific notation. Then put the number into perspective by comparing it with another number.

CHAPTER 6 SUMMARY

Definitions and Concepts	Examples

Section 6.1 Adding and Subtracting Polynomials

A polynomial is a single term or the sum of two or more terms containing variables in the numerator with whole number exponents. A monomial is a simplified polynomial with exactly one term; a binomial has exactly two terms; a trinomial has exactly three terms. The degree of a polynomial is the highest power of all the terms. The standard form of a polynomial is written in descending powers of the variable.	Polynomials Monomial: $2x^5$ $\quad$ Degree is 5. Binomial: $6x^3 + 5x$ $\quad$ Degree is 3. Trinomial: $7x + 4x^2 - 5$ $\quad$ Degree is 2.
To add polynomials, add like terms.	$(6x^3 + 5x^2 - 7x) + (-9x^3 + x^2 + 6x)$ $= (6x^3 - 9x^3) + (5x^2 + x^2) + (-7x + 6x)$ $= -3x^3 + 6x^2 - x$
The opposite, or additive inverse, of a polynomial is that polynomial with the sign of every coefficient changed. To subtract two polynomials, add the first polynomial and the opposite of the polynomial being subtracted.	$(5y^3 - 9y^2 - 4) - (3y^3 - 12y^2 - 5)$ $= (5y^3 - 9y^2 - 4) + (-3y^3 + 12y^2 + 5)$ $= (5y^3 - 3y^3) + (-9y^2 + 12y^2) + (-4 + 5)$ $= 2y^3 + 3y^2 + 1$

Definitions and Concepts	Examples

Section 6.2 Multiplying Polynomials

Properties of Exponents Product Rule: $b^m \cdot b^n = b^{m+n}$ Power Rule: $(b^m)^n = b^{mn}$ Products to Powers: $(ab)^n = a^n b^n$	$x^3 \cdot x^8 = x^{3+8} = x^{11}$ $(x^3)^8 = x^{3 \cdot 8} = x^{24}$ $(-5x^2)^3 = (-5)^3(x^2)^3 = -125x^6$
To multiply monomials, multiply coefficients and add exponents.	$(-6x^4)(3x^{10}) = -6 \cdot 3x^{4+10} = -18x^{14}$
To multiply a monomial and a polynomial, multiply each term of the polynomial by the monomial.	$2x^4(3x^2 - 6x + 5)$ $= 2x^4 \cdot 3x^2 - 2x^4 \cdot 6x + 2x^4 \cdot 5$ $= 6x^6 - 12x^5 + 10x^4$
To multiply polynomials when neither is a monomial, multiply each term of one polynomial by each term of the other polynomial. Then combine like terms.	$(2x + 3)(5x^2 - 4x + 2)$ $= 2x(5x^2 - 4x + 2) + 3(5x^2 - 4x + 2)$ $= 10x^3 - 8x^2 + 4x + 15x^2 - 12x + 6$ $= 10x^3 + 7x^2 - 8x + 6$

Section 6.3 Special Products

The FOIL method may be used when multiplying two binomials: First terms multiplied. Outside terms multiplied. Inside terms multiplied. Last terms multiplied.	F O I L $(3x + 7)(2x - 5) = 3x \cdot 2x + 3x(-5) + 7 \cdot 2x + 7(-5)$ $= 6x^2 - 15x + 14x - 35$ $= 6x^2 - x - 35$
The Product of the Sum and Difference of Two Terms $(A + B)(A - B) = A^2 - B^2$	$(4x + 7)(4x - 7) = (4x)^2 - 7^2$ $= 16x^2 - 49$
The Square of a Binomial Sum $(A + B)^2 = A^2 + 2AB + B^2$	$(x^2 + 6)^2 = (x^2)^2 + 2 \cdot x^2 \cdot 6 + 6^2$ $= x^4 + 12x^2 + 36$
The Square of a Binomial Difference $(A - B)^2 = A^2 - 2AB + B^2$	$(9x - 3)^2 = (9x)^2 - 2 \cdot 9x \cdot 3 + 3^2$ $= 81x^2 - 54x + 9$

Section 6.4 Polynomials in Several Variables

To evaluate a polynomial in several variables, substitute the given value for each variable and perform the resulting computation.	Evaluate $4x^2y + 3xy - 2x$ for $x = -1$ and $y = -3$. $4x^2y + 3xy - 2x$ $= 4(-1)^2(-3) + 3(-1)(-3) - 2(-1)$ $= 4(1)(-3) + 3(-1)(-3) - 2(-1)$ $= -12 + 9 + 2 = -1$

Definitions and Concepts	Examples

Section 6.4 Polynomials in Several Variables (continued)

For a polynomial in two variables, the degree of a term is the sum of the exponents on its variables. The degree of the polynomial is the highest degree of all its terms.

$$7x^2y + 12x^4y^3 - 17x^5 + 6$$

degree: $2 + 1 = 3$ degree: $4 + 3 = 7$ degree: 5 degree: 0

Degree of polynomial $= 7$

Polynomials in several variables are added, subtracted, and multiplied using the same rules for polynomials in one variable.

$$(5x^2y^3 - xy + 4y^2) - (8x^2y^3 - 6xy - 2y^2)$$
$$= (5x^2y^3 - xy + 4y^2) + (-8x^2y^3 + 6xy + 2y^2)$$
$$= (5x^2y^3 - 8x^2y^3) + (-xy + 6xy) + (4y^2 + 2y^2)$$
$$= -3x^2y^3 + 5xy + 6y^2$$

F   O   I   L

$$(3x - 2y)(x - y) = 3x \cdot x + 3x(-y) + (-2y)x + (-2y)(-y)$$
$$= 3x^2 - 3xy - 2xy + 2y^2$$
$$= 3x^2 - 5xy + 2y^2$$

Section 6.5 Dividing Polynomials

Additional Properties of Exponents

Quotient Rule: $\dfrac{b^m}{b^n} = b^{m-n}, \quad b \neq 0$

Zero-Exponent Rule: $b^0 = 1, \quad b \neq 0$

Quotients to Powers: $\left(\dfrac{a}{b}\right)^n = \dfrac{a^n}{b^n}, \quad b \neq 0$

$$\frac{x^{12}}{x^4} = x^{12-4} = x^8$$

$$(-3)^0 = 1 \qquad -3^0 = -(3^0) = -1$$

$$\left(\frac{y^2}{4}\right)^3 = \frac{(y^2)^3}{4^3} = \frac{y^{2 \cdot 3}}{4 \cdot 4 \cdot 4} = \frac{y^6}{64}$$

To divide monomials, divide coefficients and subtract exponents.

$$\frac{-40x^{40}}{20x^{20}} = \frac{-40}{20}x^{40-20} = -2x^{20}$$

To divide a polynomial by a monomial, divide each term of the polynomial by the monomial.

$$\frac{8x^6 - 4x^3 + 10x}{2x}$$

$$= \frac{8x^6}{2x} - \frac{4x^3}{2x} + \frac{10x}{2x}$$

$$= 4x^{6-1} - 2x^{3-1} + 5x^{1-1} = 4x^5 - 2x^2 + 5$$

Section 6.6 Dividing Polynomials by Binomials

To divide a polynomial by a binomial, begin by arranging the polynomial in descending powers of the variable. If a power of a variable is missing, add that power with a coefficient of 0. Repeat the four steps—divide, multiply, subtract, bring down the next term—until the degree of the remainder is less than the degree of the divisor.

Divide: $\dfrac{10x^2 + 13x + 8}{2x + 3}$.

$$
\begin{array}{r}
5x - 1 + \dfrac{11}{2x+3} \\[4pt]
2x+3\overline{)10x^2 + 13x + 8} \\
\underline{10x^2 + 15x} \\
-2x + 8 \\
\underline{-2x - 3} \\
11
\end{array}
$$

Definitions and Concepts	Examples

Section 6.7 Negative Exponents and Scientific Notation

Definitions and Concepts	Examples
Negative Exponents in Numerators and Denominators If $b \neq 0$, $b^{-n} = \dfrac{1}{b^n}$ and $\dfrac{1}{b^{-n}} = b^n$.	$6^{-2} = \dfrac{1}{6^2} = \dfrac{1}{36}$ $\dfrac{1}{(-2)^{-4}} = (-2)^4 = 16$ $\left(\dfrac{2}{3}\right)^{-3} = \dfrac{2^{-3}}{3^{-3}} = \dfrac{3^3}{2^3} = \dfrac{27}{8}$
An exponential expression is simplified when • No parentheses appear. • No powers are raised to powers. • Each base occurs only once. • No negative or zero exponents appear.	Simplify: $\dfrac{(2x^4)^3}{x^{18}}$. $\dfrac{(2x^4)^3}{x^{18}} = \dfrac{2^3(x^4)^3}{x^{18}} = \dfrac{8x^{4 \cdot 3}}{x^{18}} = \dfrac{8x^{12}}{x^{18}} = 8x^{12-18} = 8x^{-6} = \dfrac{8}{x^6}$
A positive number in scientific notation is expressed as $a \times 10^n$, where $1 \leq a < 10$ and n is an integer.	Write 2.9×10^{-3} in decimal notation. $2.9 \times 10^{-3} = .0029 = 0.0029$ Write 16,000 in scientific notation. $16{,}000 = 1.6 \times 10^4$
Use properties of exponents with base 10 $10^m \cdot 10^n = 10^{m+n}$, $\dfrac{10^m}{10^n} = 10^{m-n}$, and $(10^m)^n = 10^{mn}$ to perform computations with scientific notation.	$(5 \times 10^3)(4 \times 10^{-8})$ $= 5 \cdot 4 \times 10^{3-8}$ $= 20 \times 10^{-5}$ $= 2 \times 10^1 \times 10^{-5} = 2 \times 10^{-4}$

CHAPTER 6 REVIEW EXERCISES

6.1 *In Exercises 1–3, identify each polynomial as a monomial, binomial, or trinomial. Give the degree of the polynomial.*

1. $7x^4 + 9x$

2. $3x + 5x^2 - 2$

3. $16x$

In Exercises 4–8, add or subtract as indicated.

4. $(-6x^3 + 7x^2 - 9x + 3) + (14x^3 + 3x^2 - 11x - 7)$

5. $(9y^3 - 7y^2 + 5) + (4y^3 - y^2 + 7y - 10)$

6. $(5y^2 - y - 8) - (-6y^2 + 3y - 4)$

7. $(13x^4 - 8x^3 + 2x^2) - (5x^4 - 3x^3 + 2x^2 - 6)$

8. Subtract $x^4 + 7x^2 - 11x$ from $-13x^4 - 6x^2 + 5x$.

In Exercises 9–11, add or subtract as indicated.

9. Add. $7y^4 - 6y^3 + 4y^2 - 4y$
$\underline{ y^3 - y^2 + 3y - 4}$

10. Subtract. $7x^2 - 9x + 2$
$\underline{-(4x^2 - 2x - 7)}$

11. Subtract. $5x^3 - 6x^2 - 9x + 14$
$\underline{-(-5x^3 + 3x^2 - 11x + 3)}$

12. The polynomial $104.5x^2 - 1501.5x + 6016$ models the death rate per year per 100,000 men for men averaging x hours of sleep each night. Evaluate the polynomial when $x = 10$. Describe what the answer means in practical terms.

6.2 *In Exercises 13–17, simplify each expression.*

13. $x^{20} \cdot x^3$

14. $y \cdot y^5 \cdot y^8$

15. $(x^{20})^5$

16. $(10y)^2$

17. $(-4x^{10})^3$

In Exercises 18–26, find each product.

18. $(5x)(10x^3)$

19. $(-12y^7)(3y^4)$

20. $(-2x^5)(-3x^4)(5x^3)$

21. $7x(3x^2 + 9)$

22. $5x^3(4x^2 - 11x)$

23. $3y^2(-7y^2 + 3y - 6)$

24. $2y^5(8y^3 - 10y^2 + 1)$

25. $(x + 3)(x^2 - 5x + 2)$

26. $(3y - 2)(4y^2 + 3y - 5)$

In Exercises 27–28, use a vertical format to find each product.

27. $y^2 - 4y + 7$
$\underline{ 3y - 5}$

28. $4x^3 - 2x^2 - 6x - 1$
$\underline{ 2x + 3}$

6.3 *In Exercises 29–41, find each product.*

29. $(x + 6)(x + 2)$

30. $(3y - 5)(2y + 1)$

31. $(4x^2 - 2)(x^2 - 3)$

32. $(5x + 4)(5x - 4)$

33. $(7 - 2y)(7 + 2y)$

34. $(y^2 + 1)(y^2 - 1)$

35. $(x + 3)^2$

36. $(3y + 4)^2$

37. $(y - 1)^2$

38. $(5y - 2)^2$

39. $(x^2 + 4)^2$

40. $(x^2 + 4)(x^2 - 4)$

41. $(x^2 + 4)(x^2 - 5)$

42. Write a polynomial in descending powers of x that represents the area of the shaded region.

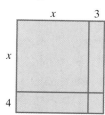

43. The parking garage shown in the figure in the next column measures 30 yards by 20 yards. The length and the width are

each increased by a fixed amount, x yards. Write a trinomial that describes the area of the expanded garage.

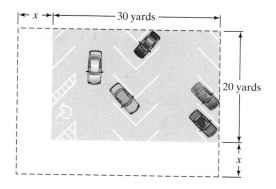

6.4

44. Evaluate $2x^3y - 4xy^2 + 5y + 6$ for $x = -1$ and $y = 2$.

45. Determine the coefficient of each term, the degree of each term, and the degree of the polynomial:
$4x^2y + 9x^3y^2 - 17x^4 - 12$.

In Exercises 46–55, perform the indicated operations.

46. $(7x^2 - 8xy + y^2) + (-8x^2 - 9xy + 4y^2)$

47. $(13x^3y^2 - 5x^2y - 9x^2) - (11x^3y^2 - 6x^2y - 3x^2 + 4)$

48. $(-7x^2y^3)(5x^4y^6)$

49. $5ab^2(3a^2b^3 - 4ab)$

50. $(x + 7y)(3x - 5y)$

51. $(4xy - 3)(9xy - 1)$

52. $(3x + 5y)^2$

53. $(xy - 7)^2$

54. $(7x + 4y)(7x - 4y)$

55. $(a - b)(a^2 + ab + b^2)$

6.5 *In Exercises 56–62, simplify each expression.*

56. $\dfrac{6^{40}}{6^{10}}$

57. $\dfrac{x^{18}}{x^3}$

58. $(-10)^0$

59. -10^0

60. $400x^0$

61. $\left(\dfrac{x^4}{2}\right)^3$

62. $\left(\dfrac{-3}{2y^6}\right)^4$

In Exercises 63–67, divide and check each answer.

63. $\dfrac{-15y^8}{3y^2}$

64. $\dfrac{40x^8y^6}{5xy^3}$

65. $\dfrac{18x^4 - 12x^2 + 36x}{6x}$

66. $\dfrac{30x^8 - 25x^7 - 40x^5}{-5x^3}$

67. $\dfrac{27x^3y^2 - 9x^2y - 18xy^2}{3xy}$

6.6 *In Exercises 68–71, divide and check each answer.*

68. $\dfrac{2x^2 + 3x - 14}{x - 2}$

69. $\dfrac{2x^3 - 5x^2 + 7x + 5}{2x + 1}$

70. $\dfrac{x^3 - 2x^2 - 33x - 7}{x - 7}$

71. $\dfrac{y^3 - 27}{y - 3}$

6.7 *In Exercises 72–76, write each expression with positive exponents only and then simplify.*

72. 7^{-2}

73. $(-4)^{-3}$

74. $2^{-1} + 4^{-1}$

75. $\dfrac{1}{5^{-2}}$

76. $\left(\dfrac{2}{5}\right)^{-3}$

In Exercises 77–85, simplify each exponential expression. Assume that variables in denominators do not equal zero.

77. $\dfrac{x^3}{x^9}$

78. $\dfrac{30y^6}{5y^8}$

79. $(5x^{-7})(6x^2)$

80. $\dfrac{x^4 \cdot x^{-2}}{x^{-6}}$

81. $\dfrac{(3y^3)^4}{y^{10}}$

82. $\dfrac{y^{-7}}{(y^4)^3}$

83. $(2x^{-1})^{-3}$

84. $\left(\dfrac{x^7}{x^4}\right)^{-2}$

85. $\dfrac{(y^3)^4}{(y^{-2})^4}$

In Exercises 86–88, write each number in decimal notation without the use of exponents.

86. 2.3×10^4

87. 1.76×10^{-3}

88. 9×10^{-1}

In Exercises 89–92, write each number in scientific notation.

89. 73,900,000

90. 0.00062

91. 0.38

92. 3.8

In Exercises 93–95, perform the indicated computation. Write the answers in scientific notation.

93. $(6 \times 10^{-3})(1.5 \times 10^6)$

94. $\dfrac{2 \times 10^2}{4 \times 10^{-3}}$

95. $(4 \times 10^{-2})^2$

96. A microsecond is 10^{-6} second and a nanosecond is 10^{-9} second. How many nanoseconds make a microsecond?

97. The world's population is approximately 6.3×10^9 people. Current projections double this population in 40 years. Write the population 40 years from now in scientific notation.

CHAPTER 6   TEST Remember to use your Chapter Test Prep Video CD to see the worked-out solutions to the test questions you want to review.

1. Identify $9x + 6x^2 - 4$ as a monomial, binomial, or trinomial. Give the degree of the polynomial.

In Exercises 2–3, add or subtract as indicated.

2. $(7x^3 + 3x^2 - 5x - 11) + (6x^3 - 2x^2 + 4x - 13)$

3. $(9x^3 - 6x^2 - 11x - 4) - (4x^3 - 8x^2 - 13x + 5)$

In Exercises 4–10, find each product.

4. $(-7x^3)(5x^8)$

5. $6x^2(8x^3 - 5x - 2)$

6. $(3x + 2)(x^2 - 4x - 3)$

7. $(3y + 7)(2y - 9)$

8. $(7x + 5)(7x - 5)$

9. $(x^2 + 3)^2$

10. $(5x - 3)^2$

11. Evaluate $4x^2y + 5xy - 6x$ for $x = -2$ and $y = 3$.

In Exercises 12–14, perform the indicated operations.

12. $(8x^2y^3 - xy + 2y^2) - (6x^2y^3 - 4xy - 10y^2)$

13. $(3a - 7b)(4a + 5b)$

14. $(2x + 3y)^2$

In Exercises 15–17, divide and check each answer.

15. $\dfrac{-25x^{16}}{5x^4}$

16. $\dfrac{15x^4 - 10x^3 + 25x^2}{5x}$

17. $\dfrac{2x^3 - 3x^2 + 4x + 4}{2x + 1}$

In Exercises 18–19, write each expression with positive exponents only and then simplify.

18. 10^{-2}

19. $\dfrac{1}{4^{-3}}$

In Exercises 20–25, simplify each expression.

20. $(-3x^2)^3$

21. $\dfrac{20x^3}{5x^8}$

22. $(-7x^{-8})(3x^2)$

23. $\dfrac{(2y^3)^4}{y^8}$

24. $(5x^{-4})^{-2}$

25. $\left(\dfrac{x^{10}}{x^5}\right)^{-3}$

26. Write 3.7×10^{-4} in decimal notation.

27. Write 7,600,000 in scientific notation.

In Exercises 28–29, perform the indicated computation. Write the answers in scientific notation.

28. $(4.1 \times 10^2)(3 \times 10^{-5})$

29. $\dfrac{8.4 \times 10^6}{4 \times 10^{-2}}$

30. Write a polynomial in descending powers of x that represents the area of the figure.

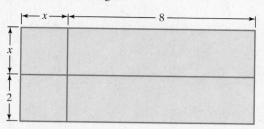

CUMULATIVE REVIEW EXERCISES (CHAPTERS 1–6)

In Exercises 1–2, perform the indicated operation or operations.

1. $(-7)(-5) \div (12 - 3)$

2. $(3 - 7)^2(9 - 11)^3$

3. What is the difference in elevation between a plane flying 14,300 feet above sea level and a submarine traveling 750 feet below sea level?

In Exercises 4–5, solve each equation.

4. $2(x + 3) + 2x = x + 4$

5. $\dfrac{x}{5} - \dfrac{1}{3} = \dfrac{x}{10} - \dfrac{1}{2}$

6. The length of a rectangular sign is 2 feet less than three times its width. If the perimeter of the sign is 28 feet, what are its dimensions?

7. Solve: $7 - 8x \le -6x - 5$. Express the solution set in set-builder notation and graph the solution set on a number line.

8. You and a friend live 72 miles apart. You decide to meet by riding your bikes directly toward each other on the same road. If you bike at 13 miles per hour and your friend bikes at 11 miles per hour, in how many hours will you meet?

9. If a 20-pound bag of fertilizer covers 5000 square feet, how many pounds are needed to cover an area of 26,000 square feet? How many bags of fertilizer are needed?

10. Graph $y = -\frac{2}{5}x + 2$ using the slope and y-intercept.

11. Graph $x - 2y = 4$ using intercepts.

12. Find the slope of the line passing through the points $(-3, 2)$ and $(2, -4)$. Is the line rising, falling, horizontal, or vertical?

13. The slope of a line is -2 and the line passes through the point $(3, -1)$. Write the line's equation in point-slope form and slope-intercept form.

In Exercises 14–15, solve each system by the method of your choice.

14. $3x + 2y = 10$
$4x - 3y = -15$

15. $2x + 3y = -6$
$y = 3x - 13$

16. You are choosing between two long-distance telephone plans. One has a monthly fee of $15 with a charge of $0.05 per minute for all long-distance calls. The other plan has a monthly fee of $5 with a charge of $0.07 per minute for all long-distance calls. For how many minutes of long-distance calls will the costs for the two plans be the same? What will be the cost for each plan?

17. Graph the solutions for the system of linear inequalities:

$$2x + 5y \le 10$$
$$x - y \ge 4.$$

18. Subtract: $(9x^5 - 3x^3 + 2x - 7) - (6x^5 + 3x^3 - 7x - 9)$.

19. Divide: $\dfrac{x^3 + 3x^2 + 5x + 3}{x + 1}$.

20. Simplify: $\dfrac{(3x^2)^4}{x^{10}}$.

Landscaped parks often form an oasis of greenery and calm in which to escape from the bustle of city life. Flowers, trees, ponds, and fountains provide a natural setting that mirrors the interest that many city dwellers have in their environment.

The role of polynomials in landscape design is explored in Exercises 84–85 in Exercise Set 7.6.

Factoring Polynomials

Have you ever thought about creating attractive and inviting home landscaping? Algebra and geometry play an important role in landscape design. In this chapter, you will see how rewriting a polynomial sum or difference in terms of multiplication can be used in the creation of horticultural masterpieces.

SECTION 7.1

THE GREATEST COMMON FACTOR AND FACTORING BY GROUPING

Objectives

1 Factor monomials.

2 Find the greatest common factor.

3 Factor out the greatest common factor of a polynomial.

4 Factor by grouping.

A two-year-old boy is asked, "Do you have a brother?" He answers, "Yes." "What is your brother's name?" "Tom." Asked if Tom has a brother, the two-year-old replies, "No." The child can go in the direction from self to brother, but he cannot reverse this direction and move from brother back to self.

As our intellects develop, we learn to reverse the direction of our thinking. Reversibility of thought is found throughout algebra. For example, we can multiply polynomials and show that

$$5x(2x + 3) = 10x^2 + 15x.$$

We can also reverse this process and express the resulting polynomial as

$$10x^2 + 15x = 5x(2x + 3).$$

Factoring a polynomial containing the sum of monomials means finding an equivalent expression that is a product.

Factoring $10x^2 + 15x$

Sum of monomials

Equivalent expression that is a product

$$10x^2 + 15x = 5x(2x + 3)$$

The factors of $10x^2 + 15x$ are $5x$ and $2x + 3$.

In this chapter, we will be factoring over the set of integers, meaning that the coefficients in the factors are integers. Polynomials that cannot be factored using integer coefficients are called **prime polynomials** over the set of integers.

Factoring Monomials Factoring a monomial means finding two monomials whose product gives the original monomial. For example, $30x^2$ can be factored in a number of different ways, such as

$$30x^2 = (5x)(6x) \qquad \text{The factors are } 5x \text{ and } 6x.$$
$$30x^2 = (15x)(2x) \qquad \text{The factors are } 15x \text{ and } 2x.$$
$$30x^2 = (10x^2)(3) \qquad \text{The factors are } 10x^2 \text{ and } 3.$$
$$30x^2 = (-6x)(-5x). \qquad \text{The factors are } -6x \text{ and } -5x.$$

Observe that each part of the factorization is called a *factor* of the given monomial.

1 Factor monomials.

DISCOVER FOR YOURSELF

Write three more ways of factoring the monomial $30x^2$.

Factoring Out the Greatest Common Factor   We use the distributive property to multiply a monomial and a polynomial of two or more terms. When we factor, we reverse this process, expressing the polynomial as a product.

Multiplication	**Factoring**
$a(b + c) = ab + ac$	$ab + ac = a(b + c)$

Here is a specific example:

Multiplication	**Factoring**
$5x(2x + 3)$	$10x^2 + 15x$
$= 5x \cdot 2x + 5x \cdot 3$	$= 5x \cdot 2x + 5x \cdot 3$
$= 10x^2 + 15x$	$= 5x(2x + 3).$

In the process of finding an equivalent expression for $10x^2 + 15x$ that is a product, we used the fact that $5x$ is a factor of both $10x^2$ and $15x$. The factoring on the right shows that $5x$ is a *common factor* for all the terms of the binomial $10x^2 + 15x$.

In any factoring problem, the first step is to look for the *greatest common factor*. The **greatest common factor**, abbreviated GCF, is an expression of the highest degree that divides each term of the polynomial. Can you see that $5x$ is the greatest common factor of $10x^2 + 15x$? 5 is the greatest integer that divides 10 and 15. Furthermore, x is the greatest expression that divides x^2 and x.

The variable part of the greatest common factor always contains the smallest power of a variable that appears in all terms of the polynomial. For example, consider the polynomial

$$10x^2 + 15x.$$

> x^1, or x, is the variable raised to the smallest exponent.

We see that x is the variable part of the greatest common factor, $5x$.

2 Find the greatest common factor.

EXAMPLE 1   Finding the Greatest Common Factor

Find the greatest common factor of each list of terms:

a. $6x^3$ and $10x^2$ **b.** $15y^5$, $-9y^4$, and $27y^3$ **c.** x^5y^3, x^4y^4, and x^3y^2.

SOLUTION   Use numerical coefficients to determine the coefficient of the GCF. Use variable factors to determine the variable factor of the GCF.

> 2 is the greatest integer that divides 6 and 10.

a. $6x^3$ and $10x^2$

> x^2 is the variable raised to the smallest exponent.

We see that 2 is the coefficient of the GCF and x^2 is the variable factor of the GCF. Thus, the GCF of $6x^3$ and $10x^2$ is $2x^2$.

> 3 is the greatest integer that divides 15, −9, and 27.

b. $15y^5$, $-9y^4$,   and $27y^3$

> y^3 is the variable raised to the smallest exponent.

We see that 3 is the coefficient of the GCF and y^3 is the variable factor of the GCF. Thus, the GCF of $15y^5$, $-9y^4$, and $27y^3$ is $3y^3$.

x^3 is the variable, x, raised to the smallest exponent.

c. $x^5y^3,\qquad x^4y^4,\qquad$ and $\qquad x^3y^2$

y^2 is the variable, y, raised to the smallest exponent.

Because all terms have coefficients of 1, 1 is the greatest integer that divides these coefficients. Thus, 1 is the coefficient of the GCF. The voice balloons show that x^3 and y^2 are the variable factors of the GCF. Thus, the GCF of x^5y^3, x^4y^4, and x^3y^2 is x^3y^2. ■

 CHECK POINT 1 Find the greatest common factor of each list of terms:

a. $18x^3$ and $15x^2$ **b.** $-20x^2$, $12x^4$, and $40x^3$

c. x^4y, x^3y^2, and x^2y.

3   Factor out the greatest common factor of a polynomial.

When we factor a monomial from a polynomial, we determine the greatest common factor of all terms in the polynomial. Sometimes there may not be a GCF other than 1. When a GCF other than 1 exists, we use the following procedure:

FACTORING A MONOMIAL FROM A POLYNOMIAL

1. Determine the greatest common factor of all terms in the polynomial.
2. Express each term as the product of the GCF and its other factor.
3. Use the distributive property to factor out the GCF.

EXAMPLE 2   Factoring Out the Greatest Common Factor

Factor: $5x^2 + 30$.

SOLUTION   The GCF of $5x^2$ and 30 is 5.

$$5x^2 + 30$$
$$= 5 \cdot x^2 + 5 \cdot 6 \qquad \text{Express each term as the product of the GCF and its other factor.}$$
$$= 5(x^2 + 6) \qquad \text{Factor out the GCF.}$$

Because factoring reverses the process of multiplication, all factoring results can be checked by multiplying.

$$5(x^2 + 6) = 5 \cdot x^2 + 5 \cdot 6 = 5x^2 + 30$$

The factoring is correct because multiplication gives us the original polynomial. ■

 CHECK POINT 2 Factor: $6x^2 + 18$.

EXAMPLE 3 Factoring Out the Greatest Common Factor

Factor: $18x^3 + 27x^2$.

SOLUTION We begin by determining the greatest common factor.

9 is the greatest integer that divides 18 and 27.

$$18x^3 \quad \text{and} \quad 27x^2$$

x^2 is the variable raised to the smallest exponent.

The GCF of the two terms in the polynomial is $9x^2$.

$$18x^3 + 27x^2$$
$$= 9x^2(2x) + 9x^2(3) \qquad \text{Express each term as the product of the GCF and its other factor.}$$
$$= 9x^2(2x + 3) \qquad \text{Factor out the GCF.}$$

We can check this factorization by multiplying $9x^2$ and $2x + 3$, obtaining the original polynomial as the answer. ■

DISCOVER FOR YOURSELF

What happens if you factor out $3x^2$ rather than $9x^2$ from $18x^3 + 27x^2$? Although $3x^2$ is a common factor of the two terms, it is not the *greatest* common factor. Remove $3x^2$ from $18x^3 + 27x^2$ and describe what happens with the second factor. Now factor again. Make the final result look like the factorization in Example 3. What is the advantage of factoring out the greatest common factor rather than just a common factor?

✔ **CHECK POINT 3** Factor: $25x^2 + 35x^3$.

EXAMPLE 4 Factoring Out the Greatest Common Factor

Factor: $16x^5 - 12x^4 + 4x^3$.

SOLUTION First, determine the greatest common factor.

4 is the greatest integer that divides 16, −12, and 4.

$$16x^5, \quad -12x^4, \quad \text{and} \quad 4x^3$$

x^3 is the variable raised to the smallest exponent.

The GCF of the three terms of the polynomial is $4x^3$.

$$16x^5 - 12x^4 + 4x^3$$
$$= 4x^3 \cdot 4x^2 - 4x^3 \cdot 3x + 4x^3 \cdot 1 \qquad \text{Express each term as the product of the GCF and its other factor.}$$
$$= 4x^3(4x^2 - 3x + 1) \qquad \text{Factor out the GCF.}$$

Don't leave out the 1.

✔ **CHECK POINT 4** Factor: $15x^5 + 12x^4 - 27x^3$.

EXAMPLE 5 Factoring Out the Greatest Common Factor

Factor: $27x^2y^3 - 9xy^2 + 81xy$.

SOLUTION First, determine the greatest common factor.

9 is the greatest integer that divides 27, −9, and 81.

$$27x^2y^3, \quad -9xy^2, \quad \text{and} \quad 81xy$$

The variables raised to the smallest exponents are x and y.

The GCF of the three terms of the polynomial is $9xy$.

$$27x^2y^3 - 9xy^2 + 81xy$$

$$= 9xy \cdot 3xy^2 - 9xy \cdot y + 9xy \cdot 9 \qquad \text{Express each term as the product of the GCF and its other factor.}$$

$$= 9xy(3xy^2 - y + 9) \qquad \text{Factor out the GCF.} \quad \blacksquare$$

 CHECK POINT 5 Factor: $8x^3y^2 - 14x^2y + 2xy$.

---

4 Factor by grouping.

Factoring by Grouping Up to now, we have factored a monomial from a polynomial. By contrast, in our next example, the greatest common factor of the polynomial is a binomial.

EXAMPLE 6 Factoring Out the Greatest Common Binomial Factor

Factor:

a. $x^2(x + 3) + 5(x + 3)$ **b.** $x(y + 1) - 2(y + 1)$.

SOLUTION Let's identify the common binomial factor in each part of the problem.

$$x^2(x + 3) \quad \text{and} \quad 5(x + 3) \qquad\qquad x(y + 1) \quad \text{and} \quad -2(y + 1)$$

The GCF, a binomial, is $x + 3$. The GCF, a binomial, is $y + 1$.

We factor out these common binomial factors as follows.

a. $x^2(x + 3) + 5(x + 3)$

$$= (x + 3)x^2 + (x + 3)5 \qquad \text{Express each term as the product of the GCF and its other factor, in that order. Hereafter, we omit this step.}$$

$$= (x + 3)(x^2 + 5) \qquad \text{Factor out the GCF, } x + 3.$$

b. $x(y + 1) - 2(y + 1) \qquad \text{The GCF is } y + 1.$

$$= (y + 1)(x - 2) \qquad \text{Factor out the GCF.} \quad \blacksquare$$

 CHECK POINT 6 Factor:

a. $x^2(x + 1) + 7(x + 1)$ **b.** $x(y + 4) - 7(y + 4)$.

Some polynomials have only a greatest common factor of 1. However, by a suitable grouping of the terms, it still may be possible to factor. This process, called **factoring by grouping**, is illustrated in Example 7.

EXAMPLE 7 Factoring by Grouping

Factor: $x^3 + 4x^2 + 3x + 12$.

SOLUTION There is no factor other than 1 common to all terms. However, we can group terms that have a common factor:

$$\boxed{x^3 + 4x^2} \; + \; \boxed{3x + 12}.$$

Common factor is x^2. Common factor is 3.

We now factor the given polynomial as follows:

$$x^3 + 4x^2 + 3x + 12$$
$$= (x^3 + 4x^2) + (3x + 12) \qquad \text{Group terms with common factors.}$$
$$= x^2(x + 4) + 3(x + 4) \qquad \text{Factor out the greatest common factor}$$

Factor out the greatest common factor from the grouped terms. The remaining two terms have $x + 4$ as a common binomial factor.

$$= (x + 4)(x^2 + 3). \qquad \text{Factor out the GCF, } x + 4.$$

Thus, $x^3 + 4x^2 + 3x + 12 = (x + 4)(x^2 + 3)$. Check the factorization by multiplying the right side of the equation using the FOIL method. Because the factorization is correct, you should obtain the original polynomial. ∎

 CHECK POINT 7 Factor: $x^3 + 5x^2 + 2x + 10$.

FACTORING BY GROUPING

1. Group terms that have a common monomial factor. There will usually be two groups. Sometimes the terms must be rearranged.
2. Factor out the common monomial factor from each group.
3. Factor out the remaining common binomial factor (if one exists).

EXAMPLE 8 Factoring by Grouping

Factor: $xy + 5x - 4y - 20$.

SOLUTION There is no factor other than 1 common to all terms. However, we can group terms that have a common factor:

$$\boxed{xy + 5x} \; + \; \boxed{-4y - 20}.$$

Common factor is x: $xy + 5x = x(y + 5)$.

Use -4, rather than 4, as the common factor: $-4y - 20 = -4(y + 5)$. In this way, the common binomial factor, $y + 5$, appears.

The voice balloons illustrate that it is sometimes necessary to factor out a negative number from a grouping to obtain a common binomial factor for the two groupings. We now factor the given polynomial as follows:

$$xy + 5x - 4y - 20$$
$$= x(y + 5) - 4(y + 5) \qquad \text{Factor } x \text{ and } -4, \text{ respectively, from each grouping.}$$
$$= (y + 5)(x - 4). \qquad \text{Factor out the GCF, } y + 5.$$

Thus, $xy + 5x - 4y - 20 = (y + 5)(x - 4)$. Using the commutative property of multiplication, the factorization can also be expressed as $(x - 4)(y + 5)$. Multiply these factors using the FOIL method to verify that, regardless of the order, these are the correct factors. ■

 CHECK POINT 8 Factor: $xy + 3x - 5y - 15$.

7.1 EXERCISE SET

Student Solutions Manual CD/Video PH Math/Tutor Center MathXL Tutorials on CD MathXL® MyMathLab Interactmath.com

Practice Exercises

In Exercises 1–6, find three factorizations for each monomial.

1. $8x^3$

2. $20x^4$

3. $-12x^5$

4. $-15x^6$

5. $36x^4$

6. $27x^5$

In Exercises 7–18, find the greatest common factor of each list of terms.

7. 4 and $8x$

8. 5 and $15x$

9. $12x^2$ and $8x$

10. $20x^2$ and $15x$

11. $-2x^4$ and $6x^3$

12. $-3x^4$ and $6x^3$

13. $9y^5, 18y^2,$ and $-3y$

14. $10y^5, 20y^2,$ and $-5y$

15. $xy, xy^2,$ and xy^3

16. $x^2y, 3x^3y,$ and $6x^2$

17. $16x^5y^4, 8x^6y^3,$ and $20x^4y^5$

18. $18x^5y^4, 6x^6y^3,$ and $12x^4y^5$

In Exercises 19–54, factor each polynomial using the greatest common factor. If there is no common factor other than 1 and the polynomial cannot be factored, so state.

19. $8x + 8$

20. $9x + 9$

21. $4y - 4$

22. $5y - 5$

23. $5x + 30$

24. $10x + 30$

25. $30x - 12$

26. $32x - 24$

27. $x^2 + 5x$

28. $x^2 + 6x$

29. $18y^2 + 12$

30. $20y^2 + 15$

31. $14x^3 + 21x^2$

32. $6x^3 + 15x^2$

33. $13y^2 - 25y$

34. $11y^2 - 30y$

35. $9y^4 + 27y^6$

36. $10y^4 + 15y^6$

37. $8x^2 - 4x^4$

38. $12x^2 - 4x^4$

39. $12y^2 + 16y - 8$

40. $15y^2 - 3y + 9$

41. $9x^4 + 18x^3 + 6x^2$

42. $32x^4 + 2x^3 + 8x^2$

43. $100y^5 - 50y^3 + 100y^2$

44. $26y^5 - 13y^3 + 39y^2$

45. $10x - 20x^2 + 5x^3$

46. $6x - 4x^2 + 2x^3$

47. $11x^2 - 23$

48. $12x^2 - 25$

49. $6x^3y^2 + 9xy$

50. $4x^2y^3 + 6xy$

51. $30x^2y^3 - 10xy^2 + 20xy$

52. $27x^2y^3 - 18xy^2 + 45x^2y$

53. $32x^3y^2 - 24x^3y - 16x^2y$

54. $18x^3y^2 - 12x^3y - 24x^2y$

In Exercises 55–66, factor each polynomial using the greatest common binomial factor.

55. $x(x + 5) + 3(x + 5)$

56. $x(x + 7) + 10(x + 7)$

57. $x(x + 2) - 4(x + 2)$

58. $x(x + 3) - 8(x + 3)$

59. $x(y + 6) - 7(y + 6)$

60. $x(y + 9) - 11(y + 9)$

61. $3x(x + y) - (x + y)$

62. $7x(x + y) - (x + y)$

63. $4x(3x + 1) + 3x + 1$

64. $5x(2x + 1) + 2x + 1$

65. $7x^2(5x + 4) + 5x + 4$

66. $9x^2(7x + 2) + 7x + 2$

In Exercises 67–84, factor by grouping.

67. $x^2 + 2x + 4x + 8$

68. $x^2 + 3x + 5x + 15$

69. $x^2 + 3x - 5x - 15$

70. $x^2 + 7x - 4x - 28$

71. $x^3 - 2x^2 + 5x - 10$

72. $x^3 - 3x^2 + 4x - 12$

73. $x^3 - x^2 + 2x - 2$

74. $x^3 + 6x^2 - 2x - 12$

75. $xy + 5x + 9y + 45$

76. $xy + 6x + 2y + 12$

77. $xy - x + 5y - 5$

78. $xy - x + 7y - 7$

79. $3x^2 - 6xy + 5xy - 10y^2$

80. $10x^2 - 12xy + 35xy - 42y^2$

81. $3x^3 - 2x^2 - 6x + 4$

82. $4x^3 - x^2 - 12x + 3$

83. $x^2 - ax - bx + ab$

84. $x^2 + ax + bx + ab$

Practice Plus

In Exercises 85–92, factor each polynomial.

85. $24x^3y^3z^3 + 30x^2y^2z + 18x^2yz^2$

86. $16x^2y^2z^2 + 32x^2yz^2 + 24x^2yz$

87. $x^3 - 4 + 3x^3y - 12y$

88. $x^3 - 5 + 2x^3y - 10y$

89. $4x^5(x + 1) - 6x^3(x + 1) - 8x^2(x + 1)$

90. $8x^5(x + 2) - 10x^3(x + 2) - 2x^2(x + 2)$

91. $3x^5 - 3x^4 + x^3 - x^2 + 5x - 5$

92. $7x^5 - 7x^4 + x^3 - x^2 + 3x - 3$

The figures for Exercises 93–94 show one or more circles drawn inside a square. Write a polynomial that represents the shaded area in each figure. Then factor the polynomial.

93.

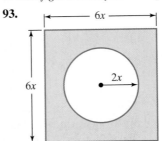

94.

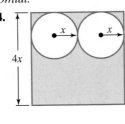

Application Exercises

95. An explosion causes debris to rise vertically with an initial velocity of 64 feet per second. The polynomial $64x - 16x^2$ describes the height of the debris above the ground, in feet, after x seconds.

   a. Find the height of the debris after 3 seconds.

   b. Factor the polynomial.

   c. Use the factored form of the polynomial in part (b) to find the height of the debris after 3 seconds. Do you get the same answer as you did in part (a)? If so, does this prove that your factorization is correct? Explain.

96. An explosion causes debris to rise vertically with an initial velocity of 72 feet per second. The polynomial $72x - 16x^2$ describes the height of the debris above the ground, in feet, after x seconds.

   a. Find the height of the debris after 4 seconds.

   b. Factor the polynomial.

   c. Use the factored form of the polynomial in part (b) to find the height of the debris after 4 seconds. Do you get the same answer as you did in part (a)? If so, does this prove that your factorization is correct? Explain.

In Exercises 97–98, write a polynomial for the length of each rectangle.

97.

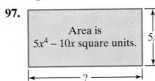

98.

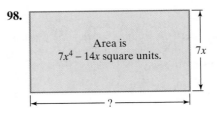

Writing in Mathematics

99. What is factoring?

100. What is a prime polynomial?

101. Explain how to find the greatest common factor of a list of terms. Give an example with your explanation.

102. Use an example and explain how to factor out the greatest common factor of a polynomial.

103. Suppose that a polynomial contains four terms and can be factored by grouping. Explain how to obtain the factorization.

104. Write a sentence that uses the word "factor" as a noun. Then write a sentence that uses the word "factor" as a verb.

Critical Thinking Exercises

105. Which one of the following is true?

  a. Because a monomial contains one term, it follows that a monomial can be factored in precisely one way.

  b. The GCF for $8x^3 - 16x^2$ is $8x$.

  c. The integers 10 and 31 have no GCF.

  d. $-4x^2 + 12x$ can be factored as $-4x(x - 3)$ or $4x(-x + 3)$.

106. Suppose you receive x dollars in January. Each month thereafter, you receive $100 more than you received the month before. Write a factored polynomial that describes the total dollar amount you receive from January through April.

In Exercises 107–108, write a polynomial that fits the given description. Do not use a polynomial that appears in this section or in the exercise set.

107. The polynomial has four terms and can be factored using a greatest common factor that has both a coefficient and a variable.

108. The polynomial has four terms and can be factored by grouping.

Technology Exercises

In Exercises 109–111, use a graphing utility to graph each side of the equation in the same viewing rectangle. Do the graphs coincide? If so, this means that the polynomial on the left side has been factored correctly. If not, factor the polynomial correctly and then use your graphing utility to verify the factorization.

109. $-3x - 6 = -3(x - 2)$

110. $x^2 - 2x + 5x - 10 = (x - 2)(x - 5)$

111. $x^2 + 2x + x + 2 = x(x + 2) + 1$

Review Exercises

112. Multiply: $(x + 7)(x + 10)$. (Section 6.3, Example 1)

113. Solve the system by graphing:
$$2x - y = -4$$
$$x - 3y = 3.$$
(Section 5.1, Example 2)

114. Write the point-slope form of a line passing through $(-7, 2)$ and $(-4, 5)$. Then use the point-slope equation to write the slope-intercept equation. (Section 4.5, Example 2)

SECTION 7.2

FACTORING TRINOMIALS WHOSE LEADING COEFFICIENT IS ONE

Objective

1 Factor trinomials of the form $x^2 + bx + c$.

Not afraid of heights and cutting-edge excitement? How about sky diving? Behind your exhilarating experience is the world of algebra. After you jump from the airplane, your height above the ground at every instant of your fall can be described by a formula involving a variable that is squared. At a height of approximately 2000 feet, you'll need to open your parachute. How can you determine when you must do so?

The answer to this critical question involves using the factoring technique presented in this section. In Section 7.6, in which applications are discussed, this technique is applied to models involving the height of any free-falling object—in this case, you.

1 Factor trinomials of the form $x^2 + bx + c$.

A Strategy for Factoring $x^2 + bx + c$ In Section 6.3, we used the FOIL method to multiply two binomials. The product was often a trinomial. The following are some examples:

Factored Form	**F**	**O**	**I**	**L**	**Trinomial Form**

$$(x + 3)(x + 4) = x^2 + 4x + 3x + 12 = x^2 + 7x + 12$$
$$(x - 3)(x - 4) = x^2 - 4x - 3x + 12 = x^2 - 7x + 12$$
$$(x + 3)(x - 5) = x^2 - 5x + 3x - 15 = x^2 - 2x - 15.$$

Observe that each trinomial is of the form $x^2 + bx + c$, where the coefficient of the squared term is 1. Our goal in this section is to start with the trinomial form and, assuming that it is factorable, return to the factored form.

The first FOIL multiplication shown above indicates that $(x + 3)(x + 4) = x^2 + 7x + 12$. Let's reverse the sides of this equation:

$$x^2 + 7x + 12 = (x + 3)(x + 4).$$

We can make several important observations about the factors on the right side.

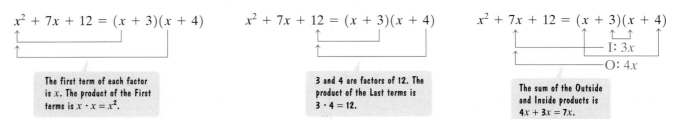

The first term of each factor is x. The product of the First terms is $x \cdot x = x^2$.

3 and 4 are factors of 12. The product of the Last terms is $3 \cdot 4 = 12$.

The sum of the Outside and Inside products is $4x + 3x = 7x$.

These observations provide us with a procedure for factoring $x^2 + bx + c$.

A STRATEGY FOR FACTORING $x^2 + bx + c$

1. Enter x as the first term of each factor.

$$(x \quad)(x \quad) = x^2 + bx + c$$

2. List pairs of factors of the constant c.

3. Try various combinations of these factors as the second term in each set of parentheses. Select the combination in which the sum of the Outside and Inside products is equal to bx.

$$(x + \square)(x + \square) = x^2 + bx + c$$

I
O
Sum of O + I

4. Check your work by multiplying the factors using the FOIL method. You should obtain the original trinomial.

If none of the possible combinations yield an Outside product and an Inside product whose sum is equal to bx, the trinomial cannot be factored using integers and is called **prime** over the set of integers.

EXAMPLE 1 Factoring a Trinomial in $x^2 + bx + c$ Form

Factor: $x^2 + 6x + 8$.

SOLUTION

Step 1. **Enter x as the first term of each factor.**

$$x^2 + 6x + 8 = (x \quad)(x \quad)$$

To find the second term of each factor, we must find two integers whose product is 8 and whose sum is 6.

Step 2. **List pairs of factors of the constant, 8.**

Factors of 8	8, 1	4, 2	−8, −1	−4, −2

Step 3. **Try various combinations of these factors.** The correct factorization of $x^2 + 6x + 8$ is the one in which the sum of the Outside and Inside products is equal to $6x$. Here is a list of the possible factorizations:

Possible Factorizations of $x^2 + 6x + 8$	Sum of Outside and Inside Products (Should Equal $6x$)
$(x + 8)(x + 1)$	$x + 8x = 9x$
$(x + 4)(x + 2)$	$2x + 4x = 6x$ ← This is the required middle term.
$(x - 8)(x - 1)$	$-x - 8x = -9x$
$(x - 4)(x - 2)$	$-2x - 4x = -6x$

Thus, $x^2 + 6x + 8 = (x + 4)(x + 2)$.

Step 4. **Check this result by multiplying the right side using the FOIL method.** You should obtain the original trinomial. Because of the commutative property, we can also say that

$$x^2 + 6x + 8 = (x + 2)(x + 4).$$

USING TECHNOLOGY

If a polynomial contains one variable, a graphing utility can be used to check its factorization. For example, the factorization in Example 1

$$x^2 + 6x + 8 = (x + 4)(x + 2)$$

can be checked graphically or numerically.

Graphic Check

Use the GRAPH feature. Graph $y_1 = x^2 + 6x + 8$ and $y_2 = (x + 4)(x + 2)$ on the same screen. Because the graphs are identical, the factorization appears to be correct.

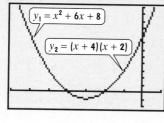

$y_1 = x^2 + 6x + 8$

$y_2 = (x + 4)(x + 2)$

$[-7, 1, 1]$ by $[-2, 12, 1]$

Numeric Check

Use the TABLE feature. Enter $y_1 = x^2 + 6x + 8$ and $y_2 = (x + 4)(x + 2)$ and press TABLE. Two columns of values are shown, one for y_1 and one for y_2. Because the corresponding values are equal regardless of how far up or down we scroll, the factorization is correct.

X	Y1	Y2
-3	-1	-1
-2	0	0
-1	3	3
0	8	8
1	15	15
2	24	24
3	35	35

X=-3

✔ CHECK POINT 1 Factor: $x^2 + 5x + 6$.

EXAMPLE 2 Factoring a Trinomial in $x^2 + bx + c$ Form

Factor: $x^2 - 5x + 6$.

SOLUTION

Step 1. **Enter x as the first term of each factor.**
$$x^2 - 5x + 6 = (x \quad)(x \quad)$$

To find the second term of each factor, we must find two integers whose product is 6 and whose sum is -5.

Step 2. **List pairs of factors of the constant, 6.**

Factors of 6	6, 1	3, 2	−6, −1	−3, −2

Step 3. **Try various combinations of these factors.** The correct factorization of $x^2 - 5x + 6$ is the one in which the sum of the Outside and Inside products is equal to $-5x$. Here is a list of the possible factorizations:

Possible Factorizations of $x^2 - 5x + 6$	Sum of Outside and Inside Products (Should Equal $-5x$)
$(x + 6)(x + 1)$	$x + 6x = 7x$
$(x + 3)(x + 2)$	$2x + 3x = 5x$
$(x - 6)(x - 1)$	$-x - 6x = -7x$
$(x - 3)(x - 2)$	$-2x - 3x = -5x$

This is the required middle term.

Thus, $x^2 - 5x + 6 = (x - 3)(x - 2)$. Verify this result using the FOIL method. ∎

In factoring a trinomial of the form $x^2 + bx + c$, you can speed things up by listing the factors of c and then finding their sums. We are interested in a sum of b. For example, in factoring $x^2 - 5x + 6$, we are interested in the factors of 6 whose sum is -5.

Factors of 6	6, 1	3, 2	−6, −1	−3, −2
Sum of Factors	7	5	−7	−5

This is the desired sum.

Thus, $x^2 - 5x + 6 = (x - 3)(x - 2)$.

✔ CHECK POINT 2 Factor: $x^2 - 6x + 8$.

EXAMPLE 3 Factoring a Trinomial in $x^2 + bx + c$ Form

Factor: $x^2 + 2x - 35$.

SOLUTION

Step 1. **Enter x as the first term of each factor.**
$$x^2 + 2x - 35 = (x \quad)(x \quad)$$

To find the second term of each factor, we must find two integers whose product is -35 and whose sum is 2.

STUDY TIP

To factor $x^2 + bx + c$ when c is positive, find two numbers with the same sign as the middle term.

$x^2 + 6x + 8 = (x + 2)(x + 4)$

Same signs

$x^2 - 5x + 6 = (x - 3)(x - 2)$

Same signs

Step 2. **List pairs of factors of the constant, −35.**

Factors of −35	−35, 1	−7, 5	35, −1	7, −5

Step 3. **Try various combinations of these factors.** We are looking for the factors whose sum is 2.

Factors of −35	−35, 1	−7, 5	35, −1	7, −5
Sum of Factors	−34	−2	34	2

This is the desired sum.

Thus, $x^2 + 2x - 35 = (x + 7)(x - 5)$.

Step 4. **Verify the factorization using the FOIL method.**

$$(x + 7)(x - 5) = x^2 - 5x + 7x - 35 = x^2 + 2x - 35$$

Because the product of the factors is the original polynomial, the factorization is correct. ∎

✔ CHECK POINT **3** Factor: $x^2 + 3x - 10$.

EXAMPLE 4 Factoring a Trinomial Whose Leading Coefficient Is One

Factor: $y^2 - 2y - 99$.

SOLUTION

Step 1. **Enter y as the first term of each factor.**

$$y^2 - 2y - 99 = (y \quad)(y \quad)$$

To find the second term of each factor, we must find two integers whose product is −99 and whose sum is −2.

Step 2. **List pairs of factors of the constant, − 99.**

Factors of − 99	−99, 1	−11, 9	−33, 3	99, −1	11, −9	33, −3

Step 3. **Try various combinations of these factors.** We are interested in factors whose sum is −2.

Factors of −99	−99, 1	−11, 9	−33, 3	99, −1	11, −9	33, −3
Sum of Factors	−98	−2	−30	98	2	30

This is the desired sum.

Thus, $y^2 - 2y - 99 = (y - 11)(y + 9)$. Verify this result using the FOIL method. ∎

✔ CHECK POINT **4** Factor: $y^2 - 6y - 27$.

STUDY TIP

To factor $x^2 + bx + c$ when c is negative, find two numbers with opposite signs whose sum is the coefficient of the middle term.

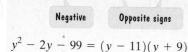

$x^2 + 2x - 35 = (x + 7)(x - 5)$

Negative   Opposite signs

$y^2 - 2y - 99 = (y - 11)(y + 9)$

Negative   Opposite signs

EXAMPLE 5 Trying to Factor a Trinomial in $x^2 + bx + c$ Form

Factor: $x^2 + x - 5$.

SOLUTION

Step 1. **Enter x as the first term of each factor.**

$$x^2 + x - 5 = (x \qquad)(x \qquad)$$

To find the second term of each factor, we must find two integers whose product is -5 and whose sum is 1.

Steps 2 and 3. **List pairs of factors of the constant, -5, and try various combinations of these factors.** We are interested in factors whose sum is 1.

Factors of -5	$-5, 1$	$5, -1$
Sum of Factors	-4	4

No pair gives the desired sum, 1.

Because neither pair has a sum of 1, $x^2 + x - 5$ cannot be factored using integers. This trinomial is prime.

✔ **CHECK POINT 5** Factor: $x^2 + x - 7$.

EXAMPLE 6 Factoring a Trinomial in Two Variables

Factor: $x^2 - 5xy + 6y^2$.

SOLUTION

Step 1. **Enter x as the first term of each factor.** Because the last term of the trinomial contains y^2, the second term of each factor must contain y.

$$x^2 - 5xy + 6y^2 = (x \ ?y)(x \ ?y)$$

The question marks indicate that we are looking for the coefficients of y in each factor. To find these coefficients, we must find two integers whose product is 6 and whose sum is -5.

Steps 2 and 3. **List pairs of factors of the coefficient of the last term, 6, and try various combinations of these factors.** We are interested in factors whose sum is -5.

Factors of 6	6, 1	3, 2	$-6, -1$	$-3, -2$
Sum of Factors	7	5	-7	-5

This is the desired sum.

Thus, $x^2 - 5xy + 6y^2 = (x - 3y)(x - 2y)$.

Step 4. **Verify the factorization using the FOIL method.**

$$(x - 3y)(x - 2y) = x^2 - 2xy - 3xy + 6y^2 = x^2 - 5xy + 6y^2$$

Because the product of the factors is the original polynomial, the factorization is correct. ■

✔ **CHECK POINT 6** Factor: $x^2 - 4xy + 3y^2$.

Some polynomials can be factored using more than one technique. **Always begin by looking for a greatest common factor** and, if there is one, factor it out. A polynomial is **factored completely** when it is written as the product of prime polynomials.

EXAMPLE 7 Factoring Completely

Factor: $3x^3 - 15x^2 - 42x$.

SOLUTION The GCF of the three terms of the polynomial is $3x$. We begin by factoring out $3x$. Then we factor the remaining trinomial by the methods of this section.

$$3x^3 - 15x^2 - 42x$$
$$= 3x(x^2 - 5x - 14) \qquad \text{Factor out the GCF.}$$
$$= 3x(x \qquad)(x \qquad) \qquad \text{Begin factoring } x^2 - 5x - 14. \text{ Find two integers whose product is } -14 \text{ and whose sum is } -5.$$
$$= 3x(x - 7)(x + 2) \qquad \text{The integers are } -7 \text{ and } 2.$$

Thus,

$$3x^3 - 15x^2 - 42x = 3x(x - 7)(x + 2).$$

> Be sure to include the GCF in the factorization.

How can we check this factorization? We will multiply the binomials using the FOIL method. Then use the distributive property and multiply each term of this product by $3x$. If the factorization is correct, we should obtain the original polynomial.

$$3x(x - 7)(x + 2) = 3x(x^2 + 2x - 7x - 14) = 3x(x^2 - 5x - 14) = 3x^3 - 15x^2 - 42x$$

> Use the FOIL method on $(x - 7)(x + 2)$.

> This is the original polynomial.

The factorization is correct.

✔ **CHECK POINT 7** Factor: $2x^3 + 6x^2 - 56x$.

7.2 EXERCISE SET

Student Solutions Manual   CD/Video   PH Math/Tutor Center   MathXL Tutorials on CD   MathXL®   MyMathLab   Interactmath.com

Practice Exercises

In Exercises 1–42, factor each trinomial, or state that the trinomial is prime. Check each factorization using FOIL multiplication.

1. $x^2 + 7x + 6$

2. $x^2 + 9x + 1$

3. $x^2 + 7x + 10$

4. $x^2 + 9x + 14$

5. $x^2 + 11x + 10$

6. $x^2 + 13x + 12$

7. $x^2 - 7x + 12$

8. $x^2 - 13x + 40$

9. $x^2 - 12x + 36$

10. $x^2 - 8x + 16$

11. $y^2 - 8y + 15$

12. $y^2 - 8y + 7$

13. $x^2 + 3x - 10$

14. $x^2 + 3x - 28$

15. $y^2 + 10y - 39$

16. $y^2 + 5y - 24$

17. $x^2 - 2x - 15$

18. $x^2 - 4x - 5$

19. $x^2 - 2x - 8$

20. $x^2 - 5x - 6$

21. $x^2 + 4x + 12$ **22.** $x^2 + 4x + 5$

23. $y^2 - 16y + 48$ **24.** $y^2 - 10y + 21$

25. $x^2 - 3x + 6$ **26.** $x^2 + 4x - 10$

27. $w^2 - 30w - 64$ **28.** $w^2 + 12w - 64$

29. $y^2 - 18y + 65$ **30.** $y^2 - 22y + 72$

31. $r^2 + 12r + 27$ **32.** $r^2 - 15r - 16$

33. $y^2 - 7y + 5$ **34.** $y^2 - 15y + 5$

35. $x^2 + 7xy + 6y^2$ **36.** $x^2 + 6xy + 8y^2$

37. $x^2 - 8xy + 15y^2$ **38.** $x^2 - 9xy + 14y^2$

39. $x^2 - 3xy - 18y^2$ **40.** $x^2 - xy - 30y^2$

41. $a^2 - 18ab + 45b^2$ **42.** $a^2 - 18ab + 80b^2$

In Exercises 43–66, factor completely.

43. $3x^2 + 15x + 18$ **44.** $3x^2 + 21x + 36$

45. $4y^2 - 4y - 8$ **46.** $3y^2 + 3y - 18$

47. $10x^2 - 40x - 600$ **48.** $2x^2 + 10x - 48$

49. $3x^2 - 33x + 54$ **50.** $2x^2 - 14x + 24$

51. $2r^3 + 6r^2 + 4r$ **52.** $2r^3 + 8r^2 + 6r$

53. $4x^3 + 12x^2 - 72x$ **54.** $3x^3 - 15x^2 + 18x$

55. $2r^3 + 8r^2 - 64r$ **56.** $3r^3 - 9r^2 - 54r$

57. $y^4 + 2y^3 - 80y^2$ **58.** $y^4 - 12y^3 + 35y^2$

59. $x^4 - 3x^3 - 10x^2$ **60.** $x^4 - 22x^3 + 120x^2$

61. $2w^4 - 26w^3 - 96w^2$ **62.** $3w^4 + 54w^3 + 135w^2$

63. $15xy^2 + 45xy - 60x$ **64.** $20x^2y - 100xy + 120y$

65. $x^5 + 3x^4y - 4x^3y^2$ **66.** $x^3y - 2x^2y^2 - 3xy^3$

Practice Plus

In Exercises 67–76, factor completely.

67. $2x^2y^2 - 32x^2yz + 30x^2z^2$

68. $2x^2y^2 - 30x^2yz + 28x^2z^2$

69. $(a + b)x^2 + (a + b)x - 20(a + b)$

70. $(a + b)x^2 - 13(a + b)x + 36(a + b)$

71. $x^2 + 0.5x + 0.06$

72. $x^2 - 0.5x - 0.06$

73. $x^2 - \dfrac{2}{5}x + \dfrac{1}{25}$

74. $x^2 + \dfrac{2}{3}x + \dfrac{1}{9}$

75. $-x^2 - 3x + 40$

76. $-x^2 - 4x + 45$

Application Exercises

77. You dive directly upward from a board that is 32 feet high. After t seconds, your height above the water is described by the polynomial

$$-16t^2 + 16t + 32.$$

   a. Factor the polynomial completely. Begin by factoring -16 from each term.

   b. Evaluate both the original polynomial and its factored form for $t = 2$. Do you get the same answer for each evaluation? Describe what this answer means.

78. You dive directly upward from a board that is 48 feet high. After t seconds, your height above the water is described by the polynomial

$$-16t^2 + 32t + 48.$$

   a. Factor the polynomial completely. Begin by factoring -16 from each term.

   b. Evaluate both the original polynomial and its factored form for $t = 3$. Do you get the same answer for each evaluation? Describe what this answer means.

Writing in Mathematics

79. Explain how to factor $x^2 + 8x + 15$.

80. Give two helpful suggestions for factoring $x^2 - 5x + 6$.

81. In factoring $x^2 + bx + c$, describe how the last terms in each factor are related to b and c.

82. Without actually factoring and without multiplying the given factors, explain why the following factorization is not correct:

$$x^2 + 46x + 513 = (x - 27)(x - 19).$$

Critical Thinking Exercises

83. Which one of the following is true?

   a. A factor of $x^2 + x + 20$ is $x + 5$.

   b. A trinomial can never have two identical factors.

   c. A factor of $y^2 + 5y - 24$ is $y - 3$.

   d. $x^2 + 4 = (x + 2)(x + 2)$

In Exercises 84–85, find all positive integers b so that the trinomial can be factored.

84. $x^2 + bx + 15$

85. $x^2 + 4x + b$

86. Factor: $x^{2n} + 20x^n + 99$.

87. Factor $x^3 + 3x^2 + 2x$. If x represents an integer, use the factorization to describe what the trinomial represents.

88. A box with no top is to be made from an 8-inch by 6-inch piece of metal by cutting identical squares from each corner and turning up the sides (see the figure). The volume of the box is modeled by the polynomial $4x^3 - 28x^2 + 48x$. Factor the polynomial completely. Then use the dimensions given on the box and show that its volume is equivalent to the factorization that you obtain.

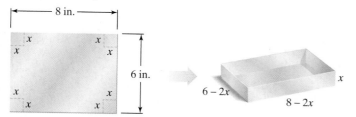

Technology Exercises

In Exercises 89–92, use the GRAPH *or* TABLE *feature of a graphing utility to determine if the polynomial on the left side of each equation has been correctly factored. If the graphs of y_1 and y_2 coincide, or if their corresponding table values are equal, this means that the polynomial on the left side has been correctly factored. If not, factor the trinomial correctly and then use your graphing utility to verify the factorization.*

89. $x^2 - 5x + 6 = (x - 2)(x - 3)$

90. $2x^2 + 2x - 12 = 2(x - 3)(x + 2)$

91. $x^2 - 2x + 1 = (x + 1)(x - 1)$

92. $2x^2 + 8x + 6 = (x + 3)(x + 1)$

Review Exercises

93. Multiply: $(2x + 3)(x - 2)$. (Section 6.3, Example 2)

94. Multiply: $(3x + 4)(3x + 1)$. (Section 6.3, Example 2)

95. Solve: $4(x - 2) = 3x + 5$. (Section 2.3, Example 2)

George Tooker, American, born 1920. "Farewell" 1966, egg tempera on gessoed masonite, 61×60.1 cm. (24.034×23.679 in.) P.967.76. Hood Museum of Art. Dartmouth College, Hanover, New Hampshire; gift of Pennington Haile, Class of 1924. ©George Tooker/DC Moore Gallery.

SECTION 7.3

Objectives

1 Factor trinomials by trial and error.

2 Factor trinomials by grouping.

FACTORING TRINOMIALS WHOSE LEADING COEFFICIENT IS NOT ONE

The special significance of the number 1 is reflected in our language. "One," "an," and "a" mean the same thing. The words "unit," "unity," "union," "unique," and "universal" are derived from the Latin word for "one." For the ancient Greeks, 1 was the indivisible unit from which all other numbers arose.

The Greeks' philosophy of 1 applies to our work in this section. Factoring trinomials whose leading coefficient is 1 is the basic technique from which other methods of factoring $ax^2 + bx + c$, where a is not equal to 1, follow.

1   Factor trinomials by trial and error.

Factoring by the Trial-and-Error Method   How do we factor a trinomial such as $3x^2 - 20x + 28$? Notice that the leading coefficient is 3. We must find two binomials whose product is $3x^2 - 20x + 28$. The product of the First terms must be $3x^2$:

$$(3x \quad)(x \quad).$$

From this point on, the factoring strategy is exactly the same as the one we use to factor trinomials whose leading coefficient is 1.

A STRATEGY FOR FACTORING $ax^2 + bx + c$ Assume, for the moment, that there is no greatest common factor.

1. Find two First terms whose product is ax^2:

$$(\Box x + \quad)(\Box x + \quad) = ax^2 + bx + c.$$

2. Find two Last terms whose product is c:

$$(\Box x + \Box)(\Box x + \Box) = ax^2 + bx + c.$$

3. By trial and error, perform steps 1 and 2 until the sum of the Outside product and the Inside product is bx:

$$(\Box x + \Box)(\Box x + \Box) = ax^2 + bx + c.$$

    I
    O
    Sum of O + I

If no such combinations exist, the polynomial is prime.

EXAMPLE 1   Factoring a Trinomial Whose Leading Coefficient Is Not One

Factor: $3x^2 - 20x + 28$.

SOLUTION

Step 1.   **Find two First terms whose product is $3x^2$.**

$$3x^2 - 20x + 28 = (3x \quad)(x \quad)$$

Step 2.   **Find two Last terms whose product is 28.** The number 28 has pairs of factors that are either both positive or both negative. Because the middle term, $-20x$, is negative, both factors must be negative. The negative factorizations of 28 are $-1(-28)$, $-2(-14)$, and $-4(-7)$.

Step 3.   **Try various combinations of these factors.** The correct factorization of $3x^2 - 20x + 28$ is the one in which the sum of the Outside and Inside products is equal to $-20x$. Here is a list of the possible factorizations:

Possible Factorizations of $3x^2 - 20x + 28$	Sum of Outside and Inside Products (Should Equal $-20x$)
$(3x - 1)(x - 28)$	$-84x - x = -85x$
$(3x - 28)(x - 1)$	$-3x - 28x = -31x$
$(3x - 2)(x - 14)$	$-42x - 2x = -44x$
$(3x - 14)(x - 2)$	$-6x - 14x = -20x$
$(3x - 4)(x - 7)$	$-21x - 4x = -25x$
$(3x - 7)(x - 4)$	$-12x - 7x = -19x$

This is the required middle term.

Thus,

$$3x^2 - 20x + 28 = (3x - 14)(x - 2) \quad \text{or} \quad (x - 2)(3x - 14).$$

Show that this factorization is correct by multiplying the factors using the FOIL method. You should obtain the original trinomial. ■

 CHECK POINT 1 Factor: $5x^2 - 14x + 8$.

EXAMPLE 2 Factoring a Trinomial Whose Leading Coefficient Is Not One

Factor: $8x^2 - 10x - 3$.

SOLUTION

Step 1. **Find two First terms whose product is $8x^2$.**

$$8x^2 - 10x - 3 \overset{?}{=} (8x \quad)(x \quad)$$
$$8x^2 - 10x - 3 \overset{?}{=} (4x \quad)(2x \quad)$$

Step 2. **Find two Last terms whose product is -3.** The possible factorizations are $1(-3)$ and $-1(3)$.

Step 3. **Try various combinations of these factors.** The correct factorization of $8x^2 - 10x - 3$ is the one in which the sum of the Outside and Inside products is equal to $-10x$. Here is a list of the possible factorizations:

Possible Factorizations of $8x^2 - 10x - 3$	Sum of Outside and Inside Products (Should Equal $-10x$)
$(8x + 1)(x - 3)$	$-24x + x = -23x$
$(8x - 3)(x + 1)$	$8x - 3x = 5x$
$(8x - 1)(x + 3)$	$24x - x = 23x$
$(8x + 3)(x - 1)$	$-8x + 3x = -5x$
$(4x + 1)(2x - 3)$	$-12x + 2x = -10x$
$(4x - 3)(2x + 1)$	$4x - 6x = -2x$
$(4x - 1)(2x + 3)$	$12x - 2x = 10x$
$(4x + 3)(2x - 1)$	$-4x + 6x = 2x$

This is the required middle term.

Thus,

$$8x^2 - 10x - 3 = (4x + 1)(2x - 3) \quad \text{or} \quad (2x - 3)(4x + 1).$$

Use FOIL multiplication to check either of these factorizations. ■

 CHECK POINT 2 Factor: $6x^2 + 19x - 7$.

EXAMPLE 3 Factoring a Trinomial in Two Variables

Factor: $2x^2 - 7xy + 3y^2$.

SOLUTION

Step 1. **Find two First terms whose product is $2x^2$.**

$$2x^2 - 7xy + 3y^2 = (2x \qquad)(x \qquad)$$

Step 2. **Find two Last terms whose product is $3y^2$.** The possible factorizations are $(y)(3y)$ and $(-y)(-3y)$.

Step 3. **Try various combinations of these factors.** The correct factorization of $2x^2 - 7xy + 3y^2$ is the one in which the sum of the Outside and Inside products is equal to $-7xy$. Here is a list of possible factorizations:

Possible Factorizations of $2x^2 - 7xy + 3y^2$	Sum of Outside and Inside Products (Should Equal $-7xy$)
$(2x + 3y)(x + y)$	$2xy + 3xy = 5xy$
$(2x + y)(x + 3y)$	$6xy + xy = 7xy$
$(2x - 3y)(x - y)$	$-2xy - 3xy = -5xy$
$(2x - y)(x - 3y)$	$-6xy - xy = -7xy$

This is the required middle term.

Thus,

$$2x^2 - 7xy + 3y^2 = (2x - y)(x - 3y) \quad \text{or} \quad (x - 3y)(2x - y).$$

Use FOIL multiplication to check either of these factorizations. ∎

 CHECK POINT 3 Factor: $3x^2 - 13xy + 4y^2$.

② Factor trinomials by grouping.

Factoring by the Grouping Method A second method for factoring $ax^2 + bx + c$, $a \neq 0$, is called the **grouping method**. The method involves both trial and error, as well as grouping. The trial and error in factoring $ax^2 + bx + c$ depends on finding two numbers, p and q, for which $p + q = b$. Then we factor $ax^2 + px + qx + c$ using grouping.

Let's see how this works by looking at our factorization in Example 2:

$$8x^2 - 10x - 3 = (2x - 3)(4x + 1).$$

If we multiply using FOIL on the right, we obtain:

$$(2x - 3)(4x + 1) = 8x^2 + 2x - 12x - 3.$$

In this case, the desired numbers, p and q, are $p = 2$ and $q = -12$. Compare these numbers to ac and b in the given polynomial:

$$8x^2 - 10x - 3.$$

$ac = 8(-3) = -24$

$a = 8 \quad b = -10 \quad c = -3$

Can you see that p and q, 2 and -12, are factors of ac, or -24? Furthermore, p and q have a sum of b, namely -10. By expressing the middle term, $-10x$, in terms of p and q, we can factor by grouping as follows:

$$8x^2 - 10x - 3$$
$$= 8x^2 + (2x - 12x) - 3 \qquad \text{Rewrite } -10x \text{ as } 2x - 12x.$$
$$= (8x^2 + 2x) + (-12x - 3) \qquad \text{Group terms.}$$
$$= 2x(4x + 1) - 3(4x + 1) \qquad \text{Factor from each group.}$$
$$= (4x + 1)(2x - 3) \qquad \text{Factor out the common binomial factor.}$$

As we obtained in Example 2,

$$8x^2 - 10x - 3 = (4x + 1)(2x - 3).$$

Generalizing from this example, here's how to factor a trinomial by grouping:

FACTORING $ax^2 + bx + c$ USING GROUPING ($a \neq 1$)

1. Multiply the leading coefficient, a, and the constant, c.
2. Find the factors of ac whose sum is b.
3. Rewrite the middle term, bx, as a sum or difference using the factors from step 2.
4. Factor by grouping.

EXAMPLE 4 Factoring by Grouping

Factor by grouping: $2x^2 - x - 6$.

SOLUTION   The trinomial is of the form $ax^2 + bx + c$.

$$2x^2 - x - 6$$

$\boxed{a = 2}$ $\boxed{b = -1}$ $\boxed{c = -6}$

Step 1. **Multiply the leading coefficient, a, and the constant, c.** Using $a = 2$ and $c = -6$,

$$ac = 2(-6) = -12.$$

Step 2. **Find the factors of ac whose sum is b.** We want the factors of -12 whose sum is b, or -1. The factors of -12 whose sum is -1 are -4 and 3.

Step 3. **Rewrite the middle term, $-x$, as a sum or difference using the factors from step 2, -4 and 3.**

$$2x^2 - x - 6 = 2x^2 - 4x + 3x - 6$$

Step 4. **Factor by grouping.**

$$= (2x^2 - 4x) + (3x - 6) \qquad \text{Group terms.}$$
$$= 2x(x - 2) + 3(x - 2) \qquad \text{Factor from each group.}$$
$$= (x - 2)(2x + 3) \qquad \text{Factor out the common binomial factor.}$$

Thus,

$$2x^2 - x - 6 = (x - 2)(2x + 3) \quad \text{or} \quad (2x + 3)(x - 2). \qquad \blacksquare$$

DISCOVER FOR YOURSELF

In step 2, we discovered that the desired numbers were -4 and 3, and we wrote $-x$ as $-4x + 3x$. What happens if we write $-x$ as $3x - 4x$? Use factoring by grouping on

$$2x^2 - x - 6$$
$$= 2x^2 + 3x - 4x - 6.$$

Describe what happens.

✔ **CHECK POINT 4** Factor by grouping: $3x^2 - x - 10$.

EXAMPLE 5 Factoring by Grouping

Factor by grouping: $8x^2 - 22x + 5$.

SOLUTION   The trinomial is of the form $ax^2 + bx + c$.

$$8x^2 - 22x + 5$$

$$a = 8 \qquad b = -22 \qquad c = 5$$

Step 1. **Multiply the leading coefficient, a, and the constant, c.** Using $a = 8$ and $c = 5$, $ac = 8 \cdot 5 = 40$.

Step 2. **Find the factors of ac whose sum is b.** We want the factors of 40 whose sum is b, or -22. The factors of 40 whose sum is -22 are -2 and -20.

Step 3. **Rewrite the middle term, $-22x$, as a sum or difference using the factors from step 2, -2 and -20.**

$$8x^2 - 22x + 5 = 8x^2 - 2x - 20x + 5$$

Step 4. **Factor by grouping.**

$$= (8x^2 - 2x) + (-20x + 5) \quad \text{Group terms.}$$
$$= 2x(4x - 1) - 5(4x - 1) \quad \text{Factor from each group.}$$
$$= (4x - 1)(2x - 5) \quad \text{Factor out the common binomial factor.}$$

Thus,

$$8x^2 - 22x + 5 = (4x - 1)(2x - 5) \quad \text{or} \quad (2x - 5)(4x - 1). \quad \blacksquare$$

 CHECK POINT 5 Factor by grouping: $8x^2 - 10x + 3$.

Factoring Completely   Always begin the process of factoring a polynomial by looking for a greatest common factor. If there is one, **factor out the GCF first**. After doing this, you should attempt to factor the remaining trinomial by one of the methods presented in this section.

EXAMPLE 6 Factoring Completely

Factor completely: $15y^4 + 26y^3 + 7y^2$.

SOLUTION   We will first factor out a common monomial factor from the polynomial and then factor the resulting trinomial by the methods of this section. The GCF of the three terms is y^2.

$$15y^4 + 26y^3 + 7y^2 = y^2(15y^2 + 26y + 7) \quad \text{Factor out the GCF.}$$
$$= y^2(5y + 7)(3y + 1) \quad \text{Factor } 15y^2 + 26y + 7 \text{ using trial and error or grouping.}$$

Thus,

$$15y^4 + 26y^3 + 7y^2 = y^2(5y + 7)(3y + 1) \quad \text{or} \quad y^2(3y + 1)(5y + 7).$$

Be sure to include the GCF, y^2, in the factorization. $\blacksquare$

 **CHECK POINT 6** Factor completely: $5y^4 + 13y^3 + 6y^2$.

7.3 EXERCISE SET

Student Solutions Manual   CD/Video   PH Math/Tutor Center   MathXL Tutorials on CD   MathXL®   MyMathLab   Interactmath.com

Practice Exercises

In Exercises 1–58, use the method of your choice to factor each trinomial, or state that the trinomial is prime. Check each factorization using FOIL multiplication.

1. $2x^2 + 5x + 3$

2. $3x^2 + 5x + 2$

3. $3x^2 + 13x + 4$

4. $2x^2 + 7x + 3$

5. $2x^2 + 11x + 12$

6. $2x^2 + 19x + 35$

7. $5y^2 - 16y + 3$

8. $5y^2 - 17y + 6$

9. $3y^2 + y - 4$

10. $3y^2 - y - 4$

11. $3x^2 + 13x - 10$

12. $3x^2 + 14x - 5$

13. $3x^2 - 22x + 7$

14. $3x^2 - 10x + 7$

15. $5y^2 - 16y + 3$

16. $5y^2 - 8y + 3$

17. $3x^2 - 17x + 10$

18. $3x^2 - 25x - 28$

19. $6w^2 - 11w + 4$

20. $6w^2 - 17w + 12$

21. $8x^2 + 33x + 4$

22. $7x^2 + 43x + 6$

23. $5x^2 + 33x - 14$

24. $3x^2 + 22x - 16$

25. $14y^2 + 15y - 9$

26. $6y^2 + 7y - 24$

27. $6x^2 - 7x + 3$

28. $9x^2 + 3x + 2$

29. $25z^2 - 30z + 9$

30. $9z^2 + 12z + 4$

31. $15y^2 - y - 2$

32. $15y^2 + 13y - 2$

33. $5x^2 + 2x + 9$

34. $3x^2 - 5x + 1$

35. $10y^2 + 43y - 9$

36. $16y^2 - 46y + 15$

37. $8x^2 - 2x - 1$

38. $8x^2 - 22x + 5$

39. $9y^2 - 9y + 2$

40. $9y^2 + 5y - 4$

41. $20x^2 + 27x - 8$

42. $15x^2 - 19x + 6$

43. $2x^2 + 3xy + y^2$

44. $3x^2 + 4xy + y^2$

45. $3x^2 + 5xy + 2y^2$

46. $3x^2 + 11xy + 6y^2$

47. $2x^2 - 9xy + 9y^2$

48. $3x^2 + 5xy - 2y^2$

49. $6x^2 - 5xy - 6y^2$

50. $6x^2 - 7xy - 5y^2$

51. $15x^2 + 11xy - 14y^2$

52. $15x^2 - 31xy + 10y^2$

53. $2a^2 + 7ab + 5b^2$

54. $2a^2 + 5ab + 2b^2$

55. $15a^2 - ab - 6b^2$

56. $3a^2 - ab - 14b^2$

57. $12x^2 - 25xy + 12y^2$

58. $12x^2 + 7xy - 12y^2$

In Exercises 59–86, factor completely.

59. $4x^2 + 26x + 30$

60. $4x^2 - 18x - 10$

61. $9x^2 - 6x - 24$

62. $12x^2 - 33x + 21$

63. $4y^2 + 2y - 30$

64. $36y^2 + 6y - 12$

65. $9y^2 + 33y - 60$

66. $16y^2 - 16y - 12$

67. $3x^3 + 4x^2 + x$

68. $3x^3 + 14x^2 + 8x$

69. $2x^3 - 3x^2 - 5x$

70. $6x^3 + 4x^2 - 10x$

71. $9y^3 - 39y^2 + 12y$

72. $10y^3 + 12y^2 + 2y$

73. $60z^3 + 40z^2 + 5z$

74. $80z^3 + 80z^2 - 60z$

75. $15x^4 - 39x^3 + 18x^2$

76. $24x^4 + 10x^3 - 4x^2$

77. $10x^5 - 17x^4 + 3x^3$

78. $15x^5 - 2x^4 - x^3$

79. $6x^2 - 3xy - 18y^2$

80. $4x^2 + 14xy + 10y^2$

81. $12x^2 + 10xy - 8y^2$

82. $24x^2 + 3xy - 27y^2$

83. $8x^2y + 34xy - 84y$

84. $6x^2y - 2xy - 60y$

85. $12a^2b - 46ab^2 + 14b^3$

86. $12a^2b - 34ab^2 + 14b^3$

Practice Plus

In Exercises 87–90, factor completely.

87. $30(y + 1)x^2 + 10(y + 1)x - 20(y + 1)$

88. $6(y + 1)x^2 + 33(y + 1)x + 15(y + 1)$

89. $-32x^2y^4 + 20xy^4 + 12y^4$

90. $-10x^2y^4 + 14xy^4 + 12y^4$

91. a. Factor $2x^2 - 5x - 3$.

  b. Use the factorization in part (a) to factor
$$2(y + 1)^2 - 5(y + 1) - 3.$$
  Then simplify each factor.

92. a. Factor $3x^2 + 5x - 2$.

  b. Use the factorization in part (a) to factor
$$3(y + 1)^2 + 5(y + 1) - 2.$$
  Then simplify each factor.

93. Divide $3x^3 - 11x^2 + 12x - 4$ by $x - 2$. Use the quotient to factor $3x^3 - 11x^2 + 12x - 4$ completely.

94. Divide $2x^3 + x^2 - 13x + 6$ by $x - 2$. Use the quotient to factor $2x^3 + x^2 - 13x + 6$ completely.

Application Exercises

It is possible to construct geometric models for factorizations so that you can see the factoring. This idea is developed in Exercises 95–96.

95. Consider the following figure.

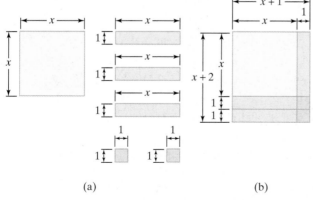

(a)   (b)

  a. Write a trinomial that expresses the sum of the areas of the six rectangular pieces shown in figure (a).

  b. Express the area of the large rectangle in figure (b) as the product of two binomials.

  c. Are the pieces in figures (a) and (b) the same? Set the expressions that you wrote in parts (a) and (b) equal to each other. What factorization is illustrated?

96. Copy the figure and cut out the six pieces. Use the pieces to create a geometric model for the factorization
$$2x^2 + 3x + 1 = (2x + 1)(x + 1)$$
by forming a large rectangle using all the pieces.

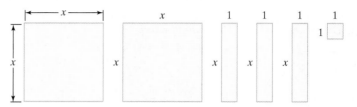

Writing in Mathematics

97. Explain how to factor $2x^2 - x - 1$.

98. Why is it a good idea to factor out the GCF first and then use other methods of factoring? Use $3x^2 - 18x + 15$ as an example. Discuss what happens if one first uses trial and error to factor as two binomials rather than first factoring out the GCF.

99. In factoring $3x^2 - 10x - 8$, a student lists $(3x - 2)(x + 4)$ as a possible factorization. Use FOIL multiplication to determine if this factorization is correct. If it is not correct, describe how the correct factorization can quickly be obtained using these factors.

100. Explain why $2x - 10$ cannot be one of the factors in the correct factorization of $6x^2 - 19x + 10$.

Critical Thinking Exercises

101. Which one of the following is true?

  a. Once a GCF is factored from $18y^2 - 6y + 6$, the remaining trinomial factor is prime.

  b. One factor of $12x^2 - 13x + 3$ is $4x + 3$.

  c. One factor of $4y^2 - 11y - 3$ is $y + 3$.

  d. The trinomial $3x^2 + 2x + 1$ has relatively small coefficients and therefore can be factored.

In Exercises 102–103, find all integers b so that the trinomial can be factored.

102. $3x^2 + bx + 2$

103. $2x^2 + bx + 3$

104. Factor: $3x^{10} - 4x^5 - 15$.

105. Factor: $2x^{2n} - 7x^n - 4$.

Review Exercises

In Exercises 106–108, perform the indicated operations.

106. $(9x + 10)(9x - 10)$ (Section 6.3, Example 4)

107. $(4x + 5y)^2$ (Section 6.3, Example 5)

108. $(x + 2)(x^2 - 2x + 4)$ (Section 6.2, Example 7)

✔ MID-CHAPTER CHECK POINT

CHAPTER 7

What You Know: We learned to factor out a polynomial's greatest common factor and to use grouping to factor polynomials with four terms. We factored polynomials with three terms, beginning with trinomials with leading coefficient 1 and moving on to $ax^2 + bx + c$, with $a \neq 1$. We saw that the factoring process should begin by looking for a GCF and, if there is one, factoring it out first.

In Exercises 1–12, factor completely, or state that the polynomial is prime.

1. $x^5 + x^4$
2. $x^2 + 7x - 18$
3. $x^2y^3 - x^2y^2 + x^2y$
4. $x^2 - 2x + 4$
5. $7x^2 - 22x + 3$
6. $x^3 + 5x^2 + 3x + 15$
7. $2x^3 - 11x^2 + 5x$
8. $xy - 7x - 4y + 28$
9. $x^2 - 17xy + 30y^2$
10. $25x^2 - 25x - 14$
11. $16x^2 - 70x + 24$
12. $3x^2 + 10xy + 7y^2$

SECTION 7.4

FACTORING SPECIAL FORMS

Objectives

1 Factor the difference of two squares.

2 Factor perfect square trinomials.

3 Factor the sum and difference of two cubes.

Do you enjoy solving puzzles? The process is a natural way to develop problem-solving skills that are important to every area of our lives. Engaging in problem solving for sheer pleasure releases chemicals in the brain that enhance our feeling of well-being. Perhaps this is why puzzles date back 12,000 years.

In this section, we develop factoring techniques by reversing the formulas for special products discussed in Chapter 6. These factorizations can be visualized by fitting pieces of a puzzle together to form rectangles.

1 Factor the difference of two squares.

Factoring the Difference of Two Squares A method for factoring the difference of two squares is obtained by reversing the special product for the sum and difference of two terms.

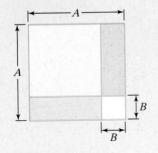
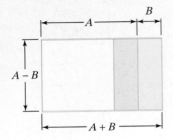
THE DIFFERENCE OF TWO SQUARES If A and B are real numbers, variables, or algebraic expressions, then
$$A^2 - B^2 = (A + B)(A - B).$$

In words: The difference of the squares of two terms factors as the product of a sum and a difference of those terms.

EXAMPLE 1 Factoring the Difference of Two Squares

Factor:

a. $x^2 - 4$ **b.** $81x^2 - 49$.

SOLUTION We must express each term as the square of some monomial. Then we use the formula for factoring $A^2 - B^2$.

a. $x^2 - 4 = x^2 - 2^2 = (x + 2)(x - 2)$

$A^2 - B^2 = (A + B)(A - B)$

b. $81x^2 - 49 = (9x)^2 - 7^2 = (9x + 7)(9x - 7)$ ∎

✔ **CHECK POINT 1** Factor:

a. $x^2 - 81$ **b.** $36x^2 - 25$.

Can $x^2 - 5$ be factored using integers and the formula for factoring $A^2 - B^2$? No. The number 5 is not the square of an integer. Thus, $x^2 - 5$ is prime over the set of integers.

EXAMPLE 2 Factoring the Difference of Two Squares

Factor:

a. $9 - 16x^{10}$ **b.** $25x^2 - 4y^2$.

SOLUTION Begin by expressing each term as the square of some monomial. Then use the formula for factoring $A^2 - B^2$.

a. $9 - 16x^{10} = 3^2 - (4x^5)^2 = (3 + 4x^5)(3 - 4x^5)$

$A^2 - B^2 = (A + B)(A - B)$

b. $25x^2 - 4y^2 = (5x)^2 - (2y)^2 = (5x + 2y)(5x - 2y)$ ∎

✔ **CHECK POINT 2** Factor:

a. $25 - 4x^{10}$ **b.** $100x^2 - 9y^2$.

When factoring, always check first for common factors. If there are common factors, factor out the GCF and then factor the resulting polynomial.

EXAMPLE 3 Factoring Out the GCF and Then Factoring the Difference of Two Squares

Factor:

a. $12x^3 - 3x$ **b.** $80 - 125x^2$.

SOLUTION

a. $12x^3 - 3x = 3x(4x^2 - 1) = 3x[(2x)^2 - 1^2] = 3x(2x + 1)(2x - 1)$

          Factor out the GCF. $A^2 - B^2 = (A + B)(A - B)$

b. $80 - 125x^2 = 5(16 - 25x^2) = 5[4^2 - (5x)^2] = 5(4 + 5x)(4 - 5x)$

 CHECK POINT 3 Factor:
    a. $18x^3 - 2x$ **b.** $72 - 18x^2$.

We have seen that a polynomial is factored completely when it is written as the product of prime polynomials. To be sure that you have factored completely, check to see whether any of the factors can be factored.

EXAMPLE 4 A Repeated Factorization

Factor completely: $x^4 - 81$.

SOLUTION

$x^4 - 81 = (x^2)^2 - 9^2$	Express as the difference of two squares.
$= (x^2 + 9)(x^2 - 9)$	The factors are the sum and the difference of the expressions being squared.
$= (x^2 + 9)(x^2 - 3^2)$	The factor $x^2 - 9$ is the difference of two squares and can be factored.
$= (x^2 + 9)(x + 3)(x - 3)$	The factors of $x^2 - 9$ are the sum and the difference of the expressions being squared.

STUDY TIP

Factoring $x^4 - 81$ as

$$(x^2 + 9)(x^2 - 9)$$

is not a complete factorization. The second factor, $x^2 - 9$, is itself a difference of two squares and can be factored.

Are you tempted to further factor $x^2 + 9$, the sum of two squares, in Example 4? Resist the temptation! **The sum of two squares, $A^2 + B^2$, with no common factor other than 1 is a prime polynomial over the integers.**

 CHECK POINT 4 Factor completely: $81x^4 - 16$.

---

2   Factor perfect square trinomials.

Factoring Perfect Square Trinomials Our next factoring technique is obtained by reversing the special products for squaring binomials. The trinomials that are factored using this technique are called **perfect square trinomials**.

Visualizing the Factoring for a Perfect Square Trinomial

Area:

$$(A + B)^2$$

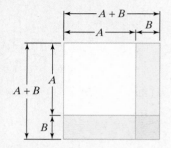

Sum of Areas:

$$A^2 + 2AB + B^2$$

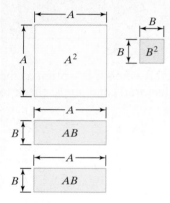

Conclusion:

$$A^2 + 2AB + B^2 = (A + B)^2$$

FACTORING PERFECT SQUARE TRINOMIALS Let A and B be real numbers, variables, or algebraic expressions.

1. $A^2 + 2AB + B^2 = (A + B)^2$

Same sign

2. $A^2 - 2AB + B^2 = (A - B)^2$

Same sign

The two items in the box show that perfect square trinomials come in two forms: one in which the middle term is positive and one in which the middle term is negative. Here's how to recognize a perfect square trinomial:

1. The first and last terms are squares of monomials or integers.

2. The middle term is twice the product of the expressions being squared in the first and last terms.

EXAMPLE 5 Factoring Perfect Square Trinomials

Factor:

a. $x^2 + 6x + 9$ **b.** $x^2 - 16x + 64$ **c.** $25x^2 - 60x + 36$.

SOLUTION

a. $x^2 + 6x + 9 = x^2 + 2 \cdot x \cdot 3 + 3^2 = (x + 3)^2$

The middle term has a positive sign.

$$A^2 + 2AB + B^2 = (A + B)^2$$

b. $x^2 - 16x + 64 = x^2 - 2 \cdot x \cdot 8 + 8^2 = (x - 8)^2$

The middle term has a negative sign.

$$A^2 - 2AB + B^2 = (A - B)^2$$

c. We suspect that $25x^2 - 60x + 36$ is a perfect square trinomial because $25x^2 = (5x)^2$ and $36 = 6^2$. The middle term can be expressed as twice the product of $5x$ and 6.

$$25x^2 - 60x + 36 = (5x)^2 - 2 \cdot 5x \cdot 6 + 6^2 = (5x - 6)^2$$

$$A^2 - 2AB + B^2 = (A - B)^2$$

■

 CHECK POINT 5 Factor:

a. $x^2 + 14x + 49$ **b.** $x^2 - 6x + 9$

c. $16x^2 - 56x + 49$.

EXAMPLE 6 Factoring a Perfect Square Trinomial in Two Variables

Factor: $16x^2 + 40xy + 25y^2$.

SOLUTION Observe that $16x^2 = (4x)^2$, $25y^2 = (5y)^2$, and $40xy$ is twice the product of $4x$ and $5y$. Thus, we have a perfect square trinomial.

$$16x^2 + 40xy + 25y^2 = (4x)^2 + 2 \cdot 4x \cdot 5y + (5y)^2 = (4x + 5y)^2$$

$$A^2 \ + \ 2AB \ + \ B^2 \ = \ (A \ + \ B)^2$$

✔ **CHECK POINT 6** Factor: $4x^2 + 12xy + 9y^2$.

Factoring the Sum and Difference of Two Cubes We can use the following formulas to factor the sum or the difference of two cubes:

3 Factor the sum and difference of two cubes.

FACTORING THE SUM AND DIFFERENCE OF TWO CUBES

1. Factoring the Sum of Two Cubes

$$A^3 + B^3 = (A + B)(A^2 - AB + B^2)$$

Same sign   Opposite signs

2. Factoring the Difference of Two Cubes

$$A^3 - B^3 = (A - B)(A^2 + AB + B^2)$$

Same sign   Opposite signs

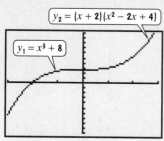
EXAMPLE 7   Factoring the Sum of Two Cubes

Factor: $x^3 + 8$.

SOLUTION We must express each term as the cube of some monomial. Then we use the formula for factoring $A^3 + B^3$.

$$x^3 + 8 = x^3 + 2^3 = (x + 2)(x^2 - x \cdot 2 + 2^2) = (x + 2)(x^2 - 2x + 4)$$

$$A^3 \ + \ B^3 \ = \ (A \ + \ B) \ (A^2 \ - \ AB \ + \ B^2)$$

✔ **CHECK POINT 7** Factor: $x^3 + 27$.

EXAMPLE 8   Factoring the Difference of Two Cubes

Factor: $27 - y^3$.

SOLUTION Express each term as the cube of some monomial. Then use the formula for factoring $A^3 - B^3$.

$$27 - y^3 = 3^3 - y^3 = (3 - y)(3^2 + 3y + y^2) = (3 - y)(9 + 3y + y^2)$$

$$A^3 - B^3 \ = \ (A \ - \ B) \ (A^2 \ + \ AB \ + \ B^2)$$

✔ **CHECK POINT 8** Factor: $1 - y^3$.

EXAMPLE 9 Factoring the Sum of Two Cubes

Factor: $64x^3 + 125$.

SOLUTION Express each term as the cube of some monomial. Then use the formula for factoring $A^3 + B^3$.

$$64x^3 + 125 = (4x)^3 + 5^3 = (4x + 5)[(4x)^2 - (4x)(5) + 5^2]$$

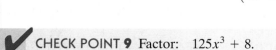

$$A^3 + B^3 = (A + B)(A^2 - AB + B^2)$$

$$= (4x + 5)(16x^2 - 20x + 25)$$

■

✔ **CHECK POINT 9** Factor: $125x^3 + 8$.

7.4 EXERCISE SET

Student Solutions Manual   CD/Video   PH Math/Tutor Center   MathXL Tutorials on CD   MathXL®   MyMathLab   Interactmath.com

Practice Exercises

In Exercises 1–26, factor each difference of two squares.

1. $x^2 - 25$ **2.** $x^2 - 16$

3. $y^2 - 1$ **4.** $y^2 - 9$

5. $4x^2 - 9$ **6.** $9x^2 - 25$

7. $25 - x^2$ **8.** $16 - x^2$

9. $1 - 49x^2$ **10.** $1 - 64x^2$

11. $9 - 25y^2$ **12.** $16 - 49y^2$

13. $x^4 - 9$ **14.** $x^4 - 25$

15. $49y^4 - 16$ **16.** $49y^4 - 25$

17. $x^{10} - 9$ **18.** $x^{10} - 1$

19. $25x^2 - 16y^2$ **20.** $9x^2 - 25y^2$

21. $x^4 - y^{10}$ **22.** $x^{14} - y^4$

23. $x^4 - 16$ **24.** $x^4 - 1$

25. $16x^4 - 81$ **26.** $81x^4 - 1$

In Exercises 27–40, factor completely, or state that the polynomial is prime.

27. $2x^2 - 18$ **28.** $5x^2 - 45$

29. $2x^3 - 72x$ **30.** $2x^3 - 8x$

31. $x^2 + 36$ **32.** $x^2 + 4$

33. $3x^3 + 27x$ **34.** $3x^3 + 15x$

35. $18 - 2y^2$ **36.** $32 - 2y^2$

37. $3y^3 - 48y$ **38.** $3y^3 - 75y$

39. $18x^3 - 2x$ **40.** $20x^3 - 5x$

In Exercises 41–62, factor any perfect square trinomials, or state that the polynomial is prime.

41. $x^2 + 2x + 1$ **42.** $x^2 + 4x + 4$

43. $x^2 - 14x + 49$ **44.** $x^2 - 10x + 25$

45. $x^2 - 2x + 1$ **46.** $x^2 - 4x + 4$

47. $x^2 + 22x + 121$ **48.** $x^2 + 24x + 144$

49. $4x^2 + 4x + 1$ **50.** $9x^2 + 6x + 1$

51. $25y^2 - 10y + 1$ **52.** $64y^2 - 16y + 1$

53. $x^2 - 10x + 100$

54. $x^2 - 7x + 49$

55. $x^2 + 14xy + 49y^2$

56. $x^2 + 16xy + 64y^2$

57. $x^2 - 12xy + 36y^2$

58. $x^2 - 18xy + 81y^2$

59. $x^2 - 8xy + 64y^2$

60. $x^2 + 9xy + 16y^2$

61. $16x^2 - 40xy + 25y^2$

62. $9x^2 + 48xy + 64y^2$

In Exercises 63–70, factor completely.

63. $12x^2 - 12x + 3$

64. $18x^2 + 24x + 8$

65. $9x^3 + 6x^2 + x$

66. $25x^3 - 10x^2 + x$

67. $2y^2 - 4y + 2$

68. $2y^2 - 40y + 200$

69. $2y^3 + 28y^2 + 98y$

70. $50y^3 + 20y^2 + 2y$

In Exercises 71–88, factor using the formula for the sum or difference of two cubes.

71. $x^3 + 1$

72. $x^3 + 64$

73. $x^3 - 27$

74. $x^3 - 64$

75. $8y^3 - 1$

76. $27y^3 - 1$

77. $27x^3 + 8$

78. $125x^3 + 8$

79. $x^3y^3 - 64$

80. $x^3y^3 - 27$

81. $27y^4 + 8y$

82. $64y - y^4$

83. $54 - 16y^3$

84. $128 - 250y^3$

85. $64x^3 + 27y^3$

86. $8x^3 + 27y^3$

87. $125x^3 - 64y^3$

88. $125x^3 - y^3$

Practice Plus

In Exercises 89–96, factor completely.

89. $25x^2 - \dfrac{4}{49}$

90. $16x^2 - \dfrac{9}{25}$

91. $y^4 - \dfrac{y}{1000}$

92. $y^4 - \dfrac{y}{8}$

93. $0.25x - x^3$

94. $0.64x - x^3$

95. $(x + 1)^2 - 25$

96. $(x + 2)^2 - 49$

97. Divide $x^3 - x^2 - 5x - 3$ by $x - 3$. Use the quotient to factor $x^3 - x^2 - 5x - 3$ completely.

98. Divide $x^3 + 4x^2 - 3x - 18$ by $x - 2$. Use the quotient to factor $x^3 + 4x^2 - 3x - 18$ completely.

Application Exercises

In Exercises 99–102, find the formula for the area of the shaded region and express it in factored form.

99.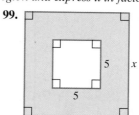

100.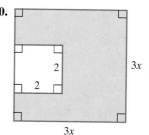

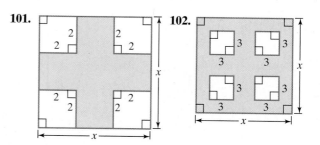

101. **102.**

Writing in Mathematics

103. Explain how to factor the difference of two squares. Provide an example with your explanation.

104. What is a perfect square trinomial and how is it factored?

105. Explain why $x^2 - 1$ is factorable, but $x^2 + 1$ is not.

106. Explain how to factor $x^3 + 1$.

Critical Thinking Exercises

107. Which one of the following is true?

a. Because $x^2 - 25 = (x + 5)(x - 5)$, then $x^2 + 25 = (x - 5)(x + 5)$.

b. All perfect square trinomials are squares of binomials.

c. Any polynomial that is the sum of two squares is prime.

d. The polynomial $16x^2 + 20x + 25$ is a perfect square trinomial.

108. Where is the error in this "proof" that $2 = 0$?

$a = b$	Suppose that a and b are any equal real numbers.
$a^2 = b^2$	Square both sides of the equation.
$a^2 - b^2 = 0$	Subtract b^2 from both sides.
$2(a^2 - b^2) = 2 \cdot 0$	Multiply both sides by 2.
$2(a^2 - b^2) = 0$	On the right side, $2 \cdot 0 = 0$.
$2(a + b)(a - b) = 0$	Factor $a^2 - b^2$.
$2(a + b) = 0$	Divide both sides by $a - b$.
$2 = 0$	Divide both sides by $a + b$.

In Exercises 109–112, factor each polynomial.

109. $x^2 - y^2 + 3x + 3y$

110. $x^{2n} - 25y^{2n}$

111. $4x^{2n} + 12x^n + 9$

112. $(x + 3)^2 - 2(x + 3) + 1$

In Exercises 113–114, find all integers k so that the trinomial is a perfect square trinomial.

113. $9x^2 + kx + 1$

114. $64x^2 - 16x + k$

Technology Exercises

In Exercises 115–118, use the GRAPH *or* TABLE *feature of a graphing utility to determine if the polynomial on the left side of each equation has been correctly factored. If the graphs of y_1 and y_2 coincide, or if their corresponding table values are equal, this means that the polynomial on the left side has been correctly factored. If not, factor the polynomial correctly and then use your graphing utility to verify the factorization.*

115. $4x^2 - 9 = (4x + 3)(4x - 3)$

116. $x^2 - 6x + 9 = (x - 3)^2$

117. $4x^2 - 4x + 1 = (4x - 1)^2$

118. $x^3 - 1 = (x - 1)(x^2 - x + 1)$

Review Exercises

119. Simplify: $(2x^2y^3)^4(5xy^2)$. (Section 6.7, Example 5)

120. Subtract: $(10x^2 - 5x + 2) - (14x^2 - 5x - 1)$. (Section 6.1, Example 3)

121. Divide: $\dfrac{6x^2 + 11x - 10}{3x - 2}$. (Section 6.6, Example 1)

A GENERAL FACTORING STRATEGY

Yogi Berra, catcher and renowned hitter for the New York Yankees (1946–1963), said it best: "If you don't know where you're going, you'll probably end up someplace else." When it comes to factoring, it's easy to know where you're going. Why? In this section, you will learn a step-by-step strategy that provides a plan and direction for solving factoring problems.

A Strategy for Factoring Polynomials It is important to practice factoring a wide variety of polynomials so that you can quickly select the appropriate technique. The polynomial is factored completely when all its polynomial factors, except possibly the monomial factors, are prime. Because of the commutative property, the order of the factors does not matter.

Here is a general strategy for factoring polynomials:

1 Recognize the appropriate method for factoring a polynomial.

A STRATEGY FOR FACTORING A POLYNOMIAL

1. If there is a common factor other than 1, factor out the GCF.

2. Determine the number of terms in the polynomial and try factoring as follows:

  a. If there are two terms, can the binomial be factored by one of the following special forms?

  Difference of two squares: $A^2 - B^2 = (A + B)(A - B)$
  Sum of two cubes: $A^3 + B^3 = (A + B)(A^2 - AB + B^2)$
  Difference of two cubes: $A^3 - B^3 = (A - B)(A^2 + AB + B^2)$

  b. If there are three terms, is the trinomial a perfect square trinomial? If so, factor by one of the following special forms:

  $$A^2 + 2AB + B^2 = (A + B)^2$$
  $$A^2 - 2AB + B^2 = (A - B)^2.$$

  If the trinomial is not a perfect square trinomial, try factoring by trial and error or grouping.

  c. If there are four or more terms, try factoring by grouping.

3. Check to see if any factors with more than one term in the factored polynomial can be factored further. If so, factor completely.

4. Check by multiplying.

2 Use a general strategy for factoring polynomials.

The following examples and those in the exercise set are similar to the previous factoring problems. One difference is that although these polynomials may be factored using the techniques we have studied in this chapter, each must be factored using at least two techniques. Also different is that these factorizations are not all of the same type. They are intentionally mixed to promote the development of a general factoring strategy.

EXAMPLE 1   Factoring a Polynomial

Factor: $4x^4 - 16x^2$.

SOLUTION

Step 1.   **If there is a common factor, factor out the GCF.** Because $4x^2$ is common to both terms, we factor it out.

$$4x^4 - 16x^2 = 4x^2(x^2 - 4) \qquad \text{Factor out the GCF.}$$

Step 2.   **Determine the number of terms and factor accordingly.** The factor $x^2 - 4$ has two terms. It is the difference of two squares: $x^2 - 2^2$. We factor using the special form for the difference of two squares and rewrite the GCF.

$$4x^4 - 16x^2 = 4x^2(x + 2)(x - 2) \qquad \text{Use } A^2 - B^2 = (A + B)(A - B)$$
$$\text{on } x^2 - 4: A = x \text{ and } B = 2.$$

Step 3.   **Check to see if any factors with more than one term can be factored further.** No factor with more than one term can be factored further, so we have factored completely.

Step 4. **Check by multiplying.**

$$4x^2(x + 2)(x - 2) = 4x^2(x^2 - 4) = 4x^4 - 16x^2$$

This is the original polynomial, so the factorization is correct.

✔ **CHECK POINT 1** Factor: $5x^4 - 45x^2$.

EXAMPLE 2 Factoring a Polynomial

Factor: $3x^2 - 6x - 45$.

SOLUTION

Step 1. **If there is a common factor, factor out the GCF.** Because 3 is common to all terms, we factor it out.

$$3x^2 - 6x - 45 = 3(x^2 - 2x - 15) \qquad \textit{Factor out the GCF.}$$

Step 2. **Determine the number of terms and factor accordingly.** The factor $x^2 - 2x - 15$ has three terms, but it is not a perfect square trinomial. We factor it using trial and error.

$$3x^2 - 6x - 45 = 3(x^2 - 2x - 15) = 3(x - 5)(x + 3)$$

Step 3. **Check to see if factors can be factored further.** In this case, they cannot, so we have factored completely.

Step 4. **Check by multiplying.**

$$3(x - 5)(x + 3) = 3(x^2 - 2x - 15) = 3x^2 - 6x - 45$$

FOIL

This is the original polynomial, so the factorization is correct.

✔ **CHECK POINT 2** Factor: $4x^2 - 16x - 48$.

EXAMPLE 3 Factoring a Polynomial

Factor: $7x^5 - 7x$.

SOLUTION

Step 1. **If there is a common factor, factor out the GCF.** Because $7x$ is common to both terms, we factor it out.

$$7x^5 - 7x = 7x(x^4 - 1) \qquad \textit{Factor out the GCF.}$$

Step 2. **Determine the number of terms and factor accordingly.** The factor $x^4 - 1$ has two terms. This binomial can be expressed as $(x^2)^2 - 1^2$, so it can be factored as the difference of two squares.

$$7x^5 - 7x = 7x(x^4 - 1) = 7x(x^2 + 1)(x^2 - 1) \qquad \textit{Use } A^2 - B^2 = (A + B)(A - B) \textit{ on } x^4 - 1: A = x^2 \textit{ and } B = 1.$$

Step 3.   **Check to see if factors can be factored further.** We note that $(x^2 - 1)$ is also the difference of two squares, $x^2 - 1^2$, so we continue factoring.

$$7x^5 - 7x = 7x(x^2 + 1)(x + 1)(x - 1)$$
Factor $x^2 - 1$ as the difference of two squares.

Step 4.   **Check by multiplying.**

$$7x(x^2 + 1)(x + 1)(x - 1) = 7x(x^2 + 1)(x^2 - 1) = 7x(x^4 - 1) = 7x^5 - 7x$$

We obtain the original polynomial, so the factorization is correct. ■

✓ CHECK POINT **3** Factor: $4x^5 - 64x$.

EXAMPLE 4   Factoring a Polynomial

Factor: $x^3 - 5x^2 - 4x + 20$.

SOLUTION

Step 1.   **If there is a common factor, factor out the GCF.** Other than 1, there is no common factor.

Step 2.   **Determine the number of terms and factor accordingly.** There are four terms. We try factoring by grouping.

$$x^3 - 5x^2 - 4x + 20$$
$$= (x^3 - 5x^2) + (-4x + 20) \qquad \text{Group terms with common factors.}$$
$$= x^2(x - 5) - 4(x - 5) \qquad \text{Factor from each group.}$$
$$= (x - 5)(x^2 - 4) \qquad \text{Factor out the common binomial factor, } x - 5.$$

Step 3.   **Check to see if factors can be factored further.** We note that $(x^2 - 4)$ is the difference of two squares, $x^2 - 2^2$, so we continue factoring.

$$x^3 - 5x^2 - 4x + 20 = (x - 5)(x + 2)(x - 2) \qquad \text{Factor } x^2 - 4 \text{ as the difference of two squares.}$$

We have factored completely because no factor with more than one term can be factored further.

Step 4.   **Check by multiplying.**

F O I L

$$(x - 5)(x + 2)(x - 2) = (x - 5)(x^2 - 4) = x^3 - 4x - 5x^2 + 20$$
$$= x^3 - 5x^2 - 4x + 20$$

We obtain the original polynomial, so the factorization is correct. ■

USING TECHNOLOGY

You can use a graphing utility to check the factorization in Example 4. Enter the given polynomial and its complete factorization:

$$y_1 = x^3 - 5x^2 - 4x + 20 \quad \text{and} \quad y_2 = (x - 5)(x + 2)(x - 2).$$

Graphic Check

The graphs are identical.

Numeric Check

No matter how far up or down we scroll, $y_1 = y_2$.

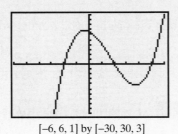

[−6, 6, 1] by [−30, 30, 3]

Caution: Keep in mind that a graphing utility cannot verify that a polynomial is factored completely. If you enter $(x - 5)(x^2 - 4)$ for y_2, which is not a complete factorization, you will still obtain identical graphs or tables with equal corresponding values.

✔ **CHECK POINT 4** Factor: $x^3 - 4x^2 - 9x + 36$.

EXAMPLE 5 Factoring a Polynomial

Factor: $2x^3 - 24x^2 + 72x$.

SOLUTION

Step 1. **If there is a common factor, factor out the GCF.** Because $2x$ is common to all terms, we factor it out.

$$2x^3 - 24x^2 + 72x = 2x(x^2 - 12x + 36) \quad \text{Factor out the GCF.}$$

Step 2. **Determine the number of terms and factor accordingly.** The factor $x^2 - 12x + 36$ has three terms. Is it a perfect square trinomial? Yes. The first term, x^2, is the square of a monomial. The last term, 36, or 6^2, is the square of an integer. The middle term involves twice the product of x and 6. We factor using $A^2 - 2AB + B^2 = (A - B)^2$.

$$2x^3 - 24x^2 + 72x = 2x(x^2 - 12x + 36)$$

$$= 2x(x^2 - 2 \cdot x \cdot 6 + 6^2) \quad \text{The second factor is a perfect square trinomial.}$$

$A^2 - 2\ A\ B + B^2$

$$= 2x(x - 6)^2 \quad A^2 - 2AB + B^2 = (A - B)^2$$

Step 3. **Check to see if factors can be factored further.** In this problem, they cannot, so we have factored completely.

Step 4.   **Check by multiplying.**

$$2x(x - 6)^2 = 2x(x^2 - 12x + 36) = 2x^3 - 24x^2 + 72x$$

We obtain the original polynomial, so the factorization is correct.   ∎

✔ CHECK POINT **5** Factor: $3x^3 - 30x^2 + 75x$.

EXAMPLE 6   Factoring a Polynomial

Factor: $3x^5 + 24x^2$.

SOLUTION

Step 1.   **If there is a common factor, factor out the GCF.** Because $3x^2$ is common to both terms, we factor it out.

$$3x^5 + 24x^2 = 3x^2(x^3 + 8) \qquad \text{Factor out the GCF.}$$

Step 2.   **Determine the number of terms and factor accordingly.** The factor $x^3 + 8$ has two terms. This binomial can be expressed as $x^3 + 2^3$, so it can be factored as the sum of two cubes.

$$3x^5 + 24x^2 = 3x^2(\underbrace{x^3 + 2^3}_{A^3 + B^3}) \qquad \begin{array}{l}\text{Express } x^3 + 8 \text{ as the sum of}\\ \text{two cubes.}\end{array}$$

$$= 3x^2\underbrace{(x + 2)(x^2 - 2x + 4)}_{(A + B)(A^2 - AB + B^2)} \qquad \text{Factor the sum of two cubes.}$$

Step 3.   **Check to see if factors can be factored further.** In this problem, they cannot, so we have factored completely.

Step 4.   **Check by multiplying.**

$$3x^2(x + 2)(x^2 - 2x + 4) = 3x^2[x(x^2 - 2x + 4) + 2(x^2 - 2x + 4)]$$

$$= 3x^2(x^3 - 2x^2 + 4x + 2x^2 - 4x + 8)$$

$$= 3x^2(x^3 + 8) = 3x^5 + 24x^2$$

We obtain the original polynomial, so the factorization is correct.   ∎

✔ CHECK POINT **6** Factor: $2x^5 + 54x^2$.

DISCOVER FOR YOURSELF

In Examples 1–6, substitute 1 for the variable in both the given polynomial and in its factored form. Evaluate each expression. What do you observe? Do this for a second value of the variable. Is this a complete check or only a partial check of the factorization? Explain.

EXAMPLE 7   Factoring a Polynomial in Two Variables

Factor: $32x^4y - 2y^5$.

SOLUTION

Step 1.   **If there is a common factor, factor out the GCF.** Because $2y$ is common to both terms, we factor it out.

$$32x^4y - 2y^5 = 2y(16x^4 - y^4) \qquad \text{Factor out the GCF.}$$

Step 2. **Determine the number of terms and factor accordingly.** The factor $16x^4 - y^4$ has two terms. It is the difference of two squares: $(4x^2)^2 - (y^2)^2$. We factor using the special form for the difference of two squares.

$$32x^4y - 2y^5 = 2y[(4x^2)^2 - (y^2)^2]$$

Express $16x^4 - y^4$ as the difference of two squares.

$$= 2y(4x^2 + y^2)(4x^2 - y^2)$$

$A^2 - B^2 = (A + B)(A - B)$

$(A + B)$ $(A - B)$

Step 3. **Check to see if factors can be factored further.** We note that the last factor, $4x^2 - y^2$, is also the difference of two squares, $(2x)^2 - y^2$, so we continue factoring.

$$32x^4y - 2y^5 = 2y(4x^2 + y^2)(2x + y)(2x - y)$$

Step 4. **Check by multiplying.** Multiply the factors in the factorization and verify that you obtain the original polynomial. ∎

✔ **CHECK POINT 7** Factor: $3x^4y - 48y^5$.

EXAMPLE 8 Factoring a Polynomial in Two Variables

Factor: $18x^3 + 48x^2y + 32xy^2$.

SOLUTION

Step 1. **If there is a common factor, factor out the GCF.** Because $2x$ is common to all terms, we factor it out.

$$18x^3 + 48x^2y + 32xy^2 = 2x(9x^2 + 24xy + 16y^2)$$

Step 2. **Determine the number of terms and factor accordingly.** The factor $9x^2 + 24xy + 16y^2$ has three terms. Is it a perfect square trinomial? Yes. The first term, $9x^2$, or $(3x)^2$, and the last term, $16y^2$, or $(4y)^2$, are squares of monomials. The middle term, $24xy$, is twice the product of $3x$ and $4y$. We factor using $A^2 + 2AB + B^2 = (A + B)^2$.

$$18x^3 + 48x^2y + 32xy^2 = 2x(9x^2 + 24xy + 16y^2)$$

$$= 2x[(3x)^2 + 2 \cdot 3x \cdot 4y + (4y)^2]$$

The second factor is a perfect square trinomial.

$A^2 + 2 \cdot A \cdot B + B^2$

$$= 2x(3x + 4y)^2$$

$A^2 + 2AB + B^2 = (A + B)^2$

Step 3. **Check to see if factors can be factored further.** In this problem, they cannot, so we have factored completely.

Step 4. **Check by multiplication.** Multiply the factors in the factorization and verify that you obtain the original polynomial.

✔ **CHECK POINT 8** Factor: $12x^3 + 36x^2y + 27xy^2$.

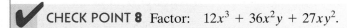

7.5 EXERCISE SET

Student Solutions Manual CD/Video PH Math/Tutor Center MathXL Tutorials on CD MathXL® MyMathLab Interactmath.com

Practice Exercises

In Exercises 1–62, factor completely, or state that the polynomial is prime. Check factorizations using multiplication or a graphing utility.

1. $5x^3 - 20x$

2. $4x^3 - 100x$

3. $7x^3 + 7x$

4. $6x^3 + 24x$

5. $5x^2 - 5x - 30$

6. $5x^2 - 15x - 50$

7. $2x^4 - 162$

8. $7x^4 - 7$

9. $x^3 + 2x^2 - 9x - 18$

10. $x^3 + 3x^2 - 25x - 75$

11. $3x^3 - 24x^2 + 48x$

12. $5x^3 - 20x^2 + 20x$

13. $2x^5 + 2x^2$

14. $2x^5 + 128x^2$

15. $6x^2 + 8x$

16. $21x^2 - 35x$

17. $2y^2 - 2y - 112$

18. $6x^2 - 6x - 12$

19. $7y^4 + 14y^3 + 7y^2$

20. $2y^4 + 28y^3 + 98y^2$

21. $y^2 + 8y - 16$

22. $y^2 - 18y - 81$

23. $16y^2 - 4y - 2$

24. $32y^2 + 4y - 6$

25. $r^2 - 25r$

26. $3r^2 - 27r$

27. $4w^2 + 8w - 5$

28. $35w^2 - 2w - 1$

29. $x^3 - 4x$

30. $9x^3 - 9x$

31. $x^2 + 64$

32. $y^2 + 36$

33. $9y^2 + 13y + 4$

34. $20y^2 + 12y + 1$

35. $y^3 + 2y^2 - 4y - 8$

36. $y^3 + 2y^2 - y - 2$

37. $16y^2 + 24y + 9$

38. $25y^2 + 20y + 4$

39. $4y^3 - 28y^2 + 40y$

40. $7y^3 - 21y^2 + 14y$

41. $y^5 - 81y$

42. $y^5 - 16y$

43. $20a^4 - 45a^2$

44. $48a^4 - 3a^2$

45. $12y^2 - 11y + 2$

46. $21x^2 - 25x - 4$

47. $9y^2 - 64$

48. $100y^2 - 49$

49. $9y^2 + 64$

50. $100y^2 + 49$

51. $2y^3 + 3y^2 - 50y - 75$

52. $12y^3 + 16y^2 - 3y - 4$

53. $2r^3 + 30r^2 - 68r$

54. $3r^3 - 27r^2 - 210r$

55. $8x^5 - 2x^3$

56. $y^9 - y^5$

57. $3x^2 + 243$

58. $27x^2 + 75$

59. $x^4 + 8x$

60. $x^4 + 27x$

61. $2y^5 - 2y^2$

62. $2y^5 - 128y^2$

Exercises 63–92 contain polynomials in several variables. Factor each polynomial completely and check using multiplication.

63. $6x^2 + 8xy$

64. $21x^2 - 35xy$

65. $xy - 7x + 3y - 21$

66. $xy - 5x + 2y - 10$

67. $x^2 - 3xy - 4y^2$

68. $x^2 - 4xy - 12y^2$

69. $72a^3b^2 + 12a^2 - 24a^4b^2$

70. $24a^4b + 60a^3b^2 + 150a^2b^3$

71. $3a^2 + 27ab + 54b^2$

72. $3a^2 + 15ab + 18b^2$

73. $48x^4y - 3x^2y$

74. $16a^3b^2 - 4ab^2$

75. $6a^2b + ab - 2b$

76. $16a^2 - 32ab + 12b^2$

77. $7x^5y - 7xy^5$

78. $3x^4y^2 - 3x^2y^2$

79. $10x^3y - 14x^2y^2 + 4xy^3$

80. $18x^3y + 57x^2y^2 + 30xy^3$

81. $2bx^2 + 44bx + 242b$

82. $3xz^2 - 72xz + 432x$

83. $15a^2 + 11ab - 14b^2$

84. $25a^2 + 25ab + 6b^2$

85. $36x^3y - 62x^2y^2 + 12xy^3$

86. $10a^4b^2 - 15a^3b^3 - 25a^2b^4$

87. $a^2y - b^2y - a^2x + b^2x$

88. $bx^2 - 4b + ax^2 - 4a$

89. $9ax^3 + 15ax^2 - 14ax$

90. $4ay^3 - 12ay^2 + 9ay$

91. $81x^4y - y^5$

92. $16x^4y - y^5$

Practice Plus

In Exercises 93–102, factor completely.

93. $10x^2(x + 1) - 7x(x + 1) - 6(x + 1)$

94. $12x^2(x - 1) - 4x(x - 1) - 5(x - 1)$

95. $6x^4 + 35x^2 - 6$

96. $7x^4 + 34x^2 - 5$

97. $(x - 7)^2 - 4a^2$

98. $(x - 6)^2 - 9a^2$

99. $x^2 + 8x + 16 - 25a^2$

100. $x^2 + 14x + 49 - 16a^2$

101. $y^7 + y$

102. $(y + 1)^3 + 1$

Application Exercises

103. A rock is dropped from the top of a 256-foot cliff. The height, in feet, of the rock above the water after *t* seconds is modeled by the polynomial $256 - 16t^2$. Factor this expression completely.

256 feet

104. The building shown in the figure has a height represented by *x* feet. The building's base is a square and its volume is $x^3 - 60x^2 + 900x$ cubic feet. Express the building's dimensions in terms of *x*.

105. Express the area of the shaded ring shown in the figure in terms of π. Then factor this expression completely.

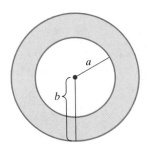

Writing in Mathematics

106. Describe a strategy that can be used to factor polynomials.

107. Describe some of the difficulties in factoring polynomials. What suggestions can you offer to overcome these difficulties?

108. You are about to take a great picture of fog rolling into San Francisco from the middle of the Golden Gate Bridge, 400 feet above the water. Whoops! You accidently lean too far over the safety rail and drop your camera. The height, in feet, of the camera after *t* seconds is modeled by the polynomial $400 - 16t^2$. The factored form of the polynomial is $16(5 + t)(5 - t)$. Describe something about your falling camera that is easier to see from the factored form, $16(5 + t)(5 - t)$, than from the form $400 - 16t^2$.

Critical Thinking Exercises

109. Which one of the following is true?

   a. $x^2 - 9 = (x - 3)^2$ for any real number *x*.

   b. The polynomial $4x^2 + 100$ is the sum of two squares and therefore cannot be factored.

   c. If the general factoring strategy is used to factor a polynomial, at least two factorizations are necessary before the given polynomial is factored completely.

   d. Once a common monomial factor is removed from $3xy^3 + 9xy^2 + 21xy$, the remaining trinomial factor cannot be factored further.

In Exercises 110–114, factor completely.

110. $3x^5 - 21x^3 - 54x$

111. $5y^5 - 5y^4 - 20y^3 + 20y^2$

112. $4x^4 - 9x^2 + 5$

113. $(x + 5)^2 - 20(x + 5) + 100$

114. $3x^{2n} - 27y^{2n}$

Technology Exercises

In Exercises 115–119, use the GRAPH *or* TABLE *feature of a graphing utility to determine if the polynomial on the left side of each equation has been correctly factored. If not, factor the polynomial correctly and then use your graphing utility to verify the factorization.*

115. $4x^2 - 12x + 9 = (4x - 3)^2$; $[-5, 5, 1]$ by $[0, 20, 1]$

116. $3x^3 - 12x^2 - 15x = 3x(x + 5)(x - 1)$; $[-5, 7, 1]$ by $[-80, 80, 10]$

117. $6x^2 + 10x - 4 = 2(3x - 1)(x + 2)$; $[-5, 5, 1]$ by $[-20, 20, 2]$

118. $x^4 - 16 = (x^2 + 4)(x + 2)(x - 2)$; $[-5, 5, 1]$ by $[-20, 20, 2]$

119. $2x^3 + 10x^2 - 2x - 10 = 2(x + 5)(x^2 + 1)$; $[-8, 4, 1]$ by $[-100, 100, 10]$

Review Exercises

120. Factor: $9x^2 - 16$. (Section 7.4, Example 1)

121. Graph using intercepts: $5x - 2y = 10$. (Section 4.2, Example 4)

122. The second angle of a triangle measures three times that of the first angle's measure. The third angle measures 80° more than the first. Find the measure of each angle. (Section 3.3, Example 6)

7.6

Objectives

1 Use the zero-product principle.

2 Solve quadratic equations by factoring.

3 Solve problems using quadratic equations.

SOLVING QUADRATIC EQUATIONS BY FACTORING

The alligator, an endangered species, was the subject of a protection program at Florida's Everglades National Park. Park rangers used the formula

$$P = -10x^2 + 475x + 3500$$

to estimate the alligator population, P, after x years of the protection program. Their goal was to bring the population up to 7250. To find out how long the program had to be continued for this to happen, we substitute 7250 for P in the formula and solve for x:

$$7250 = -10x^2 + 475x + 3500.$$

Do you see how this equation differs from a linear equation? The highest exponent on x is 2. Solving such an equation involves finding the numbers that will make the equation a true statement. In this section, we use factoring to solve equations in the form $ax^2 + bx + c = 0$. We also look at applications of these equations.

The Standard Form of a Quadratic Equation   We begin by defining a quadratic equation.

DEFINITION OF A QUADRATIC EQUATION   A **quadratic equation** in x is an equation that can be written in the **standard form**

$$ax^2 + bx + c = 0,$$

where a, b, and c are real numbers, with $a \neq 0$. A quadratic equation in x is also called a **second-degree polynomial equation** in x.

Here is an example of a quadratic equation in standard form:

$$x^2 - 7x + 10 = 0.$$

$a = 1$ $b = -7$ $c = 10$

1 Use the zero-product principle.

Solving Quadratic Equations by Factoring We can factor the left side of the quadratic equation $x^2 - 7x + 10 = 0$. We obtain $(x - 5)(x - 2) = 0$. If a quadratic equation has zero on one side and a factored expression on the other side, it can be solved using the **zero-product principle**.

THE ZERO-PRODUCT PRINCIPLE If the product of two algebraic expressions is zero, then at least one of the factors is equal to zero.

If $AB = 0$, then $A = 0$ or $B = 0$.

For example, consider the equation $(x - 5)(x - 2) = 0$. According to the zero-product principle, this product can be zero only if at least one of the factors is zero. We set each individual factor equal to zero and solve each resulting equation for x.

$$(x - 5)(x - 2) = 0$$
$$x - 5 = 0 \quad \text{or} \quad x - 2 = 0$$
$$x = 5 \qquad\qquad x = 2$$

We can check each of the proposed solutions, 5 and 2, in the original quadratic equation, $x^2 - 7x + 10 = 0$. Substitute each one separately for x in the equation.

Check 5:

$$x^2 - 7x + 10 = 0$$
$$5^2 - 7 \cdot 5 + 10 \overset{?}{=} 0$$
$$25 - 35 + 10 \overset{?}{=} 0$$
$$0 = 0, \quad \text{true}$$

Check 2:

$$x^2 - 7x + 10 = 0$$
$$2^2 - 7 \cdot 2 + 10 \overset{?}{=} 0$$
$$4 - 14 + 10 \overset{?}{=} 0$$
$$0 = 0, \quad \text{true}$$

The resulting true statements indicate that the solutions are 5 and 2. Note that with a quadratic equation, we can have two solutions, compared to the linear equation that usually had one.

EXAMPLE 1 Using the Zero-Product Principle

Solve the equation: $(3x - 1)(x + 2) = 0$.

SOLUTION The product $(3x - 1)(x + 2)$ is equal to zero. By the zero-product principle, the only way that this product can be zero is if at least one of the factors is zero. Thus,

$$3x - 1 = 0 \quad \text{or} \quad x + 2 = 0.$$
$$3x = 1 \qquad\qquad x = -2 \qquad \textit{Solve each equation for x.}$$
$$x = \frac{1}{3}$$

Because each linear equation has a solution, the original equation, $(3x - 1)(x + 2) = 0$, has two solutions, $\frac{1}{3}$ and -2. Check these solutions by substituting each one separately into the given equation. ■

✔ **CHECK POINT 1** Solve the equation: $(2x + 1)(x - 4) = 0$.

2 Solve quadratic equations by factoring.

In Example 1 and Check Point 1, the given equations were in factored form. Here is a procedure for solving a quadratic equation when we must first do the factoring.

SOLVING A QUADRATIC EQUATION BY FACTORING

1. If necessary, rewrite the equation in the standard form $ax^2 + bx + c = 0$, moving all terms to one side, thereby obtaining zero on the other side.
2. Factor.
3. Apply the zero-product principle, setting each factor equal to zero.
4. Solve the equations in step 3.
5. Check the solutions in the original equation.

EXAMPLE 2 Solving a Quadratic Equation by Factoring

Solve: $2x^2 + 7x - 4 = 0$.

SOLUTION

Step 1.   **Move all terms to one side and obtain zero on the other side.** All terms are already on the left and zero is on the other side, so we can skip this step.

Step 2.   **Factor.**

$$2x^2 + 7x - 4 = 0$$
$$(2x - 1)(x + 4) = 0$$

Steps 3 and 4. **Set each factor equal to zero and solve each resulting equation.**

$$2x - 1 = 0 \quad \text{or} \quad x + 4 = 0$$
$$2x = 1 \qquad\qquad x = -4$$
$$x = \frac{1}{2}$$

Step 5.   **Check the solutions in the original equation.**

Check $\frac{1}{2}$:

$$2x^2 + 7x - 4 = 0$$
$$2\left(\frac{1}{2}\right)^2 + 7\left(\frac{1}{2}\right) - 4 \stackrel{?}{=} 0$$
$$2\left(\frac{1}{4}\right) + 7\left(\frac{1}{2}\right) - 4 \stackrel{?}{=} 0$$
$$\frac{1}{2} + \frac{7}{2} - 4 \stackrel{?}{=} 0$$
$$4 - 4 \stackrel{?}{=} 0$$
$$0 = 0, \quad \text{true}$$

Check -4:

$$2x^2 + 7x - 4 = 0$$
$$2(-4)^2 + 7(-4) - 4 \stackrel{?}{=} 0$$
$$2(16) + 7(-4) - 4 \stackrel{?}{=} 0$$
$$32 + (-28) - 4 \stackrel{?}{=} 0$$
$$4 - 4 \stackrel{?}{=} 0$$
$$0 = 0, \quad \text{true}$$

The solutions are $\frac{1}{2}$ and -4. ∎

STUDY TIP

Do not confuse factoring a polynomial with solving a quadratic equation by factoring.

~~INCORRECT!~~
~~Factor: $2x^2 + 7x - 4$.~~
~~$(2x - 1)(x + 4)$~~
~~$2x - 1 = 0$ or $x + 4 = 0$~~
~~$x = \frac{1}{2}$ $x = -4$~~

✔ **CHECK POINT 2** Solve: $x^2 - 6x + 5 = 0$.

USING TECHNOLOGY

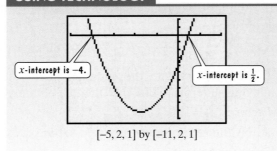

x-intercept is -4.

x-intercept is $\frac{1}{2}$.

$[-5, 2, 1]$ by $[-11, 2, 1]$

You can use a graphing utility to check the real number solutions of a quadratic equation. **The solutions of $ax^2 + bx + c = 0$ correspond to the x-intercepts for the graph of $y = ax^2 + bx + c$.** For example, to check the solutions of $2x^2 + 7x - 4 = 0$, graph $y = 2x^2 + 7x - 4$. The U-shaped, cuplike, graph is shown on the left. The x-intercepts are -4 and $\frac{1}{2}$, verifying -4 and $\frac{1}{2}$ as the solutions.

EXAMPLE 3 Solving a Quadratic Equation by Factoring

Solve: $3x^2 = 2x$.

SOLUTION

Step 1. **Move all terms to one side and obtain zero on the other side.** Subtract $2x$ from both sides and write the equation in standard form.

$$3x^2 - 2x = 2x - 2x$$
$$3x^2 - 2x = 0$$

Step 2. **Factor.** We factor out x from the two terms on the left side.

$$3x^2 - 2x = 0$$
$$x(3x - 2) = 0$$

Steps 3 and 4. Set each factor equal to zero and solve the resulting equations.

$$x = 0 \quad \text{or} \quad 3x - 2 = 0$$
$$3x = 2$$
$$x = \frac{2}{3}$$

Step 5. **Check the solutions in the original equation.**

STUDY TIP

Avoid dividing both sides of $3x^2 = 2x$ by x. You will obtain $3x = 2$ and, consequently, $x = \frac{2}{3}$. The other solution, 0, is lost. We can divide both sides of an equation by any *nonzero* real number. If x is zero, we lose the second solution.

Check 0:

$$3x^2 = 2x$$
$$3 \cdot 0^2 \stackrel{?}{=} 2 \cdot 0$$
$$0 = 0, \quad \text{true}$$

Check $\frac{2}{3}$:

$$3x^2 = 2x$$
$$3\left(\frac{2}{3}\right)^2 \stackrel{?}{=} 2\left(\frac{2}{3}\right)$$
$$3\left(\frac{4}{9}\right) \stackrel{?}{=} 2\left(\frac{2}{3}\right)$$
$$\frac{4}{3} = \frac{4}{3}, \quad \text{true}$$

The solutions are 0 and $\frac{2}{3}$.

✓ **CHECK POINT 3** Solve: $4x^2 = 2x$.

EXAMPLE 4 Solving a Quadratic Equation by Factoring

Solve: $x^2 = 6x - 9$.

SOLUTION

Step 1. **Move all terms to one side and obtain zero on the other side.** To obtain zero on the right, we subtract $6x$ and add 9 on both sides.

$$x^2 - 6x + 9 = 6x - 6x - 9 + 9$$
$$x^2 - 6x + 9 = 0$$

Step 2. **Factor.** The trinomial on the left side is a perfect square trinomial: $x^2 - 6x + 9 = x^2 - 2 \cdot x \cdot 3 + 3^2$. We factor using $A^2 - 2AB + B^2 = (A - B)^2$: $A = x$ and $B = 3$.

$$x^2 - 6x + 9 = 0$$
$$(x - 3)^2 = 0$$

Steps 3 and 4. **Set each factor equal to zero and solve the resulting equations.** Because both factors are the same, it is only necessary to set one of them equal to zero.

$$x - 3 = 0$$
$$x = 3$$

Step 5. **Check the solution in the original equation.**

Check 3:

$$x^2 = 6x - 9$$
$$3^2 \overset{?}{=} 6 \cdot 3 - 9$$
$$9 \overset{?}{=} 18 - 9$$
$$9 = 9, \quad \text{true}$$

The solution is 3. ∎

USING TECHNOLOGY

The graph of $y = x^2 - 6x + 9$ is shown below. Notice that there is only one x-intercept, namely 3, verifying that the solution of

$$x^2 - 6x + 9 = 0$$

is 3.

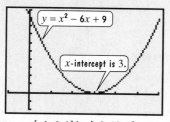

$y = x^2 - 6x + 9$

x-intercept is 3.

$[-1, 6, 1]$ by $[-2, 10, 1]$

✔ **CHECK POINT 4** Solve: $x^2 = 10x - 25$.

EXAMPLE 5 Solving a Quadratic Equation by Factoring

Solve: $9x^2 = 16$.

SOLUTION

Step 1. **Move all terms to one side and obtain zero on the other side.** Subtract 16 from both sides and write the equation in standard form.

$$9x^2 - 16 = 16 - 16$$
$$9x^2 - 16 = 0$$

Step 2. **Factor.** The binomial on the left side is the difference of two squares: $9x^2 - 16 = (3x)^2 - 4^2$. We factor using $A^2 - B^2 = (A + B)(A - B)$: $A = 3x$ and $B = 4$.

$$9x^2 - 16 = 0$$
$$(3x + 4)(3x - 4) = 0$$

Steps 3 and 4. Set each factor equal to zero and solve the resulting equations. We use the zero-product principle to solve $(3x + 4)(3x - 4) = 0$.

$$3x + 4 = 0 \quad \text{or} \quad 3x - 4 = 0$$
$$3x = -4 \qquad\qquad 3x = 4$$
$$x = -\frac{4}{3} \qquad\qquad x = \frac{4}{3}$$

Step 5. Check the solutions in the original equation. Do this now and verify that the solutions of $9x^2 = 16$ are $-\frac{4}{3}$ and $\frac{4}{3}$. ∎

 CHECK POINT 5 Solve: $16x^2 = 25$.

EXAMPLE 6 Solving a Quadratic Equation by Factoring

Solve: $(x - 2)(x + 3) = 6$.

SOLUTION

Step 1. Move all terms to one side and obtain zero on the other side. We write the equation in standard form by multiplying out the product on the left side and then subtracting 6 from both sides.

$$(x - 2)(x + 3) = 6 \qquad \text{This is the given equation.}$$
$$x^2 + 3x - 2x - 6 = 6 \qquad \text{Use the FOIL method.}$$
$$x^2 + x - 6 = 6 \qquad \text{Simplify.}$$
$$x^2 + x - 6 - 6 = 6 - 6 \qquad \text{Subtract 6 from both sides.}$$
$$x^2 + x - 12 = 0 \qquad \text{Simplify.}$$

Step 2. Factor.

$$x^2 + x - 12 = 0$$
$$(x + 4)(x - 3) = 0$$

Steps 3 and 4. Set each factor equal to zero and solve the resulting equations.

$$x + 4 = 0 \quad \text{or} \quad x - 3 = 0$$
$$x = -4 \qquad\qquad x = 3$$

Step 5. Check the solutions in the original equation. Do this now and verify that the solutions are -4 and 3. ∎

 CHECK POINT 6 Solve: $(x - 5)(x - 2) = 28$.

---

3 Solve problems using quadratic equations.

Applications of Quadratic Equations Solving quadratic equations by factoring can be used to answer questions about variables contained in mathematical models.

EXAMPLE 7 Modeling Motion

You throw a ball straight up from a rooftop 160 feet high with an initial speed of 48 feet per second. The formula

$$h = -16t^2 + 48t + 160$$

describes the ball's height above the ground, h, in feet, t seconds after you throw it. The ball misses the rooftop on its way down and eventually strikes the ground. The situation is illustrated in Figure 7.1. How long will it take for the ball to hit the ground?

SOLUTION The ball hits the ground when h, its height above the ground, is 0 feet. Thus, we substitute 0 for h in the given formula and solve for t.

$$h = -16t^2 + 48t + 160 \qquad \text{This is the formula that models the ball's height.}$$
$$0 = -16t^2 + 48t + 160 \qquad \text{Substitute 0 for } h.$$

FIGURE 7.1

It is easier to factor a trinomial with a positive leading coefficient. Thus, if a negative squared term appears in a quadratic equation, we make it positive by multiplying both sides of the equation by -1.

$$-1 \cdot 0 = -1(-16t^2 + 48t + 160)$$
$$0 = 16t^2 - 48t - 160$$

Do you see that each term on the right side of the equation changed sign? The left side of the equation remained zero. Now we continue to solve the equation.

$$16t^2 - 48t - 160 = 0 \qquad \text{Reverse the two sides of the equation. This step is optional.}$$

$$16(t^2 - 3t - 10) = 0 \qquad \text{Factor out the GCF, 16.}$$

Do not set the constant, 16, equal to zero: $16 \neq 0$.

$$16(t - 5)(t + 2) = 0 \qquad \text{Factor the trinomial.}$$

$$t - 5 = 0 \quad \text{or} \quad t + 2 = 0 \qquad \text{Set each variable factor equal to 0.}$$

$$t = 5 \qquad\qquad t = -2 \qquad \text{Solve for } t.$$

Because we begin describing the ball's height at $t = 0$, we discard the solution $t = -2$. The ball hits the ground after 5 seconds.

Figure 7.2 shows the graph of the formula $h = -16t^2 + 48t + 160$. The horizontal axis is labeled t, for the ball's time in motion. The vertical axis is labeled h, for the ball's height above the ground. Because time and height are both positive, the model is graphed in quadrant I only.

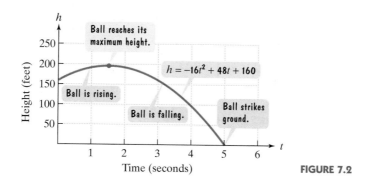

FIGURE 7.2

The graph visually shows what we discovered algebraically: The ball hits the ground after 5 seconds. The graph also reveals that the ball reaches its maximum height, nearly 200 feet, after 1.5 seconds. Then the ball begins to fall.

✔ **CHECK POINT 7** Use the formula $h = -16t^2 + 48t + 160$ to determine when the ball's height is 192 feet. Identify your solutions as points on the graph in Figure 7.2.

In our next example, we use our five-step strategy for solving word problems.

EXAMPLE 8 Solving a Problem About a Rectangle's Area

An architect is allowed no more than 15 square meters to add a small bedroom to a house. Because of the room's design in relationship to the existing structure, the width of its rectangular floor must be 7 meters less than two times the length. Find the precise length and width of the rectangular floor of maximum area that the architect is permitted.

SOLUTION

Step 1. **Let x represent one of the quantities.** We know something about the width: It must be 7 meters less than two times the length. We will let

$$x = \text{the length of the floor.}$$

Step 2. **Represent other quantities in terms of x.** Because the width must be 7 meters less than two times the length, let

$$2x - 7 = \text{the width of the floor.}$$

The problem is illustrated in Figure 7.3.

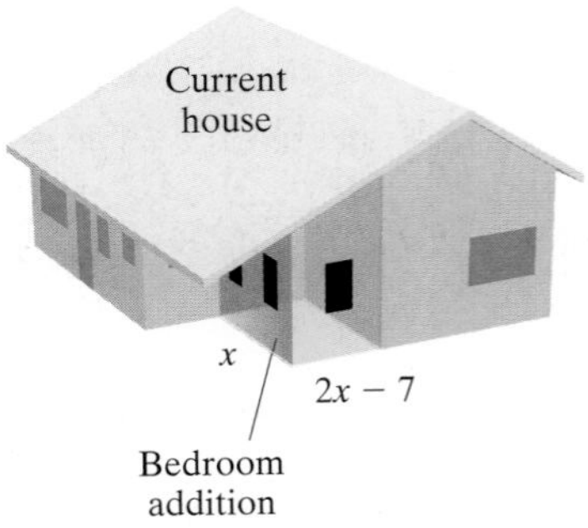

Current house

x

$2x - 7$

Bedroom addition

FIGURE 7.3

Step 3. **Write an equation that describes the conditions.** Because the architect is allowed no more than 15 square meters, an area of 15 square meters is the maximum area permitted. The area of a rectangle is the product of its length and its width.

Length of the floor	times	Width of the floor	is	the area.
x	$\cdot$	$(2x - 7)$	$=$	15

Step 4. **Solve the equation and answer the question.**

$$x(2x - 7) = 15 \qquad \text{This is the equation for the problem's conditions.}$$

$$2x^2 - 7x = 15 \qquad \text{Use the distributive property.}$$

$$2x^2 - 7x - 15 = 0 \qquad \text{Subtract 15 from both sides.}$$

$$(2x + 3)(x - 5) = 0 \qquad \text{Factor.}$$

$$2x + 3 = 0 \quad \text{or} \quad x - 5 = 0 \qquad \text{Set each factor equal to zero.}$$

$$2x = -3 \qquad\qquad x = 5 \qquad \text{Solve the resulting equations.}$$

$$x = -\frac{3}{2}$$

A rectangle cannot have a negative length. Thus,

$$\text{Length} = x = 5$$
$$\text{Width} = 2x - 7 = 2 \cdot 5 - 7 = 10 - 7 = 3.$$

The architect is permitted a room of maximum area whose length is 5 meters and whose width is 3 meters.

Step 5. **Check the proposed solution in the original wording of the problem.** The area of the floor using the dimensions that we found is

$$A = lw = (5 \text{ meters})(3 \text{ meters}) = 15 \text{ square meters.}$$

Because the problem's wording tells us that the maximum area permitted is 15 square meters, our dimensions are correct. ∎

✔ CHECK POINT 8 The length of a rectangular sign is 3 feet longer than the width. If the sign's area is 54 square feet, find its length and width.

7.6 EXERCISE SET

Student Solutions Manual   CD/Video   PH Math/Tutor Center   MathXL Tutorials on CD   MathXL®   MyMathLab   Interactmath.com

Practice Exercises

In Exercises 1–8, solve each equation using the zero-product principle.

1. $x(x + 7) = 0$

2. $x(x - 3) = 0$

3. $(x - 6)(x + 4) = 0$

4. $(x - 3)(x + 8) = 0$

5. $(x - 9)(5x + 4) = 0$

6. $(x + 7)(3x - 2) = 0$

7. $10(x - 4)(2x + 9) = 0$

8. $8(x - 5)(3x + 11) = 0$

In Exercises 9–56, use factoring to solve each quadratic equation. Check by substitution or by using a graphing utility and identifying x-intercepts.

9. $x^2 + 8x + 15 = 0$

10. $x^2 + 5x + 6 = 0$

11. $x^2 - 2x - 15 = 0$

12. $x^2 + x - 42 = 0$

13. $x^2 - 4x = 21$

14. $x^2 + 7x = 18$

15. $x^2 + 9x = -8$

16. $x^2 - 11x = -10$

17. $x^2 + 4x = 0$

18. $x^2 - 6x = 0$

19. $x^2 - 5x = 0$

20. $x^2 + 3x = 0$

21. $x^2 = 4x$

22. $x^2 = 8x$

23. $2x^2 = 5x$

24. $3x^2 = 5x$

25. $3x^2 = -5x$

26. $2x^2 = -3x$

27. $x^2 + 4x + 4 = 0$

28. $x^2 + 6x + 9 = 0$

29. $x^2 = 12x - 36$

30. $x^2 = 14x - 49$

31. $4x^2 = 12x - 9$

32. $9x^2 = 30x - 25$

33. $2x^2 = 7x + 4$

34. $3x^2 = x + 4$

35. $5x^2 = 18 - x$

36. $3x^2 = 15 + 4x$

37. $x^2 - 49 = 0$

38. $x^2 - 25 = 0$

39. $4x^2 - 25 = 0$

40. $9x^2 - 100 = 0$

41. $81x^2 = 25$

42. $25x^2 = 49$

43. $x(x - 4) = 21$

44. $x(x - 3) = 18$

45. $4x(x + 1) = 15$

46. $x(3x + 8) = -5$

47. $(x - 1)(x + 4) = 14$

48. $(x - 3)(x + 8) = -30$

49. $(x + 1)(2x + 5) = -1$

50. $(x + 3)(3x + 5) = 7$

51. $y(y + 8) = 16(y - 1)$

52. $y(y + 9) = 4(2y + 5)$

53. $4y^2 + 20y + 25 = 0$

54. $4y^2 + 44y + 121 = 0$

55. $64w^2 = 48w - 9$

56. $25w^2 = 80w - 64$

Practice Plus

In Exercises 57–66, solve each equation and check your solutions.

57. $(x - 4)(x^2 + 5x + 6) = 0$

58. $(x - 5)(x^2 - 3x + 2) = 0$

59. $x^3 - 36x = 0$

60. $x^3 - 4x = 0$

61. $y^3 + 3y^2 + 2y = 0$

62. $y^3 + 2y^2 - 3y = 0$

63. $2(x - 4)^2 + x^2 = x(x + 50) - 46x$

64. $(x - 4)(x - 5) + (2x + 3)(x - 1) = x(2x - 25) - 13$

65. $(x - 2)^2 - 5(x - 2) + 6 = 0$

66. $(x - 3)^2 + 2(x - 3) - 8 = 0$

Application Exercises

A ball is thrown straight up from a rooftop 300 feet high. The formula

$$h = -16t^2 + 20t + 300$$

describes the ball's height above the ground, h, in feet, t seconds after it was thrown. The ball misses the rooftop on its way down and eventually strikes the ground. The graph of the formula is shown, with tick marks omitted along the horizontal axis. Use the formula to solve Exercises 67–69.

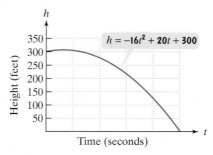

67. How long will it take for the ball to hit the ground? Use this information to provide tick marks with appropriate numbers along the horizontal axis in the figure shown.

68. When will the ball's height be 304 feet? Identify the solution as a point on the graph.

69. When will the ball's height be 276 feet? Identify the solution as a point on the graph.

An explosion causes debris to rise vertically with an initial speed of 72 feet per second. The formula

$$h = -16t^2 + 72t$$

describes the height of the debris above the ground, h, in feet, t seconds after the explosion. Use this information to solve Exercises 70–71.

70. How long will it take for the debris to hit the ground?

71. When will the debris be 32 feet above the ground?

The formula

$$N = 2x^2 + 22x + 320$$

models the number of inmates, N, in thousands, in U.S. state and federal prisons x years after 1980. The graph of the formula is shown in a $[0, 20, 1]$ by $[0, 1600, 100]$ viewing rectangle. Use the formula to solve Exercises 72–73.

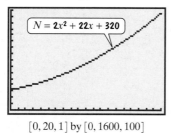

$[0, 20, 1]$ by $[0, 1600, 100]$

72. In which year were there 740 thousand inmates in U.S. state and federal prisons? Identify the solution as a point on the graph shown.

73. In which year were there 1100 thousand inmates in U.S. state and federal prisons? Identify the solution as a point on the graph shown.

The alligator, an endangered species, is the subject of a protection program. The formula

$$P = -10x^2 + 475x + 3500$$

models the alligator population, P, after x years of the protection program, where $0 \le x \le 12$. Use the formula to solve Exercises 74–75.

74. After how long is the population up to 5990?

75. After how long is the population up to 7250?

The graph of the alligator population is shown over time. Use the graph to solve Exercises 76–77.

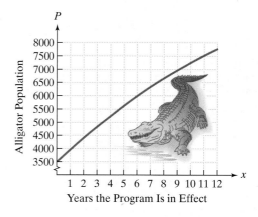

76. Identify your solution in Exercise 74 as a point on the graph.

77. Identify your solution in Exercise 75 as a point on the graph.

The formula

$$N = \frac{t^2 - t}{2}$$

describes the number of football games, N, that must be played in a league with t teams if each team is to play every other team once. Use this information to solve Exercises 78–79.

78. If a league has 36 games scheduled, how many teams belong to the league, assuming that each team plays every other team once?

79. If a league has 45 games scheduled, how many teams belong to the league, assuming that each team plays every other team once?

80. The length of a rectangular garden is 5 feet greater than the width. The area of the rectangle is 300 square feet. Find the length and the width.

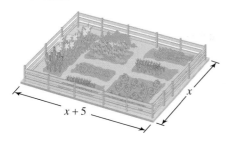

81. A rectangular parking lot has a length that is 3 yards greater than the width. The area of the parking lot is 180 square yards. Find the length and the width.

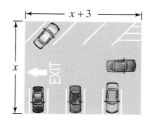

82. Each end of a glass prism is a triangle with a height that is 1 inch shorter than twice the base. If the area of the triangle is 60 square inches, how long are the base and height?

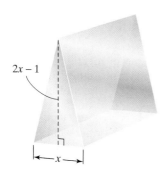

83. Great white sharks have triangular teeth with a height that is 1 centimeter longer than the base. If the area of one tooth is 15 square centimeters, find its base and height.

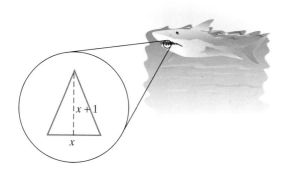

84. A vacant rectangular lot is being turned into a community vegetable garden measuring 15 meters by 12 meters. A path of uniform width is to surround the garden. If the area of the lot is 378 square meters, find the width of the path surrounding the garden.

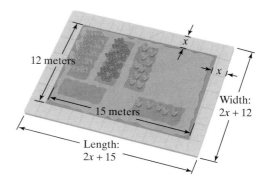

85. As part of a landscaping project, you put in a flower bed measuring 10 feet by 12 feet. You plan to surround the bed with a uniform border of low-growing plants.

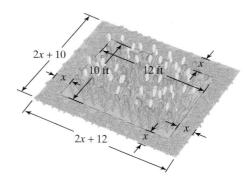

a. Write a polynomial that describes the area of the uniform border that surrounds your flower bed. (*Hint*: The area of the border is the area of the large rectangle shown in the figure minus the area of the flower bed.)

b. The low-growing plants surrounding the flower bed require 1 square foot each when mature. If you have 168 of these plants, how wide a strip around the flower bed should you prepare for the border?

Writing in Mathematics

86. What is a quadratic equation?

87. Explain how to solve $x^2 + 6x + 8 = 0$ using factoring and the zero-product principle.

88. If $(x + 2)(x - 4) = 0$ indicates that $x + 2 = 0$ or $x - 4 = 0$, explain why $(x + 2)(x - 4) = 6$ does not mean $x + 2 = 6$ or $x - 4 = 6$. Could we solve the equation using $x + 2 = 3$ and $x - 4 = 2$ because $3 \cdot 2 = 6$?

Critical Thinking Exercises

89. Which one of the following is true?

   a. If $(x + 3)(x - 4) = 2$, then $x + 3 = 0$ or $x - 4 = 0$.

   b. The solutions of the equation $4(x - 5)(x + 3) = 0$ are $4, 5,$ and -3.

   c. Equations solved by factoring always have two different solutions.

   d. Both 0 and $-\pi$ are solutions of the equation $x(x + \pi) = 0$.

90. Write a quadratic equation in standard form whose solutions are -3 and 5.

In Exercises 91–93, solve each equation.

91. $x^3 - x^2 - 16x + 16 = 0$

92. $3^{x^2 - 9x + 20} = 1$ **93.** $(x^2 - 5x + 5)^3 = 1$

In Exercises 94–97, match each equation with its graph. The graphs are labeled (a) through (d).

94. $y = x^2 - x - 2$ **95.** $y = x^2 + x - 2$

96. $y = x^2 - 4$ **97.** $y = x^2 - 4x$

 a.

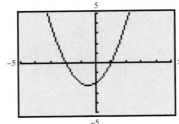

b.

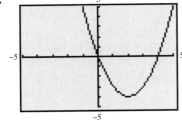

c.

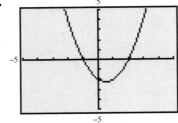

d.

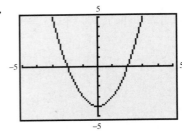

Technology Exercises

In Exercises 98–102, use the x-intercepts for the graph in a $[-10, 10, 1]$ by $[-13, 10, 1]$ viewing rectangle to solve the quadratic equation. Check by substitution.

98. Use the graph of $y = x^2 + 3x - 4$ to solve

$$x^2 + 3x - 4 = 0.$$

99. Use the graph of $y = x^2 + x - 6$ to solve

$$x^2 + x - 6 = 0.$$

100. Use the graph of $y = (x - 2)(x + 3) - 6$ to solve

$$(x - 2)(x + 3) - 6 = 0.$$

101. Use the graph of $y = x^2 - 2x + 1$ to solve

$$x^2 - 2x + 1 = 0.$$

102. Use the technique of identifying x-intercepts on a graph generated by a graphing utility to check any five equations that you solved in Exercises 9–56.

103. If you have access to a calculator that solves quadratic equations, consult the owner's manual to determine how to use this feature. Then use your calculator to solve any five of the equations in Exercises 9–56.

Review Exercises

104. Graph: $y > -\dfrac{2}{3}x + 1$. (Section 4.6, Example 2)

105. Simplify: $\left(\dfrac{8x^4}{4x^7}\right)^2$. (Section 6.7, Example 6)

106. Solve: $5x + 28 = 6 - 6x$. (Section 2.2, Example 7)

GROUP PROJECT

CHAPTER 7

Group members are on the board of a condominium association. The condominium has just installed a small 35-foot-by-30-foot pool. Your job is to choose a material to surround the pool to create a border of uniform width.

a. Begin by writing an algebraic expression for the area, in square feet, of the border around the pool. (*Hint:* The border's area is the combined area of the pool and border minus the area of the pool.)

b. You must select one of the following options for the border.

Options for the Border	Price
Cement	$6 per square foot
Outdoor carpeting	$5 per square foot plus $10 per foot to install edging around the rectangular border
Brick	$8 per square foot plus a $60 charge for delivering the bricks

Write an algebraic expression for the cost of installing the border for each of these options.

c. You would like the border to be 5 feet wide. Use the algebraic expressions in part (b) to find the cost of the border for each of the three options.

d. You would prefer not to use cement. However, the condominium association is limited by a $5000 budget. Given this limitation, approximately how wide can the border be using outdoor carpeting or brick? Which option should you select and why?

CHAPTER 7 SUMMARY

Definitions and Concepts	Examples

Section 7.1 The Greatest Common Factor and Factoring By Grouping

Factoring a polynomial containing the sum of monomials means finding an equivalent expression that is a product. The greatest common factor, GCF, is an expression that divides every term of the polynomial. The variable part of the GCF contains the smallest power of a variable that appears in all terms of the polynomial.	Find the GCF of $16x^2y$, $20x^3y^2$, and $8x^2y^3$. The GCF of 16, 20, and 8 is 4. The GCF of x^2, x^3, and x^2 is x^2. The GCF of y, y^2, and y^3 is y. $$\text{GCF} = 4 \cdot x^2 \cdot y = 4x^2y$$
To factor a monomial from a polynomial, express each term as the product of the GCF and its other factor. Then use the distributive property to factor out the GCF.	$$16x^2y + 20x^3y^2 + 8x^2y^3$$ $$= 4x^2y \cdot 4 + 4x^2y \cdot 5xy + 4x^2y \cdot 2y^2$$ $$= 4x^2y(4 + 5xy + 2y^2)$$
To factor by grouping, factor out the GCF from each group. Then factor out the remaining common factor.	$$xy + 5x - 3y - 15$$ $$= x(y + 5) - 3(y + 5)$$ $$= (y + 5)(x - 3)$$

Definitions and Concepts	Examples

Section 7.2 Factoring Trinomials Whose Leading Coefficient Is One

To factor a trinomial of the form $x^2 + bx + c$, find two numbers whose product is c and whose sum is b. The factorization is $(x + \text{one number})(x + \text{other number})$.	Factor: $x^2 + 9x + 20$. Find two numbers whose product is 20 and whose sum is 9. The numbers are 4 and 5. $x^2 + 9x + 20 = (x + 4)(x + 5)$

Section 7.3 Factoring Trinomials Whose Leading Coefficient Is Not One

To factor $ax^2 + bx + c$ by trial and error, try various combinations of factors of ax^2 and c until a middle term of bx is obtained for the sum of outside and inside products.	Factor: $3x^2 + 7x - 6$. Factors of $3x^2$: $3x, x$ Factors of -6: 1 and -6, -1 and 6, 2 and -3, -2 and 3. A possible combination of these factors is $(3x - 2)(x + 3)$. Sum of outside and inside products should equal $7x$. $9x - 2x = 7x$ Thus, $3x^2 + 7x - 6 = (3x - 2)(x + 3)$.
To factor $ax^2 + bx + c$ by grouping, find the factors of ac whose sum is b. Write bx using these factors. Then factor by grouping.	Factor: $3x^2 + 7x - 6$. Find the factors of $3(-6)$, or -18, whose sum is 7. They are 9 and -2. $3x^2 + 7x - 6$ $= 3x^2 + 9x - 2x - 6$ $= 3x(x + 3) - 2(x + 3) = (x + 3)(3x - 2)$

Section 7.4 Factoring Special Forms

The Difference of Two Squares $A^2 - B^2 = (A + B)(A - B)$	$9x^2 - 25y^2$ $= (3x)^2 - (5y)^2 = (3x + 5y)(3x - 5y)$
Perfect Square Trinomials $A^2 + 2AB + B^2 = (A + B)^2$ $A^2 - 2AB + B^2 = (A - B)^2$	$x^2 + 16x + 64 = x^2 + 2 \cdot x \cdot 8 + 8^2 = (x + 8)^2$ $25x^2 - 30x + 9 = (5x)^2 - 2 \cdot 5x \cdot 3 + 3^2 = (5x - 3)^2$
Sum and Difference of Cubes $A^3 + B^3 = (A + B)(A^2 - AB + B^2)$ $A^3 - B^3 = (A - B)(A^2 + AB + B^2)$	$8x^3 - 125 = (2x)^3 - 5^3$ $= (2x - 5)[(2x)^2 + 2x \cdot 5 + 5^2]$ $= (2x - 5)(4x^2 + 10x + 25)$

Definitions and Concepts | Examples

Section 7.5 A General Factoring Strategy

A Factoring Strategy

1. Factor out the GCF.

2. a. If two terms, try
$$A^2 - B^2 = (A + B)(A - B)$$
$$A^3 + B^3 = (A + B)(A^2 - AB + B^2)$$
$$A^3 - B^3 = (A - B)(A^2 + AB + B^2).$$

b. If three terms, try
$$A^2 + 2AB + B^2 = (A + B)^2$$
$$A^2 - 2AB + B^2 = (A - B)^2.$$

If not a perfect square trinomial, try trial and error or grouping.

c. If four terms, try factoring by grouping.

3. See if any factors can be factored further.

4. Check by multiplying.

Factor: $2x^4 + 10x^3 - 8x^2 - 40x$.
The GCF is $2x$.

$$2x^4 + 10x^3 - 8x^2 - 40x$$
$$= 2x(x^3 + 5x^2 - 4x - 20)$$

Four terms: Try grouping.

$$= 2x[x^2(x + 5) - 4(x + 5)]$$
$$= 2x(x + 5)(x^2 - 4)$$

This can be factored further.

$$= 2x(x + 5)(x + 2)(x - 2)$$

Section 7.6 Solving Quadratic Equations by Factoring

The Zero-Product Principle
If $AB = 0$, then $A = 0$ or $B = 0$.

Solve: $(x - 6)(x + 10) = 0$
$$x - 6 = 0 \quad \text{or} \quad x + 10 = 0$$
$$x = 6 \qquad\qquad x = -10$$

A quadratic equation in x is an equation that can be written in the standard form
$$ax^2 + bx + c = 0, \quad a \neq 0.$$

To solve by factoring, write the equation in standard form, factor, set each factor equal to zero, and solve each resulting equation. Check proposed solutions in the original equation.

Solve: $4x^2 + 9x = 9$.
$$4x^2 + 9x - 9 = 0$$
$$(4x - 3)(x + 3) = 0$$
$$4x - 3 = 0 \quad \text{or} \quad x + 3 = 0$$
$$x = \frac{3}{4} \qquad\qquad x = -3$$

CHAPTER 7 REVIEW EXERCISES

7.1 *In Exercises 1–5, factor each polynomial using the greatest common factor. If there is no common factor other than 1 and the polynomial cannot be factored, so state.*

1. $30x - 45$

2. $12x^3 + 16x^2 - 400x$

3. $30x^4y + 15x^3y + 5x^2y$

4. $7(x + 3) - 2(x + 3)$

5. $7x^2(x + y) - (x + y)$

In Exercises 6–9, factor by grouping.

6. $x^3 + 3x^2 + 2x + 6$

7. $xy + y + 4x + 4$

8. $x^3 + 5x + x^2 + 5$

9. $xy + 4x - 2y - 8$

7.2 *In Exercises 10–17, factor completely, or state that the trinomial is prime.*

10. $x^2 - 3x + 2$

11. $x^2 - x - 20$

12. $x^2 + 19x + 48$

13. $x^2 - 6xy + 8y^2$

14. $x^2 + 5x - 9$

15. $x^2 + 16xy - 17y^2$

16. $3x^2 + 6x - 24$

17. $3x^3 - 36x^2 + 33x$

7.3 *In Exercises 18–26, factor completely, or state that the trinomial is prime.*

18. $3x^2 + 17x + 10$

19. $5y^2 - 17y + 6$

20. $4x^2 + 4x - 15$

21. $5y^2 + 11y + 4$

22. $8x^2 + 8x - 6$

23. $2x^3 + 7x^2 - 72x$

24. $12y^3 + 28y^2 + 8y$

25. $2x^2 - 7xy + 3y^2$

26. $5x^2 - 6xy - 8y^2$

7.4 *In Exercises 27–30, factor each difference of two squares completely.*

27. $4x^2 - 1$

28. $81 - 100y^2$

29. $25a^2 - 49b^2$

30. $z^4 - 16$

In Exercises 31–34, factor completely, or state that the polynomial is prime.

31. $2x^2 - 18$

32. $x^2 + 1$

33. $9x^3 - x$

34. $18xy^2 - 8x$

In Exercises 35–41, factor any perfect square trinomials, or state that the polynomial is prime.

35. $x^2 + 22x + 121$

36. $x^2 - 16x + 64$

37. $9y^2 + 48y + 64$

38. $16x^2 - 40x + 25$

39. $25x^2 + 15x + 9$

40. $36x^2 + 60xy + 25y^2$

41. $25x^2 - 40xy + 16y^2$

In Exercises 42–45, factor using the formula for the sum or difference of two cubes.

42. $x^3 - 27$

43. $64x^3 + 1$

44. $54x^3 - 16y^3$

45. $27x^3y + 8y$

In Exercises 46–47, find the formula for the area of the shaded region and express it in factored form.

46.

47.

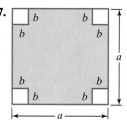

48. The figure shows a geometric interpretation of a factorization. Use the sum of the areas of the four pieces on the left and the area of the square on the right to write the factorization that is illustrated.

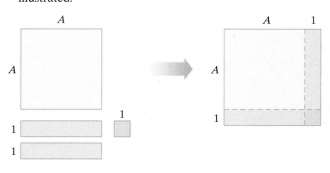

7.5 *In Exercises 49–81, factor completely, or state that the polynomial is prime.*

49. $x^3 - 8x^2 + 7x$

50. $10y^2 + 9y + 2$

51. $128 - 2y^2$

52. $9x^2 + 6x + 1$

53. $20x^7 - 36x^3$

54. $x^3 - 3x^2 - 9x + 27$

55. $y^2 + 16$

56. $2x^3 + 19x^2 + 35x$

57. $3x^3 - 30x^2 + 75x$

58. $3x^5 - 24x^2$

59. $4y^4 - 36y^2$

60. $5x^2 + 20x - 105$

61. $9x^2 + 8x - 3$

62. $10x^5 - 44x^4 + 16x^3$

63. $100y^2 - 49$

64. $9x^5 - 18x^4$

65. $x^4 - 1$

66. $2y^3 - 16$

67. $x^3 + 64$

68. $6x^2 + 11x - 10$

69. $3x^4 - 12x^2$

70. $x^2 - x - 90$

71. $25x^2 + 25xy + 6y^2$

72. $x^4 + 125x$

73. $32y^3 + 32y^2 + 6y$

74. $2y^2 - 16y + 32$

75. $x^2 - 2xy - 35y^2$

76. $x^2 + 7x + xy + 7y$

77. $9x^2 + 24xy + 16y^2$

78. $2x^4y - 2x^2y$

79. $100y^2 - 49z^2$

80. $x^2 + xy + y^2$

81. $3x^4y^2 - 12x^2y^4$

7.6 In Exercises 82–83, solve each equation using the zero-product principle.

82. $x(x - 12) = 0$

83. $3(x - 7)(4x + 9) = 0$

In Exercises 84–92, use factoring to solve each quadratic equation.

84. $x^2 + 5x - 14 = 0$

85. $5x^2 + 20x = 0$

86. $2x^2 + 15x = 8$

87. $x(x - 4) = 32$

88. $(x + 3)(x - 2) = 50$

89. $x^2 = 14x - 49$

90. $9x^2 = 100$

91. $3x^2 + 21x + 30 = 0$

92. $3x^2 = 22x - 7$

93. You dive from a board that is 32 feet above the water. The formula

$$h = -16t^2 + 16t + 32$$

describes your height above the water, h, in feet, t seconds after you dive. How long will it take you to hit the water?

94. The length of a rectangular sign is 3 feet longer than the width. If the sign has space for 40 square feet of advertising, find its length and its width.

95. The square lot shown here is being turned into a garden with a 3-meter path at one end. If the area of the garden is 88 square meters, find the dimensions of the square lot.

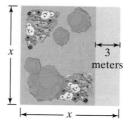

CHAPTER 7   TEST

 Remember to use your Chapter Test Prep Video CD to see the worked-out solutions to the test questions you want to review.

In Exercises 1–21, factor completely, or state that the polynomial is prime.

1. $x^2 - 9x + 18$

2. $x^2 - 14x + 49$

3. $15y^4 - 35y^3 + 10y^2$

4. $x^3 + 2x^2 + 3x + 6$

5. $x^2 - 9x$

6. $x^3 + 6x^2 - 7x$

7. $14x^2 + 64x - 30$

8. $25x^2 - 9$

9. $x^3 + 8$

10. $x^2 - 4x - 21$

11. $x^2 + 4$

12. $6y^3 + 9y^2 + 3y$

13. $4y^2 - 36$

14. $16x^2 + 48x + 36$

15. $2x^4 - 32$

16. $36x^2 - 84x + 49$

17. $7x^2 - 50x + 7$

18. $x^3 + 2x^2 - 5x - 10$

19. $12y^3 - 12y^2 - 45y$

20. $y^3 - 125$

21. $5x^2 - 5xy - 30y^2$

In Exercises 22–27, solve ech quadratic eqution.

22. $x^2 + 2x - 24 = 0$

23. $3x^2 - 5x = 2$

24. $x(x - 6) = 16$

25. $6x^2 = 21x$

26. $16x^2 = 81$

27. $(5x + 4)(x - 1) = 2$

28. Find a formula for the area of the shaded region and express it in factored form.

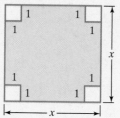

29. A model rocket is launched from a height of 96 feet. The formula

$$h = -16t^2 + 80t + 96$$

describes the rocket's height, h, in feet, t seconds after it was launched. How long will it take the rocket to reach the ground?

30. The length of a rectangular garden is 6 feet longer than its width. If the area of the garden is 55 square feet, find its length and its width.

CUMULATIVE REVIEW EXERCISES (CHAPTERS 1–7)

1. Simplify: $6[5 + 2(3 - 8) - 3]$.

2. Solve: $4(x - 2) = 2(x - 4) + 3x$.

3. Solve: $\dfrac{x}{2} - 1 = \dfrac{x}{3} + 1$.

4. Solve and express the solution set in set-builder notation. Graph the solution set on a number line.

$$5 - 5x > 2(5 - x) + 1$$

5. Find the measures of the angles of a triangle whose two base angles have equal measure and whose third angle is $10°$ less than three times the measure of a base angle.

6. A dinner for six people cost $159, including a 6% tax. What was the dinner's cost before tax?

7. Graph using the slope and y-intercept: $y = -\frac{3}{5}x + 3$.

8. Write the point-slope form of the line passing through $(2, -4)$ and $(3, 1)$. Then use the point-slope form of the equation to write the slope-intercept equation.

9. Graph: $5x - 6y > 30$.

10. Solve the system:

$$5x + 2y = 13$$
$$y = 2x - 7.$$

11. Solve the system:

$$2x + 3y = 5$$
$$3x - 2y = -4.$$

12. Subtract: $\dfrac{4}{5} - \dfrac{9}{8}$.

In Exercises 13–15, perform the indicated operations.

13. $\dfrac{6x^5 - 3x^4 + 9x^2 + 27x}{3x}$

14. $(3x - 5y)(2x + 9y)$

15. $\dfrac{6x^3 + 5x^2 - 34x + 13}{3x - 5}$

16. Write 0.0071 in scientific notation.

In Exercises 17–19, factor completely.

17. $3x^2 + 11x + 6$

18. $y^5 - 16y$

19. $4x^2 + 12x + 9$

20. The length of a rectangle is 2 feet greater than its width. If the rectangle's area is 24 square feet, find its dimensions.

At the start of the twenty-first century, we are plagued by questions about the environment. Will we run out of gas? How hot will it get? Will there be neighborhoods where the air is pristine? Can we make garbage disappear? Will there be any wilderness left? Which wild animals will become extinct? How much will it cost to clean up toxic wastes from our rivers so that they can safely provide food, recreation, and enjoyment of wildlife for the millions who live along and visit their shores?

The role of algebraic fractions in modeling environmental issues is introduced in Section 8.1 and developed in Example 6 in Section 8.6.

Rational Expressions

When making decisions on public policies dealing with the environment, two important questions are

- What are the costs?
- What are the benefits?

Algebraic fractions play an important role in modeling the costs. By learning to work with these fractional expressions, you will gain new insights into phenomena as diverse as the dosage of drugs prescribed for children, inventory costs for a business, the cost of environmental cleanup, and even the shape of our heads.

SECTION 8.1

Objectives

1 Find numbers for which a rational expression is undefined.

2 Simplify rational expressions.

3 Solve applied problems involving rational expressions.

DISCOVER FOR YOURSELF

What happens if you try substituting 100 for x in

$$\frac{250x}{100 - x}?$$

What does this tell you about the cost of cleaning up all of the river's pollutants?

RATIONAL EXPRESSIONS AND THEIR SIMPLIFICATION

How do we describe the costs of reducing environmental pollution? We often use algebraic expressions involving quotients of polynomials. For example, the algebraic expression

$$\frac{250x}{100 - x}$$

describes the cost, in millions of dollars, to remove x percent of the pollutants that are discharged into a river. Removing a modest percentage of pollutants, say 40%, is far less costly than removing a substantially greater percentage, such as 95%. We see this by evaluating the algebraic expression for $x = 40$ and $x = 95$.

Evaluating $\dfrac{250x}{100 - x}$ for

$x = 40$: $x = 95$:

Cost is $\dfrac{250(40)}{100 - 40} \approx 167.$ Cost is $\dfrac{250(95)}{100 - 95} = 4750.$

The cost increases from approximately $167 million to a possibly prohibitive $4750 million, or $4.75 billion. Costs spiral upward as the percentage of removed pollutants increases.

Many algebraic expressions that describe costs of environmental projects are examples of *rational expressions*. In this section, we introduce rational expressions and their simplification.

1 Find numbers for which a rational expression is undefined.

Excluding Numbers from Rational Expressions A **rational expression** is the quotient of two polynomials. Some examples are

$$\frac{x - 2}{4}, \quad \frac{4}{x - 2}, \quad \frac{x}{x^2 - 1}, \quad \text{and} \quad \frac{x^2 + 1}{x^2 + 2x - 3}.$$

Rational expressions indicate division and division by zero is undefined. This means that **we must exclude any value or values of the variable that make a denominator zero.** For example, consider the rational expression

$$\frac{4}{x - 2}.$$

When x is replaced with 2, the denominator is 0 and the expression is undefined.

$$\text{If } x = 2: \quad \frac{4}{x-2} = \frac{4}{2-2} = \frac{4}{0} \quad \boxed{\text{Division by zero is undefined.}}$$

Notice that if x is replaced by a number other than 2, such as 1, the expression is defined because the denominator is nonzero.

$$\text{If } x = 1: \quad \frac{4}{x-2} = \frac{4}{1-2} = \frac{4}{-1} = -4.$$

Thus, only 2 must be excluded as a replacement for x in the rational expression $\dfrac{4}{x-2}$.

USING TECHNOLOGY

We can use the $\boxed{\text{TABLE}}$ feature of a graphing utility to verify our work with $\dfrac{4}{x-2}$. Enter

$$y_1 = 4 \boxed{\div} \boxed{(} \boxed{(} \boxed{x} \boxed{-} 2 \boxed{)}$$

and press $\boxed{\text{TABLE}}$.

$$y_1 = \frac{4}{x-2}$$

X	Y1
-3	-.8
-2	-1
-1	-1.333
0	-2
1	-4
2	ERROR
3	4

X= -3

This verifies that if $x = 1$, the value of $\dfrac{4}{x-2}$ is -4.

This verifies that 2 must be excluded as a replacement for x.

EXCLUDING VALUES FROM RATIONAL EXPRESSIONS   If a variable in a rational expression is replaced by a number that causes the denominator to be 0, that number must be excluded as a replacement for the variable. The rational expression is undefined at any value that produces a denominator of 0.

How do we determine the value or values of the variable for which a rational expression is undefined? Set the denominator equal to 0 and then solve the resulting equation for the variable.

EXAMPLE 1   Determining Numbers for Which Rational Expressions Are Undefined

Find all the numbers for which the rational expression is undefined:

a. $\dfrac{6x+12}{7x-28}$ **b.** $\dfrac{2x+6}{x^2+3x-10}$.

SOLUTION   In each case, we set the denominator equal to 0 and solve.

$$\frac{6x+12}{7x-28} \qquad \boxed{\text{Exclude values of } x \text{ that make these denominators 0.}} \qquad \frac{2x+6}{x^2+3x-10}$$

a. $7x - 28 = 0$ Set the denominator of $\dfrac{6x + 12}{7x - 28}$ equal to 0.

$\qquad 7x = 28$ Add 28 to both sides.

$\qquad\quad x = 4$ Divide both sides by 7.

Thus, $\dfrac{6x + 12}{7x - 28}$ is undefined for $x = 4$.

b. $x^2 + 3x - 10 = 0$ Set the denominator of $\dfrac{2x + 6}{x^2 + 3x - 10}$ equal to 0.

$\qquad (x + 5)(x - 2) = 0$ Factor.

$\qquad x + 5 = 0 \quad \text{or} \quad x - 2 = 0$ Set each factor equal to 0.

$\qquad\quad x = -5 \qquad\qquad x = 2$ Solve the resulting equations.

Thus, $\dfrac{2x + 6}{x^2 + 3x - 10}$ is undefined for $x = -5$ and $x = 2$.

USING TECHNOLOGY

When using a graphing utility to graph an equation containing a rational expression, you might not be pleased with the quality of the display. Compare these two graphs of $y = \dfrac{6x + 12}{7x - 28}$.

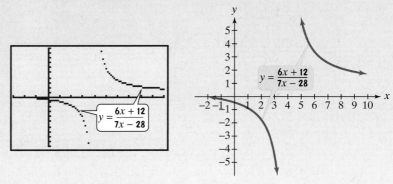

The graph on the left was obtained using the $\boxed{\text{DOT}}$ mode in a $[-3, 10, 1]$ by $[-10, 10, 1]$ viewing rectangle. Examine the behavior of the graph near $x = 4$, the number for which the rational expression is undefined. The values of the rational expression are decreasing as the values of x get closer to 4 on the left and increasing as the values of x get closer to 4 on the right. However, there is no point on the graph corresponding to $x = 4$. Would you agree that this behavior is better illustrated in the hand-drawn graph on the right?

 CHECK POINT 1 Find all the numbers for which the rational expression is undefined:

a. $\dfrac{7x - 28}{8x - 40}$ **b.** $\dfrac{8x - 40}{x^2 + 3x - 28}$.

Is every rational expression undefined for at least one number? No. Consider

$$\frac{x - 2}{4}.$$

Because the denominator is not zero for any value of x, the rational expression is defined for all real numbers. Thus, it is not necessary to exclude any values for x.

2 Simplify rational expressions.

Simplifying Rational Expressions A rational expression is **simplified** if its numerator and denominator have no common factors other than 1 or −1. The following principle is used to simplify a rational expression:

FUNDAMENTAL PRINCIPLE OF RATIONAL EXPRESSIONS If P, Q, and R are polynomials, and Q and R are not 0,

$$\frac{PR}{QR} = \frac{P}{Q}.$$

As you read the Fundamental Principle, can you see why $\dfrac{PR}{QR}$ is not simplified? The numerator and denominator have a common factor, the polynomial R. By dividing the numerator and the denominator by the common factor, R, we obtain the simplified form $\dfrac{P}{Q}$. This is often shown as follows:

$$\frac{P\overset{1}{R}}{Q\underset{1}{R}} = \frac{P}{Q}.$$

Observe that
$$\frac{PR}{QR} = \frac{P}{Q} \cdot \frac{R}{R} = \frac{P}{Q} \cdot 1 = \frac{P}{Q}.$$

The following procedure can be used to simplify rational expressions:

SIMPLIFYING RATIONAL EXPRESSIONS

1. Factor the numerator and the denominator completely.
2. Divide both the numerator and the denominator by any common factors.

EXAMPLE 2 Simplifying a Rational Expression

Simplify: $\dfrac{5x + 35}{20x}$.

SOLUTION

$$\frac{5x + 35}{20x} = \frac{5(x + 7)}{5 \cdot 4x}$$ Factor the numerator and denominator. Because the denominator is 20x, x ≠ 0.

$$= \frac{\overset{1}{5}(x + 7)}{\underset{1}{5} \cdot 4x}$$ Divide out the common factor of 5.

$$= \frac{x + 7}{4x}$$ ∎

✔ **CHECK POINT 2** Simplify: $\dfrac{7x + 28}{21x}$.

EXAMPLE 3 Simplifying a Rational Expression

Simplify: $\dfrac{x^3 + x^2}{x + 1}$.

SOLUTION

$$\frac{x^3 + x^2}{x + 1} = \frac{x^2(x + 1)}{x + 1} \qquad \text{Factor the numerator. Because the denominator is } x + 1, x \neq -1.$$

$$= \frac{x^2 \overset{1}{\cancel{(x + 1)}}}{\underset{1}{\cancel{x + 1}}} \qquad \text{Divide out the common factor of } x + 1.$$

$$= x^2 \qquad\blacksquare$$

Simplifying a rational expression can change the numbers that make it undefined. For example, we just showed that

$$\frac{x^3 + x^2}{x + 1} = x^2.$$

This is undefined for $x = -1$.

This simplified form is defined for all real numbers.

Thus, to equate the two expressions, we must restrict the values for x in the simplified expression to exclude -1. We can write

$$\frac{x^3 + x^2}{x + 1} = x^2, \quad x \neq -1.$$

Hereafter, we will assume that the simplified rational expression is equal to the original rational expression for all real numbers except those for which either denominator is 0.

USING TECHNOLOGY

A graphing utility can be used to verify that

$$\frac{x^3 + x^2}{x + 1} = x^2, \quad x \neq -1.$$

Enter $y_1 = \dfrac{x^3 + x^2}{x + 1}$ and $y_2 = x^2$.

Graphic Check

The graphs of y_1 and y_2 appear to be identical. You can use the TRACE feature to trace y_1 and show that it is undefined for $x = -1$.

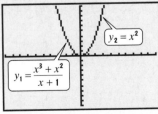

$y_2 = x^2$

$y_1 = \dfrac{x^3 + x^2}{x + 1}$

$[-10, 10, 1]$ by $[-10, 10, 1]$

Numeric Check

No matter how far up or down we scroll, if $x \neq -1$, $y_1 = y_2$. If $x = -1$, y_1 is undefined, although the value of y_2 is 1.

X	Y₁	Y₂
-3	9	9
-2	4	4
-1	ERROR	1
0	0	0
1	1	1
2	4	4
3	9	9

X = -3

✔ **CHECK POINT 3** Simplify: $\dfrac{x^3 - x^2}{7x - 7}$.

EXAMPLE 4 Simplifying a Rational Expression

Simplify: $\dfrac{x^2 + 6x + 5}{x^2 - 25}$.

SOLUTION

$$\dfrac{x^2 + 6x + 5}{x^2 - 25} = \dfrac{(x + 5)(x + 1)}{(x + 5)(x - 5)}$$

Factor the numerator and denominator. Because the denominator is $(x + 5)(x - 5)$, $x \neq -5$ and $x \neq 5$.

$$= \dfrac{\overset{1}{\cancel{(x + 5)}}(x + 1)}{\underset{1}{\cancel{(x + 5)}}(x - 5)}$$

Divide out the common factor of $x + 5$.

$$= \dfrac{x + 1}{x - 5}$$

■

✔ **CHECK POINT 4** Simplify: $\dfrac{x^2 - 1}{x^2 + 2x + 1}$.

STUDY TIP

When simplifying rational expressions, only *factors* that are common to the *entire numerator* and the *entire denominator* can be divided out. **It is incorrect to divide out common terms from the numerator and denominator.**

Incorrect!

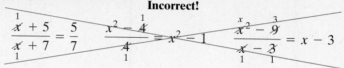

The first two expressions, $\dfrac{x + 5}{x + 7}$ and $\dfrac{x^2 - 4}{4}$, have no common factors in their numerators and denominators. Thus, these rational expressions are in simplified form. The rational expression $\dfrac{x^2 - 9}{x - 3}$ can be simplified as follows:

Correct

$$\dfrac{x^2 - 9}{x - 3} = \dfrac{(x + 3)\overset{1}{\cancel{(x - 3)}}}{\underset{1}{\cancel{x - 3}}} = x + 3.$$

> Divide out the common factor, $x - 3$.

Factors That Are Opposites How do we simplify rational expressions that contain factors in the numerator and denominator that are opposites, or additive inverses? Here is an example of such an expression:

$$\dfrac{x - 3}{3 - x}.$$

> The numerator and denominator are opposites. They differ only in their signs.

Factor out -1 from either the numerator or the denominator. Then divide out the common factor.

$$\dfrac{x - 3}{3 - x} = \dfrac{-1(-x + 3)}{3 - x}$$

Factor -1 from the numerator. Notice how the sign of each term in the polynomial $x - 3$ changes.

$$= \dfrac{-1(3 - x)}{3 - x}$$

In the numerator, use the commutative property to rewrite $-x + 3$ as $3 - x$.

$$= \dfrac{-1\overset{1}{\cancel{(3 - x)}}}{\underset{1}{\cancel{3 - x}}}$$

Divide out the common factor of $3 - x$.

$$= -1$$

Our result, -1, suggests a useful property that is stated at the top of the next page.

SIMPLIFYING RATIONAL EXPRESSIONS WITH OPPOSITE FACTORS IN THE NUMERATOR AND DENOMINATOR The quotient of two polynomials that have opposite signs and are additive inverses is −1.

EXAMPLE 5 Simplifying a Rational Expression

Simplify: $\dfrac{4x^2 - 25}{15 - 6x}$.

SOLUTION

$$\frac{4x^2 - 25}{15 - 6x} = \frac{(2x + 5)(2x - 5)}{3(5 - 2x)}$$ Factor the numerator and denominator.

$$= \frac{(2x + 5)\overset{-1}{\cancel{(2x - 5)}}}{3\cancel{(5 - 2x)}}$$ The quotient of polynomials with opposite signs is −1.

$$= \frac{-(2x + 5)}{3} \quad \text{or} \quad -\frac{2x + 5}{3} \quad \text{or} \quad \frac{-2x - 5}{3}$$

Each of these forms is an acceptable answer. ∎

✔ **CHECK POINT 5** Simplify: $\dfrac{9x^2 - 49}{28 - 12x}$.

---

3 Solve applied problems involving rational expressions.

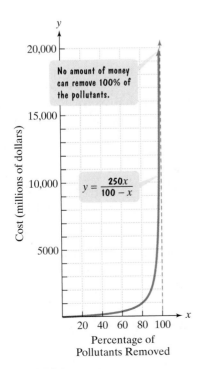

FIGURE 8.1

Applications The equation

$$y = \frac{250x}{100 - x}$$

models the cost, in millions of dollars, to remove x percent of the pollutants that are discharged into a river. This equation contains the rational expression that we looked at in the opening to this section. Do you remember how costs were spiraling upward as the percentage of removed pollutants increased?

Is it possible to clean up the river completely? To do this, we must remove 100% of the pollutants. The problem is that the rational expression is undefined for $x = 100$.

$$y = \frac{250x}{100 - x}$$ If $x = 100$, the value of the denominator is 0.

Notice how the graph of $y = \dfrac{250x}{100 - x}$, shown in Figure 8.1, approaches but never touches the dashed vertical line $x = 100$, our undefined value. The graph continues to rise more and more steeply, visually showing the escalating costs. By never touching the dashed vertical line, the graph illustrates that no amount of money will be enough to remove all pollutants from the river.

8.1 EXERCISE SET

Student Solutions Manual CD/Video PH Math/Tutor Center MathXL Tutorials on CD MathXL® MyMathLab Interactmath.com

Practice Exercises

In Exercises 1–20, find all numbers for which each rational expression is undefined. If the rational expression is defined for all real numbers, so state.

1. $\dfrac{5}{2x}$

2. $\dfrac{11}{3x}$

3. $\dfrac{x}{x-8}$

4. $\dfrac{x}{x-6}$

5. $\dfrac{13}{5x-20}$

6. $\dfrac{17}{6x-30}$

7. $\dfrac{x+3}{(x+9)(x-2)}$

8. $\dfrac{x+5}{(x+7)(x-9)}$

9. $\dfrac{4x}{(3x-17)(x+3)}$

10. $\dfrac{8x}{(4x-19)(x+2)}$

11. $\dfrac{x+5}{x^2+x-12}$

12. $\dfrac{7x-14}{x^2-9x+20}$

13. $\dfrac{x+5}{5}$

14. $\dfrac{x+7}{7}$

15. $\dfrac{y+3}{4y^2+y-3}$

16. $\dfrac{y+8}{6y^2-y-2}$

17. $\dfrac{y+5}{y^2-25}$

18. $\dfrac{y+7}{y^2-49}$

19. $\dfrac{5}{x^2+1}$

20. $\dfrac{8}{x^2+4}$

In Exercises 21–76, simplify each rational expression. If the rational expression cannot be simplified, so state.

21. $\dfrac{14x^2}{7x}$

22. $\dfrac{9x^2}{6x}$

23. $\dfrac{5x-15}{25}$

24. $\dfrac{7x+21}{49}$

25. $\dfrac{2x-8}{4x}$

26. $\dfrac{3x-9}{6x}$

27. $\dfrac{3}{3x-9}$

28. $\dfrac{12}{6x-18}$

29. $\dfrac{-15}{3x-9}$

30. $\dfrac{-21}{7x-14}$

31. $\dfrac{3x+9}{x+3}$

32. $\dfrac{5x-10}{x-2}$

33. $\dfrac{x+5}{x^2-25}$

34. $\dfrac{x+4}{x^2-16}$

35. $\dfrac{2y-10}{3y-15}$

36. $\dfrac{6y+18}{11y+33}$

37. $\dfrac{x+1}{x^2-2x-3}$

38. $\dfrac{x+2}{x^2-x-6}$

39. $\dfrac{4x-8}{x^2-4x+4}$

40. $\dfrac{x^2-12x+36}{4x-24}$

41. $\dfrac{y^2-3y+2}{y^2+7y-18}$

42. $\dfrac{y^2+5y+4}{y^2-4y-5}$

43. $\dfrac{2y^2-7y+3}{2y^2-5y+2}$

44. $\dfrac{3y^2+4y-4}{6y^2-y-2}$

45. $\dfrac{2x+3}{2x+5}$

46. $\dfrac{3x+7}{3x+10}$

47. $\dfrac{x^2+12x+36}{x^2-36}$

48. $\dfrac{x^2-14x+49}{x^2-49}$

49. $\dfrac{x^3-2x^2+x-2}{x-2}$

50. $\dfrac{x^3+4x^2-3x-12}{x+4}$

51. $\dfrac{x^3-8}{x-2}$

52. $\dfrac{x^3-125}{x^2-25}$

53. $\dfrac{(x-4)^2}{x^2-16}$

54. $\dfrac{(x+5)^2}{x^2-25}$

55. $\dfrac{x}{x+1}$

56. $\dfrac{x}{x+7}$

57. $\dfrac{x+4}{x^2+16}$

58. $\dfrac{x+5}{x^2+25}$

59. $\dfrac{x-5}{5-x}$

60. $\dfrac{x-7}{7-x}$

61. $\dfrac{2x-3}{3-2x}$

62. $\dfrac{5x-4}{4-5x}$

63. $\dfrac{x-5}{x+5}$

64. $\dfrac{x-7}{x+7}$

65. $\dfrac{4x-6}{3-2x}$

66. $\dfrac{9x-15}{5-3x}$

67. $\dfrac{4-6x}{3x^2-2x}$

68. $\dfrac{9-15x}{5x^2-3x}$

69. $\dfrac{x^2-1}{1-x}$

70. $\dfrac{x^2-4}{2-x}$

71. $\dfrac{y^2 - y - 12}{4 - y}$

72. $\dfrac{y^2 - 7y + 12}{3 - y}$

73. $\dfrac{x^2y - x^2}{x^3 - x^3y}$

74. $\dfrac{xy - 2x}{3y - 6}$

75. $\dfrac{x^2 + 2xy - 3y^2}{2x^2 + 5xy - 3y^2}$

76. $\dfrac{x^2 + 3xy - 10y^2}{3x^2 - 7xy + 2y^2}$

Practice Plus

In Exercises 77–84, simplify each rational expression.

77. $\dfrac{x^2 - 9x + 18}{x^3 - 27}$

78. $\dfrac{x^3 - 8}{x^2 + 2x - 8}$

79. $\dfrac{9 - y^2}{y^2 - 3(2y - 3)}$

80. $\dfrac{16 - y^2}{y(y - 8) + 16}$

81. $\dfrac{xy + 2y + 3x + 6}{x^2 + 5x + 6}$

82. $\dfrac{xy + 4y - 7x - 28}{x^2 + 11x + 28}$

83. $\dfrac{8x^2 + 4x + 2}{1 - 8x^3}$

84. $\dfrac{x^3 - 3x^2 + 9x}{x^3 + 27}$

Application Exercises

85. The rational expression

$$\frac{130x}{100 - x}$$

describes the cost, in millions of dollars, to inoculate x percent of the population against a particular strain of flu.

   a. Evaluate the expression for $x = 40$, $x = 80$, and $x = 90$. Describe the meaning of each evaluation in terms of percentage inoculated and cost.

   b. For what value of x is the expression undefined?

   c. What happens to the cost as x approaches 100%? How can you interpret this observation?

86. The rational expression

$$\frac{60{,}000x}{100 - x}$$

describes the cost, in dollars, to remove x percent of the air pollutants in the smokestack emission of a utility company that burns coal to generate electricity.

   a. Evaluate the expression for $x = 20$, $x = 50$, and $x = 80$. Describe the meaning of each evaluation in terms of percentage of pollutants removed and cost.

   b. For what value of x is the expression undefined?

   c. What happens to the cost as x approaches 100%? How can you interpret this observation?

Doctors use the rational expression

$$\frac{DA}{A + 12}$$

to determine the dosage of a drug prescribed for children. In this expression, A = the child's age and D = the adult dosage. Use the expression to solve Exercises 87–88.

87. If the normal adult dosage of medication is 1000 milligrams, what dosage should an 8-year-old child receive?

88. If the normal adult dosage of medication is 1000 milligrams, what dosage should a 4-year-old child receive?

89. A company that manufactures bicycles has costs given by the equation

$$C = \frac{100x + 100{,}000}{x}$$

in which x is the number of bicycles manufactured and C is the cost to manufacture each bicycle.

   a. Find the cost per bicycle when manufacturing 500 bicycles.

   b. Find the cost per bicycle when manufacturing 4000 bicycles.

   c. Does the cost per bicycle increase or decrease as more bicycles are manufactured? Explain why this happens.

90. A company that manufactures small canoes has costs given by the equation

$$C = \frac{20x + 20{,}000}{x}$$

in which x is the number of canoes manufactured and C is the cost to manufacture each canoe.

   a. Find the cost per canoe when manufacturing 100 canoes.

   b. Find the cost per canoe when manufacturing 10,000 canoes.

   c. Does the cost per canoe increase or decrease as more canoes are manufactured? Explain why this happens.

A drug is injected into a patient and the concentration of the drug in the bloodstream is monitored. The drug's concentration, y, in milligrams per liter, after x hours is modeled by

$$y = \frac{5x}{x^2 + 1}.$$

The graph of this equation, obtained with a graphing utility, is shown in the figure in a [0, 10, 1] by [0, 3, 1] viewing rectangle. Use this information to solve Exercises 91–92.

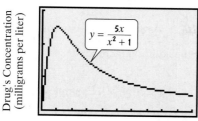

Hours after Injection

[0, 10, 1] by [0, 3, 1]

91. Use the equation to find the drug's concentration after 3 hours. Then identify the point on the equation's graph that conveys this information.

92. Use the graph of the equation to find after how many hours the drug reaches its maximum concentration. Then use the equation to find the drug's concentration at this time.

Body-mass index takes both weight and height into account when assessing whether an individual is underweight or overweight. The formula for body-mass index, BMI, is

$$\text{BMI} = \frac{703w}{h^2},$$

where w is weight, in pounds, and h is height, in inches. In adults, normal values for the BMI are between 20 and 25, inclusive. Values below 20 indicate that an individual is underweight and values above 30 indicate that an individual is obese. Use this information to solve Exercises 93–94.

93. Calculate the BMI, to the nearest tenth, for a 145-pound person who is 5 feet 10 inches tall. Is this person underweight?

94. Calculate the BMI, to the nearest tenth, for a 150-pound person who is 5 feet 6 inches tall. Is this person overweight?

95. The bar graph shows the total number of crimes in the United States, in millions, from 1995 through 2001.

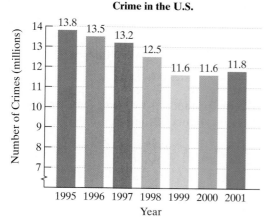

Crime in the U.S.

Source: FBI

The polynomial $3.7t + 257.4$ describes the U.S. population, in millions, t years after 1994. The polynomial $-0.4t + 14.2$ describes the number of crimes in the United States, in millions, t years after 1994.

a. Write a rational expression that describes the crime rate in the United States t years after 1994.

b. According to the rational expression in part (a), what was the crime rate in 2001? Round to two decimal places. How many crimes does this indicate per 100,000 inhabitants?

c. According to the FBI, there were 4161 crimes per 100,000 U.S. inhabitants in 2001. How well does the rational expression that you evaluated in part (b) model this number?

Writing in Mathematics

96. What is a rational expression? Give an example with your explanation.

97. Explain how to find the number or numbers, if any, for which a rational expression is undefined.

98. Explain how to simplify a rational expression.

99. Explain how to simplify a rational expression with opposite factors in the numerator and denominator.

100. A politician claims that each year the crime rate in the United States is decreasing. Explain how to use the polynomials in Exercise 95 to verify this claim.

101. Use the graph shown for Exercises 91–92 to write a description of the drug's concentration over time. In your description, try to convey as much information as possible that is displayed visually by the graph.

Critical Thinking Exercises

102. Which one of the following is true?

a. $\dfrac{x + 5}{x} = 5$ **b.** $\dfrac{x^2 + 3}{3} = x^2 + 1$

c. $\dfrac{3x + 9}{3x + 13} = \dfrac{9}{13}$

d. The expression $\dfrac{-3y - 6}{y + 2}$ reduces to the consecutive integer that follows -4.

103. Write a rational expression that cannot be simplified.

104. Write a rational expression that is undefined for $x = -4$.

105. Write a rational expression with $x^2 - x - 6$ in the numerator that can be simplified to $x - 3$.

Technology Exercises

In Exercises 106–109, use the GRAPH *or* TABLE *feature of a graphing utility to determine if the rational expression has been correctly simplified. If the simplification is wrong, correct it and then verify your answer using the graphing utility.*

106. $\dfrac{3x + 15}{x + 5} = 3, \quad x \neq -5$

107. $\dfrac{2x^2 - x - 1}{x - 1} = 2x^2 - 1, \ x \neq 1$

108. $\dfrac{x^2 - x}{x} = x^2 - 1, \ x \neq 0$

109. Use a graphing utility to verify the graph in Figure 8.1 on page 460. [TRACE] along the graph as x approaches 100. What do you observe?

Review Exercises

110. Multiply: $\dfrac{5}{6} \cdot \dfrac{9}{25}$. (Section 1.1, Example 5)

111. Divide: $\dfrac{2}{3} \div 4$. (Section 1.1, Example 6)

112. Solve by the addition method:

$2x - 5y = -2$

$3x + 4y = 20$. (Section 5.3, Example 3)

8.2

Objectives

1 Multiply rational expressions.

2 Divide rational expressions.

MULTIPLYING AND DIVIDING RATIONAL EXPRESSIONS

Highbrow wit in conjunction with lowbrow comedy characterize Stephen Sondheim's *A Funny Thing Happened on the Way to the Forum*. The musical is based on the plays of Plautus, comic dramatist of ancient Rome.

Your psychology class is learning various techniques to double what we remember over time. At the beginning of the course, students memorize 40 words in Latin, a language with which they are not familiar. The rational expression

$$\frac{5t + 30}{t}$$

models the class average for the number of words remembered after t days, where $t \geq 1$. If the techniques are successful, what will be the new memory model?

The new model can be found by multiplying the given rational expression by 2. In this section, you will see that we multiply rational expressions in the same way that we multiply rational numbers. Thus, we multiply numerators and multiply denominators. The rational expression for doubling what the class remembers over time is

$$\frac{2}{1} \cdot \frac{5t + 30}{t} = \frac{2(5t + 30)}{1 \cdot t} = \frac{2 \cdot 5t + 2 \cdot 30}{t} = \frac{10t + 60}{t}.$$

Multiplying Rational Expressions The product of two rational expressions is the product of their numerators divided by the product of their denominators.

1 Multiply rational expressions.

> **MULTIPLYING RATIONAL EXPRESSIONS** If P, Q, R, and S are polynomials, where $Q \neq 0$ and $S \neq 0$, then
>
> $$\frac{P}{Q} \cdot \frac{R}{S} = \frac{PR}{QS}.$$

EXAMPLE 1 Multiplying Rational Expressions

Multiply: $\dfrac{7}{x+3} \cdot \dfrac{x-2}{5}$.

SOLUTION

$$\dfrac{7}{x+3} \cdot \dfrac{x-2}{5} = \dfrac{7(x-2)}{(x+3)5}$$ Multiply numerators. Multiply denominators. $(x \neq -3)$

$$= \dfrac{7x-14}{5x+15}$$ ■

✔ **CHECK POINT 1** Multiply: $\dfrac{9}{x+4} \cdot \dfrac{x-5}{2}$.

Here is a step-by-step procedure for multiplying rational expressions. Before multiplying, divide out any factors common to both a numerator and a denominator.

MULTIPLYING RATIONAL EXPRESSIONS

1. Factor all numerators and denominators completely.
2. Divide numerators and denominators by common factors.
3. Multiply the remaining factors in the numerators and multiply the remaining factors in the denominators.

EXAMPLE 2 Multiplying Rational Expressions

Multiply: $\dfrac{x-3}{x+5} \cdot \dfrac{10x+50}{7x-21}$.

SOLUTION

$$\dfrac{x-3}{x+5} \cdot \dfrac{10x+50}{7x-21}$$

$$= \dfrac{x-3}{x+5} \cdot \dfrac{10(x+5)}{7(x-3)}$$ Factor as many numerators and denominators as possible.

$$= \dfrac{\overset{1}{\cancel{x-3}}}{\underset{1}{\cancel{x+5}}} \cdot \dfrac{10\overset{1}{\cancel{(x+5)}}}{7\underset{1}{\cancel{(x-3)}}}$$ Divide numerators and denominators by common factors.

$$= \dfrac{10}{7}$$ Multiply the remaining factors in the numerators and denominators. ■

✔ **CHECK POINT 2** Multiply: $\dfrac{x+4}{x-7} \cdot \dfrac{3x-21}{8x+32}$.

EXAMPLE 3 Multiplying Rational Expressions

Multiply: $\dfrac{x-7}{x-1} \cdot \dfrac{x^2-1}{3x-21}$.

SOLUTION

$$\dfrac{x-7}{x-1} \cdot \dfrac{x^2-1}{3x-21}$$

$$= \dfrac{x-7}{x-1} \cdot \dfrac{(x+1)(x-1)}{3(x-7)}$$ Factor as many numerators and denominators as possible.

$$= \dfrac{\cancel{x-7}^{1}}{\cancel{x-1}_{1}} \cdot \dfrac{(x+1)\cancel{(x-1)}^{1}}{3\cancel{(x-7)}_{1}}$$ Divide numerators and denominators by common factors.

$$= \dfrac{x+1}{3}$$ Multiply the remaining factors in the numerators and denominators. ∎

 CHECK POINT 3 Multiply: $\dfrac{x-5}{x-2} \cdot \dfrac{x^2-4}{9x-45}$.

EXAMPLE 4 Multiplying Rational Expressions

Multiply: $\dfrac{4x+8}{6x-3x^2} \cdot \dfrac{3x^2-4x-4}{9x^2-4}$.

SOLUTION

$$\dfrac{4x+8}{6x-3x^2} \cdot \dfrac{3x^2-4x-4}{9x^2-4}$$

$$= \dfrac{4(x+2)}{3x(2-x)} \cdot \dfrac{(3x+2)(x-2)}{(3x+2)(3x-2)}$$ Factor as many numerators and denominators as possible.

$$= \dfrac{4(x+2)}{3x\cancel{(2-x)}_{1}} \cdot \dfrac{\cancel{(3x+2)}^{1}\cancel{(x-2)}^{-1}}{\cancel{(3x+2)}_{1}(3x-2)}$$ Divide numerators and denominators by common factors. Because $2-x$ and $x-2$ have opposite signs, their quotient is -1.

$$= \dfrac{-4(x+2)}{3x(3x-2)} \quad \text{or} \quad -\dfrac{4(x+2)}{3x(3x-2)}$$ Multiply the remaining factors in the numerators and denominators.

> It is not necessary to carry out these multiplications.

∎

 CHECK POINT 4 Multiply: $\dfrac{5x+5}{7x-7x^2} \cdot \dfrac{2x^2+x-3}{4x^2-9}$.

2 Divide rational expressions.

Dividing Rational Expressions   The quotient of two rational expressions is the product of the first expression and the multiplicative inverse, or reciprocal, of the second. The reciprocal is found by interchanging the numerator and the denominator.

DIVIDING RATIONAL EXPRESSIONS If P, Q, R, and S are polynomials, where $Q \neq 0$, $R \neq 0$, and $S \neq 0$, then

$$\frac{P}{Q} \div \frac{R}{S} = \frac{P}{Q} \cdot \frac{S}{R} = \frac{PS}{QR}.$$

Change division to multiplication.

Replace $\frac{R}{S}$ with its reciprocal by interchanging numerator and denominator.

Thus, **we find the quotient of two rational expressions by inverting the divisor and multiplying.** For example,

$$\frac{x}{7} \div \frac{6}{y} = \frac{x}{7} \cdot \frac{y}{6} = \frac{xy}{42}.$$

Change the division to multiplication.

Replace $\frac{6}{y}$ with its reciprocal by interchanging numerator and denominator.

EXAMPLE 5 Dividing Rational Expressions

Divide: $(x + 5) \div \dfrac{x - 2}{x + 9}$.

SOLUTION

$$(x + 5) \div \frac{x - 2}{x + 9} = \frac{x + 5}{1} \cdot \frac{x + 9}{x - 2}$$

Invert the divisor and multiply.

$$= \frac{(x + 5)(x + 9)}{x - 2}$$

Multiply the factors in the numerators and denominators. We need not carry out the multiplication in the numerator.

CHECK POINT 5 Divide: $(x + 3) \div \dfrac{x - 4}{x + 7}$.

EXAMPLE 6 Dividing Rational Expressions

Divide: $\dfrac{x^2 - 2x - 8}{x^2 - 9} \div \dfrac{x - 4}{x + 3}$.

SOLUTION

$$\frac{x^2 - 2x - 8}{x^2 - 9} \div \frac{x - 4}{x + 3}$$

$$= \frac{x^2 - 2x - 8}{x^2 - 9} \cdot \frac{x + 3}{x - 4}$$

Invert the divisor and multiply.

$$= \frac{(x - 4)(x + 2)}{(x + 3)(x - 3)} \cdot \frac{x + 3}{x - 4}$$

Factor as many numerators and denominators as possible.

$$= \frac{\overset{1}{\cancel{(x - 4)}}(x + 2)}{\cancel{(x + 3)}(x - 3)} \cdot \frac{\overset{1}{\cancel{(x + 3)}}}{\cancel{(x - 4)}}$$

Divide numerators and denominators by common factors.

$$= \frac{x + 2}{x - 3}$$

Multiply the remaining factors in the numerators and the denominators.

✔ **CHECK POINT 6** Divide: $\dfrac{x^2 + 5x + 6}{x^2 - 25} \div \dfrac{x + 2}{x + 5}$.

EXAMPLE 7 Dividing Rational Expressions

Divide: $\dfrac{y^2 + 7y + 12}{y^2 + 9} \div (7y^2 + 21y)$.

SOLUTION

$$\dfrac{y^2 + 7y + 12}{y^2 + 9} \div \dfrac{7y^2 + 21y}{1}$$
It is helpful to write the divisor with a denominator of 1.

$$= \dfrac{y^2 + 7y + 12}{y^2 + 9} \cdot \dfrac{1}{7y^2 + 21y}$$
Invert the divisor and multiply.

$$= \dfrac{(y + 4)(y + 3)}{y^2 + 9} \cdot \dfrac{1}{7y(y + 3)}$$
Factor as many numerators and denominators as possible.

$$= \dfrac{(y + 4)\cancel{(y + 3)}^{1}}{y^2 + 9} \cdot \dfrac{1}{7y\cancel{(y + 3)}_{1}}$$
Divide numerators and denominators by common factors.

$$= \dfrac{y + 4}{7y(y^2 + 9)}$$
Multiply the remaining factors in the numerators and the denominators.

✔ **CHECK POINT 7** Divide: $\dfrac{y^2 + 3y + 2}{y^2 + 1} \div (5y^2 + 10y)$.

8.2 EXERCISE SET

Student Solutions Manual   CD/Video   PH Math/Tutor Center   MathXL Tutorials on CD   MathXL®   MyMathLab   Interactmath.com

Practice Exercises

In Exercises 1–32, multiply as indicated.

1. $\dfrac{4}{x + 3} \cdot \dfrac{x - 5}{9}$

2. $\dfrac{8}{x - 2} \cdot \dfrac{x + 5}{3}$

3. $\dfrac{x}{3} \cdot \dfrac{12}{x + 5}$

4. $\dfrac{x}{5} \cdot \dfrac{30}{x - 4}$

5. $\dfrac{3}{x} \cdot \dfrac{4x}{15}$

6. $\dfrac{7}{x} \cdot \dfrac{5x}{35}$

7. $\dfrac{x - 3}{x + 5} \cdot \dfrac{4x + 20}{9x - 27}$

8. $\dfrac{x - 2}{x + 9} \cdot \dfrac{5x + 45}{2x - 4}$

9. $\dfrac{x^2 + 9x + 14}{x + 7} \cdot \dfrac{1}{x + 2}$

10. $\dfrac{x^2 + 9x + 18}{x + 6} \cdot \dfrac{1}{x + 3}$

11. $\dfrac{x^2 - 25}{x^2 - 3x - 10} \cdot \dfrac{x + 2}{x}$

12. $\dfrac{x^2 - 49}{x^2 - 4x - 21} \cdot \dfrac{x + 3}{x}$

13. $\dfrac{4y + 30}{y^2 - 3y} \cdot \dfrac{y - 3}{2y + 15}$

14. $\dfrac{9y + 21}{y^2 - 2y} \cdot \dfrac{y - 2}{3y + 7}$

15. $\dfrac{y^2 - 7y - 30}{y^2 - 6y - 40} \cdot \dfrac{2y^2 + 5y + 2}{2y^2 + 7y + 3}$

16. $\dfrac{3y^2 + 17y + 10}{3y^2 - 22y - 16} \cdot \dfrac{y^2 - 4y - 32}{y^2 - 8y - 48}$

17. $(y^2 - 9) \cdot \dfrac{4}{y - 3}$

18. $(y^2 - 16) \cdot \dfrac{3}{y - 4}$

19. $\dfrac{x^2 - 5x + 6}{x^2 - 2x - 3} \cdot \dfrac{x^2 - 1}{x^2 - 4}$

20. $\dfrac{x^2 + 5x + 6}{x^2 + x - 6} \cdot \dfrac{x^2 - 9}{x^2 - x - 6}$

21. $\dfrac{x^3 - 8}{x^2 - 4} \cdot \dfrac{x + 2}{3x}$

22. $\dfrac{x^2 + 6x + 9}{x^3 + 27} \cdot \dfrac{1}{x + 3}$

23. $\dfrac{(x - 2)^3}{(x - 1)^3} \cdot \dfrac{x^2 - 2x + 1}{x^2 - 4x + 4}$

24. $\dfrac{(x + 4)^3}{(x + 2)^3} \cdot \dfrac{x^2 + 4x + 4}{x^2 + 8x + 16}$

25. $\dfrac{6x + 2}{x^2 - 1} \cdot \dfrac{1 - x}{3x^2 + x}$

26. $\dfrac{8x + 2}{x^2 - 9} \cdot \dfrac{3 - x}{4x^2 + x}$

27. $\dfrac{25 - y^2}{y^2 - 2y - 35} \cdot \dfrac{y^2 - 8y - 20}{y^2 - 3y - 10}$

28. $\dfrac{2y}{3y - y^2} \cdot \dfrac{2y^2 - 9y + 9}{8y - 12}$

29. $\dfrac{x^2 - y^2}{x} \cdot \dfrac{x^2 + xy}{x + y}$

30. $\dfrac{4x - 4y}{x} \cdot \dfrac{x^2 + xy}{x^2 - y^2}$

31. $\dfrac{x^2 + 2xy + y^2}{x^2 - 2xy + y^2} \cdot \dfrac{4x - 4y}{3x + 3y}$

32. $\dfrac{x^2 - y^2}{x + y} \cdot \dfrac{x + 2y}{2x^2 - xy - y^2}$

In Exercises 33–64, divide as indicated.

33. $\dfrac{x}{7} \div \dfrac{5}{3}$

34. $\dfrac{x}{3} \div \dfrac{3}{8}$

35. $\dfrac{3}{x} \div \dfrac{12}{x}$

36. $\dfrac{x}{5} \div \dfrac{20}{x}$

37. $\dfrac{15}{x} \div \dfrac{3}{2x}$

38. $\dfrac{9}{x} \div \dfrac{3}{4x}$

39. $\dfrac{x + 1}{3} \div \dfrac{3x + 3}{7}$

40. $\dfrac{x + 5}{7} \div \dfrac{4x + 20}{9}$

41. $\dfrac{7}{x - 5} \div \dfrac{28}{3x - 15}$

42. $\dfrac{4}{x - 6} \div \dfrac{40}{7x - 42}$

43. $\dfrac{x^2 - 4}{x} \div \dfrac{x + 2}{x - 2}$

44. $\dfrac{x^2 - 4}{x - 2} \div \dfrac{x + 2}{4x - 8}$

45. $(y^2 - 16) \div \dfrac{y^2 + 3y - 4}{y^2 + 4}$

46. $(y^2 + 4y - 5) \div \dfrac{y^2 - 25}{y + 7}$

47. $\dfrac{y^2 - y}{15} \div \dfrac{y - 1}{5}$

48. $\dfrac{y^2 - 2y}{15} \div \dfrac{y - 2}{5}$

49. $\dfrac{4x^2 + 10}{x - 3} \div \dfrac{6x^2 + 15}{x^2 - 9}$

50. $\dfrac{x^2 + x}{x^2 - 4} \div \dfrac{x^2 - 1}{x^2 + 5x + 6}$

51. $\dfrac{x^2 - 25}{2x - 2} \div \dfrac{x^2 + 10x + 25}{x^2 + 4x - 5}$

52. $\dfrac{x^2 - 4}{x^2 + 3x - 10} \div \dfrac{x^2 + 5x + 6}{x^2 + 8x + 15}$

53. $\dfrac{y^3 + y}{y^2 - y} \div \dfrac{y^3 - y^2}{y^2 - 2y + 1}$

54. $\dfrac{3y^2 - 12}{y^2 + 4y + 4} \div \dfrac{y^3 - 2y^2}{y^2 + 2y}$

55. $\dfrac{y^2 + 5y + 4}{y^2 + 12y + 32} \div \dfrac{y^2 - 12y + 35}{y^2 + 3y - 40}$

56. $\dfrac{y^2 + 4y - 21}{y^2 + 3y - 28} \div \dfrac{y^2 + 14y + 48}{y^2 + 4y - 32}$

57. $\dfrac{2y^2 - 128}{y^2 + 16y + 64} \div \dfrac{y^2 - 6y - 16}{3y^2 + 30y + 48}$

58. $\dfrac{3y + 12}{y^2 + 3y} \div \dfrac{y^2 + y - 12}{9y - y^3}$

59. $\dfrac{2x + 2y}{3} \div \dfrac{x^2 - y^2}{x - y}$

60. $\dfrac{5x + 5y}{7} \div \dfrac{x^2 - y^2}{x - y}$

61. $\dfrac{x^2 - y^2}{8x^2 - 16xy + 8y^2} \div \dfrac{4x - 4y}{x + y}$

62. $\dfrac{4x^2 - y^2}{x^2 + 4xy + 4y^2} \div \dfrac{4x - 2y}{3x + 6y}$

63. $\dfrac{xy - y^2}{x^2 + 2x + 1} \div \dfrac{2x^2 + xy - 3y^2}{2x^2 + 5xy + 3y^2}$

64. $\dfrac{x^2 - 4y^2}{x^2 + 3xy + 2y^2} \div \dfrac{x^2 - 4xy + 4y^2}{x + y}$

Practice Plus

In Exercises 65–72, perform the indicated operation or operations.

65. $\left(\dfrac{y - 2}{y^2 - 9y + 18} \cdot \dfrac{y^2 - 4y - 12}{y + 2}\right) \div \dfrac{y^2 - 4}{y^2 + 5y + 6}$

66. $\left(\dfrac{6y^2 + 31y + 18}{3y^2 - 20y + 12} \cdot \dfrac{2y^2 - 15y + 18}{6y^2 + 35y + 36}\right) \div \dfrac{2y^2 - 13y + 15}{9y^2 + 15y + 4}$

67. $\dfrac{3x^2 + 3x - 60}{2x - 8} \div \left(\dfrac{30x^2}{x^2 - 7x + 10} \cdot \dfrac{x^3 + 3x^2 - 10x}{25x^3}\right)$

68. $\dfrac{5x^2 - x}{3x + 2} \div \left(\dfrac{6x^2 + x - 2}{10x^2 + 3x - 1} \cdot \dfrac{2x^2 - x - 1}{2x^2 - x}\right)$

69. $\dfrac{x^2 + xz + xy + yz}{x - y} \div \dfrac{x + z}{x + y}$

70. $\dfrac{x^2 - xz + xy - yz}{x - y} \div \dfrac{x - z}{y - x}$

71. $\dfrac{3xy + ay + 3xb + ab}{9x^2 - a^2} \div \dfrac{y^3 + b^3}{6x - 2a}$

72. $\dfrac{5xy - ay - 5xb + ab}{25x^2 - a^2} \div \dfrac{y^3 - b^3}{15x + 3a}$

Application Exercises

73. In the Section 8.1 opener, we used

$$\dfrac{250x}{100 - x}$$

to describe the cost, in millions of dollars, to remove x percent of the pollutants that are discharged into the river. We were wrong. The cost will be half of what we originally anticipated. Write a rational expression that represents the reduced cost.

74. We originally thought that the cost, in dollars, to manufacture each of x bicycles was

$$\frac{100x + 100,000}{x}.$$

We were wrong. We can manufacture each bicycle at half of what we originally anticipated. Write a rational expression that represents the reduced cost.

Writing in Mathematics

75. Explain how to multiply rational expressions.

76. Explain how to divide rational expressions.

77. In dividing polynomials

$$\frac{P}{Q} \div \frac{R}{S},$$

why is it necessary to state that polynomial R is not equal to 0?

Critical Thinking Exercises

78. Which one of the following is true?

a. $5 \div x = \frac{1}{5} \cdot x$ for any nonzero number x.

b. $\frac{4}{x} \div \frac{x-2}{x} = \frac{4}{x-2}$ if $x \neq 0$ and $x \neq 2$.

c. $\frac{x-5}{6} \cdot \frac{3}{5-x} = \frac{1}{2}$ for any value of x except 5.

d. The quotient of two rational expressions can be found by dividing their numerators and dividing their denominators.

79. Find the missing polynomials: $\dfrac{}{} \cdot \dfrac{3x - 12}{2x} = \dfrac{3}{2}$.

80. Find the missing polynomials: $-\dfrac{1}{2x - 3} \div \dfrac{}{} = \dfrac{1}{3}$.

81. Divide:

$$\frac{9x^2 - y^2 + 15x - 5y}{3x^2 + xy + 5x} \div \frac{3x + y}{9x^3 + 6x^2y + xy^2}.$$

Technology Exercises

In Exercises 82–85, use the GRAPH *or* TABLE *feature of a graphing utility to determine if the multiplication or division has been performed correctly. If the answer is wrong, correct it and then verify your correction using the graphing utility.*

82. $\dfrac{x^2 + x}{3x} \cdot \dfrac{6x}{x + 1} = 2x$

83. $\dfrac{x^3 - 25x}{x^2 - 3x - 10} \cdot \dfrac{x + 2}{x} = x + 5$

84. $\dfrac{x^2 - 9}{x + 4} \div \dfrac{x - 3}{x + 4} = x - 3$

85. $(x - 5) \div \dfrac{2x^2 - 11x + 5}{4x^2 - 1} = 2x - 1$

Review Exercises

86. Solve: $2x + 3 < 3(x - 5)$. (Section 2.6, Example 6)

87. Factor completely: $3x^2 - 15x - 42$. (Section 7.5, Example 2)

88. Solve: $x(2x + 9) = 5$. (Section 7.6, Example 6)

SECTION **8.3**

ADDING AND SUBTRACTING RATIONAL EXPRESSIONS WITH THE SAME DENOMINATOR

Objectives

1 Add rational expressions with the same denominator.

2 Subtract rational expressions with the same denominator.

3 Add and subtract rational expressions with opposite denominators.

Are you long, medium, or round? Your skull, that is? The varying shapes of the human skull create glorious diversity in the human species. By learning to add and subtract rational expressions with the same denominator, you will obtain an expression that models this diversity.

1 Add rational expressions with the same denominator.

Addition when Denominators Are the Same To add rational numbers having the same denominators, such as $\frac{2}{9}$ and $\frac{5}{9}$, we add the numerators and place the sum over the common denominator:

$$\frac{2}{9} + \frac{5}{9} = \frac{2+5}{9} = \frac{7}{9}.$$

We add rational expressions with the same denominator in an identical manner.

> **ADDING RATIONAL EXPRESSIONS WITH COMMON DENOMINATORS**
>
> If $\dfrac{P}{R}$ and $\dfrac{Q}{R}$ are rational expressions, then
>
> $$\frac{P}{R} + \frac{Q}{R} = \frac{P+Q}{R}.$$
>
> To add rational expressions with the same denominator, add numerators and place the sum over the common denominator. If possible, simplify the result.

USING TECHNOLOGY

The graphs of

$$y_1 = \frac{2x-1}{3} + \frac{x+4}{3}$$

and

$$y_2 = x+1$$

are the same line. Thus,

$$\frac{2x-1}{3} + \frac{x+4}{3} = x+1.$$

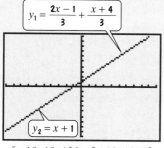

$[-10, 10, 1]$ by $[-10, 10, 1]$

EXAMPLE 1 Adding Rational Expressions when Denominators Are the Same

Add: $\dfrac{2x-1}{3} + \dfrac{x+4}{3}$.

SOLUTION

$$\frac{2x-1}{3} + \frac{x+4}{3} = \frac{2x-1+x+4}{3} \qquad \text{Add numerators. Place this sum over the common denominator.}$$

$$= \frac{3x+3}{3} \qquad \text{Combine like terms.}$$

$$= \frac{\overset{1}{\cancel{3}}(x+1)}{\underset{1}{\cancel{3}}} \qquad \text{Factor and simplify.}$$

$$= x+1 \qquad\qquad \blacksquare$$

✔ **CHECK POINT 1** Add: $\dfrac{3x-2}{5} + \dfrac{2x+12}{5}$.

EXAMPLE 2 Adding Rational Expressions when Denominators Are the Same

Add: $\dfrac{x^2}{x^2-9} + \dfrac{9-6x}{x^2-9}$.

SOLUTION

$$\frac{x^2}{x^2-9} + \frac{9-6x}{x^2-9} = \frac{x^2+9-6x}{x^2-9} \qquad \text{Add numerators. Place this sum over the common denominator.}$$

$$= \frac{x^2-6x+9}{x^2-9} \qquad \text{Write the numerator in descending powers of } x.$$

$$= \frac{(x-3)\overset{1}{\cancel{(x-3)}}}{(x+3)\underset{1}{\cancel{(x-3)}}} \qquad \text{Factor and simplify. What values of } x \text{ are not permitted?}$$

$$= \frac{x-3}{x+3} \qquad\qquad\qquad\qquad \blacksquare$$

✔ CHECK POINT 2 Add: $\dfrac{x^2}{x^2 - 25} + \dfrac{25 - 10x}{x^2 - 25}$.

2 Subtract rational expressions with the same denominator.

Subtraction when Denominators Are the Same The following box shows how to subtract rational expressions with the same denominator:

SUBTRACTING RATIONAL EXPRESSIONS WITH COMMON DENOMINATORS

If $\dfrac{P}{R}$ and $\dfrac{Q}{R}$ are rational expressions, then

$$\frac{P}{R} - \frac{Q}{R} = \frac{P - Q}{R}.$$

To subtract rational expressions with the same denominator, subtract numerators and place the difference over the common denominator. If possible, simplify the result.

EXAMPLE 3 Subtracting Rational Expressions when Denominators Are the Same

Subtract:

a. $\dfrac{2x + 3}{x + 1} - \dfrac{x}{x + 1}$ **b.** $\dfrac{5x + 1}{x^2 - 9} - \dfrac{4x - 2}{x^2 - 9}$.

SOLUTION

a. $\dfrac{2x + 3}{x + 1} - \dfrac{x}{x + 1} = \dfrac{2x + 3 - x}{x + 1}$ Subtract numerators. Place this difference over the common denominator.

$= \dfrac{x + 3}{x + 1}$ Combine like terms.

b. $\dfrac{5x + 1}{x^2 - 9} - \dfrac{4x - 2}{x^2 - 9} = \dfrac{5x + 1 - (4x - 2)}{x^2 - 9}$ Subtract numerators and include parentheses to indicate that both terms are subtracted. Place this difference over the common denominator.

$= \dfrac{5x + 1 - 4x + 2}{x^2 - 9}$ Remove parentheses and then change the sign of each term.

$= \dfrac{x + 3}{x^2 - 9}$ Combine like terms.

$= \dfrac{\overset{1}{\cancel{x + 3}}}{\cancel{(x + 3)}(x - 3)}$ Factor and simplify ($x \neq -3$ and $x \neq 3$).

$= \dfrac{1}{x - 3}$

✔ CHECK POINT 3 Subtract:

a. $\dfrac{4x + 5}{x + 7} - \dfrac{x}{x + 7}$ **b.** $\dfrac{3x^2 + 4x}{x - 1} - \dfrac{11x - 4}{x - 1}$.

USING TECHNOLOGY

To check Example 3(b) numerically, enter

$$y_1 = \frac{5x + 1}{x^2 - 9} - \frac{4x - 2}{x^2 - 9}$$

$$y_2 = \frac{1}{x - 3}$$

and use the TABLE feature. If $x \neq -3$ and $x \neq 3$, no matter how far up or down we scroll, $y_1 = y_2$.

X	Y1	Y2
-3	ERROR	-.1667
-2	-.2	-.2
-1	-.25	-.25
0	-.3333	-.3333
1	-.5	-.5
2	-1	-1
3	ERROR	ERROR

X= -3

STUDY TIP

When a numerator is being subtracted, be sure to **subtract every term in that expression.**

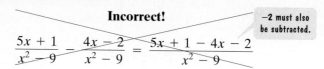

The − sign applies to the entire numerator, $4x - 2$.

Insert parentheses to indicate this.

The sign of every term of $4x - 2$ changes.

$$\frac{5x + 1}{x^2 - 9} - \frac{4x - 2}{x^2 - 9} = \frac{5x + 1 - (4x - 2)}{x^2 - 9} = \frac{5x + 1 - 4x + 2}{x^2 - 9}$$

The entire numerator of the second rational expression must be subtracted. Avoid the common error of subtracting only the first term.

Incorrect!

−2 must also be subtracted.

$$\frac{5x + 1}{x^2 - 9} - \frac{4x - 2}{x^2 - 9} = \frac{5x + 1 - 4x - 2}{x^2 - 9}$$

EXAMPLE 4 Subtracting Rational Expressions when Denominators Are the Same

Subtract: $\dfrac{20y^2 + 5y + 1}{6y^2 + y - 2} - \dfrac{8y^2 - 12y - 5}{6y^2 + y - 2}.$

SOLUTION

$$\frac{20y^2 + 5y + 1}{6y^2 + y - 2} - \frac{8y^2 - 12y - 5}{6y^2 + y - 2}$$

Don't forget the parentheses.

$$= \frac{20y^2 + 5y + 1 - (8y^2 - 12y - 5)}{6y^2 + y - 2}$$

Subtract numerators. Place this difference over the common denominator.

$$= \frac{20y^2 + 5y + 1 - 8y^2 + 12y + 5}{6y^2 + y - 2}$$

Remove parentheses and then change the sign of each term.

$$= \frac{(20y^2 - 8y^2) + (5y + 12y) + (1 + 5)}{6y^2 + y - 2}$$

Group like terms. This step is usually performed mentally.

$$= \frac{12y^2 + 17y + 6}{6y^2 + y - 2}$$

Combine like terms.

$$= \frac{\overset{1}{\cancel{(3y + 2)}}(4y + 3)}{\underset{1}{\cancel{(3y + 2)}}(2y - 1)}$$

Factor and simplify.

$$= \frac{4y + 3}{2y - 1}$$

■

✔ **CHECK POINT 4** Subtract: $\dfrac{y^2 + 3y - 6}{y^2 - 5y + 4} - \dfrac{4y - 4 - 2y^2}{y^2 - 5y + 4}.$

3 Add and subtract rational expressions with opposite denominators.

Addition and Subtraction when Denominators Are Opposites How do we add or subtract rational expressions when denominators are opposites, or additive inverses? Here is an example of this type of addition problem:

$$\frac{x^2}{x-5} + \frac{4x+5}{5-x}.$$

These denominators are opposites. The differ only in their signs.

Multiply the numerator and the denominator of either of the rational expressions by −1. Then they will both have the same denominator.

EXAMPLE 5　Adding Rational Expressions when Denominators Are Opposites

Add: $\dfrac{x^2}{x-5} + \dfrac{4x+5}{5-x}$.

SOLUTION

$$\frac{x^2}{x-5} + \frac{4x+5}{5-x}$$

$$= \frac{x^2}{x-5} + \frac{(-1)}{(-1)}\cdot\frac{4x+5}{5-x}$$
Multiply the numerator and denominator of the second rational expression by −1.

$$= \frac{x^2}{x-5} + \frac{-4x-5}{-5+x}$$
Perform the multiplications by −1 by changing every term's sign.

$$= \frac{x^2}{x-5} + \frac{-4x-5}{x-5}$$
Rewrite −5 + x as x − 5. Both rational expressions have the same denominator.

$$= \frac{x^2 + (-4x-5)}{x-5}$$
Add numerators. Place this sum over the common denominator.

$$= \frac{x^2 - 4x - 5}{x-5}$$
Remove parentheses.

$$= \frac{\overset{1}{\cancel{(x-5)}}(x+1)}{\underset{1}{\cancel{x-5}}}$$
Factor and simplify.

$$= x + 1$$　∎

✔ **CHECK POINT 5** Add: $\dfrac{x^2}{x-7} + \dfrac{4x+21}{7-x}$.

ADDING AND SUBTRACTING RATIONAL EXPRESSIONS WITH OPPOSITE DENOMINATORS When one denominator is the additive inverse of the other, first multiply either rational expression by $\frac{-1}{-1}$ to obtain a common denominator.

EXAMPLE 6 Subtracting Rational Expressions when Denominators Are Opposites

Subtract: $\dfrac{5x - x^2}{x^2 - 4x - 3} - \dfrac{3x - x^2}{3 + 4x - x^2}$.

SOLUTION We note that $x^2 - 4x - 3$ and $3 + 4x - x^2$ are opposites. We multiply the second rational expression by $\frac{-1}{-1}$.

$$\frac{(-1)}{(-1)} \cdot \frac{3x - x^2}{3 + 4x - x^2} = \frac{-3x + x^2}{-3 - 4x + x^2}$$
Multiply the numerator and denominator by -1 by changing every term's sign.

$$= \frac{x^2 - 3x}{x^2 - 4x - 3}$$
Write the numerator and the denominator in descending powers of x.

We now return to the original subtraction problem.

$$\frac{5x - x^2}{x^2 - 4x - 3} - \frac{3x - x^2}{3 + 4x - x^2}$$
This is the given problem.

$$= \frac{5x - x^2}{x^2 - 4x - 3} - \frac{x^2 - 3x}{x^2 - 4x - 3}$$
Replace the second rational expression by the form obtained through multiplication by $\frac{-1}{-1}$.

$$= \frac{5x - x^2 - (x^2 - 3x)}{x^2 - 4x - 3}$$
Subtract numerators. Place this difference over the common denominator. Don't forget parentheses!

$$= \frac{5x - x^2 - x^2 + 3x}{x^2 - 4x - 3}$$
Remove parentheses and then change the sign of each term.

$$= \frac{-2x^2 + 8x}{x^2 - 4x - 3}$$
Combine like terms in the numerator. Although the numerator can be factored, further simplification is not possible. ∎

✔ **CHECK POINT 6** Subtract: $\dfrac{7x - x^2}{x^2 - 2x - 9} - \dfrac{5x - 3x^2}{9 + 2x - x^2}$.

8.3 EXERCISE SET

     Math XL *MyMathLab*

Student Solutions Manual   CD/Video   PH Math/Tutor Center   MathXL Tutorials on CD   MathXL®   MyMathLab   Interactmath.com

Practice Exercises

In Exercises 1–38, add or subtract as indicated. Simplify the result, if possible.

1. $\dfrac{7x}{13} + \dfrac{2x}{13}$

2. $\dfrac{3x}{17} + \dfrac{8x}{17}$

3. $\dfrac{8x}{15} + \dfrac{x}{15}$

4. $\dfrac{9x}{24} + \dfrac{x}{24}$

5. $\dfrac{x - 3}{12} + \dfrac{5x + 21}{12}$

6. $\dfrac{x + 4}{9} + \dfrac{2x - 25}{9}$

7. $\dfrac{4}{x} + \dfrac{2}{x}$

8. $\dfrac{5}{x} + \dfrac{13}{x}$

9. $\dfrac{8}{9x} + \dfrac{13}{9x}$

10. $\dfrac{4}{9x} + \dfrac{11}{9x}$

11. $\dfrac{5}{x + 3} + \dfrac{4}{x + 3}$

12. $\dfrac{8}{x + 6} + \dfrac{10}{x + 6}$

13. $\dfrac{x}{x - 3} + \dfrac{4x + 5}{x - 3}$

14. $\dfrac{x}{x - 4} + \dfrac{9x + 7}{x - 4}$

15. $\dfrac{4x + 1}{6x + 5} + \dfrac{8x + 9}{6x + 5}$

16. $\dfrac{3x + 2}{3x + 4} + \dfrac{3x + 6}{3x + 4}$

17. $\dfrac{y^2 + 7y}{y^2 - 5y} + \dfrac{y^2 - 4y}{y^2 - 5y}$

18. $\dfrac{y^2 - 2y}{y^2 + 3y} + \dfrac{y^2 + y}{y^2 + 3y}$

19. $\dfrac{4y - 1}{5y^2} + \dfrac{3y + 1}{5y^2}$

20. $\dfrac{y + 2}{6y^3} + \dfrac{3y - 2}{6y^3}$

21. $\dfrac{x^2 - 2}{x^2 + x - 2} + \dfrac{2x - x^2}{x^2 + x - 2}$

22. $\dfrac{x^2 + 9x}{4x^2 - 11x - 3} + \dfrac{3x - 5x^2}{4x^2 - 11x - 3}$

23. $\dfrac{x^2 - 4x}{x^2 - x - 6} + \dfrac{4x - 4}{x^2 - x - 6}$

24. $\dfrac{x}{2x + 7} - \dfrac{2}{2x + 7}$

25. $\dfrac{3x}{5x - 4} - \dfrac{4}{5x - 4}$

26. $\dfrac{x}{x - 1} - \dfrac{1}{x - 1}$

27. $\dfrac{4x}{4x - 3} - \dfrac{3}{4x - 3}$

28. $\dfrac{2y + 1}{3y - 7} - \dfrac{y + 8}{3y - 7}$

29. $\dfrac{14y}{7y + 2} - \dfrac{7y - 2}{7y + 2}$

30. $\dfrac{2x + 3}{3x - 6} - \dfrac{3 - x}{3x - 6}$

31. $\dfrac{3x + 1}{4x - 2} - \dfrac{x + 1}{4x - 2}$

32. $\dfrac{x^3 - 3}{2x^4} - \dfrac{7x^3 - 3}{2x^4}$

33. $\dfrac{3y^2 - 1}{3y^3} - \dfrac{6y^2 - 1}{3y^3}$

34. $\dfrac{y^2 + 3y}{y^2 + y - 12} - \dfrac{y^2 - 12}{y^2 + y - 12}$

35. $\dfrac{4y^2 + 5}{9y^2 - 64} - \dfrac{y^2 - y + 29}{9y^2 - 64}$

36. $\dfrac{2y^2 + 6y + 8}{y^2 - 16} - \dfrac{y^2 - 3y - 12}{y^2 - 16}$

37. $\dfrac{6y^2 + y}{2y^2 - 9y + 9} - \dfrac{2y + 9}{2y^2 - 9y + 9} - \dfrac{4y - 3}{2y^2 - 9y + 9}$

38. $\dfrac{3y^2 - 2}{3y^2 + 10y - 8} - \dfrac{y + 10}{3y^2 + 10y - 8} - \dfrac{y^2 - 6y}{3y^2 + 10y - 8}$

In Exercises 39–64, denominators are additive inverses. Add or subtract as indicated. Simplify the result, if possible.

39. $\dfrac{4}{x - 3} + \dfrac{2}{3 - x}$

40. $\dfrac{6}{x - 5} + \dfrac{2}{5 - x}$

41. $\dfrac{6x + 7}{x - 6} + \dfrac{3x}{6 - x}$

42. $\dfrac{6x + 5}{x - 2} + \dfrac{4x}{2 - x}$

43. $\dfrac{5x - 2}{3x - 4} + \dfrac{2x - 3}{4 - 3x}$

44. $\dfrac{9x - 1}{7x - 3} + \dfrac{6x - 2}{3 - 7x}$

45. $\dfrac{x^2}{x - 2} + \dfrac{4}{2 - x}$

46. $\dfrac{x^2}{x - 3} + \dfrac{9}{3 - x}$

47. $\dfrac{y - 3}{y^2 - 25} + \dfrac{y - 3}{25 - y^2}$

48. $\dfrac{y - 7}{y^2 - 16} + \dfrac{7 - y}{16 - y^2}$

49. $\dfrac{6}{x - 1} - \dfrac{5}{1 - x}$

50. $\dfrac{10}{x - 2} - \dfrac{6}{2 - x}$

51. $\dfrac{10}{x + 3} - \dfrac{2}{-x - 3}$

52. $\dfrac{11}{x + 7} - \dfrac{5}{-x - 7}$

53. $\dfrac{y}{y - 1} - \dfrac{1}{1 - y}$

54. $\dfrac{y}{y - 4} - \dfrac{4}{4 - y}$

55. $\dfrac{3 - x}{x - 7} - \dfrac{2x - 5}{7 - x}$

56. $\dfrac{4 - x}{x - 9} - \dfrac{3x - 8}{9 - x}$

57. $\dfrac{x - 2}{x^2 - 25} - \dfrac{x - 2}{25 - x^2}$

58. $\dfrac{x - 8}{x^2 - 16} - \dfrac{x - 8}{16 - x^2}$

59. $\dfrac{x}{x - y} + \dfrac{y}{y - x}$

60. $\dfrac{2x - y}{x - y} + \dfrac{x - 2y}{y - x}$

61. $\dfrac{2x}{x^2 - y^2} + \dfrac{2y}{y^2 - x^2}$

62. $\dfrac{2y}{x^2 - y^2} + \dfrac{2x}{y^2 - x^2}$

63. $\dfrac{x^2 - 2}{x^2 + 6x - 7} + \dfrac{19 - 4x}{7 - 6x - x^2}$

64. $\dfrac{2x + 3}{x^2 - x - 30} + \dfrac{x - 2}{30 + x - x^2}$

Practice Plus

In Exercises 65–72, perform the indicated operation or operations. Simplify the result, if possible.

65. $\dfrac{6b^2 - 10b}{16b^2 - 48b + 27} + \dfrac{7b^2 - 20b}{16b^2 - 48b + 27} - \dfrac{6b - 3b^2}{16b^2 - 48b + 27}$

66. $\dfrac{22b + 15}{12b^2 + 52b - 9} + \dfrac{30b - 20}{12b^2 + 52b - 9} - \dfrac{4 - 2b}{12b^2 + 52b - 9}$

67. $\dfrac{2y}{y - 5} - \left(\dfrac{2}{y - 5} + \dfrac{y - 2}{y - 5} \right)$

68. $\dfrac{3x}{(x + 1)^2} - \left[\dfrac{5x + 1}{(x + 1)^2} - \dfrac{3x + 2}{(x + 1)^2} \right]$

69. $\dfrac{b}{ac + ad - bc - bd} - \dfrac{a}{ac + ad - bc - bd}$

70. $\dfrac{y}{ax + bx - ay - by} - \dfrac{x}{ax + bx - ay - by}$

71. $\dfrac{(y-3)(y+2)}{(y+1)(y-4)} - \dfrac{(y+2)(y+3)}{(y+1)(4-y)} - \dfrac{(y+5)(y-1)}{(y+1)(4-y)}$

72. $\dfrac{(y+1)(2y-1)}{(y-2)(y-3)} + \dfrac{(y+2)(y-1)}{(y-2)(y-3)} - \dfrac{(y+5)(2y+1)}{(3-y)(2-y)}$

Application Exercises

73. Anthropologists and forensic scientists classify skulls using

$$\frac{L+60W}{L} - \frac{L-40W}{L},$$

where L is the skull's length and W is its width.

a. Express the classification as a single rational expression.

b. If the value of the rational expression in part (a) is less than 75, a skull is classified as long. A medium skull has a value between 75 and 80, and a round skull has a value over 80. Use your rational expression from part (a) to classify a skull that is 5 inches wide and 6 inches long.

74. The temperature, in degrees Fahrenheit, of a dessert placed in a freezer for t hours is modeled by

$$\frac{t+30}{t^2+4t+1} - \frac{t-50}{t^2+4t+1}.$$

a. Express the temperature as a single rational expression.

b. Use your rational expression from part (a) to find the temperature of the dessert, to the nearest hundredth of a degree, after 1 hour and after 2 hours.

In Exercises 75–76, find the perimeter of each rectangle.

75.

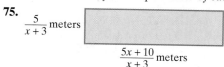

$\dfrac{5}{x+3}$ meters

$\dfrac{5x+10}{x+3}$ meters

76.

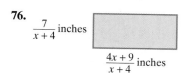

$\dfrac{7}{x+4}$ inches

$\dfrac{4x+9}{x+4}$ inches

Writing in Mathematics

77. Explain how to add rational expressions when denominators are the same. Give an example with your explanation.

78. Explain how to subtract rational expressions when denominators are the same. Give an example with your explanation.

79. Describe two similarities between the following problems:

$$\frac{3}{8} + \frac{1}{8} \quad \text{and} \quad \frac{x}{x^2-1} + \frac{1}{x^2-1}.$$

80. Explain how to add rational expressions when denominators are opposites. Use an example to support your explanation.

Critical Thinking Exercises

81. Which one of the following is true?

a. The sum of two rational expressions with the same denominator can be found by adding numerators, adding denominators, and then simplifying.

b. $\dfrac{4}{b} - \dfrac{2}{-b} = -\dfrac{2}{b}$

c. The difference between two rational expressions with the same denominator can always be simplified.

d. $\dfrac{2x+1}{x-7} + \dfrac{3x+1}{x-7} - \dfrac{5x+2}{x-7} = 0$

In Exercises 82–83, perform the indicated operations. Simplify the result if possible.

82. $\left(\dfrac{3x-1}{x^2+5x-6} - \dfrac{2x-7}{x^2+5x-6} \right) \div \dfrac{x+2}{x^2-1}$

83. $\left(\dfrac{3x^2-4x+4}{3x^2+7x+2} - \dfrac{10x+9}{3x^2+7x+2} \right) \div \dfrac{x-5}{x^2-4}$

In Exercises 84–88, find the missing expression.

84. $\dfrac{2x}{x+3} + \dfrac{\rule{1cm}{0.4pt}}{x+3} = \dfrac{4x+1}{x+3}$

85. $\dfrac{3x}{x+2} - \dfrac{\rule{1cm}{0.4pt}}{x+2} = \dfrac{6-17x}{x+2}$

86. $\dfrac{6}{x-2} + \dfrac{\rule{1cm}{0.4pt}}{2-x} = \dfrac{13}{x-2}$

87. $\dfrac{a^2}{a-4} - \dfrac{\rule{1cm}{0.4pt}}{a-4} = a+3$

88. $\dfrac{3x}{x-5} + \dfrac{\rule{1cm}{0.4pt}}{5-x} = \dfrac{7x+1}{x-5}$

Technology Exercises

In Exercises 89–91, use the GRAPH *or* TABLE *feature of a graphing utility to determine if the subtraction has been performed correctly. If the answer is wrong, correct it and then verify your correction using the graphing utility.*

89. $\dfrac{3x + 6}{2} - \dfrac{x}{2} = x + 3$

90. $\dfrac{x^2 + 4x + 3}{x + 2} - \dfrac{5x + 9}{x + 2} = x - 2, x \neq -2$

91. $\dfrac{x^2 - 13}{x + 4} - \dfrac{3}{x + 4} = x + 4, x \neq -4$

Review Exercises

92. Subtract: $\dfrac{13}{15} - \dfrac{8}{45}$. (Section 1.1, Example 9)

93. Factor completely: $81x^4 - 1$. (Section 7.4, Example 4)

94. Divide: $\dfrac{3x^3 + 2x^2 - 26x - 15}{x + 3}$. (Section 6.6, Example 2)

SECTION

8.4

Objectives

1 Find the least common denominator.

2 Add and subtract rational expressions with different denominators.

ADDING AND SUBTRACTING RATIONAL EXPRESSIONS WITH DIFFERENT DENOMINATORS

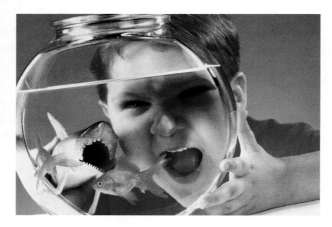

When my aunt asked how I liked my five-year-old nephew, I replied "medium rare." Unfortunately, my little joke did not get me out of baby sitting for the Dennis the Menace of our family. Now the little squirt doesn't want to go to bed because his head hurts. Does my aunt have any aspirin? What is the proper dosage for a child his age?

In this section's exercise set, you will use two formulas that model drug dosage for children. Before working with these models, we continue drawing on your experience from arithmetic to add and subtract rational expressions that have different denominators.

1 Find the least common denominator.

Finding the Least Common Denominator We can gain insight into adding rational expressions with different denominators by looking closely at what we do when adding fractions with different denominators. For example, suppose that we want to add $\frac{1}{2}$ and $\frac{2}{3}$. We must first write the fractions with the same denominator. We look for the smallest number that contains both 2 and 3 as factors. This number, 6, is then used as the *least common denominator*, or LCD.

The **least common denominator** of several rational expressions is a polynomial consisting of the product of all prime factors in the denominators, with each factor raised to the greatest power of its occurrence in any denominator.

FINDING THE LEAST COMMON DENOMINATOR
1. Factor each denominator completely.
2. List the factors of the first denominator.
3. Add to the list in step 2 any factors of the second denominator that do not appear in the list.
4. Form the product of each different factor from the list in step 3. This product is the least common denominator.

EXAMPLE 1 Finding the Least Common Denominator

Find the LCD of $\dfrac{7}{6x^2}$ and $\dfrac{2}{9x}$.

SOLUTION

Step 1. **Factor each denominator completely.**
$$6x^2 = 3 \cdot 2x^2 \quad (\text{or } 3 \cdot 2 \cdot x \cdot x)$$
$$9x = 3 \cdot 3x$$

Step 2. **List the factors of the first denominator.**
$$3, 2, x^2 \quad (\text{or } 3, 2, x, x)$$

Step 3. **Add any unlisted factors from the second denominator.** Two factors from $3 \cdot 3x$ are already in our list. These factors include x and one factor of 3. We add the other factor of 3 to our list. We have
$$3, 3, 2, x^2.$$

Step 4. **The least common denominator is the product of all factors in the final list.** Thus,
$$3 \cdot 3 \cdot 2x^2$$
or $18x^2$ is the least common denominator. ∎

 CHECK POINT 1 Find the LCD of $\dfrac{3}{10x^2}$ and $\dfrac{7}{15x}$.

EXAMPLE 2 Finding the Least Common Denominator

Find the LCD of $\dfrac{3}{x + 1}$ and $\dfrac{5}{x - 1}$.

SOLUTION

Step 1. **Factor each denominator completely.**
$$x + 1 = 1(x + 1)$$
$$x - 1 = 1(x - 1)$$

Step 2. **List the factors of the first denominator.**
$$1, x + 1$$

Step 3. **Add any unlisted factors from the second denominator.** We listed 1 and $x + 1$ as factors of the first denominator, $1(x + 1)$. The factors of the second denominator, $1(x - 1)$, include 1 and $x - 1$. One factor, 1, is already in our list, but the other factor, $x - 1$, is not. We add $x - 1$ to the list. We have

$$1, x + 1, x - 1.$$

Step 4. **The least common denominator is the product of all factors in the final list.** Thus,

$$1(x + 1)(x - 1)$$

or $(x + 1)(x - 1)$ is the least common denominator of $\dfrac{3}{x + 1}$ and $\dfrac{5}{x - 1}$. ■

 CHECK POINT 2 Find the LCD of $\dfrac{2}{x + 3}$ and $\dfrac{4}{x - 3}$.

EXAMPLE 3 Finding the Least Common Denominator

Find the LCD of

$$\frac{7}{5x^2 + 15x} \quad \text{and} \quad \frac{9}{x^2 + 6x + 9}.$$

SOLUTION

Step 1. **Factor each denominator completely.**

$$5x^2 + 15x = 5x(x + 3)$$
$$x^2 + 6x + 9 = (x + 3)^2$$

Step 2. **List the factors of the first denominator.**

$$5, x, (x + 3)$$

Step 3. **Add any unlisted factors from the second denominator.** The second denominator is $(x + 3)^2$ or $(x + 3)(x + 3)$. One factor of $x + 3$ is already in our list, but the other factor is not. We add $x + 3$ to the list. We have

$$5, x, (x + 3), (x + 3).$$

Step 4. **The least common denominator is the product of all factors in the final list.** Thus,

$$5x(x + 3)(x + 3) \quad \text{or} \quad 5x(x + 3)^2$$

is the least common denominator. ■

 CHECK POINT 3 Find the LCD of $\dfrac{9}{7x^2 + 28x}$ and $\dfrac{11}{x^2 + 8x + 16}$.

---

2 Add and subtract rational expressions with different denominators.

Adding and Subtracting Rational Expressions with Different Denominators
Finding the least common denominator for two (or more) rational expressions is the first step needed to add or subtract the expressions. For example, to add $\frac{1}{2}$ and $\frac{2}{3}$, we first determine that the LCD is 6. Then we write each fraction in terms of the LCD.

$$\frac{1}{2} + \frac{2}{3} = \frac{1}{2} \cdot \frac{3}{3} + \frac{2}{3} \cdot \frac{2}{2}$$

Multiply the numerator and denominator of each fraction by whatever extra factors are required to form 6, the LCD.

$\frac{3}{3} = 1$ and $\frac{2}{2} = 1$. Multiplying by 1 does not change a fraction's value.

$$= \frac{3}{6} + \frac{4}{6}$$

$$= \frac{3 + 4}{6}$$

Add numerators. Place this sum over the LCD.

$$= \frac{7}{6}$$

We follow the same steps in adding or subtracting rational expressions with different denominators.

ADDING AND SUBTRACTING RATIONAL EXPRESSIONS THAT HAVE DIFFERENT DENOMINATORS

1. Find the LCD of the rational expressions.

2. Rewrite each rational expression as an equivalent expression whose denominator is the LCD. To do so, multiply the numerator and the denominator of each rational expression by any factor(s) needed to convert the denominator into the LCD.

3. Add or subtract numerators, placing the resulting expression over the LCD.

4. If possible, simplify the resulting rational expression.

EXAMPLE 4 Adding Rational Expressions with Different Denominators

Add: $\dfrac{7}{6x^2} + \dfrac{2}{9x}$.

SOLUTION

Step 1. **Find the least common denominator.** In Example 1, we found that the LCD for these rational expressions is $18x^2$.

Step 2. **Write equivalent expressions with the LCD as denominators.** We must rewrite each rational expression with a denominator of $18x^2$.

$$\frac{7}{6x^2} \cdot \frac{3}{3} = \frac{21}{18x^2} \qquad\qquad \frac{2}{9x} \cdot \frac{2x}{2x} = \frac{4x}{18x^2}$$

Multiply the numerator and denominator by 3 to get $18x^2$, the LCD.

Multiply the numerator and denominator by $2x$ to get $18x^2$, the LCD.

Because $\dfrac{3}{3} = 1$ and $\dfrac{2x}{2x} = 1$, we are not changing the value of either rational expression, only its appearance.

STUDY TIP

It is incorrect to add rational expressions by adding numerators and adding denominators. Avoid this common error.

Incorrect!

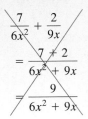

Now we are ready to perform the indicated addition.

$$\frac{7}{6x^2} + \frac{2}{9x}$$ This is the given problem. The LCD is $18x^2$.

$$= \frac{7}{6x^2} \cdot \frac{3}{3} + \frac{2}{9x} \cdot \frac{2x}{2x}$$ Write equivalent expressions with the LCD.

$$= \frac{21}{18x^2} + \frac{4x}{18x^2}$$

Steps 3 and 4. **Add numerators, putting this sum over the LCD. Simplify if possible.**

$$= \frac{21 + 4x}{18x^2} \quad \text{or} \quad \frac{4x + 21}{18x^2}$$

The numerator is prime and further simplification is not possible. ■

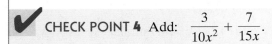 **CHECK POINT 4** Add: $\dfrac{3}{10x^2} + \dfrac{7}{15x}$.

EXAMPLE 5 Adding Rational Expressions with Different Denominators

Add: $\dfrac{3}{x + 1} + \dfrac{5}{x - 1}$.

SOLUTION

Step 1. **Find the least common denominator.** The factors of the denominators are $x + 1$ and $x - 1$. In Example 2, we found that the LCD is $(x + 1)(x - 1)$.

Step 2. **Write equivalent expressions with the LCD as denominators.**

$$\frac{3}{x + 1} + \frac{5}{x - 1}$$

$$= \frac{3(x - 1)}{(x + 1)(x - 1)} + \frac{5(x + 1)}{(x + 1)(x - 1)}$$ Multiply each numerator and denominator by the extra factor required to form $(x + 1)(x - 1)$, the LCD.

Steps 3 and 4. **Add numerators, putting this sum over the LCD. Simplify if possible.**

$$= \frac{3(x - 1) + 5(x + 1)}{(x + 1)(x - 1)}$$

$$= \frac{3x - 3 + 5x + 5}{(x + 1)(x - 1)}$$ Use the distributive property to multiply and remove grouping symbols.

$$= \frac{8x + 2}{(x + 1)(x - 1)}$$ Combine like terms: $3x + 5x = 8x$ and $-3 + 5 = 2$. ■

We can factor 2 from the numerator of the answer in Example 5 to obtain

$$\frac{2(4x + 1)}{(x + 1)(x - 1)}.$$

Because the numerator and denominator do not have any common factors, further simplification is not possible. In this section, unless there is a common factor in the numerator and denominator, we will leave an answer's numerator in unfactored form and the denominator in factored form.

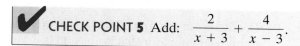 **CHECK POINT 5** Add: $\dfrac{2}{x + 3} + \dfrac{4}{x - 3}$.

EXAMPLE 6 Subtracting Rational Expressions with Different Denominators

Subtract: $\dfrac{x}{x + 3} - 1$.

SOLUTION

Step 1. Find the least common denominator. We know that 1 means $\frac{1}{1}$. The factor of the first denominator is $x + 3$. Adding the factor of the second denominator, 1, the LCD is $1(x + 3)$ or $x + 3$.

Step 2. Write equivalent expressions with the LCD as denominators.

$$\frac{x}{x + 3} - 1$$

$$= \frac{x}{x + 3} - \frac{1}{1} \qquad \text{Write 1 as } \tfrac{1}{1}.$$

$$= \frac{x}{x + 3} - \frac{1(x + 3)}{1(x + 3)} \qquad \text{Multiply the numerator and denominator of } \tfrac{1}{1} \text{ by the extra factor required to form } x + 3, \text{ the LCD.}$$

Steps 3 and 4. Subtract numerators, putting this difference over the LCD. Simplify if possible.

$$= \frac{x - (x + 3)}{x + 3}$$

$$= \frac{x - x - 3}{x + 3} \qquad \text{Remove parentheses and then change the sign of each term.}$$

$$= \frac{-3}{x + 3} \quad \text{or} \quad -\frac{3}{x + 3} \qquad \text{Simplify.} \qquad \blacksquare$$

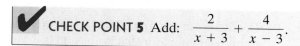 **CHECK POINT 6** Subtract: $\dfrac{x}{x + 5} - 1$.

EXAMPLE 7 Subtracting Rational Expressions with Different Denominators

Subtract: $\dfrac{y + 2}{4y + 16} - \dfrac{2}{y^2 + 4y}$.

SOLUTION

Step 1. Find the least common denominator. Start by factoring the denominators.

$$4y + 16 = 4(y + 4)$$
$$y^2 + 4y = y(y + 4)$$

The factors of the first denominator are 4 and $y + 4$. The only factor from the second denominator that is unlisted is y. Thus, the least common denominator is $4y(y + 4)$.

Step 2. Write equivalent expressions with the LCD as denominators.

$$\frac{y + 2}{4y + 16} - \frac{2}{y^2 + 4y}$$

$$= \frac{y + 2}{4(y + 4)} - \frac{2}{y(y + 4)} \qquad \begin{array}{l}\text{Factor denominators.}\\ \text{The LCD is } 4y(y + 4).\end{array}$$

$$= \frac{(y + 2)y}{4y(y + 4)} - \frac{2 \cdot 4}{4y(y + 4)} \qquad \begin{array}{l}\text{Multiply each numerator and}\\ \text{denominator by the extra factor}\\ \text{required to form } 4y(y + 4), \text{the LCD.}\end{array}$$

Steps 3 and 4. Subtract numerators, putting this difference over the LCD. Simplify if possible.

$$= \frac{(y + 2)y - 2 \cdot 4}{4y(y + 4)}$$

$$= \frac{y^2 + 2y - 8}{4y(y + 4)} \qquad \begin{array}{l}\text{Use the distributive property:}\\ (y + 2)y = y^2 + 2y. \text{ Multiply: } 2 \cdot 4 = 8.\end{array}$$

$$= \frac{\overset{1}{\cancel{(y + 4)}}(y - 2)}{4y\underset{1}{\cancel{(y + 4)}}} \qquad \text{Factor and simplify.}$$

$$= \frac{y - 2}{4y}$$

■

✔ **CHECK POINT 7** Subtract: $\dfrac{5}{y^2 - 5y} - \dfrac{y}{5y - 25}$.

In some situations, after factoring denominators, a factor in one denominator is the opposite of a factor in the other denominator. When this happens, we can use the following procedure:

ADDING AND SUBTRACTING RATIONAL EXPRESSIONS WHEN DENOMINATORS CONTAIN OPPOSITE FACTORS When one denominator contains the opposite factor of the other, first multiply either rational expression by $\frac{-1}{-1}$. Then apply the procedure for adding or subtracting rational expressions that have different denominators to the rewritten problem.

EXAMPLE 8 Adding Rational Expressions with Opposite Factors in the Denominators

Add: $\dfrac{x^2 - 2}{2x^2 - x - 3} + \dfrac{x - 2}{3 - 2x}$.

SOLUTION

Step 1. **Find the least common denominator.** Start by factoring the denominators.

$$2x^2 - x - 3 = (2x - 3)(x + 1)$$
$$3 - 2x = 1(3 - 2x)$$

Do you see that $2x - 3$ and $3 - 2x$ are opposite factors? Thus, we multiply either rational expression by $\dfrac{-1}{-1}$. We will use the second rational expression, resulting in $2x - 3$ in the denominator.

$$\dfrac{x^2 - 2}{2x^2 - x - 3} + \dfrac{x - 2}{3 - 2x}$$

$$= \dfrac{x^2 - 2}{(2x - 3)(x + 1)} + \dfrac{(-1)}{(-1)} \cdot \dfrac{x - 2}{3 - 2x}$$

> Factor the first denominator. Multiply the second rational expression by $\frac{-1}{-1}$.

$$= \dfrac{x^2 - 2}{(2x - 3)(x + 1)} + \dfrac{-x + 2}{-3 + 2x}$$

> Perform the multiplications by -1 by changing every term's sign.

$$= \dfrac{x^2 - 2}{(2x - 3)(x + 1)} + \dfrac{2 - x}{2x - 3}$$

The LCD of our rewritten addition problem is $(2x - 3)(x + 1)$.

Step 2. **Write equivalent expressions with the LCD as denominators.**

$$= \dfrac{x^2 - 2}{(2x - 3)(x + 1)} + \dfrac{(2 - x)(x + 1)}{(2x - 3)(x + 1)}$$

> Multiply the numerator and denominator of the second rational expression by the extra factor required to form $(2x - 3)(x + 1)$, the LCD.

Steps 3 and 4. **Add numerators, putting this sum over the LCD. Simplify if possible.**

$$= \dfrac{x^2 - 2 + (2 - x)(x + 1)}{(2x - 3)(x + 1)}$$

$$= \dfrac{x^2 - 2 + 2x + 2 - x^2 - x}{(2x - 3)(x + 1)}$$

> Use the FOIL method to multiply $(2 - x)(x + 1)$.

$$= \dfrac{(x^2 - x^2) + (2x - x) + (-2 + 2)}{(2x - 3)(x + 1)}$$

> Group like terms.

$$= \dfrac{x}{(2x - 3)(x + 1)}$$

> Combine like terms. ∎

DISCOVER FOR YOURSELF

In Example 8, the denominators can be factored as follows:

$$2x^2 - x - 3 = (2x - 3)(x + 1)$$
$$3 - 2x = -1(2x - 3).$$

Using these factorizations, what is the LCD? Solve Example 8 by obtaining this LCD in each rational expression. Then combine the expressions. How does your solution compare with the one shown on the right?

✔ CHECK POINT **8** Add: $\dfrac{4x}{x^2 - 25} + \dfrac{3}{5 - x}$.

8.4 EXERCISE SET

Student Solutions Manual   CD/Video   PH Math/Tutor Center   MathXL Tutorials on CD   MathXL®   MyMathLab   Interactmath.com

Practice Exercises

In Exercises 1–16, find the least common denominator of the rational expressions.

1. $\dfrac{7}{15x^2}$ and $\dfrac{13}{24x}$

2. $\dfrac{11}{25x^2}$ and $\dfrac{17}{35x}$

3. $\dfrac{8}{15x^2}$ and $\dfrac{5}{6x^5}$

4. $\dfrac{7}{15x^2}$ and $\dfrac{11}{24x^5}$

5. $\dfrac{4}{x-3}$ and $\dfrac{7}{x+1}$

6. $\dfrac{2}{x-5}$ and $\dfrac{3}{x+7}$

7. $\dfrac{5}{7(y+2)}$ and $\dfrac{10}{y}$

8. $\dfrac{8}{11(y+5)}$ and $\dfrac{12}{y}$

9. $\dfrac{17}{x+4}$ and $\dfrac{18}{x^2-16}$

10. $\dfrac{3}{x-6}$ and $\dfrac{4}{x^2-36}$

11. $\dfrac{8}{y^2-9}$ and $\dfrac{14}{y(y+3)}$

12. $\dfrac{14}{y^2-49}$ and $\dfrac{12}{y(y-7)}$

13. $\dfrac{7}{y^2-1}$ and $\dfrac{y}{y^2-2y+1}$

14. $\dfrac{9}{y^2-25}$ and $\dfrac{y}{y^2-10y+25}$

15. $\dfrac{3}{x^2-x-20}$ and $\dfrac{x}{2x^2+7x-4}$

16. $\dfrac{7}{x^2-5x-6}$ and $\dfrac{x}{x^2-4x-5}$

In Exercises 17–82, add or subtract as indicated. Simplify the result, if possible.

17. $\dfrac{3}{x}+\dfrac{5}{x^2}$

18. $\dfrac{4}{x}+\dfrac{8}{x^2}$

19. $\dfrac{2}{9x}+\dfrac{11}{6x}$

20. $\dfrac{5}{6x}+\dfrac{7}{8x}$

21. $\dfrac{4}{x}+\dfrac{7}{2x^2}$

22. $\dfrac{10}{x}+\dfrac{3}{5x^2}$

23. $6+\dfrac{1}{x}$

24. $3+\dfrac{1}{x}$

25. $\dfrac{2}{x}+9$

26. $\dfrac{7}{x}+4$

27. $\dfrac{x-1}{6}+\dfrac{x+2}{3}$

28. $\dfrac{x+3}{2}+\dfrac{x+5}{4}$

29. $\dfrac{4}{x}+\dfrac{3}{x-5}$

30. $\dfrac{3}{x}+\dfrac{4}{x-6}$

31. $\dfrac{2}{x-1}+\dfrac{3}{x+2}$

32. $\dfrac{3}{x-2}+\dfrac{4}{x+3}$

33. $\dfrac{2}{y+5}+\dfrac{3}{4y}$

34. $\dfrac{3}{y+1}+\dfrac{2}{3y}$

35. $\dfrac{x}{x+7}-1$

36. $\dfrac{x}{x+6}-1$

37. $\dfrac{7}{x+5}-\dfrac{4}{x-5}$

38. $\dfrac{8}{x+6}-\dfrac{2}{x-6}$

39. $\dfrac{2x}{x^2-16}+\dfrac{x}{x-4}$

40. $\dfrac{4x}{x^2-25}+\dfrac{x}{x+5}$

41. $\dfrac{5y}{y^2-9}-\dfrac{4}{y+3}$

42. $\dfrac{8y}{y^2-16}-\dfrac{5}{y+4}$

43. $\dfrac{7}{x-1}-\dfrac{3}{(x-1)^2}$

44. $\dfrac{5}{x+3}-\dfrac{2}{(x+3)^2}$

45. $\dfrac{3y}{4y-20}+\dfrac{9y}{6y-30}$

46. $\dfrac{4y}{5y-10}+\dfrac{3y}{10y-20}$

47. $\dfrac{y+4}{y}-\dfrac{y}{y+4}$

48. $\dfrac{y}{y-5}-\dfrac{y-5}{y}$

49. $\dfrac{2x+9}{x^2-7x+12}-\dfrac{2}{x-3}$

50. $\dfrac{3x+7}{x^2-5x+6}-\dfrac{3}{x-3}$

51. $\dfrac{3}{x^2-1}+\dfrac{4}{(x+1)^2}$

52. $\dfrac{6}{x^2 - 4} + \dfrac{2}{(x + 2)^2}$

53. $\dfrac{3x}{x^2 + 3x - 10} - \dfrac{2x}{x^2 + x - 6}$

54. $\dfrac{x}{x^2 - 2x - 24} - \dfrac{x}{x^2 - 7x + 6}$

55. $\dfrac{y}{y^2 + 2y + 1} + \dfrac{4}{y^2 + 5y + 4}$

56. $\dfrac{y}{y^2 + 5y + 6} + \dfrac{4}{y^2 - y - 6}$

57. $\dfrac{x - 5}{x + 3} + \dfrac{x + 3}{x - 5}$

58. $\dfrac{x - 7}{x + 4} + \dfrac{x + 4}{x - 7}$

59. $\dfrac{5}{2y^2 - 2y} - \dfrac{3}{2y - 2}$

60. $\dfrac{7}{5y^2 - 5y} - \dfrac{2}{5y - 5}$

61. $\dfrac{4x + 3}{x^2 - 9} - \dfrac{x + 1}{x - 3}$

62. $\dfrac{2x - 1}{x + 6} - \dfrac{6 - 5x}{x^2 - 36}$

63. $\dfrac{y^2 - 39}{y^2 + 3y - 10} - \dfrac{y - 7}{y - 2}$

64. $\dfrac{y^2 - 6}{y^2 + 9y + 18} - \dfrac{y - 4}{y + 6}$

65. $4 + \dfrac{1}{x - 3}$ **66.** $7 + \dfrac{1}{x - 5}$

67. $3 - \dfrac{3y}{y + 1}$ **68.** $7 - \dfrac{4y}{y + 5}$

69. $\dfrac{9x + 3}{x^2 - x - 6} + \dfrac{x}{3 - x}$

70. $\dfrac{x^2 + 9x}{x^2 - 2x - 3} + \dfrac{5}{3 - x}$

71. $\dfrac{x + 3}{x^2 + x - 2} - \dfrac{2}{x^2 - 1}$

72. $\dfrac{x}{x^2 - 10x + 25} - \dfrac{x - 4}{2x - 10}$

73. $\dfrac{y + 3}{5y^2} - \dfrac{y - 5}{15y}$

74. $\dfrac{y - 7}{3y^2} - \dfrac{y - 2}{12y}$

75. $\dfrac{x + 3}{3x + 6} + \dfrac{x}{4 - x^2}$

76. $\dfrac{x + 7}{4x + 12} + \dfrac{x}{9 - x^2}$

77. $\dfrac{y}{y^2 - 1} + \dfrac{2y}{y - y^2}$

78. $\dfrac{y}{y^2 - 1} + \dfrac{5y}{y - y^2}$

79. $\dfrac{x - 1}{x} + \dfrac{y + 1}{y}$

80. $\dfrac{x + 2}{y} + \dfrac{y - 2}{x}$

81. $\dfrac{3x}{x^2 - y^2} - \dfrac{2}{y - x}$

82. $\dfrac{7x}{x^2 - y^2} - \dfrac{3}{y - x}$

Practice Plus

In Exercises 83–92, perform the indicated operation or operations. Simplify the result, if possible.

83. $\dfrac{x + 6}{x^2 - 4} - \dfrac{x + 3}{x + 2} + \dfrac{x - 3}{x - 2}$

84. $\dfrac{x + 8}{x^2 - 9} - \dfrac{x + 2}{x + 3} + \dfrac{x - 2}{x - 3}$

85. $\dfrac{5}{x^2 - 25} + \dfrac{4}{x^2 - 11x + 30} - \dfrac{3}{x^2 - x - 30}$

86. $\dfrac{3}{x^2 - 49} + \dfrac{2}{x^2 - 15x + 56} - \dfrac{5}{x^2 - x - 56}$

87. $\dfrac{x + 6}{x^3 - 27} - \dfrac{x}{x^3 + 3x^2 + 9x}$

88. $\dfrac{x + 8}{x^3 - 8} - \dfrac{x}{x^3 + 2x^2 + 4x}$

89. $\dfrac{9y + 3}{y^2 - y - 6} + \dfrac{y}{3 - y} + \dfrac{y - 1}{y + 2}$

90. $\dfrac{7y - 2}{y^2 - y - 12} + \dfrac{2y}{4 - y} + \dfrac{y + 1}{y + 3}$

91. $\dfrac{3}{x^2 + 4xy + 3y^2} - \dfrac{5}{x^2 - 2xy - 3y^2} + \dfrac{2}{x^2 - 9y^2}$

92. $\dfrac{5}{x^2 + 3xy + 2y^2} - \dfrac{7}{x^2 - xy - 2y^2} + \dfrac{4}{x^2 - 4y^2}$

Application Exercises

Two formulas that approximate the dosage of a drug prescribed for children are

$$\text{Young's Rule: } C = \frac{DA}{A + 12}$$

$$\text{and Cowling's Rule: } C = \frac{D(A + 1)}{24}.$$

In each formula, A = the child's age, in years, D = an adult dosage, and C = the proper child's dosage. The formulas apply for ages 2 through 13, inclusive. Use the formulas to solve Exercises 93–96.

93. Use Young's rule to find the difference in a child's dosage for an 8-year-old child and a 3-year-old child. Express the answer as a single rational expression in terms of D. Then describe what your answer means in terms of the variables in the model.

94. Use Young's rule to find the difference in a child's dosage for a 10-year-old child and a 3-year-old child. Express the answer as a single rational expression in terms of D. Then describe what your answer means in terms of the variables in the model.

95. For a 12-year-old child, what is the difference in the dosage given by Cowling's rule and Young's rule? Express the answer as a single rational expression in terms of D. Then describe what your answer means in terms of the variables in the models.

96. Use Cowling's rule to find the difference in a child's dosage for a 12-year-old child and a 10-year-old child. Express the answer as a single rational expression in terms of D. Then describe what your answer means in terms of the variables in the model.

The graphs illustrate Young's rule and Cowling's rule when the dosage of a drug prescribed for an adult is 1000 milligrams. Use the graphs to solve Exercises 97–100.

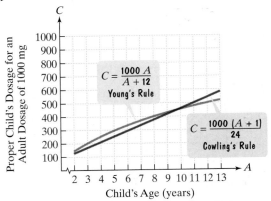

97. Does either formula consistently give a smaller dosage than the other? If so, which one?

98. Is there an age at which the dosage given by one formula becomes greater than the dosage given by the other? If so, what is a reasonable estimate of that age?

99. For what age under 11 is the difference in dosage given by the two formulas the greatest?

100. For what age over 11 is the difference in dosage given by the two formulas the greatest?

In Exercises 101–102, express the perimeter of each rectangle as a single rational expression.

101. $\dfrac{x}{x+3}$, $\dfrac{x}{x+4}$

102. $\dfrac{x}{x+5}$, $\dfrac{x}{x+6}$

Writing in Mathematics

103. Explain how to find the least common denominator for denominators of $x^2 - 100$ and $x^2 - 20x + 100$.

104. Explain how to add rational expressions that have different denominators. Use $\dfrac{3}{x+5} + \dfrac{7}{x+2}$ in your explanation.

Explain the error in Exercises 105–106. Then rewrite the right side of the equation to correct the error that now exists.

105. $\dfrac{1}{x} + \dfrac{2}{5} = \dfrac{3}{x+5}$

106. $\dfrac{1}{x} + 7 = \dfrac{1}{x+7}$

107. The formulas in Exercises 93–96 relate the dosage of a drug prescribed for children to the child's age. Describe another factor that might be used when determining a child's dosage. Is this factor more or less important than age? Explain why.

Critical Thinking Exercises

108. Which one of the following is true?

a. $x - \dfrac{1}{5} = \dfrac{4}{5}x$

b. The LCD of $\dfrac{1}{x}$ and $\dfrac{2x}{x-1}$ is $x^2 - 1$.

c. $\dfrac{1}{x} + \dfrac{x}{1} = \dfrac{1}{\frac{x}{1}} + \dfrac{\frac{1}{x}}{1} = 1 + 1 = 2$

d. $\dfrac{2}{x} + 1 = \dfrac{2 + x}{x}, x \neq 0$

In Exercises 109–110, perform the indicated operations. Simplify the result, if possible.

109. $\dfrac{y^2 + 5y + 4}{y^2 + 2y - 3} \cdot \dfrac{y^2 + y - 6}{y^2 + 2y - 3} - \dfrac{2}{y - 1}$

110. $\left(\dfrac{1}{x + h} - \dfrac{1}{x}\right) \div h$

In Exercises 111–112, find the missing rational expression.

111. $\dfrac{2}{x - 1} + \underline{\hspace{1cm}} = \dfrac{2x^2 + 3x - 1}{x^2(x - 1)}$

112. $\dfrac{4}{x - 2} - \underline{\hspace{1cm}} = \dfrac{2x + 8}{(x - 2)(x + 1)}$

Review Exercises

113. Multiply: $(3x + 5)(2x - 7)$. (Section 6.3, Example 2)

114. Graph: $3x - y < 3$. (Section 4.6, Example 2)

115. Write the slope-intercept form of the equation of the line passing through $(-3, -4)$ and $(1, 0)$. (Section 4.5, Example 2)

✓ MID-CHAPTER CHECK POINT

CHAPTER 8

What You Know: We learned that it is necessary to exclude any value or values of a variable that make the denominator of a rational expression zero. We learned to simplify rational expressions by dividing the numerator and the denominator by common factors. We performed a variety of operations with rational expressions, including multiplication, division, addition, and subtraction.

1. Find all numbers for which $\dfrac{x^2 - 4}{x^2 - 2x - 8}$ is undefined.

In Exercises 2–4, simplify each rational expression.

2. $\dfrac{3x^2 - 7x + 2}{6x^2 + x - 1}$

3. $\dfrac{9 - 3y}{y^2 - 5y + 6}$

4. $\dfrac{16w^3 - 24w^2}{8w^4 - 12w^3}$

In Exercises 5–20, perform the indicated operations. Simplify the result, if possible.

5. $\dfrac{7x - 3}{x^2 + 3x - 4} - \dfrac{3x + 1}{x^2 + 3x - 4}$

6. $\dfrac{x + 2}{2x - 4} \cdot \dfrac{8}{x^2 - 4}$

7. $1 + \dfrac{7}{x - 2}$

8. $\dfrac{2x^2 + x - 1}{2x^2 - 7x + 3} \div \dfrac{x^2 - 3x - 4}{x^2 - x - 6}$

9. $\dfrac{1}{x^2 + 2x - 3} + \dfrac{1}{x^2 + 5x + 6}$

10. $\dfrac{17}{x - 5} + \dfrac{x + 8}{5 - x}$

11. $\dfrac{4y^2 - 1}{9y - 3y^2} \cdot \dfrac{y^2 - 7y + 12}{2y^2 - 7y - 4}$

12. $\dfrac{y}{y + 1} - \dfrac{2y}{y + 2}$

13. $\dfrac{w^2 + 6w + 5}{7w^2 - 63} \div \dfrac{w^2 + 10w + 25}{7w + 21}$

14. $\dfrac{2z}{z^2 - 9} - \dfrac{5}{z^2 + 4z + 3}$

15. $\dfrac{z + 2}{3z - 1} + \dfrac{5}{(3z - 1)^2}$

16. $\dfrac{8}{x^2 + 4x - 21} + \dfrac{3}{x + 7}$

17. $\dfrac{x^4 - 27x}{x^2 - 9} \cdot \dfrac{x + 3}{x^2 + 3x + 9}$

18. $\dfrac{x - 1}{x^2 - x - 2} - \dfrac{x + 2}{x^2 + 4x + 3}$

19. $\dfrac{x^2 - 2xy + y^2}{x + y} \div \dfrac{x^2 - xy}{5x + 5y}$

20. $\dfrac{5}{x + 5} + \dfrac{x}{x - 4} - \dfrac{11x - 8}{x^2 + x - 20}$

SECTION
8.5

COMPLEX RATIONAL EXPRESSIONS

Objectives

1 Simplify complex rational expressions by dividing.

2 Simplify complex rational expressions by multiplying by the LCD.

Do you drive to and from campus each day? If the one-way distance of your round-trip commute is d, then your average rate, or speed, is given by the expression

$$\frac{2d}{\dfrac{d}{r_1} + \dfrac{d}{r_2}}$$

in which r_1 and r_2 are your average rates on the outgoing and return trips, respectively. Do you notice anything unusual about this expression? It has two separate rational expressions in its denominator.

Numerator
$$\frac{2d}{\dfrac{d}{r_1} + \dfrac{d}{r_2}}$$
Main fraction bar
Denominator

Separate rational expressions occur in the denominator.

Complex rational expressions, also called **complex fractions**, have numerators or denominators containing one or more rational expressions. Here is another example of such an expression:

Numerator
$$\frac{1 + \dfrac{1}{x}}{1 - \dfrac{1}{x}}.$$
Main fraction bar
Denominator

Separate rational expressions occur in the numerator and denominator.

In this section, we study two methods for simplifying complex rational expressions.

1 Simplify complex rational expressions by dividing.

Simplifying by Rewriting Complex Rational Expressions as a Quotient of Two Rational Expressions One method for simplifying a complex rational expression is to combine its numerator into a single expression and combine its denominator into a single expression. Then perform the division by inverting the denominator and multiplying.

SIMPLIFYING A COMPLEX RATIONAL EXPRESSION BY DIVIDING

1. If necessary, add or subtract to get a single rational expression in the numerator.
2. If necessary, add or subtract to get a single rational expression in the denominator.
3. Perform the division indicated by the main fraction bar: Invert the denominator of the complex rational expression and multiply.
4. If possible, simplify.

The following examples illustrate the use of this first method.

EXAMPLE 1 Simplifying a Complex Rational Expression

Simplify:

$$\frac{\dfrac{1}{3} + \dfrac{2}{5}}{\dfrac{2}{5} - \dfrac{1}{3}}.$$

SOLUTION Let's first identify the parts of this complex rational expression.

Numerator

$$\frac{\dfrac{1}{3} + \dfrac{2}{5}}{\dfrac{2}{5} - \dfrac{1}{3}}$$

Main fraction bar

Denominator

Step 1. **Add to get a single rational expression in the numerator.**

$$\frac{1}{3} + \frac{2}{5} = \frac{1 \cdot 5}{3 \cdot 5} + \frac{2 \cdot 3}{5 \cdot 3} = \frac{5}{15} + \frac{6}{15} = \frac{11}{15}$$

The LCD is 3 · 5, or 15.

Step 2. **Subtract to get a single rational expression in the denominator.**

$$\frac{2}{5} - \frac{1}{3} = \frac{2 \cdot 3}{5 \cdot 3} - \frac{1 \cdot 5}{3 \cdot 5} = \frac{6}{15} - \frac{5}{15} = \frac{1}{15}$$

The LCD is 15.

Steps 3 and 4. **Perform the division indicated by the main fraction bar: Invert and multiply. If possible, simplify.**

$$\frac{\dfrac{1}{3} + \dfrac{2}{5}}{\dfrac{2}{5} - \dfrac{1}{3}} = \frac{\dfrac{11}{15}}{\dfrac{1}{15}} = \frac{11}{15} \cdot \frac{15}{1} = \frac{11}{\overset{1}{\cancel{15}}} \cdot \frac{\overset{1}{\cancel{15}}}{1} = 11$$

Invert and multiply.

✔ **CHECK POINT 1** Simplify:

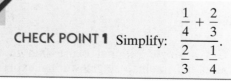

$$\frac{\dfrac{1}{4} + \dfrac{2}{3}}{\dfrac{2}{3} - \dfrac{1}{4}}.$$

EXAMPLE 2 Simplifying a Complex Rational Expression

Simplify:

$$\frac{1 + \dfrac{1}{x}}{1 - \dfrac{1}{x}}.$$

SOLUTION

Step 1. Add to get a single rational expression in the numerator.

$$1 + \frac{1}{x} = \frac{1}{1} + \frac{1}{x} = \frac{1 \cdot x}{1 \cdot x} + \frac{1}{x} = \frac{x}{x} + \frac{1}{x} = \frac{x + 1}{x}$$

> The LCD is 1 · x, or x.

Step 2. Subtract to get a single rational expression in the denominator.

$$1 - \frac{1}{x} = \frac{1}{1} - \frac{1}{x} = \frac{1 \cdot x}{1 \cdot x} - \frac{1}{x} = \frac{x}{x} - \frac{1}{x} = \frac{x - 1}{x}$$

> The LCD is 1 · x, or x.

Steps 3 and 4. Perform the division indicated by the main fraction bar: Invert and multiply. If possible, simplify.

$$\frac{1 + \dfrac{1}{x}}{1 - \dfrac{1}{x}} = \frac{\dfrac{x + 1}{x}}{\dfrac{x - 1}{x}} = \frac{x + 1}{x} \cdot \frac{x}{x - 1} = \frac{x + 1}{\overset{1}{x}} \cdot \frac{\overset{1}{x}}{x - 1} = \frac{x + 1}{x - 1}$$

> Invert and multiply.

✔ **CHECK POINT 2** Simplify: $\dfrac{2 - \dfrac{1}{x}}{2 + \dfrac{1}{x}}.$

EXAMPLE 3 Simplifying a Complex Rational Expression

Simplify:

$$\frac{\dfrac{1}{xy}}{\dfrac{1}{x} + \dfrac{1}{y}}.$$

SOLUTION

Step 1. Get a single rational expression in the numerator. The numerator, $\dfrac{1}{xy}$, already contains a single rational expression, so we can skip this step.

Step 2. **Add to get a single rational expression in the denominator.**

$$\frac{1}{x} + \frac{1}{y} = \frac{1 \cdot y}{x \cdot y} + \frac{1 \cdot x}{y \cdot x} = \frac{y}{xy} + \frac{x}{xy} = \frac{y + x}{xy}$$

The LCD is xy.

Steps 3 and 4. **Perform the division indicated by the main fraction bar: Invert and multiply. If possible, simplify.**

$$\frac{\dfrac{1}{xy}}{\dfrac{1}{x} + \dfrac{1}{y}} = \frac{\dfrac{1}{xy}}{\dfrac{y+x}{xy}} = \frac{1}{xy} \cdot \frac{xy}{y+x} = \frac{1}{\overset{1}{\cancel{xy}}} \cdot \frac{\overset{1}{\cancel{xy}}}{y+x} = \frac{1}{y+x}$$

Invert and multiply.

■

✔ **CHECK POINT 3** Simplify: $\dfrac{\dfrac{1}{x} - \dfrac{1}{y}}{\dfrac{1}{xy}}$.

2 Simplify complex rational expressions by multiplying by the LCD.

Simplifying Complex Rational Expressions by Multiplying by the LCD A second method for simplifying a complex rational expression is to find the least common denominator of all the rational expressions in its numerator and denominator. Then multiply each term in its numerator and denominator by this least common denominator. Because we are multiplying by a form of 1, we will obtain an equivalent expression that does not contain fractions in its numerator or denominator.

SIMPLIFYING A COMPLEX RATIONAL EXPRESSION BY MULTIPLYING BY THE LCD

1. Find the LCD of all rational expressions within the complex rational expression.

2. Multiply both the numerator and the denominator of the complex rational expression by this LCD.

3. Use the distributive property and multiply each term in the numerator and denominator by this LCD. Simplify. No fractional expressions should remain in the numerator and denominator.

4. If possible, factor and simplify.

We now rework Examples 1, 2, and 3 using the method of multiplying by the LCD. Compare the two simplification methods to see if there is one method that you prefer.

EXAMPLE 4 Simplifying a Complex Rational Expression by the LCD Method

Simplify:

$$\frac{\dfrac{1}{3} + \dfrac{2}{5}}{\dfrac{2}{5} - \dfrac{1}{3}}.$$

SOLUTION   The denominators in the complex rational expression are 3, 5, 5, and 3. The LCD is $3 \cdot 5$, or 15. Multiply both the numerator and denominator of the complex rational expression by 15.

$$\frac{\dfrac{1}{3} + \dfrac{2}{5}}{\dfrac{2}{5} - \dfrac{1}{3}} = \frac{15}{15} \cdot \frac{\left(\dfrac{1}{3} + \dfrac{2}{5}\right)}{\left(\dfrac{2}{5} - \dfrac{1}{3}\right)} = \frac{15 \cdot \dfrac{1}{3} + 15 \cdot \dfrac{2}{5}}{15 \cdot \dfrac{2}{5} - 15 \cdot \dfrac{1}{3}} = \frac{5 + 6}{6 - 5} = \frac{11}{1} = 11$$

$\frac{15}{15} = 1$, so we are not changing the complex fraction's value.

✔ **CHECK POINT 4** Simplify by the LCD method: $\dfrac{\dfrac{1}{4} + \dfrac{2}{3}}{\dfrac{2}{3} - \dfrac{1}{4}}$.

EXAMPLE 5   **Simplifying a Complex Rational Expression by the LCD Method**

Simplify:

$$\frac{1 + \dfrac{1}{x}}{1 - \dfrac{1}{x}}.$$

SOLUTION   The denominators in the complex rational expression are $1, x, 1$, and x.

$$\frac{1 + \dfrac{1}{x}}{1 - \dfrac{1}{x}} = \frac{\dfrac{1}{1} + \dfrac{1}{x}}{\dfrac{1}{1} - \dfrac{1}{x}} \quad \text{Denominators}$$

Denominators

The LCD is $1 \cdot x$, or x. Multiply both the numerator and denominator of the complex rational expression by x.

$$\frac{1 + \dfrac{1}{x}}{1 - \dfrac{1}{x}} = \frac{x}{x} \cdot \frac{\left(1 + \dfrac{1}{x}\right)}{\left(1 - \dfrac{1}{x}\right)} = \frac{x \cdot 1 + x \cdot \dfrac{1}{x}}{x \cdot 1 - x \cdot \dfrac{1}{x}} = \frac{x + 1}{x - 1}$$

✔ **CHECK POINT 5** Simplify by the LCD method: $\dfrac{2 - \dfrac{1}{x}}{2 + \dfrac{1}{x}}$.

EXAMPLE 6 Simplifying a Complex Rational Expression by the LCD Method

Simplify:

$$\frac{\dfrac{1}{xy}}{\dfrac{1}{x}+\dfrac{1}{y}}.$$

SOLUTION The denominators in the complex rational expression are xy, x, and y. The LCD is xy. Multiply both the numerator and denominator of the complex rational expression by xy.

$$\frac{\dfrac{1}{xy}}{\dfrac{1}{x}+\dfrac{1}{y}}=\frac{xy}{xy}\cdot\frac{\left(\dfrac{1}{xy}\right)}{\left(\dfrac{1}{x}+\dfrac{1}{y}\right)}=\frac{xy\cdot\dfrac{1}{xy}}{xy\cdot\dfrac{1}{x}+xy\cdot\dfrac{1}{y}}=\frac{1}{y+x}.$$

■

✔ **CHECK POINT 6** Simplify by the LCD method: $\dfrac{\dfrac{1}{x}-\dfrac{1}{y}}{\dfrac{1}{xy}}.$

8.5 EXERCISE SET

Student Solutions Manual   CD/Video   PH Math/Tutor Center   MathXL Tutorials on CD   MathXL®   MyMathLab   Interactmath.com

Practice Exercises

In Exercises 1–40, simplify each complex rational expression by the method of your choice.

1. $\dfrac{\dfrac{1}{2}+\dfrac{1}{4}}{\dfrac{1}{2}+\dfrac{1}{3}}$

2. $\dfrac{\dfrac{1}{3}+\dfrac{1}{4}}{\dfrac{1}{3}+\dfrac{1}{6}}$

3. $\dfrac{5+\dfrac{2}{5}}{7-\dfrac{1}{10}}$

4. $\dfrac{1+\dfrac{3}{5}}{2-\dfrac{1}{4}}$

5. $\dfrac{\dfrac{2}{5}-\dfrac{1}{3}}{\dfrac{2}{3}-\dfrac{3}{4}}$

6. $\dfrac{\dfrac{1}{2}-\dfrac{1}{4}}{\dfrac{3}{8}+\dfrac{1}{16}}$

7. $\dfrac{\dfrac{3}{4}-x}{\dfrac{3}{4}+x}$

8. $\dfrac{\dfrac{2}{3}-x}{\dfrac{2}{3}+x}$

9. $\dfrac{7-\dfrac{2}{x}}{5+\dfrac{1}{x}}$

10. $\dfrac{8+\dfrac{3}{x}}{1-\dfrac{7}{x}}$

11. $\dfrac{2+\dfrac{3}{y}}{1-\dfrac{7}{y}}$

12. $\dfrac{4-\dfrac{7}{y}}{3-\dfrac{2}{y}}$

13. $\dfrac{\dfrac{1}{y}-\dfrac{3}{2}}{\dfrac{1}{y}+\dfrac{3}{4}}$

14. $\dfrac{\dfrac{1}{y}-\dfrac{3}{4}}{\dfrac{1}{y}+\dfrac{2}{3}}$

15. $\dfrac{\dfrac{x}{5}-\dfrac{5}{x}}{\dfrac{1}{5}+\dfrac{1}{x}}$

16. $\dfrac{\dfrac{3}{x}+\dfrac{x}{3}}{\dfrac{x}{3}-\dfrac{3}{x}}$

17. $\dfrac{1+\dfrac{1}{x}}{1-\dfrac{1}{x^2}}$

18. $\dfrac{1+\dfrac{2}{x}}{1-\dfrac{4}{x^2}}$

19. $\dfrac{\dfrac{1}{7}-\dfrac{1}{y}}{7-y}$

20. $\dfrac{\dfrac{1}{9}-\dfrac{1}{y}}{9-y}$

21. $\dfrac{x+\dfrac{2}{y}}{\dfrac{x}{y}}$

22. $\dfrac{x-\dfrac{2}{y}}{\dfrac{x}{y}}$

23. $\dfrac{\dfrac{1}{x} + \dfrac{1}{y}}{xy}$

24. $\dfrac{\dfrac{1}{x} + \dfrac{1}{y}}{x + y}$

25. $\dfrac{\dfrac{x}{y} + \dfrac{1}{x}}{\dfrac{y}{x} + \dfrac{1}{x}}$

26. $\dfrac{\dfrac{1}{x} + \dfrac{1}{y}}{\dfrac{1}{x} - \dfrac{1}{y}}$

27. $\dfrac{\dfrac{1}{y} + \dfrac{2}{y^2}}{\dfrac{2}{y} + 1}$

28. $\dfrac{\dfrac{1}{y} + \dfrac{3}{y^2}}{\dfrac{3}{y} + 1}$

29. $\dfrac{\dfrac{12}{x^2} - \dfrac{3}{x}}{\dfrac{15}{x} - \dfrac{9}{x^2}}$

30. $\dfrac{\dfrac{8}{x^2} - \dfrac{2}{x}}{\dfrac{10}{x} - \dfrac{6}{x^2}}$

31. $\dfrac{2 + \dfrac{6}{y}}{1 - \dfrac{9}{y^2}}$

32. $\dfrac{3 + \dfrac{12}{y}}{1 - \dfrac{16}{y^2}}$

33. $\dfrac{\dfrac{1}{x + 2}}{1 + \dfrac{1}{x + 2}}$

34. $\dfrac{\dfrac{1}{x - 2}}{1 - \dfrac{1}{x - 2}}$

35. $\dfrac{x - 5 + \dfrac{3}{x}}{x - 7 + \dfrac{2}{x}}$

36. $\dfrac{x + 9 - \dfrac{7}{x}}{x - 6 + \dfrac{4}{x}}$

37. $\dfrac{\dfrac{3}{xy^2} + \dfrac{2}{x^2y}}{\dfrac{1}{x^2y} + \dfrac{2}{xy^3}}$

38. $\dfrac{\dfrac{2}{x^3y} + \dfrac{5}{xy^4}}{\dfrac{5}{x^3y} - \dfrac{3}{xy}}$

39. $\dfrac{\dfrac{3}{x + 1} - \dfrac{3}{x - 1}}{\dfrac{5}{x^2 - 1}}$

40. $\dfrac{\dfrac{3}{x + 2} - \dfrac{3}{x - 2}}{\dfrac{5}{x^2 - 4}}$

Practice Plus

In Exercises 41–48, simplify each complex rational expression.

41. $\dfrac{\dfrac{6}{x^2 + 2x - 15} - \dfrac{1}{x - 3}}{\dfrac{1}{x + 5} + 1}$

42. $\dfrac{\dfrac{1}{x - 2} - \dfrac{6}{x^2 + 3x - 10}}{1 + \dfrac{1}{x - 2}}$

43. $\dfrac{y^{-1} - (y + 5)^{-1}}{5}$

44. $\dfrac{y^{-1} - (y + 2)^{-1}}{2}$

45. $\dfrac{\dfrac{1}{1 - \dfrac{1}{x}} - 1}{}$

46. $\dfrac{\dfrac{1}{1 - \dfrac{1}{x + 1}} - 1}{}$

47. $\dfrac{1}{1 + \dfrac{1}{1 + \dfrac{1}{x}}}$

48. $\dfrac{1}{1 + \dfrac{1}{1 + \dfrac{1}{2}}}$

Application Exercises

49. The average rate on a round-trip commute having a one-way distance d is given by the complex rational expression

$$\dfrac{2d}{\dfrac{d}{r_1} + \dfrac{d}{r_2}}$$

in which r_1 and r_2 are the average rates on the outgoing and return trips, respectively. Simplify the expression. Then find your average rate if you drive to campus averaging 40 miles per hour and return home on the same route averaging 30 miles per hour.

50. If two electrical resistors with resistances R_1 and R_2 are connected in parallel (see the figure), then the total resistance in the circuit is given by the complex rational expression

$$\dfrac{1}{\dfrac{1}{R_1} + \dfrac{1}{R_2}}.$$

Simplify the expression. Then find the total resistance if $R_1 = 10$ ohms and $R_2 = 20$ ohms.

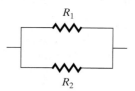

Writing in Mathematics

51. What is a complex rational expression? Give an example with your explanation.

52. Describe two ways to simplify $\dfrac{\dfrac{3}{x} + \dfrac{2}{x^2}}{\dfrac{1}{x^2} + \dfrac{2}{x}}$.

53. Which method do you prefer for simplifying complex rational expressions? Why?

Critical Thinking Exercises

54. Which one of the following is true?

a. The fraction $\dfrac{31,729,546}{72,578,112}$ is a complex rational expression.

b. $\dfrac{y - \dfrac{1}{2}}{y + \dfrac{3}{4}} = \dfrac{4y - 2}{4y + 3}$ for any value of y except $-\dfrac{3}{4}$.

c. $\dfrac{\dfrac{1}{4} - \dfrac{1}{3}}{\dfrac{1}{3} + \dfrac{1}{6}} = \dfrac{1}{12} \div \dfrac{3}{6} = \dfrac{1}{6}$

d. Some complex rational expressions cannot be simplified by both methods discussed in this section.

55. In one short sentence, five words or less, explain what

$$\frac{\dfrac{1}{x} + \dfrac{1}{x^2} + \dfrac{1}{x^3}}{\dfrac{1}{x^4} + \dfrac{1}{x^5} + \dfrac{1}{x^6}}$$

does to each number x.

In Exercises 56–57, simplify completely.

56. $\dfrac{2y}{2 + \dfrac{2}{y}} + \dfrac{y}{1 + \dfrac{1}{y}}$

57. $\dfrac{1 + \dfrac{1}{y} - \dfrac{6}{y^2}}{1 - \dfrac{5}{y} + \dfrac{6}{y^2}} - \dfrac{1 - \dfrac{1}{y}}{1 - \dfrac{2}{y} - \dfrac{3}{y^2}}$

Technology Exercises

In Exercises 58–60, use the GRAPH *or* TABLE *feature of a graphing utility to determine if the simplification is correct. If the answer is wrong, correct it and then verify your corrected simplification using the graphing utility.*

58. $\dfrac{x - \dfrac{1}{2x + 1}}{1 - \dfrac{x}{2x + 1}} = 2x - 1$

59. $\dfrac{\dfrac{1}{x} + 1}{\dfrac{1}{x}} = 2$

60. $\dfrac{\dfrac{1}{x} + \dfrac{1}{3}}{\dfrac{1}{3x}} = x + \dfrac{1}{3}$

Review Exercises

61. Factor completely: $2x^3 - 20x^2 + 50x$. (Section 7.5, Example 2)

62. Solve: $2 - 3(x - 2) = 5(x + 5) - 1$. (Section 2.3, Example 3)

63. Multiply: $(x + y)(x^2 - xy + y^2)$. (Section 6.2, Example 7)

SECTION 8.6

SOLVING RATIONAL EQUATIONS

Objectives

1 Solve rational equations.

2 Solve problems involving formulas with rational expressions.

The time has come to clean up the river. Suppose that the government has committed $375 million for this project. We know that

$$y = \frac{250x}{100 - x}$$

models the cost, in millions of dollars, to remove x percent of the river's pollutants. What percentage of pollutants can be removed for $375 million?

In order to determine the percentage, we use the given model. The government has committed $375 million, so substitute 375 for y:

$$375 = \frac{250x}{100 - x} \quad \text{or} \quad \frac{250x}{100 - x} = 375.$$

> The equation contains a rational expression.

Now we need to solve the equation and find the value for x. This variable represents the percentage of pollutants that can be removed for $375 million.

A **rational**, or **fractional**, **equation** is an equation containing one or more rational expressions. The preceding equation is an example of a rational equation. Do you see that there is a variable in a denominator? This is a characteristic of many rational equations. In this section, you will learn a procedure for solving such equations.

Solving Rational Equations We have seen that the LCD is used to add and subtract rational expressions. By contrast, when solving rational equations, **the LCD is used as a multiplier that clears an equation of fractions**.

1 Solve rational equations.

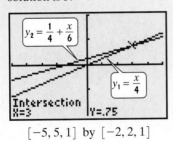
EXAMPLE 1 Solving a Rational Equation

Solve: $\dfrac{x}{4} = \dfrac{1}{4} + \dfrac{x}{6}$.

SOLUTION The LCD of 4, 4, and 6 is 12. To clear the equation of fractions, we multiply both sides by 12.

$$\frac{x}{4} = \frac{1}{4} + \frac{x}{6} \qquad \text{This is the given equation.}$$

$$12\left(\frac{x}{4}\right) = 12\left(\frac{1}{4} + \frac{x}{6}\right) \qquad \begin{array}{l}\text{Multiply both sides by 12, the LCD of all}\\\text{the fractions in the equation.}\end{array}$$

$$12 \cdot \frac{x}{4} = 12 \cdot \frac{1}{4} + 12 \cdot \frac{x}{6} \qquad \text{Use the distributive property on the right side.}$$

$$3x = 3 + 2x \qquad \text{Simplify: } \overset{3}{\cancel{12}} \cdot \frac{x}{\cancel{4}} = 3x; \; \overset{3}{\cancel{12}} \cdot \frac{1}{\cancel{4}} = 3; \; \overset{2}{\cancel{12}} \cdot \frac{x}{\cancel{6}} = 2x.$$

$$x = 3 \qquad \text{Subtract 2x from both sides.}$$

Substitute 3 for x in the original equation. You should obtain the true statement $\dfrac{3}{4} = \dfrac{3}{4}$. This verifies that the solution is 3. ∎

✔ **CHECK POINT 1** Solve: $\dfrac{x}{6} = \dfrac{1}{6} + \dfrac{x}{8}$.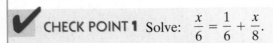

In Example 1, we solved a rational equation with constants in the denominators. Now, let's consider an equation such as

$$\frac{1}{x} = \frac{1}{5} + \frac{3}{2x}.$$

Can you see how this equation differs from the rational equation that we solved earlier? The variable, x, appears in two of the denominators. The procedure for solving this equation still involves multiplying each side by the least common denominator. However, we must avoid any values of the variable that make a denominator zero. For example, examine the denominators in the equation:

$$\frac{1}{x} = \frac{1}{5} + \frac{3}{2x}.$$

| This denominator would equal zero if $x = 0$. | This denominator would equal zero if $x = 0$. |

We see that x cannot equal zero. With this in mind, let's solve the equation.

EXAMPLE 2 Solving a Rational Equation

Solve: $\dfrac{1}{x} = \dfrac{1}{5} + \dfrac{3}{2x}.$

SOLUTION The denominators are $x, 5,$ and $2x$. The least common denominator is $10x$. We begin by multiplying both sides of the equation by $10x$. We will also write the restriction that x cannot equal zero to the right of the equation.

$$\frac{1}{x} = \frac{1}{5} + \frac{3}{2x}, \quad x \neq 0 \qquad \text{This is the given equation.}$$

$$10x \cdot \frac{1}{x} = 10x\left(\frac{1}{5} + \frac{3}{2x}\right) \qquad \text{Multiply both sides by 10x.}$$

$$10x \cdot \frac{1}{x} = 10x \cdot \frac{1}{5} + 10x \cdot \frac{3}{2x} \qquad \text{Use the distributive property. Be sure to multiply all terms by 10x.}$$

$$10\cancel{x} \cdot \frac{1}{\cancel{x}} = \overset{2}{\cancel{10}}x \cdot \frac{1}{\cancel{5}} + \overset{5}{\cancel{10}}\cancel{x} \cdot \frac{3}{\cancel{2x}} \qquad \text{Divide out common factors in the multiplications.}$$

$$10 = 2x + 15 \qquad \text{Simplify.}$$

Observe that the resulting equation,

$$10 = 2x + 15$$

is now cleared of fractions. With the variable term, $2x$, already on the right, we will collect constant terms on the left by subtracting 15 from both sides.

$$-5 = 2x \qquad \text{Subtract 15 from both sides.}$$

$$-\frac{5}{2} = x \qquad \text{Divide both sides by 2.}$$

We check our solution by substituting $-\frac{5}{2}$ into the original equation or by using a calculator. With a calculator, evaluate each side of the equation for $x = -\frac{5}{2}$, or for $x = -2.5$. Note that the original restriction that $x \neq 0$ is met. The solution is $-\frac{5}{2}$. ∎

 CHECK POINT 2 Solve: $\dfrac{5}{2x} = \dfrac{17}{18} - \dfrac{1}{3x}.$

The following steps may be used to solve a rational equation:

SOLVING RATIONAL EQUATIONS

1. List restrictions on the variable. Avoid any values of the variable that make a denominator zero.
2. Clear the equation of fractions by multiplying both sides by the LCD of all rational expressions in the equation.
3. Solve the resulting equation.
4. Reject any proposed solution that is in the list of restrictions on the variable. Check other proposed solutions in the original equation.

EXAMPLE 3   Solving a Rational Equation

Solve: $x + \dfrac{1}{x} = \dfrac{5}{2}$.

SOLUTION

Step 1.   List restrictions on the variable.

$$x + \frac{1}{x} = \frac{5}{2}$$

> This denominator would equal 0 if $x = 0$.

The restriction is $x \neq 0$.

Step 2.   Multiply both sides by the LCD. The denominators are x and 2. Thus, the LCD is $2x$. We multiply both sides by $2x$.

$$x + \frac{1}{x} = \frac{5}{2}, \quad x \neq 0 \qquad \text{This is the given equation.}$$

$$2x\left(x + \frac{1}{x}\right) = 2x\left(\frac{5}{2}\right) \qquad \text{Multiply both sides by the LCD.}$$

$$2x \cdot x + 2x \cdot \frac{1}{x} = 2x \cdot \frac{5}{2} \qquad \text{Use the distributive property on the left side.}$$

$$2x^2 + 2 = 5x \qquad \text{Simplify.}$$

Step 3.   Solve the resulting equation. Can you see that we have a quadratic equation? Write the equation in standard form and solve for x.

$$2x^2 - 5x + 2 = 0 \qquad \text{Subtract 5x from both sides.}$$
$$(2x - 1)(x - 2) = 0 \qquad \text{Factor.}$$
$$2x - 1 = 0 \quad \text{or} \quad x - 2 = 0 \qquad \text{Set each factor equal to 0.}$$
$$2x = 1 \qquad\qquad x = 2 \qquad \text{Solve the resulting equations.}$$
$$x = \frac{1}{2}$$

Step 4.   Check proposed solutions in the original equation. The proposed solutions, $\frac{1}{2}$ and 2, are not part of the restriction that $x \neq 0$. Neither makes a denominator in the original equation equal to zero.

Check $\frac{1}{2}$:   **Check 2:**

$$x + \frac{1}{x} = \frac{5}{2} \qquad\qquad\qquad x + \frac{1}{x} = \frac{5}{2}$$

$$\frac{1}{2} + \frac{1}{\frac{1}{2}} \stackrel{?}{=} \frac{5}{2} \qquad\qquad\qquad 2 + \frac{1}{2} \stackrel{?}{=} \frac{5}{2}$$

$$\frac{1}{2} + 2 \stackrel{?}{=} \frac{5}{2} \qquad\qquad\qquad \frac{4}{2} + \frac{1}{2} \stackrel{?}{=} \frac{5}{2}$$

$$\frac{1}{2} + \frac{4}{2} \stackrel{?}{=} \frac{5}{2} \qquad\qquad\qquad \frac{5}{2} = \frac{5}{2}, \text{true}$$

$$\frac{5}{2} = \frac{5}{2}, \text{true}$$

The solutions are $\frac{1}{2}$ and 2. ∎

✔ **CHECK POINT 3** Solve: $x + \dfrac{6}{x} = -5$.

EXAMPLE 4   Solving a Rational Equation

Solve: $\dfrac{3x}{x^2 - 9} + \dfrac{1}{x - 3} = \dfrac{3}{x + 3}$.

SOLUTION

Step 1.   **List restrictions on the variable.** By factoring denominators, it makes it easier to see values that make denominators zero.

$$\frac{3x}{(x + 3)(x - 3)} + \frac{1}{x - 3} = \frac{3}{x + 3}$$

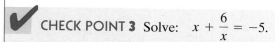

This denominator is zero if $x = -3$ or $x = 3$.   This denominator is zero if $x = 3$.   This denominator is zero if $x = -3$.

The restrictions are $x \neq -3$ and $x \neq 3$.

Step 2.   **Multiply both sides by the LCD.** The LCD is $(x + 3)(x - 3)$.

$$\frac{3x}{(x + 3)(x - 3)} + \frac{1}{x - 3} = \frac{3}{x + 3}, \quad x \neq -3, x \neq 3$$

This is the given equation with a denominator factored.

$$(x + 3)(x - 3)\left[\frac{3x}{(x + 3)(x - 3)} + \frac{1}{x - 3}\right] = (x + 3)(x - 3) \cdot \frac{3}{x + 3}$$

Multiply both sides by the LCD.

$$\cancel{(x + 3)}\,\cancel{(x - 3)} \cdot \frac{3x}{\cancel{(x + 3)}\,\cancel{(x - 3)}} + (x + 3)\cancel{(x - 3)} \cdot \frac{1}{\cancel{x - 3}}$$

$$= \cancel{(x + 3)}(x - 3) \cdot \frac{3}{\cancel{x + 3}}$$

Use the distributive property on the left side.

$$3x + (x + 3) = 3(x - 3)$$

Simplify.

Step 3. **Solve the resulting equation.**

$$3x + (x + 3) = 3(x - 3)$$ This is the equation cleared of fractions.

$$4x + 3 = 3x - 9$$ Combine like terms on the left side.
Use the distributive property on the right side.

$$x + 3 = -9$$ Subtract 3x from both sides.

$$x = -12$$ Subtract 3 from both sides.

Step 4. **Check proposed solutions in the original equation.** The proposed solution, -12, is not part of the restriction that $x \neq -3$ and $x \neq 3$. Substitute -12 for x in the given equation and show that -12 is the solution. ■

✔ **CHECK POINT 4** Solve: $\dfrac{11}{x^2 - 25} + \dfrac{4}{x + 5} = \dfrac{3}{x - 5}$.

EXAMPLE 5 Solving a Rational Equation

Solve: $\dfrac{8x}{x + 1} = 4 - \dfrac{8}{x + 1}$.

SOLUTION

Step 1. **List restrictions on the variable.**

$$\frac{8x}{x + 1} = 4 - \frac{8}{x + 1}$$

These denominators are zero if $x = -1$.

The restriction is $x \neq -1$.

Step 2. **Multiply both sides by the LCD.** The LCD is $x + 1$.

$$\frac{8x}{x + 1} = 4 - \frac{8}{x + 1}, \quad x \neq -1$$ This is the given equation.

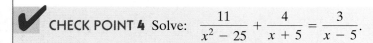

$$(x + 1) \cdot \frac{8x}{x + 1} = (x + 1)\left[4 - \frac{8}{x + 1}\right]$$ Multiply both sides by the LCD.

$$(x + 1) \cdot \frac{8x}{x + 1} = (x + 1) \cdot 4 - (x + 1) \cdot \frac{8}{x + 1}$$ Use the distributive property on the right side.

$$8x = 4(x + 1) - 8$$ Simplify.

Step 3. **Solve the resulting equation.**

$$8x = 4(x + 1) - 8$$ This is the equation cleared of fractions.

$$8x = 4x + 4 - 8$$ Use the distributive property on the right side.

$$8x = 4x - 4$$ Simplify.

$$4x = -4$$ Subtract 4x from both sides.

$$x = -1$$ Divide both sides by 4.

STUDY TIP

Reject any proposed solution that causes any denominator in a rational equation to equal 0.

Step 4. **Check proposed solutions.** The proposed solution, -1, is *not* a solution because of the restriction that $x \neq -1$. Notice that -1 makes both of the denominators zero in the original equation. There is *no solution to this equation.* ■

✔ **CHECK POINT 5** Solve: $\dfrac{x}{x-3} = \dfrac{3}{x-3} + 9$.

STUDY TIP

It is important to distinguish between adding and subtracting rational expressions and solving rational equations. We *simplify* sums and differences of terms. On the other hand, we *solve* equations. This is shown in the following two problems, both with an LCD of $3x$.

Adding Rational Expressions

Simplify:

$\dfrac{5}{3x} + \dfrac{3}{x}$.

$= \dfrac{5}{3x} + \dfrac{3}{x} \cdot \dfrac{3}{3}$

$= \dfrac{5}{3x} + \dfrac{9}{3x}$

$= \dfrac{5+9}{3x}$

$= \dfrac{14}{3x}$

Solving Rational Equations

Solve:

$\dfrac{5}{3x} + \dfrac{3}{x} = 1$.

$3x\left(\dfrac{5}{3x} + \dfrac{3}{x}\right) = 3x \cdot 1$

$3x \cdot \dfrac{5}{3x} + 3x \cdot \dfrac{3}{x} = 3x$

$5 + 9 = 3x$

$14 = 3x$

$\dfrac{14}{3} = x$

---

2 Solve problems involving formulas with rational expressions.

Applications of Rational Equations Rational equations can be solved to answer questions about variables contained in mathematical models.

EXAMPLE 6 A Government-Funded Cleanup

The formula

$$y = \frac{250x}{100 - x}$$

models the cost, *y*, in millions of dollars, to remove *x* percent of a river's pollutants. If the government commits $375 million for this project, what percentage of pollutants can be removed?

SOLUTION Substitute 375 for *y* and solve the resulting rational equation for *x*.

$375 = \dfrac{250x}{100 - x}$ The LCD is $100 - x$.

$(100 - x)375 = \cancel{(100 - x)} \cdot \dfrac{250x}{\cancel{100 - x}}$ Multiply both sides by the LCD.

$375(100 - x) = 250x$ Simplify.

$37{,}500 - 375x = 250x$ Use the distributive property on the left side.

$37{,}500 = 625x$ Add 375x to both sides.

$\dfrac{37{,}500}{625} = \dfrac{625x}{625}$ Divide both sides by 625.

$60 = x$ Simplify.

If the government spends $375 million, 60% of the river's pollutants can be removed. ■

✔ **CHECK POINT 6** Use the model in Example 6 to answer this question: If government funding is increased to $750 million, what percentage of pollutants can be removed?

8.6 EXERCISE SET

Student Solutions Manual CD/Video PH Math/Tutor Center MathXL Tutorials on CD MathXL® MyMathLab Interactmath.com

Practice Exercises

In Exercises 1–44, solve each rational equation. If an equation has no solution, so state.

1. $\dfrac{x}{3} = \dfrac{x}{2} - 2$

2. $\dfrac{x}{5} = \dfrac{x}{6} + 1$

3. $\dfrac{4x}{3} = \dfrac{x}{18} - \dfrac{x}{6}$

4. $\dfrac{5x}{4} = \dfrac{x}{12} - \dfrac{x}{2}$

5. $2 - \dfrac{8}{x} = 6$

6. $1 - \dfrac{9}{x} = 4$

7. $\dfrac{2}{x} + \dfrac{1}{3} = \dfrac{4}{x}$

8. $\dfrac{5}{x} + \dfrac{1}{3} = \dfrac{6}{x}$

9. $\dfrac{2}{x} + 3 = \dfrac{5}{2x} + \dfrac{13}{4}$

10. $\dfrac{7}{2x} = \dfrac{5}{3x} + \dfrac{22}{3}$

11. $\dfrac{2}{3x} + \dfrac{1}{4} = \dfrac{11}{6x} - \dfrac{1}{3}$

12. $\dfrac{5}{2x} - \dfrac{8}{9} = \dfrac{1}{18} - \dfrac{1}{3x}$

13. $\dfrac{6}{x + 3} = \dfrac{4}{x - 3}$

14. $\dfrac{7}{x + 1} = \dfrac{5}{x - 3}$

15. $\dfrac{x - 2}{2x} + 1 = \dfrac{x + 1}{x}$

16. $\dfrac{7x - 4}{5x} = \dfrac{9}{5} - \dfrac{4}{x}$

17. $x + \dfrac{6}{x} = -7$

18. $x + \dfrac{7}{x} = -8$

19. $\dfrac{x}{5} - \dfrac{5}{x} = 0$

20. $\dfrac{x}{4} - \dfrac{4}{x} = 0$

21. $x + \dfrac{3}{x} = \dfrac{12}{x}$

22. $x + \dfrac{3}{x} = \dfrac{19}{x}$

23. $\dfrac{4}{y} - \dfrac{y}{2} = \dfrac{7}{2}$

24. $\dfrac{4}{3y} - \dfrac{1}{3} = y$

25. $\dfrac{x - 4}{x} = \dfrac{15}{x + 4}$

26. $\dfrac{x - 1}{2x + 3} = \dfrac{6}{x - 2}$

27. $\dfrac{1}{x - 1} + 5 = \dfrac{11}{x - 1}$

28. $\dfrac{3}{x + 4} - 7 = \dfrac{-4}{x + 4}$

29. $\dfrac{8y}{y + 1} = 4 - \dfrac{8}{y + 1}$

30. $\dfrac{2}{y - 2} = \dfrac{y}{y - 2} - 2$

31. $\dfrac{3}{x - 1} + \dfrac{8}{x} = 3$

32. $\dfrac{2}{x - 2} + \dfrac{4}{x} = 2$

33. $\dfrac{3y}{y - 4} - 5 = \dfrac{12}{y - 4}$

34. $\dfrac{10}{y + 2} = 3 - \dfrac{5y}{y + 2}$

35. $\dfrac{1}{x} + \dfrac{1}{x - 3} = \dfrac{x - 2}{x - 3}$

36. $\dfrac{1}{x - 1} + \dfrac{2}{x} = \dfrac{x}{x - 1}$

37. $\dfrac{x + 1}{3x + 9} + \dfrac{x}{2x + 6} = \dfrac{2}{4x + 12}$

38. $\dfrac{3}{2y - 2} + \dfrac{1}{2} = \dfrac{2}{y - 1}$

39. $\dfrac{4y}{y^2 - 25} + \dfrac{2}{y - 5} = \dfrac{1}{y + 5}$

40. $\dfrac{1}{x + 4} + \dfrac{1}{x - 4} = \dfrac{22}{x^2 - 16}$

41. $\dfrac{1}{x - 4} - \dfrac{5}{x + 2} = \dfrac{6}{x^2 - 2x - 8}$

42. $\dfrac{6}{x + 3} - \dfrac{5}{x - 2} = \dfrac{-20}{x^2 + x - 6}$

43. $\dfrac{2}{x + 3} - \dfrac{2x + 3}{x - 1} = \dfrac{6x - 5}{x^2 + 2x - 3}$

44. $\dfrac{x - 3}{x - 2} + \dfrac{x + 1}{x + 3} = \dfrac{2x^2 - 15}{x^2 + x - 6}$

Practice Plus

In Exercises 45–52, solve or simplify, whichever is appropriate.

45. $\dfrac{x^2 - 10}{x^2 - x - 20} = 1 + \dfrac{7}{x - 5}$

46. $\dfrac{x^2 + 4x - 2}{x^2 - 2x - 8} = 1 + \dfrac{4}{x - 4}$

47. $\dfrac{x^2 - 10}{x^2 - x - 20} - 1 - \dfrac{7}{x - 5}$

48. $\dfrac{x^2 + 4x - 2}{x^2 - 2x - 8} - 1 - \dfrac{4}{x - 4}$

49. $5y^{-2} + 1 = 6y^{-1}$

50. $3y^{-2} + 1 = 4y^{-1}$

51. $\dfrac{3}{y + 1} - \dfrac{1}{1 - y} = \dfrac{10}{y^2 - 1}$

52. $\dfrac{4}{y - 2} - \dfrac{1}{2 - y} = \dfrac{25}{y + 6}$

Application Exercises

A company that manufactures wheelchairs has fixed costs of $500,000. The average cost per wheelchair, C, for the company to manufacture x wheelchairs per month is modeled by the formula

$$C = \frac{400x + 500,000}{x}.$$

Use this mathematical model to solve Exercises 53–54.

53. How many wheelchairs per month can be produced at an average cost of $450 per wheelchair?

54. How many wheelchairs per month can be produced at an average cost of $405 per wheelchair?

In Palo Alto, California, a government agency ordered computer-related companies to contribute to a pool of money to clean up underground water supplies. (The companies had stored toxic chemicals in leaking underground containers.) The formula

$$C = \frac{2x}{100 - x}$$

models the cost, C, in millions of dollars, for removing x percent of the contaminants. Use this mathematical model to solve Exercises 55–56.

55. What percentage of the contaminants can be removed for $2 million?

56. What percentage of the contaminants can be removed for $8 million?

We have seen that Young's rule

$$C = \frac{DA}{A + 12}$$

can be used to approximate the dosage of a drug prescribed for children. In this formula, A = the child's age, in years, D = an adult dosage, and C = the proper child's dosage. Use this formula to solve Exercises 57–58.

57. When the adult dosage is 1000 milligrams, a child is given 300 milligrams. What is that child's age? Round to the nearest year.

58. When the adult dosage is 1000 milligrams, a child is given 500 milligrams. What is that child's age?

A grocery store sells 4000 cases of canned soup per year. By averaging costs to purchase soup and to pay storage costs, the owner has determined that if x cases are ordered at a time, the yearly inventory cost, C, can be modeled by

$$C = \frac{10,000}{x} + 3x.$$

The graph of this model is shown at the top of the next column. Use this information to solve Exercises 59–60.

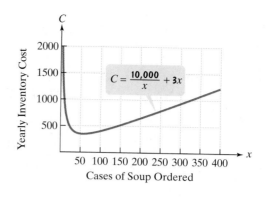

59. How many cases should be ordered at a time for yearly inventory costs to be $350? Identify your solutions as points on the graph.

60. How many cases should be ordered at a time for yearly inventory costs to be $790? Identify your solutions as points on the graph.

In baseball, a player's batting average is the total number of hits divided by the total number of times at bat. Use this information to solve Exercises 61–62.

61. A player has 12 hits after 40 times at bat. How many additional consecutive times must the player hit the ball to achieve a batting average of 0.440?

62. A player has eight hits after 50 times at bat. How many additional consecutive times must the player hit the ball to achieve a batting average of 0.250?

Writing in Mathematics

63. What is a rational equation?

64. Explain how to solve a rational equation.

65. Explain how to find restrictions on the variable in a rational equation.

66. Why should restrictions on the variable in a rational equation be listed before you begin solving the equation?

67. Describe similarities and differences between the procedures needed to solve the following problems:

$$\text{Add: } \frac{2}{x} + \frac{3}{4} \qquad \text{Solve: } \frac{2}{x} + \frac{3}{4} = 1.$$

68. The equation

$$P = \frac{72,900}{100x^2 + 729}$$

models the percentage of people in the United States, P, who have x years of education and are unemployed. Use this model to write a problem that can be solved with a rational equation. It is not necessary to solve the problem.

Critical Thinking Exercises

69. Which one of the following is true?

a. $\dfrac{1}{x} + \dfrac{1}{6} = 6x\left(\dfrac{1}{x} + \dfrac{1}{6}\right) = 6 + x$

b. If a is any real number, the equation $\dfrac{a}{x} + 1 = \dfrac{a}{x}$ has no solution.

c. All real numbers satisfy the equation $\dfrac{3}{x} - \dfrac{1}{x} = \dfrac{2}{x}$.

d. To solve $\dfrac{5}{3x} + \dfrac{3}{x} = 1$, we must first add the rational expressions on the left side.

70. Solve for f: $\dfrac{1}{p} + \dfrac{1}{q} = \dfrac{1}{f}$.

71. Solve for f_2: $f = \dfrac{f_1 f_2}{f_1 + f_2}$.

In Exercises 72–73, solve each rational equation.

72. $\dfrac{x + 1}{2x^2 - 11x + 5} = \dfrac{x - 7}{2x^2 + 9x - 5} - \dfrac{2x - 6}{x^2 - 25}$

73. $\left(\dfrac{x + 1}{x + 7}\right)^2 \div \left(\dfrac{x + 1}{x + 7}\right)^4 = 0.$

74. Find b so that the solution of

$$\dfrac{7x + 4}{b} + 13 = x$$

is -6.

Technology Exercises

In Exercises 75–77, use a graphing utility to solve each rational equation. Graph each side of the equation in the given viewing rectangle. The solution is the first coordinate of the point(s) of intersection. Check by direct substitution.

75. $\dfrac{x}{2} + \dfrac{x}{4} = 6$

$[-5, 10, 1]$ by $[-5, 10, 1]$

76. $\dfrac{50}{x} = 2x$

$[-10, 10, 1]$ by $[-20, 20, 2]$

77. $x + \dfrac{6}{x} = -5$

$[-10, 10, 1]$ by $[-10, 10, 1]$

Review Exercises

78. Factor completely: $x^4 + 2x^3 - 3x - 6$. (Section 7.1, Example 7)

79. Simplify: $(3x^2)(-4x^{-10})$. (Section 6.7, Example 3)

80. Simplify: $-5[4(x - 2) - 3]$. (Section 1.8, Example 11)

SECTION

8.7

APPLICATIONS USING RATIONAL EQUATIONS AND VARIATION

Objectives

1 Solve problems involving motion.

2 Solve problems involving work.

3 Solve problems involving similar triangles.

4 Solve problems involving variation.

Look around. Everything you see will probably be replaced or thrown away someday. Skates, clothes, the computer, the refrigerator, furniture—they may break or wear out, or you may get tired of them and want new ones sooner or later. The amount of garbage produced in urban areas varies directly as the population. How does this affect the environment? And what does the phrase "*varies directly*" mean? In this section, we provide an answer to the latter question, as we turn to applications using rational equations.

1 Solve problems involving motion.

Problems Involving Motion We have seen that the distance, d, covered by any moving body is the product of its average rate, r, and its time in motion, t: $d = rt$. Rational expressions appear in motion problems when the conditions of the problem involve the time traveled. We can obtain an expression for t, the time traveled, by dividing both sides of $d = rt$ by r.

$$d = rt \qquad \text{Distance equals rate times time.}$$

$$\frac{d}{r} = \frac{rt}{r} \qquad \text{Divide both sides by } r.$$

$$\frac{d}{r} = t \qquad \text{Simplify.}$$

TIME IN MOTION

$$t = \frac{d}{r}$$

$$\text{Time traveled} = \frac{\text{Distance traveled}}{\text{Rate of travel}}$$

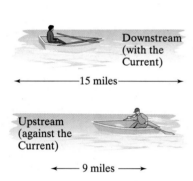

Downstream (with the Current)

←——— 15 miles ———→

Upstream (against the Current)

←——— 9 miles ———→

EXAMPLE 1 A Motion Problem Involving Time

In still water, your small boat averages 8 miles per hour. It takes you the same amount of time to travel 15 miles downstream, with the current, as 9 miles upstream, against the current. What is the rate of the water's current?

SOLUTION

Step 1. **Let x represent one of the quantities.** Let

$$x = \text{the rate of the current.}$$

Step 2. **Represent other quantities in terms of x.** We still need expressions for the rate of your boat with the current and the rate against the current. Traveling with the current, the boat's rate in still water, 8 miles per hour, is increased by the current's rate, x miles per hour. Thus,

$$8 + x = \text{the boat's rate with the current.}$$

Traveling against the current, the boat's rate in still water, 8 miles per hour, is decreased by the current's rate, x miles per hour. Thus,

$$8 - x = \text{the boat's rate against the current.}$$

Step 3. **Write an equation that describes the conditions.** By reading the problem again, we discover that the crucial idea is that the time spent going 15 miles with the current equals the time spent going 9 miles against the current. This information is summarized in the following table.

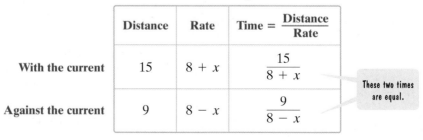

	Distance	Rate	Time = $\dfrac{\text{Distance}}{\text{Rate}}$
With the current	15	$8 + x$	$\dfrac{15}{8 + x}$
Against the current	9	$8 - x$	$\dfrac{9}{8 - x}$

These two times are equal.

We are now ready to write an equation that describes the problem's conditions.

The time spent going 15 miles with the current	equals	the time spent going 9 miles against the current.

$$\frac{15}{8 + x} = \frac{9}{8 - x}$$

Step 4. **Solve the equation and answer the question.**

$$\frac{15}{8 + x} = \frac{9}{8 - x}$$
This is the equation for the problem's conditions.

$$(8 + x)(8 - x) \cdot \frac{15}{8 + x} = (8 + x)(8 - x) \cdot \frac{9}{8 - x}$$
Multiply both sides by the LCD, $(8 + x)(8 - x)$.

$$15(8 - x) = 9(8 + x)$$
Simplify.

$$120 - 15x = 72 + 9x$$
Use the distributive property.

$$120 = 72 + 24x$$
Add 15x to both sides.

$$48 = 24x$$
Subtract 72 from both sides.

$$2 = x$$
Divide both sides by 24.

The rate of the water's current is 2 miles per hour.

Step 5. **Check the proposed solution in the original wording of the problem.** Does it take you the same amount of time to travel 15 miles downstream as 9 miles upstream if the current is 2 miles per hour? Keep in mind that your rate in still water is 8 miles per hour.

Time required to travel 15 miles with the current $= \dfrac{\text{Distance}}{\text{Rate}} = \dfrac{15}{8 + 2} = \dfrac{15}{10} = 1\frac{1}{2}$ hours

Time required to travel 9 miles against the current $= \dfrac{\text{Distance}}{\text{Rate}} = \dfrac{9}{8 - 2} = \dfrac{9}{6} = 1\frac{1}{2}$ hours

These times are the same, which checks with the original conditions of the problem. ∎

 CHECK POINT 1 Forget the small boat! This time we have you canoeing on the Colorado River. In still water, your average canoeing rate is 3 miles per hour. It takes you the same amount of time to travel 10 miles downstream, with the current, as 2 miles upstream, against the current. What is the rate of the water's current?

2 Solve problems involving work.

Problems Involving Work You are thinking of designing your own Web site. You estimate that it will take 30 hours to do the job. In 1 hour, $\frac{1}{30}$ of the job is completed. In 2 hours, $\frac{2}{30}$, or $\frac{1}{15}$, of the job is completed. In 3 hours, the fractional part of the job done is $\frac{3}{30}$, or $\frac{1}{10}$. In x hours, the fractional part of the job that you can complete is $\frac{x}{30}$.

Your friend, who has experience developing Web sites, took 20 hours working on his own to design an impressive site. You wonder about the possibility of working together. How long would it take both of you to design your Web site?

Problems involving work usually have two people working together to complete a job. The amount of time it takes each person to do the job working alone is frequently known. The question deals with how long it will take both people working together to complete the job.

In work problems, **the number 1 represents one whole job completed**. For example, the completion of your Web site is represented by 1. Equations in work problems are based on the following condition:

$$\underset{\substack{\text{Fractional part of}\\\text{the job done by the}\\\text{first person}}}{} + \underset{\substack{\text{fractional part of}\\\text{the job done by the}\\\text{second person}}}{} = \underset{\substack{\text{1 (one whole}\\\text{job completed).}}}{}$$

EXAMPLE 2 — Solving a Problem Involving Work

You can design a Web site in 30 hours. Your friend can design the same site in 20 hours. How long will it take to design the Web site if you both work together?

SOLUTION

Step 1. **Let x represent one of the quantities.** Let x = the time, in hours, for you and your friend to design the Web site together.

Step 2. **Represent other quantities in terms of x.** Because there are no other unknown quantities, we can skip this step.

Step 3. **Write an equation that describes the conditions.** We construct a table to help find the fractional part of the task completed by you and your friend in x hours.

	Fractional part of job completed in 1 hour	Time working together	Fractional part of job completed in x hours
You — *You can design the site in 30 hours.*	$\dfrac{1}{30}$	x	$\dfrac{x}{30}$
Your friend — *Your friend can design the site in 20 hours.*	$\dfrac{1}{20}$	x	$\dfrac{x}{20}$

$$\underset{\substack{\text{Fractional part of}\\\text{the job done by you}}}{\dfrac{x}{30}} + \underset{\substack{\text{fractional part of the}\\\text{job done by your friend}}}{\dfrac{x}{20}} = \underset{\substack{\text{one whole}\\\text{job.}}}{1}$$

Step 4. **Solve the equation and answer the question.**

$$\frac{x}{30} + \frac{x}{20} = 1 \qquad \text{This is the equation for the problem's conditions.}$$

$$60\left(\frac{x}{30} + \frac{x}{20}\right) = 60 \cdot 1 \qquad \text{Multiply both sides by 60, the LCD.}$$

$$60 \cdot \frac{x}{30} + 60 \cdot \frac{x}{20} = 60$$ Use the distributive property on the left side.

$$2x + 3x = 60$$ Simplify: $\frac{\overset{2}{60}}{1} \cdot \frac{x}{\underset{1}{30}} = 2x$ and $\frac{\overset{3}{60}}{1} \cdot \frac{x}{\underset{1}{20}} = 3x$.

$$5x = 60$$ Combine like terms.

$$x = 12$$ Divide both sides by 5.

If you both work together, you can design your Web site in 12 hours.

Step 5. **Check the proposed solution in the original wording of the problem.** Will you both complete the job in 12 hours? Because you can design the site in 30 hours, in 12 hours, you can complete $\frac{12}{30}$, or $\frac{2}{5}$, of the job. Because your friend can design the site in 20 hours, in 12 hours, he can complete $\frac{12}{20}$, or $\frac{3}{5}$, of the job. Notice that $\frac{2}{5} + \frac{3}{5} = 1$, which represents the completion of the entire job, or one whole job. ∎

STUDY TIP

Let

     $a =$ the time it takes person A to do a job working alone

     $b =$ the time it takes person B to do the same job working alone.

If x represents the time it takes for A and B to complete the entire job working together, then the situation can be modeled by the rational equation

$$\frac{x}{a} + \frac{x}{b} = 1.$$

✔ **CHECK POINT 2** One person can paint the outside of a house in 8 hours. A second person can do it in 4 hours. How long will it take them to do the job if they work together?

 Solve problems involving similar triangles.

Similar Triangles Shown in the margin is an international road sign. This sign is shaped just like the actual sign, although its size is smaller. Figures that have the same shape, but not the same size, are used in **scale drawings**. A scale drawing always pictures the exact shape of the object that the drawing represents. Architects, engineers, landscape gardeners, and interior decorators use scale drawings in planning their work.

     Figures that have the same shape, but not necessarily the same size, are called **similar figures**. In Figure 8.2, triangles ABC and DEF are similar. Angles A and D measure the same number of degrees and are called **corresponding angles**. Angles C and F are corresponding angles, as are angles B and E. Angles with the same number of tick marks in Figure 8.2 are the corresponding angles.

Pedestrian Crossing

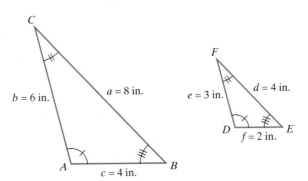

FIGURE 8.2

The sides opposite the corresponding angles are called **corresponding sides**. Although the measures of corresponding angles are equal, corresponding sides may or may not be the same length. For the triangles in Figure 8.2, each side in the smaller triangle is half the length of the corresponding side in the larger triangle.

The triangles in Figure 8.2 illustrate what it means to be **similar triangles**. **Corresponding angles have the same measure and the ratios of the lengths of the corresponding sides are equal.** For the triangles in Figure 8.2, each of these ratios is equal to 2:

$$\frac{a}{d} = \frac{8}{4} = 2 \qquad \frac{b}{e} = \frac{6}{3} = 2 \qquad \frac{c}{f} = \frac{4}{2} = 2.$$

In similar triangles, the lengths of the corresponding sides are proportional. Thus,

$$\frac{a}{d} = \frac{b}{e} = \frac{c}{f}.$$

If we know that two triangles are similar, we can set up a proportion to solve for the length of an unknown side.

EXAMPLE 3   Using Similar Triangles

The triangles in Figure 8.3 are similar. Find the missing length, x.

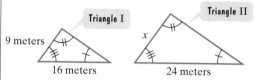

Triangle I

Triangle II

9 meters

x

16 meters

24 meters

FIGURE 8.3

SOLUTION   Because the triangles are similar, their corresponding sides are proportional.

Left side of △ I. $\dfrac{9}{x} = \dfrac{16}{24}$ Bottom side of △ I.

Corresponding side on left of △ II.   Corresponding side on bottom of △ II.

We solve this rational equation by multiplying both sides by the LCD, 24x. (You can also apply the cross-products principle for solving proportions.)

$$24x \cdot \frac{9}{x} = 24x \cdot \frac{16}{24} \qquad \text{Multiply both sides by the LCD, 24x.}$$

$$24 \cdot 9 = 16x \qquad \text{Simplify.}$$

$$216 = 16x \qquad \text{Multiply: } 24 \cdot 9 = 216.$$

$$13.5 = x \qquad \text{Divide both sides by 16.}$$

The missing length, x, is 13.5 meters.   ■

CHECK POINT 3 The similar triangles in the figure are positioned so that they have the same orientation. Find the missing length, x.

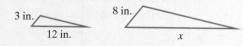

3 in.   8 in.

12 in. x

How can we quickly determine if two triangles are similar? **If the measures of two angles of one triangle are equal to those of two angles of a second triangle, then the two triangles are similar.** If the triangles are similar, then their corresponding sides are proportional.

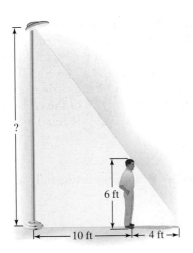

FIGURE 8.4

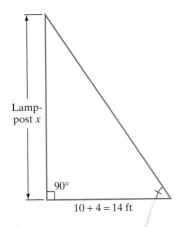

FIGURE 8.5

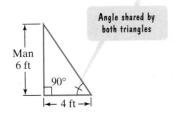

4 Solve problems involving variation.

EXAMPLE 4 Problem Solving Using Similar Triangles

A man who is 6 feet tall is standing 10 feet from the base of a lamppost (see Figure 8.4). The man's shadow has a length of 4 feet. How tall is the lamppost?

SOLUTION The drawing in Figure 8.5 makes the similarity of the triangles easier to see. The large triangle with the lamppost on the left and the small triangle with the man on the left both contain 90° angles. They also share an angle. Thus, two angles of the large triangle are equal in measure to two angles of the small triangle. This means that the triangles are similar and their corresponding sides are proportional. We begin by letting x represent the height of the lamppost, in feet. Because corresponding sides of similar triangles are proportional,

Left side of big △. Bottom side of big △.

$$\frac{x}{6} = \frac{14}{4}.$$

Corresponding side on left of small △. Corresponding side on bottom of small △.

We solve for x by multiplying both sides by the LCD, 12.

$$12 \cdot \frac{x}{6} = 12 \cdot \frac{14}{4} \qquad \text{Multiply both sides by the LCD, 12.}$$

$$2x = 42 \qquad \text{Simplify: } \frac{\overset{2}{\cancel{12}}}{1} \cdot \frac{x}{\cancel{6}} = 2x \text{ and } \frac{\overset{3}{\cancel{12}}}{1} \cdot \frac{14}{\cancel{4}} = 42.$$

$$x = 21 \qquad \text{Divide both sides by 2.}$$

The lamppost is 21 feet tall. ∎

✔ **CHECK POINT 4** Find the height of the lookout tower using the figure that lines up the top of the tower with the top of a stick that is 2 yards long.

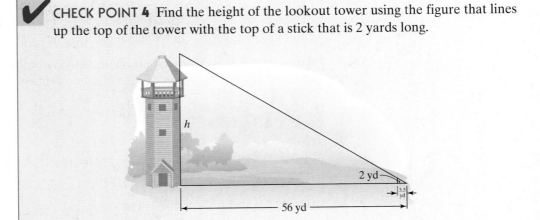

Variation Certain formulas occur so frequently in applied situations that they are given special names. Variation formulas show how one quantity changes in relation to another quantity. Quantities can vary directly or inversely.

Direct Variation Because light travels faster than sound, during a thunderstorm we see lightning before we hear thunder. The formula

$$d = 1080t$$

describes your distance, d, in feet, from the storm's center if it takes you t seconds to hear thunder after seeing lightning. Thus,

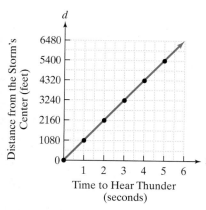

The graph of $d = 1080t$. Distance to a storm's center varies directly as the time it takes to hear thunder.

If $t = 1$, $d = 1080 \cdot 1 = 1080$:   If it takes 1 second to hear thunder, the storm's center is 1080 feet away.

If $t = 2$, $d = 1080 \cdot 2 = 2160$:   If it takes 2 seconds to hear thunder, the storm's center is 2160 feet away.

If $t = 3$, $d = 1080 \cdot 3 = 3240$:   If it takes 3 seconds to hear thunder, the storm's center is 3240 feet away.

As the formula $d = 1080t$ illustrates, the distance to the storm's center is a constant multiple of how long it takes to hear the thunder. When the time is doubled, the storm's distance is doubled; when the time is tripled, the storm's distance is tripled; and so on. Because of this, the distance is said to **vary directly** as the time. The **equation of variation** is

$$d = 1080t.$$

Generalizing, we obtain the following statement:

DIRECT VARIATION   If a situation is described by an equation in the form

$$y = kx,$$

where k is a constant, we say that y **varies directly as x**. The number k is called the **constant of variation**.

EXAMPLE 5 **Writing a Direct Variation Equation**

A person's salary, S, varies directly as the number of hours worked, h.

a. Write an equation that expresses this relationship.

b. Margarita earns $18 per hour. Substitute 18 for k, the constant of variation, in the equation in part (a) and write the equation for Margarita's salary.

SOLUTION

a. We know that y varies directly as x is expressed as

$$y = kx.$$

By changing letters, we can write an equation that describes the following English statement: Salary, S, varies directly as the number of hours worked, h.

$$S = kh$$

b. Substituting 18 for k in the direct variation equation gives

$$S = 18h.$$

This equation describes Margarita's salary in terms of the number of hours she works. For example, if she works 10 hours, we can substitute 10 for h and determine her salary:

$$S = 18(10) = 180.$$

Her salary for working 10 hours is $180. Notice that, as the number of hours worked increases, the salary increases. ∎

CHECK POINT 5 A person's hair length, *L*, in inches, varies directly as the number of years it has been growing, *N*.

   a. Write an equation that expresses this relationship.

   b. The longest moustache on record was grown by Kalyan Sain of India. His moustache grew 4 inches each year. Substitute 4 for *k*, the constant of variation, in the equation in part (a) and write the equation for the length of Sain's moustache.

   c. Sain grew his moustache for 17 years. Substitute 17 for *N* in the equation from part (b) and find its length.

In Example 5 and Check Point 5, the constants of variation were given. If the constant of variation is not given, we can find it by substituting given values in the variation formula and solving for *k*. Example 6 shows how this is done.

EXAMPLE 6   Finding *k*, the Constant of Variation

Height, *H*, varies directly as foot length, *F*.

   a. Write an equation that expresses this relationship.

   b. Photographs of large footprints were published in 1951. Some speculated that these footprints were made by the Abominable Snowman. Each footprint was 23 inches long. The Abominable Snowman's height was determined to be 154.1 inches. (This is 12 feet, 10.1 inches, so it might not be a pleasant experience to run into this critter on a mellow hike through the woods!) Use $H = 154.1$ and $F = 23$ to find the constant of variation.

SOLUTION

   a. We know that *y* varies directly as *x* is expressed as

$$y = kx.$$

By changing letters, we can write an equation that describes the following English statement: Height, *H*, varies directly as foot length, *F*.

$$H = kF$$

   b. The Abominable Snowman's height is 154.1 inches, and foot length is 23 inches. Substitute 154.1 for *H* and 23 for *F* in the direct variation equation.

$$H = kF$$
$$154.1 = k \cdot 23$$

Solve for *k*, the constant of variation, by dividing both sides of the equation by 23:

$$\frac{154.1}{23} = \frac{k \cdot 23}{23}$$
$$6.7 = k.$$

Thus, the constant of variation is 6.7. ∎

In Example 6, now that we know the constant of variation ($k = 6.7$), we can rewrite $H = kF$ using this constant. The equation of variation is

$$H = 6.7F.$$

We can use this equation to find other values. For example, if your foot length is 10 inches, your height is

$$H = 6.7(10) = 67,$$

or approximately 67 inches.

CHECK POINT **6** The weight, W, of an aluminum canoe varies directly as its length, L.

a. Write an equation that expresses this relationship.

b. A 6-foot canoe weighs 75 pounds. Substitute 75 for W and 6 for L in the equation from part (a) and find k, the constant of variation.

c. Substitute the value of k into your equation in part (a) and write the equation that describes the weight of this type of canoe in terms of its length.

d. Use the equation from part (c) to find the weight of a 16-foot canoe of this type.

Our work up to this point provides a step-by-step procedure for solving variation problems. This procedure applies to direct variation problems, as well as to the other kind of variation problem that we will discuss.

SOLVING VARIATION PROBLEMS

1. Write an equation that describes the given English statement.
2. Substitute the given pair of values into the equation in step 1 and find the value of k.
3. Substitute the value of k into the equation in step 1.
4. Use the equation from step 3 to answer the problem's question.

EXAMPLE 7   Solving a Direct Variation Problem

The amount of garbage, G, created in a geographic area varies directly as the area's population, P. Dallas, Texas, has a population of 1.2 million and creates 38.4 million pounds of garbage each week. Find the number of pounds of garbage per week produced by New York City with a population of 8 million.

SOLUTION

Step 1. **Write an equation.** We know that y varies directly as x is expressed as

$$y = kx.$$

By changing letters, we can write an equation that describes the following English statement: Garbage production, G, varies directly as the population, P.

$$G = kP$$

Step 2. **Use the given values to find k.** Dallas has a population of 1.2 million and creates 38.4 million pounds of garbage weekly. Substitute 38.4 for G and 1.2 for P in the direct variation equation. Then solve for k.

$G = kP$	Garbage production varies directly as the population.
$38.4 = k \cdot 1.2$	Substitute 38.4 for G and 1.2 for P.
$\dfrac{38.4}{1.2} = \dfrac{k \cdot 1.2}{1.2}$	Divide both sides by 1.2.
$32 = k$	Simplify.

Step 3. **Substitute the value of k into the equation.**

$G = kP$	Use the equation from step 1.
$G = 32P$	Replace k, the constant of variation, with 32.

Step 4. **Answer the problem's question.** New York City has a population of 8 million. To find its weekly garbage production, substitute 8 for P in $G = 32P$ and solve for G.

$$G = 32P \qquad \text{Use the equation from step 3.}$$
$$G = 32(8) \qquad \text{Substitute 8 for } P.$$
$$= 256 \qquad \text{Multiply.}$$

The number of pounds of garbage per week produced by New York City is 256 million pounds. ■

 CHECK POINT 7 The pressure, P, of water on an object below the surface varies directly as its distance, D, below the surface. If a submarine experiences a pressure of 25 pounds per square inch 60 feet below the surface, how much pressure will it experience 330 feet below the surface?

Inverse Variation The distance from Atlanta, Georgia, to Orlando, Florida, is 450 miles. The time that it takes to drive from Atlanta to Orlando depends on the rate at which one drives and is given by

$$\text{Time} = \frac{450}{\text{Rate}}. \qquad \boxed{\text{Time} = \frac{\text{Distance}}{\text{Rate}}}$$

For example, if you average 45 miles per hour, the time for the drive is

$$\text{Time} = \frac{450}{45} = 10$$

or 10 hours. If you average 75 miles per hour, the time for the drive is

$$\text{Time} = \frac{450}{75} = 6$$

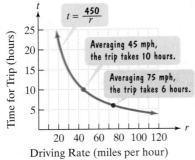

FIGURE 8.6

or 6 hours. As your rate (or speed) increases, the time for the trip decreases, and vice versa. This is illustrated by the graph in Figure 8.6. Observe the graph's shape. It shows that as the rate increases, the time for the trip decreases quite rapidly.

We can express the time for the Atlanta–Orlando trip using t for time and r for rate:

$$t = \frac{450}{r}.$$

This equation is an example of an **inverse variation** equation. Time, t, **varies inversely** as rate, r. When two quantities vary inversely, one quantity increases as the other decreases, and vice versa.

Generalizing, we obtain the following statement:

INVERSE VARIATION If a situation is described by an equation in the form

$$y = \frac{k}{x}$$

where k is a constant, we say that y **varies inversely as x**. The number k is called the **constant of variation**.

We use the same procedure to solve inverse variation problems as we did to solve direct variation problems. Example 8 illustrates this procedure.

EXAMPLE 8 Solving an Inverse Variation Problem

Figure 8.7 shows the price per barrel for crude oil from 1990 through 2004. The price, P, of oil varies inversely as the supply, S. An OPEC nation sells oil for $19.50 per barrel when its daily production level is 4 million barrels. At what price will it sell oil if the daily production level is decreased to 3 million barrels?

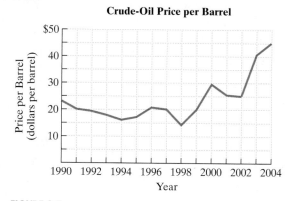

FIGURE 8.7

Source: U.S. Department of Commerce

SOLUTION

Step 1.   **Write an equation.** We know that y varies inversely as x is expressed as

$$y = \frac{k}{x}.$$

By changing letters, we can write an equation that describes the following English statement: Price, P, varies inversely as supply, S.

$$P = \frac{k}{S}$$

Step 2.   **Use the given values to find k.** Oil is sold for $19.50 at a production level of 4 million barrels. Substitute $19.50 for P and 4 for S in the inverse variation equation. Then solve for k.

$$P = \frac{k}{S} \qquad \text{Price varies inversely as supply.}$$

$$19.50 = \frac{k}{4} \qquad \text{Oil is sold for \$19.50 at a production level of 4 million barrels.}$$

$$(19.50)(4) = \frac{k}{4} \cdot 4 \qquad \text{Multiply both sides by 4.}$$

$$78 = k \qquad \text{Simplify.}$$

Step 3.   **Substitute the value of k into the equation.**

$$P = \frac{k}{S} \qquad \text{Use the equation from step 1.}$$

$$P = \frac{78}{S} \qquad \text{Replace } k, \text{ the constant of variation, with 78.}$$

Step 4.   **Answer the problem's question.** We need to find at what price the OPEC nation will sell oil if it reduces daily production to 3 million barrels. Substitute 3 for S in the equation and solve for P.

$$P = \frac{78}{S} = \frac{78}{3} = 26$$

The price will be be $26 per barrel.

Doubling the pressure halves the volume.

CHECK POINT 8 When you use a spray can and press the valve at the top, you decrease the pressure of the gas in the can. This decrease of pressure causes the volume of the gas in the can to increase. Because the gas needs more room than is provided in the can, it expands in spray form through the small hole near the valve. In general, if the temperature is constant, the pressure, P, of a gas in a container varies inversely as the volume, V, of the container. The pressure of a gas sample in a container whose volume is 8 cubic inches is 12 pounds per square inch. If the sample expands to a volume of 22 cubic inches, what is the new pressure of the gas?

8.7 EERCISE SET

Student Solutions Manual   CD/Video   PH Math/Tutor Center   MathXL Tutorials on CD   MathXL®   MyMathLab   Interactmath.com

Practice and Application Exercises

Use rational equations to solve Exercises 1–10. Each exercise is a problem involving motion.

1. How bad is the heavy traffic? You can walk 10 miles in the same time that it takes to travel 15 miles by car. If the car's rate is 3 miles per hour faster than your walking rate, find the average rate of each.

	Distance	Rate	Time = $\dfrac{\text{Distance}}{\text{Rate}}$
Walking	10	x	$\dfrac{10}{x}$
Car in Heavy Traffic	15	$x + 3$	$\dfrac{15}{x + 3}$

2. You can travel 40 miles on motorcycle in the same time that it takes to travel 15 miles on bicycle. If your motorcycle's rate is 20 miles per hour faster than your bicycle's, find the average rate for each.

	Distance	Rate	Time = $\dfrac{\text{Distance}}{\text{Rate}}$
Motorcycle	40	$x + 20$	$\dfrac{40}{x + 20}$
Bicycle	15	x	$\dfrac{15}{x}$

3. A jogger runs 4 miles per hour faster downhill than uphill. If the jogger can run 5 miles downhill in the same time that it takes to run 3 miles uphill, find the jogging rate in each direction.

4. A truck can travel 120 miles in the same time that it takes a car to travel 180 miles. If the truck's rate is 20 miles per hour slower than the car's, find the average rate for each.

5. In still water, a boat averages 15 miles per hour. It takes the same amount of time to travel 20 miles downstream, with the current, as 10 miles upstream, against the current. What is the rate of the water's current?

6. In still water, a boat averages 18 miles per hour. It takes the same amount of time to travel 33 miles downstream, with the current, as 21 miles upstream, against the current. What is the rate of the water's current?

7. As part of an exercise regimen, you walk 2 miles on an indoor track. Then you jog at twice your walking speed for another 2 miles. If the total time spent walking and jogging is 1 hour, find the walking and jogging rates.

8. The joys of the Pacific Coast! You drive 90 miles along the Pacific Coast Highway and then take a 5-mile run along a hiking trail in Point Reyes National Seashore. Your driving rate is nine times that of your running rate. If the total time for driving and running is 3 hours, find the average rate driving and the average rate running.

9. The water's current is 2 miles per hour. A boat can travel 6 miles downstream, with the current, in the same amount of time it travels 4 miles upstream, against the current. What is the boat's average rate in still water?

10. The water's current is 2 miles per hour. A canoe can travel 6 miles downstream, with the current, in the same amount of time it travels 2 miles upstream, against the current. What is the canoe's average rate in still water?

Use a rational equation to solve Exercises 11–16. Each exercise is a problem involving work.

11. You must leave for campus in 10 minutes or you will be late for class. Unfortunately, you are snowed in. You can shovel the driveway in 20 minutes and your brother claims he can do it in 15 minutes. If you shovel together, how long will it take to clear the driveway? Will this give you enough time before you have to leave?

12. You promised your parents that you would wash the family car. You have not started the job and they are due home in 16 minutes. You can wash the car in 40 minutes and your sister claims she can do it in 30 minutes. If you work together, how long will it take to do the job? Will this give you enough time before your parents return?

13. The MTV crew will arrive in one week and begin filming the city for *The Real World Kalamazoo*. The mayor is desperate to clean the city streets before filming begins. Two teams are available, one that requires 400 hours and one that requires 300 hours. If the teams work together, how long will it take to clean all of Kalamazoo's streets? Is this enough time before the cameras begin rolling?

14. A hurricane strikes and a rural area is without food or water. Three crews arrive. One can dispense needed supplies in 10 hours, a second in 15 hours, and a third in 20 hours. How long will it take all three crews working together to dispense food and water?

15. A pool can be filled by one pipe in 4 hours and by a second pipe in 6 hours. How long will it take using both pipes to fill the pool?

16. A pool can be filled by one pipe in 3 hours and by a second pipe in 6 hours. How long will it take using both pipes to fill the pool?

In Exercises 17–22, use similar triangles and the fact that corresponding sides are proportional to find the length of the side marked with an x.

17.

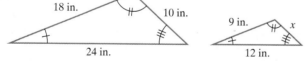

18.

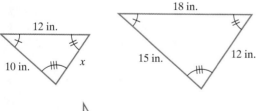

19.

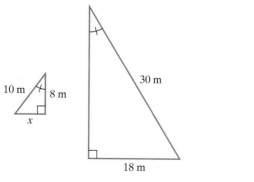

20.

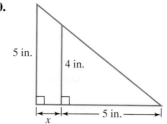

21.

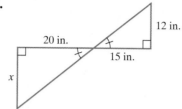

22.

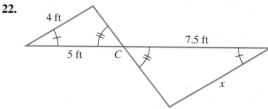

Use similar triangles to solve Exercises 23–24.

23. A tree casts a shadow 12 feet long. At the same time, a vertical rod 8 feet high casts a shadow 6 feet long. How tall is the tree?

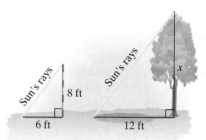

24. A person who is 5 feet tall is standing 80 feet from the base of a tree. The tree casts an 86-foot shadow. The person's shadow is 6 feet in length. What is the tree's height?

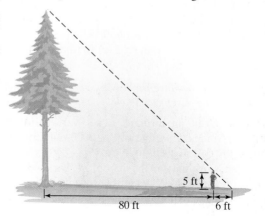

In Exercises 25–28, write an equation that expresses each relationship. Use k as the constant of variation.

25. g varies directly as h.   **26.** v varies directly as r.

27. w varies inversely as v.   **28.** a varies inversely as b.

In Exercises 29–32, determine the constant of variation for each stated condition.

29. y varies directly as x, and y = 80 when x = 4.
30. y varies directly as x, and y = 108 when x = 12.
31. W varies inversely as r, and W = 600 when r = 10.
32. T varies inversely as n, and T = 4 when n = 24.

Use the four-step procedure for solving variation problems given on page 515 to solve Exercises 33–36.

33. y varies directly as x. y = 35 when x = 5. Find y when x = 12.
34. y varies directly as x. y = 55 when x = 5. Find y when x = 13.
35. y varies inversely as x. y = 10 when x = 5. Find y when x = 2.
36. y varies inversely as x. y = 5 when x = 3. Find y when x = 9.

37. A person's fingernail growth, G, in inches, varies directly as the number of weeks it has been growing, W.
   a. Write an equation that expresses this relationship.
   b. Fingernails grow at a rate of about 0.02 inch per week. Substitute 0.02 for k, the constant of variation, in the equation in part (a) and write the equation for fingernail growth.
   c. Substitute 52 for W to determine your fingernail growth at the end of one year if for some bizarre reason you decided not to cut them and they did not break.

38. A person's salary, S, varies directly as the number of hours worked, h.
   a. Write an equation that expresses this relationship.
   b. For a 40-hour work week, Gloria earned $1400. Substitute 1400 for S and 40 for h in the equation from part (a) and find k, the constant of variation.
   c. Substitute the value of k into your equation in part (a) and write the equation that describes Gloria's salary in terms of the number of hours she works.
   d. Use the equation from part (c) to find Gloria's salary for 25 hours of work.

Use the four-step procedure for solving variation problems given on page 515 to solve Exercises 39–48.

39. The cost, C, of an airplane ticket varies directly as the number of miles, M, in the trip. A 3000-mile trip costs $400. What is the cost of a 450-mile trip?
40. An object's Weight on the moon, M, varies directly as its weight on Earth, E. A person who weighs 55 kilograms on Earth weighs 8.8 kilograms on the moon. What is the moon weight of a person who weighs 90 kilograms on Earth?

41. The Mach number is a measurement of speed named after the man who suggested it, Ernst Mach (1838–1916). The speed of an aircraft varies directly as its Mach number. Shown here are two aircraft. Use the figures for the Concorde to determine the Blackbird's speed.

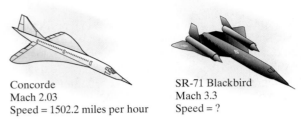

Concorde
Mach 2.03
Speed = 1502.2 miles per hour

SR-71 Blackbird
Mach 3.3
Speed = ?

42. A golf ball's bounce height, B, in inches, varies directly as its drop height, d, in inches. A golf ball bounces 36 inches when dropped from a height of 40 inches. What is the ball's bounce height if the drop height is increased to 50 inches?
43. The time that it takes to get to campus varies inversely as your driving rate. Averaging 20 miles per hour in terrible traffic, it takes you 1.5 hours to get to campus. How long would the trip take averaging 60 miles per hour?
44. The weight that can be supported by a 2-inch by 4-inch piece of pine (called a 2-by-4) varies inversely as its length. A 10-foot 2-by-4 can support 500 pounds. What weight can be supported by a 5-foot 2-by-4?
45. The volume of a gas in a container at a constant temperature varies inversely as the pressure. If the volume is 32 cubic centimeters at a pressure of 8 pounds per square centimeter, find the pressure when the volume is 40 cubic centimeters.
46. The current in a circuit varies inversely as the resistance. The current is 20 amperes when the resistance is 5 ohms. Find the current for a resistance of 16 ohms.
47. The number of pens sold varies inversely as the price per pen. If 4000 pens are sold at a price of $1.50 each, find the number of pens sold at a price of $1.20 each.
48. The time required to accomplish a task varies inversely as the number of people working on the task. It takes 6 hours for 20 people to put a new roof on a porch. How long would it take 30 people to do the job?

Writing in Mathematics

49. What is the relationship among time traveled, distance traveled, and rate of travel?
50. If you know how many hours it takes for you to do a job, explain how to find the fractional part of the job you can complete in x hours.
51. If you can do a job in 6 hours and your friend can do the same job in 3 hours, explain how to find how long it takes to complete the job working together. It is not necessary to solve the problem.
52. When two people work together to complete a job, describe one factor that can result in more or less time than the time given by the rational equations we have been using.

53. What are similar triangles?

54. If the ratio of the corresponding sides of two similar triangles is 1 to 1 $\left(\frac{1}{1}\right)$, what must be true about the triangles?

55. What are corresponding angles in similar triangles?

56. Describe how to identify the corresponding sides in similar triangles.

57. What does it mean if two quantities vary directly?

58. In your own words, explain how to solve a variation problem.

59. What does it mean if two quantities vary inversely?

60. Explain the meaning of this statement: A company's monthly sales vary directly as its advertising budget.

61. Explain the meaning of this statement: A company's monthly sales vary inversely as the price of its product.

Critical Thinking Exercises

62. Two skiers begin skiing along a trail at the same time. The faster skier averages 9 miles per hour and the slower skier averages 6 miles per hour. The faster skier completes the trail $\frac{1}{4}$ hour before the slower skier. How long is the trail?

63. A snowstorm causes a bus driver to decrease the usual average rate along a 60-mile route by 15 miles per hour. As a result, the bus takes two hours longer than usual to complete the route. At what average rate does the bus usually cover the 60-mile route?

64. One pipe can fill a swimming pool in 2 hours, a second can fill the pool in 3 hours, and a third pipe can fill the pool in 4 hours. How many minutes, to the nearest minute, would it take to fill the pool with all three pipes operating?

65. Ben can prepare a company report in 3 hours. Shane can prepare the report in 4.2 hours. How long will it take them, working together, to prepare *four* company reports?

66. An experienced carpenter can panel a room 3 times faster than an apprentice can. Working together, they can panel the room in 6 hours. How long would it take each person working alone to do the job?

67. It normally takes 2 hours to fill a swimming pool. The pool has developed a slow leak. If the pool were full, it would take 10 hours for all the water to leak out. If the pool is empty, how long will it take to fill it?

68. The intensity of radiation from a machine used to treat tumors varies inversely as the square of the distance from the machine. If the intensity is 140.5 milliroentgens per hour at 2 meters, what is the intensity at a distance of 3 meters?

69. Two investments have interest rates that differ by 1%. An investment for 1 year at the lower rate earns $175. The same principal invested for a year at the higher rate earns $200. What are the two interest rates?

Review Exercises

70. Factor: $25x^2 - 81$. (Section 7.4, Example 1)

71. Solve: $x^2 - 12x + 36 = 0$. (Section 7.6, Example 4)

72. Graph: $y = -\frac{2}{3}x + 4$. (Section 4.4, Example 3)

GROUP PROJECT

CHAPTER 8

Group members make up the sales team for a company that makes computer video games. It has been determined that the formula

$$y = \frac{200x}{x^2 + 100}$$

models the monthly sales, y, in thousands of games, of a new video game x months after the game is introduced. The figure shows the graph of the formula. What are the team's recommendations to the company in terms of how long the video game should be on the market before another new video game is introduced? What other factors might members want to take into account in terms of the recommendations? What will eventually happen to sales, and how is this indicated by the graph? What could the company do to change the behavior of this model and continue generating sales? Would this be cost effective?

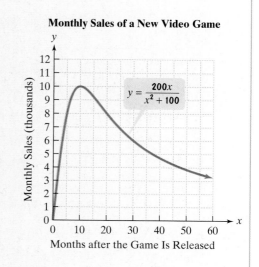

Monthly Sales of a New Video Game

$y = \dfrac{200x}{x^2 + 100}$

Monthly Sales (thousands)

Months after the Game Is Released

CHAPTER 8 SUMMARY

Definitions and Concepts	Examples

Section 8.1 Rational Expressions and their Simplification

A rational expression is the quotient of two polynomials. To find values for which a rational expression is undefined, set the denominator equal to 0 and solve.

Find all numbers for which

$$\frac{7x}{x^2 - 3x - 4}$$

is undefined.

$$x^2 - 3x - 4 = 0$$

$$(x - 4)(x + 1) = 0$$

$$x - 4 = 0 \quad \text{or} \quad x + 1 = 0$$

$$x = 4 \qquad x = -1$$

Undefined at 4 and −1

To simplify a rational expression:

1. Factor the numerator and the denominator completely.

2. Divide the numerator and the denominator by any common factors.

If factors in the numerator and denominator are opposites, their quotient is −1.

Simplify: $\dfrac{3x + 18}{x^2 - 36}$.

$$\frac{3x + 18}{x^2 - 36} = \frac{3\overset{1}{\cancel{(x + 6)}}}{\underset{1}{\cancel{(x + 6)}}(x - 6)} = \frac{3}{x - 6}$$

Section 8.2 Multiplying and Dividing Rational Expressions

Multiplying Rational Expressions

1. Factor completely.

2. Divide numerators and denominators by common factors.

3. Multiply remaining factors in the numerators and multiply the remaining factors in the denominators.

$$\frac{x^2 + 3x - 10}{x^2 - 2x} \cdot \frac{x^2}{x^2 - 25}$$

$$= \frac{\overset{1}{\cancel{(x + 5)}}\overset{1}{\cancel{(x - 2)}}}{\underset{1}{\cancel{x}}\underset{1}{\cancel{(x - 2)}}} \cdot \frac{\overset{1}{x} \cdot x}{\underset{1}{\cancel{(x + 5)}}(x - 5)}$$

$$= \frac{x}{x - 5}$$

Dividing Rational Expressions
Invert the divisor and multiply.

$$\frac{3y + 3}{(y + 2)^2} \div \frac{y^2 - 1}{y + 2}$$

$$= \frac{3y + 3}{(y + 2)^2} \cdot \frac{y + 2}{y^2 - 1}$$

$$= \frac{3\overset{1}{\cancel{(y + 1)}}}{(y + 2)\underset{1}{\cancel{(y + 2)}}} \cdot \frac{\overset{1}{\cancel{(y + 2)}}}{\underset{1}{\cancel{(y + 1)}}(y - 1)}$$

$$= \frac{3}{(y + 2)(y - 1)}$$

Definitions and Concepts	Examples

Section 8.3 Adding and Subtracting Rational Expressions with the Same Denominator

To add or subtract rational expressions with the same denominator, add or subtract the numerators and place the result over the common denominator. If possible, simplify the resulting expression.

$$\frac{y^2 - 3y + 4}{y^2 + 8y + 15} - \frac{y^2 - 5y - 2}{y^2 + 8y + 15}$$

$$= \frac{y^2 - 3y + 4 - (y^2 - 5y - 2)}{y^2 + 8y + 15}$$

$$= \frac{y^2 - 3y + 4 - y^2 + 5y + 2}{y^2 + 8y + 15}$$

$$= \frac{2y + 6}{(y + 5)(y + 3)}$$

$$= \frac{2\overset{1}{\cancel{(y + 3)}}}{(y + 5)\underset{1}{\cancel{(y + 3)}}} = \frac{2}{y + 5}$$

To add or subtract rational expressions with opposite denominators, multiply either rational expression by $\frac{-1}{-1}$ to obtain a common denominator.

$$\frac{7}{x - 6} + \frac{x + 4}{6 - x}$$

$$= \frac{7}{x - 6} + \frac{(-1)}{(-1)} \cdot \frac{x + 4}{6 - x}$$

$$= \frac{7}{x + 6} + \frac{-x - 4}{x - 6}$$

$$= \frac{7 - x - 4}{x - 6} = \frac{3 - x}{x - 6}$$

Section 8.4 Adding and Subtracting Rational Expressions with Different Denominators

Finding the Least Common Denominator (LCD)

1. Factor denominators completely.
2. List factors of the first denominator.
3. Add to the list factors of the second denominator that are not already in the list.
4. The LCD is the product of factors in step 3.

Find the LCD of

$$\frac{x + 1}{2x - 2} \quad \text{and} \quad \frac{2x}{x^2 + 2x - 3}.$$

$$2x - 2 = 2(x - 1)$$

$$x^2 + 2x - 3 = (x - 1)(x + 3)$$

Factors of first denominator: $2, x - 1$
Factors of second denominator not in the list: $x + 3$
LCD: $2(x - 1)(x + 3)$

Adding and Subtracting Rational Expressions with Different Denominators

1. Find the LCD.
2. Rewrite each rational expression as an equivalent expression with the LCD.
3. Add or subtract numerators, placing the resulting expression over the LCD.
4. If possible, simplify.

$$\frac{x + 1}{2x - 2} - \frac{2x}{x^2 + 2x - 3}$$

$$= \frac{x + 1}{2(x - 1)} - \frac{2x}{(x - 1)(x + 3)}$$

LCD is $2(x - 1)(x + 3)$.

$$= \frac{(x + 1)(x + 3)}{2(x - 1)(x + 3)} - \frac{2x \cdot 2}{2(x - 1)(x + 3)}$$

$$= \frac{x^2 + 4x + 3 - 4x}{2(x - 1)(x + 3)}$$

$$= \frac{x^2 + 3}{2(x - 1)(x + 3)}$$

Definitions and Concepts	Examples

Section 8.5 Complex Rational Expressions

Complex rational expressions have numerators or denominators containing one or more rational expressions. Complex rational expressions can be simplified by obtaining single expressions in the numerator and denominator and then dividing. They can also be simplified by multiplying the numerator and denominator by the LCD of all rational expressions within the complex rational expression.

Simplify by dividing: $\dfrac{\dfrac{1}{x}+5}{\dfrac{1}{x}-\dfrac{1}{3}}$.

$$= \frac{\dfrac{1}{x}+\dfrac{5x}{x}}{\dfrac{3}{3x}-\dfrac{x}{3x}} = \frac{\dfrac{1+5x}{x}}{\dfrac{3-x}{3x}} = \frac{1+5x}{\overset{1}{\cancel{x}}} \cdot \frac{\overset{1}{\cancel{3x}}}{3-x}$$

$$= \frac{3(1+5x)}{3-x} \quad \text{or} \quad \frac{3+15x}{3-x}$$

Simplify by the LCD method: $\dfrac{\dfrac{1}{x}+5}{\dfrac{1}{x}-\dfrac{1}{3}}$.

LCD is $3x$.

$$\frac{3x}{3x} \cdot \frac{\left(\dfrac{1}{x}+5\right)}{\left(\dfrac{1}{x}-\dfrac{1}{3}\right)} = \frac{3x \cdot \dfrac{1}{x} + 3x \cdot 5}{3x \cdot \dfrac{1}{x} - 3x \cdot \dfrac{1}{3}}$$

$$= \frac{3+15x}{3-x}$$

Section 8.6 Solving Rational Equations

A rational equation is an equation containing one or more rational expressions.

Solving Rational Equations

1. List restrictions on the variable.

2. Clear fractions by multiplying both sides by the LCD.

3. Solve the resulting equation.

4. Reject any proposed solution in the list of restrictions. Check other proposed solutions in the original equation.

Solve: $\dfrac{7x}{x^2-4} + \dfrac{5}{x-2} = \dfrac{2x}{x^2-4}$

$$\frac{7x}{(x+2)(x-2)} + \frac{5}{x-2} = \frac{2x}{(x+2)(x-2)}$$

Denominators would equal 0 if $x=-2$ or $x=2$.
Restrictions: $x \neq -2$ and $x \neq 2$.

LCD is $(x+2)(x-2)$.

$$(x+2)(x-2)\left[\frac{7x}{(x+2)(x-2)} + \frac{5}{x-2}\right]$$

$$= (x+2)(x-2) \cdot \frac{2x}{(x+2)(x-2)}$$

$$7x + 5(x+2) = 2x$$
$$7x + 5x + 10 = 2x$$
$$12x + 10 = 2x$$
$$10 = -10x$$
$$-1 = x$$

The proposed solution, -1, is not part of the restriction $x \neq -2$ and $x \neq 2$. It checks. The solution is -1.

Definitions and Concepts	Examples

Section 8.7 Applications Using Rational Equations and Variation

Motion problems involving time are solved using

$$t = \frac{d}{r}.$$

$$\text{Time traveled} = \frac{\text{Distance traveled}}{\text{Rate of travel}}$$

It takes a cyclist who averages 16 miles per hour in still air the same time to travel 48 miles with the wind as 16 miles against the wind. What is the wind's rate?

$$x = \text{wind's rate}$$

$$16 + x = \text{cyclist's rate with wind}$$

$$16 - x = \text{cyclist's rate against wind}$$

	Distance	Rate	Time = $\dfrac{\text{Distance}}{\text{Rate}}$
With wind	48	$16 + x$	$\dfrac{48}{16 + x}$
Against wind	16	$16 - x$	$\dfrac{9}{16 - x}$

Two times are equal

$$\frac{48}{16 + x} = \frac{16}{16 - x}$$

$$(16 + x)(16 - x) \cdot \frac{48}{16 + x} = \frac{16}{16 - x} \cdot (16 + x)(16 - x)$$

$$48(16 - x) = 16(16 + x)$$

Solving this equation, $x = 8$.
The wind's rate is 8 miles per hour.

Work problems are solved using the following condition:

Fraction of job done by the first	+	fraction of job done by the second	=	1.

One pipe fills a pool in 20 hours and a second pipe in 15 hours. How long will it take to fill the pool using both pipes?

$$x = \text{time using both pipes}$$

Fraction of pool filled by pipe 1 in x hours	+	fraction of pool filled by pipe 2 in x hours	=	1.

$$\frac{x}{20} + \frac{x}{15} = 1$$

$$60\left(\frac{x}{20} + \frac{x}{15}\right) = 60 \cdot 1$$

$$3x + 4x = 60$$

$$7x = 60$$

$$x = \frac{60}{7} = 8\frac{4}{7} \text{ hours}$$

It will take $8\frac{4}{7}$ hours for both pipes to fill the pool.

Definitions and Concepts	Examples

Section 8.7 Applications Using Rational Equations and Variation (continued)

Similar triangles have the same shape, but not necessarily the same size. Corresponding angles have the same measure, and corresponding sides are proportional. If the measures of two angles of one triangle are equal to those of two angles of a second triangle, then the two triangles are similar.

Find x for these similar triangles.

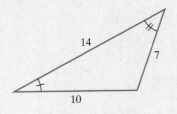

Corresponding sides are proportional:

$$\frac{7}{x} = \frac{10}{5}. \quad \left(or \ \frac{7}{x} = \frac{14}{7} \right)$$

$$5x \cdot \frac{7}{x} = \frac{10}{5} \cdot 5x$$

$$35 = 10x$$

$$x = \frac{35}{10} = 3.5$$

Variation

$$y \text{ varies directly as } x: \quad y = kx.$$

$$y \text{ varies inversely as } x: \quad y = \frac{k}{x}. \quad \text{constant of variation}$$

Solving a Variation Problem

1. Write the variation equation.
2. Substitute the pair of values and find k.
3. Substitute the value of k into the variation equation.
4. Use the variation equation with k to answer the problem's question.

The time that it takes you to get to drive a certain distance varies inversely as your driving rate. Averaging 30 miles per hour, it takes you 10 hours to drive the distance. How long would the trip take averaging 50 miles per hour?

1. $t = \dfrac{k}{r}$ Time, t, varies inversely as rate, r.

2. It takes 10 hours at 30 miles per hour.

$$10 = \frac{k}{30}$$

$$k = 10(30) = 300$$

3. $t = \dfrac{300}{r}$

4. How long at 50 miles per hour? Substitute 50 for r.

$$t = \frac{300}{50} = 6$$

It takes 6 hours at 50 miles per hour.

CHAPTER 8 REVIEW EXERCISES

8.1 *In Exercises 1–4, find all numbers for which each rational expression is undefined. If the rational expression is defined for all real numbers, so state.*

1. $\dfrac{5x}{6x - 24}$

2. $\dfrac{x + 3}{(x - 2)(x + 5)}$

3. $\dfrac{x^2 + 3}{x^2 - 3x + 2}$

4. $\dfrac{7}{x^2 + 81}$

In Exercises 5–12, simplify each rational expression. If the rational expression cannot be simplified, so state.

5. $\dfrac{16x^2}{12x}$

6. $\dfrac{x^2 - 4}{x - 2}$

7. $\dfrac{x^3 + 2x^2}{x + 2}$

8. $\dfrac{x^2 + 3x - 18}{x^2 - 36}$

9. $\dfrac{x^2 - 4x - 5}{x^2 + 8x + 7}$

10. $\dfrac{y^2 + 2y}{y^2 + 4y + 4}$

11. $\dfrac{x^2}{x^2 + 4}$

12. $\dfrac{2x^2 - 18y^2}{3y - x}$

8.2 *In Exercises 13–17, multiply as indicated.*

13. $\dfrac{x^2 - 4}{12x} \cdot \dfrac{3x}{x + 2}$

14. $\dfrac{5x + 5}{6} \cdot \dfrac{3x}{x^2 + x}$

15. $\dfrac{x^2 + 6x + 9}{x^2 - 4} \cdot \dfrac{x - 2}{x + 3}$

16. $\dfrac{y^2 - 2y + 1}{y^2 - 1} \cdot \dfrac{2y^2 + y - 1}{5y - 5}$

17. $\dfrac{2y^2 + y - 3}{4y^2 - 9} \cdot \dfrac{3y + 3}{5y - 5y^2}$

In Exercises 18–22, divide as indicated.

18. $\dfrac{x^2 + x - 2}{10} \div \dfrac{2x + 4}{5}$

19. $\dfrac{6x + 2}{x^2 - 1} \div \dfrac{3x^2 + x}{x - 1}$

20. $\dfrac{1}{y^2 + 8y + 15} \div \dfrac{7}{y + 5}$

21. $\dfrac{y^2 + y - 42}{y - 3} \div \dfrac{y + 7}{(y - 3)^2}$

22. $\dfrac{8x + 8y}{x^2} \div \dfrac{x^2 - y^2}{x^2}$

8.3 *In Exercises 23–28, add or subtract as indicated. Simplify the result, if possible.*

23. $\dfrac{4x}{x + 5} + \dfrac{20}{x + 5}$

24. $\dfrac{8x - 5}{3x - 1} + \dfrac{4x + 1}{3x - 1}$

25. $\dfrac{3x^2 + 2x}{x - 1} - \dfrac{10x - 5}{x - 1}$

26. $\dfrac{6y^2 - 4y}{2y - 3} - \dfrac{12 - 3y}{2y - 3}$

27. $\dfrac{x}{x - 2} + \dfrac{x - 4}{2 - x}$

28. $\dfrac{x + 5}{x - 3} - \dfrac{x}{3 - x}$

8.4 *In Exercises 29–31, find the least common denominator of the rational expressions.*

29. $\dfrac{7}{9x^3}$ and $\dfrac{5}{12x}$

30. $\dfrac{3}{x^2(x - 1)}$ and $\dfrac{11}{x(x - 1)^2}$

31. $\dfrac{x}{x^2 + 4x + 3}$ and $\dfrac{17}{x^2 + 10x + 21}$

In Exercises 32–42, add or subtract as indicated. Simplify the result, if possible.

32. $\dfrac{7}{3x} + \dfrac{5}{2x^2}$

33. $\dfrac{5}{x + 1} + \dfrac{2}{x}$

34. $\dfrac{7}{x + 3} + \dfrac{4}{(x + 3)^2}$

35. $\dfrac{6y}{y^2 - 4} - \dfrac{3}{y + 2}$

36. $\dfrac{y - 1}{y^2 - 2y + 1} - \dfrac{y + 1}{y - 1}$

37. $\dfrac{x + y}{y} - \dfrac{x - y}{x}$

38. $\dfrac{2x}{x^2 + 2x + 1} + \dfrac{x}{x^2 - 1}$

39. $\dfrac{5x}{x + 1} - \dfrac{2x}{1 - x^2}$

40. $\dfrac{4}{x^2 - x - 6} - \dfrac{4}{x^2 - 4}$

41. $\dfrac{7}{x + 3} + 2$

42. $\dfrac{2y - 5}{6y + 9} - \dfrac{4}{2y^2 + 3y}$

8.5 *In Exercises 43–47, simplify each complex rational expression.*

43. $\dfrac{\dfrac{1}{2} + \dfrac{3}{8}}{\dfrac{3}{4} - \dfrac{1}{2}}$

44. $\dfrac{\dfrac{1}{x}}{1 - \dfrac{1}{x}}$

45. $\dfrac{\dfrac{1}{x} + \dfrac{1}{y}}{\dfrac{1}{xy}}$

46. $\dfrac{\dfrac{1}{x} - \dfrac{1}{2}}{\dfrac{1}{3} - \dfrac{x}{6}}$

47. $\dfrac{3 + \dfrac{12}{x}}{1 - \dfrac{16}{x^2}}$

8.6 *In Exercises 48–55, solve each rational equation. If an equation has no solution, so state.*

48. $\dfrac{3}{x} - \dfrac{1}{6} = \dfrac{1}{x}$

49. $\dfrac{3}{4x} = \dfrac{1}{x} + \dfrac{1}{4}$

50. $x + 5 = \dfrac{6}{x}$

51. $4 - \dfrac{x}{x+5} = \dfrac{5}{x+5}$

52. $\dfrac{2}{x-3} = \dfrac{4}{x+3} + \dfrac{8}{x^2-9}$

53. $\dfrac{2}{x} = \dfrac{2}{3} + \dfrac{x}{6}$

54. $\dfrac{13}{y-1} - 3 = \dfrac{1}{y-1}$

55. $\dfrac{1}{x+3} - \dfrac{1}{x-1} = \dfrac{x+1}{x^2+2x-3}$

56. Park rangers introduce 50 elk into a wildlife preserve. The formula

$$P = \dfrac{250(3t + 5)}{t + 25}$$

models the elk population, P, after t years. How many years will it take for the population to increase to 125 elk?

57. The formula

$$S = \dfrac{C}{1 - r}$$

describes the selling price, S, of a product in terms of its cost to the retailer, C, and its markup, r, usually expressed as a percent. A small television cost a retailer $140 and was sold for $200. Find the markup. Express the answer as a percent.

8.7

58. In still water, a paddle boat averages 20 miles per hour. It takes the boat the same amount of time to travel 72 miles downstream, with the current, as 48 miles upstream, against the current. What is the rate of the water's current?

59. A car travels 60 miles in the same time that a car traveling 10 miles per hour faster travels 90 miles. What is the rate of each car?

60. A painter can paint a fence around a house in 6 hours. Working alone, the painter's apprentice can paint the same fence in 12 hours. How many hours would it take them to do the job if they worked together?

61. The triangles shown in the figure are similar. Find the length of the side marked with an x.

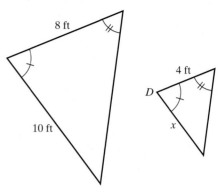

62. Find the height of the lamppost in the figure.

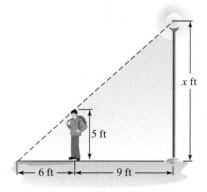

63. An electric bill varies directly as the amount of electricity used. The bill for 1400 kilowatts of electricity is $98. What is the bill for 2200 kilowatts of electricity?

64. The time it takes to drive a certain distance varies inversely as the rate of travel. If it takes 4 hours at 50 miles per hour to drive the distance, how long will it take at 40 miles per hour?

CHAPTER 8   TEST   Remember to use your Chapter Test Prep Video CD to see the worked-out solutions to the test questions you want to review.

1. Find all numbers for which

$$\dfrac{x + 7}{x^2 + 5x - 36}$$

is undefined.

In Exercises 2–3, simplify each rational expression.

2. $\dfrac{x^2 + 2x - 3}{x^2 - 3x + 2}$

3. $\dfrac{4x^2 - 20x}{x^2 - 4x - 5}$

In Exercises 4–16, perform the indicated operations. Simplify the result, if possible.

4. $\dfrac{x^2 - 16}{10} \cdot \dfrac{5}{x + 4}$

5. $\dfrac{x^2 - 7x + 12}{x^2 - 4x} \cdot \dfrac{x^2}{x^2 - 9}$

6. $\dfrac{2x + 8}{x - 3} \div \dfrac{x^2 + 5x + 4}{x^2 - 9}$

7. $\dfrac{5y + 5}{(y - 3)^2} \div \dfrac{y^2 - 1}{y - 3}$

8. $\dfrac{2y^2 + 5}{y + 3} + \dfrac{6y - 5}{y + 3}$

9. $\dfrac{y^2 - 2y + 3}{y^2 + 7y + 12} - \dfrac{y^2 - 4y - 5}{y^2 + 7y + 12}$

10. $\dfrac{x}{x + 3} + \dfrac{5}{x - 3}$

11. $\dfrac{2}{x^2 - 4x + 3} + \dfrac{6}{x^2 + x - 2}$

12. $\dfrac{4}{x - 3} + \dfrac{x + 5}{3 - x}$

13. $1 + \dfrac{3}{x - 1}$

14. $\dfrac{2x + 3}{x^2 - 7x + 12} - \dfrac{2}{x - 3}$

15. $\dfrac{8y}{y^2 - 16} - \dfrac{4}{y - 4}$

16. $\dfrac{(x - y)^2}{x + y} \div \dfrac{x^2 - xy}{3x + 3y}$

In Exercises 17–18, simplify each complex rational expression.

17. $\dfrac{5 + \dfrac{5}{x}}{2 + \dfrac{1}{x}}$

18. $\dfrac{\dfrac{1}{x} - \dfrac{1}{y}}{\dfrac{1}{x}}$

In Exercises 19–21, solve each rational equation.

19. $\dfrac{5}{x} + \dfrac{2}{3} = 2 - \dfrac{2}{x} - \dfrac{1}{6}$

20. $\dfrac{3}{y + 5} - 1 = \dfrac{4 - y}{2y + 10}$

21. $\dfrac{2}{x - 1} = \dfrac{3}{x^2 - 1} + 1$

22. In still water, a boat averages 30 miles per hour. It takes the boat the same amount of time to travel 16 miles downstream, with the current, as 14 miles upstream, against the current. What is the rate of the water's current?

23. One pipe can fill a hot tub in 20 minutes and a second pipe can fill it in 30 minutes. If the hot tub is empty, how long will it take both pipes to fill it?

24. The triangles in the figure are similar. Find the length of the side marked with an x.

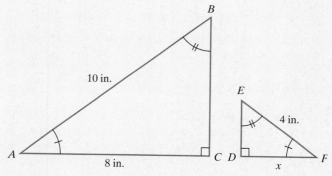

25. The amount of current flowing in an electrical circuit varies inversely as the resistance in the circuit. When the resistance in a particular circuit is 5 ohms, the current is 42 amperes. What is the current when the resistance is 4 ohms?

CUMULATIVE REVIEW EXERCISES (CHAPTERS 1–8)

In Exercises 1–6, solve each equation, inequality, or system of equations.

1. $2(x - 3) + 5x = 8(x - 1)$

2. $-3(2x - 4) > 2(6x - 12)$

3. $x^2 + 3x = 18$

4. $\dfrac{2x}{x^2 - 4} + \dfrac{1}{x - 2} = \dfrac{2}{x + 2}$

5. $y = 2x - 3$
$x + 2y = 9$

6. $3x + 2y = -2$
$-4x + 5y = 18$

In Exercises 7–9, graph each equation or inequality in a rectangular coordinate system.

7. $3x - 2y = 6$

8. $y > -2x + 3$

9. $y = -3$

In Exercises 10–12, simplify each expression.

10. $-21 - 16 - 3(2 - 8)$

11. $\left(\dfrac{4x^5}{2x^2}\right)^3$

12. $\dfrac{\dfrac{1}{x} - 2}{4 - \dfrac{1}{x}}$

In Exercises 13–15, factor completely.

13. $4x^2 - 13x + 3$

14. $4x^2 - 20x + 25$

15. $3x^2 - 75$

In Exercises 16–18, perform the indicated operations.

16. $(4x^2 - 3x + 2) - (5x^2 - 7x - 6)$

17. $\dfrac{-8x^6 + 12x^4 - 4x^2}{4x^2}$

18. $\dfrac{x + 6}{x - 2} + \dfrac{2x + 1}{x + 3}$

19. You invested $4000, part at 5% and the remainder at 9% annual interest. At the end of the year, the total interest from these investments was $311. How much was invested at each rate?

20. A 68-inch board is to be cut into two pieces. If one piece must be three times as long as the other, find the length of each piece.

CHAPTER 9

Roots and Radicals

How rapidly is cyberspace reshaping our lives? Data indicate that the percentage of U.S. households online increased quite rapidly toward the end of the twentieth century. However, now this growth is beginning to slow down. In this chapter, you will see why models with square roots are used to describe phenomena that are continuing to grow, but whose growth is leveling off. By learning about roots and radicals, you will have new algebraic tools for describing your world.

SECTION

9.1

Objectives

1 Find square roots.

2 Evaluate models containing square roots.

3 Use a calculator to find decimal approximations for irrational square roots.

4 Find higher roots.

FINDING ROOTS

What is the maximum speed at which a racing cyclist can turn a corner without tipping over? The answer, in miles per hour, is given by the algebraic expression $4\sqrt{x}$, where x is the radius of the corner, in feet. Algebraic expressions containing roots describe phenomena as diverse as a wild animal's territorial area, evaporation on a lake's surface, and Albert Einstein's bizarre concept of how an astronaut moving close to the speed of light would barely age relative to friends watching from Earth. No description of your world can be complete without roots and radicals. In this section, we develop the basics of radical expressions through a notation that takes us from a number raised to a power back to the number itself.

| **1** | Find square roots. |

Square Roots From our earlier work with exponents, we are aware that the square of both 5 and −5 is 25:

$$5^2 = 25 \quad \text{and} \quad (-5)^2 = 25.$$

The reverse operation of squaring a number is finding the *square root* of the number. For example,

- A square root of 25 is 5 because $5^2 = 25$.
- A square root of 25 is also −5 because $(-5)^2 = 25$.

In general, **if $b^2 = a$, then b is a square root of a.**

The symbol $\sqrt{\ \ }$ is used to denote the nonnegative or *principal square root* of a number. For example,

- $\sqrt{25} = 5$ because $5^2 = 25$ and 5 is positive.
- $\sqrt{100} = 10$ because $10^2 = 100$ and 10 is positive.

The symbol $\sqrt{\ \ }$ that we use to denote the principal square root is called a **radical sign**. The number under the radical sign is called the **radicand**. Together we refer to the radical sign and its radicand as a **radical**.

Radical sign $\sqrt{a}$ Radicand

Radical

> **DEFINITION OF THE PRINCIPAL SQUARE ROOT** If a is a nonnegative real number, the nonnegative number b such that $b^2 = a$, denoted by $b = \sqrt{a}$, is the **principal square root** of a.

The symbol $-\sqrt{}$ is used to denote the negative square root of a number. For example,

- $-\sqrt{25} = -5$ because $(-5)^2 = 25$ and -5 is negative.
- $-\sqrt{100} = -10$ because $(-10)^2 = 100$ and -10 is negative.

EXAMPLE 1 Evaluating Expressions Containing Radicals

Evaluate:

a. $\sqrt{64}$ **b.** $-\sqrt{49}$ **c.** $\sqrt{\dfrac{1}{4}}$ **d.** $\sqrt{9 + 16}$ **e.** $\sqrt{9} + \sqrt{16}$.

SOLUTION

STUDY TIP
In Example 1, parts (d) and (e), observe that $\sqrt{9 + 16}$ is not equal to $\sqrt{9} + \sqrt{16}$. In general, $$\sqrt{a + b} \neq \sqrt{a} + \sqrt{b}$$ and $$\sqrt{a - b} \neq \sqrt{a} - \sqrt{b}.$$

a. $\sqrt{64} = 8$ The principal square root of 64 is 8. Check: $8^2 = 64$.

b. $-\sqrt{49} = -7$ The negative square root of 49 is -7. Check: $(-7)^2 = 49$.

c. $\sqrt{\dfrac{1}{4}} = \dfrac{1}{2}$ The principal square root of $\frac{1}{4}$ is $\frac{1}{2}$. Check: $\left(\frac{1}{2}\right)^2 = \frac{1}{4}$.

d. $\sqrt{9 + 16} = \sqrt{25}$ First simplify the expression under the radical sign.
$\qquad\qquad\quad = 5$ Then take the principal square root of 25, which is 5.

e. $\sqrt{9} + \sqrt{16} = 3 + 4$ $\sqrt{9} = 3$ because $3^2 = 9$. $\sqrt{16} = 4$ because $4^2 = 16$.
$\qquad\qquad\qquad = 7$

✔ **CHECK POINT 1** Evaluate:

a. $\sqrt{81}$ **b.** $-\sqrt{9}$ **c.** $\sqrt{\dfrac{1}{25}}$

d. $\sqrt{36 + 64}$ **e.** $\sqrt{36} + \sqrt{64}$.

2 Evaluate models containing square roots.

EXAMPLE 2 An Application: A Mathematical Model Containing a Radical

The amount of evaporation, in inches per day, of a large body of water can be modeled by the formula

$$E = \frac{w}{20\sqrt{a}}$$

where a = surface area of the water, in square miles

w = average wind speed of the air over the water, in miles per hour

E = evaporation, in inches per day.

Determine the evaporation on a lake whose surface area is 9 square miles on a day when the wind speed over the water is 10 miles per hour.

Some square roots occur so frequently that you may want to memorize them.

$$\sqrt{1} = 1 \qquad \sqrt{4} = 2$$
$$\sqrt{9} = 3 \qquad \sqrt{16} = 4$$
$$\sqrt{25} = 5 \qquad \sqrt{36} = 6$$
$$\sqrt{49} = 7 \qquad \sqrt{64} = 8$$
$$\sqrt{81} = 9 \qquad \sqrt{100} = 10$$
$$\sqrt{121} = 11 \qquad \sqrt{144} = 12$$

SOLUTION

$$E = \frac{w}{20\sqrt{a}} \qquad \text{This is the given formula.}$$

$$E = \frac{10}{20\sqrt{9}} \qquad \begin{array}{l}\text{Substitute the given values:} \\ w \text{ (wind speed)} = 10 \text{ miles per hour and} \\ a \text{ (area)} = 9 \text{ square miles.}\end{array}$$

$$E = \frac{10}{20 \cdot 3} \qquad \sqrt{9} = 3 \text{ because } 3^2 = 9.$$

$$= \frac{1}{6} \qquad \frac{\overset{1}{\cancel{10}}}{\underset{2}{\cancel{20}} \cdot 3} = \frac{1}{6}$$

The evaporation is $\frac{1}{6}$ of an inch on that day. ■

 CHECK POINT 2 Use the model in Example 2 to solve this problem. Determine the evaporation on a lake whose surface area is 4 square miles on a day when the wind speed is 20 miles per hour.

Modeling Data with Square Roots   Many real-world phenomena start with rapid growth, and then the growth begins to level off. This type of behavior can be modeled by equations involving square roots. For example, due to heightened interest in the Internet and the World Wide Web, the number of U.S. households online is increasing. Table 9.1 gives the percentage of U.S. households online from 1999 through 2003. The data are displayed as a set of five points in the scatter plot in Figure 9.1.

Table 9.1

Year	Percentage of U.S. Households Online
1999	28
2000	44
2001	51
2002	55
2003	56

Source: Forrester Research

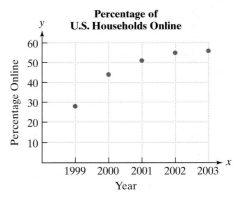

FIGURE 9.1

Can you see that the growth in the percentage of online households is slowing down in 2001 through 2003? Let's see why square roots can be used to model the data by looking at the shape of the graph of $y = \sqrt{x}$.

We graph $y = \sqrt{x}$ by first selecting integers for x. It is easiest to choose **perfect squares**, numbers that have rational square roots. Table 9.2 shows five ordered pairs that are solutions of $y = \sqrt{x}$. We plot these ordered pairs as points in the rectangular coordinate system and connect the points with a smooth curve. The graph of $y = \sqrt{x}$ is shown in Figure 9.2.

Table 9.2

x	$y = \sqrt{x}$	(x, y)
0	$y = \sqrt{0} = 0$	$(0, 0)$
1	$y = \sqrt{1} = 1$	$(1, 1)$
4	$y = \sqrt{4} = 2$	$(4, 2)$
9	$y = \sqrt{9} = 3$	$(9, 3)$
16	$y = \sqrt{16} = 4$	$(16, 4)$

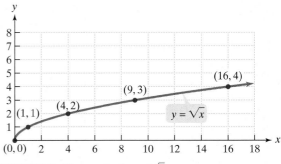

FIGURE 9.2   The graph of $y = \sqrt{x}$

The graph in Figure 9.2 is increasing from left to right. However, the rate of increase is slowing down as the graph moves to the right. This is why equations with square roots are often used to model growing phenomena with growth that is leveling off.

Is it possible to choose values of x in Table 9.2 that are not perfect squares? Yes. For example, we can let $x = 3$. Thus, $y = \sqrt{x} = \sqrt{3}$. Because 3 is not a perfect square, $\sqrt{3}$ is an irrational number, one that cannot be expressed as a quotient of integers. We can use a calculator to find a decimal approximation for $\sqrt{3}$.

Most Scientific Calculators

3 $\boxed{\sqrt{}}$

Most Graphing Calculators

$\boxed{\sqrt{}}$ 3 $\boxed{\text{ENTER}}$

Rounding the displayed number to two decimal places, $\sqrt{3} \approx 1.73$. This information is shown visually as a point, approximately $(3, 1.73)$, on the graph of $y = \sqrt{x}$ in Figure 9.3.

3 Use a calculator to find decimal approximations for irrational square roots.

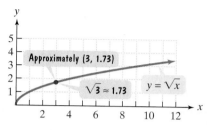

FIGURE 9.3 Visualizing $\sqrt{3} \approx 1.73$

USING TECHNOLOGY

You can use your calculator to approximate

$$14\sqrt{3} + 30$$

as follows:

Scientific Calculator

14 $\boxed{\times}$ 3 $\boxed{\sqrt{}}$ $\boxed{+}$ 30 $\boxed{=}$

Graphing Calculator

14 $\boxed{\times}$ $\boxed{\sqrt{}}$ 3 $\boxed{+}$ 30 $\boxed{\text{ENTER}}$

Some calculators display an open parenthesis after $\boxed{\sqrt{}}$ is pressed. In this case, use a parenthesis after entering the radicand.

4 Find higher roots.

EXAMPLE 3 Modeling Online Households with a Square Root

The formula

$$P = 14\sqrt{t} + 30$$

models the percentage, P, of U.S. households online t years after 1999. Find the percentage online in 2002.

SOLUTION Because 2002 is 3 years after 1999, we substitute 3 for t in the given formula. Then we use a calculator to find P, the percentage of online households in 2002.

$$P = 14\sqrt{t} + 30 \qquad \text{Use the given formula.}$$
$$P = 14\sqrt{3} + 30 \qquad \text{Substitute 3 for } t.$$
$$\approx 54.25 \qquad \text{Use a calculator.}$$

Rounding to the nearest percent, approximately 54% of U.S. households were online in 2002. This is close to the the actual percentage, 55%, shown in Table 9.1. ∎

 CHECK POINT 3 Use the formula in Example 3 to find the percentage of U.S. households online in 2001. Round to the nearest percent. How well does your answer model the actual data from Table 9.1?

Did you notice that we graphed $y = \sqrt{x}$ in quadrant I, where $x \geq 0$? What happens if x is negative? Is the square root of a negative number a real number? For example, consider $\sqrt{-25}$. Is there a real number whose square is -25? No. Thus, $\sqrt{-25}$ is not a real number. In general, **a square root of a negative number is not a real number**.

Roots Greater Than Square Roots Finding the square root of a number reverses the process of squaring a number. Similarly, the cube root of a number reverses the process of cubing a number. For example, $2^3 = 8$, and so the cube root of 8 is 2. The notation that we use is $\sqrt[3]{8} = 2$.

We define the **principal nth root** of a real number a, symbolized by $\sqrt[n]{a}$, as follows:

DEFINITION OF THE PRINCIPAL nTH ROOT OF A REAL NUMBER

$$\sqrt[n]{a} = b \text{ means that } b^n = a.$$

The natural number n is called the **index**. The index, 2, for square roots is usually omitted.

For example,

$$\sqrt[3]{64} = 4 \text{ because } 4^3 = 64 \text{ and } \sqrt[5]{-32} = -2 \text{ because } (-2)^5 = -32.$$

The same vocabulary that we learned for square roots applies to *n*th roots. The symbol $\sqrt[n]{a}$ is called a **radical** and *a* is called the **radicand**.

Table 9.3 shows how various roots reverse raising numbers to powers.

Table 9.3 **Reversing *n*th Powers with *n*th Roots**

	Powers	Roots	Vocabulary
Cube Roots	$4^3 = 64$	$\sqrt[3]{64} = 4$	3 is the index of the radical.
	$(-2)^3 = -8$	$\sqrt[3]{-8} = -2$	
	$5^3 = 125$	$\sqrt[3]{125} = 5$	
Fourth Roots	$1^4 = 1$	$\sqrt[4]{1} = 1$	Index = 4
	$3^4 = 81$	$\sqrt[4]{81} = 3$	
Fifth Roots	$2^5 = 32$	$\sqrt[5]{32} = 2$	Index = 5
nth Roots	$b^n = a$	$\sqrt[n]{a} = b$	Index = n

One of the entries in the table involves a negative radicand: $\sqrt[3]{-8} = -2$. By contrast to the square root of a negative number, the cube root of a negative number is a real number. In general, **if the index is even**, such as $\sqrt{\ }$, $\sqrt[4]{\ }$, $\sqrt[6]{\ }$, and so on, **the radicand must be nonnegative for the root to be a real number**. For example,

$$\sqrt[4]{81} = 3, \text{ but } \sqrt[4]{-81} \text{ is not a real number.}$$
$$\sqrt[6]{64} = 2, \text{ but } \sqrt[6]{-64} \text{ is not a real number.}$$

Furthermore, if the index is even, the principal *n*th root is nonnegative.

STUDY TIP

Some higher roots occur so frequently that you may want to memorize them.

Cube Roots	Fourth Roots
$\sqrt[3]{1} = 1$	$\sqrt[4]{1} = 1$
$\sqrt[3]{8} = 2$	$\sqrt[4]{16} = 2$
$\sqrt[3]{27} = 3$	$\sqrt[4]{81} = 3$
$\sqrt[3]{64} = 4$	$\sqrt[4]{256} = 4$
$\sqrt[3]{125} = 5$	$\sqrt[4]{625} = 5$
$\sqrt[3]{216} = 6$	

Fifth Roots
$\sqrt[5]{1} = 1$
$\sqrt[5]{32} = 2$
$\sqrt[5]{243} = 3$

EXAMPLE 4 Finding Cube Roots

Find the cube roots:

a. $\sqrt[3]{27}$ **b.** $\sqrt[3]{-1}$ **c.** $\sqrt[3]{\dfrac{1}{8}}$.

SOLUTION

a. $\sqrt[3]{27} = 3$ because $3^3 = 27$.
b. $\sqrt[3]{-1} = -1$ because $(-1)^3 = -1$.
c. $\sqrt[3]{\dfrac{1}{8}} = \dfrac{1}{2}$ because $\left(\dfrac{1}{2}\right)^3 = \dfrac{1}{8}$.

✔ **CHECK POINT 4** Find the cube roots:

a. $\sqrt[3]{1}$ **b.** $\sqrt[3]{-27}$ **c.** $\sqrt[3]{\dfrac{1}{125}}$.

EXAMPLE 5 Finding Higher Roots

Find the indicated root, or state that the expression is not a real number:

a. $\sqrt[4]{16}$ **b.** $\sqrt[4]{-16}$ **c.** $-\sqrt[4]{16}$ **d.** $\sqrt[5]{-32}$.

SOLUTION

a. $\sqrt[4]{16} = 2$ because $2^4 = 16$.

b. $\sqrt[4]{-16}$ is not a real number because the index, 4, is even and the radicand, -16, is negative. No real number equals $\sqrt[4]{-16}$ because any real number raised to the fourth power gives a nonnegative result.

c.
$$-\sqrt[4]{16} = -2$$

> Copy the negative sign and use the fact that $\sqrt[4]{16} = 2$.

d. $\sqrt[5]{-32} = -2$ because $(-2)^5 = -32$.   ■

CHECK POINT 5 Find the indicated root, or state that the expression is not a real number:

a. $\sqrt[4]{81}$

b. $\sqrt[4]{-81}$

c. $-\sqrt[4]{81}$

d. $\sqrt[5]{-\dfrac{1}{32}}$.

9.1 EXERCISE SET

Student Solutions Manual   CD/Video   PH Math/Tutor Center   MathXL Tutorials on CD   MathXL®   MyMathLab   Interactmath.com

Practice Exercises

In Exercises 1–26, evaluate each expression, or state that the expression is not a real number.

1. $\sqrt{36}$

2. $\sqrt{16}$

3. $-\sqrt{36}$

4. $-\sqrt{16}$

5. $\sqrt{-36}$

6. $\sqrt{-16}$

7. $\sqrt{\dfrac{1}{9}}$

8. $\sqrt{\dfrac{1}{49}}$

9. $\sqrt{\dfrac{1}{100}}$

10. $\sqrt{\dfrac{1}{81}}$

11. $-\sqrt{\dfrac{1}{36}}$

12. $-\sqrt{\dfrac{1}{121}}$

13. $\sqrt{-\dfrac{1}{36}}$

14. $\sqrt{-\dfrac{1}{121}}$

15. $\sqrt{0.04}$

16. $\sqrt{0.64}$

17. $\sqrt{33 - 8}$

18. $\sqrt{51 + 13}$

19. $\sqrt{2 \cdot 32}$

20. $\sqrt{\dfrac{75}{3}}$

21. $\sqrt{144 + 25}$

22. $\sqrt{25 - 16}$

23. $\sqrt{144} + \sqrt{25}$

24. $\sqrt{25} - \sqrt{16}$

25. $\sqrt{25 - 144}$

26. $\sqrt{16 - 25}$

In Exercises 27–28, graph each equation. Begin by filling in the table and finding five solutions of the equation. Then plot these ordered pairs as points in the rectangular coordinate system and connect the points with a smooth curve.

27. $y = \sqrt{x - 1}$

> Because the radicand cannot be negative, the graph begins at the point with this x-coordinate.

x	$y = \sqrt{x - 1}$	(x, y)
1		
2		
5		
10		
17		

28. $y = \sqrt{x + 2}$

> Because the radicand cannot be negative, the graph begins at the point with this x-coordinate.

x	$y = \sqrt{x + 2}$	(x, y)
-2		
-1		
2		
7		
14		

29. Describe one similarity and one difference between your graph in Exercise 27 and the graph of $y = \sqrt{x}$, shown in Figure 9.2 on page 534.

30. Describe one similarity and one difference between your graph in Exercise 28 and the graph of $y = \sqrt{x}$ shown in Figure 9.2 on page 534.

In Exercises 31–36, use a calculator to approximate each square root. Round to three decimal places.

31. $\sqrt{7}$

32. $\sqrt{11}$

33. $\sqrt{23}$

34. $\sqrt{97}$

35. $-\sqrt{65}$

36. $-\sqrt{83}$

In Exercises 37–46, use a calculator to approximate each expression. Round to three decimal places. If the expression is not a real number and an approximation is not possible, so state.

37. $12 + \sqrt{11}$

38. $14 + \sqrt{13}$

39. $\dfrac{12 + \sqrt{11}}{2}$

40. $\dfrac{14 + \sqrt{13}}{2}$

41. $\dfrac{-5 + \sqrt{321}}{6}$

42. $\dfrac{-7 + \sqrt{839}}{5}$

43. $\sqrt{13 - 5}$

44. $\sqrt{21 - 4}$

45. $\sqrt{5 - 13}$

46. $\sqrt{4 - 21}$

In Exercises 47–56, find each cube root.

47. $\sqrt[3]{64}$

48. $\sqrt[3]{27}$

49. $\sqrt[3]{-27}$

50. $\sqrt[3]{-64}$

51. $-\sqrt[3]{8}$

52. $-\sqrt[3]{27}$

53. $\sqrt[3]{\dfrac{1}{125}}$

54. $\sqrt[3]{\dfrac{1}{1000}}$

55. $\sqrt[3]{-1000}$

56. $\sqrt[3]{-125}$

In Exercises 57–74, find the indicated root, or state that the expression is not a real number.

57. $\sqrt[4]{1}$

58. $\sqrt[5]{1}$

59. $\sqrt[4]{16}$

60. $\sqrt[4]{81}$

61. $-\sqrt[4]{16}$

62. $-\sqrt[4]{81}$

63. $\sqrt[4]{-16}$

64. $\sqrt[4]{-81}$

65. $\sqrt[5]{-1}$

66. $\sqrt[7]{-1}$

67. $\sqrt[6]{-1}$

68. $\sqrt[8]{-1}$

69. $-\sqrt[4]{256}$

70. $-\sqrt[4]{10,000}$

71. $\sqrt[6]{64}$

72. $\sqrt[5]{32}$

73. $-\sqrt[5]{32}$

74. $-\sqrt[6]{64}$

Practice Plus

In Exercises 75–84, for which values of x is each radical expression a real number? Express your answer as an inequality or write "all real numbers."

75. $\sqrt{2x}$

76. $\sqrt{3x}$

77. $\sqrt{x - 2}$

78. $\sqrt{x - 3}$

79. $\sqrt{2 - x}$

80. $\sqrt{3 - x}$

81. $\sqrt{x^2 + 2}$

82. $\sqrt{x^2 + 3}$

83. $\sqrt{12 - 2x}$

84. $\sqrt{8 - 2x}$

Application Exercises

85. Racing cyclists use the formula $v = 4\sqrt{r}$ to determine the maximum velocity, v, in miles per hour, to turn a corner with radius r, in feet, without tipping over. What is the maximum velocity that a cyclist should travel around a corner of radius 9 feet without tipping over?

86. The formula $v = \sqrt{2.5r}$ models the maximum safe speed, v, in miles per hour, at which a car can travel on a curved road with radius of curvature r, in feet. A highway crew measures the radius of curvature at an exit ramp on a highway as 360 feet. What is the maximum safe speed?

Police use the formula $v = \sqrt{20L}$ to estimate the speed of a car, v, in miles per hour, based on the length, L, in feet, of its skid marks upon sudden braking on a dry asphalt road. Use the formula to solve Exercises 87–88.

87. A motorist is involved in an accident. A police officer measures the car's skid marks to be 245 feet long. Estimate the speed at which the motorist was traveling before braking. If the posted speed limit is 50 miles per hour and the motorist tells the officer he was not speeding, should the officer believe him? Explain.

88. A motorist is involved in an accident. A police officer measures the car's skid marks to be 45 feet long. Estimate the speed at which the motorist was traveling before braking, If the posted speed limit is 35 miles per hour and the motorist tells the officer she was not speeding, should the officer believe her? Explain.

Autism is a neurological disorder that impedes language and derails social and emotional development. New findings suggest that the condition is not a sudden calamity that strikes children at the age of 2 or 3, but a developmental problem linked to abnormally rapid brain growth during infancy. The graphs show that the heads of severely autistic children start out smaller than average and then go through a period of explosive growth. Exercises 89–90 involve mathematical models for the data shown by the graphs.

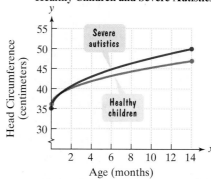

Developmental Differences between Healthy Children and Severe Autistics

Source: The Journal of the American Medical Association

89. The data for one of the two groups shown by the graphs can be modeled by

$$y = 2.9\sqrt{x} + 36,$$

where y is the head circumference, in centimeters, at age x months, $0 \le x \le 14$.

a. According to the model, what is the head circumference at birth?

b. According to the model, what is the head circumference at 9 months?

c. According to the model, what is the head circumference at 14 months? Use a calculator and round to the nearest tenth of a centimeter.

d. Use the values that you obtained in parts (a) through (c) and the graphs shown on the previous page to determine if the given model describes healthy children or severe autistics.

90. The data for one of the two groups shown by the graphs on the previous page can be modeled by

$$y = 4\sqrt{x} + 35,$$

where y is the head circumference, in centimeters, at age x months, $0 \le x \le 14$.

a. According to the model, what is the head circumference at birth?

b. According to the model, what is the head circumference at 9 months?

c. According to the model, what is the head circumference at 14 months? Use a calculator and round to the nearest centimeter.

d. Use the values that you obtained in parts (a) through (c) and the graphs shown on the previous page to determine if the given model describes healthy children or severe autistics.

91. The bar graph shows the cumulative number of AIDS cases diagnosed in the United States from 1991 through 2001. The data can be modeled by an equation in the form

$$y = a\sqrt{x} + b,$$

where y is the cumulative number of AIDS cases, in thousands, x years after 1991. In this exercise, you will use the data from 1991 and 2000 to find values for a and b. This will give you an equation of a square root model for AIDS cases in the United States.

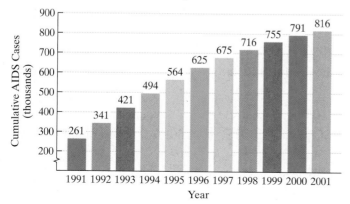

Cumulative Number of AIDS Cases Diagnosed in the U.S.

Source: Centers for Disease Control

a. In 1991, there were 261 thousand cumulative AIDS cases. Substitute 0 for x (1991 is 0 years after 1991) and 261 for y in $y = a\sqrt{x} + b$. Find the value for b and rewrite the model using this value.

b. In 2000, there were 791 thousand cumulative AIDS cases. Substitute 9 for x (2000 is 9 years after 1991) and 791 for y in your model from part (a). Find the value for a, rounded to the nearest whole number, and rewrite the model using this value.

c. According to your model from part (b), how many cumulative AIDS cases were there in 2001? Use a calculator and round to the nearest whole number, meaning the nearest thousands of cases. How well does the model describe the actual data number given for 2001 in the bar graph?

d. According to your model from part (b), how many cumulative AIDS cases will there be in the United States in 2007?

Writing in Mathematics

92. What are the square roots of 36? Explain why each of these numbers is a square root.

93. What does the symbol $\sqrt{}$ denote? Which of your answers in Exercise 92 is given by this symbol? Write the symbol needed to obtain the other answer.

94. Explain why $\sqrt{-1}$ is not a real number.

95. Explain why $\sqrt[3]{8}$ is 2. Then describe what is meant by $\sqrt[n]{a} = b$.

96. Explain the meaning of the words "radical," "radicand," and "index." Give an example with your explanation.

97. Describe the trend in the cumulative number of AIDS cases diagnosed in the United States shown in the bar graph for Exercise 91. Why is a square root model useful for describing the data?

98. Suppose that a motorist first lightly applies the brakes and then forcefully applies the brakes after a few seconds. Does the model given in Exercises 87–88 underestimate or overestimate the speed at which the motorist was traveling before braking? Explain your answer.

99. The formula $d = \sqrt{1.5h}$ models the distance, d, in miles, that can be seen to the horizon from a height h, in feet. Write a word problem using the formula. Then solve the problem.

Critical Thinking Exercises

100. Which one of the following is true?

a. $\sqrt{9} + \sqrt{16} = \sqrt{25}$

b. $\dfrac{\sqrt{64}}{2} = \sqrt{32}$

c. $\sqrt[3]{-27}$ is not a real number.

d. $\sqrt{\dfrac{1}{4}} + \sqrt{\dfrac{1}{9}} = \sqrt{\dfrac{25}{36}}$

In Exercises 101–103, simplify each expression.

101. $\sqrt{\sqrt[3]{64}}$

102. $\sqrt[3]{-\sqrt{1}}$

103. $\sqrt{\sqrt{16}} - \sqrt[3]{\sqrt{64}}$

104. Between which two consecutive integers is $-\sqrt{47}$?

Technology Exercises

105. Use a graphing utility to graph $y_1 = \sqrt{x}$, $y_2 = \sqrt{x + 4}$ and $y_3 = \sqrt{x - 3}$ in the same $[-5, 10, 1]$ by $[0, 6, 1]$ viewing rectangle. Describe one similarity and one difference that you observe among the graphs. Use the word "shift" in your response.

106. Use a graphing utility to graph $y = \sqrt{x}$, $y = \sqrt{x} + 4$, and $y = \sqrt{x} - 3$ in the same $[-1, 10, 1]$ by $[-10, 10, 1]$ viewing rectangle. Describe one similarity and one difference that you observe among the graphs.

Review Exercises

107. Graph: $4x - 5y = 20$. (Section 4.2, Example 4)

108. Solve and graph the solution on a number line: $2(x - 3) > 4x + 10$. (Section 2.6, Example 7)

109. Divide: $\dfrac{1}{x^2 - 17x + 30} \div \dfrac{1}{x^2 + 7x - 18}$. (Section 8.2, Example 6)

SECTION 9.2

Objectives

1 Multiply square roots.

2 Simplify square roots.

3 Use the quotient rule for square roots.

4 Use the product and quotient rules for other roots.

MULTIPLYING AND DIVIDING RADICALS

Hi, mom. I'm back from my two-year journey traveling in space. You've changed. Do you know that you're getting to look just like Norman Bates's mom? What's going on?

What is going on, indeed! In this section, in addition to learning to multiply and divide radicals, you will see how radicals model your return to a futuristic world from which friends and loved ones have long departed.

1 Multiply square roots.

The Product Rule for Square Roots A rule for multiplying square roots can be generalized by comparing $\sqrt{25} \cdot \sqrt{4}$ and $\sqrt{25 \cdot 4}$. Notice that

$$\sqrt{25} \cdot \sqrt{4} = 5 \cdot 2 = 10 \quad \text{and} \quad \sqrt{25 \cdot 4} = \sqrt{100} = 10.$$

Because we obtain 10 in both situations, the original radical expressions must be equal. That is,

$$\sqrt{25} \cdot \sqrt{4} = \sqrt{25 \cdot 4}.$$

This result is a special case of the **product rule for square roots** that can be generalized as follows:

> **THE PRODUCT RULE FOR SQUARE ROOTS** If a and b represent nonnegative real numbers, then
> $$\sqrt{ab} = \sqrt{a} \cdot \sqrt{b} \quad \text{and} \quad \sqrt{a} \cdot \sqrt{b} = \sqrt{ab}.$$
> The square root of a product is the product of the square roots.

EXAMPLE 1 Using the Product Rule to Multiply Square Roots

Use the product rule for square roots to find each product:

a. $\sqrt{2} \cdot \sqrt{5}$ **b.** $\sqrt{7x} \cdot \sqrt{11y}$ **c.** $\sqrt{7} \cdot \sqrt{7}$ **d.** $\sqrt{\dfrac{2}{5}} \cdot \sqrt{\dfrac{3}{7}}$.

SOLUTION

a. $\sqrt{2} \cdot \sqrt{5} = \sqrt{2 \cdot 5} = \sqrt{10}$
b. $\sqrt{7x} \cdot \sqrt{11y} = \sqrt{7x \cdot 11y} = \sqrt{77xy}$ Assume that $x \geq 0$ and $y \geq 0$.
c. $\sqrt{7} \cdot \sqrt{7} = \sqrt{7 \cdot 7} = \sqrt{49} = 7$
d. $\sqrt{\dfrac{2}{5}} \cdot \sqrt{\dfrac{3}{7}} = \sqrt{\dfrac{2}{5} \cdot \dfrac{3}{7}} = \sqrt{\dfrac{6}{35}}$

■

✔ **CHECK POINT 1** Use the product rule for square roots to find each product:

a. $\sqrt{3} \cdot \sqrt{10}$ **b.** $\sqrt{2x} \cdot \sqrt{13y}$

c. $\sqrt{5} \cdot \sqrt{5}$ **d.** $\sqrt{\dfrac{3}{2}} \cdot \sqrt{\dfrac{5}{11}}$.

 2 Simplify square roots.

Using the Product Rule to Simplify Square Roots We have seen that a number that is the square of a rational number is called a **perfect square**. For example, 100 is a perfect square because $100 = 10^2$. Thus, $\sqrt{100} = 10$.

A square root is **simplified** when its radicand has no factors other than 1 that are perfect squares. For example, $\sqrt{500}$ is not simplified because it can be expressed as $\sqrt{100 \cdot 5}$ and 100 is a perfect square. We can use the product rule in the form

$$\sqrt{ab} = \sqrt{a}\,\sqrt{b}$$

to simplify $\sqrt{500}$. We factor 500 so that one of its factors is the largest perfect square possible.

$$\sqrt{500} = \sqrt{100 \cdot 5}$$ Factor 500. 100 is the largest perfect square factor.

$$= \sqrt{100}\,\sqrt{5}$$ Use the product rule: $\sqrt{ab} = \sqrt{a}\,\sqrt{b}$.

$$= 10\sqrt{5}$$ Write $\sqrt{100}$ as 10. We read $10\sqrt{5}$ as "ten times the square root of 5."

STUDY TIP

When simplifying square roots, always look for the *greatest* perfect square factor possible. The following factorization will lead to further simplification:

$$\sqrt{500} = \sqrt{25 \cdot 20} = \sqrt{25}\sqrt{20} = 5\sqrt{20}.$$

25 is a perfect square factor of 500, but not the greatest perfect square factor.

Because 20 contains a perfect square factor, 4, the simplification is not complete.

$$5\sqrt{20} = 5\sqrt{4 \cdot 5} = 5\sqrt{4}\,\sqrt{5} = 5 \cdot 2\sqrt{5} = 10\sqrt{5}$$

Although the result checks with our simplification using $\sqrt{500} = \sqrt{100 \cdot 5}$, more work is required when the greatest perfect square factor is not used.

EXAMPLE 2 Using the Product Rule to Simplify Square Roots

Simplify:

a. $\sqrt{75}$ b. $\sqrt{80}$ c. $\sqrt{38}$.

SOLUTION In each case, try to factor the radicand as a product of the greatest perfect square factor and another factor.

a. $\sqrt{75} = \sqrt{25 \cdot 3}$ 25 is the greatest perfect square that is a factor of 75.

$\qquad = \sqrt{25} \sqrt{3}$ $\sqrt{ab} = \sqrt{a}\,\sqrt{b}$

$\qquad = 5\sqrt{3}$ Write $\sqrt{25}$ as 5.

b. $\sqrt{80} = \sqrt{16 \cdot 5}$ 16 is the greatest perfect square that is a factor of 80.

$\qquad = \sqrt{16} \sqrt{5}$ $\sqrt{ab} = \sqrt{a}\,\sqrt{b}$

$\qquad = 4\sqrt{5}$ Write $\sqrt{16}$ as 4.

c. Although the radicand, 38, can be factored as $2 \cdot 19$, neither of these factors is a perfect square. Because 38 has no perfect square factors other than 1, $\sqrt{38}$ cannot be simplified.

USING TECHNOLOGY

You can use a calculator to provide numerical support that your simplified answer is correct. For example, to support $\sqrt{75} = 5\sqrt{3}$, find decimal approximations for $\sqrt{75}$ and $5\sqrt{3}$. These approximations should be equal.

	$\sqrt{75}$:	$5\sqrt{3}$:
Scientific Calculator	75 $\sqrt{\ }$ $\boxed{\ }$	5×3 $\sqrt{\ }$ $\boxed{\ }$ $=$
Graphing Calculator	$\sqrt{\ }$ $\boxed{\ }$ 75 $\boxed{\text{ENTER}}$	5 $\sqrt{\ }$ $\boxed{\ }$ 3 $\boxed{\text{ENTER}}$

Correct to two decimal places, $\sqrt{75} \approx 8.66$ and $5\sqrt{3} \approx 8.66$. Use this technique to support the numerical results for the answers in this section. Caution: **A simplified square root does not mean a decimal approximation.**

 CHECK POINT 2 Simplify, if possible:

a. $\sqrt{12}$ b. $\sqrt{60}$ c. $\sqrt{55}$.

Simplifying Square Roots Containing Variables Square roots can also contain variables. Here are three examples:

$$\sqrt{x^2} \qquad \sqrt{x^4} \qquad \sqrt{x^6}.$$

We know that the square root of a negative number is not a real number. Thus, we want to avoid values for variables in the radicand that would make the radicand negative. To avoid negative radicands, we assume throughout this chapter that **if a variable appears in the radicand of a radical expression, it represents nonnegative numbers only**.

Do you see that we can simplify each of the three square root expressions just shown?

$$\sqrt{x^2} = x \quad \underset{(x)^2 = x^2}{\text{Because}} \qquad \sqrt{x^4} = x^2 \quad \underset{(x^2)^2 = x^4}{\text{Because}} \qquad \sqrt{x^6} = x^3 \quad \underset{(x^3)^2 = x^6}{\text{Because}}$$

Each simplified form is the variable in the radicand raised to one-half the power in the radicand.

To simplify square roots when the radicand contains x to an even power, we can use the following rule:

> **SIMPLIFYING SQUARE ROOTS WITH VARIABLES TO EVEN POWERS**
>
> $$\sqrt{x^{2n}} = x^n$$
>
> The square root of a variable raised to an even power equals the variable raised to one-half that power.

For example,

$$\sqrt{x^{10}} = x^5 \qquad \text{and} \qquad \sqrt{y^{34}} = y^{17}.$$

$\frac{1}{2} \cdot 10 = 5$ $\qquad$ $\frac{1}{2} \cdot 34 = 17$

ENRICHMENT ESSAY

A Radical Idea: Time Is Relative

What does travel in space have to do with radicals? Imagine that in the future we will be able to travel at velocities approaching the speed of light (approximately 186,000 miles per second). According to Einstein's theory of relativity, time would pass more quickly on earth than it would in the moving spaceship. The expression

$$R_f\sqrt{1 - \left(\frac{v}{c}\right)^2}$$

gives the aging rate of an astronaut relative to the aging rate, R_f, of a friend on earth. In the expression, v is the astronaut's speed and c is the speed of light. As the astronaut's speed approaches the speed of light, we can substitute c for v:

$$R_f\sqrt{1 - \left(\frac{v}{c}\right)^2} \quad \text{Let } v = c.$$

$$= R_f\sqrt{1 - 1^2}$$

$$= R_f\sqrt{0} = 0$$

Close to the speed of light, the astronaut's aging rate relative to a friend on earth is nearly 0. What does this mean? As we age here on earth, the space traveler would barely get older. The space traveler would return to a futuristic world in which friends and loved ones would be long dead.

EXAMPLE 3 **Simplifying Square Roots with Variables to Even Powers**

Simplify: $\sqrt{72x^{14}}$.

SOLUTION We write the radicand as the product of the greatest perfect square factor and another factor. Variables to even powers are part of the perfect square factor.

$$\sqrt{72x^{14}} = \sqrt{36x^{14} \cdot 2} \qquad 36x^{14} \text{ is the greatest perfect square factor of } 72x^{14}.$$

$$= \sqrt{36x^{14}}\sqrt{2} \qquad \text{Use the product rule: } \sqrt{ab} = \sqrt{a}\,\sqrt{b}.$$

$$= 6x^7\sqrt{2} \text{ or } 6\sqrt{2}x^7 \qquad \sqrt{36} = 6 \text{ and, using } \sqrt{x^{2n}} = x^n, \sqrt{x^{14}} = x^7. \blacksquare$$

✔ **CHECK POINT 3** Simplify: $\sqrt{40x^{16}}$.

How do we simplify the square root of a radicand containing a variable raised to an odd power, such as $\sqrt{x^5}$? Express the variable as the product of two factors, one of which has an even power. Then use the product rule to simplify. For example,

$$\sqrt{x^5} = \sqrt{x^4 \cdot x} = \sqrt{x^4}\sqrt{x} = x^2\sqrt{x}.$$

$\sqrt{x^{2n}} = x^n$

EXAMPLE 4 **Simplifying Square Roots with Variables to Odd Powers**

Simplify: $\sqrt{50x^7}$.

SOLUTION

$$\sqrt{50x^7} = \sqrt{25x^6 \cdot 2x} \qquad 25x^6 \text{ is the greatest perfect square factor of } 50x^7.$$

$$= \sqrt{25x^6}\sqrt{2x} \qquad \text{Use the product rule: } \sqrt{ab} = \sqrt{a}\,\sqrt{b}.$$

$$= 5x^3\sqrt{2x} \qquad \sqrt{25} = 5 \text{ and, using } \sqrt{x^{2n}} = x^n, \sqrt{x^6} = x^3. \blacksquare$$

✔ **CHECK POINT 4** Simplify: $\sqrt{27x^9}$.

Now we look at an example where we use the product rule to multiply radicals and then simplify.

EXAMPLE 5 Multiplying and Simplifying Square Roots

Multiply and then simplify:

$$\sqrt{6x^8} \cdot \sqrt{2x^3}.$$

SOLUTION

$$\sqrt{6x^8} \cdot \sqrt{2x^3} = \sqrt{6x^8 \cdot 2x^3}$$ Use the product rule: $\sqrt{a}\,\sqrt{b} = \sqrt{ab}$.

$$= \sqrt{12x^{11}}$$ Multiply in the radicand: Multiply coefficients and add exponents.

$$= \sqrt{4x^{10} \cdot 3x}$$ Simplify. $4x^{10}$ is the greatest perfect square factor of $12x^{11}$.

$$= \sqrt{4x^{10}}\sqrt{3x}$$ Use the product rule: $\sqrt{ab} = \sqrt{a}\,\sqrt{b}$.

$$= 2x^5\sqrt{3x}$$ $\sqrt{4} = 2$ and, using $\sqrt{x^{2n}} = x^n$, $\sqrt{x^{10}} = x^5$. ∎

 CHECK POINT 5 Multiply and then simplify:

$$\sqrt{15x^6} \cdot \sqrt{3x^7}.$$

---

3 Use the quotient rule for square roots.

The Quotient Rule for Square Roots A rule for dividing square roots can be generalized by comparing

$$\sqrt{\frac{64}{4}} \quad \text{and} \quad \frac{\sqrt{64}}{\sqrt{4}}.$$

Note that

$$\sqrt{\frac{64}{4}} = \sqrt{16} = 4 \quad \text{and} \quad \frac{\sqrt{64}}{\sqrt{4}} = \frac{8}{2} = 4.$$

Because we obtain 4 in both situations, the original radical expressions must be equal:

$$\sqrt{\frac{64}{4}} = \frac{\sqrt{64}}{\sqrt{4}}.$$

This result is a special case of the **quotient rule for square roots** that can be generalized as follows:

THE QUOTIENT RULE FOR SQUARE ROOTS If a and b represent nonnegative real numbers and $b \neq 0$, then

$$\frac{\sqrt{a}}{\sqrt{b}} = \sqrt{\frac{a}{b}} \quad \text{and} \quad \sqrt{\frac{a}{b}} = \frac{\sqrt{a}}{\sqrt{b}}.$$

The square root of a quotient is the quotient of the square roots.

EXAMPLE 6 Using the Quotient Rule to Simplify Square Roots

Simplify:

a. $\sqrt{\dfrac{100}{9}}$ **b.** $\dfrac{\sqrt{48x^3}}{\sqrt{6x}}$.

SOLUTION

a. $\sqrt{\dfrac{100}{9}} = \dfrac{\sqrt{100}}{\sqrt{9}} = \dfrac{10}{3}$

b. $\dfrac{\sqrt{48x^3}}{\sqrt{6x}} = \sqrt{\dfrac{48x^3}{6x}} = \sqrt{8x^2} = \sqrt{4x^2}\sqrt{2} = 2x\sqrt{2}$ ∎

 CHECK POINT 6 Simplify:

a. $\sqrt{\dfrac{49}{25}}$

b. $\dfrac{\sqrt{48x^5}}{\sqrt{3x}}$.

---

4 Use the product and quotient rules for other roots.

The Product and Quotient Rules for Other Roots The product and quotient rules apply to cube roots, fourth roots, and all higher roots.

> **THE PRODUCT AND QUOTIENT RULES FOR nTH ROOTS** For all real numbers, where the indicated roots represent real numbers,
>
> $$\sqrt[n]{a} \cdot \sqrt[n]{b} = \sqrt[n]{ab} \quad \text{and} \quad \dfrac{\sqrt[n]{a}}{\sqrt[n]{b}} = \sqrt[n]{\dfrac{a}{b}}, b \neq 0.$$

EXAMPLE 7 Simplifying, Multiplying, and Dividing Higher Roots

Simplify:

a. $\sqrt[3]{24}$ b. $\sqrt[4]{8} \cdot \sqrt[4]{4}$ c. $\sqrt[4]{\dfrac{81}{16}}$.

SOLUTION

a. $\sqrt[3]{24} = \sqrt[3]{8 \cdot 3}$ *Find the greatest perfect cube that is a factor of 24. $2^3 = 8$, so 8 is a perfect cube and is the greatest perfect cube factor of 24.*

$= \sqrt[3]{8} \cdot \sqrt[3]{3}$ $\sqrt[n]{ab} = \sqrt[n]{a}\,\sqrt[n]{b}$

$= 2\sqrt[3]{3}$ $\sqrt[3]{8} = 2$

b. $\sqrt[4]{8} \cdot \sqrt[4]{4} = \sqrt[4]{8 \cdot 4}$ $\sqrt[n]{a} \cdot \sqrt[n]{b} = \sqrt[n]{ab}$

$= \sqrt[4]{32}$ *Find the greatest perfect fourth power that is a factor of 32.*

$= \sqrt[4]{16 \cdot 2}$ $2^4 = 16$, *so 16 is a perfect fourth power and is the largest perfect fourth power that is a factor of 32.*

$= \sqrt[4]{16} \cdot \sqrt[4]{2}$ $\sqrt[n]{ab} = \sqrt[n]{a} \cdot \sqrt[n]{b}$

$= 2\sqrt[4]{2}$ $\sqrt[4]{16} = 2$

c. $\sqrt[4]{\dfrac{81}{16}} = \dfrac{\sqrt[4]{81}}{\sqrt[4]{16}}$ $\sqrt[n]{\dfrac{a}{b}} = \dfrac{\sqrt[n]{a}}{\sqrt[n]{b}}$

$= \dfrac{3}{2}$ $\sqrt[4]{81} = 3$ *because* $3^4 = 81$, *and* $\sqrt[4]{16} = 2$ *because* $2^4 = 16$. ∎

 CHECK POINT 7 Simplify:

a. $\sqrt[3]{40}$ b. $\sqrt[5]{8} \cdot \sqrt[5]{8}$ c. $\sqrt[3]{\dfrac{125}{27}}$.

9.2 EXERCISE SET

    Math XL MyMathLab

Student Solutions Manual   CD/Video   PH Math/Tutor Center   MathXL Tutorials on CD   MathXL®   MyMathLab   Interactmath.com

Practice Exercises

Remember that throughout this chapter, variable expressions in radicands represent nonnegative real numbers.

In Exercises 1–14, use the product rule for square roots to find each product.

1. $\sqrt{2} \cdot \sqrt{7}$

2. $\sqrt{2} \cdot \sqrt{17}$

3. $\sqrt{3x} \cdot \sqrt{5y}$

4. $\sqrt{7x} \cdot \sqrt{11y}$

5. $\sqrt{5} \cdot \sqrt{5}$

6. $\sqrt{10} \cdot \sqrt{10}$

7. $\sqrt{\dfrac{2}{3}} \cdot \sqrt{\dfrac{5}{7}}$

8. $\sqrt{\dfrac{3}{5}} \cdot \sqrt{\dfrac{5}{7}}$

9. $\sqrt{0.1x} \cdot \sqrt{5y}$

10. $\sqrt{0.2x} \cdot \sqrt{3y}$

11. $\sqrt{\dfrac{1}{5}a} \cdot \sqrt{\dfrac{1}{5}b}$

12. $\sqrt{\dfrac{1}{7}a} \cdot \sqrt{\dfrac{1}{7}b}$

13. $\sqrt{\dfrac{2x}{9}} \cdot \sqrt{\dfrac{9}{2}}$

14. $\sqrt{\dfrac{5x}{11}} \cdot \sqrt{\dfrac{11}{5}}$

In Exercises 15–54, simplify each expression. If the expression cannot be simplified, so state.

15. $\sqrt{50}$

16. $\sqrt{27}$

17. $\sqrt{45}$

18. $\sqrt{28}$

19. $\sqrt{200}$

20. $\sqrt{300}$

21. $\sqrt{75x}$

22. $\sqrt{40x}$

23. $\sqrt{9x}$

24. $\sqrt{25x}$

25. $\sqrt{35}$

26. $\sqrt{22}$

27. $\sqrt{y^2}$

28. $\sqrt{z^2}$

29. $\sqrt{64x^2}$

30. $\sqrt{36x^2}$

31. $\sqrt{11x^2}$

32. $\sqrt{17x^2}$

33. $\sqrt{8x^2}$

34. $\sqrt{12x^2}$

35. $\sqrt{x^{20}}$

36. $\sqrt{x^{30}}$

37. $\sqrt{25y^{10}}$

38. $\sqrt{36y^{10}}$

39. $\sqrt{20x^6}$

40. $\sqrt{24x^8}$

41. $\sqrt{72y^{100}}$

42. $\sqrt{32y^{200}}$

43. $\sqrt{x^3}$

44. $\sqrt{y^3}$

45. $\sqrt{x^7}$

46. $\sqrt{x^9}$

47. $\sqrt{y^{17}}$

48. $\sqrt{y^{19}}$

49. $\sqrt{25x^5}$

50. $\sqrt{49x^5}$

51. $\sqrt{8x^{17}}$

52. $\sqrt{12x^7}$

53. $\sqrt{90y^{19}}$

54. $\sqrt{600y^{23}}$

In Exercises 55–68, multiply and, if possible, simplify.

55. $\sqrt{3} \cdot \sqrt{15}$

56. $\sqrt{3} \cdot \sqrt{6}$

57. $\sqrt{5x} \cdot \sqrt{10y}$

58. $\sqrt{8x} \cdot \sqrt{10y}$

59. $\sqrt{12x} \cdot \sqrt{3x}$

60. $\sqrt{20x} \cdot \sqrt{5x}$

61. $\sqrt{15x^2} \cdot \sqrt{3x}$

62. $\sqrt{2x^2} \cdot \sqrt{10x}$

63. $\sqrt{15x^4} \cdot \sqrt{5x^9}$

64. $\sqrt{50x^9} \cdot \sqrt{4x^4}$

65. $\sqrt{7x} \cdot \sqrt{3y}$

66. $\sqrt{5x} \cdot \sqrt{11y}$

67. $\sqrt{50xy} \cdot \sqrt{4xy^2}$

68. $\sqrt{5xy} \cdot \sqrt{10xy^2}$

In Exercises 69–92, simplify using the quotient rule for square roots.

69. $\sqrt{\dfrac{49}{16}}$

70. $\sqrt{\dfrac{100}{9}}$

71. $\sqrt{\dfrac{3}{4}}$

72. $\sqrt{\dfrac{7}{25}}$

73. $\sqrt{\dfrac{x^2}{36}}$

74. $\sqrt{\dfrac{x^2}{49}}$

75. $\sqrt{\dfrac{7}{x^4}}$

76. $\sqrt{\dfrac{13}{x^6}}$

77. $\sqrt{\dfrac{72}{y^{20}}}$

78. $\sqrt{\dfrac{300}{y^{30}}}$

79. $\dfrac{\sqrt{54}}{\sqrt{6}}$

80. $\dfrac{\sqrt{75}}{\sqrt{3}}$

81. $\dfrac{\sqrt{24}}{\sqrt{3}}$

82. $\dfrac{\sqrt{60}}{\sqrt{5}}$

83. $\dfrac{\sqrt{75}}{\sqrt{15}}$

84. $\dfrac{\sqrt{21}}{\sqrt{3}}$

85. $\dfrac{\sqrt{48x}}{\sqrt{3x}}$

86. $\dfrac{\sqrt{27x}}{\sqrt{3x}}$

87. $\dfrac{\sqrt{32x^3}}{\sqrt{8x}}$

88. $\dfrac{\sqrt{75x^3}}{\sqrt{3x}}$

89. $\dfrac{\sqrt{150x^4}}{\sqrt{3x}}$

90. $\dfrac{\sqrt{400x^4}}{\sqrt{2x}}$

91. $\dfrac{\sqrt{400x^{10}}}{\sqrt{10x^3}}$

92. $\dfrac{\sqrt{800x^{12}}}{\sqrt{10x^3}}$

In Exercises 93–108, simplify each radical expression.

93. $\sqrt[3]{16}$

94. $\sqrt[3]{32}$

95. $\sqrt[3]{54}$

96. $\sqrt[3]{250}$

97. $\sqrt[4]{32}$

98. $\sqrt[4]{48}$

99. $\sqrt[3]{4} \cdot \sqrt[3]{2}$

100. $\sqrt[3]{3} \cdot \sqrt[3]{9}$

101. $\sqrt[3]{9} \cdot \sqrt[3]{6}$

102. $\sqrt[3]{12} \cdot \sqrt[3]{4}$

103. $\sqrt[4]{4} \cdot \sqrt[4]{8}$

104. $\sqrt[5]{16} \cdot \sqrt[5]{4}$

105. $\sqrt[3]{\dfrac{27}{8}}$

106. $\sqrt[3]{\dfrac{125}{8}}$

107. $\sqrt[3]{\dfrac{3}{8}}$

108. $\sqrt[3]{\dfrac{5}{27}}$

Practice Plus

In Exercises 109–114, simplify each expression.

109. $\sqrt{90(x+4)^3}$

110. $\sqrt{150(x+8)^3}$

111. $\sqrt{x^2 - 6x + 9}$

112. $\sqrt{x^2 - 10x + 25}$

113. $\sqrt{2^{43}x^{104}y^{13}}$

114. $\sqrt{3^{41}x^{102}y^{17}}$

In Exercises 115–116, use the fact that $\sqrt[3]{x^3} = x$ (why?) to simplify each expression.

115. $\sqrt[3]{24x^5}$

116. $\sqrt[3]{16x^4}$

Application Exercises

117. The algebraic expression $2\sqrt{5L}$ is used to estimate the speed of a car, in miles per hour, prior to an accident based on the length of its skid marks L, in feet. Find the speed of a car that left skid marks 40 feet long and write the answer in simplified radical form.

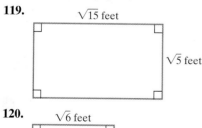

118. The time, in seconds, that it takes an object to fall a distance d, in feet, is given by the algebraic expression $\sqrt{\dfrac{d}{16}}$. Find how long it will take a ball dropped from the top of a building 320 feet tall to hit the ground. Write the answer in simplified radical form.

In Exercises 119–120, express the area of each rectangle as a square root in simplified form.

119.

$\sqrt{15}$ feet

$\sqrt{5}$ feet

120.

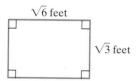

$\sqrt{6}$ feet

$\sqrt{3}$ feet

Writing in Mathematics

121. Use words to state the product rule for square roots. Give an example with your description.

122. Explain why $\sqrt{50}$ is not simplified. What do we mean when we say that a square root is simplified?

123. Explain how to simplify square roots with variables to even powers.

124. Explain how to simplify square roots with variables to odd powers.

125. Explain how to simplify $\sqrt{10} \cdot \sqrt{5}$.

126. Use words to state the quotient rule for square roots. Give an example with your description.

127. Read the essay on page 543. The future is now: You have the opportunity to explore the cosmos in a starship traveling near the speed of light. The experience will enable you to understand the mysteries of the universe in deeply personal ways, transporting you to unimagined levels of knowing and being. The down side: You return from your two-year journey to a futuristic world in which friends and loved ones are long dead. Do you explore space or stay here on earth? What are the reasons for your choice?

Critical Thinking Exercises

128. Which one of the following is true?

   a. $\sqrt{20} = 4\sqrt{5}$

   b. If $y \geq 0$, $\sqrt{y^9} = y^3$.

   c. $\sqrt{2x}\,\sqrt{6y} = 2\sqrt{3xy}$ if x and y are nonnegative real numbers.

   d. $\sqrt{2} \cdot \sqrt{8} = 16$

In Exercises 129–130, fill in the missing coefficients and exponents to make each statement true.

129. $\sqrt{\boxed{}\,x^{\boxed{}}} = 5x^7$

130. $\sqrt{15x^{\boxed{}}} \cdot \sqrt{\boxed{}\,x^5} = 3x^3\sqrt{10x}$

Technology Exercises

131. Use a calculator to provide numerical support for your simplifications in Exercises 15–20. In each case, find a decimal approximation for the given expression. Then find a decimal approximation for your simplified expression. The approximations should be the same.

In Exercises 132–134, determine if each simplification is correct by graphing each side of the equation with your graphing utility. Use the given viewing rectangle. The graphs should be the same. If they are not, correct the right side of the equation and then use your graphing utility to verify the simplification.

132. $\sqrt{x^4} = x^2$ $[0, 5, 1]$ by $[0, 20, 1]$

133. $\sqrt{8x^2} = 4x\sqrt{2}$ $[0, 5, 1]$ by $[0, 20, 1]$

134. $\sqrt{x^3} = x\sqrt{x}$ $[0, 5, 1]$ by $[0, 10, 1]$

Review Exercises

135. Solve the system:
$$4x + 3y = 18$$
$$5x - 9y = 48.$$

(Section 5.3, Example 3)

136. Subtract: $\dfrac{6x}{x^2 - 4} - \dfrac{3}{x + 2}$. (Section 8.4, Example 7)

137. Factor completely: $2x^3 - 16x^2 + 30x$. (Section 7.5, Example 2)

SECTION 9.3

Objectives

1 Add and subtract radicals.

2 Multiply radical expressions with more than one term.

3 Multiply conjugates.

OPERATIONS WITH RADICALS

I need to call the Weather Channel to let them know I just shoveled three feet of "partly cloudy" off my driveway. What can radicals tell me about how I'm perceiving the temperature on this cold day? In this section, we provide an answer to this question as we focus our attention on adding, subtracting, and multiplying radicals.

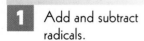

1 Add and subtract radicals.

Adding and Subtracting Like Radicals Two or more square roots can be combined using the distributive property provided that they have the same radicand. Such radicals are called **like radicals**. For example,

$$7\sqrt{11} + 6\sqrt{11} = (7 + 6)\sqrt{11} = 13\sqrt{11}.$$

7 square roots of 11 plus 6 square roots of 11 result in 13 square roots of 11.

EXAMPLE 1 Adding and Subtracting Like Radicals

Add or subtract as indicated:

a. $7\sqrt{2} + 5\sqrt{2}$ **b.** $\sqrt{5x} - 7\sqrt{5x}$.

SOLUTION

a. $7\sqrt{2} + 5\sqrt{2} = (7 + 5)\sqrt{2}$ *Apply the distributive property.*

$\qquad\qquad\quad = 12\sqrt{2}$ *Simplify.*

b. $\sqrt{5x} - 7\sqrt{5x} = 1\sqrt{5x} - 7\sqrt{5x}$ *Write $\sqrt{5x}$ as $1\sqrt{5x}$.*

$\qquad\qquad\quad = (1 - 7)\sqrt{5x}$ *Apply the distributive property.*

$\qquad\qquad\quad = -6\sqrt{5x}$ *Simplify.*

✔ **CHECK POINT 1** Add or subtract as indicated:

a. $8\sqrt{13} + 9\sqrt{13}$ **b.** $\sqrt{17x} - 20\sqrt{17x}$.

In some cases, radicals can be combined once they have been simplified. For example, to add $\sqrt{2}$ and $\sqrt{8}$, we can write $\sqrt{8}$ as $\sqrt{4 \cdot 2}$ because 4 is a perfect square factor of 8.

$$\sqrt{2} + \sqrt{8} = \sqrt{2} + \sqrt{4 \cdot 2} = 1\sqrt{2} + 2\sqrt{2} = (1 + 2)\sqrt{2} = 3\sqrt{2}$$

EXAMPLE 2 Combining Radicals That First Require Simplification

Add or subtract as indicated:

a. $7\sqrt{3} + \sqrt{12}$ **b.** $4\sqrt{50x} - 6\sqrt{32x}$.

SOLUTION

a. $7\sqrt{3} + \sqrt{12}$

$= 7\sqrt{3} + \sqrt{4 \cdot 3}$ Split 12 into two factors such that one is a perfect square.

$= 7\sqrt{3} + 2\sqrt{3}$ $\sqrt{4 \cdot 3} = \sqrt{4}\sqrt{3} = 2\sqrt{3}$

$= (7 + 2)\sqrt{3}$ Apply the distributive property. You will find that this step is usually done mentally.

$= 9\sqrt{3}$ Simplify.

b. $4\sqrt{50x} - 6\sqrt{32x}$

$= 4\sqrt{25 \cdot 2x} - 6\sqrt{16 \cdot 2x}$ 25 is the greatest perfect square factor of 50x and 16 is the greatest perfect square factor of 32x.

$= 4 \cdot 5\sqrt{2x} - 6 \cdot 4\sqrt{2x}$ $\sqrt{25 \cdot 2x} = \sqrt{25}\sqrt{2x} = 5\sqrt{2x}$ and $\sqrt{16 \cdot 2x} = \sqrt{16}\sqrt{2x} = 4\sqrt{2x}$.

$= 20\sqrt{2x} - 24\sqrt{2x}$ Multiply: $4 \cdot 5 = 20$ and $6 \cdot 4 = 24$.

$= (20 - 24)\sqrt{2x}$ Apply the distributive property.

$= -4\sqrt{2x}$ Simplify.   ∎

✔ **CHECK POINT 2** Add or subtract as indicated:

a. $5\sqrt{27} + \sqrt{12}$ **b.** $6\sqrt{18x} - 4\sqrt{8x}$.

2 Multiply radical expressions with more than one term.

Multiplying Radical Expressions with More Than One Term Radical expressions with more than one term are multiplied in much the same way that polynomials with more than one term are multiplied. Example 3 uses the distributive property and the FOIL method to perform multiplications.

EXAMPLE 3 Multiplying Radicals

Multiply:

a. $\sqrt{3}(\sqrt{7} + \sqrt{5})$ **b.** $(3 + \sqrt{2})(5 + \sqrt{2})$ **c.** $(5 + \sqrt{7})(6 - 3\sqrt{7})$.

SOLUTION

a. $\sqrt{3}(\sqrt{7} + \sqrt{5}) = \sqrt{3} \cdot \sqrt{7} + \sqrt{3} \cdot \sqrt{5}$ Use the distributive property.

$= \sqrt{21} + \sqrt{15}$ Use the product rule: $\sqrt{a}\sqrt{b} = \sqrt{ab}$.

b.

$$(3 + \sqrt{2})(5 + \sqrt{2}) = 3 \cdot 5 + 3 \cdot \sqrt{2} + 5 \cdot \sqrt{2} + \sqrt{2} \cdot \sqrt{2}$$

$$= 15 + 3\sqrt{2} + 5\sqrt{2} + 2 \qquad \text{Multiply.}$$

$$= (15 + 2) + (3\sqrt{2} + 5\sqrt{2}) \qquad \text{Group terms to be added. This step can be done mentally.}$$

$$= 17 + 8\sqrt{2} \qquad \text{Add like terms.}$$

c.

$$(5 + \sqrt{7})(6 - 3\sqrt{7}) = 5 \cdot 6 + 5(-3\sqrt{7}) + 6 \cdot \sqrt{7} + (\sqrt{7})(-3\sqrt{7})$$

$$= 30 - 15\sqrt{7} + 6\sqrt{7} - 21 \qquad \text{Multiply. For L, or Last terms:} \\ (\sqrt{7})(-3\sqrt{7}) = -3\sqrt{49} = -3 \cdot 7 = -21.$$

$$= (30 - 21) + (-15\sqrt{7} + 6\sqrt{7}) \qquad \text{Group like terms.}$$

$$= 9 - 9\sqrt{7} \qquad \text{Combine like terms.} \blacksquare$$

✔ **CHECK POINT 3** Multiply:

a. $\sqrt{2}(\sqrt{5} + \sqrt{11})$
b. $(4 + \sqrt{3})(2 + \sqrt{3})$
c. $(3 + \sqrt{5})(8 - 4\sqrt{5})$.

3 Multiply conjugates.

Radical expressions that involve the sum and difference of the same two terms are called **conjugates**. For example,

$$5 + \sqrt{2} \quad \text{and} \quad 5 - \sqrt{2}$$

are conjugates of each other. Although you can multiply these conjugates using the FOIL method, there is an even faster way. Use the special-product formula for the sum and difference of the same two terms:

$$(A + B)(A - B) = A^2 - B^2.$$

Thus,

$$(A + B)(A - B) = A^2 - B^2$$

$$(5 + \sqrt{2})(5 - \sqrt{2}) = 5^2 - (\sqrt{2})^2 = 25 - 2 = 23.$$

EXAMPLE 4 Multiplying Conjugates

Multiply:

a. $(2 + \sqrt{7})(2 - \sqrt{7})$ **b.** $(\sqrt{5} - \sqrt{3})(\sqrt{5} + \sqrt{3})$.

SOLUTION Use the special-product formula shown.

$$(A + B)(A - B) = A^2 - B^2$$

First term squared − second term squared = product

a. $(2 + \sqrt{7})(2 - \sqrt{7}) = 2^2 - (\sqrt{7})^2 = 4 - 7 = -3$

b. $(\sqrt{5} - \sqrt{3})(\sqrt{5} + \sqrt{3}) = (\sqrt{5})^2 - (\sqrt{3})^2 = 5 - 3 = 2$

In the next section, we will use conjugates to simplify quotients.

 CHECK POINT 4 Multiply:

a. $\left(3 + \sqrt{11}\right)\left(3 - \sqrt{11}\right)$ **b.** $\left(\sqrt{7} - \sqrt{2}\right)\left(\sqrt{7} + \sqrt{2}\right)$.

ENRICHMENT ESSAY

Radicals and Windchill

The way that we perceive the temperature on a cold day depends on both air temperature and wind speed. The windchill temperature is what the air temperature would have to be with no wind to achieve the same chilling effect on the skin. The formula that describes windchill temperature, W, in terms of the velocity of the wind, v, in miles per hour, and the actual air temperature, t, in degrees Fahrenheit, is

$$W = 35.74 + 0.6215t - 35.75 \sqrt[25]{v^4} + 0.4275t \sqrt[25]{v^4}.$$

Use your calculator to describe how cold the air temperature feels (that is, the windchill temperature) when the temperature is 15° Fahrenheit and the wind is 5 miles per hour. Contrast this with a temperature of 40° Fahrenheit and a wind blowing at 50 miles per hour.

9.3 EXERCISE SET

Student Solutions Manual | CD/Video | PH Math/Tutor Center | MathXL Tutorials on CD | MathXL® | MyMathLab | Interactmath.com

Practice Exercises

In Exercises 1–22, add or subtract as indicated. If terms are not like radicals and cannot be combined, so state.

1. $8\sqrt{3} + 5\sqrt{3}$ **2.** $9\sqrt{5} + 6\sqrt{5}$

3. $17\sqrt{6} - 2\sqrt{6}$ **4.** $19\sqrt{7} - 2\sqrt{7}$

5. $3\sqrt{13} - 8\sqrt{13}$ **6.** $5\sqrt{17} - 11\sqrt{17}$

7. $12\sqrt{x} + 3\sqrt{x}$ **8.** $9\sqrt{x} + 2\sqrt{x}$

9. $70\sqrt{y} - 76\sqrt{y}$ **10.** $8\sqrt{y} - 28\sqrt{y}$

11. $7\sqrt{10x} + 2\sqrt{10x}$ **12.** $4\sqrt{6x} + 3\sqrt{6x}$

13. $7\sqrt{5y} - \sqrt{5y}$ **14.** $8\sqrt{3y} - \sqrt{3y}$

15. $\sqrt{5} + \sqrt{5}$ **16.** $\sqrt{3} + \sqrt{3}$

17. $4\sqrt{2} + 3\sqrt{2} + 5\sqrt{2}$

18. $6\sqrt{3} + 2\sqrt{3} + 5\sqrt{3}$

19. $4\sqrt{7} - 5\sqrt{7} + 8\sqrt{7}$

20. $5\sqrt{7} - 6\sqrt{7} + 10\sqrt{7}$

21. $4\sqrt{11} - 6\sqrt{11} + 2\sqrt{11}$

22. $7\sqrt{17} - 10\sqrt{17} + 3\sqrt{17}$

In Exercises 23–42, add or subtract as indicated. You will need to simplify terms before they can be combined. If terms cannot be simplified so that they can be combined, so state.

23. $\sqrt{5} + \sqrt{20}$ **24.** $\sqrt{3} + \sqrt{27}$

25. $\sqrt{8} - \sqrt{2}$ **26.** $\sqrt{50} - \sqrt{2}$

27. $\sqrt{50} + \sqrt{18}$ **28.** $\sqrt{28} + \sqrt{63}$

29. $7\sqrt{12} + \sqrt{75}$ **30.** $5\sqrt{12} + \sqrt{75}$

31. $3\sqrt{27} - 2\sqrt{18}$

32. $5\sqrt{27} - 3\sqrt{18}$

33. $2\sqrt{45x} - 2\sqrt{20x}$

34. $2\sqrt{50x} - 2\sqrt{18x}$

35. $\sqrt{8} + \sqrt{16} + \sqrt{18} + \sqrt{25}$

36. $\sqrt{6} + \sqrt{9} + \sqrt{24} + \sqrt{25}$

37. $\sqrt{2} + \sqrt{11}$

38. $\sqrt{3} + \sqrt{13}$

39. $2\sqrt{80} + 3\sqrt{75}$

40. $2\sqrt{75} + 3\sqrt{125}$

41. $3\sqrt{54} - 2\sqrt{20} + 4\sqrt{45} - \sqrt{24}$

42. $4\sqrt{8} - \sqrt{128} + 2\sqrt{48} + 3\sqrt{18}$

In Exercises 43–78, multiply as indicated. If possible, simplify any square roots that appear in the product.

43. $\sqrt{2}\left(\sqrt{3} + \sqrt{5}\right)$

44. $\sqrt{5}\left(\sqrt{3} + \sqrt{6}\right)$

45. $\sqrt{7}\left(\sqrt{6} - \sqrt{10}\right)$

46. $\sqrt{7}\left(\sqrt{5} - \sqrt{11}\right)$

47. $\sqrt{3}\left(5 + \sqrt{3}\right)$

48. $\sqrt{6}\left(7 + \sqrt{6}\right)$

49. $\sqrt{3}\left(\sqrt{6} - \sqrt{3}\right)$

50. $\sqrt{6}\left(\sqrt{6} - \sqrt{2}\right)$

51. $\left(5 + \sqrt{2}\right)\left(6 + \sqrt{2}\right)$

52. $\left(7 + \sqrt{2}\right)\left(8 + \sqrt{2}\right)$

53. $\left(4 + \sqrt{5}\right)\left(10 - 3\sqrt{5}\right)$

54. $\left(6 + \sqrt{5}\right)\left(9 - 4\sqrt{5}\right)$

55. $\left(6 - 3\sqrt{7}\right)\left(2 - 5\sqrt{7}\right)$

56. $\left(7 - 2\sqrt{7}\right)\left(5 - 3\sqrt{7}\right)$

57. $\left(\sqrt{10} - 3\right)\left(\sqrt{10} - 5\right)$

58. $\left(\sqrt{10} - 4\right)\left(\sqrt{10} - 6\right)$

59. $\left(\sqrt{3} + \sqrt{6}\right)\left(\sqrt{3} + 2\sqrt{6}\right)$

60. $\left(\sqrt{6} + \sqrt{3}\right)\left(\sqrt{6} + 5\sqrt{3}\right)$

61. $\left(\sqrt{2} + 1\right)\left(\sqrt{3} - 6\right)$

62. $\left(\sqrt{5} + 3\right)\left(\sqrt{2} - 8\right)$

63. $\left(3 + \sqrt{5}\right)\left(3 - \sqrt{5}\right)$

64. $\left(4 + \sqrt{7}\right)\left(4 - \sqrt{7}\right)$

65. $\left(1 - \sqrt{6}\right)\left(1 + \sqrt{6}\right)$

66. $\left(1 - \sqrt{5}\right)\left(1 + \sqrt{5}\right)$

67. $\left(\sqrt{11} + 5\right)\left(\sqrt{11} - 5\right)$

68. $\left(\sqrt{11} + 6\right)\left(\sqrt{11} - 6\right)$

69. $\left(\sqrt{7} - \sqrt{5}\right)\left(\sqrt{7} + \sqrt{5}\right)$

70. $\left(\sqrt{10} - \sqrt{7}\right)\left(\sqrt{10} + \sqrt{7}\right)$

71. $\left(2\sqrt{3} + 7\right)\left(2\sqrt{3} - 7\right)$

72. $\left(5\sqrt{3} + 6\right)\left(5\sqrt{3} - 6\right)$

73. $\left(2\sqrt{3} + \sqrt{5}\right)\left(2\sqrt{3} - \sqrt{5}\right)$

74. $\left(4\sqrt{5} + \sqrt{2}\right)\left(4\sqrt{5} - \sqrt{2}\right)$

75. $\left(\sqrt{2} + \sqrt{3}\right)^2$

76. $\left(\sqrt{3} + \sqrt{5}\right)^2$

77. $\left(\sqrt{x} - \sqrt{10}\right)^2$

78. $\left(\sqrt{x} - \sqrt{11}\right)^2$

Practice Plus

In Exercises 79–86, add or subtract as indicated. You will need to simplify terms before they can be combined. Assume all variables represent nonnegative real numbers.

79. $5\sqrt{27x^3} - 3x\sqrt{12x}$

80. $7\sqrt{32x^3} - 3x\sqrt{50x}$

81. $6y^2\sqrt{x^5y} + 2x^2\sqrt{xy^5}$

82. $9y^2\sqrt{x^5y} + x^2\sqrt{xy^5}$

83. $3\sqrt[3]{54} - 4\sqrt[3]{16}$

84. $7\sqrt[3]{24} - 5\sqrt[3]{81}$

85. $x\sqrt[3]{32x} + 9\sqrt[3]{4x^4}$ Hint: $\sqrt[3]{x^3} = x$

86. $x\sqrt[3]{48x} + 11\sqrt[3]{6x^4}$ Hint: $\sqrt[3]{x^3} = x$

Application Exercises

In Exercises 87–92, write expressions for the perimeter and area of each figure. Then simplify these expressions. Assume that all measures are given in inches.

87.

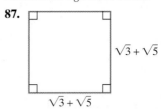

$\sqrt{3} + \sqrt{5}$

$\sqrt{3} + \sqrt{5}$

88.

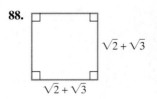

$\sqrt{2} + \sqrt{3}$

$\sqrt{2} + \sqrt{3}$

89.

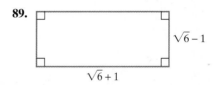

$\sqrt{6} - 1$

$\sqrt{6} + 1$

90.

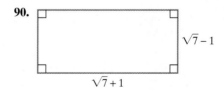

$\sqrt{7} - 1$

$\sqrt{7} + 1$

91.

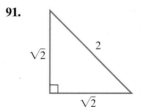

$\sqrt{2}$ 2

$\sqrt{2}$

92.

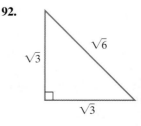

$\sqrt{3}$ $\sqrt{6}$

$\sqrt{3}$

There is a formula for adding $\sqrt{a}$ *and* $\sqrt{b}$*. The formula is* $\sqrt{a} + \sqrt{b} = \sqrt{(a+b) + 2\sqrt{ab}}$*. Use this formula to add the radicals in Exercises 93–94. Then work the problem again by the methods discussed in this section. Which method do you prefer? Why?*

93. $\sqrt{2} + \sqrt{8}$

94. $\sqrt{5} + \sqrt{20}$

Writing in Mathematics

95. What are like radicals? Give an example with your explanation.

96. Explain how to add like radicals. Give an example with your explanation.

97. If only like radicals can be added, why is it possible to add $\sqrt{2}$ and $\sqrt{8}$?

98. Explain how to perform this multiplication: $\sqrt{2}(\sqrt{7} + \sqrt{10})$.

99. Explain how to perform this multiplication: $(2 + \sqrt{3})(4 + \sqrt{3})$.

100. What are conjugates? Give an example with your explanation.

101. Describe how to multiply conjugates.

Critical Thinking Exercises

102. Which one of the following is true?

a. $\sqrt{2} + \sqrt{8} = \sqrt{10}$ **b.** $(\sqrt{5} + \sqrt{3})^2 = 5 + 3$

c. $4\sqrt{3} + 5\sqrt{3} = 9\sqrt{6}$ **d.** None of the above is true.

103. Simplify: $\sqrt{5} \cdot \sqrt{15} + 6\sqrt{3}$.

104. Multiply: $(\sqrt[3]{4} + 1)(\sqrt[3]{2} - 3)$.

105. Fill in the boxes to make the statement true:
$(5 + \sqrt{})(5 - \sqrt{}) = 22$.

106. Multiply: $(4\sqrt{3x} + \sqrt{2y})(4\sqrt{3x} - \sqrt{2y})$.

Technology Exercises

In Exercises 107–110, determine if each operation is performed correctly by graphing each side of the equation with your graphing utility. Use the given viewing rectangle. The graphs should be the same. If they are not, correct the right side of the equation and then use your graphing utility to verify the correction.

107. $\sqrt{4x} + \sqrt{9x} = 5\sqrt{x}$
$[0, 5, 1]$ by $[0, 10, 1]$

108. $\sqrt{16x} - \sqrt{9x} = \sqrt{7x}$
$[0, 5, 1]$ by $[0, 5, 1]$

109. $(\sqrt{x} - 1)(\sqrt{x} - 1) = x + 1$
$[0, 5, 1]$ by $[-1, 2, 1]$

110. $(\sqrt{x} + 2)(\sqrt{x} - 2) = x^2 - 4$
$[-10, 10, 1]$ by $[-10, 10, 1]$

Review Exercises

111. Multiply: $(5x + 3)(5x - 3)$. (Section 6.3, Example 4)

112. Factor completely: $64x^3 - x$. (Section 7.5, Example 3)

113. Graph: $y = -\dfrac{1}{4}x + 3$. (Section 4.4, Example 3)

✔ **MID-CHAPTER CHECK POINT**

CHAPTER 9

What You Know: We learned to find roots of numbers. We saw that if an index is even, the radicand must be nonnegative for a root to be a real number. We learned to simplify radicals. We performed various operations with radicals including multiplication, division, addition, and subtraction.

In Exercises 1–20, simplify the given expression or perform the indicated operation and, if possible, simplify. Assume that all variables represent positive real numbers.

1. $\sqrt{50} \cdot \sqrt{6}$

2. $\sqrt{6} + 9\sqrt{6}$

3. $\sqrt{96x^3}$

4. $\sqrt[5]{\dfrac{4}{32}}$

5. $\sqrt{27} + 3\sqrt{12}$

6. $(\sqrt{10} + \sqrt{3})(\sqrt{10} - \sqrt{3})$

7. $\dfrac{\sqrt{5}}{2}(4\sqrt{3} - 6\sqrt{20})$

8. $-\sqrt{32x^{21}}$

9. $\sqrt{6x^3} \cdot \sqrt{2x^4}$

10. $\dfrac{\sqrt[3]{32}}{\sqrt[3]{2}}$

11. $-3\sqrt{90} - 5\sqrt{40}$

12. $(2 - \sqrt{3})(5 + 2\sqrt{3})$

13. $\dfrac{\sqrt{56x^5}}{\sqrt{7x^3}}$

14. $-\sqrt[4]{32}$

15. $(\sqrt{2} + \sqrt{7})^2$

16. $\sqrt[3]{\dfrac{1}{2}} \cdot \sqrt[3]{32}$

17. $\sqrt{5} + \sqrt{20} + \sqrt{45}$

18. $\dfrac{1}{3}\sqrt{\dfrac{90}{16}}$

19. $3\sqrt{2}(\sqrt{2} + \sqrt{5})$

20. $(5 - \sqrt{2})(5 + \sqrt{2})$

9.4

Objectives

1 Rationalize denominators containing one term.

2 Rationalize denominators containing two terms.

RATIONALIZING THE DENOMINATOR

They say a fool and his money are soon parted and the rest of us just wait to be taxed. Of course, there's more complaining when you're not expecting a refund and more taxes must be paid. In most tax years, approximately 59% of all taxpayers receive a tax refund, whereas 41% must pay more taxes than were withheld.

In this section's exercise set, you will use a formula that models the percentage of taxpayers who must pay more taxes based on the taxpayer's age. In the section itself, you will learn how to rewrite the formula without a radical in its denominator using a technique called *rationalizing the denominator*.

1 Rationalize denominators containing one term.

Rationalizing Denominators Containing One Term You can use a calculator to compare the approximate values for $\dfrac{1}{\sqrt{3}}$ and $\dfrac{\sqrt{3}}{3}$. The two approximations are the same. This is not a coincidence:

$$\frac{1}{\sqrt{3}} = \frac{1}{\sqrt{3}} \cdot \boxed{\frac{\sqrt{3}}{\sqrt{3}}} = \frac{\sqrt{3}}{\sqrt{9}} = \frac{\sqrt{3}}{3}.$$

Any number divided by itself is 1.
Multiplication by 1 does not change the value of $\frac{1}{\sqrt{3}}$.

This process involves rewriting a radical expression as an equivalent expression in which the denominator no longer contains any radicals. The process is called **rationalizing the denominator**. If the denominator contains the square root of a natural number that is not a perfect square, **multiply the numerator and the denominator by the smallest number that produces the square root of a perfect square in the denominator**.

EXAMPLE 1 Rationalizing Denominators

Rationalize the denominator:

a. $\dfrac{15}{\sqrt{6}}$ **b.** $\sqrt{\dfrac{3}{5}}$.

SOLUTION

a. If we multiply the numerator and denominator of $\dfrac{15}{\sqrt{6}}$ by $\sqrt{6}$, the denominator

becomes $\sqrt{6} \cdot \sqrt{6} = \sqrt{36} = 6$. Therefore, we multiply by 1, choosing $\dfrac{\sqrt{6}}{\sqrt{6}}$ for 1.

$$\frac{15}{\sqrt{6}} = \frac{15}{\sqrt{6}} \cdot \frac{\sqrt{6}}{\sqrt{6}}$$ Multiply the numerator and denominator by $\sqrt{6}$ to remove the square root in the denominator.

$$= \frac{15\sqrt{6}}{6}$$ $\sqrt{6} \cdot \sqrt{6} = \sqrt{36} = 6$

$$= \frac{5\sqrt{6}}{2}$$ Simplify, dividing numerator and denominator by 3.

b. $\sqrt{\dfrac{3}{5}} = \dfrac{\sqrt{3}}{\sqrt{5}}$ The square root of a quotient is the quotient of the square roots.

$$= \frac{\sqrt{3}}{\sqrt{5}} \cdot \frac{\sqrt{5}}{\sqrt{5}}$$ Because $\sqrt{5}$ is the smallest factor that will produce a perfect square in the denominator, multiply by 1, choosing $\dfrac{\sqrt{5}}{\sqrt{5}}$ for 1.

$$= \frac{\sqrt{15}}{5}$$ $\sqrt{5} \cdot \sqrt{5} = \sqrt{25} = 5$

 CHECK POINT 1 Rationalize the denominator:

a. $\dfrac{25}{\sqrt{10}}$ **b.** $\sqrt{\dfrac{2}{7}}$.

It is a good idea to simplify a radical expression before attempting to rationalize the denominator.

EXAMPLE 2 Simplifying and Then Rationalizing Denominators

Rationalize the denominator:

a. $\dfrac{12}{\sqrt{8}}$ **b.** $\sqrt{\dfrac{7x}{75}}$.

SOLUTION

a. We begin by simplifying $\sqrt{8}$.

$$\frac{12}{\sqrt{8}} = \frac{12}{\sqrt{4 \cdot 2}}$$ 4 is the greatest perfect square factor of 8.

$$= \frac{12}{2\sqrt{2}}$$ $\sqrt{4 \cdot 2} = \sqrt{4}\,\sqrt{2} = 2\sqrt{2}$

$$= \frac{6}{\sqrt{2}}$$ Simplify, dividing the numerator and denominator by 2.

$$= \frac{6}{\sqrt{2}} \cdot \frac{\sqrt{2}}{\sqrt{2}}$$ Rationalize the denominator.

$$= \frac{6\sqrt{2}}{2}$$ $\sqrt{2} \cdot \sqrt{2} = \sqrt{4} = 2$

$$= 3\sqrt{2}$$ Simplify.

DISCOVER FOR YOURSELF

Rationalize the denominator in Example 2(a) without first simplifying. Multiply the numerator and the denominator by $\sqrt{8}$. Do this again by multiplying by $\dfrac{\sqrt{2}}{\sqrt{2}}$. Which method do you prefer?

b. $\sqrt{\dfrac{7x}{75}} = \dfrac{\sqrt{7x}}{\sqrt{75}}$ The square root of a quotient is the quotient of the square roots.

$$= \dfrac{\sqrt{7x}}{\sqrt{25 \cdot 3}}$$ Simplify the denominator. 25 is the greatest perfect square factor of 75.

$$= \dfrac{\sqrt{7x}}{5\sqrt{3}}$$ $\sqrt{25 \cdot 3} = \sqrt{25}\,\sqrt{3} = 5\sqrt{3}$

$$= \dfrac{\sqrt{7x}}{5\sqrt{3}} \cdot \dfrac{\sqrt{3}}{\sqrt{3}}$$ Rationalize the denominator, choosing $\dfrac{\sqrt{3}}{\sqrt{3}}$ for 1.

$$= \dfrac{\sqrt{21x}}{5 \cdot 3}$$ $\sqrt{3} \cdot \sqrt{3} = \sqrt{9} = 3$

$$= \dfrac{\sqrt{21x}}{15}$$

✔ **CHECK POINT 2** Rationalize the denominator:

a. $\dfrac{15}{\sqrt{18}}$ **b.** $\sqrt{\dfrac{7x}{20}}.$

2   Rationalize denominators containing two terms.

Rationalizing Denominators Containing Two Terms   How can we rationalize a denominator if the denominator contains two terms with one or more square roots? **Multiply the numerator and the denominator by the conjugate of the denominator.** Here are two examples of such expressions:

$$\dfrac{7}{5 + \sqrt{3}} \qquad\qquad \dfrac{2}{\sqrt{6} - \sqrt{2}}.$$

The conjugate of the denominator is $5 - \sqrt{3}$.   The conjugate of the denominator is $\sqrt{6} + \sqrt{2}$.

The product of the denominator and its conjugate is found using the formula

$$(A + B)(A - B) = A^2 - B^2.$$

The simplified product will not contain a radical.

EXAMPLE 3   Rationalizing a Denominator Containing Two Terms

Rationalize the denominator: $\dfrac{7}{5 + \sqrt{3}}.$

SOLUTION   The conjugate of the denominator is $5 - \sqrt{3}$. If we multiply the numerator and the denominator by $5 - \sqrt{3}$, the denominator will not contain a radical. Therefore, we multiply by 1, choosing $\dfrac{5 - \sqrt{3}}{5 - \sqrt{3}}$ for 1.

$$\frac{7}{5 + \sqrt{3}} = \frac{7}{5 + \sqrt{3}} \cdot \frac{5 - \sqrt{3}}{5 - \sqrt{3}}$$

Multiply by 1.

$$= \frac{7(5 - \sqrt{3})}{5^2 - (\sqrt{3})^2}$$

$(A + B)(A - B) = A^2 - B^2$

$$= \frac{7(5 - \sqrt{3})}{25 - 3}$$

$(\sqrt{3})^2 = \sqrt{3} \cdot \sqrt{3} = \sqrt{9} = 3$

$$= \frac{7(5 - \sqrt{3})}{22} \quad \text{or} \quad \frac{35 - 7\sqrt{3}}{22}$$

■

✔ **CHECK POINT 3** Rationalize the denominator: $\dfrac{8}{4 + \sqrt{5}}$.

EXAMPLE 4 Rationalizing a Denominator Containing Two Terms

Rationalize the denominator: $\dfrac{2}{\sqrt{6} - \sqrt{2}}$.

SOLUTION   The conjugate of the denominator is $\sqrt{6} + \sqrt{2}$. If we multiply the numerator and the denominator by $\sqrt{6} + \sqrt{2}$, the denominator will not contain a radical. Therefore, we multiply by 1, choosing $\dfrac{\sqrt{6} + \sqrt{2}}{\sqrt{6} + \sqrt{2}}$ for 1.

$$\frac{2}{\sqrt{6} - \sqrt{2}} = \frac{2}{\sqrt{6} - \sqrt{2}} \cdot \frac{\sqrt{6} + \sqrt{2}}{\sqrt{6} + \sqrt{2}}$$

Multiply by 1.

$$= \frac{2(\sqrt{6} + \sqrt{2})}{(\sqrt{6})^2 - (\sqrt{2})^2}$$

$(A - B)(A + B) = A^2 - B^2$

$$= \frac{2(\sqrt{6} + \sqrt{2})}{6 - 2}$$

$(\sqrt{6})^2 = \sqrt{6} \cdot \sqrt{6} = \sqrt{36} = 6$
and $(\sqrt{2})^2 = \sqrt{2} \cdot \sqrt{2} = \sqrt{4} = 2$

$$= \frac{2(\sqrt{6} + \sqrt{2})}{4}$$

$$= \frac{\overset{1}{2}(\sqrt{6} + \sqrt{2})}{\underset{2}{4}}$$

Divide the numerator and denominator by the common factor, 2.

$$= \frac{\sqrt{6} + \sqrt{2}}{2}$$

■

✔ **CHECK POINT 4** Rationalize the denominator: $\dfrac{8}{\sqrt{7} - \sqrt{3}}$.

9.4 EXERCISE SET

 Student Solutions Manual CD/Video PH Math/Tutor Center MathXL Tutorials on CD Math XL MathXL® MyMathLab MyMathLab Interactmath.com

Practice Exercises

In Exercises 1–24, rationalize each denominator. If possible, simplify the rationalized expression by dividing the numerator and denominator by the greatest common factor.

1. $\dfrac{1}{\sqrt{10}}$

2. $\dfrac{1}{\sqrt{2}}$

3. $\dfrac{5}{\sqrt{5}}$

4. $\dfrac{7}{\sqrt{7}}$

5. $\dfrac{2}{\sqrt{6}}$

6. $\dfrac{4}{\sqrt{6}}$

7. $\dfrac{28}{\sqrt{7}}$

8. $\dfrac{40}{\sqrt{5}}$

9. $\sqrt{\dfrac{3}{5}}$

10. $\sqrt{\dfrac{3}{7}}$

11. $\sqrt{\dfrac{7}{3}}$

12. $\sqrt{\dfrac{5}{2}}$

13. $\sqrt{\dfrac{x^2}{3}}$

14. $\sqrt{\dfrac{x^2}{7}}$

15. $\sqrt{\dfrac{11}{x}}$

16. $\sqrt{\dfrac{6}{x}}$

17. $\sqrt{\dfrac{x}{y}}$

18. $\sqrt{\dfrac{a}{b}}$

19. $\sqrt{\dfrac{x^4}{2}}$

20. $\sqrt{\dfrac{x^4}{3}}$

21. $\dfrac{\sqrt{7}}{\sqrt{5}}$

22. $\dfrac{\sqrt{11}}{\sqrt{5}}$

23. $\dfrac{\sqrt{3x}}{\sqrt{14}}$

24. $\dfrac{\sqrt{2x}}{\sqrt{17}}$

In Exercises 25–52, begin by simplifying the expression. Then rationalize the denominator using the simplified expression.

25. $\dfrac{1}{\sqrt{20}}$

26. $\dfrac{1}{\sqrt{18}}$

27. $\dfrac{12}{\sqrt{32}}$

28. $\dfrac{15}{\sqrt{50}}$

29. $\dfrac{15}{\sqrt{12}}$

30. $\dfrac{13}{\sqrt{40}}$

31. $\sqrt{\dfrac{5}{18}}$

32. $\sqrt{\dfrac{7}{12}}$

33. $\sqrt{\dfrac{x}{32}}$

34. $\sqrt{\dfrac{x}{40}}$

35. $\sqrt{\dfrac{1}{45}}$

36. $\sqrt{\dfrac{1}{54}}$

37. $\dfrac{\sqrt{7}}{\sqrt{12}}$

38. $\dfrac{\sqrt{5}}{\sqrt{18}}$

39. $\dfrac{8x}{\sqrt{8}}$

40. $\dfrac{27x}{\sqrt{27}}$

41. $\dfrac{\sqrt{7y}}{\sqrt{8}}$

42. $\dfrac{\sqrt{3y}}{\sqrt{125}}$

43. $\sqrt{\dfrac{7x}{12}}$

44. $\sqrt{\dfrac{11x}{18}}$

45. $\sqrt{\dfrac{45}{x}}$

46. $\sqrt{\dfrac{27}{x}}$

47. $\dfrac{5}{\sqrt{x^3}}$

48. $\dfrac{11}{\sqrt{x^3}}$

49. $\sqrt{\dfrac{27}{y^3}}$

50. $\sqrt{\dfrac{45}{y^3}}$

51. $\dfrac{\sqrt{50x^2}}{\sqrt{12y^3}}$

52. $\dfrac{\sqrt{27x^2}}{\sqrt{12y^3}}$

In Exercises 53–74, rationalize each denominator. Simplify, if possible.

53. $\dfrac{1}{4 + \sqrt{3}}$

54. $\dfrac{1}{5 + \sqrt{2}}$

55. $\dfrac{9}{2 - \sqrt{7}}$

56. $\dfrac{12}{2 - \sqrt{7}}$

57. $\dfrac{16}{\sqrt{11} + 3}$

58. $\dfrac{15}{\sqrt{7} + 2}$

59. $\dfrac{18}{3 - \sqrt{3}}$

60. $\dfrac{40}{5 - \sqrt{5}}$

61. $\dfrac{\sqrt{2}}{\sqrt{2} + 1}$

62. $\dfrac{\sqrt{3}}{\sqrt{3} + 1}$

63. $\dfrac{\sqrt{10}}{\sqrt{10} - \sqrt{7}}$

64. $\dfrac{\sqrt{5}}{\sqrt{5} - \sqrt{3}}$

65. $\dfrac{6}{\sqrt{6} + \sqrt{3}}$

66. $\dfrac{8}{\sqrt{7} + \sqrt{3}}$

67. $\dfrac{2}{\sqrt{5} - \sqrt{3}}$

68. $\dfrac{5}{\sqrt{7} - \sqrt{2}}$

69. $\dfrac{2}{4 + \sqrt{x}}$

70. $\dfrac{6}{4 - \sqrt{x}}$

71. $\dfrac{2\sqrt{3}}{\sqrt{15} + 2}$

72. $\dfrac{3\sqrt{2}}{\sqrt{10} + 2}$

73. $\dfrac{\sqrt{5} + \sqrt{2}}{\sqrt{5} - \sqrt{2}}$

74. $\dfrac{\sqrt{5} + \sqrt{3}}{\sqrt{5} - \sqrt{3}}$

Practice Plus

In Exercises 75–82, rationalize each denominator. Simplify, if possible.

75. $\dfrac{\sqrt{36x^2 y^5}}{\sqrt{2x^3 y}}$

76. $\dfrac{\sqrt{100x^5 y^2}}{\sqrt{2xy^3}}$

77. $\dfrac{2}{\sqrt{x + 2} - \sqrt{x}}$

78. $\dfrac{3}{\sqrt{x + 3} - \sqrt{x}}$

79. $\dfrac{\sqrt{2}}{\sqrt{3}} + \dfrac{\sqrt{3}}{\sqrt{2}}$

80. $\dfrac{\sqrt{2}}{\sqrt{7}} + \dfrac{\sqrt{7}}{\sqrt{2}}$

81. $\dfrac{2x + 4 - 2h}{\sqrt{x + 2} - h}$

82. $\dfrac{4x + 12 - 4h}{\sqrt{x + 3} - h}$

Application Exercises

83. Do you expect to pay more taxes than were withheld? Would you be surprised to know that the percentage of taxpayers who receive a refund and the percentage of taxpayers who pay more taxes vary according to age? The formula

$$P = \frac{x(13 + \sqrt{x})}{5\sqrt{x}}$$

models the percentage, P, of taxpayers who are x years old who must pay more taxes.

a. What percentage of 25-year-olds must pay more taxes?

b. Rewrite the formula by rationalizing the denominator.

c. Use the rationalized form of the formula from part (b) to find the percentage of 25-year-olds who must pay more taxes. Do you get the same answer as you did in part (a)? If so, does this prove that you correctly rationalized the denominator? Explain.

84. In the Peanuts cartoon, Woodstock appears to be working steps mentally. Fill in the missing steps that show how to go from $\dfrac{7\sqrt{2 \cdot 2 \cdot 3}}{6}$ to $\dfrac{7}{3}\sqrt{3}$.

PEANUTS reprinted by permission of United Feature Syndicate, Inc.

85. The early Greeks believed that the most pleasing of all rectangles were *golden rectangles*, whose ratio of width to height is

$$\frac{w}{h} = \frac{2}{\sqrt{5} - 1}.$$

Rationalize the denominator for this ratio and then use a calculator to approximate the answer correct to the nearest hundredth.

Writing in Mathematics

86. Describe what it means to rationalize a denominator. Use both $\dfrac{1}{\sqrt{5}}$ and $\dfrac{1}{5 + \sqrt{5}}$ in your explanation.

87. When a radical expression has its denominator rationalized, we change the denominator so that it no longer contains a radical. Doesn't this change the value of the radical expression? Explain.

88. Square the real number $\dfrac{2}{\sqrt{3}}$. Observe that the radical is eliminated from the denominator. Explain whether this process is equivalent to rationalizing the denominator.

89. Use the model in Exercise 83 and a calculator to find the percentage of taxpayers ages 30, 40, and 50 who expect to pay more taxes. Describe the trend that you observe. What explanation can you offer for this trend?

Critical Thinking Exercises

90. Which one of the following is true?

a. $\dfrac{4 + 8\sqrt{3}}{4} = 1 + 8\sqrt{3}$

b. $\dfrac{3\sqrt{x}}{x\sqrt{6}} = \dfrac{\sqrt{6x}}{2x}$ for $x > 0$

c. Conjugates are used to rationalize the denominator of $\dfrac{2 - \sqrt{5}}{\sqrt{3}}$.

d. Radical expressions with rationalized denominators require less space to write than before they are rationalized.

91. Rationalize the denominator: $\dfrac{1}{\sqrt[3]{2}}$.

92. Simplify: $\sqrt{2} + \sqrt{\dfrac{1}{2}}$.

93. Simplify: $\sqrt{13 + \sqrt{2} + \dfrac{7}{3 + \sqrt{2}}}$.

94. Fill in the box to make the statement true: $\dfrac{4}{2 + \sqrt{\boxed{}}} = 8 - 4\sqrt{3}$.

Review Exercises

95. Solve: $2x - 1 = x^2 - 4x + 4$. (Section 7.6, Example 4)

96. Simplify: $(2x^2)^{-3}$. (Section 6.7, Example 6)

97. Multiply: $\dfrac{x^2 - 6x + 9}{12} \cdot \dfrac{3}{x^2 - 9}$. (Section 8.2, Example 3)

Objectives

1 Solve radical equations

2 Solve problems involving square-root models.

RADICAL EQUATIONS

"A breakthrough is something that changes the behavior of hundreds of millions of people where, if you took it away from them, they'd say, 'You can't take that away from me.' Breakthroughs are critical for us." Bill Gates

Computers have become faster and more powerful. Moore's Law, which states that every 18 months computer power doubles at no extra cost, is still going strong. But customers are not trading up for new models the way they used to, despite being lured by cheaper and cheaper prices. The bar graph in Figure 9.4 shows the number of personal computers sold in the United States each year from 1996 through 2002. Does the trend shown by the graph suggest a new idea afoot in the land, a philosophy of "good enough" when it comes to high tech? Or will a new wave of innovation reverse this trend?

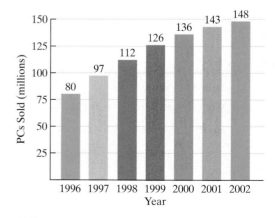

Millions of Personal Computers Sold in the U. S.

FIGURE 9.4

Source: Newsweek

The bar graph indicates that PC sales are increasing for the period shown. However, the rate of increase is slowing down. For this reason, a square-root formula is appropriate for describing the data. The formula

$$P = 28\sqrt{t} + 80$$

models the number of personal computers sold in the United States, P, in millions, t years after 1996.

If trends from 1996 through 2002 continue and there are no new breakthroughs to spur sales, in what year will 164 million personal computers be sold? Substitute 164 for P in the formula and solve for t:

$$164 = 28\sqrt{t} + 80.$$

The resulting equation contains a variable in the radicand and is called a *radical equation*. A **radical equation** is an equation in which the variable occurs in a square root, cube root, or any higher root.

$$164 = 28\sqrt{t} + 80$$

The variable occurs in a square root.

In this section, you will learn how to solve radical equations with square roots. Solving such equations will enable you to solve new kinds of problems using square-root models, such as the model for the number of PCs sold in the United States. We will return to this model at the end of the section.

1 Solve radical equations.

Radical Equations with Square Roots Consider the following radical equation:

$$\sqrt{x} = 5.$$

We solve the equation by squaring both sides:

Squaring both sides eliminates the square root.

$$\left(\sqrt{x}\right)^2 = 5^2$$
$$x = 25.$$

The proposed solution, 25, can be checked in the original equation, $\sqrt{x} = 5$. Because $\sqrt{25} = 5$, the solution is 25.

In general, we solve radical equations with square roots by squaring both sides of the equation. Unfortunately, all the solutions of the squared equation may not be solutions of the original equation. Consider, for example, the equation

$$x = 4.$$

If we square both sides, we obtain

$$x^2 = 16.$$

This new equation has two solutions, -4 and 4. By contrast, only 4 is a solution of the original equation, $x = 4$. For this reason, we must **always check proposed solutions of radical equations in the original equation**.

Here is a general method for solving radical equations with square roots:

SOLVING RADICAL EQUATIONS CONTAINING SQUARE ROOTS

1. If necessary, arrange terms so that one radical is isolated on one side of the equation.

2. Square both sides of the equation to eliminate the square root.

3. Solve the resulting equation.

4. Check all proposed solutions in the original equation.

EXAMPLE 1 Solving a Radical Equation

Solve: $\sqrt{3x + 4} = 8$.

SOLUTION

Step 1. **Isolate the radical.** The radical, $\sqrt{3x + 4}$, is already isolated on the left side of the equation, so we can skip this step.

Step 2. **Square both sides.**

$$\sqrt{3x + 4} = 8 \qquad \text{This is the given equation.}$$
$$\left(\sqrt{3x + 4}\right)^2 = 8^2 \qquad \text{Square both sides to eliminate the radical.}$$
$$3x + 4 = 64 \qquad \text{Simplify.}$$

Step 3. **Solve the resulting equation.**

$$3x + 4 = 64 \qquad \text{This resulting equation is a linear equation.}$$
$$3x = 60 \qquad \text{Subtract 4 from both sides.}$$
$$x = 20 \qquad \text{Divide both sides by 3.}$$

Step 4. **Check the proposed solution in the original equation.**

Check 20:

$$\sqrt{3x + 4} = 8$$
$$\sqrt{3 \cdot 20 + 4} \overset{?}{=} 8$$
$$\sqrt{60 + 4} \overset{?}{=} 8$$
$$\sqrt{64} \overset{?}{=} 8$$
$$8 = 8, \qquad \text{true}$$

The solution is 20. ∎

USING TECHNOLOGY

We can use a graphing utility to verify the solution to the equation in Example 1:

$$\sqrt{3x + 4} = 8.$$

Graph each side of the equation:

$$y_1 = \sqrt{3x + 4}$$
$$y_2 = 8.$$

TRACE along the curves or use the utility's intersection feature. The solution, as shown below, is the first coordinate of the intersection point. Thus, the solution is 20.

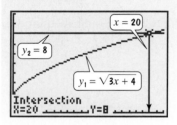

$[0, 22, 1]$ by $[0, 10, 1]$

✔ **CHECK POINT 1** Solve: $\sqrt{2x + 3} = 5$.

EXAMPLE 2 Solving a Radical Equation

Solve: $\sqrt{3x + 9} - 2\sqrt{x} = 0$.

SOLUTION

Step 1. **Isolate each radical.** We can get each radical by itself on one side of the equation by adding $2\sqrt{x}$ to both sides.

$$\sqrt{3x + 9} - 2\sqrt{x} = 0 \qquad \text{This is the given equation.}$$
$$\sqrt{3x + 9} - 2\sqrt{x} + 2\sqrt{x} = 0 + 2\sqrt{x} \qquad \text{Add } 2\sqrt{x} \text{ to both sides.}$$
$$\sqrt{3x + 9} = 2\sqrt{x} \qquad \text{Simplify.}$$

Step 2. **Square both sides.**

$$\left(\sqrt{3x + 9}\right)^2 = (2\sqrt{x})^2$$
$$3x + 9 = 4x \qquad \text{Simplify.}$$

Step 3. **Solve the resulting equation.**

$$3x + 9 = 4x \qquad \text{The resulting equation is a linear equation.}$$
$$9 = x \qquad \text{Subtract } 3x \text{ from both sides.}$$

Step 4. **Check the proposed solution in the original equation.**

Check 9:

$$\sqrt{3x + 9} - 2\sqrt{x} = 0$$
$$\sqrt{3 \cdot 9 + 9} - 2\sqrt{9} \overset{?}{=} 0$$
$$\sqrt{36} - 2\sqrt{9} \overset{?}{=} 0$$
$$6 - 2 \cdot 3 \overset{?}{=} 0$$
$$6 - 6 \overset{?}{=} 0$$
$$0 = 0, \quad \text{true}$$

The solution is 9. ∎

✔ **CHECK POINT 2** Solve: $\sqrt{x + 32} - 3\sqrt{x} = 0.$

EXAMPLE 3 Solving a Radical Equation

Solve: $\sqrt{x} + 5 = 0.$

SOLUTION

Step 1. **Isolate the radical.** We isolate the radical, $\sqrt{x}$, on the left side by subtracting 5 from both sides.

$$\sqrt{x} + 5 = 0 \qquad \text{This is the given equation.}$$
$$\sqrt{x} = -5 \qquad \text{Subtract 5 from both sides.}$$

A principal square root cannot be negative. This equation has no solution. Let's continue the solution procedure to see what happens.

Step 2. **Square both sides.**

$$(\sqrt{x})^2 = (-5)^2$$
$$x = 25 \qquad \text{Simplify.}$$

Step 3. **Solve the resulting equation.** We immediately see that 25 is the proposed solution.

Step 4. **Check the proposed solution in the original equation.**

Check 25:

$$\sqrt{x} + 5 = 0$$
$$\sqrt{25} + 5 \overset{?}{=} 0$$
$$5 + 5 \overset{?}{=} 0$$
$$10 = 0, \quad \text{false}$$

This false statement indicates that 25 is not a solution. Thus, the equation has no solution. ∎

USING TECHNOLOGY

The graphs of

$$y_1 = \sqrt{x}$$

and

$$y_2 = -5$$

do not intersect, as shown below. Thus, $\sqrt{x} = -5$ has no real number solution.

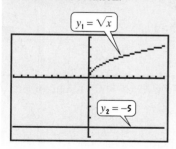

$[-10, 10, 1]$ by $[-6, 4, 1]$

Example 3 illustrates that extra solutions may be introduced when you raise both sides of a radical equation to an even power. Such solutions, which are not solutions of the given equation, are called **extraneous solutions**. Thus, 25 is an extraneous solution of the equation $\sqrt{x} + 5 = 0.$

✔ **CHECK POINT 3** Solve: $\sqrt{x} + 1 = 0.$

EXAMPLE 4 Solving a Radical Equation

Solve: $\sqrt{2x - 1} + 2 = x$.

SOLUTION

Step 1. **Isolate the radical.** We isolate the radical, $\sqrt{2x - 1}$, by subtracting 2 from both sides.

$$\sqrt{2x - 1} + 2 = x \qquad \text{This is the given equation.}$$
$$\sqrt{2x - 1} = x - 2 \qquad \text{Subtract 2 from both sides.}$$

Step 2. **Square both sides.**

$$\left(\sqrt{2x - 1}\right)^2 = (x - 2)^2$$

$$2x - 1 = x^2 - 4x + 4 \qquad \begin{array}{l}\text{Simplify. Use the formula}\\ (A - B)^2 = A^2 - 2AB + B^2 \\ \text{on the right side.}\end{array}$$

Step 3. **Solve the resulting equation.** Because of the x^2-term, the resulting equation is a quadratic equation. We can obtain 0 on the left side by subtracting $2x$ and adding 1 on both sides.

$$2x - 1 = x^2 - 4x + 4 \qquad \text{The resulting equation is quadratic.}$$
$$0 = x^2 - 6x + 5 \qquad \begin{array}{l}\text{Write in standard form, subtracting}\\ 2x \text{ and adding 1 on both sides.}\end{array}$$
$$0 = (x - 1)(x - 5) \qquad \text{Factor.}$$
$$x - 1 = 0 \quad \text{or} \quad x - 5 = 0 \qquad \text{Set each factor equal to 0.}$$
$$x = 1 \qquad\qquad x = 5 \qquad \text{Solve the resulting equations.}$$

Step 4. **Check the proposed solutions in the original equation.**

Check 1:	**Check 5:**
$\sqrt{2x - 1} + 2 = x$	$\sqrt{2x - 1} + 2 = x$
$\sqrt{2 \cdot 1 - 1} + 2 \stackrel{?}{=} 1$	$\sqrt{2 \cdot 5 - 1} + 2 \stackrel{?}{=} 5$
$\sqrt{1} + 2 \stackrel{?}{=} 1$	$\sqrt{9} + 2 \stackrel{?}{=} 5$
$1 + 2 \stackrel{?}{=} 1$	$3 + 2 \stackrel{?}{=} 5$
$3 = 1,$ false	$5 = 5,$ true

Thus, 1 is an extraneous solution. The only solution is 5.

USING TECHNOLOGY

A graphing utility's TABLE feature provides a numeric check that 1 is not a solution and 5 is a solution of $\sqrt{2x - 1} + 2 = x$.

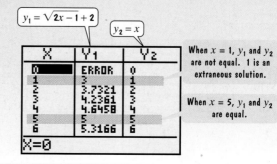

$y_1 = \sqrt{2x - 1} + 2$

$y_2 = x$

When $x = 1$, y_1 and y_2 are not equal. 1 is an extraneous solution.

When $x = 5$, y_1 and y_2 are equal.

✔ CHECK POINT 4 Solve: $\sqrt{x + 3} + 3 = x$.

2 Solve problems involving square-root models.

Applications of Radical Equations Radical equations can be solved to answer questions about variables contained in square-root models.

EXAMPLE 5 Using a Square-Root Model

The formula

$$P = 28\sqrt{t} + 80$$

models the number of personal computers sold in the United States, P, in millions, t years after 1996. If trends indicated by this model continue, in what year will 164 million personal computers be sold?

SOLUTION Because we are interested in finding when 164 million PCs will be sold, we substitute 164 for P in the given formula. Then we solve for t, the number of years after 1996.

$P = 28\sqrt{t} + 80$	This is the given formula.
$164 = 28\sqrt{t} + 80$	Substitute 164 for P.
$84 = 28\sqrt{t}$	Subtract 80 from both sides.
$3 = \sqrt{t}$	Divide both sides by 28: $\frac{84}{28} = 3$.
$3^2 = (\sqrt{t})^2$	Square both sides.
$9 = t$	Simplify.

The model indicates that 164 million personal computers will be sold in the United States 9 years after 1996. Because $1996 + 9 = 2005$, this will occur in 2005. ∎

✔ CHECK POINT 5 Use the formula in Example 5 to predict when 192 million personal computers will be sold in the United States.

9.5 EXERCISE SET

Student Solutions Manual CD/Video PH Math/Tutor Center MathXL Tutorials on CD MathXL® MyMathLab Interactmath.com

Practice Exercises

In Exercises 1–44, solve each radical equation. If the equation has no solution, so state.

1. $\sqrt{x} = 5$
2. $\sqrt{x} = 6$
3. $\sqrt{x} - 4 = 0$
4. $\sqrt{x} - 9 = 0$
5. $\sqrt{x + 2} = 3$
6. $\sqrt{x - 6} = 5$
7. $\sqrt{x - 3} - 11 = 0$
8. $\sqrt{x + 5} - 8 = 0$
9. $\sqrt{3x - 5} = 4$
10. $\sqrt{5x - 6} = 8$
11. $\sqrt{x + 5} + 2 = 5$
12. $\sqrt{x + 6} + 1 = 3$
13. $\sqrt{x + 3} = \sqrt{4x - 3}$
14. $\sqrt{x + 8} = \sqrt{5x - 4}$
15. $\sqrt{6x - 2} = \sqrt{4x + 4}$
16. $\sqrt{5x - 4} = \sqrt{3x + 6}$

17. $11 = 6 + \sqrt{x + 1}$
18. $7 = 5 + \sqrt{x + 1}$
19. $\sqrt{x} + 10 = 0$
20. $\sqrt{x} + 8 = 0$
21. $\sqrt{x - 1} = -3$
22. $\sqrt{x - 2} = -5$
23. $3\sqrt{x} = \sqrt{8x + 16}$
24. $3\sqrt{x - 2} = \sqrt{7x + 4}$
25. $\sqrt{2x - 3} + 5 = 0$
26. $\sqrt{2x - 8} + 4 = 0$
27. $\sqrt{3x + 4} - 2 = 3$
28. $\sqrt{5x - 4} - 2 = 4$
29. $3\sqrt{x - 1} = \sqrt{3x + 3}$
30. $2\sqrt{3x + 4} = \sqrt{5x + 9}$
31. $\sqrt{x + 7} = x + 5$
32. $\sqrt{x + 10} = x - 2$
33. $\sqrt{2x + 13} = x + 7$
34. $\sqrt{6x + 1} = x - 1$

35. $\sqrt{9x^2 + 2x - 4} = 3x$ **36.** $\sqrt{9x^2 - 2x + 8} = 3x$

37. $x = \sqrt{2x - 2} + 1$ **38.** $x = \sqrt{3x + 7} - 3$

39. $x = \sqrt{8 - 7x} + 2$ **40.** $x = \sqrt{1 - 8x} + 2$

41. $\sqrt{3x + 10} = x + 4$ **42.** $\sqrt{x - 3} = x - 9$

43. $3\sqrt{x} + 5 = 2$ **44.** $3\sqrt{x} + 8 = 5$

Practice Plus

45. Two added to the square root of the product of 4 and a number is equal to 10. Find the number.

46. Five added to the square root of the product of 6 and a number is equal to 8. Find the number.

47. A number is 4 more than the principal square root of twice the number. Find the number.

48. A number is 6 more than the principal square root of 3 times the number. Find the number.

49. Solve for h: $v = \sqrt{2gh}$.

50. Solve for l: $t = \dfrac{\pi}{2}\sqrt{\dfrac{l}{2}}$.

Solve the equations in Exercises 51–54. You will need to square both sides of each equation twice.

51. $\sqrt{x} + 2 = \sqrt{x + 8}$ **52.** $\sqrt{x} + 6 = \sqrt{x + 72}$

53. $\sqrt{x - 8} = \sqrt{x} - 2$ **54.** $\sqrt{x - 4} = \sqrt{x} - 2$

Application Exercises

55. The bar graph shows the obesity rates for adults in the United States for six selected years from 1991 through 2001. The formula $y = 3.2\sqrt{x} + 11$ models the percentage of obese U.S. adults, y, x years after 1991.

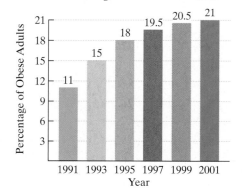

Percentage of Obese U. S. Adults

Source: Centers for Disease Control

a. According to the formula, what percentage of U.S. adults were obese in 1995? How well does the formula model the data shown in the bar graph for that year?

b. If trends indicated by the data continue, use the formula to determine when 25% of U.S. adults will be obese. Use a calculator and round to the nearest year.

56. By 2010, India could become the world's most HIV-afflicted country. The bar graph shows the increase in the country's HIV infections from 1998 through 2001. The formula $y = 0.3\sqrt{x} + 3.4$ models the number of HIV infections in India, y, in millions, x years after 1998.

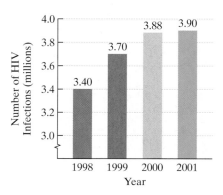

AIDS in India

Source: UNAIDS

a. According to the formula, how many Indians were infected with HIV in 2001? Use a calculator and round to two decimal places. How well does the formula model the data shown in the bar graph for that year?

b. If trends indicated by the data continue, use the formula to determine when the number of HIV infections in India will reach 4.3 million.

57. Out of a group of 50,000 births, the number of people, N, surviving to age x is modeled by the formula $N = 5000\sqrt{100 - x}$. To what age will 25,000 people in the group survive?

58. The time, t, in seconds for a free-falling object to fall d feet is modeled by the formula $t = \sqrt{\dfrac{d}{16}}$. If a worker accidentally drops a hammer from a building and it hits the ground after 4 seconds, from what height was the hammer dropped?

59. The distance, d, in kilometers, that one can see to the horizon from an altitude of h meters is modeled by the formula $d = 3.5\sqrt{h}$. A plane flying at an altitude of 8 kilometers loses altitude so that the pilot can see a distance of 200 kilometers to the horizon. How much altitude did the plane lose? Use a calculator and round to the nearest kilometer.

60. The figure shows a grandfather clock whose pendulum length is l feet. The time, t, in seconds, it takes the pendulum of the clock to swing through one complete cycle is described by

$$t = \frac{11}{7}\sqrt{\frac{l}{2}}.$$

Determine how long the pendulum must be for one complete cycle to take 2 seconds. Round to the nearest hundredth of a foot.

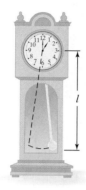

61. Two tractors are removing a tree stump from the ground. If two forces, A and B, pull at right angles to each other, the size of the resulting force, R, is given by the formula $R = \sqrt{A^2 + B^2}$. Tractor A exerts 300 pounds of force. If the resulting force is 500 pounds, how much force is tractor B exerting in the removal of the stump?

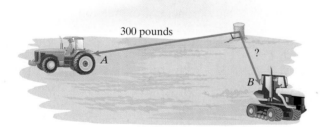

300 pounds

A

?

B

Writing in Mathematics

62. What is a radical equation?

63. In solving $\sqrt{2x-1} + 2 = x$, why is it a good idea to isolate the radical term? What if we don't do this and simply square each side? Describe what happens.

64. What is an extraneous solution of a radical equation?

65. Explain why $\sqrt{x} = -1$ has no solution.

66. Some commentators are proclaiming the end of the computer world's glory days. Describe two new technological innovations that might reverse the trend shown by the bar graph in Figure 9.4 on page 560. Assuming that these innovations cause the rate of PC sales to increase, will a square-root model be appropriate for describing the new data? Explain your answer.

Critical Thinking Exercises

67. Which one of the following is true?
  a. The equation $y^2 = 25$ has the same solutions as the equation $y = 5$.
  b. The equation $\sqrt{x^2 + 2x} = -1$ has no real number solution.
  c. The first step in solving $\sqrt{x} + 3 = 4$ is to take the square root of each side.
  d. When an extraneous root is substituted into an equation with radicals, a denominator of zero results.

68. The square root of the sum of two consecutive integers is one less than the smaller integer. Find the integers.

69. If $w = 2$, find x, y, and z if $y = \sqrt{x-2}+2$, $z = \sqrt{y-2}+2$, and $w = \sqrt{z-2}+2$.

Technology Exercises

In Exercises 70–74, use a graphing utility to solve each radical equation. Graph each side of the equation in the given viewing rectangle. The equation's solution is given by the x-coordinate of the point(s) of intersection. Check by substitution.

70. $\sqrt{2x+2} = \sqrt{3x-5}$
  $[-1, 10, 1]$ by $[-1, 5, 1]$

71. $\sqrt{x} + 3 = 5$
  $[-1, 6, 1]$ by $[-1, 6, 1]$

72. $\sqrt{x^2+3} = x + 1$
  $[-1, 6, 1]$ by $[-1, 6, 1]$

73. $4\sqrt{x} = x + 3$
  $[-1, 10, 1]$ by $[-1, 13, 1]$

74. $\sqrt{x} + 4 = 2$
  $[-2, 18, 1]$ by $[0, 10, 1]$

75. Use a graphing utility's $\boxed{\text{TABLE}}$ feature to provide a numeric check for the solutions you obtained in any three radical equations from Exercises 29–38.

Review Exercises

76. A total of $9000 was invested for 1 year, part at 6% and the remainder at 4% simple interest. At the end of the year, the investments earned $500 in interest How much was invested at each rate? (Section 3.1, Example 2)

77. Producers of *The Lord of the Rings*, the musical, are a bit worried about their basic concept and decide to sell tickets for previews at cut-rate prices. If four orchestra and two mezzanine seats sell for $22, while two orchestra and three mezzanine seats sell for $16, what is the price of an orchestra seat? (Section 5.4, Example 1)

78. Solve by graphing:

$$2x + y = -4$$
$$x + y = -3.$$

(Section 5.1, Example 2)

9.6

RATIONAL EXPONENTS

Objectives

1 Evaluate expressions with rational exponents.

2 Solve problems using models with rational exponents.

Animals in the wild have regions to which they confine their movement, called their territorial area. Territorial area, in square miles, is related to an animal's body weight. If an animal weighs W pounds, its territorial area is

$$W^{\frac{141}{100}}$$

square miles.

W to the *what* power?! How can we interpret the information given by this algebraic expression? In this section, we turn our attention to rational exponents such as $\frac{141}{100}$ and their relationship to roots of real numbers.

1 Evaluate expressions with rational exponents.

Defining Rational Exponents We define rational exponents so that their properties are the same as the properties for integer exponents. For example, we know that exponents are multiplied when an exponential expression is raised to a power. For this to be true,

$$\left(7^{\frac{1}{2}}\right)^2 = 7^{\frac{1}{2} \cdot 2} = 7^1 = 7.$$

We also know that

$$\left(\sqrt{7}\right)^2 = \sqrt{7} \cdot \sqrt{7} = \sqrt{49} = 7.$$

Can you see that the square of both $7^{\frac{1}{2}}$ and $\sqrt{7}$ is 7? It is reasonable to conclude that

$$7^{\frac{1}{2}} \text{ means } \sqrt{7}.$$

We can generalize this idea with the following definition:

THE DEFINITION OF $a^{\frac{1}{n}}$ If $\sqrt[n]{a}$ represents a real number and $n \geq 2$ is an integer, then

$$a^{\frac{1}{n}} = \sqrt[n]{a}.$$

The denominator of the rational exponent is the radical's index.

EXAMPLE 1 Using the Definition of $a^{\frac{1}{n}}$

Simplify:

a. $64^{\frac{1}{2}}$ **b.** $125^{\frac{1}{3}}$ **c.** $-16^{\frac{1}{4}}$ **d.** $(-27)^{\frac{1}{3}}$.

SOLUTION

a. $64^{\frac{1}{2}} = \sqrt{64} = 8$

b. $125^{\frac{1}{3}} = \sqrt[3]{125} = 5$

> The denominator is the index.

c. $-16^{\frac{1}{4}} = -(\sqrt[4]{16}) = -2$

> The base is 16 and the negative sign is not affected by the exponent.

d. $(-27)^{\frac{1}{3}} = \sqrt[3]{-27} = -3$

> Parentheses show that the base is −27 and that the negative sign is affected by the exponent.

✔ **CHECK POINT 1** Simplify:

a. $25^{\frac{1}{2}}$ **b.** $8^{\frac{1}{3}}$ **c.** $-81^{\frac{1}{4}}$ **d.** $(-8)^{\frac{1}{3}}$.

In Example 1 and Check Point 1, each rational exponent had a numerator of 1. If the numerator is some other integer, we still want to multiply exponents when raising a power to a power. For this reason,

$$a^{\frac{2}{3}} = \left(a^{\frac{1}{3}}\right)^2 \text{ and } a^{\frac{2}{3}} = \left(a^2\right)^{\frac{1}{3}}.$$

> This means $(\sqrt[3]{a})^2$.   This means $\sqrt[3]{a^2}$.

Thus,

$$a^{\frac{2}{3}} = (\sqrt[3]{a})^2 = \sqrt[3]{a^2}.$$

Do you see that the denominator, 3, of the rational exponent is the same as the index of the radical? The numerator, 2, of the rational exponent serves as an exponent in each of the two radical forms. We generalize these ideas with the following definition:

THE DEFINITION OF $a^{\frac{m}{n}}$ If $\sqrt[n]{a}$ represents a real number, and $\frac{m}{n}$ is a positive rational number, $n \geq 2$, then

$$a^{\frac{m}{n}} = (\sqrt[n]{a})^m.$$

Also,

$$a^{\frac{m}{n}} = \sqrt[n]{a^m}.$$

The first form of the definition, shown again below, involves taking the root first. This form is often preferable because smaller numbers are involved. Notice that the rational exponent consists of two parts, indicated by the following voice balloons:

The numerator is the exponent.

$$a^{\frac{m}{n}} = \left(\sqrt[n]{a}\right)^m.$$

The denominator is the radical's index

USING TECHNOLOGY

Here are the calculator keystroke sequences for $27^{\frac{2}{3}}$:

Many Scientific Calculators

27 $\boxed{y^x}$ $\boxed{(}$ 2 $\boxed{\div}$ 3 $\boxed{)}$ $\boxed{=}$

Many Graphing Calculators

27 $\boxed{\wedge}$ $\boxed{(}$ 2 $\boxed{\div}$ 3 $\boxed{)}$ $\boxed{\text{ENTER}}$.

EXAMPLE 2 Using the Definition of $a^{\frac{m}{n}}$

Simplify:

a. $27^{\frac{2}{3}}$ **b.** $9^{\frac{3}{2}}$ **c.** $-32^{\frac{4}{5}}$.

SOLUTION

a. $27^{\frac{2}{3}} = \left(\sqrt[3]{27}\right)^2 = 3^2 = 9$

b. $9^{\frac{3}{2}} = \left(\sqrt{9}\right)^3 = 3^3 = 27$

c. $-32^{\frac{4}{5}} = -\left(\sqrt[5]{32}\right)^4 = -2^4 = -16$

The base is 32 and the negative sign is not affected by the expontent.

✔ **CHECK POINT 2** Simplify:

a. $27^{\frac{4}{3}}$ **b.** $4^{\frac{3}{2}}$ **c.** $-16^{\frac{3}{4}}$.

Can a rational exponent be negative? Yes. The way that negative rational exponents are defined is similar to the way that negative integer exponents are defined:

THE DEFINITION OF $a^{-\frac{m}{n}}$ If $a^{-\frac{m}{n}}$ is a nonzero real number, then

$$a^{-\frac{m}{n}} = \frac{1}{a^{\frac{m}{n}}}.$$

EXAMPLE 3 Using the Definition of $a^{-\frac{m}{n}}$

Simplify:

a. $100^{-\frac{1}{2}}$ **b.** $27^{-\frac{1}{3}}$ **c.** $81^{-\frac{3}{4}}$.

SOLUTION

a. $100^{-\frac{1}{2}} = \frac{1}{100^{\frac{1}{2}}} = \frac{1}{\sqrt{100}} = \frac{1}{10}$

b. $27^{-\frac{1}{3}} = \frac{1}{27^{\frac{1}{3}}} = \frac{1}{\sqrt[3]{27}} = \frac{1}{3}$

USING TECHNOLOGY

Here are the calculator keystroke sequences for $81^{-\frac{3}{4}}$:

Many Scientific Calculators

81 $\boxed{y^x}$ $\boxed{(}$ 3 $\boxed{+/-}$ $\boxed{\div}$ 4 $\boxed{)}$ $\boxed{=}$

Many Graphing Calculators

81 $\boxed{\wedge}$ $\boxed{(}$ $\boxed{(-)}$ 3 $\boxed{\div}$ 4 $\boxed{)}$ $\boxed{\text{ENTER}}$.

c. $81^{-\frac{3}{4}} = \frac{1}{81^{\frac{3}{4}}} = \frac{1}{\left(\sqrt[4]{81}\right)^3} = \frac{1}{3^3} = \frac{1}{27}$ ∎

 CHECK POINT 3 Simplify:

a. $25^{-\frac{1}{2}}$ **b.** $64^{-\frac{1}{3}}$ **c.** $32^{-\frac{4}{5}}$.

2 Solve problems using models with rational exponents.

Applications Now that you know the meaning of rational exponents, you can work with mathematical models that contain these exponents.

EXAMPLE 4 Filing Tax Returns Early

The bar graph in Figure 9.5 shows the percentage of U.S. taxpayers who file their tax returns at least one month before the April 15 deadline. The formula

$$P = 152x^{-\frac{1}{5}}$$

models the percentage, P, of taxpayers who are x years old who file early. What percentage of 32-year-olds are expected to file early?

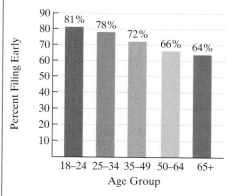

Percentage of Taxpayers Filing at Least One Month before the Deadline

FIGURE 9.5

Source: Bruskin/Goldring Research

SOLUTION Because we are interested in taxpayers of age 32, we substitute 32 for x in the given formula. Then we find P, the percentage of 32-year-olds expected to file at least one month before the tax deadline.

$$P = 152x^{-\frac{1}{5}}$$ *This is the given formula.*

$$P = 152 \cdot 32^{-\frac{1}{5}}$$ *Substitute 32 for x.*

$$P = \frac{152}{32^{\frac{1}{5}}} = \frac{152}{\sqrt[5]{32}} = \frac{152}{2} = 76$$

We see that 76% of 32-year-olds are expected to file early. ∎

 CHECK POINT 4 The formula $S = 63.25x^{\frac{1}{4}}$ models the average sale price, S, in thousands of dollars, of a single-family home in the U.S. Midwest x years after 1981. What was the average sale price in 1997?

9.6 EXERCISE SET

Student Solutions Manual   CD/Video   PH Math/Tutor Center   MathXL Tutorials on CD   MathXL®   MyMathLab   Interactmath.com

In Exercises 1–48, simplify by first writing the expression in radical form. If applicable, use a calculator to verify your answer.

1. $49^{\frac{1}{2}}$

2. $100^{\frac{1}{2}}$

3. $121^{\frac{1}{2}}$

4. $1^{\frac{1}{2}}$

5. $27^{\frac{1}{3}}$

6. $64^{\frac{1}{3}}$

7. $-125^{\frac{1}{3}}$

8. $-27^{\frac{1}{3}}$

9. $16^{\frac{1}{4}}$

10. $81^{\frac{1}{4}}$

11. $-32^{\frac{1}{5}}$

12. $-243^{\frac{1}{5}}$

13. $\left(\frac{1}{9}\right)^{\frac{1}{2}}$

14. $\left(\frac{1}{25}\right)^{\frac{1}{2}}$

15. $\left(\frac{27}{64}\right)^{\frac{1}{3}}$

16. $\left(\frac{64}{125}\right)^{\frac{1}{3}}$

17. $81^{\frac{3}{2}}$

18. $25^{\frac{3}{2}}$

19. $125^{\frac{2}{3}}$

20. $1000^{\frac{2}{3}}$

21. $9^{\frac{3}{2}}$

22. $16^{\frac{3}{2}}$

23. $(-32)^{\frac{3}{5}}$

24. $(-27)^{\frac{2}{3}}$

25. $9^{-\frac{1}{2}}$

26. $49^{-\frac{1}{2}}$

27. $125^{-\frac{1}{3}}$

28. $27^{-\frac{1}{3}}$

29. $32^{-\frac{1}{5}}$

30. $243^{-\frac{1}{5}}$

31. $\left(\frac{1}{4}\right)^{-\frac{1}{2}}$

32. $\left(\frac{1}{9}\right)^{-\frac{1}{2}}$

33. $16^{-\frac{3}{4}}$

34. $625^{-\frac{3}{4}}$

35. $81^{-\frac{5}{4}}$

36. $32^{-\frac{4}{5}}$

37. $8^{-\frac{2}{3}}$

38. $625^{-\frac{5}{4}}$

39. $\left(\frac{4}{25}\right)^{-\frac{1}{2}}$

40. $\left(\frac{8}{27}\right)^{-\frac{1}{3}}$

41. $\left(\frac{8}{125}\right)^{-\frac{1}{3}}$

42. $\left(\frac{9}{100}\right)^{-\frac{1}{2}}$

43. $(-8)^{-\frac{2}{3}}$

44. $(-64)^{-\frac{2}{3}}$

45. $27^{\frac{2}{3}} + 16^{\frac{3}{4}}$

46. $4^{\frac{5}{2}} - 8^{\frac{2}{3}}$

47. $25^{\frac{3}{2}} \cdot 81^{\frac{1}{4}}$

48. $16^{-\frac{3}{4}} \cdot 16^{\frac{3}{2}}$

Practice Plus

In Exercises 49–56, simplify each expression. Write answers in exponential form with positive exponents only. Assume that all variables represent positive real numbers.

49. $x^{\frac{1}{3}} \cdot x^{\frac{1}{4}}$

50. $x^{\frac{1}{4}} \cdot x^{\frac{1}{5}}$

51. $\dfrac{x^{\frac{1}{6}}}{x^{\frac{5}{6}}}$

52. $\dfrac{x^{\frac{1}{4}}}{x^{\frac{3}{4}}}$

53. $\left(x^{\frac{1}{4}}y^3\right)^{\frac{2}{3}}$

54. $\left(x^{\frac{1}{6}}y^{15}\right)^{\frac{3}{5}}$

55. $\left(\dfrac{x^{\frac{2}{5}}}{x^{\frac{6}{5}} \cdot x^{\frac{3}{5}}}\right)^{5}$

56. $\left(\dfrac{x^{\frac{4}{7}}}{x^{\frac{3}{7}} \cdot x^{\frac{2}{7}}}\right)^{49}$

Application Exercises

57. The maximum velocity v, in miles per hour, that an automobile can travel around a curve with radius r feet without skidding is modeled by the formula

$$v = \left(\frac{5r}{2}\right)^{\frac{1}{2}}.$$

If the curve has a radius of 250 feet, find the maximum velocity a car can travel around it without skidding.

58. The formula

$$v = \left(\frac{p}{0.015}\right)^{\frac{1}{3}}$$

models the wind speed, v, in miles per hour, needed to produce p watts of power from a windmill. How fast must the wind be blowing to produce 120 watts of power?

According to the American Management Association, the percentage of potential employees testing positive for illegal drugs is on the decline. The formula

$$P = \frac{73t^{\frac{1}{3}} - 28t^{\frac{2}{3}}}{t}$$

models the percentage, P, of people applying for jobs who tested positive t years after 1985. Use this model to solve Exercises 59–60.

59. What percentage of people applying for jobs tested positive for illegal drugs in 1993?

60. What percentage of people applying for jobs tested positive for illegal drugs in 2001? Use a calculator and round to the nearest hundredth of a percent.

The formula $V = 194.8t^{\frac{1}{6}}$ models the number of visitors, V, in millions, to America's national parks t years after 1985. The formula $E = 339.6t^{\frac{2}{3}}$ models the expenditures, E, in millions of dollars, for the National Park Service t years after 1985. Use these models to solve Exercises 61–62.

61. Express the models for E and V in radical form.

62. Use $\dfrac{E}{V}$ to find the cost per visitor for the National Park Service in 2001. Round to the nearest cent.

Writing in Mathematics

63. What is the meaning of $a^{\frac{1}{n}}$? Give an example to support your explanation.

64. What is the meaning of $a^{\frac{m}{n}}$? Give an example.

65. What is the meaning of $a^{-\frac{m}{n}}$? Give an example.

66. Explain why $a^{\frac{1}{n}}$ is negative when n is odd and a is negative. What happens if n is even and a is negative? Why?

67. In simplifying $36^{\frac{3}{2}}$, is it better to use $a^{\frac{m}{n}} = \sqrt[n]{a^m}$ or $a^{\frac{m}{n}} = (\sqrt[n]{a})^m$? Explain.

68. In Exercises 59–60, you used a model for the percentage of people applying for jobs who tested positive for illegal drugs. Drug testing is a controversial issue. Is testing of this kind a violation of privacy rights or an appropriate procedure for companies to screen potential new hires? Explain your position.

Critical Thinking Exercises

69. Which one of the following is true?

a. $2^{\frac{1}{2}} \cdot 2^{\frac{1}{2}} = 4^{\frac{1}{2}}$

b. $8^{-\frac{1}{2}} = \frac{1}{4}$

c. $25^{-\frac{1}{2}} = -5$

d. $-3^{-2} = \frac{1}{9}$

70. Which one of the following is true?

a. $2^{\frac{1}{2}} \cdot 2^{\frac{3}{2}} = \left(\frac{1}{4}\right)^{-1}$

b. $16^{-\frac{1}{4}} = -2$

c. The result of $81^{\frac{1}{4}} \cdot 125^{\frac{1}{3}}$ is not an integer.

d. $-8^{\frac{1}{3}}$ and $(-8)^{\frac{1}{3}}$ do not result in the same answer.

Without using a calculator, simplify the expressions in Exercises 71–72 completely.

71. $25^{\frac{1}{4}} \cdot 25^{-\frac{3}{4}}$

72. $\dfrac{3^{-1} \cdot 3^{\frac{1}{2}}}{3^{-\frac{3}{2}}}$

Technology Exercises

73. The territorial area of an animal in the wild is the area of the region to which the animal confines its movements. The formula

$$T = W^{1.41} = W^{\frac{141}{100}} = \sqrt[100]{W^{141}}$$

models the territorial area, T, in square miles, in terms of an animal's body weight, W, in pounds.

a. Use a calculator to fill in the table of values, rounding T to the nearest whole square mile.

W	0	25	50	150	200	250	300
$T = W^{1.41}$							

b. Use the table of values to graph $T = W^{1.41}$. What does the shape of the graph indicate about the relationship between body weight and territorial area?

c. Verify your hand-drawn graph by using a graphing utility to graph the model

74. If A is the surface area of a cube and V is its volume, then $A = 6V^{\frac{2}{3}}$.

a. Graph the equation $\left(y = 6x^{\frac{2}{3}}\right)$ relating a cube's surface area and volume using a graphing utility and a $[0, 30, 3]$ by $[0, 60, 3]$ viewing rectangle.

b. $\boxed{\text{TRACE}}$ or $\boxed{\text{ZOOM IN}}$ along the curve and verify the numbers shown in the figure below. In particular, show that a cube whose volume is 27 cubic units has a surface area of 54 square units.

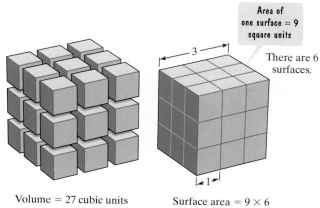

Volume = 27 cubic units

Area of one surface = 9 square units

There are 6 surfaces.

Surface area = 9×6 = 54 square units

c. $\boxed{\text{TRACE}}$ or $\boxed{\text{ZOOM IN}}$ along the curve and find the surface area of a cube whose volume is 8 cubic units.

Review Exercises

75. Solve the system:

$$7x - 3y = -14$$
$$y = 3x + 6.$$

(Section 5.2, Example 1)

76. Graph the solutions of the system:

$$-3x + 4y \leq 12$$
$$x \geq 2.$$

(Section 5.5, Example 2)

77. Simplify: $\dfrac{(2x)^5}{x^3}$. (Section 6.7, Example 5)

GROUP PROJECT

CHAPTER 9

The following topics related to irrational roots are appropriate for a group research project. A group report should be given to the class on the researched topic.

a. A History of How Irrational Numbers Developed
b. Proving that $\sqrt{2}$ Is Irrational
c. Golden Rectangles in Art and Architecture (See Exercise Set 9.4, Exercise 85.)
d. Golden Ratios in Proportions of the Human Body (See Exercise Set 9.4, Exercise 85.)
e. Radicals in Nature

CHAPTER 9 SUMMARY

Definitions and Concepts	Examples
Section 9.1 Finding Roots	
If $b^2 = a$, then b is a square root of a. The principal square root of a, designated $\sqrt{a}$, is the nonnegative number satisfying $b^2 = a$. The negative square root of a is written $-\sqrt{a}$. A square root of a negative number is not a real number.	• $\sqrt{100} = 10$ because $10^2 = 100$. • $-\sqrt{100} = -10$ • $\sqrt{-100}$ is not a real number.
The principal nth root of a real number a, symbolized by $\sqrt[n]{a}$, is defined as follows: $$\sqrt[n]{a} = b \text{ means that } b^n = a.$$ The natural number n is called the index, the symbol $\sqrt[n]{}$ is called a radical sign, and the expression under the radical sign is called the radicand. If the index is even, the radicand must be nonnegative for the root to be a real number and the principal nth root is nonnegative.	• $\sqrt[3]{8} = 2$ because $2^3 = 8$. • $\sqrt[3]{-125} = -5$ because $(-5)^3 = -125$. • $\sqrt[4]{81} = 3$ because $3^4 = 81$. • $\sqrt[4]{-81}$ is not a real number.
Section 9.2 Multiplying and Dividing Radicals	
The Product Rule for Roots $\sqrt[n]{a} \cdot \sqrt[n]{b} = \sqrt[n]{ab}$ (The roots represent real numbers.)	• $\sqrt{11} \cdot \sqrt{7} = \sqrt{11 \cdot 7} = \sqrt{77}$ • $\sqrt[3]{4} \cdot \sqrt[3]{2} = \sqrt[3]{4 \cdot 2} = \sqrt[3]{8} = 2$
A square root is simplified when its radicand has no factors other than 1 that are perfect squares. To simplify a square root, factor the radicand so that one of its factors is a perfect square. Simplify square roots with a variable to an even power using $\sqrt{x^{2n}} = x^n$. Simplify square roots with a variable to an odd power by expressing the variable as the product of two factors, one of which has an even power.	• $\sqrt{63} = \sqrt{9 \cdot 7} = \sqrt{9} \cdot \sqrt{7} = 3\sqrt{7}$ • $\sqrt{x^{26}} = x^{13}$ • $\sqrt{x^{27}} = \sqrt{x^{26} \cdot x}$ $\quad = \sqrt{x^{26}} \cdot \sqrt{x} = x^{13}\sqrt{x}$ • $\sqrt{45x^3} = \sqrt{9x^2 \cdot 5x}$ $\quad = \sqrt{9x^2} \cdot \sqrt{5x} = 3x\sqrt{5x}$ • $\sqrt{8} \cdot \sqrt{3} = \sqrt{24} = \sqrt{4 \cdot 6}$ $\quad = \sqrt{4} \cdot \sqrt{6} = 2\sqrt{6}$

Definitions and Concepts	Examples

Section 9.2 Multiplying and Dividing Radicals (continued)

The Quotient Rule for Roots

$$\frac{\sqrt[n]{a}}{\sqrt[n]{b}} = \sqrt[n]{\frac{a}{b}}$$

(The roots represent real numbers and no denominators are 0.)

• $\sqrt{\dfrac{36}{x^2}} = \dfrac{\sqrt{36}}{\sqrt{x^2}} = \dfrac{6}{x}$

• $\sqrt[3]{\dfrac{2}{125}} = \dfrac{\sqrt[3]{2}}{\sqrt[3]{125}} = \dfrac{\sqrt[3]{2}}{5}$

• $\dfrac{\sqrt{50x^4}}{\sqrt{2x}} = \sqrt{\dfrac{50x^4}{2x}} = \sqrt{25x^3}$

$$= \sqrt{25x^2 \cdot x} = 5x\sqrt{x}$$

Section 9.3 Operations with Radicals

Square roots with the same radicand are like radicals. Like radicals can be added or subtracted using the distributive property. In some cases, radicals can be combined once they have been simplified.

• $7\sqrt{2} + 9\sqrt{2} = (7 + 9)\sqrt{2}$

$$= 16\sqrt{2}$$

• $6\sqrt{8} - \sqrt{2} = 6\sqrt{4 \cdot 2} - \sqrt{2}$

$$= 6\sqrt{4}\,\sqrt{2} - \sqrt{2}$$
$$= 6 \cdot 2\sqrt{2} - \sqrt{2}$$
$$= 12\sqrt{2} - \sqrt{2}$$
$$= (12 - 1)\sqrt{2}$$
$$= 11\sqrt{2}$$

Radical expressions with more than one term are multiplied in much the same way that polynomials with more than one term are multiplied.

• $\sqrt{5}\left(\sqrt{2} + \sqrt{5}\right) = \sqrt{10} + \sqrt{25}$

$$= \sqrt{10} + 5$$

• $\left(2 + \sqrt{3}\right)\left(4 + 5\sqrt{3}\right)$

F   O   I   L

$$= 2 \cdot 4 + 2(5\sqrt{3}) + 4\sqrt{3} + \sqrt{3}(5\sqrt{3})$$
$$= 8 + 10\sqrt{3} + 4\sqrt{3} + 15$$
$$= 23 + 14\sqrt{3}$$

Radical expressions that involve the sum and difference of the same two terms are called conjugates. Use the special product

$$(A + B)(A - B) = A^2 - B^2$$

to multiply conjugates.

$\left(\sqrt{11} + 5\right)\left(\sqrt{11} - 5\right)$
$$= \left(\sqrt{11}\right)^2 - 5^2 = 11 - 25 = -14$$

Section 9.4 Rationalizing the Denominator

The process of rewriting a radical expression as an equivalent expression in which the denominator no longer contains any radicals is called rationalizing the denominator. If the denominator contains the square root of a natural number that is not a perfect square, multiply the numerator and the denominator by the smallest number that produces the square root of a perfect square in the denominator.

• Rationalize the denominator:

$$\frac{4}{\sqrt{7}}.$$

$$\frac{4}{\sqrt{7}} = \frac{4}{\sqrt{7}} \cdot \frac{\sqrt{7}}{\sqrt{7}} = \frac{4\sqrt{7}}{7}$$

Definitions and Concepts	Examples

Section 9.4 Rationalizing the Denominator (continued)

If the denominator contains two terms, rationalize the denominator by multiplying the numerator and the denominator by the conjugate of the denominator.

$$\frac{9}{7 - \sqrt{5}} = \frac{9}{7 - \sqrt{5}} \cdot \frac{7 + \sqrt{5}}{7 + \sqrt{5}}$$

$$= \frac{9(7 + \sqrt{5})}{7^2 - (\sqrt{5})^2}$$

$$= \frac{9(7 + \sqrt{5})}{49 - 5} = \frac{9(7 + \sqrt{5})}{44}$$

Section 9.5 Radical Equations

A radical equation is an equation in which the variable occurs in a radicand.

Solving Radical Equations Containing Square Roots

1. Isolate the radical.
2. Square both sides.
3. Solve the resulting equation.
4. Check proposed solutions in the original equation. Solutions of the squared equation, but not the original equation, are called extraneous solutions.

Solve: $\sqrt{2x + 1} - x = -7$.

$\sqrt{2x + 1} = x - 7$ Isolate the radical.

$(\sqrt{2x + 1})^2 = (x - 7)^2$ Square both sides.

$2x + 1 = x^2 - 14x + 49$ Use $(A - B)^2 = A^2 - 2AB + B^2$.

$0 = x^2 - 16x + 48$ Write in standard form.

$0 = (x - 12)(x - 4)$ Factor.

$x - 12 = 0$ or $x - 4 = 0$ Set each factor equal to 0.

$x = 12$ $x = 4$ Solve the resulting equations.

Check both proposed solutions. 12 checks, but 4 is extraneous. The only solution is 12.

Section 9.6 Rational Exponents

- $a^{\frac{1}{n}} = \sqrt[n]{a}$

- $a^{\frac{m}{n}} = (\sqrt[n]{a})^m$

- $a^{-\frac{m}{n}} = \frac{1}{a^{\frac{m}{n}}}$

- $16^{\frac{1}{2}} = \sqrt{16} = 4$

- $8^{\frac{5}{3}} = (\sqrt[3]{8})^5 = 2^5 = 32$

- $81^{-\frac{3}{4}} = \frac{1}{81^{\frac{3}{4}}} = \frac{1}{(\sqrt[4]{81})^3} = \frac{1}{3^3} = \frac{1}{27}$

- $27^{\frac{1}{3}} = \sqrt[3]{27} = 3$

CHAPTER 9   REVIEW EXERCISES

9.1 *In Exercises 1–6, find the indicated root, or state that the expression is not a real number.*

1. $\sqrt{121}$

2. $-\sqrt{121}$

3. $\sqrt{-121}$

4. $\sqrt[3]{\frac{8}{125}}$

5. $\sqrt[5]{-32}$

6. $-\sqrt[4]{81}$

In Exercises 7–8, use a calculator to approximate each square root. Round to three decimal places.

7. $\sqrt{75}$

8. $\sqrt{398 - 5}$

9. The formula $P = 26.5\sqrt{t}$ models the thousands of people over age 85, P, in Arizona t years after 1990. Find the over-85 population in 1999.

10. Use the model in Exercise 9 to describe what is happening to Arizona's over-85 population over time.

11. The formula

$$d = \sqrt{\frac{3h}{2}}$$

models the distance, d, in miles, that you can see to the horizon at a height of h feet. A 1575-foot skyscraper that is being built in Hong Kong will be the world's tallest building. How far to the horizon will visitors be able to see from the top of the building? Use a calculator and round to the nearest mile.

9.2 *In Exercises 12–19, simplify each expression.*

12. $\sqrt{54}$

13. $6\sqrt{20}$

14. $\sqrt{63x^2}$

15. $\sqrt{48x^3}$

16. $\sqrt{x^8}$

17. $\sqrt{75x^9}$

18. $\sqrt{45x^{23}}$

19. $\sqrt[3]{24}$

In Exercises 20–25, multiply and, if possible, simplify.

20. $\sqrt{7} \cdot \sqrt{11}$

21. $\sqrt{3} \cdot \sqrt{12}$

22. $\sqrt{5x} \cdot \sqrt{10x}$

23. $\sqrt{3x^2} \cdot \sqrt{4x^3}$

24. $\sqrt[3]{6} \cdot \sqrt[3]{9}$

25. $\sqrt{\dfrac{5}{2}} \cdot \sqrt{\dfrac{3}{8}}$

In Exercises 26–33, simplify using the quotient rule.

26. $\sqrt{\dfrac{121}{4}}$

27. $\sqrt{\dfrac{7x}{25}}$

28. $\sqrt{\dfrac{18}{x^2}}$

29. $\dfrac{\sqrt{200}}{\sqrt{2}}$

30. $\dfrac{\sqrt{96}}{\sqrt{3}}$

31. $\dfrac{\sqrt{72x^8}}{\sqrt{x^3}}$

32. $\sqrt[3]{\dfrac{5}{64}}$

33. $\sqrt[3]{\dfrac{40}{27}}$

9.3 *In Exercises 34–39, add or subtract as indicated.*

34. $7\sqrt{5} + 13\sqrt{5}$

35. $\sqrt{8} + \sqrt{50}$

36. $\sqrt{75} - \sqrt{48}$

37. $2\sqrt{80} + 3\sqrt{45}$

38. $4\sqrt{72} - 2\sqrt{48}$

39. $2\sqrt{18} + 3\sqrt{27} - \sqrt{12}$

In Exercises 40–47, multiply as indicated and, if possible, simplify.

40. $\sqrt{10}\left(\sqrt{5} + \sqrt{6}\right)$

41. $\sqrt{3}\left(\sqrt{6} - \sqrt{12}\right)$

42. $\left(9 + \sqrt{2}\right)\left(10 + \sqrt{2}\right)$

43. $\left(1 + 3\sqrt{7}\right)\left(4 - \sqrt{7}\right)$

44. $\left(\sqrt{3} + 2\right)\left(\sqrt{6} - 4\right)$

45. $\left(2 + \sqrt{7}\right)\left(2 - \sqrt{7}\right)$

46. $\left(\sqrt{11} - \sqrt{5}\right)\left(\sqrt{11} + \sqrt{5}\right)$

47. $\left(1 + \sqrt{2}\right)^2$

9.4 *In Exercises 48–56, rationalize each denominator and, if possible, simplify.*

48. $\dfrac{30}{\sqrt{5}}$

49. $\dfrac{13}{\sqrt{50}}$

50. $\sqrt{\dfrac{2}{3}}$

51. $\sqrt{\dfrac{3}{8}}$

52. $\sqrt{\dfrac{17}{x}}$

53. $\dfrac{11}{\sqrt{5} + 2}$

54. $\dfrac{21}{4 - \sqrt{3}}$

55. $\dfrac{12}{\sqrt{5} + \sqrt{3}}$

56. $\dfrac{7\sqrt{2}}{\sqrt{2} - 4}$

9.5 *In Exercises 57–63, solve each radical equation. If the equation has no solution, so state.*

57. $\sqrt{x + 3} = 4$

58. $\sqrt{2x + 3} = 5$

59. $3\sqrt{x} = \sqrt{6x + 15}$

60. $\sqrt{5x + 1} = x + 1$

61. $\sqrt{x + 1} + 5 = x$

62. $\sqrt{x - 2} + 5 = 1$

63. $x = \sqrt{x^2 + 4x + 4}$

64. The time, t, in seconds, for a free-falling object to fall d feet is modeled by the formula

$$t = \sqrt{\dfrac{d}{16}}.$$

A rock that is dropped from a bridge takes 3 seconds to hit the water. How far above the water is the bridge?

65. The distance to the horizon that you can see, D, in miles, on the top of a mountain H feet high is modeled by the formula $D = \sqrt{2H}$. You've hiked to the top of a mountain with views extending 50 miles to the horizon. How high is the mountain?

9.6 *In Exercises 66–71, simplify by first writing the expression in radical form.*

66. $16^{\frac{1}{2}}$

67. $125^{\frac{1}{3}}$

68. $64^{\frac{2}{3}}$

69. $25^{-\frac{1}{2}}$

70. $27^{-\frac{1}{3}}$

71. $(-8)^{-\frac{4}{3}}$

72. The formula $S = 28.6A^{\frac{1}{3}}$ models the number of plant species, S, on the various islands of the Galápagos chain in terms of the area, A, in square miles, of a particular island. Approximately how many species of plants are there on a Galápagos island whose area is 8 square miles?

CHAPTER 9   TEST

Remember to use your Chapter Test Prep Video CD to help you study and view solutions to the test questions you need help with.

In Exercises 1–2, find the indicated root, or state that the expression is not a real number.

1. $-\sqrt{64}$

2. $\sqrt[3]{64}$

In Exercises 3–8, simplify each expression.

3. $\sqrt{48}$

4. $\sqrt{72x^3}$

5. $\sqrt{x^{29}}$

6. $\sqrt{\dfrac{25}{x^2}}$

7. $\sqrt{\dfrac{75}{27}}$

8. $\sqrt[3]{\dfrac{5}{8}}$

In Exercises 9–18, perform the indicated operation and, if possible, simplify.

9. $\dfrac{\sqrt{80x^4}}{\sqrt{2x^2}}$

10. $\sqrt{10} \cdot \sqrt{5}$

11. $\sqrt{6x} \cdot \sqrt{6y}$

12. $\sqrt{10x^2} \cdot \sqrt{2x^3}$

13. $\sqrt{24} + 3\sqrt{54}$

14. $7\sqrt{8} - 2\sqrt{32}$

15. $\sqrt{3}\left(\sqrt{10} + \sqrt{3}\right)$

16. $\left(7 - \sqrt{5}\right)\left(10 + 3\sqrt{5}\right)$

17. $\left(\sqrt{6} + 2\right)\left(\sqrt{6} - 2\right)$

18. $\left(3 + \sqrt{7}\right)^2$

In Exercises 19–20, rationalize each denominator and, if possible, simplify.

19. $\dfrac{4}{\sqrt{5}}$

20. $\dfrac{5}{4 + \sqrt{3}}$

In Exercises 21–22, solve each radical equation. If the equation has no solution, so state.

21. $\sqrt{3x} + 5 = 11$

22. $\sqrt{2x - 1} = x - 2$

23. The time, t, in seconds, for a free-falling object to fall d feet is modeled by the formula

$$t = \sqrt{\dfrac{d}{16}}.$$

How many feet will a skydiver fall in 10 seconds?

In Exercises 24–25, simplify by first writing the expression in radical form.

24. $8^{\frac{2}{3}}$

25. $9^{-\frac{1}{2}}$

CUMULATIVE REVIEW EXERCISES (CHAPTERS 1–9)

In Exercises 1–6, solve each equation or system of equations.

1. $2x + 3x - 5 + 7 = 10x + 3 - 6x - 4$

2. $2x^2 + 5x = 12$

3. $8x - 5y = -4$
$2x + 15y = -66$

4. $\dfrac{15}{x} - 4 = \dfrac{6}{x} + 3$

5. $-3x - 7 = 8$

6. $\sqrt{2x - 1} - x = -2$

In Exercises 7–11, simplify each expression.

7. $\dfrac{8x^3}{-4x^7}$

8. $6\sqrt{75} - 4\sqrt{12}$

9. $\dfrac{\dfrac{1}{x} - \dfrac{1}{2}}{\dfrac{1}{3} - \dfrac{x}{6}}$

10. $\dfrac{4 - x^2}{3x^2 - 5x - 2}$

11. $-5 - (-8) - (4 - 6)$

In Exercises 12–13, factor completely.

12. $x^2 - 18x + 77$

13. $x^3 - 25x$

In Exercises 14–18, perform the indicated operations. If possible, simplify the answer.

14. $\dfrac{6x^3 - 19x^2 + 16x - 4}{x - 2}$

15. $(2x - 3)(4x^2 + 6x + 9)$

16. $\dfrac{3x}{x^2 + x - 2} - \dfrac{2}{x + 2}$

17. $\dfrac{5x^2 - 6x + 1}{x^2 - 1} \div \dfrac{16x^2 - 9}{4x^2 + 7x + 3}$

18. $\sqrt{12} - 4\sqrt{75}$

In Exercises 19–21, graph each equation or inequality in a rectangular coordinate system.

19. $2x - y = 4$

20. $y = -\dfrac{2}{3}x$

21. $x \geq -1$

22. Find the slope of the line through $(-1, 5)$ and $(2, -3)$.

23. Write the point-slope form of the equation of the line with slope 5, passing through $(-2, -3)$. Then use the point-slope equation to write the slope-intercept form of the line's equation.

24. Seven subtracted from five times a number is 208. Find the number.

25. Park rangers catch, tag, and then release 318 deer back into a state park. Two weeks later, they select a sample of 168 deer, 56 of which are tagged. Assuming the ratio of tagged deer in the sample holds for all deer in the park, approximately how many deer are in the park?

This is the most spectacular display of fireworks you've ever seen. You can hear the collective oohs and ahs coming from the massive throng of people mesmerized by the presentation. How do the people launching the fireworks know when they should be set off so they can be viewed at the greatest possible height?

This problem appears as Exercise 49 in Exercise Set 10.5.

Quadratic Equations and Functions

When did the number of people receiving food stamps reach a maximum? When were women's earnings as a percentage of men's at the lowest? What is the youngest average age at which American men were first married and when did this occur? Applications in which a quantity is to have a maximum or minimum value can often be modeled by equations of the form

$$y = ax^2 + bx + c.$$

In this chapter, you will learn techniques for solving quadratic equations,

$$ax^2 + bx + c = 0,$$

that will give you insight into these models. You will also learn to graph the models so that you can visualize their highest or lowest point.

10.1

Objectives

1 Solve quadratic equations using the square root property.

2 Solve problems using the Pythagorean Theorem.

3 Find the distance between two points.

SOLVING QUADRATIC EQUATIONS BY THE SQUARE ROOT PROPERTY

Shown here is Renaissance artist Raphael Sanzio's (1483–1520) image of Pythagoras from *The School of Athens* mural. Detail of left side.

Stanza della Segnatura, Vatican Palace, Vatican State. Scala/Art Resource, NY.

Pythagoras

For the followers of the Greek mathematician Pythagoras in the sixth century B.C., numbers took on a life-and-death importance. The "Pythagorean Brotherhood" was a secret group whose members were convinced that properties of whole numbers were the key to understanding the universe. Members of the Brotherhood (which admitted women) thought that all numbers that were not whole numbers could be represented as the ratio of whole numbers. A crisis occurred for the Pythagoreans when they discovered the existence of a number that was not rational. Because the Pythagoreans viewed numbers with reverence and awe, the punishment for speaking about this irrational number was death. However, a member of the Brotherhood revealed the secret of the irrational number's existence. When he later died in a shipwreck, his death was viewed as punishment from the gods.

In this section, you will work with the triangle that led the Pythagoreans to the discovery of irrational numbers. You will find the lengths of one of the triangle's sides using a property other than factoring that can be used to solve quadratic equations.

1 Solve quadratic equations using the square root property.

The Square Root Property Let's begin with a relatively simple quadratic equation:

$$x^2 = 9.$$

The value of x must be a number whose square is 9. There are two numbers whose square is 9:

$$x = \sqrt{9} = 3 \quad \text{or} \quad x = -\sqrt{9} = -3.$$

Thus, the solutions of $x^2 = 9$ are 3 and -3. This is an example of the **square root property**.

> **THE SQUARE ROOT PROPERTY** If u is an algebraic expression and d is a positive real number, then $u^2 = d$ has exactly two solutions:
>
> If $u^2 = d$, then $u = \sqrt{d}$ or $u = -\sqrt{d}$.
>
> Equivalently,
>
> If $u^2 = d$, then $u = \pm\sqrt{d}$.

Notice that $u = \pm\sqrt{d}$ is a shorthand notation to indicate that $u = \sqrt{d}$ or $u = -\sqrt{d}$. Although we usually read $u = \pm\sqrt{d}$ as "u equals plus or minus the square root of d," we actually mean that u is the positive square root of d or the negative square root of d.

EXAMPLE 1 Solving Quadratic Equations by the Square Root Property

Solve by the square root property:

    a. $x^2 = 49$ **b.** $4x^2 = 20$ **c.** $2x^2 - 5 = 0$.

SOLUTION

    a. $x^2 = 49$ *This is the original equation.*

        $x = \sqrt{49}$ or $x = -\sqrt{49}$ *Apply the square root property. You can also write $x = \pm\sqrt{49}$.*

        $x = 7$ or $x = -7$ *In abbreviated notation, $x = \pm 7$.*

Substitute both values into the original equation and confirm that the solutions are 7 and -7.

    b. To apply the square root property, we need a squared expression by itself on one side of the equation.

$$4x^2 = 20$$

We want x^2 by itself.

We can get x^2 by itself if we divide both sides by 4.

        $4x^2 = 20$ *This is the original equation.*

        $\dfrac{4x^2}{4} = \dfrac{20}{4}$ *Divide both sides by 4.*

        $x^2 = 5$ *Simplify.*

    $x = \sqrt{5}$ or $x = -\sqrt{5}$ *Apply the square root property.*

Now let's check these proposed solutions in the original equation.

        Check $\sqrt{5}$:   **Check $-\sqrt{5}$:**

        $4x^2 = 20$ $4x^2 = 20$

    $4\left(\sqrt{5}\right)^2 \stackrel{?}{=} 20$ $4\left(-\sqrt{5}\right)^2 \stackrel{?}{=} 20$

      $4 \cdot 5 \stackrel{?}{=} 20$ $4 \cdot 5 \stackrel{?}{=} 20$

     $20 = 20$,   *true* $20 = 20$,   *true*

The solutions are $\sqrt{5}$ and $-\sqrt{5}$.

c. To solve $2x^2 - 5 = 0$ by the square root property, we must isolate the squared expression by itself on one side of the equation.

$$2x^2 - 5 = 0$$

We want x^2 by itself.

$$2x^2 - 5 = 0 \qquad \text{This is the original equation.}$$
$$2x^2 = 5 \qquad \text{Add 5 to both sides.}$$
$$x^2 = \frac{5}{2} \qquad \text{Divide both sides by 2.}$$
$$x = \sqrt{\frac{5}{2}} \quad \text{or} \quad x = -\sqrt{\frac{5}{2}} \qquad \text{Apply the square root property.}$$

Substitute $\sqrt{\frac{5}{2}}$ and $-\sqrt{\frac{5}{2}}$ into the original equation and verify that these numbers are solutions. In this section, we will express irrational solutions in simplified radical form, rationalizing denominators when possible. Rationalizing denominators, the solutions are $\dfrac{\sqrt{10}}{2}$ and $-\dfrac{\sqrt{10}}{2}$. ∎

USING TECHNOLOGY

The graph of $y = (x - 5)^2 - 16$ has x-intercepts at 1 and 9. The solutions of $(x - 5)^2 - 16 = 0$, or $(x - 5)^2 = 16$, are 1 and 9.

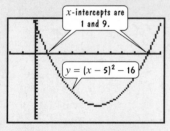

x-intercepts are 1 and 9.

$y = (x - 5)^2 - 16$

$[-2, 10, 1]$ by $[-20, 10, 1]$

Another option is to graph each side of the equation and find the x-coordinates of the intersection points. Graphing

$$y_1 = (x - 5)^2$$

and

$$y_2 = 16,$$

the x-coordinates of the intersection points are 1 and 9.

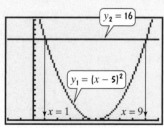

$y_2 = 16$

$y_1 = (x - 5)^2$

$x = 1$ $x = 9$

$[-2, 10, 1]$ by $[0, 20, 1]$

 **CHECK POINT 1** Solve by the square root property:

a. $x^2 = 36$ **b.** $5x^2 = 15$

c. $2x^2 - 7 = 0.$

Can we solve an equation such as $(x - 5)^2 = 16$ using the square root property? Yes. The equation is in the form $u^2 = d$, where u^2, the squared expression, is by itself on the left side.

$$(x - 5)^2 = 16$$

This is u^2 in $u^2 = d$ with $u = x - 5$.

This is d in $u^2 = d$ with $d = 16$.

EXAMPLE 2 Solving a Quadratic Equation by the Square Root Property

Solve by the square root property: $(x - 5)^2 = 16$.

SOLUTION

$$(x - 5)^2 = 16 \qquad \text{This is the original equation.}$$
$$x - 5 = \sqrt{16} \quad \text{or} \quad x - 5 = -\sqrt{16} \qquad \text{Apply the square root property.}$$
$$x - 5 = 4 \quad \text{or} \quad x - 5 = -4 \qquad \text{Simplify.}$$
$$x = 9 \qquad\qquad x = 1 \qquad \text{Add 5 to both sides in each equation.}$$

Substitute both values into the original equation and confirm that the solutions are 9 and 1. The solutions are visually confirmed in the Using Technology box. ∎

 CHECK POINT 2 Solve by the square root property: $(x - 3)^2 = 25$.

> ### EXAMPLE 3 Solving a Quadratic Equation by the Square Root Property
>
> Solve by the square root property: $(x - 1)^2 = 5$.
>
> **SOLUTION**
>
> $$(x - 1)^2 = 5 \qquad \text{This is the original equation.}$$
> $$x - 1 = \sqrt{5} \quad \text{or} \quad x - 1 = -\sqrt{5} \qquad \text{Apply the square root property.}$$
> $$x = 1 + \sqrt{5} \qquad x = 1 - \sqrt{5} \qquad \text{Add 1 to both sides in each equation.}$$
>
> **Check** $1 + \sqrt{5}$: | **Check** $1 - \sqrt{5}$:
>
> $$(x - 1)^2 = 5 \qquad\qquad (x - 1)^2 = 5$$
> $$\left(1 + \sqrt{5} - 1\right)^2 \stackrel{?}{=} 5 \qquad\qquad \left(1 - \sqrt{5} - 1\right)^2 \stackrel{?}{=} 5$$
> $$\left(\sqrt{5}\right)^2 \stackrel{?}{=} 5 \qquad\qquad \left(-\sqrt{5}\right)^2 \stackrel{?}{=} 5$$
> $$5 = 5, \quad \text{true} \qquad\qquad 5 = 5, \quad \text{true}$$
>
> The solutions are $1 + \sqrt{5}$ and $1 - \sqrt{5}$, expressed in abbreviated notation as $1 \pm \sqrt{5}$. ∎
>
> ✔ **CHECK POINT 3** Solve by the square root property: $(x - 2)^2 = 7$.

2 Solve problems using the Pythagorean Theorem.

The Pythagorean Theorem and the Square Root Property The ancient Greek philosopher and mathematician Pythagoras (approximately 582–500 B.C.) founded a school whose motto was "All is number." Pythagoras is best remembered for his work with the **right triangle**, a triangle with one angle measuring 90°. The side opposite the 90° angle is called the **hypotenuse**. The other sides are called **legs**. Pythagoras found that if he constructed squares on each of the legs, as well as a larger square on the hypotenuse, the sum of the areas of the smaller squares is equal to the area of the larger square. This is illustrated in Figure 10.1.

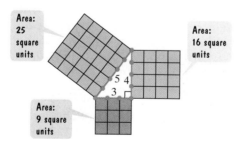

Area: 25 square units

Area: 16 square units

Area: 9 square units

FIGURE 10.1 The area of the large square equals the sum of the areas of the smaller squares.

This relationship is usually stated in terms of the lengths of the three sides of a right triangle and is called the **Pythagorean Theorem**.

THE PYTHAGOREAN THEOREM The sum of the squares of the lengths of the legs of a right triangle equals the square of the length of the hypotenuse.

If the legs have lengths a and b, and the hypotenuse has length c, then

$$a^2 + b^2 = c^2.$$

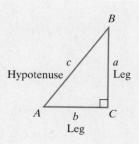

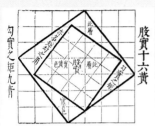
EXAMPLE 4 Using the Pythagorean Theorem

In a 25-inch television set, the length of the screen's diagonal is 25 inches. If the screen's height is 15 inches, what is its width?

SOLUTION   Figure 10.2 shows a right triangle that is formed by the height, width, and diagonal. We can find w, the screen's width, using the Pythagorean Theorem.

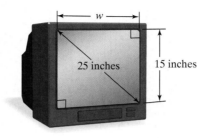

FIGURE 10.2 A right triangle is formed by the television's height, width, and diagonal.

(Leg)²	plus	(Leg)²	equals	(Hypotenuse)².
w^2	$+$	15^2	$=$	25^2

This is the equation resulting from the Pythagorean Theorem.

The equation $w^2 + 15^2 = 25^2$ can be solved by the square root property.

$$w^2 + 15^2 = 25^2$$ This is the equation that models the verbal conditions.

$$w^2 + 225 = 625$$ Square 15 and 25.

$$w^2 = 400$$ Isolate w^2 by subtracting 225 from both sides.

$$w = \sqrt{400} \quad \text{or} \quad w = -\sqrt{400}$$ Apply the square root property.

$$w = 20 \qquad\qquad w = -20$$ Simplify.

Because w represents the width of the television's screen, this dimension must be positive. We reject -20. Thus, the width of the television is 20 inches. ∎

 CHECK POINT 4 What is the width in a 15-inch television set whose height is 9 inches?

3 Find the distance between two points.

The Distance Formula   Using the Pythagorean Theorem, we can find the distance between the two points $P_1(x_1, y_1)$ and $P_2(x_2, y_2)$ in the rectangular coordinate system. The two points are illustrated in Figure 10.3.

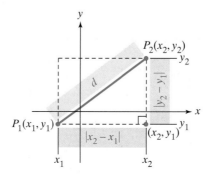

FIGURE 10.3

The distance that we need to find is represented by d and shown in blue. Notice that the distance between the two points on the dashed horizontal line is the absolute value of the difference between the x-coordinates of the two points. This distance, $|x_2 - x_1|$, is shown in pink. Similarly, the distance between the two points on the dashed vertical line is the absolute value of the difference between the y-coordinates of the two points. This distance, $|y_2 - y_1|$, is also shown in pink.

Because the dashed lines are horizontal and vertical, a right triangle is formed. Thus, we can use the Pythagorean Theorem to find the distance d. Squaring the lengths of the triangle's sides results in positive numbers, so absolute value notation is not necessary.

$$d^2 = (x_2 - x_1)^2 + (y_2 - y_1)^2 \qquad \text{This is the equation resulting from the Pythagorean Theorem.}$$

$$d = \pm\sqrt{(x_2 - x_1)^2 + (y_2 - y_1)^2} \qquad \text{Apply the square root property.}$$

$$d = \sqrt{(x_2 - x_1)^2 + (y_2 - y_1)^2} \qquad \text{Because distance is nonnegative, write only the principal square root.}$$

This result is called the **distance formula**.

> **THE DISTANCE FORMULA** The distance, d, between the points (x_1, y_1) and (x_2, y_2) in the rectangular coordinate system is
> $$d = \sqrt{(x_2 - x_1)^2 + (y_2 - y_1)^2}.$$

When using the distance formula, it does not matter which point you call (x_1, y_1) and which you call (x_2, y_2).

EXAMPLE 5 Using the Distance Formula

Find the distance between $(-4, -3)$ and $(6, 2)$.

SOLUTION We will let $(x_1, y_1) = (-4, -3)$ and $(x_2, y_2) = (6, 2)$.

$$d = \sqrt{(x_2 - x_1)^2 + (y_2 - y_1)^2} \qquad \text{Use the distance formula.}$$

$$= \sqrt{[6 - (-4)]^2 + [2 - (-3)]^2} \qquad \text{Substitute the given values.}$$

$$= \sqrt{10^2 + 5^2} \qquad \text{Perform operations within grouping symbols: } 6 - (-4) = 6 + 4 = 10 \text{ and } 2 - (-3) = 2 + 3 = 5.$$

$$= \sqrt{100 + 25} \qquad \text{Square 10 and 5.}$$

Caution! This is **not** equal to $\sqrt{100} + \sqrt{25}$.

$$= \sqrt{125} \qquad \text{Add.}$$

$$= 5\sqrt{5} \approx 11.18 \qquad \sqrt{125} = \sqrt{25 \cdot 5} = \sqrt{25}\,\sqrt{5} = 5\sqrt{5}$$

The distance between the given points is $5\sqrt{5}$ units, or approximately 11.18 units. The situation is illustrated in Figure 10.4. ∎

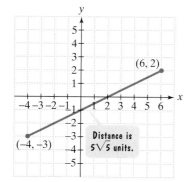

(6, 2)

(-4, -3)

Distance is $5\sqrt{5}$ units.

FIGURE 10.4 Finding the distance between two points

✔ **CHECK POINT 5** Find the distance between $(-4, 9)$ and $(1, -3)$.

10.1 EXERCISE SET

Student Solutions Manual   CD/Video   PH Math/Tutor Center   MathXL Tutorials on CD   MathXL®   MyMathLab   Interactmath.com

Practice Exercises

In Exercises 1–30, solve each quadratic equation by the square root property. If possible, simplify radicals or rationalize denominators.

1. $x^2 = 16$

2. $x^2 = 100$

3. $y^2 = 81$

4. $y^2 = 144$

5. $x^2 = 7$

6. $x^2 = 13$

7. $x^2 = 50$

8. $x^2 = 27$

9. $5x^2 = 20$

10. $3x^2 = 75$

11. $4y^2 = 49$

12. $16y^2 = 25$

13. $2x^2 + 1 = 51$

14. $3x^2 - 1 = 47$

15. $3x^2 - 2 = 0$

16. $3x^2 - 5 = 0$

17. $5z^2 - 7 = 0$

18. $5z^2 - 2 = 0$

19. $(x - 3)^2 = 16$

20. $(x - 2)^2 = 25$

21. $(x + 5)^2 = 121$

22. $(x + 6)^2 = 144$

23. $(3x + 2)^2 = 9$

24. $(2x + 1)^2 = 49$

25. $(x - 5)^2 = 3$

26. $(x - 3)^2 = 15$

27. $(y + 8)^2 = 11$

28. $(y + 7)^2 = 5$

29. $(z - 4)^2 = 18$

30. $(z - 6)^2 = 12$

In Exercises 31–40, solve each quadratic equation by first factoring the perfect square trinomial on the left side. Then apply the square root property. Simplify radicals, if possible.

31. $x^2 + 4x + 4 = 16$

32. $x^2 + 4x + 4 = 25$

33. $x^2 - 6x + 9 = 36$

34. $x^2 - 6x + 9 = 49$

35. $x^2 - 10x + 25 = 2$

36. $x^2 - 10x + 25 = 3$

37. $x^2 + 2x + 1 = 5$

38. $x^2 + 2x + 1 = 7$

39. $y^2 - 14y + 49 = 12$

40. $y^2 - 14y + 49 = 18$

In Exercises 41–48, use the Pythagorean Theorem to find the missing length in each right triangle. Express the answer in radical form and simplify, if possible.

41.

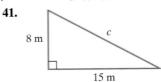

8 m, c, 15 m

42.

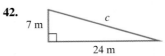

7 m, c, 24 m

43.

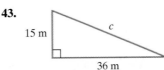

15 m, c, 36 m

44.

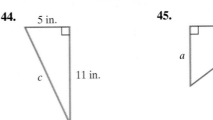

5 in., c, 11 in.

45.

16 cm, a, 20 cm

46.

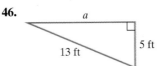

a, 13 ft, 5 ft

47.

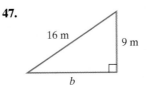

16 m, 9 m, b

48.

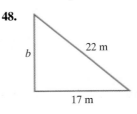

b, 22 m, 17 m

In Exercises 49–58, find the distance between each pair of points. Express answers in simplest radical form and, if necessary, round to two decimal places.

49. $(3, 5)$ and $(4, 1)$

50. $(1, 5)$ and $(6, 2)$

51. $(-4, 2)$ and $(4, 17)$

52. $(2, -2)$ and $(5, 2)$

53. $(6, -1)$ and $(9, 5)$

54. $(-4, -1)$ and $(2, -3)$

55. $(-7, -5)$ and $(-2, -1)$

56. $(-8, -4)$ and $(-3, -8)$

57. $\left(-2\sqrt{7}, 10\right)$ and $\left(4\sqrt{7}, 8\right)$

58. $\left(-\sqrt{3}, 4\sqrt{6}\right)$ and $\left(2\sqrt{3}, \sqrt{6}\right)$

Practice Plus

59. The square of the difference between a number and 3 is 25. Find the number(s).

60. The square of the difference between a number and 7 is 16. Find the number(s).

61. If 3 times a number is increased by 2 and this sum is squared, the result is 49. Find the number(s).

62. If 4 times a number is decreased by 3 and this difference is squared, the result is 9. Find the number(s).

In Exercises 63–66, solve the formula for the specified variable. Because each variable is nonnegative, list only the principal square root. If possible, simplify radicals or rationalize denominators.

63. $A = \pi r^2$ for r

64. $ax^2 - b = 0$ for x

65. $I = \dfrac{k}{d^2}$ for d

66. $A = p(1 + r)^2$ for r

Application Exercises

Use the Pythagorean Theorem to solve Exercises 67–72. Express the answer in radical form and simplify, if possible.

67. Find the length of the ladder.

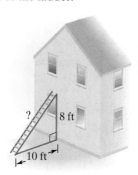

68. How high is the airplane above the ground?

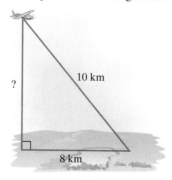

69. A baseball diamond is actually a square with 90-foot sides. What is the distance from home plate to second base?

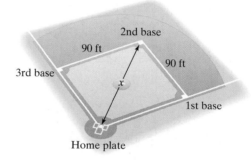

70. The base of a 20-foot ladder is 15 feet from the house. How far up the house does the ladder reach?

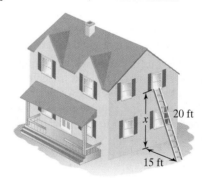

71. In a 25-inch square television set, the length of the screen's diagonal is 25 inches. Find the measure of the side of the screen.

72. In a 27-inch square television set, the length of the screen's diagonal is 27 inches. Find the measure of the side of the screen.

Use the formula for the area of a circle, $A = \pi r^2$, to solve Exercises 73–74.

73. If the area of a circle is 36π square inches, find its radius.

74. If the area of a circle is 49π square inches, find its radius.

The weight of a human fetus is modeled by the formula $W = 3t^2$, where W is the weight, in grams, and t is the time, in weeks, with $0 \leq t \leq 39$. Use this formula to solve Exercises 75–76.

75. After how many weeks does the fetus weigh 108 grams?

76. After how many weeks does the fetus weigh 192 grams?

The distance, d, in feet, that an object falls in t seconds is modeled by the formula $d = 16t^2$. Use this formula to solve Exercises 77–78.

77. If you drop a rock from a cliff 400 feet above the water, how long will it take for the rock to hit the water?

78. If you drop a rock from a cliff 576 feet above the water, how long will it take for the rock to hit the water?

79. A square flower bed is to be enlarged by adding 2 meters on each side. If the larger square has an area of 144 square meters, what is the length of the original square?

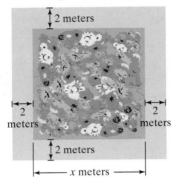

80. A square flower bed is to be enlarged by adding 3 feet on each side. If the larger square has an area of 169 square feet, what is the length of the original square?

81. A machine produces open boxes using square sheets of metal. The figure illustrates that the machine cuts equal-sized squares measuring 2 inches on a side from the corners and then shapes the metal into an open box by turning up the sides. If each box must have a volume of 200 cubic inches, find the size of the length and width of the open box.

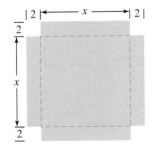
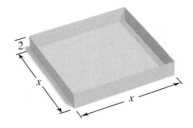

82. A machine produces open boxes using square sheets of metal. The machine cuts equal-sized squares measuring 3 inches on a side from the corners and then shapes the metal into an open box by turning up the sides. If each box must have a volume of 75 cubic inches, find the size of the length and width of the open box.

Writing in Mathematics

83. What is the square root property?

84. Explain how to solve $(x - 1)^2 = 16$ using the square root property.

85. In your own words, state the Pythagorean Theorem.

86. In the 1939 movie *The Wizard of Oz*, upon being presented with a Th.D. (Doctor of Thinkology), the Scarecrow proudly exclaims, "The sum of the square roots of any two sides of an isosceles triangle is equal to the square root of the remaining side." Did the Scarecrow get the Pythagorean Theorem right? In particular, describe four errors in the Scarecrow's statement.

Critical Thinking Exercises

87. Which one of the following is true?

a. The equation $(x + 5)^2 = 8$ is equivalent to $x + 5 = 2\sqrt{2}$.

b. The equation $x^2 = 0$ has no solution.

c. The equation $x^2 = -1$ has no solutions that are real numbers.

d. The solutions of $3x^2 - 5 = 0$ are $\dfrac{\sqrt{5}}{3}$ and $-\dfrac{\sqrt{5}}{3}$.

88. Find the value(s) of x if the distance between $(-3, -2)$ and $(x, -5)$ is 5 units.

Technology Exercises

89. Use a graphing utility to solve $4 - (x + 1)^2 = 0$. Graph $y = 4 - (x + 1)^2$ in a $[-5, 5, 1]$ by $[-5, 5, 1]$ viewing rectangle. The equation's solutions are the graph's x-intercepts. Check by substitution in the given equation.

90. Use a graphing utility to solve $(x - 1)^2 - 9 = 0$. Graph $y = (x - 1)^2 - 9$ in a $[-5, 5, 1]$ by $[-9, 3, 1]$ viewing rectangle. The equation's solutions are the graph's x-intercepts. Check by substitution in the given equation.

Review Exercises

91. Factor completely: $12x^2 + 14x - 6$. (Section 7.5, Example 2)

92. Divide: $\dfrac{x^2 - x - 6}{3x - 3} \div \dfrac{x^2 - 4}{x - 1}$. (Section 8.2, Example 6)

93. Solve: $4(x - 5) = 22 + 2(6x + 3)$. (Section 2.3, Example 3)

SOLVING QUADRATIC EQUATIONS BY COMPLETING THE SQUARE

SECTION

10.2

Objectives

1 Complete the square of a binomial.

2 Solve quadratic equations by completing the square.

There is a lack of completion in both the Escher image and the unfinished square on the left. Completion for the geometric figure can be obtained by adding a small square to its upper-right-hand corner. Understanding this process algebraically will give you a new method, appropriately called *completing the square*, for solving quadratic equations.

1 Complete the square of a binomial.

Completing the Square   How do we solve a quadratic equation, $ax^2 + bx + c = 0$, if the trinomial $ax^2 + bx + c$ cannot be factored? We can convert the equation into an equivalent equation that can be solved using the square root property. This is accomplished by **completing the square**.

COMPLETING THE SQUARE   If $x^2 + bx$ is a binomial, then by adding $\left(\dfrac{b}{2}\right)^2$, which is the square of half the coefficient of x, a perfect square trinomial will result.

That is,

$$x^2 + bx + \left(\frac{b}{2}\right)^2 = \left(x + \frac{b}{2}\right)^2.$$

EXAMPLE 1   Completing the Square

Complete the square for each binomial. Then factor the resulting perfect square trinomial:

   a. $x^2 + 8x$ **b.** $x^2 - 14x$ **c.** $x^2 + 5x$.

SOLUTION   To complete the square, we must add a term to each binomial. The term that should be added is the square of half the coefficient of x.

$$x^2 + 8x \qquad x^2 - 14x \qquad x^2 + 5x$$

Add $\left(\frac{8}{2}\right)^2 = 4^2$.
Add 16 to complete the square.

Add $\left(\frac{-14}{2}\right)^2 = (-7)^2$.
Add 49 to complete the square.

Add $\left(\frac{5}{2}\right)^2$, or $\frac{25}{4}$, to complete the square.

a. The coefficient of the x-term of $x^2 + 8x$ is 8. Half of 8 is 4, and $4^2 = 16$. Add 16.

$$x^2 + 8x + 16 = (x + 4)^2$$

b. The coefficient of the x-term of $x^2 - 14x$ is -14. Half of -14 is -7 and $(-7)^2 = 49$. Add 49.

$$x^2 - 14x + 49 = (x - 7)^2$$

c. The coefficient of the x-term of $x^2 + 5x$ is 5. Half of 5 is $\frac{5}{2}$, and $\left(\frac{5}{2}\right)^2 = \frac{25}{4}$. Add $\frac{25}{4}$.

$$x^2 + 5x + \frac{25}{4} = \left(x + \frac{5}{2}\right)^2$$

ENRICHMENT ESSAY

Visualizing Completeing the Square

This figure, with area $x^2 + 8x$, is not a complete square. The bottom-right corner is missing.

Add 16 square units to the missing portion and you literally complete the square.

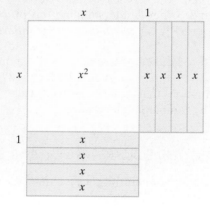

Area: $x^2 + 8x$

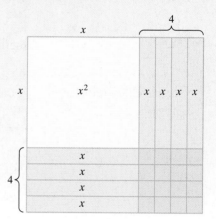

Area: $x^2 + 8x + 16 = (x + 4)^2$

 CHECK POINT 1 Complete the square for each binomial. Then factor the resulting perfect square trinomial:

a. $x^2 + 10x$

b. $x^2 - 6x$

c. $x^2 + 3x$.

---

2 Solve quadratic equations by completing the square.

Solving Quadratic Equations by Completing the Square We can solve any quadratic equation by completing the square. If the coefficient of the x^2-term is one, we add the square of half the coefficient of x to both sides of the equation. **When you add a constant term to one side of the equation to complete the square, be certain to add the same constant to the other side of the equation.** These ideas are illustrated in Example 2.

EXAMPLE 2 Solving Quadratic Equations by Completing the Square

Solve by completing the square:

a. $x^2 + 8x = -15$ **b.** $x^2 - 6x + 2 = 0$.

SOLUTION

a. To complete the square on the binomial $x^2 + 8x$, we take half of 8, which is 4, and square 4, giving 16. We add 16 to both sides of the equation. This makes the left side a perfect square trinomial.

$$x^2 + 8x = -15$$ This is the given equation.

$$x^2 + 8x + 16 = -15 + 16$$ Add 16 to both sides to complete the square.

$$(x + 4)^2 = 1$$ Factor and simplify.

$$x + 4 = \sqrt{1} \quad \text{or} \quad x + 4 = -\sqrt{1}$$ Apply the square root property.

$$x + 4 = 1 \qquad\qquad x + 4 = -1$$ Simplify.

$$x = -3 \qquad\qquad x = -5$$ Subtract 4 from both sides in each equation.

The solutions are -3 and -5.

b. To solve $x^2 - 6x + 2 = 0$ by completing the square, we first subtract 2 from both sides. This is done to isolate the binomial $x^2 - 6x$ so that we can complete the square.

$$x^2 - 6x + 2 = 0$$ This is the original equation.

$$x^2 - 6x = -2$$ Subtract 2 from both sides.

Next, we complete the square. Find half the coefficient of the x-term and square it. The coefficient of the x-term is -6. Half of -6 is -3 and $(-3)^2 = 9$. Thus, we add 9 to both sides of the equation.

$$x^2 - 6x + 9 = -2 + 9$$ Add 9 to both sides to complete the square.

$$(x - 3)^2 = 7$$ Factor and simplify.

$$x - 3 = \sqrt{7} \quad \text{or} \quad x - 3 = -\sqrt{7}$$ Apply the square root property.

$$x = 3 + \sqrt{7} \qquad x = 3 - \sqrt{7}$$ Add 3 to both sides in each equation.

The solutions are $3 + \sqrt{7}$ and $3 - \sqrt{7}$, expressed in abbreviated notation as $3 \pm \sqrt{7}$.

DISCOVER FOR YOURSELF

Try to solve the equations in Example 2 by factoring. Which equation can be solved by factoring and which one cannot? Which equation has rational solutions and which one has irrational solutions?

Write a statement about the kinds of real solutions a quadratic equation can have, and relate this statement to whether or not the equation can be solved by factoring.

If you solve a quadratic equation by completing the square and the solutions are rational numbers, the equation can also be solved by factoring. By contrast, quadratic equations with irrational solutions cannot be solved by factoring. However, all quadratic equations can be solved by completing the square.

 CHECK POINT 2 Solve by completing the square:

a. $x^2 + 6x = 7$ **b.** $x^2 - 10x + 18 = 0$.

If the coefficient of the x^2-term in a quadratic equation is not 1, you must divide each side of the equation by this coefficient before completing the square. For example, to solve $2x^2 + 5x - 4 = 0$ by completing the square, first divide every term by 2:

$$\frac{2x^2}{2} + \frac{5x}{2} - \frac{4}{2} = \frac{0}{2}$$

$$x^2 + \frac{5}{2}x - 2 = 0.$$

Now that the coefficient of the x^2-term is 1, we can solve by completing the square.

EXAMPLE 3   Solving a Quadratic Equation by Completing the Square

Solve by completing the square: $2x^2 + 5x - 4 = 0$.

SOLUTION

$2x^2 + 5x - 4 = 0$	This is the original equation.
$x^2 + \dfrac{5}{2}x - 2 = 0$	Divide both sides by 2.
$x^2 + \dfrac{5}{2}x = 2$	Add 2 to both sides and isolate the binomial.
$x^2 + \dfrac{5}{2}x + \dfrac{25}{16} = 2 + \dfrac{25}{16}$	Complete the square: Half of $\frac{5}{2}$ is $\frac{5}{4}$ and $\left(\frac{5}{4}\right)^2 = \frac{25}{16}$.
$\left(x + \dfrac{5}{4}\right)^2 = \dfrac{57}{16}$	Factor and simplify. On the right: $2 + \frac{25}{16} = \frac{32}{16} + \frac{25}{16} = \frac{57}{16}$.
$x + \dfrac{5}{4} = \sqrt{\dfrac{57}{16}}$ or $x + \dfrac{5}{4} = -\sqrt{\dfrac{57}{16}}$	Apply the square root property.
$x + \dfrac{5}{4} = \dfrac{\sqrt{57}}{4}$ $x + \dfrac{5}{4} = -\dfrac{\sqrt{57}}{4}$	$\sqrt{\dfrac{57}{16}} = \dfrac{\sqrt{57}}{\sqrt{16}} = \dfrac{\sqrt{57}}{4}$
$x = -\dfrac{5}{4} + \dfrac{\sqrt{57}}{4}$ $x = -\dfrac{5}{4} - \dfrac{\sqrt{57}}{4}$	Solve the equations, subtracting $\frac{5}{4}$ from both sides.
$x = \dfrac{-5 + \sqrt{57}}{4}$ $x = \dfrac{-5 - \sqrt{57}}{4}$	Express solutions with a common denominator.

The solutions are $\dfrac{-5 \pm \sqrt{57}}{4}$.   ∎

USING TECHNOLOGY

Obtain a decimal approximation for each solution of $2x^2 + 5x - 4 = 0$ in Example 3:

$$\frac{-5 + \sqrt{57}}{4} \approx 0.6$$

$$\frac{-5 - \sqrt{57}}{4} \approx -3.1.$$

The x-intercepts of $y = 2x^2 + 5x - 4$ verify these solutions.

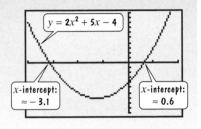

$[-4, 2, 1]$ by $[-10, 10, 1]$

 CHECK POINT 3 Solve by completing the square: $2x^2 - 10x - 1 = 0$.

10.2 EXERCISE SET

Student Solutions Manual   CD/Video   PH Math/Tutor Center   MathXL Tutorials on CD   MathXL®   MyMathLab   Interactmath.com

Practice Exercises

In Exercises 1–12, complete the square for each binomial.
Then factor the resulting perfect square trinomial.

1. $x^2 + 10x$

2. $x^2 + 12x$

3. $x^2 - 2x$

4. $x^2 - 4x$

5. $x^2 + 5x$

6. $x^2 + 3x$

7. $x^2 - 7x$

8. $x^2 - x$

9. $x^2 + \frac{1}{2}x$

10. $x^2 + \frac{1}{3}x$

11. $x^2 - \frac{4}{3}x$

12. $x^2 - \frac{4}{5}x$

In Exercises 13–34, solve each quadratic equation by completing
the square.

13. $x^2 + 4x = 5$

14. $x^2 + 6x = -8$

15. $x^2 - 10x = -24$

16. $x^2 - 2x = 8$

17. $x^2 - 2x = 5$

18. $x^2 - 4x = -2$

19. $x^2 + 4x + 1 = 0$

20. $x^2 + 6x - 5 = 0$

21. $x^2 - 3x = 28$

22. $x^2 - 5x = -6$

23. $x^2 + 3x - 1 = 0$

24. $x^2 - 3x - 5 = 0$

25. $x^2 = 7x - 3$

26. $x^2 = 5x - 3$

27. $2x^2 - 2x - 6 = 0$

28. $2x^2 - 4x - 2 = 0$

29. $2x^2 - 3x + 1 = 0$

30. $2x^2 - x - 1 = 0$

31. $2x^2 + 10x + 11 = 0$

32. $2x^2 + 8x + 5 = 0$

33. $4x^2 - 2x - 3 = 0$

34. $3x^2 - 2x - 4 = 0$

Practice Plus

In Exercises 35–40, solve each quadratic equation by completing
the square.

35. $\frac{x^2}{6} - \frac{x}{3} - 1 = 0$

36. $\frac{x^2}{6} + x - \frac{3}{2} = 0$

37. $(x + 2)(x - 3) = 1$

38. $(x - 5)(x - 3) = 1$

39. $x^2 + 4bx = 5b^2$

40. $x^2 + 6bx = 7b^2$

Writing in Mathematics

41. Explain how to complete the square for a binomial. Use $x^2 + 6x$ to illustrate your explanation.

42. Explain how to solve $x^2 + 6x + 8 = 0$ by completing the square.

Critical Thinking Exercises

43. Which one of the following is true?

   a. Completing the square is a method for finding the area and perimeter of a square.

   b. The trinomial $x^2 - 3x + 9$ is a perfect square trinomial.

   c. Although not every quadratic equation can be solved by completing the square, they can all be solved using factoring.

   d. In completing the square for $x^2 - 7x = 5$, we should add $\frac{49}{4}$ to both sides.

44. Write a perfect square trinomial whose x-term is $-20x$.

45. Solve by completing the square: $x^2 + x + c = 0$.

46. Solve by completing the square: $x^2 + bx + c = 0$.

Technology Exercise

47. Use the technique shown in the technology box on page 592 to verify the solutions of any two quadratic equations in Exercises 17–20.

Review Exercises

In Exercises 48–49, perform the indicated operations. If possible, simplify the answer.

48. $\dfrac{2x+3}{x^2-7x+12} - \dfrac{2}{x-3}$ (Section 8.4, Example 7)

49. $\dfrac{x - \dfrac{1}{3}}{3 - \dfrac{1}{x}}$ (Section 8.5, Example 6)

50. Solve: $\sqrt{2x+3} = 2x - 3$. (Section 9.5, Example 4)

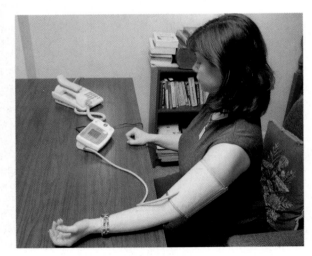

SECTION 10.3

THE QUADRATIC FORMULA

Objectives

1 Solve quadratic equations using the quadratic formula.

2 Determine the most efficient method to use when solving a quadratic equation.

3 Solve problems using quadratic equations.

Until fairly recently, many doctors believed that your blood pressure was theirs to know and yours to worry about. Today, however, people are encouraged to find out their blood pressure. That pumped-up cuff that squeezes against your upper arm measures blood pressure in millimeters (mm) of mercury (Hg). Blood pressure is given in two numbers: systolic pressure over diastolic pressure, such as 120 over 80. Systolic pressure is the pressure of blood against the artery walls when the heart contracts. Diastolic pressure is the pressure of blood against the artery walls when the heart is at rest.

There is a tendency for systolic pressure to increase with age as the arteries become less elastic. In this section, we will use quadratic equations to model normal systolic pressure based on age and gender. We begin by deriving a formula that will enable you to solve quadratic equations more quickly than using the method of completing the square.

Solve quadratic equations using the quadratic formula.

Solving Quadratic Equations Using the Quadratic Formula We can use the method of completing the square to derive a formula that can be used to solve all quadratic equations. The derivation given here also shows a particular quadratic equation, $3x^2 - 2x - 4 = 0$, to specifically illustrate each of the steps.

Deriving the Quadratic Formula

Standard Form of a Quadratic Equation	Comment	A Specific Example
$ax^2 + bx + c = 0, a > 0$	This is the given equation.	$3x^2 - 2x - 4 = 0$
$x^2 + \dfrac{b}{a}x + \dfrac{c}{a} = 0$	Divide both sides by a so that the coefficient of x^2 is 1.	$x^2 - \dfrac{2}{3}x - \dfrac{4}{3} = 0$
$x^2 + \dfrac{b}{a}x = -\dfrac{c}{a}$	Isolate the binomial by adding $-\dfrac{c}{a}$ on both sides of the equation.	$x^2 - \dfrac{2}{3}x = \dfrac{4}{3}$
$x^2 + \underbrace{\dfrac{b}{a}}x + \left(\dfrac{b}{2a}\right)^2 = -\dfrac{c}{a} + \left(\dfrac{b}{2a}\right)^2$ (half)^2	Complete the square. Add the square of half the coefficient of x to both sides.	$x^2 - \underbrace{\dfrac{2}{3}}x + \left(-\dfrac{1}{3}\right)^2 = \dfrac{4}{3} + \left(-\dfrac{1}{3}\right)^2$ (half)^2
$x^2 + \dfrac{b}{a}x + \dfrac{b^2}{4a^2} = -\dfrac{c}{a} + \dfrac{b^2}{4a^2}$		$x^2 - \dfrac{2}{3}x + \dfrac{1}{9} = \dfrac{4}{3} + \dfrac{1}{9}$
$\left(x + \dfrac{b}{2a}\right)^2 = -\dfrac{c}{a} \cdot \dfrac{4a}{4a} + \dfrac{b^2}{4a^2}$	Factor on the left side and obtain a common denominator on the right side.	$\left(x - \dfrac{1}{3}\right)^2 = \dfrac{4}{3} \cdot \dfrac{3}{3} + \dfrac{1}{9}$
$\left(x + \dfrac{b}{2a}\right)^2 = \dfrac{-4ac + b^2}{4a^2}$	Add fractions on the right side.	$\left(x - \dfrac{1}{3}\right)^2 = \dfrac{12 + 1}{9}$
$\left(x + \dfrac{b}{2a}\right)^2 = \dfrac{b^2 - 4ac}{4a^2}$		$\left(x - \dfrac{1}{3}\right)^2 = \dfrac{13}{9}$
$x + \dfrac{b}{2a} = \pm\sqrt{\dfrac{b^2 - 4ac}{4a^2}}$	Apply the square root property.	$x - \dfrac{1}{3} = \pm\sqrt{\dfrac{13}{9}}$
$x + \dfrac{b}{2a} = \pm\dfrac{\sqrt{b^2 - 4ac}}{2a}$	Take the square root of the quotient, simplifying the denominator.	$x - \dfrac{1}{3} = \pm\dfrac{\sqrt{13}}{3}$
$x = \dfrac{-b}{2a} \pm \dfrac{\sqrt{b^2 - 4ac}}{2a}$	Solve for x by subtracting $\dfrac{b}{2a}$ from both sides.	$x = \dfrac{1}{3} \pm \dfrac{\sqrt{13}}{3}$
$x = \dfrac{-b \pm \sqrt{b^2 - 4ac}}{2a}$	Combine fractions on the right side.	$x = \dfrac{1 \pm \sqrt{13}}{3}$

The formula shown at the bottom of the left column is called the *quadratic formula*. A similar proof shows that the same formula can be used to solve quadratic equations if a, the coefficient of the x^2-term, is negative.

THE QUADRATIC FORMULA The solutions of a quadratic equation in standard form $ax^2 + bx + c = 0$, with $a \neq 0$, are given by the **quadratic formula**

$$x = \frac{-b \pm \sqrt{b^2 - 4ac}}{2a}.$$

x equals negative *b* plus or minus the square root of $b^2 - 4ac$, all divided by $2a$.

To use the quadratic formula, write the quadratic equation in standard form if necessary. Then determine the numerical values for a (the coefficient of the x^2-term), b (the coefficient of the x-term), and c (the constant term). Substitute the values of a, b, and c into the quadratic formula and evaluate the expression. The $\pm$ sign indicates that there are two solutions of the equation.

EXAMPLE 1	Solving a Quadratic Equation Using the Quadratic Formula

Solve using the quadratic formula: $2x^2 + 9x - 5 = 0$.

SOLUTION The given equation is in standard form. Begin by identifying the values for a, b, and c.

$$2x^2 + 9x - 5 = 0$$

$$\boxed{a = 2} \quad \boxed{b = 9} \quad \boxed{c = -5}$$

Substituting these values into the quadratic formula and simplifying gives the equation's solutions.

$$x = \frac{-b \pm \sqrt{b^2 - 4ac}}{2a}$$
Use the quadratic formula.

$$x = \frac{-9 \pm \sqrt{9^2 - 4(2)(-5)}}{2(2)}$$
Substitute the values for a, b, and c: $a = 2$, $b = 9$, and $c = -5$.

$$= \frac{-9 \pm \sqrt{81 + 40}}{4}$$
$9^2 - 4(2)(-5) = 81 - (-40) = 81 + 40$

$$= \frac{-9 \pm \sqrt{121}}{4}$$
Add under the radical sign.

$$= \frac{-9 \pm 11}{4}$$
$\sqrt{121} = 11$

Now we will evaluate this expression in two different ways to obtain the two solutions. At the left, we will *add* 11 to -9. At the right, we will *subtract* 11 from -9.

$$x = \frac{-9 + 11}{4} \quad \text{or} \quad x = \frac{-9 - 11}{4}$$

$$= \frac{2}{4} = \frac{1}{2} \qquad\qquad = \frac{-20}{4} = -5$$

The solutions are $\frac{1}{2}$ and -5. ■

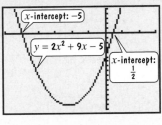
In Example 1, the solutions of $2x^2 + 9x - 5 = 0$ are rational numbers. This means that the equation can also be solved using factoring. The reason that the solutions are rational numbers is that $b^2 - 4ac$, the radicand in the quadratic formula, is 121, which is a perfect square.

To Die at Twenty

Can the equations

$$7x^5 + 12x^3 - 9x + 4 = 0$$

and

$$8x^6 - 7x^5 + 4x^3 - 19 = 0$$

be solved using a formula similar to the quadratic formula? The first equation has five solutions and the second has six solutions, but they cannot be found using a formula. How do we know? In 1832, a 20-year-old Frenchman, Evariste Galois, wrote down a proof showing that there is no general formula to solve equations when the exponent on the variable is 5 or greater.

Galois was jailed as a political activist several times while still a teenager. The day after his brilliant proof, he fought a duel over a woman. The duel was a political setup. As he lay dying, Galois told his brother, Alfred, of the manuscript that contained his proof: "Mathematical manuscripts are in my room. On the table. Take care of my work. Make it known. Important. Don't cry, Alfred. I need all my courage—to die at twenty."

(Our source is Leopold Infeld's biography of Galois, *Whom the Gods Love.* Some historians, however, dispute the story of Galois's ironic death the very day after his algebraic proof. Mathematical truths seem more reliable than historical ones!)

 CHECK POINT 1 Solve using the quadratic formula: $8x^2 + 2x - 1 = 0.$

EXAMPLE 2 Solving a Quadratic Equation Using the Quadratic Formula

Solve using the quadratic formula: $2x^2 = 6x - 1.$

SOLUTION   The quadratic equation must be in standard form to identify the values for a, b, and c. We need to move all terms to one side and obtain zero on the other side. To obtain zero on the right, we subtract $6x$ and add 1 on both sides. Then we can identify the values for a, b, and c.

$$2x^2 = 6x - 1$$
This is the given equation.

$$2x^2 - 6x + 1 = 6x - 6x - 1 + 1$$
This step is usually performed mentally.

$$2x^2 - 6x + 1 = 0$$
Identify a, the x^2-coefficient, b, the x-coefficient, and c, the constant.

$a = 2$ $b = -6$ $c = 1$

Substituting these values into the quadratic formula and simplifying gives the equation's solutions.

$$x = \frac{-b \pm \sqrt{b^2 - 4ac}}{2a}$$
Use the quadratic formula.

$$x = \frac{-(-6) \pm \sqrt{(-6)^2 - 4(2)(1)}}{2 \cdot 2}$$
Substitute the values for a, b, and c: $a = 2, b = -6$, and $c = 1$.

$$= \frac{6 \pm \sqrt{36 - 8}}{4}$$
$-(-6) = 6$ and $(-6)^2 = (-6)(-6) = 36$.

$$= \frac{6 \pm \sqrt{28}}{4}$$
Complete the subtraction under the radical.

$$= \frac{6 \pm 2\sqrt{7}}{4}$$
$\sqrt{28} = \sqrt{4 \cdot 7} = \sqrt{4}\sqrt{7} = 2\sqrt{7}$

$$= \frac{2(3 \pm \sqrt{7})}{4}$$
Factor out 2 from the numerator.

$$= \frac{3 \pm \sqrt{7}}{2}$$
Divide the numerator and denominator by 2.

The solutions are $\dfrac{3 + \sqrt{7}}{2}$ and $\dfrac{3 - \sqrt{7}}{2}$, abbreviated $\dfrac{3 \pm \sqrt{7}}{2}$.

In Example 2, the solutions of $2x^2 = 6x - 1$ are irrational numbers. This means that the equation cannot be solved using factoring. The reason that the solutions are irrational numbers is that $b^2 - 4ac$, the radicand in the quadratic formula, is 28, which is not a perfect square.

STUDY TIP

Many students use the quadratic formula correctly until the last step, where they make an error in simplifying the solutions. Be sure to factor the numerator before dividing the numerator and denominator by the greatest common factor:

$$\frac{6 \pm 2\sqrt{7}}{4} = \underbrace{\frac{2(3 \pm \sqrt{7})}{4}}_{\text{Factor first}} = \underbrace{\frac{\overset{1}{\cancel{2}}(3 \pm \sqrt{7})}{\underset{2}{\cancel{4}}}}_{\text{Then divide by the GCF.}} = \frac{3 \pm \sqrt{7}}{2}.$$

You cannot divide just one term in the numerator and the denominator by their greatest common factor.

Incorrect!

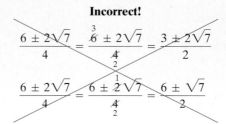

Can all irrational solutions of quadratic equations be simplified? No. The following solutions cannot be simplified.

$$\frac{5 \pm 2\sqrt{7}}{2} \qquad \boxed{\text{Other than 1, terms in each numerator have no common factor.}} \qquad \frac{-4 \pm 3\sqrt{7}}{2}$$

 CHECK POINT 2 Solve using the quadratic formula: $x^2 = 6x - 4$.

---

2 Determine the most efficient method to use when solving a quadratic equation.

Determining Which Method to Use All quadratic equations can be solved by the quadratic formula. However, if an equation is in the form $u^2 = d$, such as $x^2 = 5$ or $(2x + 3)^2 = 8$, it is faster to use the square root property, taking the square root of both sides. If the equation is not in the form $u^2 = d$, write the quadratic equation in standard form $(ax^2 + bx + c = 0)$. Try to solve the equation by factoring. If $ax^2 + bx + c$ cannot be factored, then solve the quadratic equation by using the quadratic formula.

Because we used the method of completing the square to derive the quadratic formula, we no longer need it for solving quadratic equations. However, you will use completing the square in more advanced algebra courses to help graph certain kinds of equations.

Table 10.1 summarizes our observations about which technique to use when solving a quadratic equation.

Table 10.1 Determining the Most Efficient Technique to Use When Solving a Quadratic Equation

Description and Form of the Quadratic Equation	Most Efficient Solution Method	Example
$ax^2 + bx + c = 0$ and $ax^2 + bx + c$ can be factored easily.	Factor and use the zero-product principle.	$3x^2 + 5x - 2 = 0$ $(3x - 1)(x + 2) = 0$ $3x - 1 = 0$ or $x + 2 = 0$ $x = \dfrac{1}{3}$ $x = -2$
$ax^2 + c = 0$ The quadratic equation has no x-term. ($b = 0$)	Solve for x^2 and apply the square root property.	$4x^2 - 7 = 0$ $4x^2 = 7$ $x^2 = \dfrac{7}{4}$ $x = \pm\dfrac{\sqrt{7}}{2}$
$u^2 = d$; u is a first-degree polynomial.	Use the square root property.	$(x + 4)^2 = 5$ $x + 4 = \pm\sqrt{5}$ $x = -4 \pm \sqrt{5}$
$ax^2 + bx + c = 0$ and $ax^2 + bx + c$ cannot be factored or the factoring is too difficult.	Use the quadratic formula: $x = \dfrac{-b \pm \sqrt{b^2 - 4ac}}{2a}.$	$x^2 - 2x - 6 = 0$ $(a = 1, b = -2, c = -6)$ $x = \dfrac{2 \pm \sqrt{4 - 4(1)(-6)}}{2(1)}$ $= \dfrac{2 \pm \sqrt{28}}{2} = \dfrac{2 \pm \sqrt{4}\sqrt{7}}{2}$ $= \dfrac{2 \pm 2\sqrt{7}}{2} = \dfrac{2(1 \pm \sqrt{7})}{2}$ $= 1 \pm \sqrt{7}$

3 Solve problems using quadratic equations.

Applications Quadratic equations can be solved by any efficient method to answer questions about variables contained in mathematical models.

EXAMPLE 3 Blood Pressure and Age

The graphs in Figure 10.5 illustrate that a person's normal systolic blood pressure, measured in millimeters of mercury (mm Hg), depends on his or her age.

ENRICHMENT ESSAY

Classifying Blood Pressure

Category	Systolic		Diastolic
Optimal	<120	and	<80
Normal	<130	and	<85
High Normal	130–139	or	85–89
Hypertension			
Stage 1	140–159	or	90–99
Stage 2	160–179	or	100–109
Stage 3	≥180	or	≥100

Source: National Institutes of Health

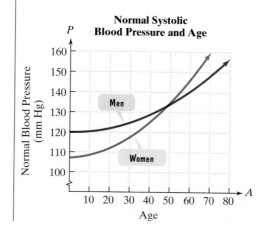

Normal Systolic Blood Pressure and Age

Normal Blood Pressure (mm Hg)

Age

FIGURE 10.5

Normal systolic blood pressure is given by the following formulas:

$$\text{Men:} \quad P = 0.006A^2 - 0.02A + 120$$

$$\text{Women:} \quad P = 0.01A^2 + 0.05A + 107.$$

In each formula, P is the normal systolic blood pressure, in millimeters of mercury, at age A.

a. Find the age, to the nearest year, of a man whose normal systolic blood pressure is 125 mm Hg.

b. Use the graphs in Figure 10.5 to describe the differences between the normal systolic blood pressures of men and women as they age.

SOLUTION

a. We are interested in the age of a man with a normal systolic blood pressure of 125 millimeters of mercury. Thus, we substitute 125 for P in the given formula for men. Then we solve for A, the man's age.

$$P = 0.006A^2 - 0.02A + 120 \qquad \text{This is the given formula for men.}$$

$$125 = 0.006A^2 - 0.02A + 120 \qquad \text{Substitute 125 for } P.$$

$$0 = 0.006A^2 - 0.02A - 5 \qquad \text{Subtract 125 from both sides}$$

$\boxed{a = 0.006} \quad \boxed{b = -0.02} \quad \boxed{c = -5}$ and write the quadratic equation in standard form.

Because the trinomial on the right side of the equation is prime, we solve using the quadratic formula.

> Notice that the variable is A, rather than the usual x.

$$A = \frac{-b \pm \sqrt{b^2 - 4ac}}{2a} \qquad \text{Use the quadratic formula.}$$

$$= \frac{-(-0.02) \pm \sqrt{(-0.02)^2 - 4(0.006)(-5)}}{2(0.006)} \qquad \begin{array}{l}\text{Substitute the values for } a, b, \text{ and } c: \\ a = 0.006, \\ b = -0.02, \text{ and} \\ c = -5.\end{array}$$

$$= \frac{0.02 \pm \sqrt{0.1204}}{0.012} \qquad \begin{array}{l}\text{Use a calculator to} \\ \text{simplify the radicand.}\end{array}$$

$$\approx \frac{0.02 \pm 0.347}{0.012} \qquad \begin{array}{l}\text{Use a calculator:} \\ \sqrt{0.1204} \approx 0.347.\end{array}$$

$$A \approx \frac{0.02 + 0.347}{0.012} \quad \text{or} \quad A \approx \frac{0.02 - 0.347}{0.012}$$

$$A \approx 31 \qquad\qquad \text{or } A \approx -27 \qquad \begin{array}{l}\text{Use a calculator} \\ \text{and round to the} \\ \text{nearest integer.}\end{array}$$

> Reject this solution. Age cannot be negative.

The positive solution indicates that 31 is the approximate age of a man whose normal systolic blood pressure is 125 mm Hg. The solution can be visualized as the point $(31, 125)$ on the red graph representing men in Figure 10.5. Take a moment to locate this point on the graph.

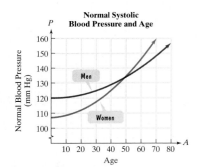

Normal Systolic Blood Pressure and Age

FIGURE 10.5 (repeated)

b. Take a second look at the graphs in Figure 10.5. The blue graph representing women's normal systolic blood pressure is narrower than the red graph representing men's normal systolic blood pressure. Up to approximately age 50, women's normal systolic blood pressure is lower than men's, although it is increasing at a faster rate. After age 50, women's normal systolic blood pressure is higher than men's. ■

 CHECK POINT 3 Using the appropriate formula in Example 3, find the age, to the nearest year, of a woman whose normal systolic blood pressure is 115 mm Hg.

10.3 EXERCISE SET

Student Solutions Manual   CD/Video   PH Math/Tutor Center   MathXL Tutorials on CD   MathXL®   MyMathLab   Interactmath.com

Practice Exercises

In Exercises 1–22, solve each equation using the quadratic formula. Simplify irrational solutions, if possible.

1. $x^2 + 5x + 6 = 0$

2. $x^2 + 7x + 10 = 0$

3. $x^2 + 5x + 3 = 0$

4. $x^2 + 5x + 2 = 0$

5. $x^2 + 4x - 6 = 0$

6. $x^2 + 2x - 4 = 0$

7. $x^2 + 4x - 7 = 0$

8. $x^2 + 4x + 1 = 0$

9. $x^2 - 3x - 18 = 0$

10. $x^2 - 3x - 10 = 0$

11. $6x^2 - 5x - 6 = 0$

12. $9x^2 - 12x - 5 = 0$

13. $x^2 - 2x - 10 = 0$

14. $x^2 + 6x - 10 = 0$

15. $x^2 - x = 14$

16. $x^2 - 5x = 10$

17. $6x^2 + 6x + 1 = 0$

18. $3x^2 - 5x + 1 = 0$

19. $9x^2 - 12x + 4 = 0$

20. $4x^2 + 12x + 9 = 0$

21. $4x^2 = 2x + 7$

22. $3x^2 = 6x - 1$

In Exercises 23–44, solve each equation by the method of your choice. Simplify irrational solutions, if possible.

23. $2x^2 - x = 1$

24. $3x^2 - 4x = 4$

25. $5x^2 + 2 = 11x$

26. $5x^2 = 6 - 13x$

27. $3x^2 = 60$

28. $2x^2 = 250$

29. $x^2 - 2x = 1$

30. $2x^2 + 3x = 1$

31. $(2x + 3)(x + 4) = 1$

32. $(2x - 5)(x + 1) = 2$

33. $(3x - 4)^2 = 16$

34. $(2x + 7)^2 = 25$

35. $3x^2 - 12x + 12 = 0$

36. $9 - 6x + x^2 = 0$

37. $4x^2 - 16 = 0$

38. $3x^2 - 27 = 0$

39. $x^2 + 9x = 0$

40. $x^2 - 6x = 0$

41. $\dfrac{3}{4}x^2 - \dfrac{5}{2}x - 2 = 0$

42. $\dfrac{1}{3}x^2 - \dfrac{1}{2}x - \dfrac{3}{2} = 0$

43. $(3x - 2)^2 = 10$

44. $(4x - 1)^2 = 15$

Practice Plus

In Exercises 45–52, solve each equation by the method of your choice. Simplify irrational solutions, if possible.

45. $\dfrac{x^2}{x + 7} - \dfrac{3}{x + 7} = 0$

46. $\dfrac{x^2}{x + 9} - \dfrac{11}{x + 9} = 0$

47. $(x + 2)^2 + x(x + 1) = 4$

48. $(x - 1)(3x + 2) = -7(x - 1)$

49. $2x^2 - 9x - 3 = 9 - 9x$

50. $3x^2 - 6x - 3 = 12 - 6x$

51. $\dfrac{1}{x} + \dfrac{1}{x + 3} = \dfrac{1}{4}$

52. $\dfrac{1}{x} + \dfrac{2}{x + 3} = \dfrac{1}{4}$

Application Exercises

53. A football is kicked straight up from a height of 4 feet with an initial speed of 60 feet per second. The formula

$$h = -16t^2 + 60t + 4$$

describes the ball's height above the ground, h, in feet, t seconds after it is kicked. How long will it take for the football to hit the ground? Use a calculator and round to the nearest tenth of a second.

54. Standing on a platform 50 feet high, a person accidentally fires a gun straight into the air. The formula

$$h = -16t^2 + 100t + 50$$

describes the bullet's height above the ground, h, in feet, t seconds after the gun is fired. How long will it take for the bullet to hit the ground? Use a calculator and round to the nearest tenth of a second.

The height of the bridge shown in the figure is modeled by

$$h = -0.05x^2 + 27,$$

where x is the distance, in feet, from the center of the arch. Use this formula to solve Exercises 55–56.

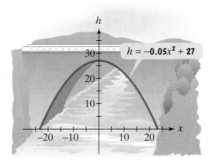

55. How far to the right of the center is the height 22 feet?

56. How far to the right of the center is the height 7 feet?

The bar graph shows the number of property crimes, in millions, in the United States for six selected years. The data can be modeled by the formula

$$N = -0.02x^2 + 0.5x + 10.2,$$

where N is the number of property crimes, in millions, x years after 1975. Use the formula to solve Exercises 57–58.

**Number of Property Crimes
in the U. S. Reported to the Police**

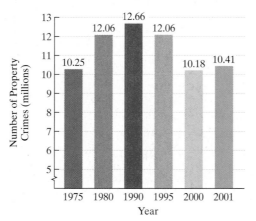

57. In which year are there 7.2 million property crimes?

58. In which year are there 6.48 million property crimes?

59. The length of a rectangle is 3 meters longer than the width. If the area is 36 square meters, find the rectangle's dimensions. Round to the nearest tenth of a meter.

60. The length of a rectangle is 2 meters longer than the width. If the area is 10 square meters, find the rectangle's dimensions. Round to the nearest tenth of a meter.

61. The hypotenuse of a right triangle is 4 feet long. One leg is 1 foot longer than the other. Find the lengths of the legs. Round to the nearest tenth of a foot.

62. The hypotenuse of a right triangle is 6 feet long. One leg is 1 foot shorter than the other. Find the lengths of the legs. Round to the nearest tenth of a foot.

63. In Exercise Set 8.4, we considered two formulas that approximate the dosage of a drug prescribed for children:

$$\text{Young's Rule:} \quad C = \frac{DA}{A + 12}$$

$$\text{Cowling's Rule:} \quad C = \frac{D(A + 1)}{24}.$$

In each formula, A = the child's age, in years, D = an adult dosage, and C = the proper child's dosage. The formulas apply for ages 2 through 13, inclusive. At which age, to the nearest tenth of a year, do the two formulas give the same dosage?

Writing in Mathematics

64. What is the quadratic formula and why is it useful?

65. Without going into specific details for each step, describe how the quadratic formula is derived.

66. Explain how to solve $x^2 + 6x + 8 = 0$ using the quadratic formula.

67. If you are given a quadratic equation, how do you determine which method to use to solve it?

Critical Thinking Exercises

68. Which one of the following is true?

  a. When using the quadratic formula to solve the equation $x^2 - x + 3 = 0$, we have $a = 1$, $b = -x$, and $c = 3$.

  b. The quadratic formula can be expressed as

  $$x = -b \pm \frac{\sqrt{b^2 - 4ac}}{2a}.$$

  c. The solutions $\dfrac{4 \pm \sqrt{3}}{2}$ can be simplified to $2 \pm \sqrt{3}$.

  d. For the quadratic equation $-2x^2 + 3x = 0$, we have $a = -2$, $b = 3$, and $c = 0$.

69. The radicand of the quadratic formula, $b^2 - 4ac$, can be used to determine whether $ax^2 + bx + c = 0$ has solutions that are rational, irrational, or not real numbers. Explain how this works. Is it possible to determine the kinds of answers that one will obtain to a quadratic equation without actually solving the equation? Explain.

70. Solve: $x^2 + 2\sqrt{3}x - 9 = 0$.

71. A rectangular vegetable garden is 5 feet wide and 9 feet long. The garden is to be surrounded by a tile border of uniform width. If there are 40 square feet of tile for the border, how wide, to the nearest tenth of a foot, should it be?

Technology Exercises

72. Graph the formula in Exercise 53

$$y = -16x^2 + 60x + 4$$

in a $[0, 4, 1]$ by $[0, 65, 5]$ viewing rectangle. Use the graph to verify your solution to the exercise.

73. Graph the formula in Exercises 57

$$y = -0.02x^2 + 0.5x + 10.2$$

in a $[0, 30, 1]$ by $[0, 14, 1]$ viewing rectangle. Use the graph to verify your solution to the exercise.

Review Exercises

74. Evaluate: $125^{-\frac{2}{3}}$. (Section 9.6, Example 3)

75. Rationalize the denominator: $\dfrac{12}{3 + \sqrt{5}}$. (Section 9.4, Example 3)

76. Multiply: $(x - y)(x^2 + xy + y^2)$. (Section 6.2, Example 7)

✔ **MID-CHAPTER CHECK POINT**

CHAPTER

10

What You Know: We saw that not all quadratic equations can be solved by factoring. We learned three new methods for solving these equations– the square root property, completing the square, and the quadratic formula. We also learned to determine the most efficient technique to use when solving a quadratic equation.

In Exercises 1–12, solve each equation by the method of your choice. Simplify irrational solutions, if possible.

1. $(3x - 2)^2 = 100$

2. $15x^2 = 5x$

3. $x^2 - 2x - 10 = 0$

4. $x^2 - 8x + 16 = 7$

5. $3x^2 - x - 2 = 0$

6. $6x^2 = 10x - 3$

7. $x^2 + (x + 1)^2 = 25$

8. $(x + 5)^2 = 40$

9. $2(x^2 - 8) = 11 - x^2$

10. $2x^2 + 5x + 1 = 0$

11. $(x - 8)(2x - 3) = 34$

12. $x + \dfrac{16}{x} = 8$

13. Solve by completing the square: $x^2 + 14x - 32 = 0$.

14. Find the missing length in the right triangle. Express the answer in simplified radical form.

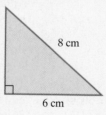

15. Find the distance between $(-3, 2)$ and $(9, -3)$.

16. The figure shows a right triangle whose hypotenuse measures 20 inches and whose leg measurements are in the ratio $3:4$. Find the length of each leg.

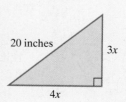

Objectives

1 Express square roots of negative numbers in terms of *i*.

2 Solve quadratic equations with imaginary solutions.

IMAGINARY NUMBERS AS SOLUTIONS OF QUADRATIC EQUATIONS

Who is this kid warning us about our eyeballs turning black if we attempt to find the square root of -9? Don't believe what you hear on the street. Although square roots of negative numbers are not real numbers, they do play a significant role in algebra. In this section, we move beyond the real numbers and discuss square roots with negative radicands.

1 Express square roots of negative numbers in terms of *i*.

The Imaginary Unit *i* Throughout this chapter, we have avoided quadratic equations that have no real numbers as solutions. A fairly simple example is the equation

$$x^2 = -1.$$

Because the square of a real number is never negative, there is no real number x such that $x^2 = -1$. To provide a setting in which such equations have solutions, mathematicians invented an expanded system of numbers, the *complex numbers*. The *imaginary number i*, defined to be a solution to the equation $x^2 = -1$, is the basis of this new set.

THE IMAGINARY UNIT *i* The **imaginary unit** *i* is defined as

$$i = \sqrt{-1}, \quad \text{where} \quad i^2 = -1.$$

Using the imaginary unit i, we can express the square root of any negative number as a real multiple of i. For example,

$$\sqrt{-25} = \sqrt{25(-1)} = \sqrt{25}\sqrt{-1} = 5i.$$

We can check this result by squaring $5i$ and obtaining -25.

$$(5i)^2 = 5^2i^2 = 25(-1) = -25$$

EXAMPLE 1 Expressing Square Roots of Negative Numbers as Multiples of i

Write as a multiple of i:

a. $\sqrt{-9}$ **b.** $\sqrt{-7}$ **c.** $\sqrt{-8}$.

SOLUTION

a. $\sqrt{-9} = \sqrt{9(-1)} = \sqrt{9}\sqrt{-1} = 3i$

b. $\sqrt{-7} = \sqrt{7(-1)} = \sqrt{7}\sqrt{-1} = \sqrt{7}i$

c. $\sqrt{-8} = \sqrt{8(-1)} = \sqrt{8}\sqrt{-1} = \sqrt{4 \cdot 2}\sqrt{-1} = 2\sqrt{2}i$

> Be sure not to write i under the radical.

✓ **CHECK POINT 1** Write as a multiple of i:

a. $\sqrt{-16}$ **b.** $\sqrt{-5}$ **c.** $\sqrt{-50}$.

A new system of numbers, called *complex numbers*, is based on adding multiples of i, such as $5i$, to the real numbers.

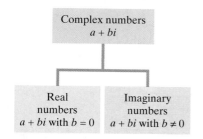

Complex numbers
$a + bi$

Real numbers
$a + bi$ with $b = 0$

Imaginary numbers
$a + bi$ with $b \neq 0$

FIGURE 10.6 The complex number system

COMPLEX NUMBERS AND IMAGINARY NUMBERS The set of all numbers in the form

$$a + bi$$

with real numbers a and b, and i, the imaginary unit, is called the set of **complex numbers**. The real number a is called the **real part** and the real number b is called the **imaginary part** of the complex number $a + bi$. If $b \neq 0$, then the complex number is called an **imaginary number** (see Figure 10.6).

Here are some examples of complex numbers. Each number can be written in the form $a + bi$.

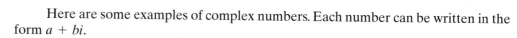

$-4 + 6i$ $2i = 0 + 2i$ $3 = 3 + 0i$

a, the real part, is -4. b, the imaginary part, is 6. a, the real part, is 0. b, the imaginary part, is 2. a, the real part, is 3. b, the imaginary part, is 0.

Can you see that b, the imaginary part, is not zero in the first two complex numbers? Because $b \neq 0$, these complex numbers are imaginary numbers. By contrast, the imaginary part of the complex number on the right, $3 + 0i$, is zero. This complex number is not an imaginary number. The number 3, or $3 + 0i$, is a real number.

2 Solve quadratic equations with imaginary solutions.

Solving Quadratic Equations with Imaginary Solutions The equation $x^2 = -25$ has no real solutions, but it does have imaginary solutions.

$$x^2 = -25 \qquad \text{No real number squared results in a negative number.}$$

$$x = \pm\sqrt{-25} \qquad \text{Apply the square root property.}$$

$$x = \pm 5i \qquad \sqrt{-25} = \sqrt{25(-1)} = \sqrt{25}\sqrt{-1} = 5i$$

The solutions are $5i$ and $-5i$. The next examples involve quadratic equations that have no real solutions, but do have imaginary solutions.

EXAMPLE 2 Solving a Quadratic Equation Using the Square Root Property

Solve: $(x + 4)^2 = -36$.

SOLUTION

$$(x + 4)^2 = -36 \qquad \text{This is the given equation.}$$

$$x + 4 = \sqrt{-36} \quad \text{or} \quad x + 4 = -\sqrt{-36} \qquad \text{Apply the square root property.}$$
$$\text{Equivalently, } x + 4 = \pm\sqrt{-36}.$$

$$x + 4 = 6i \qquad\qquad x + 4 = -6i \qquad \sqrt{-36} = \sqrt{36(-1)} = \sqrt{36}\sqrt{-1} = 6i$$

$$x = -4 + 6i \qquad\qquad x = -4 - 6i \qquad \text{Solve the equations by subtracting 4 from both sides.}$$

Check $-4 + 6i$:

$$(x + 4)^2 = -36$$
$$(-4 + 6i + 4)^2 \overset{?}{=} -36$$
$$(6i)^2 \overset{?}{=} -36$$
$$36i^2 \overset{?}{=} -36$$
$$36(-1) \overset{?}{=} -36$$
$$-36 = -36, \quad \text{true}$$

Check $-4 - 6i$:

$$(x + 4)^2 = -36$$
$$(-4 - 6i + 4)^2 \overset{?}{=} -36$$
$$(-6i)^2 \overset{?}{=} -36$$
$$36i^2 \overset{?}{=} -36$$
$$36(-1) \overset{?}{=} -36$$
$$-36 = -36, \quad \text{true}$$

The imaginary solutions are $-4 + 6i$ and $-4 - 6i$. ∎

✔ **CHECK POINT 2** Solve: $(x + 2)^2 = -25$.

EXAMPLE 3 Solving a Quadratic Equation Using the Quadratic Formula

Solve: $x^2 - 2x + 2 = 0$.

SOLUTION Because the trinomial on the left side is prime, we solve using the quadratic formula.

$$x^2 - 2x + 2 = 0$$

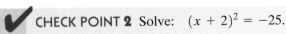

$a = 1 \qquad b = -2 \qquad c = 2$

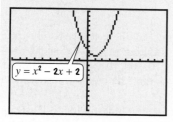
$$x = \frac{-b \pm \sqrt{b^2 - 4ac}}{2a}$$ Use the quadratic formula.

$$x = \frac{-(-2) \pm \sqrt{(-2)^2 - 4(1)(2)}}{2(1)}$$ Substitute the values for a, b, and c: $a = 1$, $b = -2$, and $c = 2$.

$$= \frac{2 \pm \sqrt{-4}}{2}$$ $(-2)^2 - 4(1)(2) = 4 - 8 = -4$

$$= \frac{2 \pm \sqrt{4(-1)}}{2}$$

$$= \frac{2 \pm 2i}{2}$$ $\sqrt{4(-1)} = \sqrt{4}\sqrt{-1} = 2i$

$$= \frac{2(1 \pm i)}{2}$$ Factor out 2 from the numerator.

$$= 1 \pm i$$ Divide the numerator and denominator by 2.

The imaginary solutions are $1 + i$ and $1 - i$.   ■

 CHECK POINT 3 Solve: $x^2 + 6x + 13 = 0$.

10.4 EXERCISE SET

Student Solutions Manual   CD/Video   PH Math/Tutor Center   MathXL Tutorials on CD   MathXL®   MyMathLab   Interactmath.com

Practice Exercises

In Exercises 1–16, express each number in terms of i.

1. $\sqrt{-36}$ **2.** $\sqrt{-49}$

3. $\sqrt{-13}$ **4.** $\sqrt{-19}$

5. $\sqrt{-50}$ **6.** $\sqrt{-12}$

7. $\sqrt{-20}$ **8.** $\sqrt{-300}$

9. $-\sqrt{-28}$ **10.** $-\sqrt{-150}$

11. $7 + \sqrt{-16}$ **12.** $9 + \sqrt{-4}$

13. $10 + \sqrt{-3}$ **14.** $5 + \sqrt{-5}$

15. $6 - \sqrt{-98}$ **16.** $6 - \sqrt{-18}$

In Exercises 17–24, solve each quadratic equation using the square root property. Express imaginary solutions in $a + bi$ form.

17. $(x - 3)^2 = -9$ **18.** $(x - 5)^2 = -36$

19. $(x + 7)^2 = -64$ **20.** $(x + 12)^2 = -100$

21. $(x - 2)^2 = -7$ **22.** $(x - 1)^2 = -13$

23. $(y + 3)^2 = -18$

24. $(y + 4)^2 = -48$

In Exercises 25–36, solve each quadratic equation using the quadratic formula.

25. $x^2 + 4x + 5 = 0$ **26.** $x^2 + 2x + 2 = 0$

27. $x^2 - 6x + 13 = 0$ **28.** $x^2 - 6x + 10 = 0$

29. $x^2 - 12x + 40 = 0$ **30.** $x^2 - 4x + 29 = 0$

31. $x^2 = 10x - 27$ **32.** $x^2 = 4x - 7$

33. $5x^2 = 2x - 3$ **34.** $6x^2 = -2x - 1$

35. $2y^2 = 4y - 5$ **36.** $5y^2 = 6y - 7$

Practice Plus

In Exercises 37–42, solve each equation by the method of your choice.

37. $12x^2 + 35 = 8x^2 + 15$ **38.** $8x^2 - 9 = 5x^2 - 30$

39. $\dfrac{x + 3}{5} = \dfrac{x - 2}{x}$ **40.** $\dfrac{x + 4}{4} = \dfrac{x - 5}{x - 2}$

41. $\dfrac{1}{x + 1} - \dfrac{1}{2} = \dfrac{1}{x}$ **42.** $\dfrac{4}{x} = \dfrac{8}{x^2} + 1$

Application Exercises

43. The personnel manager of a roller skate company knows that the company's weekly revenue, R, in thousands of dollars, can be modeled by the formula

$$R = -2x^2 + 36x,$$

where x is the price of a pair of skates, in dollars. A job applicant promises the personnel manager an advertising campaign guaranteed to generate \$200,000 in weekly revenue. Substitute 200 for R in the given formula and solve the equation. Are the solutions real numbers? Explain why the applicant will or will not be hired in the advertising department.

44. A football is kicked straight up from a height of 4 feet with an initial speed of 60 feet per second. The formula

$$h = -16t^2 + 60t + 4$$

describes the ball's height above the ground, h, in feet, t seconds after it is kicked. Will the ball reach a height of 80 feet? Substitute 80 for h in the given formula and solve the equation. Are the solutions real numbers? Explain why the ball will or will not reach 80 feet.

Writing in Mathematics

45. What is the imaginary unit i?

46. Explain how to write $\sqrt{-64}$ as a multiple of i.

47. What is a complex number?

48. What is an imaginary number?

49. Why is every real number also a complex number?

50. Explain each of the three jokes in the cartoon on page 604.

Critical Thinking Exercises

51. Which one of the following is true?

   a. $-\sqrt{-9} = -(-3) = 3$

   b. The complex number $a + 0i$ is the real number a.

   c. $2 + \sqrt{-4} = 2 - 2i$

   d. $\dfrac{2 \pm 4i}{2} = 1 \pm 4i$

52. Show that $1 + i$ is a solution of $x^2 - 2x + 2 = 0$ by substituting $1 + i$ for x. You should obtain

$$(1 + i)^2 - 2(1 + i) + 2.$$

Square $1 + i$ as you would a binomial. Distribute -2 as indicated. Then simplify the resulting expression by combining like terms and replacing i^2 with -1. You should obtain 0. Use this procedure to show that $1 - i$ is the equation's other solution.

53. Prove that there is no real number such that when twice the number is subtracted from its square, the difference is -5.

Technology Exercises

54. Reread Exercise 44. Use your graphing utility to illustrate the problem's solution by graphing $y = -16x^2 + 60x + 4$ and $y = 80$ in a $[0, 4, 1]$ by $[0, 100, 10]$ viewing rectangle. Explain how the graphs show that the ball will not reach a height of 80 feet.

55. Reread Exercise 43. Use your graphing utility to illustrate the problem's solution by graphing $y = -2x^2 + 36x$ and $y = 200$ in a $[0, 20, 1]$ by $[0, 210, 30]$ viewing rectangle. Explain how the two graphs show that weekly revenue will not reach \$200,000 and, therefore, the applicant will not be hired.

Review Exercises

In Exercises 56–58, graph each equation in a rectangular coordinate system.

56. $y = \dfrac{1}{3}x - 2$ (Section 4.4, Example 3)

57. $2x - 3y = 6$ (Section 4.2, Example 4)

58. $x = -2$ (Section 4.2, Example 8)

SECTION 10.5

GRAPHS OF QUADRATIC EQUATIONS

Objectives

1 Understand the characteristics of graphs of quadratic equations.

2 Find a parabola's intercepts.

3 Find a parabola's vertex.

4 Graph quadratic equations.

5 Solve problems using a parabola's vertex.

The Food Stamp Program is the first line of defense against hunger for millions of American families. The program provides benefits for eligible participants to purchase food items at approved food stores. Over half of all participants are children; one out of six is a low-income older adult. In this section, you will learn to use graphs of quadratic equations to gain a visual understanding of quantities that have a maximum or a minimum value, such as the number of participants in the Food Stamp Program.

1 Understand the characteristics of graphs of quadratic equations.

Graphs of Quadratic Equations The graph of the quadratic equation

$$y = ax^2 + bx + c, \quad a \neq 0,$$

is called a **parabola**. Parabolas are shaped like cups, as shown in Figure 10.7. If the coefficient of x^2 (the value of a in $ax^2 + bx + c$) is positive, the parabola opens upward. If the coefficient of x^2 is negative, the parabola opens downward. The **vertex** (or turning point) of the parabola is the lowest point, or minimum point, on the graph when it opens upward and the highest point, or maximum point, on the graph when it opens downward.

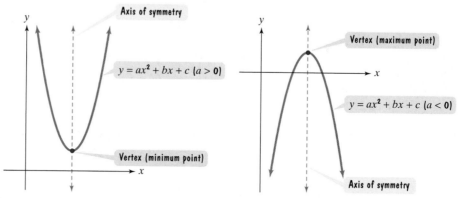

FIGURE 10.7 Characteristics of graphs of quadratic equations

$a > 0$: Parabola opens upward. $a < 0$: Parabola opens downward.

Look at the unusual image of the word "mirror" shown on the right. The artist, Scott Kim, has created the image so that the two halves of the whole are mirror images of each other. A parabola shares this kind of symmetry, in which a line through the vertex divides the figure in half. Parabolas are symmetric with respect to this line, called the **axis of symmetry**. If a parabola is folded along its axis of symmetry, the two halves match exactly.

EXAMPLE 1 Using Point Plotting to Graph a Parabola

Consider the equation $y = x^2 + 4x + 3$.

a. Is the graph a parabola that opens upward or downward?

b. Use point plotting to graph the parabola. Select integers from -5 to 1, inclusive, for x.

SOLUTION

a. To determine whether a parabola opens upward or downward, we begin by identifying a, the coefficient of x^2. The following voice balloons show the values for a, b, and c in $y = x^2 + 4x + 3$. Notice that we wrote x^2 as $1x^2$.

$$y = 1x^2 + 4x + 3$$

| a, the coefficient of x^2, is 1. | b, the coefficient of x, is 4. | c, the constant term, is 3. |

When a is greater than 0, a parabola opens upward. When a is less than 0, a parabola opens downward. Because $a = 1$, which is greater than 0, the parabola opens upward.

b. To use point plotting to graph the parabola, we first make a table of x- and y-coordinates.

x	$y = x^2 + 4x + 3$	(x, y)
-5	$y = (-5)^2 + 4(-5) + 3 = 8$	$(-5, 8)$
-4	$y = (-4)^2 + 4(-4) + 3 = 3$	$(-4, 3)$
-3	$y = (-3)^2 + 4(-3) + 3 = 0$	$(-3, 0)$
-2	$y = (-2)^2 + 4(-2) + 3 = -1$	$(-2, -1)$
-1	$y = (-1)^2 + 4(-1) + 3 = 0$	$(-1, 0)$
0	$y = 0^2 + 4(0) + 3 = 3$	$(0, 3)$
1	$y = 1^2 + 4(1) + 3 = 8$	$(1, 8)$

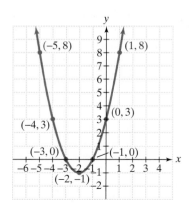

FIGURE 10.8 The graph of $y = x^2 + 4x + 3$

Then we plot the points and connect them with a smooth curve. The graph of $y = x^2 + 4x + 3$ is shown in Figure 10.8. ■

 CHECK POINT 1 Consider the equation $y = x^2 - 6x + 8$.

a. Is the graph a parabola that opens upward or downward?

b. Use point plotting to graph the parabola. Select integers from 0 to 6, inclusive, for x.

Several points are important when graphing a quadratic equation. These points, labeled in Figure 10.9, are the x-intercepts (although not every parabola has two x-intercepts), the y-intercept, and the vertex. Let's see how we can locate these points.

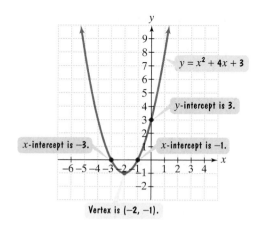

$y = x^2 + 4x + 3$

y-intercept is 3.

x-intercept is -3.

x-intercept is -1.

Vertex is $(-2, -1)$.

FIGURE 10.9 Useful points in graphing a parabola

2 Find a parabola's intercepts.

Finding a Parabola's x-Intercepts At each point where a parabola crosses the x-axis, the value of y equals 0. Thus, the x-intercepts can be found by replacing y with 0 in $y = ax^2 + bx + c$. Use factoring or the quadratic formula to solve the resulting quadratic equation for x.

EXAMPLE 2 Finding a Parabola's x-Intercepts

Find the x-intercepts for the parabola whose equation is $y = x^2 + 4x + 3$.

SOLUTION Replace y with 0 in $y = x^2 + 4x + 3$. We obtain $0 = x^2 + 4x + 3$, or $x^2 + 4x + 3 = 0$. We can solve this equation by factoring.

$$x^2 + 4x + 3 = 0$$
$$(x + 3)(x + 1) = 0$$
$$x + 3 = 0 \quad \text{or} \quad x + 1 = 0$$
$$x = -3 \qquad\qquad x = -1$$

Thus, the x-intercepts are -3 and -1. The parabola passes through $(-3, 0)$ and $(-1, 0)$, as shown in Figure 10.9. ∎

 CHECK POINT 2 Find the x-intercepts for the parabola whose equation is $y = x^2 - 6x + 8$.

Finding a Parabola's y-Intercept At the point where a parabola crosses the y-axis, the value of x equals 0. Thus, the y-intercept can be found by replacing x with 0 in $y = ax^2 + bx + c$. Simple arithmetic will produce a value for y, which is the y-intercept.

EXAMPLE 3 Finding a Parabola's y-Intercept

Find the y-intercept for the parabola whose equation is $y = x^2 + 4x + 3$.

SOLUTION Replace x with 0 in $y = x^2 + 4x + 3$.
$$y = 0^2 + 4 \cdot 0 + 3 = 0 + 0 + 3 = 3$$
The y-intercept is 3. The parabola passes through $(0, 3)$, as shown in Figure 10.9. ∎

 CHECK POINT 3 Find the y-intercept for the parabola whose equation is $y = x^2 - 6x + 8$.

3 Find a parabola's vertex.

Finding a Parabola's Vertex Keep in mind that a parabola's vertex is its turning point. The x-coordinate of the vertex for the parabola in Figure 10.9, -2, is midway between the x-intercepts, -3 and -1. If a parabola has two x-intercepts, they are found by solving $ax^2 + bx + c = 0$. The solutions of this equation,

$$x = \frac{-b - \sqrt{b^2 - 4ac}}{2a} \quad \text{and} \quad x = \frac{-b + \sqrt{b^2 - 4ac}}{2a},$$

are the x-intercepts. The value of x midway between these intercepts is $x = \dfrac{-b}{2a}$.

This equation can be used to find the x-coordinate of the vertex even when no x-intercepts exist.

> **THE VERTEX OF A PARABOLA** For a parabola whose equation is $y = ax^2 + bx + c$,
>
> **1.** The x-coordinate of the vertex is $\dfrac{-b}{2a}$.
>
> **2.** The y-coordinate of the vertex is found by substituting the x-coordinate into the parabola's equation and evaluating.

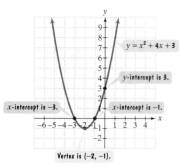

FIGURE 10.9 (repeated) Useful points in graphing a parabola

EXAMPLE 4 Finding a Parabola's Vertex

Find the vertex for the parabola whose equation is $y = x^2 + 4x + 3$.

SOLUTION In the equation $y = x^2 + 4x + 3$, $a = 1$ and $b = 4$.

$$x\text{-coordinate of vertex} = \frac{-b}{2a} = \frac{-4}{2 \cdot 1} = \frac{-4}{2} = -2$$

To find the y-coordinate of the vertex, we substitute -2 for x in $y = x^2 + 4x + 3$ and then evaluate.

$$y\text{-coordinate of vertex} = (-2)^2 + 4(-2) + 3 = 4 + (-8) + 3 = -1$$

The vertex is $(-2, -1)$, as shown in Figure 10.9. ∎

✔ **CHECK POINT 4** Find the vertex for the parabola whose equation is $y = x^2 - 6x + 8$.

---

4 Graph quadratic equations.

A Strategy for Graphing Quadratic Equations Here is a procedure to sketch the graph of the quadratic equation, $y = ax^2 + bx + c$:

> **GRAPHING QUADRATIC EQUATIONS** The graph of $y = ax^2 + bx + c$, called a parabola, can be graphed using the following steps:
>
> 1. Determine whether the parabola opens upward or downward. If $a > 0$, it opens upward. If $a < 0$, it opens downward.
> 2. Determine the vertex of the parabola. The x-coordinate is $\dfrac{-b}{2a}$.
>
>   The y-coordinate is found by substituting the x-coordinate into the parabola's equation and evaluating y.
> 3. Find any x-intercepts by replacing y with 0. Solve the resulting quadratic equation for x.
> 4. Find the y-intercept by replacing x with 0.
> 5. Plot the intercepts and the vertex.
> 6. Connect these points with a smooth curve.

EXAMPLE 5 Graphing a Parabola

Graph the quadratic equation: $y = x^2 - 2x - 3$.

SOLUTION We can graph this equation by following the steps in the box.

Step 1. **Determine how the parabola opens.** Note that a, the coefficient of x^2, is 1. Thus, $a > 0$; this positive value tells us that the parabola opens upward.

Step 2. **Find the vertex.** We know that the x-coordinate of the vertex is $\dfrac{-b}{2a}$. Let's identify the numbers a, b, and c in the given equation, which is in the form $y = ax^2 + bx + c$.

$$y = x^2 - 2x - 3$$

$a = 1 \qquad b = -2 \qquad c = -3$

Now we substitute the values of a and b into the equation for the x-coordinate:

$$x\text{-coordinate of vertex} = \frac{-b}{2a} = \frac{-(-2)}{2(1)} = \frac{2}{2} = 1.$$

The x-coordinate of the vertex is equal to 1. We can substitute 1 for x into the equation $y = x^2 - 2x - 3$ to find the y-coordinate:

$$y\text{-coordinate of vertex} = 1^2 - 2 \cdot 1 - 3 = 1 - 2 - 3 = -4.$$

The vertex is $(1, -4)$.

Step 3. **Find the x-intercepts.** Replace y with 0 in $y = x^2 - 2x - 3$. We obtain $0 = x^2 - 2x - 3$ or $x^2 - 2x - 3 = 0$. We can solve this equation by factoring.

$$x^2 - 2x - 3 = 0$$

$$(x - 3)(x + 1) = 0$$

$$x - 3 = 0 \quad \text{or} \quad x + 1 = 0$$

$$x = 3 \qquad\qquad x = -1$$

The x-intercepts are 3 and -1. The parabola passes through $(3, 0)$ and $(-1, 0)$.

Step 4. **Find the y-intercept.** Replace x with 0 in $y = x^2 - 2x - 3$:

$$y = 0^2 - 2 \cdot 0 - 3 = 0 - 0 - 3 = -3.$$

The y-intercept is -3. The parabola passes through $(0, -3)$.

Steps 5 and 6. **Plot the intercepts and the vertex. Connect these points with a smooth curve.** The intercepts and the vertex are shown as the four labeled points in Figure 10.10. Also shown is the graph of the quadratic equation, obtained by connecting the points with a smooth curve.

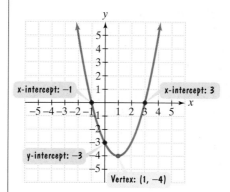

FIGURE 10.10 The graph of $y = x^2 - 2x - 3$

 CHECK POINT **5** Graph the quadratic equation: $y = x^2 + 6x + 5$.

Parabola

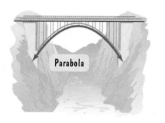

Parabola

Cables hung between structures to form suspension bridges often form parabolas. Arches constructed of steel and concrete, whose main purpose is strength, are usually parabolic in shape.

EXAMPLE 6 Graphing a Parabola

Graph the quadratic equation: $y = -x^2 + 4x - 1$.

SOLUTION

Step 1. **Determine how the parabola opens.** Note that a, the coefficient of x^2, is -1. Thus, $a < 0$; this negative value tells us that the parabola opens downward.

Step 2. **Find the vertex.** The x-coordinate of the vertex is $\dfrac{-b}{2a}$.

$$y = -x^2 + 4x - 1$$

$$a = -1 \qquad b = 4 \qquad c = -1$$

$$x\text{-coordinate of vertex} = \frac{-b}{2a} = \frac{-4}{2(-1)} = \frac{-4}{-2} = 2$$

Substitute 2 for x in $y = -x^2 + 4x - 1$ to find the y-coordinate:

$$y\text{-coordinate of vertex} = -2^2 + 4 \cdot 2 - 1 = -4 + 8 - 1 = 3.$$

The vertex is $(2, 3)$.

Step 3. **Find the x-intercepts.** Replace y with 0 in $y = -x^2 + 4x - 1$. We obtain $0 = -x^2 + 4x - 1$ or $-x^2 + 4x - 1 = 0$. This equation cannot be solved by factoring. We will use the quadratic formula to solve it.

$$a = -1, \qquad b = 4, \qquad c = -1$$

$$x = \frac{-b \pm \sqrt{b^2 - 4ac}}{2a} = \frac{-4 \pm \sqrt{4^2 - 4(-1)(-1)}}{2(-1)} = \frac{-4 \pm \sqrt{16 - 4}}{-2}$$

$$x = \frac{-4 + \sqrt{12}}{-2} \approx 0.3 \quad \text{or} \quad x = \frac{-4 - \sqrt{12}}{-2} \approx 3.7$$

The x-intercepts are approximately 0.3 and 3.7. The parabola passes through $(0.3, 0)$ and $(3.7, 0)$.

Step 4. **Find the y-intercept.** Replace x with 0 in $y = -x^2 + 4x - 1$:

$$y = -0^2 + 4 \cdot 0 - 1 = -1.$$

The y-intercept is -1. The parabola passes through $(0, -1)$.

Steps 5 and 6. **Plot the intercepts and the vertex. Connect these points with a smooth curve.** The intercepts and the vertex are shown as the four labeled points in Figure 10.11. Also shown is the graph of the quadratic equation, obtained by connecting the points with a smooth curve.

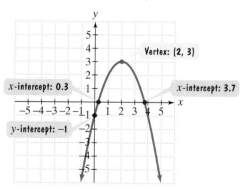

FIGURE 10.11 The graph of $y = -x^2 + 4x - 1$ ■

 CHECK POINT 6 Graph the quadratic equation: $y = -x^2 - 2x + 5$.

5 Solve problems using a parabola's vertex

Applications When did the maximum number of Americans participate in the food stamp program? How do people launching fireworks know when they should explode to be viewed at the greatest possible height? What is the age of a driver having the least number of car accidents? The answers to these questions involve finding the maximum or minimum value of equations in the form $y = ax^2 + bx + c$. The vertex of the graph is the point of interest. If $a < 0$, the parabola opens downward and the vertex is its highest point. If $a > 0$, the parabola opens upward and the vertex is its lowest point.

EXAMPLE 7 The Food Stamp Program

The formula

$$y = -0.5x^2 + 4x + 19$$

models the number of people, y, in millions, receiving food stamps x years after 1990. The graph of the resulting parabola is shown in Figure 10.12. Find the coordinates of the vertex and describe what this represents in practical terms.

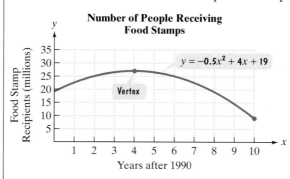

Number of People Receiving Food Stamps

$y = -0.5x^2 + 4x + 19$

Vertex

Food Stamp Recipients (millions)

Years after 1990

FIGURE 10.12

Source: New York Times

SOLUTION We begin by identifying the numbers a, b, and c in the given model.

$$y = -0.5x^2 + 4x + 19$$

$a = -0.5$ $b = 4$ $c = 19$

$$x\text{-coordinate of vertex} = \frac{-b}{2a} = \frac{-4}{2(-0.5)} = \frac{-4}{-1} = 4$$

Substitute 4 for x in $y = -0.5x^2 + 4x + 19$ to find the y-coordinate.

$$y\text{-coordinate of vertex} = -0.5(4)^2 + 4 \cdot 4 + 19 = -8 + 16 + 19 = 27.$$

The vertex is $(4, 27)$. Because the coefficient of x^2, -0.5, is negative, the parabola opens downward and the vertex is the highest point on the graph. This means that 4 years after 1990, in 1994, the number of people receiving food stamps reached a maximum of 27 million. In the years before and after 1994, the number of recipients was less than 27 million. ■

 CHECK POINT 7 The formula

$$y = 0.4x^2 - 36x + 1000$$

models the number of accidents, y, per 50 million miles driven, for drivers x years old, where $16 \leq x \leq 74$. Find the coordinates of the graph's vertex and describe what this represents in practical terms.

10.5 EXERCISE SET

Student Solutions Manual   CD/Video   PH Math/Tutor Center   MathXL Tutorials on CD   MathXL®   MyMathLab   Interactmath.com

Practice Exercises

In Exercises 1–4 determine if the parabola whose equation is given opens upward or downward.

1. $y = x^2 - 4x + 3$

2. $y = x^2 - 6x + 5$

3. $y = -2x^2 + x + 6$

4. $y = -2x^2 - 4x + 6$

In Exercises 5–10, find the x-intercepts for the parabola whose equation is given. If the x-intercepts are irrational numbers, round your answers to the nearest tenth.

5. $y = x^2 - 4x + 3$

6. $y = x^2 - 6x + 5$

7. $y = -x^2 + 8x - 12$

8. $y = -x^2 - 2x + 3$

9. $y = x^2 + 2x - 4$

10. $y = x^2 + 8x + 14$

In Exercises 11–18, find the y-intercept for the parabola whose equation is given.

11. $y = x^2 - 4x + 3$

12. $y = x^2 - 6x + 5$

13. $y = -x^2 + 8x - 12$

14. $y = -x^2 - 2x + 3$

15. $y = x^2 + 2x - 4$

16. $y = x^2 + 8x + 14$

17. $y = x^2 + 6x$

18. $y = x^2 + 8x$

In Exercises 19–24, find the vertex for the parabola whose equation is given.

19. $y = x^2 - 4x + 3$

20. $y = x^2 - 6x + 5$

21. $y = 2x^2 + 4x - 6$

22. $y = -2x^2 - 4x - 2$

23. $y = x^2 + 6x$

24. $y = x^2 + 8x$

In Exercises 25–36, graph the parabola whose equation is given.

25. $y = x^2 + 8x + 7$

26. $y = x^2 + 10x + 9$

27. $y = x^2 - 2x - 8$

28. $y = x^2 + 4x - 5$

29. $y = -x^2 + 4x - 3$

30. $y = -x^2 + 2x + 3$

31. $y = x^2 - 1$

32. $y = x^2 - 4$

33. $y = x^2 + 2x + 1$

34. $y = x^2 - 2x + 1$

35. $y = -2x^2 + 4x + 5$

36. $y = -3x^2 + 6x - 2$

Practice Plus

In Exercises 37–44, find the vertex for the parabola whose equation is given by writing the equation in the form $y = ax^2 + bx + c$.

37. $y = (x - 3)^2 + 2$

38. $y = (x - 4)^2 + 3$

39. $y = (x + 5)^2 - 4$

40. $y = (x + 6)^2 - 5$

41. $y = 2(x - 1)^2 - 3$

42. $y = 2(x - 1)^2 - 4$

43. $y = -3(x + 2)^2 + 5$

44. $y = -3(x + 4)^2 + 6$

45. Generalize your work in Exercises 37–44 and complete the following statement: For a parabola whose equation is $y = a(x - h)^2 + k$, the vertex is the point _____.

Application Exercises

46. In the United States, HCV, or hepatitis C virus, is four times as widespread as HIV. Few of the nation's three to four million carriers have any idea they are infected. The graph shows the projected mortality, in thousands, from the virus.

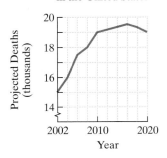

Projected Hepatitis C Deaths in the United States

Source: American Journal of Public Health

Suppose that a quadratic equation is used to model the data shown with ordered pairs representing (number of years after 2002, thousands of hepatitis C deaths). Determine, without obtaining an actual quadratic equation that models the data, the approximate coordinates of the vertex for the equation's graph. Describe what this means in practical terms. Use the word *maximum* in your description.

47. The equation $y = -0.02x^2 + x + 1$ models the number of inches, y, that a young redwood tree grows per year when the annual rainfall is x inches. The graph of the resulting parabola was obtained using a graphing utility and is shown below. Find the coordinates of the vertex. Then describe what the vertex represents in practical terms.

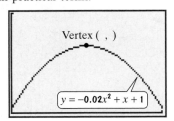

Vertex (,)

$y = -0.02x^2 + x + 1$

48. One would think that the more avocado trees planted per acre, the higher the yield of avocados. However, this is not the case. When more than a certain number of trees are present per acre, they tend to crowd one another and the yield per tree drops. The formula $y = -0.01x^2 + 0.8x$ models the yield, y, in bushels of avocados per tree, when x trees are planted per acre. The graph of the resulting parabola was obtained with a graphing utility and is shown below. Find the coordinates of the vertex. Then describe what the vertex represents in practical terms.

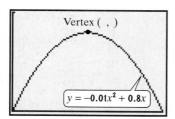

Use the coordinates of the vertex to solve Exercises 49–52.

49. Fireworks are launched into the air. The formula

$$y = -16x^2 + 200x + 4$$

models the fireworks' height, y, in feet, x seconds after they are launched. When should the fireworks explode so that they go off at the greatest height? What is that height?

50. A football is thrown by a quarterback to a receiver 40 yards away. The formula

$$y = -0.025x^2 + x + 5$$

models the football's height above the ground, y, in feet, when it is x yards from the quarterback. How many yards from the quarterback does the football reach its greatest height? What is that height?

51. You have 120 feet of fencing to enclose a rectangular plot that borders on a river. If you do not fence the side along the river, find the length and width of the plot that will maximize the area. What is the largest area that can be enclosed?

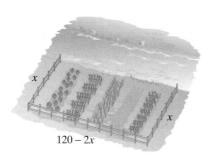

52. The figure shown indicates that you have 100 yards of fencing to enclose a rectangular area. Find the dimensions of the rectangle that maximize the enclosed area. What is the maximum area?

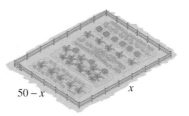

Writing in Mathematics

53. What is a parabola? Describe its shape.

54. Explain how to decide whether a parabola opens upward or downward.

55. If a parabola has two x-intercepts, explain how to find them.

56. Explain how to find a parabola's y-intercept.

57. Describe how to find a parabola's vertex.

58. A parabola that opens upward has its vertex at $(1, 2)$. Describe as much as you can about the parabola based on this information. Include in your discussion the number of x-intercepts, if any, for the parabola.

Critical Thinking Exercises

59. Which one of the following is true?

  a. The x-coordinate of the vertex of the parabola whose equation is $y = ax^2 + bx + c$ is $\dfrac{b}{2a}$.

  b. If a parabola has only one x-intercept, then the x-intercept is also the vertex.

  c. There is no relationship between the graph of $y = ax^2 + bx + c$ and the number of real solutions of the equation $ax^2 + bx + c = 0$.

  d. If $y = 4x^2 - 40x + 4$, then the vertex is the highest point on the graph.

60. Find two numbers whose sum is 200 and whose product is a maximum.

61. Graph $y = 2x^2 - 8$ and $y = -2x^2 + 8$ in the same rectangular coordinate system. What are the coordinates of the points of intersection?

62. A parabola has x-intercepts at 3 and 7, a y-intercept at -21, and $(5, 4)$ for its vertex. Write the parabola's equation.

Technology Exercises

63. Use a graphing utility to verify any five of your hand-drawn graphs in Exercises 25–36.

64. a. Use a graphing utility to graph $y = 2x^2 - 82x + 720$ in a standard viewing rectangle. What do you observe?

b. Find the coordinates of the vertex for the given quadratic equation.

c. The answer to part (b) is $(20.5, -120.5)$. Because the leading coefficient, 2 of $y = 2x^2 - 82x + 720$ is positive, the vertex is a minimum point on the graph. Use this fact to help find a viewing rectangle that will give a relatively complete picture of the parabola. With an axis of symmetry at $x = 20.5$, the setting for x should extend past this, so try Xmin = 0 and Xmax = 30. The setting for y should include (and probably go below) the y-coordinate of the graph's minimum point, so try Ymin = -130. Experiment with Ymax until your utility shows the parabola's major features.

d. In general, explain how knowing the coordinates of a parabola's vertex can help determine a reasonable viewing rectangle on a graphing utility for obtaining a complete picture of the parabola.

In Exercises 65–68, find the vertex for each parabola. Then determine a reasonable viewing rectangle on your graphing utility and use it to graph the parabola.

65. $y = -0.25x^2 + 40x$

66. $y = -4x^2 + 20x + 160$

67. $y = 5x^2 + 40x + 600$

68. $y = 0.01x^2 + 0.6x + 100$

Review Exercises

In Exercises 69–71, solve each equation or system of equations.

69. $7(x - 2) = 10 - 2(x + 3)$ (Section 2.3, Example 3)

70. $\dfrac{7}{x + 2} + \dfrac{2}{x + 3} = \dfrac{1}{x^2 + 5x + 6}$ (Section 8.6, Example 4)

71. $5x - 3y = -13$
$x = 2 - 4y$ (Section 5.2, Example 1)

10.6

INTRODUCTION TO FUNCTIONS

Objectives

1 Find the domain and range of a relation.

2 Determine whether a relation is a function.

3 Evaluate a function.

4 Use the vertical line test to identify functions.

5 Interpret function values for functions that model data.

The number of calories that you burn per hour depends on the activity in which you are engaged. The percentage of the U.S. population with a college degree depends on the year in which this percent is computed. In both of these situations, the relationship between variables can be illustrated with the notion of a *function*. Understanding this concept will give you a new perspective on many ordinary situations. Much of your work in subsequent algebra courses will be devoted to the important topic of functions and how they model your world.

1 Find the domain and range of a relation.

Relations Studies show that exercise can promote good long-term health no matter how much you weigh. A brisk half-hour walk each day is enough to get the benefits. Combined with a healthy diet, it also helps to stave off obesity. How many calories does your workout burn? The graph in Figure 10.13 shows the calories burned per hour in six activities.

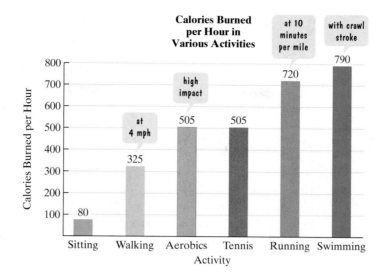

FIGURE 10.13 Counting calories
Source: FITRESOURCE.COM

The information shown in the bar graph indicates a correspondence between the activities and calories burned per hour. We can write this correspondence using a set of ordered pairs:

{(sitting, 80), (walking, 325), (aerobics, 505), (tennis, 505),
(running, 720), (swimming, 790)}.

These braces indicate that we are representing a set.

The mathematical term for a set of ordered pairs is a **relation**.

DEFINITION OF A RELATION   A **relation** is any set of ordered pairs. The set of all first components of the ordered pairs is called the **domain** of the relation and the set of all second components is called the **range** of the relation.

EXAMPLE 1   Finding the Domain and Range of a Relation

Find the domain and range of the relation:

{(sitting, 80), (walking, 325), (aerobics, 505), (tennis, 505),
(running, 720), (swimming, 790)}.

SOLUTION   The domain is the set of all first components. Thus, the domain is

{sitting, walking, aerobics, tennis, running, swimming}.

Parentheses and square brackets are not used to represent sets.

The range is the set of all second components. Thus the range is

{80, 325, 505, 720, 790}.

Although both aerobics and tennis burn 505 calories per hour, it is not necessary to list 505 twice.

✔ CHECK POINT 1 The following set shows calories burned per hour in activities not included in Figure 10.13. Find the domain and range of the relation:

{(golf, 250), (lawn mowing, 325), (water skiing, 430),
(hiking, 430), (bicycling, 720)}.

with cart

at 15 mph

2 Determine whether a relation is a function.

Activity	Calories Burned per hour
Sitting	80
Walking	325
Aerobics	505
Tennis	505
Running	720
Swimming	790

Functions Shown in the margin are the calories burned per hour for the activities in the bar graph in Figure 10.13. We've used this information to define two relations. Figure 10.14(a) shows a correspondence between activities and calories burned. Figure 10.14(b) shows a correspondence between calories burned and activities.

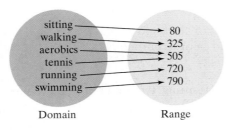

FIGURE 10.14 (a) Activities correspond to calories burned.

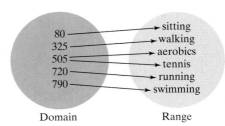

FIGURE 10.14 (b) Calories burned correspond to activities.

A relation in which each member of the domain corresponds to exactly one member of the range is a **function**. Can you see that the relation in Figure 10.14(a) is a function? Each activity in the domain corresponds to exactly one number representing calories burned per hour in the range. If we know the activity, we know the calories burned per hour. Notice that more than one element in the domain can correspond to the same element in the range: Aerobics and tennis both burn 505 calories per hour.

Is the relation in Figure 10.14(b) a function? Does each member of the domain correspond to precisely one member of the range? This relation is not a function because there is a member of the domain that corresponds to two members of the range:

(505, aerobics)   (505, tennis).

The member of the domain, 505, corresponds to both aerobics and tennis. If we know the calories burned per hour, 505, we cannot be sure of the activity. Because **a function is a relation in which no two ordered pairs have the same first component and different second components**, the ordered pairs (505, aerobics) and (505, tennis) are not ordered pairs of a function.

Same first component

(505, aerobics)   (505, tennis)

Different second components

> **DEFINITION OF A FUNCTION** A **function** is a relation in which each member of the domain corresponds to exactly one member of the range. No two ordered pairs of a function can have the same first component and different second components.

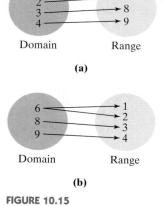

Domain   Range

(a)

Domain   Range

(b)

FIGURE 10.15

EXAMPLE 2   Determining Whether a Relation Is a Function

Determine whether each relation is a function:

a. $\{(1, 6), (2, 6), (3, 8), (4, 9)\}$ **b.** $\{(6, 1), (6, 2), (8, 3), (9, 4)\}$.

SOLUTION We begin by making a figure for each relation that shows the domain and the range (Figure 10.15).

a. Figure 10.15(a) shows that every element in the domain corresponds to exactly one element in the range. The element 1 in the domain corresponds to the element 6 in the range. Furthermore, 2 corresponds to 6, 3 corresponds to 8, and 4 corresponds to 9. No two ordered pairs in the given relation have the same first component and different second components. Thus, the relation is a function.

b. Figure 10.15(b) shows that 6 corresponds to both 1 and 2. If any element in the domain corresponds to more than one element in the range, the relation is not a function. This relation is not a function; two ordered pairs have the same first component and different second components.

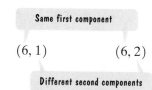

Look at Figure 10.15(a) again. The fact that 1 and 2 in the domain correspond to the same number, 6, in the range does not violate the definition of a function. **A function can have two different first components with the same second component.** By contrast, a relation is not a function when two different ordered pairs have the same first component and different second components. Thus, the relation in Figure 10.15(b) is not a function.

> ✔ **CHECK POINT 2** Determine whether each relation is a function:
>
> **a.** $\{(1, 2), (3, 4), (5, 6), (5, 8)\}$
> **b.** $\{(1, 2), (3, 4), (6, 5), (8, 5)\}$.

---

3   Evaluate a function.

Functions as Equations and Function Notation Functions are usually given in terms of equations rather than as sets of ordered pairs. For example, here is an equation that shows a woman's cholesterol level as a function of her age:

$$y = 1.17x + 147.15.$$

The variable x represents a woman's age, in years. The variable y represents her cholesterol level. For each age, x, the equation gives one and only one cholesterol level, y. The variable y is a function of the variable x.

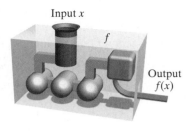

Input x

f

Output $f(x)$

FIGURE 10.16 A function as a machine with inputs and outputs

When an equation represents a function, the function is often named by a letter such as f, g, h, F, G, or H. Any letter can be used to name a function. Suppose that f names a function. Think of the domain as the set of the function's inputs and the range as the set of the function's outputs. As shown in Figure 10.16, the input is represented by x and the output by $f(x)$. The special notation $f(x)$, read "f of x" or "f at x," represents the **value of the function at the number x**.

Let's make this clearer by considering a specific example. We know the equation

$$y = 1.17x + 147.15$$

defines y as a function of x. We'll name the function f. Now, we can apply our new function notation.

Input	Output	Equation
x	$f(x)$	$f(x) = 1.17x + 147.15$

We read this equation as "f of x equals $1.17x + 147.15$."

Suppose we are interested in finding $f(30)$, the function's output when the input is 30. To find the value of the function at 30, we substitute 30 for x. We are **evaluating the function** at 30.

STUDY TIP

The notation $f(x)$ does *not* mean "f times x." The notation describes the value of the function at x.

$f(x) = 1.17x + 147.15$ This is the given function.

$f(30) = 1.17(30) + 147.15$ The input is 30.

$= 35.1 + 147.15$ Multiply.

$= 182.25$ Add.

Input
$x = 30$ $f(x) = 1.17x + 147.15$

$1.17(30) + 147.15$

Output
$f(30) = 182.25$

FIGURE 10.17 A function machine at work

The statement $f(30) = 182.25$, read "f of 30 equals 182.25," tells us that the value of the function at 30 is 182.25. When the function's input is 30, its output is 182.25. The model gives us exactly one cholesterol level for a 30-year-old woman, namely 182.25. Figure 10.17 illustrates the input and output in terms of a function machine.

EXAMPLE 3 Evaluating a Function

If $f(x) = 3x + 1$, find each of the following:

   a. $f(5)$ **b.** $f(-4)$ **c.** $f(0)$.

SOLUTION   We substitute 5, −4, and 0, respectively, for x in the function's equation, $f(x) = 3x + 1$.

   a. $f(5)\ \ = 3 \cdot 5 + 1 = 15 + 1 = 16$ f of 5 equals 16.

   b. $f(-4) = 3(-4) + 1 = -12 + 1 = -11$ f of −4 equals −11.

   c. $f(0)\ \ = 3 \cdot 0 + 1 = 0 + 1 = 1$ f of 0 equals 1. ■

The function $f(x) = 3x + 1$ is an example of a *linear function*. **Linear functions** have equations of the form $f(x) = mx + b$.

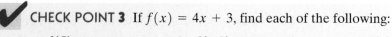

CHECK POINT 3 If $f(x) = 4x + 3$, find each of the following:

   a. $f(5)$ **b.** $f(-2)$ **c.** $f(0)$.

EXAMPLE 4 Evaluating a Function

If $g(x) = x^2 + 3x + 5$, find each of the following:

    a. $g(2)$ **b.** $g(-3)$ **c.** $g(0)$.

SOLUTION We substitute 2, −3, and 0, respectively, for x in the function's equation, $g(x) = x^2 + 3x + 5$.

    a. $g(2) = 2^2 + 3 \cdot 2 + 5 = 4 + 6 + 5 = 15$ *g of 2 is 15.*

    b. $g(-3) = (-3)^2 + 3(-3) + 5 = 9 + (-9) + 5 = 5$ *g of −3 is 5.*

    c. $g(0) = 0^2 + 3 \cdot 0 + 5 = 0 + 0 + 5 = 5$ *g of 0 is 5.* ∎

The function in Example 4 is a *quadratic function*. **Quadratic functions** have equations of the form $f(x) = ax^2 + bx + c, a \neq 0$.

 CHECK POINT 4 If $g(x) = x^2 + 4x + 3$, find each of the following:

    a. $g(5)$ **b.** $g(-4)$ **c.** $g(0)$.

4 Use the vertical line test to identify functions.

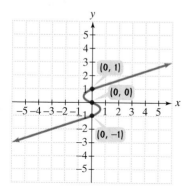

FIGURE 10.18 y is not a function of x because 0 is paired with three values of y, namely, 1, 0, and −1.

Graphs of Functions and the Vertical Line Test The **graph of a function** is the graph of its ordered pairs. For example, the graph of $f(x) = 3x + 1$ is the set of points (x, y) in the rectangular coordinate system satisfying the equation $y = 3x + 1$. Thus, the graph of f is a line with slope 3 and y-intercept 1. Similarly, the graph of $f(x) = x^2 + 3x + 5$ is the set of points (x, y) in the rectangular coordinate system satisfying the equation $y = x^2 + 3x + 5$. Thus, the graph of g is a parabola.

    Not every graph in the rectangular coordinate system is the graph of a function. The definition of a function specifies that no value of x can be paired with two or more different values of y. Consequently, if a graph contains two or more different points with the same first coordinate, the graph cannot represent a function. This is illustrated in Figure 10.18. Observe that points sharing a common first coordinate are vertically above or below each other.

    This observation is the basis of a useful test for determining whether a graph defines y as a function of x. The test is called the **vertical line test**.

THE VERTICAL LINE TEST FOR FUNCTIONS If any vertical line intersects a graph in more than one point, the graph does not define y as a function of x.

EXAMPLE 5 Using the Vertical Line Test

Use the vertical line test to identify graphs in which y is a function of x.

a.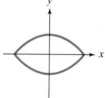   **b.**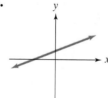   **c.**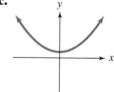   **d.**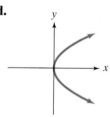

SOLUTION y is a function of x for the graphs in (b) and (c).

a.
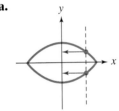
y **is not a function** of x.
Two values of y
correspond to an x-value.

b.
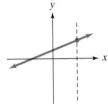
y **is a function** of x.

c.
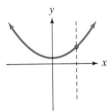
y **is a function** of x.

d.
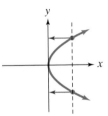
y **is not a function** of x.
Two values of y
correspond to an x-value.

CHECK POINT 5 Use the vertical line test to identify graphs in which y is a function of x.

a. y

b. y
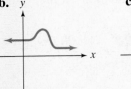

c. y
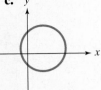

5 Interpret function values for functions that model data.

Applications Like formulas, functions can be obtained from verbal conditions or from actual data. Throughout your next algebra course, you'll have lots of practice doing this. For now, let's make sure that we can find and interpret functions values for functions that were obtained from modeling data.

EXAMPLE 6 Evaluating a Function and Interpreting the Result

The bar graph in Figure 10.19 shows the percentage of the U.S. population ages 25 or older with a college degree for six selected years from 1980 through 2002. The function

$$f(x) = 0.005x^2 + 0.338x + 17.014$$

models the percentage of the U.S. population ages 25 or older with a college degree x years after 1980. Find and interpret $f(22)$.

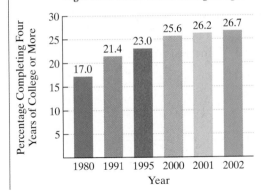

**Percentage of the U.S. Population
Ages 25 or Older with a College Degree**

FIGURE 10.19

Source: U.S. Census Bureau

SOLUTION

$$f(x) = 0.005x^2 + 0.338x + 17.014 \qquad \textit{This is the given function.}$$
$$f(22) = 0.005(22)^2 + 0.338(22) + 17.014 \qquad \textit{Replace each occurrence of x with 22.}$$
$$= 26.87 \qquad \textit{Use a calculator.}$$

We see that $f(22) = 26.87$. Because 22 represents the number of years after 1980, this means that in 2002, 26.87% of the U.S. population ages 25 or older had a college degree. Figure 10.19 shows the actual percent to be 26.7%, so the function models the data for 2002 fairly well. ∎

 CHECK POINT 6 Use the function in Example 6 to find and interpret $f(0)$.

USING TECHNOLOGY

Graphing utilities can be used to evaluate functions. The screens below show the evaluation of

$$f(x) = 0.005x^2 + 0.338x + 7.014$$

at 22 on a T1-83 graphing calculator. The function f is named Y_1.

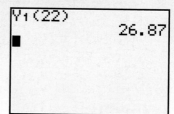

10.6 EXERCISE SET

Practice Exercises

In Exercises 1–8, determine whether each relation is a function. Give the domain and range for each relation.

1. $\{(1, 2), (3, 4), (5, 5)\}$

2. $\{(4, 5), (6, 7), (8, 8)\}$

3. $\{(3, 4), (3, 5), (4, 4), (4, 5)\}$

4. $\{(5, 6), (5, 7), (6, 6), (6, 7)\}$

5. $\{(-3, -3), (-2, -2), (-1, -1), (0, 0)\}$

6. $\{(-7, -7), (-5, -5), (-3, -3), (0, 0)\}$

7. $\{(1, 4), (1, 5), (1, 6)\}$

8. $\{(4, 1), (5, 1), (6, 1)\}$

In Exercises 9–24, evaluate each function at the given values.

9. $f(x) = x + 5$
   a. $f(7)$ **b.** $f(-6)$ **c.** $f(0)$

10. $f(x) = x + 6$
   a. $f(4)$ **b.** $f(-8)$ **c.** $f(0)$

11. $f(x) = 7x$
   a. $f(10)$ **b.** $f(-4)$ **c.** $f(0)$

12. $f(x) = 9x$
   a. $f(10)$ **b.** $f(-5)$ **c.** $f(0)$

13. $f(x) = 8x - 3$
    a. $f(12)$ **b.** $f\left(-\frac{1}{2}\right)$ **c.** $f(0)$

14. $f(x) = 6x - 5$
    a. $f(12)$ **b.** $f\left(-\frac{1}{2}\right)$ **c.** $f(0)$

15. $g(x) = x^2 + 3x$
    a. $g(2)$ **b.** $g(-2)$ **c.** $g(0)$

16. $g(x) = x^2 + 7x$
    a. $g(2)$ **b.** $g(-2)$ **c.** $g(0)$

17. $h(x) = x^2 - 2x + 3$
    a. $h(4)$ **b.** $h(-4)$ **c.** $h(0)$

18. $h(x) = x^2 - 4x + 5$
    a. $h(4)$ **b.** $h(-4)$ **c.** $h(0)$

19. $f(x) = 5$
    a. $f(9)$ **b.** $f(-9)$ **c.** $f(0)$

20. $f(x) = 7$
    a. $f(10)$ **b.** $f(-10)$ **c.** $f(0)$

21. $f(r) = \sqrt{r + 6} + 3$
    a. $f(-6)$ **b.** $f(10)$

22. $f(r) = \sqrt{25 - r} - 6$
    a. $f(16)$ **b.** $f(-24)$

23. $f(x) = \dfrac{x}{|x|}$
    a. $f(6)$ **b.** $f(-6)$

24. $f(x) = \dfrac{x}{|x|}$
    a. $f(5)$ **b.** $f(-5)$

In Exercises 25–32, use the vertical line test to identify graphs in which y is a function of x.

25.
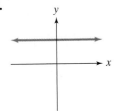

26.
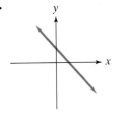

27.
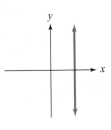

28.
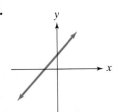

29.
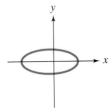

30.
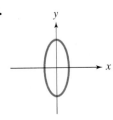

31.
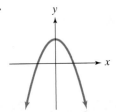

32.
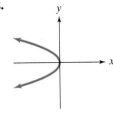

Practice Plus

In Exercises 33–36, express each function as a set of ordered pairs.

33. $f(x) = 2x + 3$; domain: $\{-1, 0, 1\}$

34. $f(x) = 3x + 5$; domain: $\{-1, 0, 1\}$

35. $g(x) = x - x^2$; domain: the set of integers from -2 to 2, inclusive

36. $g(x) = x - |x|$; domain: the set of integers from -2 to 2, inclusive

In Exercises 37–40, find

$$\frac{f(x) - f(h)}{x - h}$$

and simplify.

37. $f(x) = 6x + 7$

38. $f(x) = 8x + 9$

39. $f(x) = x^2 - 1$

40. $f(x) = x^3 - 1$

Application Exercises

41. The bar graph at the top of the next page shows the percentage of children in the world's leading industrial countries who daydream about being rich.

    a. Write a set of six ordered pairs in which countries correspond to the percentage of children daydreaming about being rich. Each ordered pair should be in the form

        (country, percent).

    b. Is the relation in part (a) a function? Explain your answer.

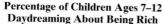

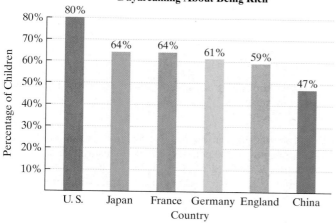

Percentage of Children Ages 7–12 Daydreaming About Being Rich

Source: Roper Starch Worldwide for A.B.C. Research

c. Write a set of six ordered pairs in which the percentage of children daydreaming about being rich corresponds to countries. Each ordered pair should be in the form

(percent, country).

d. Is the relation in part (c) a function? Explain your answer.

42. The bar graph shows the number of days in a school year for the six countries with the longest school years. (To put these numbers in perspective, a school year in the United States is 180 days.)

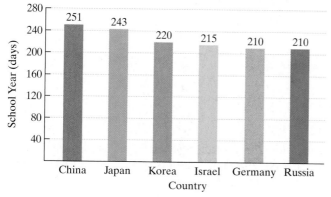

Countries with the Longest School Years

Source: UNESCO

a. Write a set of six ordered pairs in which countries correspond to the number of days in the school year. Each ordered pair should be in the form

(country, days in school year).

b. Is the relation in part (a) a function? Explain your answer.

c. Write a set of six ordered pairs in which the number of days in the school year corresponds to a country. Each ordered pair should be in the form

(days in school year, country).

d. Is the relation in part (c) a function? Explain your answer.

The function $f(x) = 0.76x + 171.4$ models the cholesterol level of an American man as a function of his age, x, in years. Use the function to solve Exercises 43–44.

43. Find and interpret $f(20)$.

44. Find and interpret $f(50)$.

The function $W(h) = 2.95h - 57.32$ models the ideal weight, $W(h)$, in pounds, of a woman with a medium frame as a function of her height, h, in inches. Use the function to solve Exercises 45–46.

45. Find and interpret $W(64)$. Round your answer to the nearest whole number.

46. Find and interpret $W(68)$. Round your answer to the nearest whole number.

The function

$$W(h) = 0.077h^2 - 7.61h + 310.11$$

models the ideal weight, in pounds, of a man with a medium frame as a function of his height, h, in inches. Use the function to solve Exercises 47–48.

47. Find and interpret $W(70)$. Round your answer to the nearest whole number.

48. Find and interpret $W(76)$. Round your answer to the nearest whole number.

The function

$$V(a) = -0.02a^2 + 1.86a + 9.90$$

models the percentage of Americans at age a who say they do volunteer work. Use the function to solve Exercises 49–50.

49. Find and interpret $V(20)$.

50. Find and interpret $V(50)$.

Writing in Mathematics

51. If a relation is represented by a set of ordered pairs, explain how to determine whether the relation is a function.

52. Your friend heard that functions are studied in intermediate and college algebra courses. He asks you what a function is. Provide him with a clear, relatively concise response.

53. Does $f(x)$ mean f times x when referring to a function f? If not, what does $f(x)$ mean? Provide an example with your explanation.

54. Explain how the vertical line test is used to determine whether a graph is a function.

55. For people filing a single return, federal income tax is a function of adjusted gross income because for each value of adjusted gross income there is a specific tax to be paid. By contrast, the price of a house is not a function of the lot size on which the house sits because houses on same-sized lots can sell for many different prices.

   a. Describe an everyday situation between variables that is a function.

   b. Describe an everyday situation between variables that is not a function.

Critical Thinking Exercises

56. Which one of the following is true?

   a. All relations are functions.

   b. No two ordered pairs of a function can have the same second component and different first components.

   c. The graph of every line is a function.

   d. A horizontal line can intersect the graph of a function at more than one point.

57. Write a linear function, $f(x) = mx + b$, satisfying the following conditions:

$$f(0) = 7 \quad \text{and} \quad f(1) = 10.$$

58. If $f(x) = ax^2 + bx + c$ and $r = \dfrac{-b + \sqrt{b^2 - 4ac}}{2a}$, find $f(r)$ without doing any algebra and explain how you arrived at your result.

59. A car was purchased for $22,500. The value of the car decreases by $3200 per year for the first seven years. Write a function V that describes the value of the car after x years, where $0 \le x \le 7$. Then find and interpret $V(3)$.

Review Exercises

60. Write 0.00397 in scientific notation. (Section 6.7, Example 8)

61. Divide: $\dfrac{x^3 + 7x^2 - 2x + 3}{x - 2}$. (Section 6.6, Example 2)

62. Solve:

$$3x + 2y = 6$$
$$8x - 3y = 1.$$

(Section 5.3, Example 5)

GROUP PROJECT

CHAPTER 10

The bar graphs in Exercises 41 and 42 in Exercise Set 10.6 (see page 627) illustrate that if a relation is a function, reversing the components in each of its ordered pairs may result in a relation that is no longer a function. Group members should find examples of bar graphs, like the ones in Exercises 41 and 42, that illustrate this idea. Consult almanacs, newspapers, magazines, or the Internet. The group should select the graph with the most intriguing data. For the graph selected, write and solve a problem with four parts similar to Exercise 41 or 42.

CHAPTER 10 SUMMARY

Definitions and Concepts	Examples

Section 10.1 Solving Quadratic Equations by the Square Root Property

The Square Root Property

If u is an algebraic expression and d is a positive real number, then

$$\text{If } u^2 = d, \text{ then } u = \sqrt{d} \text{ or } u = -\sqrt{d}.$$

Equivalently,

$$\text{If } u^2 = d, \text{ then } u = \pm\sqrt{d}.$$

Solve:
$$(x - 1)^2 = 7.$$
$$x - 1 = \pm\sqrt{7}$$
$$x = 1 \pm \sqrt{7}$$

The Pythagorean Theorem

The sum of the squares of the lengths of the legs of a right triangle equals the square of the length of the hypotenuse.

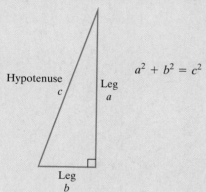

$$a^2 + b^2 = c^2$$

Find a.

$c = 6$ a $b = 2$

$$a^2 + b^2 = c^2$$
$$a^2 + 2^2 = 6^2$$
$$a^2 + 4 = 36$$
$$a^2 = 32$$

a must be positive, so do not use $a = -\sqrt{32}$. $a = \sqrt{32} = \sqrt{16 \cdot 2} = 4\sqrt{2}$

The Distance Formula

The distance, d, between the points (x_1, y_1) and (x_2, y_2) is given by

$$d = \sqrt{(x_2 - x_1)^2 + (y_2 - y_1)^2}.$$

Find the distance between $(-3, -5)$ and $(6, -2)$.

$$d = \sqrt{[6 - (-3)]^2 + [-2 - (-5)]^2}$$
$$= \sqrt{9^2 + 3^2} = \sqrt{81 + 9} = \sqrt{90} = \sqrt{9 \cdot 10} = 3\sqrt{10} \approx 9.49$$

Section 10.2 Solving Quadratic Equations by Completing the Square

Completing the Square

If $x^2 + bx$ is a binomial, then by adding $\left(\dfrac{b}{2}\right)^2$, the square of half the coefficient of x, you will obtain a perfect square trinomial. That is,

$$x^2 + bx + \left(\frac{b}{2}\right)^2 = \left(x + \frac{b}{2}\right)^2.$$

Complete the square: $x^2 + 10x$.

Add $\left(\dfrac{10}{2}\right)^2 = 5^2$, or 25.

$$x^2 + 10x + 25 = (x + 5)^2$$

Definitions and Concepts	Examples

Section 10.2 Solving Quadratic Equations by Completing the Square (continued)

Solving Quadratic Equations by Completing the Square

1. If the coefficient of x^2 is not 1, divide both sides by this coefficient.
2. Isolate variable terms on one side.
3. Complete the square by adding the square of half the coefficient of x to both sides.
4. Factor the perfect square trinomial.
5. Solve by applying the square root property.

Solve by completing the square:

$$2x^2 + 12x - 4 = 0.$$

$$\frac{2x^2}{2} + \frac{12x}{2} - \frac{4}{2} = \frac{0}{2} \qquad \text{Divide by 2.}$$

$$x^2 + 6x - 2 = 0 \qquad \text{Simplify.}$$

$$x^2 + 6x = 2 \qquad \text{Add 2.}$$

The coefficient of x is 6. Half of 6 is 3 and $3^2 = 9$. Add 9 to both sides.

$$x^2 + 6x + 9 = 2 + 9$$

$$(x + 3)^2 = 11$$

$$x + 3 = \pm\sqrt{11}$$

$$x = -3 \pm \sqrt{11}$$

Section 10.3 The Quadratic Formula

The solutions of a quadratic equation in standard form

$$ax^2 + bx + c = 0, \quad a \neq 0,$$

are given by the quadratic formula

$$x = \frac{-b \pm \sqrt{b^2 - 4ac}}{2a}.$$

Solve by the quadratic formula:

$$2x^2 + 4x = 5.$$

First write in standard form by subtracting 5 from both sides.

$$2x^2 + 4x - 5 = 0$$

$$a = 2 \quad b = 4 \quad c = -5$$

$$x = \frac{-4 \pm \sqrt{4^2 - 4 \cdot 2(-5)}}{2 \cdot 2}$$

$$= \frac{-4 \pm \sqrt{16 - (-40)}}{4}$$

$$= \frac{-4 \pm \sqrt{56}}{4} = \frac{-4 \pm \sqrt{4 \cdot 14}}{4}$$

$$= \frac{-4 \pm 2\sqrt{14}}{4}$$

$$= \frac{2(-2 \pm \sqrt{14})}{2 \cdot 2} = \frac{-2 \pm \sqrt{14}}{2}$$

Section 10.4 Imaginary Numbers as Solutions of Quadratic Equations

The imaginary unit i is defined as

$$i = \sqrt{-1}, \quad \text{where} \quad i^2 = -1.$$

The set of numbers in the form $a + bi$ is called the set of complex numbers. If $b = 0$, the complex number is a real number. If $b \neq 0$, the complex number is called an imaginary number.

- $\sqrt{-36} = \sqrt{36(-1)} = \sqrt{36}\sqrt{-1} = 6i$
- $\sqrt{-50} = \sqrt{50(-1)} = \sqrt{25 \cdot 2}\sqrt{-1} = 5\sqrt{2}i$

Definitions and Concepts	Examples

Section 10.4 Imaginary Numbers as Solutions of Quadratic Equations (continued)

Some quadratic equations have complex solutions that are imaginary numbers.

Solve: $x^2 - 2x + 2 = 0$.

$$a = 1 \quad b = -2 \quad c = 2$$

$$x = \frac{-b \pm \sqrt{b^2 - 4ac}}{2a} = \frac{-(-2) \pm \sqrt{(-2)^2 - 4 \cdot 1 \cdot 2}}{2 \cdot 1}$$

$$= \frac{2 \pm \sqrt{4 - 8}}{2} = \frac{2 \pm \sqrt{-4}}{2} = \frac{2 \pm \sqrt{4(-1)}}{2}$$

$$= \frac{2 \pm 2i}{2} = \frac{2(1 \pm i)}{2} = 1 \pm i$$

Section 10.5 Graphs of Quadratic Equations

The graph of $y = ax^2 + bx + c$, called a parabola, can be graphed using the following steps:

1. If $a > 0$, the parabola opens upward. If $a < 0$, it opens downward.

2. Find the vertex, the lowest point if the parabola opens upward and the highest point if it opens downward.
   The x-coordinate of the vertex is $\frac{-b}{2a}$. Substitute this value into the parabola's equation to find the y-coordinate.

3. Find any x-intercepts by letting $y = 0$ and solving the resulting equation.

4. Find the y-intercept by letting $x = 0$.

5. Plot the intercepts and the vertex.

6. Connect these points with a smooth curve.

Graph: $y = x^2 - 2x - 8$.

$$a = 1 \quad b = -2 \quad c = -8$$

- $a > 0$, so the parabola opens upward.

- Vertex: x-coordinate $= \dfrac{-b}{2a} = \dfrac{-(-2)}{2 \cdot 1} = 1$

  y-coordinate $= 1^2 - 2 \cdot 1 - 8 = -9$

  Vertex is $(1, -9)$.

- x-intercepts: Let $y = 0$.

$$x^2 - 2x - 8 = 0$$
$$(x - 4)(x + 2) = 0$$
$$x - 4 = 0 \quad \text{or} \quad x + 2 = 0$$
$$x = 4 \qquad\qquad x = -2$$

The parabola passes through $(4, 0)$ and $(-2, 0)$.

- y-intercept: Let $x = 0$.

$$y = 0^2 - 2 \cdot 0 - 8 = 0 - 0 - 8 = -8$$

The parabola passes through $(0, -8)$.

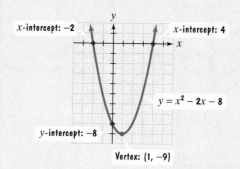

Definitions and Concepts	Examples

Section 10.6 Introduction to Functions

A relation is any set of ordered pairs. The set of first components is the domain and the set of second components is the range. A function is a relation in which each member of the domain corresponds to exactly one member of the range. No two ordered pairs of a function can have the same first component and different second components.

The domain of the relation $\{(1, 2), (3, 4), (3, 7)\}$ is $\{1, 3\}$. The range is $\{2, 4, 7\}$. The relation is not a function: 3, in the domain, corresponds to both 4 and 7 in the range.

If a function is defined as an equation, the notation $f(x)$, read "f of x" or "f at x," describes the value of the function at the number x.

If $f(x) = x^2 - 5x + 4$, find $f(3)$.

$$f(3) = 3^2 - 5 \cdot 3 + 4 = 9 - 15 + 4 = -2$$

Thus, f of 3 is -2.

The Vertical Line Test for Functions
If any vertical line intersects a graph in more than one point, the graph does not define y as a function of x.

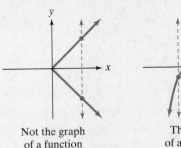

Not the graph of a function | The graph of a function

CHAPTER 10 REVIEW EXERCISES

10.1 *In Exercises 1–8, solve each quadratic equation by the square root property. If possible, simplify radicals or rationalize denominators.*

1. $x^2 = 64$

2. $x^2 = 17$

3. $2x^2 = 150$

4. $(x - 3)^2 = 9$

5. $(y + 4)^2 = 5$

6. $3y^2 - 5 = 0$

7. $(2x - 7)^2 = 25$

8. $(x + 5)^2 = 12$

In Exercises 9–11, use the Pythagorean Theorem to find the missing length in each right triangle. Express the answer in radical form and simplify, if possible.

9.

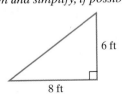

6 ft

8 ft

10.

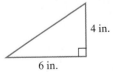

4 in.

6 in.

11.

15 cm

11 cm

Use the Pythagorean Theorem to solve Exercises 12–13.

12. How far away from the building shown in the figure is the bottom of the ladder?

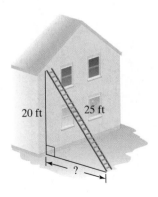

20 ft   25 ft

?

13. A vertical pole is to be supported by three wires. Each wire is 13 yards long and is anchored 5 yards from the base of the pole. How far up the pole will the wires be attached?

14. The weight of a human fetus is modeled by the formula $W = 3t^2$, where W is the weight, in grams, and t is the time, in weeks, $0 \leq t \leq 39$. After how many weeks does the fetus weigh 1200 grams?

15. The distance, d, in feet, that an object falls in t seconds is modeled by the formula $d = 16t^2$. If you dive from a height of 100 feet, how long will it take to hit the water?

In Exercises 16–17, find the distance between each pair of points. Express answers in simplest radical form and, if necessary, round to two decimal places.

16. $(-3, -2)$ and $(1, -5)$

17. $(3, 8)$ and $(5, 4)$

10.2 *In Exercises 18–21, complete the square for each binomial. Then factor the resulting perfect square trinomial.*

18. $x^2 + 16x$

19. $x^2 - 6x$

20. $x^2 + 3x$

21. $x^2 - 5x$

In Exercises 22–24, solve each quadratic equation by completing the square.

22. $x^2 - 12x + 27 = 0$

23. $x^2 - 6x + 4 = 0$

24. $3x^2 - 12x + 11 = 0$

10.3 *In Exercises 25–27, solve each equation using the quadratic formula. Simplify irrational solutions, if possible.*

25. $2x^2 + 5x - 3 = 0$

26. $x^2 = 2x + 4$

27. $3x^2 + 5 = 9x$

In Exercises 28–32, solve each equation by the method of your choice. Simplify irrational solutions, if possible.

28. $2x^2 - 11x + 5 = 0$

29. $(3x + 5)(x - 3) = 5$

30. $3x^2 - 7x + 1 = 0$

31. $x^2 - 9 = 0$

32. $(2x - 3)^2 = 5$

33. As the use of the Internet increases, so has the number of computer infections from viruses. The formula

$$N = 0.2x^2 - 1.2x + 2$$

models the number of infections per month, N, for every 1000 PCs x years after 1990. In which year was the infection rate 29 infections per month for every 1000 PCs?

34. A baseball is hit by a batter. The formula

$$h = -16t^2 + 140t + 3$$

describes the ball's height above the ground, h, in feet, t seconds after it is hit. How long will it take the ball to strike the ground? Round to the nearest tenth of a second.

10.4 *In Exercises 35–38, express each number in terms of i.*

35. $\sqrt{-81}$

36. $\sqrt{-23}$

37. $\sqrt{-48}$

38. $3 + \sqrt{-49}$

Exercises 39–43 involve quadratic equations with imaginary solutions. Solve each equation.

39. $x^2 = -100$

40. $5x^2 = -125$

41. $(2x + 1)^2 = -8$

42. $x^2 - 4x + 13 = 0$

43. $3x^2 - x + 2 = 0$

10.5 *In Exercises 44–47,*

  a. *Determine if the parabola whose equation is given opens upward or downward.*

  b. *Find the parabola's x-intercepts. If they are irrational, round to the nearest tenth.*

  c. *Find the parabola's y-intercept.*

  d. *Find the parabola's vertex.*

  e. *Graph the parabola.*

44. $y = x^2 - 6x - 7$

45. $y = -x^2 - 2x + 3$

46. $y = -3x^2 + 6x + 1$

47. $y = x^2 - 4x$

48. A batter hits a baseball into the air. The formula

$$y = -16x^2 + 96x + 3$$

models the baseball's height above the ground, y, in feet, x seconds after it is hit. When does the baseball reach its maximum height? What is that height?

10.6 *In Exercises 49–51, determine whether each relation is a function. Give the domain and range for each relation.*

49. $\{(2, 7), (3, 7), (5, 7)\}$

50. $\{(1, 10), (2, 500), (3, \pi)\}$

51. $\{(12, 13), (14, 15), (12, 19)\}$

In Exercises 52–53, evaluate each function at the given values.

52. $f(x) = 3x - 4$

   a. $f(-5)$ **b.** $f(6)$ **c.** $f(0)$

53. $g(x) = x^2 - 5x + 2$

   a. $g(-4)$ **b.** $g(3)$ **c.** $g(0)$

In Exercises 54–57, use the vertical line test to identify graphs in which y is a function of x.

54.

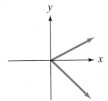

55.

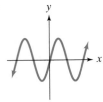

56.

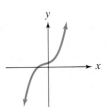

57.

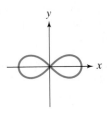

58. The function
$$f(t) = -14t^2 + 42t + 1980$$
models the average price of a computer, $f(t)$, in dollars, t years after 1993. Find and interpret $f(10)$.

CHAPTER 10 TEST

Remember to use your Chapter Test Prep Video CD to see the worked-out solutions to the test questions you want to review.

In Exercises 1–2, solve by the square root property.

1. $3x^2 = 48$

2. $(x - 3)^2 = 5$

3. To find the distance across a lake, a surveyor inserts poles at P and Q, measuring the respective distances to point R, as shown in the figure. Use the surveyor's measurements given in the figure to find the distance PQ across the lake in simplified radical form.

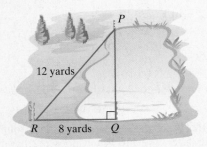

4. Find the distance between $(3, -2)$ and $(-4, 1)$. Express the answer in radical form and then round to two decimal places.

5. Solve by completing the square: $x^2 + 4x - 3 = 0$.

In Exercises 6–10, solve each equation by the method of your choice. Simplify irrational solutions, if possible.

6. $3x^2 + 5x + 1 = 0$

7. $(3x - 5)(x + 2) = -6$

8. $(2x + 1)^2 = 36$

9. $2x^2 = 6x - 1$

10. $2x^2 + 9x = 5$

In Exercises 11–12, express each number in terms of i.

11. $\sqrt{-121}$

12. $\sqrt{-75}$

In Exercises 13–15, solve each quadratic equation. Express imaginary solutions in $a + bi$ form.

13. $x^2 + 36 = 0$

14. $(x - 5)^2 = -25$

15. $x^2 - 2x + 5 = 0$

In Exercises 16–17, graph each parabola whose equation is given. Label the x-intercepts, the y-intercept, and the vertex.

16. $y = x^2 + 2x - 8$ **17.** $y = -2x^2 + 16x - 24$

A batter hits a baseball into the air. The formula

$$y = -16x^2 + 64x + 5$$

models the baseball's height above the ground, y, in feet, x seconds after it is hit. Use the formula to solve Exercises 18–19.

18. When does the baseball reach its maximum height? What is that height?

19. After how many seconds does the baseball hit the ground? Round to the nearest tenth of a second.

In Exercises 20–21, determine whether each relation is a function. Give the domain and range for each relation.

20. $\{(1, 2), (3, 4), (5, 6), (6, 6)\}$

21. $\{(2, 1), (4, 3), (6, 5), (6, 6)\}$

22. If $f(x) = 7x - 3$, find $f(10)$.

23. If $g(x) = x^2 - 3x + 7$, find $g(-2)$.

In Exercises 24–25, identify the graph or graphs in which y is a function of x.

24.

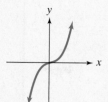

25.

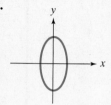

26. The function

$$f(x) = -0.5x^2 + 4x + 19$$

models the millions of people receiving food stamps x years after 1990. Find and interpret $f(10)$.

CUMULATIVE REVIEW EXERCISES (CHAPTERS 1–10)

In Exercises 1–10, solve each equation, inequality, or system of equations.

1. $2 - 4(x + 2) = 5 - 3(2x + 1)$

2. $\dfrac{x}{2} - 3 = \dfrac{x}{5}$

3. $3x + 9 \geq 5(x - 1)$

4. $2x + 3y = 6$
   $x + 2y = 5$

5. $3x - 2y = 1$
   $y = 10 - 2x$

6. $\dfrac{3}{x + 5} - 1 = \dfrac{4 - x}{2x + 10}$

7. $x + \dfrac{6}{x} = -5$

8. $x - 5 = \sqrt{x + 7}$

9. $(x - 2)^2 = 20$

10. $3x^2 - 6x + 2 = 0$

11. Solve for t: $A = \dfrac{5r + 2}{t}$.

In Exercises 12–24, perform the indicated operations. If possible, simplify the answer.

12. $\dfrac{12x^3}{3x^{12}}$ **13.** $4 \cdot 6 \div 2 \cdot 3 + (-5)$

14. $(6x^2 - 8x + 3) - (-4x^2 + x - 1)$

15. $(7x + 4)(3x - 5)$

16. $(5x - 2)^2$

17. $(x + y)(x^2 - xy + y^2)$

18. $\dfrac{x^2 + 6x + 8}{x^2} \div (3x^2 + 6x)$

19. $\dfrac{x}{x^2 + 2x - 3} - \dfrac{x}{x^2 - 5x + 4}$

20. $\dfrac{x - \dfrac{1}{5}}{5 - \dfrac{1}{x}}$

21. $3\sqrt{20} + 2\sqrt{45}$

22. $\sqrt{3x} \cdot \sqrt{6x}$

23. $\dfrac{2}{\sqrt{3}}$

24. $\dfrac{8}{3 - \sqrt{5}}$

In Exercises 25–30, factor completely.

25. $4x^2 - 49$

26. $x^3 + 3x^2 - x - 3$

27. $2x^2 + 8x - 42$

28. $x^5 - 16x$

29. $x^3 - 10x^2 + 25x$

30. $x^3 - 8$

31. Evaluate: $8^{-\frac{2}{3}}$.

In Exercises 32–37, graph each equation, inequality, or system of inequalities in a rectangular coordinate system.

32. $y = \dfrac{1}{3}x - 1$

33. $3x + 2y = -6$

34. $y = -2$

35. $3x - 4y \le 12$

36. $y = x^2 - 2x - 3$

37. $2x + y < 4$
  $x > 2$

38. Find the slope of the line passing through the points $(-1, 3)$ and $(2, -3)$.

39. Write the point-slope form of the equation of the line passing through the points $(1, 2)$ and $(3, 6)$. Then use the point-slope equation to write the slope-intercept form of the line's equation.

In Exercises 40–50, use an equation or a system of equations to solve each problem.

40. Seven subtracted from five times a number is 208. Find the number.

41. After a 20% reduction, a digital camera sold for $256. What was the price before the reduction?

42. A rectangular field is three times as long as it is wide. If the perimeter of the field is 400 yards, what are the field's dimensions?

43. You invested $20,000 in two accounts paying 7% and 9% annual interest, respectively. If the total interest earned for the year is $1550, how much was invested at each rate?

44. A chemist needs to mix a 40% acid solution with a 70% acid solution to obtain 12 liters of a 50% acid solution. How many liters of each solution should be used?

45. Two boats started at the same time from the same port. One traveled due east at 13 miles per hour and the other traveled due west at 19 miles per hour. After how many hours will the boats be 232 miles apart?

46. A university with 176 people on the faculty wants to maintain a student-to-faculty ratio of $23 : 2$. How many students should they enroll to maintain that ratio?

47. A sailboat has a triangular sail with an area of 120 square feet and a base that is 15 feet long. Find the height of the sail.

48. In a triangle, the measure of the first angle is $10°$ more than the measure of the second angle. The measure of the third angle is $20°$ more than four times that of the second angle. What is the measure of each angle?

49. A salesperson works in the TV and stereo department of an electronics store. One day she sold 3 TVs and 4 stereos for $2530. The next day, she sold 4 of the same TVs and 3 of the same stereos for $2510. Find the price of a TV and a stereo.

50. The length of a rectangle is 6 meters more than the width. The area is 55 square meters. Find the rectangle's dimensions.

Answers to Selected Exercises

CHAPTER 1

Section 1.1

Check Point Exercises

1. $\frac{21}{8}$ **2.** $1\frac{2}{3}$ **3.** $2 \cdot 2 \cdot 3 \cdot 3$ **4. a.** $\frac{2}{3}$ **b.** $\frac{7}{4}$ **c.** $\frac{13}{15}$ **d.** $\frac{1}{5}$ **5. a.** $\frac{8}{33}$ **b.** $\frac{18}{5}$ or $3\frac{3}{5}$ **c.** $\frac{2}{7}$ **d.** $\frac{51}{10}$ or $5\frac{1}{10}$ **6. a.** $\frac{10}{3}$ or $3\frac{1}{3}$
b. $\frac{2}{9}$ **c.** $\frac{3}{2}$ or $1\frac{1}{2}$ **7. a.** $\frac{5}{11}$ **b.** $\frac{2}{3}$ **c.** $\frac{9}{4}$ or $2\frac{1}{4}$ **8.** $\frac{14}{21}$ **9. a.** $\frac{11}{10}$ or $1\frac{1}{10}$ **b.** $\frac{7}{12}$ **c.** $\frac{5}{4}$ or $1\frac{1}{4}$ **10.** $\frac{53}{60}$ **11.** $\frac{29}{96}$ **12.** $\frac{1}{4}$

Exercise Set 1.1

1. $\frac{19}{8}$ **3.** $\frac{38}{5}$ **5.** $\frac{135}{16}$ **7.** $4\frac{3}{5}$ **9.** $8\frac{4}{9}$ **11.** $35\frac{11}{20}$ **13.** $2 \cdot 11$ **15.** $2 \cdot 2 \cdot 5$ **17.** prime   **19.** $2 \cdot 2 \cdot 3 \cdot 3$ **21.** $2 \cdot 2 \cdot 5 \cdot 7$
23. prime   **25.** $3 \cdot 3 \cdot 3 \cdot 3$ **27.** $2 \cdot 2 \cdot 2 \cdot 2 \cdot 3 \cdot 5$ **29.** $\frac{5}{8}$ **31.** $\frac{5}{6}$ **33.** $\frac{7}{10}$ **35.** $\frac{2}{5}$ **37.** $\frac{22}{25}$ **39.** $\frac{60}{43}$ **41.** $\frac{2}{15}$ **43.** $\frac{21}{88}$ **45.** $\frac{36}{7}$
47. $\frac{1}{12}$ **49.** $\frac{15}{14}$ **51.** 6   **53.** $\frac{15}{16}$ **55.** $\frac{9}{5}$ **57.** $\frac{5}{9}$ **59.** 3   **61.** $\frac{7}{10}$ **63.** $\frac{1}{2}$ **65.** 6   **67.** $\frac{6}{11}$ **69.** $\frac{2}{3}$ **71.** $\frac{5}{4}$ **73.** $\frac{1}{6}$
75. 2   **77.** $\frac{7}{10}$ **79.** $\frac{9}{10}$ **81.** $\frac{19}{24}$ **83.** $\frac{7}{18}$ **85.** $\frac{7}{12}$ **87.** $\frac{41}{80}$ **89.** $1\frac{5}{12}$ or $\frac{17}{12}$ **91.** $\frac{3a}{20}$ **93.** $\frac{20}{x}$ **95.** $\frac{4}{15}$ **97. a.** $\frac{11}{20}$ **b.** 240
99. $\frac{2}{5}$ **101.** $1\frac{1}{5}$ cups   **103.** $\frac{23}{20}$ mi; $\frac{7}{20}$ mi farther   **105.** 38 mi   **117.** d

Section 1.2

Check Point Exercises

1. a. -500 **b.** -282 **2.**   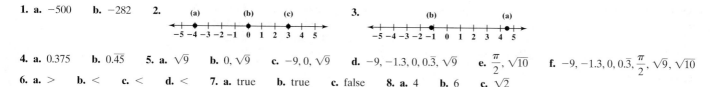   **3.**

4. a. 0.375   **b.** $0.\overline{45}$ **5. a.** $\sqrt{9}$ **b.** $0, \sqrt{9}$ **c.** $-9, 0, \sqrt{9}$ **d.** $-9, -1.3, 0, 0.\overline{3}, \sqrt{9}$ **e.** $\frac{\pi}{2}, \sqrt{10}$ **f.** $-9, -1.3, 0, 0.\overline{3}, \frac{\pi}{2}, \sqrt{9}, \sqrt{10}$
6. a. $>$ **b.** $<$ **c.** $<$ **d.** $<$ **7. a.** true   **b.** true   **c.** false   **8. a.** 4   **b.** 6   **c.** $\sqrt{2}$

Exercise Set 1.2

1. -20 **3.** 8   **5.** -3000 **7.** -4 billion   **9–19.**   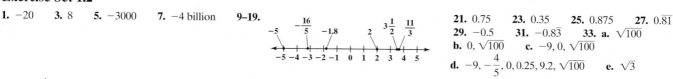   **21.** 0.75   **23.** 0.35   **25.** 0.875   **27.** $0.\overline{81}$
29. -0.5 **31.** $-0.8\overline{3}$ **33. a.** $\sqrt{100}$
b. $0, \sqrt{100}$ **c.** $-9, 0, \sqrt{100}$
d. $-9, -\frac{4}{5}, 0, 0.25, 9.2, \sqrt{100}$ **e.** $\sqrt{3}$
f. $-9, -\frac{4}{5}, 0, 0.25, \sqrt{3}, 9.2, \sqrt{100}$ **35. a.** $\sqrt{64}$ **b.** $0, \sqrt{64}$ **c.** $-11, 0, \sqrt{64}$ **d.** $-11, -\frac{5}{6}, 0, 0.75, \sqrt{64}$ **e.** $\sqrt{5}, \pi$
f. $-11, -\frac{5}{6}, 0, 0.75, \sqrt{5}, \pi, \sqrt{64}$ **37.** 0   **39.** Answers will vary; $\frac{1}{2}$ is an example.   **41.** Answers will vary; 6 is an example.   **43.** Answers will
vary; π is an example.   **45.** $<$ **47.** $>$ **49.** $>$ **51.** $<$ **53.** $>$ **55.** $<$ **57.** $<$ **59.** $>$ **61.** $>$ **63.** true   **65.** true
67. true   **69.** false   **71.** 6   **73.** 7   **75.** $\frac{5}{6}$ **77.** $\sqrt{11}$ **79.** $>$ **81.** $=$ **83.** $<$ **85.** $=$ **87.** -2 **89.** 1997, 2002; budget
deficit   **103.** c   **105.** -7 and -6 **107.** 1.732; 1 and 2   **109.** -0.414; -1 and 0

Section 1.3

Check Point Exercises

1.   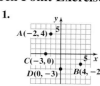   **2.** $E(-4, -2)$, $F(-2, 0)$, $G(6, 0)$ **3.** $B(8, 200)$; After 8 sec, the watermelon is 200 ft above ground.
4. $D(8.8, 0)$; After approximately 8.8 sec, the watermelon is 0 ft above ground. Equivalently, the watermelon
splatters on the ground after approximately 8.8 sec.   **5.** 30 thousand   **6. a.** 22   **b.** United States,
France, Canada, Sweden

Exercise Set 1.3

1. *(graph showing (3, 5))*   **3.** *(graph showing (−5, 1))*   **5.** *(graph showing (−3, −1))*   **7.** *(graph showing (6, −3.5))*

9–23. *(graph showing points $\left(-5, \frac{3}{2}\right)$, $(0, 2)$, $(0, 0)$, $(-2, 0)$, $(-3, -3)$, $(0, -3)$, $\left(0, -\frac{5}{2}\right)$, $\left(\frac{5}{2}, \frac{7}{2}\right)$)*

25. $(5, 2)$ **27.** $(-6, 5)$ **29.** $(-2, -3)$ **31.** $(5, -3)$ **33.** I and II   **35.** I and III

37. Answers will vary; examples are $(-2, 2), (-1, 1), (1, 1), (2, 2)$. *(graph showing $(-2, 2)$, $(2, 2)$, $(-1, 1)$, $(1, 1)$)*

39. $4\frac{1}{2}$ or $\frac{9}{2}$ **41.** $(2, 7)$; The football is 7 ft above ground when it is 2 yd from the quarterback.   **43.** $(6, 9.25)$ **45.** 12 ft; 15 yd

47. $(91, 125)$; In 1991, 125 thousand acres were used for cultivation.

49. 2001; 25 thousand acres   **51.** 1991 and 1992   **53.** 7.1 million barrels per day   **55.** 1990; 7.2 million barrels per day

57. 1970; 9.8 million barrels per day   **59.** 1.7 million barrels per day   **61.** $120 million   **63.** *No Way Out, The Postman, Dragonfly*

65. 48 yr old   **67.** approx 17 yr   **69.** $22 thousand   **71.** $7 thousand   **73.** 270; 14   **75.** 2002   **77.** 1972, 1982, 1992   **79.** $\frac{1}{20}$

89. a. *(graph showing $(-3, 9)$, $(3, 9)$, $(-2, 4)$, $(2, 4)$, $(-1, 1)$, $(1, 1)$, $(0, 0)$ and $(4, 2)$, $(9, 3)$, $(1, 1)$, $(4, -2)$, $(0, 0)$, $(9, -3)$, $(1, -1)$)*

b. A; B; Points line up vertically when x-coordinates are associated with more than one y-coordinate.   **91.** $\frac{23}{20}$

92. $<$ **93.** 5.83

Section 1.4

Check Point Exercises

1. 226.5; In 1980, the population of the United States was 226.5 million.   **2. a.** 3 terms   **b.** 6   **c.** 11   **d.** $6x$ and $2x$
3. a. $14 + x$ **b.** $y7$ **4. a.** $17 + 5x$ **b.** $x5 + 17$ **5. a.** $20 + x$ or $x + 20$ **b.** $30x$ **6.** $12 + x$ or $x + 12$ **7.** $5x + 15$
8. $24y + 42$ **9. a.** $10x$ **b.** $5a$ **10. a.** $18x + 10$ **b.** $14x + 8y$ **11.** $25x + 21$ **12.** $38x + 23y$ **13.** 108 beats/min

Exercise Set 1.4

1. 15   **3.** 80   **5.** 46   **7.** 35   **9.** 25   **11. a.** 2   **b.** 3   **c.** 5   **d.** no   **13. a.** 3   **b.** 1   **c.** 2   **d.** yes; x and $5x$
15. a. 3   **b.** 4   **c.** 1   **d.** no   **17.** $4 + y$ **19.** $3x + 5$ **21.** $5y + 4x$ **23.** $5(3 + x)$ **25.** $x9$ **27.** $x + 6y$ **29.** $x7 + 23$
31. $(x + 3)5$ **33.** $(7 + 5) + x = 12 + x$ **35.** $(7 \cdot 4)x = 28x$ **37.** $3x + 15$ **39.** $16x + 24$ **41.** $4 + 2r$ **43.** $5x + 5y$
45. $3x - 6$ **47.** $8x - 10$ **49.** $\frac{5}{2}x - 6$ **51.** $8x + 28$ **53.** $6x + 18 + 12y$ **55.** $15x - 10 + 20y$ **57.** $17x$ **59.** $8a$ **61.** $14 + x$
63. $11y + 3$ **65.** $9x + 1$ **67.** $14a + 14$ **69.** $15x + 6$ **71.** $15x + 2$ **73.** $41a + 24b$ **75.** Commutative property of addition
77. Associative property of addition   **79.** false   **81.** true   **83.** 300; You can stay in the sun for 300 min without burning with a number 15 lotion.
85. 8805; In 2002, credit card debt per U.S. household was $8805; fairly well.   **87. a.** $0.38x + 0.02$ **b.** 1.92 million; very well
101. c   **103. a.** $50.50; $5.50; $1.00   **b.** No; When producing 2000 clocks, the average cost is $3. This is more than the $1.50 maximum price at which the manufacturer can sell the clocks.

104. $0.\overline{4}$ **105.** *(graph showing $(-3, -1)$)*   **106.** $\frac{1}{5}$

Mid-Chapter 1 Check Point

1. $24a + 1$ **2.** $\frac{1}{6}$ **3.** $\frac{1}{2}$ **4.** $65x + 16$ **5.** $\frac{25}{66}$ **6.** $\frac{2}{3}$ **7.** $\frac{1}{5}$ **8.** $\frac{25}{27}$ **9.** $\frac{37}{18}$ or $2\frac{1}{18}$ **10.** IV

11. $(x + 3)5$ **12.** $5(3 + x)$ **13.** $5x + 15$ **14.** 7%   **15.** 2002, 2003, 2004   **16.** $<$

17. $-11, -\frac{3}{7}, 0, 0.45, \sqrt{25}$ **18.** 13   **19.** 4

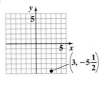

20. $(67; 20)$; In 1967, 20% of grades of undergraduate college students were A's.   **21.** 1970; 23%   **22.** 40%   **23.** 19.3
24. $0.\overline{09}$ **25.** $56x - 80 + 24y$

Section 1.5

Check Point Exercises

1. $4 + (-7) = -3$

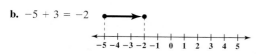

2. a. $-1 + (-3) = -4$

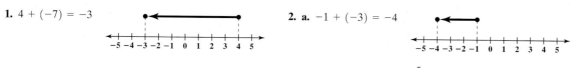

b. $-5 + 3 = -2$

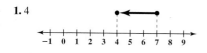

3. a. -35 **b.** -1.5 **c.** $-\dfrac{5}{6}$ **4. a.** -13 **b.** 1.2   **c.** $-\dfrac{1}{2}$ **5. a.** $-17x$
b. $-7y + 6z$ **c.** $20 - 20x$ **6.** The water level is down 3 ft at the end of 5 months.

Exercise Set 1.5

1. 4   **3.** -7 **5.** -4

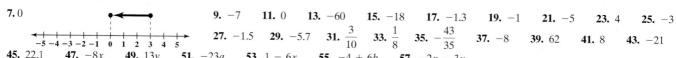

7. 0

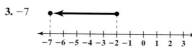   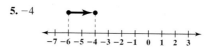

9. -7 **11.** 0   **13.** -60 **15.** -18 **17.** -1.3 **19.** -1 **21.** -5 **23.** 4   **25.** -3
27. -1.5 **29.** -5.7 **31.** $\dfrac{3}{10}$ **33.** $\dfrac{1}{8}$ **35.** $-\dfrac{43}{35}$ **37.** -8 **39.** 62   **41.** 8   **43.** -21
45. 22.1   **47.** $-8x$ **49.** $13y$ **51.** $-23a$ **53.** $1 - 6x$ **55.** $-4 + 6b$ **57.** $-2x - 3y$
59. $20x - 6$ **61.** $24 + 3y$ **63.** $47 - 33a$ **65.** 12   **67.** -30 **69.** $>$ **71.** The high temperature was 44°.
73. The elevation is 600 ft below sea level.   **75.** The temperature at 4:00 P.M. was 3°F.   **77.** The ball was at the 25-yard line at the end of the
fourth play.   **79.** \$455 billion deficit   **89.** d   **91.** $-18y$ **94.** true   **95. a.** $\sqrt{4}$ **b.** $0, \sqrt{4}$ **c.** $-6, 0, \sqrt{4}$
d. $-6, 0, 0.\overline{7}, \sqrt{4}$ **e.** $-\pi, \sqrt{3}$ **f.** $-6, -\pi, 0, 0.\overline{7}, \sqrt{3}, \sqrt{4}$ **96.** IV

Section 1.6

Check Point Exercises

1. a. -8 **b.** 9   **c.** -5 **2. a.** 9.2   **b.** $-\dfrac{14}{15}$ **c.** 7π **3.** 15   **4.** $-6, 4a, -7ab$ **5. a.** $4 - 7x$ **b.** $-9x + 4y$ **6.** 19,763 m

Exercise Set 1.6

1. a. -12 **b.** $5 + (-12)$ **3. a.** 7   **b.** $5 + 7$ **5.** 6   **7.** -6 **9.** 23   **11.** 11   **13.** -11 **15.** -38 **17.** 0   **19.** 0   **21.** 26
23. -13 **25.** 13   **27.** $-\dfrac{2}{7}$ **29.** $\dfrac{4}{5}$ **31.** -1 **33.** $-\dfrac{3}{5}$ **35.** $\dfrac{3}{4}$ **37.** $\dfrac{1}{4}$ **39.** 7.6   **41.** -2 **43.** 2.6   **45.** 0   **47.** 3π
49. 13π **51.** 19   **53.** -3 **55.** -15 **57.** 0   **59.** -52 **61.** -187 **63.** $\dfrac{7}{6}$ **65.** -4.49 **67.** $-\dfrac{3}{8}$ **69.** $-3x, -8y$
71. $12x, -5xy, -4$ **73.** $-6x$ **75.** $4 - 10y$ **77.** $5 - 7a$ **79.** $-4 - 9b$ **81.** $24 + 11x$ **83.** $3y - 8x$ **85.** 9   **87.** $\dfrac{7}{8}$
89. -2 **91.** 19,757 ft   **93.** 1039 thousand jobs   **95.** 268 thousand jobs   **97.** 21°F   **99.** 3°F
101. The maximum concentration is 0.05 and it occurs during the 3rd hour.   **103.** 0.015   **105.** from 0 to 3 hr   **113.** 711 yr
117.

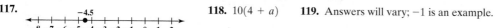

118. $10(4 + a)$ **119.** Answers will vary; -1 is an example.

Section 1.7

Check Point Exercises

1. a. -40 **b.** $-\dfrac{4}{21}$ **c.** 36   **d.** 5.5   **e.** 0   **2. a.** 24   **b.** -30 **3. a.** $\dfrac{1}{7}$ **b.** 8   **c.** $-\dfrac{1}{6}$ **d.** $-\dfrac{13}{7}$ **4. a.** -4 **b.** 8
5. a. 8   **b.** $-\dfrac{8}{15}$ **c.** -7.3 **d.** 0   **6. a.** $-20x$ **b.** $10x$ **c.** $-b$ **d.** $-21x + 28$ **e.** $-7y + 6$ **7.** $-y - 26$ **8. a.** \$330
b. \$60   **c.** \$33

Exercise Set 1.7

1. -45 **3.** 24   **5.** -21 **7.** 19   **9.** 0   **11.** -12 **13.** 9   **15.** $\dfrac{12}{35}$ **17.** $-\dfrac{14}{27}$ **19.** -3.6 **21.** 0.12   **23.** 30   **25.** -72

27. 24   **29.** -27 **31.** 90   **33.** 0   **35.** $\dfrac{1}{4}$ **37.** 5   **39.** $-\dfrac{1}{10}$ **41.** $-\dfrac{5}{2}$ **43. a.** $-32\cdot\left(\dfrac{1}{4}\right)$ **b.** -8 **45. a.** $-60\cdot\left(-\dfrac{1}{5}\right)$ **b.** 12

47. -3 **49.** -7 **51.** 30   **53.** 0   **55.** undefined   **57.** -5 **59.** -12 **61.** 6   **63.** 0   **65.** undefined   **67.** -4.3 **69.** $\dfrac{5}{6}$

71. $-\dfrac{16}{9}$ **73.** -1 **75.** -15 **77.** $-10x$ **79.** $3y$ **81.** $9x$ **83.** $-4x$ **85.** $-b$ **87.** $3y$ **89.** $-8x+12$ **91.** $6x-12$

93. $-2y+5$ **95.** $y-14$ **97.** $4(-10)+8=-32$ **99.** $(-9)(-3)-(-2)=29$ **101.** $\dfrac{-18}{-15+12}=6$ **103.** $-6-\left(\dfrac{12}{-4}\right)=-3$

105. 6.3 million; fairly well, but slight overestimation   **107. a.** 11 Latin words   **b.** 11 Latin words; it is the same.

109. a. \$2,000,000 (or \$200 tens of thousands)   **b.** \$8,000,000 (or \$800 tens of thousands)   **c.** Cost increases.   **119.** b   **121.** $5x$ **123.** $\dfrac{x}{12}$

127. $1.144x+2.5$ **129.** -9 **130.** -3 **131.** 2

Section 1.8

Check Point Exercises

1. a. 36   **b.** -64 **c.** 1   **d.** -1 **2. a.** $21x^2$ **b.** $8x^3$ **c.** cannot simplify   **3.** 15   **4.** 32   **5. a.** 36   **b.** 12   **6.** $\dfrac{3}{4}$ **7.** -40

8. -31 **9.** $\dfrac{5}{7}$ **10.** -5 **11.** $7x^2+15$ **12.** For 40-year-old drivers, there are 200 accidents per 50 million miles driven.   **13.** $30°C$

Exercise Set 1.8

1. 81   **3.** 64   **5.** 16   **7.** -64 **9.** 625   **11.** -625 **13.** -100 **15.** $19x^2$ **17.** $15x^3$ **19.** $9x^4$ **21.** $-x^2$ **23.** x^3
25. cannot be simplified   **27.** 0   **29.** 25   **31.** 27   **33.** 12   **35.** 5   **37.** 45   **39.** -24 **41.** 300   **43.** 0   **45.** -32 **47.** 64

49. 30   **51.** $\dfrac{4}{3}$ **53.** 2   **55.** 2   **57.** 3   **59.** 88   **61.** -60 **63.** -36 **65.** 14   **67.** $-\dfrac{3}{4}$ **69.** $-\dfrac{9}{40}$ **71.** $-\dfrac{37}{36}$ or $-1\dfrac{1}{36}$

73. 24   **75.** 28   **77.** 9   **79.** -7 **81.** $15x-27$ **83.** $15-3y$ **85.** $16y-25$ **87.** $-2x^2-9$ **89.** $-10-(-2)^3=-2$

91. $[2(7-10)]^2=36$ **93.** $x-(5x+8)=-4x-8$ **95.** $5(x^3-4)=5x^3-20$ **97.** 135 beats/min; (40,135) on blue graph

99. \$287.4 billion; fairly well   **101.** 65 thousand; fairly well, but slight overestimation   **103.** $20°C$ **105.** $-30°C$ **113.** $-\dfrac{79}{4}$

115. $\left(2\cdot5-\dfrac{1}{2}\cdot10\right)\cdot9=45$ **117.** Answers will vary.   **119.** 6   **120.** -24 **121.** Answers will vary; -3 is an example.

Review Exercises

1. $\dfrac{23}{7}$ **2.** $\dfrac{64}{11}$ **3.** $1\dfrac{8}{9}$ **4.** $5\dfrac{2}{5}$ **5.** $2\cdot2\cdot3\cdot5$ **6.** $3\cdot3\cdot7$ **7.** prime   **8.** $\dfrac{5}{11}$ **9.** $\dfrac{8}{15}$ **10.** $\dfrac{21}{50}$ **11.** $\dfrac{8}{3}$ **12.** $\dfrac{1}{4}$ **13.** $\dfrac{2}{3}$

14. $\dfrac{29}{18}$ **15.** $\dfrac{37}{60}$ **16.** $\dfrac{5}{12}$ **17.**   **18.**   **19.** 0.625   **20.** $0.\overline{27}$

21. a. $\sqrt{81}$ **b.** $0,\sqrt{81}$ **c.** $-17,0,\sqrt{81}$ **d.** $-17,-\dfrac{9}{13},0,0.75,\sqrt{81}$ **e.** $\sqrt{2},\pi$ **f.** $-17,-\dfrac{9}{13},0,0.75,\sqrt{2},\pi,\sqrt{81}$

22. Answers will vary; -2 is an example.   **23.** Answers will vary; $\dfrac{1}{2}$ is an example.   **24.** Answers will vary; π is an example.   **25.** $<$ **26.** $>$

27. $>$ **28.** $<$ **29.** false   **30.** true   **31.** 58   **32.** 2.75

33. IV   **34.** IV   **35.** I   **36.** II

37. $A(5,6)$; $B(-3,0)$; $C(-5,2)$; $D(-4,-2)$; $E(0,-5)$; $F(3,-1)$ **38.** 65   **39.** 80   **40.** 1989–1993   **41.** 1981–1985   **42.** 1977   **43.** 85

44. Finland, Germany, Austria   **45.** 73   **46.** 40   **47.** $13y+7$ **48.** $(x+7)9$ **49.** $(6+4)+y=10+y$ **50.** $(7\cdot10)x=70x$

51. $24x-12+30y$ **52.** $7a+2$ **53.** $28x+19$ **54.** 1800; The sale price at 25% off is \$1800 for a \$2400 computer.

55. 2   **56.** -3 **57.** $-\dfrac{11}{20}$ **58.** -7 **59.** $5y-4x$ **60.** $40-2y$ **61.** 800 ft below sea level

62. The reservoir's level at the end of 5 months is 23 ft.   **63.** $9 + (-13)$ **64.** 4   **65.** $-\dfrac{6}{5}$ **66.** -1.5 **67.** -7 **68.** -3
69. $-5 - 8a$ **70.** 27,150 ft   **71.** 84   **72.** $-\dfrac{3}{11}$ **73.** -120 **74.** -9 **75.** undefined   **76.** 2   **77.** $3x$ **78.** $-x - 1$ **79.** 36
80. -36 **81.** -32 **82.** $6x^3$ **83.** cannot be simplified   **84.** -16 **85.** -16 **86.** 10   **87.** -2 **88.** 17   **89.** -88 **90.** 14
91. $-\dfrac{20}{3}$ **92.** $-\dfrac{2}{5}$ **93.** 6   **94.** 10   **95.** $28a - 20$ **96.** $6y - 12$ **97.** 13 lb; (4, 13)   **98.** 16 lb; (6, 16)
99. 805 million; fairly well, but slight overestimation   **100.** Sales increase from 1998 to 2000 and decrease from 2000 to 2002.

Chapter 1 Test

1. 4   **2.** -11 **3.** -51 **4.** $\dfrac{1}{5}$ **5.** $-\dfrac{35}{6}$ or $-5\dfrac{5}{6}$ **6.** -5 **7.** 1   **8.** -4 **9.** -32 **10.** 1   **11.** $4x + 4$ **12.** $13x - 19y$
13. $10 - 6x$ **14.** $-7, -\dfrac{4}{5}, 0, 0.25, \sqrt{4}, \dfrac{22}{7}$ **15.** $>$ **16.** 12.8   **17.** II;

18. $(-5, -2)$ **19.** -15 **20.** 150
21. $2(3 + x)$ **22.** $(-6 \cdot 4)x = -24x$
23. $35x - 7 + 14y$ **24.** (30, 200); After 30 yr, there are 200 elk.   **25.** 50

26. 725 million   **27.** 16 sec   **28.** 17,030 ft   **29.** \$135 thousand   **30.** \$134.7 thousand

CHAPTER 2

Section 2.1

Check Point Exercises

1. a. not a solution   **b.** solution   **2.** 17   **3.** 2.29   **4.** $\dfrac{1}{4}$ **5.** 13   **6.** 12   **7.** 11   **8.** 2100 words

Exercise Set 2.1

1. 23   **3.** -20 **5.** -16 **7.** -12 **9.** 4   **11.** -11 **13.** 2   **15.** $-\dfrac{17}{12}$ **17.** $\dfrac{21}{4}$ **19.** $-\dfrac{11}{20}$ **21.** 4.3   **23.** $-\dfrac{21}{4}$
25. 18   **27.** $\dfrac{9}{10}$ **29.** -310 **31.** 4.3   **33.** 0   **35.** 11   **37.** 5   **39.** -13 **41.** 6   **43.** -12 **45.** $x = \triangle + \square$ **47.** $\triangle - \square = x$
49. $x - 12 = -2; 10$ **51.** $\dfrac{2}{5}x - 8 = \dfrac{7}{5}x; -8$ **53.** \$1700   **55.** 525,000 deaths   **57.** \$17 billion
65. Answers will vary; example: $x - 100 = -101$.   **67.** 2.7529   **68.** II   **69.** -12 **70.** $6 - 9x$

Section 2.2

Check Point Exercises

1. 36   **2. a.** 21   **b.** -4 **c.** -3.1 **3. a.** 24   **b.** -16 **4. a.** -5 **b.** 3   **5.** 6   **6.** -10 **7.** 6   **8.** 2020

Exercise Set 2.2

1. 30   **3.** -33 **5.** 7   **7.** -9 **9.** $-\dfrac{7}{2}$ **11.** 6   **13.** $-\dfrac{3}{4}$ **15.** 0   **17.** 18   **19.** -8 **21.** -17 **23.** 47   **25.** 45   **27.** -5
29. 5   **31.** 6   **33.** -1 **35.** -2 **37.** $\dfrac{9}{4}$ **39.** -6 **41.** -3 **43.** -3 **45.** 4   **47.** $-\dfrac{3}{2}$ **49.** 2   **51.** -4 **53.** -6
55. $x = \square \cdot \triangle$ **57.** $-\triangle = x$ **59.** $6x = 10; \dfrac{5}{3}$ **61.** $\dfrac{x}{-9} = 5; -45$ **63.** 10 sec   **65.** 1502.2 mph   **67.** 60 yd
73. d   **75.** Answers will vary; example: $\dfrac{5}{4}x = -20$.   **77.** 6.5   **78.** 100   **79.** -100 **80.** 3

Section 2.3

Check Point Exercises

1. 6   **2.** 2   **3.** 5   **4.** -2 **5.** inconsistent, no solution   **6.** all real numbers   **7.** 124 lb

Exercise Set 2.3

1. 3   **3.** −1   **5.** 4   **7.** 4   **9.** $\frac{7}{2}$ **11.** −3   **13.** 6   **15.** 8   **17.** 4   **19.** 1   **21.** −4   **23.** 5   **25.** 6   **27.** 1   **29.** −57
31. −10   **33.** 18   **35.** $\frac{7}{4}$ **37.** 1   **39.** 24   **41.** −6   **43.** 20   **45.** −7   **47.** no solution   **49.** all real numbers   **51.** $\frac{2}{3}$
53. all real numbers   **55.** no solution   **57.** 0   **59.** no solution   **61.** 0   **63.** $\frac{4}{3}$ **65.** $x = \square\$ - \square\triangle$ **67.** 240   **69.** $\frac{x}{5} + \frac{x}{3} = 16; 30$
71. $\frac{3x}{4} - 3 = \frac{x}{2}; 12$ **73.** 85 mph   **75.** 3.7; point (3.7, 10) on low-humor graph   **77.** 409.2 ft   **83.** c   **85.** 3   **87.** −4.2   **89.** <
90. <   **91.** −10

Section 2.4

Check Point Exercises

1. $l = \frac{A}{w}$ **2.** $l = \frac{P - 2w}{2}$ **3.** $m = \frac{T - D}{p}$ **4.** $x = 15 + 12y$ **5.** 2.3%   **6. a.** 0.67   **b.** 2.5   **7.** 4.5   **8.** 15   **9.** 36%
10. 35%   **11. a.** $1152   **b.** 4% decrease

Exercise Set 2.4

1. $r = \frac{d}{t}$ **3.** $P = \frac{I}{rt}$ **5.** $r = \frac{C}{2\pi}$ **7.** $m = \frac{E}{c^2}$ **9.** $m = \frac{y - b}{x}$ **11.** $p = \frac{T - D}{m}$ **13.** $b = \frac{2A}{h}$ **15.** $n = 5M$ **17.** $c = 4F - 160$
19. $a = 2A - b$ **21.** $r = \frac{S - P}{Pt}$ **23.** $b = \frac{2A}{h} - a$ **25.** $x = \frac{C - By}{A}$ **27.** 89%   **29.** 0.2%   **31.** 478%   **33.** 10,000%
35. 0.27   **37.** 0.634   **39.** 1.7   **41.** 0.03   **43.** 0.005   **45.** 6   **47.** 7.2   **49.** 5   **51.** 170   **53.** 20%   **55.** 12%   **57.** 60%
59. 75%   **61.** $x = \frac{y}{a + b}$ **63.** $x = \frac{y - 5}{a - b}$ **65.** $x = \frac{y}{c + d}$ **67.** $x = \frac{y + C}{A + B}$ **69. a.** $z = 3A - x - y$ **b.** 96%   **71. a.** $t = \frac{d}{r}$
b. 2.5 hr   **73.** 408   **75.** 2,369,200   **77.** 59%   **79.** 12.5%   **81.** $9   **83. a.** $1008   **b.** $17,808   **85. a.** $103.20   **b.** $756.80
87. 15%   **89.** no; 2% loss   **97.** d   **99.** 1.5 sec and 60 ft   **100.** 12   **101.** 20   **102.** 0.7x

Mid-Chapter 2 Check Point

1. 16   **2.** −3   **3.** $C = \frac{825H}{E}$ **4.** 8.4   **5.** 30   **6.** $-\frac{8}{9}$ **7.** $r = \frac{S}{2\pi h}$ **8.** 40   **9.** 20   **10.** −1   **11.** $x = \frac{By + C}{A}$
12. no solution   **13.** 2008   **14.** 40   **15.** 12.5%   **16.** $-\frac{6}{5}$ **17.** 25%   **18.** all real numbers

Section 2.5

Check Point Exercises

1. a. $4x + 6$ **b.** $\frac{x - 4}{9}$ **2.** 12   **3.** MRI scan: $575; Acupuncture: $112   **4.** pages 96 and 97   **5.** 32; 4 mi
6. 40 ft wide and 120 ft long   **7.** $940

Exercise Set 2.5

1. $x + 7$ **3.** $25 - x$ **5.** $9 - 4x$ **7.** $\frac{83}{x}$ **9.** $2x + 40$ **11.** $9x - 93$ **13.** $8(x + 14)$ **15.** $x + 60 = 410; 350$
17. $x - 23 = 214; 237$ **19.** $7x = 126; 18$ **21.** $\frac{x}{19} = 5; 95$ **23.** $4 + 2x = 56; 26$ **25.** $5x - 7 = 178; 37$ **27.** $x + 5 = 2x; 5$
29. $2(x + 4) = 36; 14$ **31.** $9x = 3x + 30; 5$ **33.** $\frac{3x}{5} + 4 = 34; 50$ **35.** *Titanic*: $200 million; *Waterworld*: $175 million
37. Liberals: 39.6%; Conservatives: 17.6%   **39.** pages 314 and 315   **41.** Rolling Stones: $121 million; Springsteen: $120 million   **43.** 32 and 34
45. 800 mi   **47.** 6 months   **49.** 50 yd wide and 200 yd long   **51.** 160 ft wide and 360 ft long   **53.** 12 ft long and 4 ft high   **55.** $400
57. $67,000   **59.** $14,500   **61.** 11 hr   **67.** d   **69.** 5 ft 7 in.   **71.** The uncle is 60 years old and the woman is 20 years old.
73. −20   **74.** 0   **75.** $w = \frac{3V}{lh}$

Section 2.6

Check Point Exercises

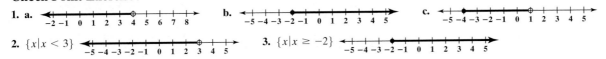

1. a.   **b.**   **c.**

2. $\{x | x < 3\}$ **3.** $\{x | x \geq -2\}$

4. a. $\{x|x < 8\}$ **b.** $\{x|x > -3\}$

5. $\{y|y \ge 4\}$ **6.** $\{x|x \ge 1\}$

7. $\{x|x \ge 1\}$ **8.** no solution; $\varnothing$ **9.** $\{x|x \text{ is a real number}\}$ **10.** at least 83%

Exercise Set 2.6

1.   **3.**   **5.**

7.   **9.**   **11.**

13. $\{x|x > -2\}$ **15.** $\{x|x \ge 4\}$ **17.** $\{x|x \ge 3\}$

19. $\{x|x > 7\}$ **21.** $\{x|x \le 6\}$

23. $\{y|y < 2\}$ **25.** $\{x|x \le 3\}$

27. $\{x|x < 16\}$ **29.** $\{x|x > 4\}$

31. $\left\{x\middle|x > \dfrac{7}{6}\right\}$ **33.** $\left\{y\middle|y \le -\dfrac{3}{8}\right\}$

35. $\{y|y > 0\}$ **37.** $\{x|x < 8\}$

39. $\{x|x > -6\}$ **41.** $\{x|x < 5\}$

43. $\{x|x \ge -7\}$ **45.** $\{x|x > -5\}$

47. $\{x|x \le -5\}$ **49.** $\{x|x < 3\}$

51. $\left\{y\middle|y \ge -\dfrac{1}{8}\right\}$ **53.** $\{x|x > -4\}$

55. $\{x|x > 5\}$ **57.** $\{x|x < 5\}$

59. $\{x|x \ge -2\}$ **61.** $\{x|x > -3\}$

63. $\{x|x \ge 4\}$ **65.** $\left\{x\middle|x > \dfrac{11}{3}\right\}$

67. $\{y|y > 2\}$ **69.** $\{y|y < 2\}$

71. $\{x|x < 3\}$ **73.** $\left\{x\middle|x > \dfrac{5}{3}\right\}$

75. $\{x|x \ge 9\}$ **77.** $\{x|x < -6\}$

79. no solution; $\varnothing$ **81.** $\{x|x \text{ is a real number}\}$ **83.** no solution; $\varnothing$ **85.** $\{x|x \text{ is a real number}\}$ **87.** $\{x|x \le 0\}$ **89.** $x > \dfrac{b-a}{3}$

91. $\dfrac{y-b}{m} \ge x$ **93.** x is between -2 and 2; $|x| < 2$ **95.** x is greater than 2 or less than -2; $|x| > 2$ **97.** Denmark, Netherlands, Norway

99. Japan, Mexico   **101.** Netherlands, Norway, Canada, U.S.   **103.** 20 yr; from 2009 onward   **105. a.** at least 96   **b.** if you get less than 66 on the final   **107.** up to 1280 mi   **109.** up to 29 bags of cement   **111.** An open dot indicates an endpoint that is not a solution and a closed dot indicates an endpoint that is a solution.   **117.** more than 720 mi   **119.** $\{x|x < 0.4\}$ **121.** 20   **122.** length: 11 in.; width: 6 in.   **123.** 4

Review Exercises

1. 32   **2.** −22   **3.** 12   **4.** −22   **5.** 5   **6.** 80   **7.** −56   **8.** 11   **9.** 4   **10.** −15   **11.** −12   **12.** −25   **13.** 10   **14.** 6
15. −5   **16.** −10   **17.** 2   **18.** 1   **19.** 16 yr; 2006   **20.** −1   **21.** 12   **22.** −13   **23.** −3   **24.** −10   **25.** 2   **26.** 2
27. no solution   **28.** all real numbers   **29.** 20 yr old   **30.** $r = \dfrac{I}{P}$ **31.** $h = \dfrac{3V}{B}$ **32.** $w = \dfrac{P - 2l}{2}$ **33.** $B = 2A - C$
34. $m = \dfrac{T - D}{P}$ **35.** 72%   **36.** 0.35%   **37.** 0.65   **38.** 1.5   **39.** 0.03   **40.** 9.6   **41.** 200   **42.** 48%   **43.** 100%   **44.** 40%
45. 12.5%   **46.** no; 1%   **47. a.** $h = 7r$ **b.** 5 ft 3 in.   **48.** 350 gallons   **49.** 10   **50.** New York: 55 days; Los Angeles: 213 days
51. pages 46 and 47   **52.** Streisand: 49 albums; Madonna: 47 albums   **53.** 9 yr; 2012   **54.** 18 checks   **55.** length: 150 yd; width: 50 yd   **56.** $240
57. ←|—|—|—|—|—|—|—|—|—|—|→   **58.** ←|—|—|—|—|—|—|—|—|—|→   **59.** $\{x\,|\,x > 4\}$ **60.** $\{x\,|\,x \leq -3\}$
     −5 −4 −3 −2 −1 0 1 2 3 4 5   −5 −4 −3 −2 −1 0 1 2 3 4 5
61. $\{x\,|\,x < 4\}$ ←|—|—|—|—|—|○|—|—|—|→   **62.** $\{x\,|\,x > -8\}$ ←|—|—|○|—|—|—|—|—|—|—|→
          −2 −1 0 1 2 3 4 5 6 7 8   −10 −9 −8 −7 −6 −5 −4 −3 −2 −1 0
63. $\{x\,|\,x \geq -3\}$ ←|—|—|●|—|—|—|—|—|→   **64.** $\{x\,|\,x > 6\}$ ←|—|—|—|—|—|—|—|○|—|→
          −5 −4 −3 −2 −1 0 1 2 3 4 5   −2 −1 0 1 2 3 4 5 6 7 8
65. $\{x\,|\,x \geq 4\}$ ←|—|—|—|—|—|●|—|—|→   **66.** $\{x\,|\,x \leq 2\}$ ←|—|—|—|—|—|—|●|—|—|→   **67.** $\{x\,|\,x$ is a real number$\}$
          −2 −1 0 1 2 3 4 5 6 7 8   −5 −4 −3 −2 −1 0 1 2 3 4 5
68. no solution; ∅   **69.** at least 64   **70.** no more than 99 min

Chapter 2 Test

1. $\dfrac{9}{2}$ **2.** −5   **3.** $-\dfrac{4}{3}$ **4.** 2   **5.** −20   **6.** $-\dfrac{5}{3}$ **7.** 60 yr; 2020   **8.** $h = \dfrac{V}{\pi r^2}$ **9.** $w = \dfrac{P - 2l}{2}$ **10.** 8.4   **11.** 150   **12.** 5%
13. 63   **14.** fitness trainer: $50,950; teacher: $28,080   **15.** 600 min   **16.** length: 150 yd; width: 75 yd   **17.** $35
18. ←|—|—|○|—|—|—|—|—|→   **19.** ←|—|—|—|—|—|—|○|—|—|→   **20.** $\{x\,|\,x \leq -1\}$
     −5 −4 −3 −2 −1 0 1 2 3 4 5   −8 −7 −6 −5 −4 −3 −2 −1 0 1 2
21. $\{x\,|\,x < -6\}$ ←|—|○|—|—|—|—|—|—|—|→   **22.** $\{x\,|\,x \leq -3\}$ ←|—|●|—|—|—|—|—|—|→
          −8 −7 −6 −5 −4 −3 −2 −1 0 1 2   −5 −4 −3 −2 −1 0 1 2 3 4 5
23. $\{x\,|\,x > 7\}$ ←|—|—|—|—|—|—|—|—|○|→   **24.** at least 92   **25.** widths greater than 8 inches
          −2 −1 0 1 2 3 4 5 6 7 8

Cumulative Review Exercises (Chapters 1–2)

1. −4   **2.** −2   **3.** −128   **4.** $-103 - 20x$ **5.** $-4, -\dfrac{1}{3}, 0, \sqrt{4}, 1063$
6. III;

7. <   **8.** $24x - 6 - 30y$ **9.** 2000; 4%   **10.** 1992; 7.8%   **11.** 1   **12.** −15   **13.** $A = \dfrac{3V}{h}$ **14.** 160
15. length: 130 yd; width: 70 yd   **16.** 75,000 gal   **17.** ←|—|—|○|—|—|—|—|●|—|→
                                                                          −5 −4 −3 −2 −1 0 1 2 3 4 5
18. $\{x\,|\,x < -3\}$ ←|—|○|—|—|—|—|—|—|—|→   **19.** $\{x\,|\,x \geq -6\}$ ←|—|●|—|—|—|—|—|—|→
          −5 −4 −3 −2 −1 0 1 2 3 4 5   −8 −7 −6 −5 −4 −3 −2 −1 0 1 2
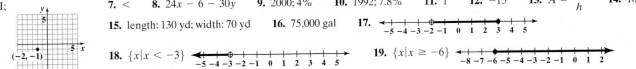
20. more than $47,500

CHAPTER 3

Section 3.1

Check Point Exercises

1. $150   **2.** $15,000 at 9% and $10,000 at 12%   **3.** 27 ml of acid   **4.** 30 ml of 10% and 20 ml of 60% acid solution   **5.** 9 mph   **6.** 4 hr

Enrichment Essay: Trick Questions

1. 12   **2.** 12   **3.** sister and brother   **4.** the match

Exercise Set 3.1

1. $160   **3.** $0.08(20,000 - x)$; $8000 at 7% and $12,000 at 8%   **5.** $120,000 - x$; $0.08x$; $0.18(120,000 - x)$; $116,000 at 8% and $4000 at 18%
7. $3600 at 6% and $2400 at 9%   **9.** $31,250 at 15% and $18,750 at 7%   **11.** 6 ml   **13.** $0.10(50 - x)$; $0.08(50)$; 20 l of 5% and 30 l of 10%
15. 6 oz   **17.** The north had 600 students and the south had 400 students.   **19.** $10x$; $12x$; 3 hr   **21.** $5(x + 5)$; $5x$; faster truck: 62.5 mph; slower
truck: 57.5 mph   **23.** 2.4 hr (or 2 hr and 24 min)   **25.** 2 hr at 50 mph and 3 hr at 40 mph   **27.** 12 lb of grade A and 8 lb of grade B   **29.** 12 lb
31. 20 dimes; 12 quarters   **33.** $10 bills: 5; $5 bills: 30   **35.** 225 adult tickets; 80 children tickets   **43.** Yes; Answers will vary.   **45.** 25 ml
47. −71   **48.** $-100, 0, \sqrt{16}$ **49.** −3

Section 3.2

Check Point Exercises

1. a. 1:16, or 1 to 16   **b.** 16:17, or 16 to 17   **2. a.** 15   **b.** −27   **3.** $1500   **4.** 720 deer

Exercise Set 3.2

1. $\frac{1}{2}$ **3.** $\frac{12}{5}$ **5.** $\frac{3}{4}$ **7.** $\frac{2}{1}$; 2:1 **9.** $\frac{1}{3}$; 1:3 **11.** 12 **13.** 7.5 **15.** -9 **17.** 34 **19.** -49 **21.** $-\frac{149}{6}$

23. $\dfrac{\text{price of 3 boxes}}{3\text{ boxes}} = \dfrac{\text{price of 8 boxes}}{8\text{ boxes}}$; $\dfrac{3\text{ boxes}}{8\text{ boxes}} = \dfrac{\text{price of 3 boxes}}{\text{price of 8 boxes}}$; $\dfrac{8\text{ boxes}}{3\text{ boxes}} = \dfrac{\text{price of 8 boxes}}{\text{price of 3 boxes}}$ **25.** $x = \dfrac{ab}{c}$ **27.** $x = \dfrac{ad + bd}{c}$

29. $x = \dfrac{ab}{c}$ **31.** $\dfrac{15}{11}$; 15:11 **33.** Answers will vary. **35.** \$1800 **37.** 20,489 fur seal pups **39.** \$950 **41.** 154.1 in.

43. $\dfrac{62}{123}$ (or 62:123); $\dfrac{10}{9}$ (or 10:9) **51.** b **53.** 5 consecutive pitches **55.** approximately 66.34 mi

56. $\{x\,|\,x \geq -2\}$ **57.** 280 **58.** $19 - 8x$

Mid-Chapter 3 Check Point

1. 8.05 days **2.** 31 **3.** $-\dfrac{3}{4}$ **4.** \$11,500 at 8% and \$13,500 at 9% **5.** $\dfrac{5}{12}$; 5:12, or 5 to 12 **6.** 375 mph and 475 mph

7. 800 elk **8.** 8 qt **9.** -3 **10.** \$2500 at 4% and \$1500 at 3% **11.** 4 hours by plane and 2 hours by car **12.** $5\dfrac{1}{3}$ in.

Section 3.3

Check Point Exercises

1. 12 ft **2.** 400π ft$^2 \approx 1256$ ft^2; 40π ft ≈ 126 ft **3.** large pizza **4.** 2 times **5.** No; About 32 more cubic inches are needed.
6. $120°, 40°, 20°$ **7.** $60°$

Exercise Set 3.3

1. 18 m; 18 m^2 **3.** 56 in.^{2} **5.** 91 m^2 **7.** 50 ft **9.** 8 ft **11.** 50 cm **13.** 16π cm$^2 \approx 50$ cm^2; 8π cm ≈ 25 cm
15. 36π yd$^2 \approx 113$ yd^2; 12π yd ≈ 38 yd **17.** 7 in.; 14 in. **19.** 36 in.^{3} **21.** 150π cm$^3 \approx 471$ cm^3 **23.** 972π cm$^3 \approx 3054$ cm^3

25. 48π m$^3 \approx 151$ m^3 **27.** $h = \dfrac{V}{\pi r^2}$ **29.** 9 times **31.** $50°, 50°, 80°$ **33.** $4x = 76; 3x + 4 = 61; 2x + 5 = 43; 76°, 61°, 43°$

35. $40°, 80°, 60°$ **37.** $32°$ **39.** $2°$ **41.** $48°$ **43.** $90°$ **45.** $75°$ **47.** $135°$ **49.** $50°$ **51.** 72 m^2 **53.** 70.5 cm^2 **55.** 168 cm^3
57. \$698.18 **59.** large pizza **61.** \$2262 **63.** approx 19.7 ft **65.** 21,000 yd^3 **67.** the can with diameter of 6 in. and height of 5 in.
69. Yes, the water tank is a little over one cubic foot too small. **81.** 2.25 times **83.** Volume increases 8 times. **85.** $35°$

86. $s = \dfrac{P - b}{2}$ **87.** 8 **88.** 0

Review Exercises

1. \$3000 at 8% and \$7000 at 10% **2.** \$8000 at 10% and \$14,000 at 12% **3.** 4 gal of 75% and 6 gal of 50% **4.** No, it is not possible.
5. North had 60 students and south had 90 students. **6.** 3 hr **7.** 30 mph and 40 mph **8.** \$5 bills: 6; \$10 bills: 9 **9.** exactly enough information
10. not enough information **11.** exactly enough information **12.** $\dfrac{2}{5}$; 2:5, or 2 to 5 **13.** $\dfrac{15}{14}$; 15:14, or 15 to 14 **14.** $\dfrac{4}{1}$; 4:1, or 4 to 1 **15.** 5
16. -65 **17.** $\dfrac{2}{5}$ **18.** 3 **19.** 324 teachers **20.** 287 trout **21.** 19,200 skates **22.** 32.5 ft^2 **23.** 50 cm^2 **24.** 135 yd^2
25. 20π m ≈ 63 m; 100π m$^2 \approx 314$ m^2 **26.** 6 ft **27.** 156 ft^2 **28.** \$1890 **29.** medium pizza **30.** 60 cm^3 **31.** 128π yd$^3 \approx 402$ yd^3
32. 288π m$^3 \approx 905$ m^3 **33.** 4800 m^3 **34.** 16 fish **35.** $x = 30, 3x = 90, 2x = 60; 30°, 60°, 90°$ **36.** $85°, 35°, 60°$ **37.** $33°$ **38.** $105°$
39. $57.5°$ **40.** $45°$ and $135°$

Chapter 3 Test

1. \$4000 at 9% and \$2000 at 6% **2.** 40 ml of 50% and 60 ml of 80% **3.** 5 hr **4.** $\dfrac{4}{7}$; 4:7, or 4 to 7 **5.** $-\dfrac{15}{2}$ **6.** 6000 tule elk
7. 137.5 lb/in.^{2} **8.** 517 m^2 **9.** 525 in.^{2} **10.** 18 in.^{3} **11.** 175π cm$^3 \approx 550$ cm^3 **12.** \$650 **13.** 14 ft **14.** $126°, 42°, 12°$ **15.** $53°$

Cumulative Review Exercises (Chapters 1–3)

1. $-\dfrac{27}{68}$ **2.** $-1 - 6x$ **3.** $-3, 0, \sqrt{9}$ **4.** IV; **5.** 39 **6.** $<$ **7.** $6°$F **8.** 8 **9.** $-\dfrac{1}{2}$ **10.** $m = 2A - n$
11. 2015 **12.** 240 **13.** width: 53 m; length: 120 m **14.** 200 lb
15. $\{x\,|\,x \leq 0\}$;

16. $\{x\,|\,x < 1\}$; **17.** $1\dfrac{1}{2}$ hr **18.** $-\dfrac{21}{4}$ **19.** Reagan: 69 years old; Buchanan: 65 years old
20. No, the percent increase is 225%, which is not more than 300%.

CHAPTER 4

Section 4.1

Check Point Exercises

1. a. solution   **b.** not a solution   **2.** $(-2, -4), (-1, -1), (0, 2), (1, 5),$ and $(2, 8)$

3.   **4.**   **5.**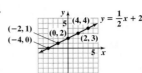

6.   **7. a.**

x	$y = 2x$ (x, y)	x	$y = 10 + x$ (x, y)
0	$(0, 0)$	0	$(0, 10)$
2	$(2, 4)$	2	$(2, 12)$
4	$(4, 8)$	4	$(4, 14)$
6	$(6, 12)$	6	$(6, 16)$
8	$(8, 16)$	8	$(8, 18)$
10	$(10, 20)$	10	$(10, 20)$
12	$(12, 24)$	12	$(12, 22)$

b. 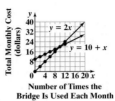   **c.** $(10, 20)$; If the bridge is used 10 times in a month, the total monthly cost without the coupon book is the same as the monthly cost with the coupon book, namely $20.

Exercise Set 4.1

1. $(2, 3)$ and $(3, 2)$ are not solutions; $(-4, -12)$ is a solution.   **3.** $(-5, -20)$ is not a solution; $(0, 0)$ and $(9, -36)$ are solutions.

5. $(2, -2)$ is not a solution; $(0, 6)$ and $(-3, 0)$ are solutions.   **7.** $(0, 5)$ is not a solution; $(-5, 6)$ and $(10, -3)$ are solutions.

9. $\left(1, \dfrac{1}{3}\right)$ is not a solution; $(0, 0)$ and $\left(2, -\dfrac{2}{3}\right)$ are solutions.   **11.** $(3, 4)$ and $(0, -4)$ are not solutions; $(4, 7)$ is a solution.

13.

x	(x, y)
-2	$(-2, -24)$
-1	$(-1, -12)$
0	$(0, 0)$
1	$(1, 12)$
2	$(2, 24)$

15.

x	(x, y)
-2	$(-2, 20)$
-1	$(-1, 10)$
0	$(0, 0)$
1	$(1, -10)$
2	$(2, -20)$

17.

x	(x, y)
-2	$(-2, -21)$
-1	$(-1, -13)$
0	$(0, -5)$
1	$(1, 3)$
2	$(2, 11)$

19.

x	(x, y)
-2	$(-2, 13)$
-1	$(-1, 10)$
0	$(0, 7)$
1	$(1, 4)$
2	$(2, 1)$

21.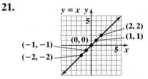

23. 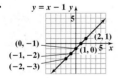   **25.**   **27.** 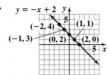   **29.**   **31.**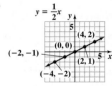

33. 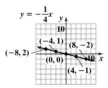   **35.** 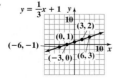   **37.** 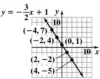   **39.** 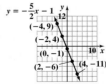   **41.**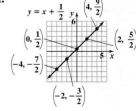

43. 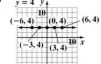   **45.**   **47.** 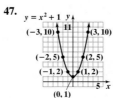   **49.** 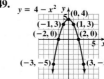   **51.**

53. $y = 2x + 5$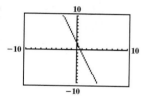

55. a. $8x + 6y = 14.50$ **b.** $1.25

57. a.

African Americans	
x	(x, y)
0	(0, 5.2)
10	(10, 7.6)
20	(20, 10)
30	(30, 12.4)

b.

Whites	
x	(x, y)
0	(0, 3.4)
10	(10, 5.6)
20	(20, 7.8)
30	(30, 10)

c.

Hispanics	
x	(x, y)
0	(0, 4.2)
10	(10, 5.5)
20	(20, 6.8)
30	(30, 8.1)

59. a. 40; Both rental companies have the same cost when the truck is used for 40 miles.   **b.** 55
c. 54; Both rental companies have the same cost of $54 when the truck is used for 40 miles.

61. a.

x	(x, y)
0	(0, 30,000)
10	(10, 30,500)
20	(20, 31,000)
30	(30, 31,500)
40	(40, 32,000)

b.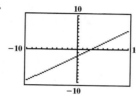

67.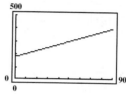

69. a. $(0, 0.6), (1, 0.3), (2, 0.2), (3, 0.3), (4, 0.6), (5, 1.1)$ **b.** between 10:00 A.M. and noon

71.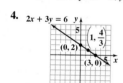

Answers will vary.

73.

Answers will vary.

75.

U.S. population increases.

76. 2   **77.** 1   **78.** $h = \dfrac{3V}{A}$

Section 4.2

Check Point Exercises

1. a. x-intercept is -3; y-intercept is 5.   **b.** y-intercept is 4; no x-intercept.   **c.** x-intercept is 0; y-intercept is 0.   **2.** 3   **3.** -4

4. $2x + 3y = 6$
5. $x - 2y = 4$
6. $x + 3y = 0$
7. $y = 3$
8. $x = -2$

  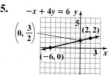

Exercise Set 4.2

1. a. 3   **b.** 4   **3. a.** -4 **b.** -2 **5. a.** 0   **b.** 0   **7. a.** no x-intercept   **b.** -2 **9.** x-intercept is 10; y-intercept is 4

11. x-intercept is $\dfrac{15}{2}$, or $7\dfrac{1}{2}$; y-intercept is -5.   **13.** x-intercept is 8; y-intercept is $-\dfrac{8}{3}$, or $-2\dfrac{2}{3}$.   **15.** x-intercept is 0; y-intercept is 0.

17. x-intercept is $-\dfrac{11}{2}$, or $-5\dfrac{1}{2}$; y-intercept is $\dfrac{11}{3}$, or $3\dfrac{2}{3}$

19. $x + y = 5$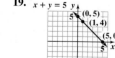
21. $x + 3y = 6$
23. $6x - 9y = 18$
25. $-x + 4y = 6$
27. $2x - y = 7$
 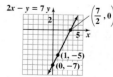

29. $3x = 5y - 15$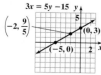
31. $25y = 100 - 50x$
33. $2x - 8y = 12$
35. $x + 2y = 0$
37. $y - 3x = 0$
  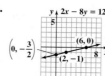 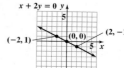

39.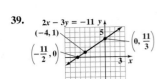

41. $y = 3$ **43.** $x = -3$ **45.** $y = 0$ **47.**

49.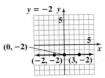

51.   **53.**   **55.** 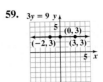   **57.**   **59.**

61.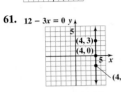

63. Exercise 4   **65.** Exercise 7   **67.** Exercise 1   **69.**
71. from 3 to 12 sec   **73.** 45; The vulture was 45 m
above the ground when the observation started.
75. 12, 13, 14, 15, 16; The vulture is on the ground at this time.

77. 8:00 A.M. to 11:00 A.M.   **79.** 11:00 A.M. to 1:00 P.M.   **81. a.** 80; In 1994, carbonated beverages had 80% of the market share.   **b.** very well
c. 140; In 2134, there will be no consumption of carbonated beverages. Model breakdown: model extended too far into the future.   **93.** $y = 6$
95. $-6; 3$ **97.** 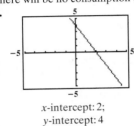   **99.**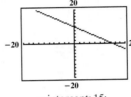

x-intercept: 2;
y-intercept: 4

x-intercept: 15;
y-intercept: 10

101. 13.4   **102.** $4x + 5$

103.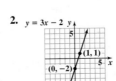

Section 4.3

Check Point Exercises

1. a. 6   **b.** $-\dfrac{7}{5}$ **2. a.** 0   **b.** undefined   **3.** Both slopes equal 2, so the lines are parallel.   **4.** $\dfrac{2}{15} \approx 0.13$; The slope indicates that the
number of U.S. men living alone is projected to increase by 0.13 million each year. The rate of change is 0.13 million men per year.

Exercise Set 4.3

1. $\dfrac{3}{4}$; rises   **3.** $\dfrac{1}{4}$; rises   **5.** 0; horizontal   **7.** -5; falls   **9.** undefined; vertical   **11.** $\dfrac{1}{2}$ **13.** $-\dfrac{1}{3}$

15. $-\dfrac{1}{2}$ **17.** $-\dfrac{2}{3}$ **19.** 0   **21.** undefined   **23.** parallel   **25.** not parallel

27.   **29.** Slopes of corresponding opposite sides are equal: $-\dfrac{2}{5}$ and $\dfrac{4}{3}$.   **31.** 2   **33.** collinear
35. 250; The amount spent online per U.S. online household increased by $250 each year from 1999 to 2001.
37. -0.4; The percentage of men is decreasing by 0.4% per year.   **39.** 2900; Mean income is increasing $2900 per year.
41. 0.40; Cost increases by $0.40 per mile driven.   **43.** pitch $= \dfrac{1}{3}$ **45.** 8.3%   **53.** b

55. b_2, b_1, b_4, b_3 **57.** -3 **59.** $\dfrac{3}{4}$ **61.** 12 in. and 24 in.   **62.** -42 **63.** $\{x \mid x \le 4\}$;

Section 4.4

Check Point Exercises

1. a. $5; -3$ **b.** $\dfrac{2}{3}; 4$ **c.** $-7; 6$ **2.** $y = 3x - 2$ **3.** $y = \dfrac{3}{5}x + 1$ **4.** $3x + 4y = 0$

Exercise Set 4.4

1. $3; 2$ **3.** $3; -5$ **5.** $-\dfrac{1}{2}; 5$ **7.** $7; 0$ **9.** $0; 10$ **11.** $-1; 4$ **13.** $y = 5x + 7; 5; 7$ **15.** $y = -x + 6; -1; 6$ **17.** $y = -6x; -6; 0$

19. $y = 2x; 2; 0$ **21.** $y = -\dfrac{2}{7}x; -\dfrac{2}{7}; 0$ **23.** $y = -\dfrac{3}{2}x + \dfrac{3}{2}; -\dfrac{3}{2}; \dfrac{3}{2}$ **25.** $y = \dfrac{3}{4}x - 3; \dfrac{3}{4}; -3$

27.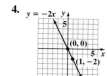

29.

31.

33.

35.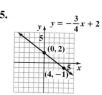

37.

39. a. $y = -3x$ **b.** $-3; 0$ **41. a.** $y = \dfrac{4}{3}x$ **b.** $\dfrac{4}{3}; 0$ **43. a.** $y = -2x + 3$ **b.** $-2; 3$

c.

45. a. $y = -\dfrac{7}{2}x + 7$ **b.** $-\dfrac{7}{2}; 7$ **c.**

47.

49.

parallel; The slopes are equal.   not parallel; The slopes are not equal.

51.

parallel; The slopes are equal.

53. $y = -3x + 5$ **55.** $y = -x + 2$ **57.** $y = x$ **59. a.** $38\%; 37.6\%; 37.2\%; 36.8\%; 34\%; 30\%$
b. -0.4; The percentage of U.S. men smoking is decreasing by 0.4% per year.   **c.** 38; In 1980, 38% of U.S. men were smoking.   **61. a.** 21; In 1997, 21 million people in sub-Saharan Africa were living with AIDS.
b. 1.6; Number of people living with AIDS in sub-Saharan Africa is increasing by 1.6 million per year.
c. $y = 1.6x + 21$ **d.** 35.4 million **67.** $F = \dfrac{9}{5}C + 32$ **69.** 8 **70.** 0 **71.** 56

Mid-Chapter 4 Check Point

1. a. 4 **b.** 2 **c.** $-\dfrac{1}{2}$ **2. a.** -5 **b.** no y-intercept **c.** undefined slope **3. a.** 0 **b.** 0 **c.** $\dfrac{3}{5}$

4.

5.

6.

7.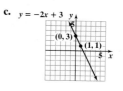

8.

9.

10.

11.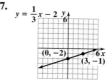

12.

13.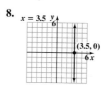

14.

15.

16. $m = \dfrac{5}{2}$; y-intercept is -5. **17.** parallel; Both have slopes of $\dfrac{4}{5}$.

18. a. 33 **b.** In 1995, 33% of U.S. colleges offered distance learning. **c.** 7.8
d. For the years 1995 through 2002, the percentage of colleges that offered distance learning increased at a rate of 7.8% per year.

Section 4.5

Check Point Exercises

1. $y + 5 = 6(x - 2); y = 6x - 17$ **2. a.** $y + 1 = -5(x + 2),$ or $y + 6 = -5(x + 1)$ **b.** $y = -5x - 11$ **3.** $y = 0.28x + 27.2; 41.2$

Exercise Set 4.5

1. $y - 5 = 3(x - 2); y = 3x - 1$ **3.** $y - 6 = 5(x + 2); y = 5x + 16$ **5.** $y + 2 = -8(x + 3); y = -8x - 26$

7. $y - 0 = -12(x + 8); y = -12x - 96$ **9.** $y + 2 = -1\left(x + \frac{1}{2}\right); y = -x - \frac{5}{2}$ **11.** $y - 0 = \frac{1}{2}(x - 0); y = \frac{1}{2}x$

13. $y + 2 = -\frac{2}{3}(x - 6); y = -\frac{2}{3}x + 2$ **15.** $y - 2 = 2(x - 1),$ or $y - 10 = 2(x - 5); y = 2x$

17. $y - 0 = 1(x + 3),$ or $y - 3 = 1(x - 0); y = x + 3$ **19.** $y + 1 = 1(x + 3),$ or $y - 4 = 1(x - 2); y = x + 2$

21. $y + 1 = \frac{5}{7}(x + 4),$ or $y - 4 = \frac{5}{7}(x - 3); y = \frac{5}{7}x + \frac{13}{7}$ **23.** $y + 1 = 0(x + 3),$ or $y + 1 = 0(x - 4); y = -1$

25. $y - 4 = 1(x - 2),$ or $y - 0 = 1(x + 2); y = x + 2$ **27.** $y - 0 = 8\left(x + \frac{1}{2}\right),$ or $y - 4 = 8(x - 0); y = 8x + 4$ **29.** $y = 4x + 14$

31. $y = -3x - 8$ **33.** $y = 3x - 2$ **35.** $y = -5x - 20$ **37. a.** $y - 162 = 1(x - 2),$ or $y - 168 = 1(x - 8)$ **b.** $y = x + 160$ **c.** 180 lb

39. Answers will vary. Example: $y - 3 = -\frac{2}{3}(x - 12); y = -\frac{2}{3}x + 11; 6.3.$

41. a–d. Answers will vary.

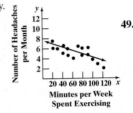

45. c

49. a.

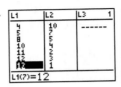

b.

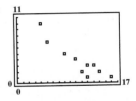

c. $a = -0.6867924528; b = 11.01132075; r = -0.9214983162;$

d.

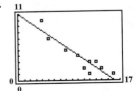

50. at most 12 sheets of paper **51.** $1, \sqrt{4}$ **52.**

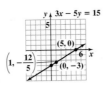

Section 4.6

Check Point Exercises

1. a. solution **b.** not a solution **2.**

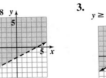

3.

4. a. $B(60, 20)$; A region that has an average annual temperature of 60°F and an average annual precipitation of 20 inches is a grassland.
b. $5(60) - 7(20) \geq 70, 160 \geq 70$ true; $3(60) - 35(20) \leq -140, -520 \leq -140$ true

Exercise Set 4.6

1. no; yes; yes **3.** yes; yes; no **5.** yes; yes; no **7.** yes; no; yes **9.**

11.

13. $x + 2y > 4$

15. $3x - y \le 6$

17. $3x - 2y \le 8$ $\frac{8}{3}$

19. $4x + 3y > 15$ $\frac{15}{4}$

21. $5x - y < -7$ $\frac{7}{5}$

23. $y \le \frac{1}{3}x$

25. $y > 2x$

27. $y > 3x + 2$

29. $y < \frac{3}{4}x - 3$

31. $x \le 1$

33. $y > 1$

35. $x \ge 0$

37. $x + y \ge 2$

39. $5x - 2y \le 10$

41. $y \ge \frac{1}{2}x$

43. $y \le -1$

45. a. $20x + 10y \le 80,000$ **b.** Answers will vary; $(1000, 2000)$ is an example. The plane can carry 1000 bottles of water and 2000 medical kits.

47. a. $50x + 150y > 2000$ **b.** $50x + 150y > 2000$ **c.** Answers will vary; $(20, 15)$ is an example. The elevator cannot carry 20 children and 15 adults. **49. a.** 27.1 **b.** borderline overweight **59.** $x + y \ge 3$

61.

63.

65. $h = \dfrac{V}{lw}$ **66.** $-\dfrac{8}{15}$ **67.** 5

Review Exercises

1. $(-3, 3)$ is not a solution; $(0, 6)$ and $(1, 9)$ are solutions. **2.** $(0, 4)$ and $(-1, 15)$ are not solutions; $(4, 0)$ is a solution.

3. a.

x	(x, y)
-2	$(-2, -7)$
-1	$(-1, -5)$
0	$(0, -3)$
1	$(1, -1)$
2	$(2, 1)$

b. $y = 2x - 3$

4. a.

x	(x, y)
-2	$(-2, 0)$
-1	$\left(-1, \frac{1}{2}\right)$
0	$(0, 1)$
1	$\left(1, \frac{3}{2}\right)$
2	$(2, 2)$

b. $y = \frac{1}{2}x + 1$

5. $y = x^2 - 3$

6. a.

x	(x, y)
10	$(10, 9)$
12	$(12, 19)$
14	$(14, 29)$
16	$(16, 39)$

b. Answers will vary.

7. a. -2 **b.** -4

8. a. no x-intercept **b.** 2

9. a. 0 **b.** 0

10. $2x + y = 4$

11. $3x - 2y = 12$

12. $3x = 6 - 2y$

13. $3x - y = 0$

14. $x = 3$

15. $y = -5$

16. $y + 3 = 5$

17. $2x = -8$

18. a. 5:00 P.M.; −4°F   **b.** 8:00 P.M.; 16°F   **c.** 4 and 6; At 4:00 P.M. and 6:00 P.M, the temperature was 0°F.   **d.** 12; At noon, the temperature was 12°F.

e. The temperature stayed the same, 12°F.   **19.** $-\frac{1}{2}$; falls   **20.** 3; rises   **21.** 0; horizontal   **22.** undefined; vertical   **23.** $\frac{3}{5}$

24. undefined   **25.** $-\frac{1}{3}$ **26.** 0   **27.** not parallel   **28.** parallel   **29. a.** $m = -44$; The number of new AIDS diagnoses decreased at a rate

of 44 each year from 1999 to 2001.   **b.** $m = 909$; The number of new AIDS diagnoses increased at a rate of 909 each year from 2001 to 2003.

c. $m = 432.5$; yes; Answers will vary.   **30.** 5; −7   **31.** −4; 6   **32.** 0; 3   **33.** $-\frac{2}{3}$; 2

34.    **35.** 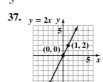   **36.** 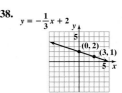   **37.**   **38.**

39. 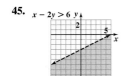   **40. a.** 25; In 1990, the average age of U.S. Hispanics was 25.
b. 0.3; The average age for U.S. whites increased at a rate of about 0.3 each year from 1990 to 2000.
c. $y = 0.3x + 35$ **d.** 41 years old
41. $y - 7 = 6(x + 4)$; $y = 6x + 31$
42. $y - 4 = 3(x - 3)$, or $y - 1 = 3(x - 2)$; $y = 3x - 5$
43. a. $y - 1.5 = 0.95(x - 1)$, or $y - 3.4 = 0.95(x - 3)$ **b.** $y = 0.95x + 0.55$ **c.** $10.05 billion
44. $(0, 0)$ and $(-3, 4)$ are not solutions; $(3, -6)$ and $(-2, -5)$ are solutions.

Yes, they are parallel since

both have slopes of $-\frac{1}{2}$ and different y-intercepts.

45. 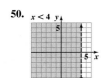   **46.**   **47.**   **48.**   **49.**

50.   **51.**   **52.**

Chapter 4 Test

1. $(-2, 1)$ is not a solution; $(0, -5)$ and $(4, 3)$ are solutions.

2.

x	(x, y)
−2	(−2, −5)
−1	(−1, −2)
0	(0, 1)
1	(1, 4)
2	(2, 7)

3.   **4. a.** 2   **b.** −3   **5.**   **6.**

7. 3; rises   **8.** undefined; vertical   **9.** $\frac{3}{2}$ **10.** parallel   **11.** −1; 10   **12.** −2; 6

13.   **14.**   **15.** $y - 4 = -2(x + 1)$;   **16.** $y - 1 = 3(x - 2)$, or $y + 8 = 3(x + 1)$;

$y = -2x + 2$ $y = 3x - 5$

17. $3x - 2y < 6$ **18.** $y \geq 2x - 2$ **19.** $x > -1$ **20.** 106; Spending per pupil increased at a rate of about $106 each year.

Cumulative Review Exercises (Chapters 1–4)

1. 2   **2.** $4 - 6x$ **3.** $\sqrt{5}$ **4.** 19   **5.** $\dfrac{5}{4}$ **6.** $x = \dfrac{y - b}{m}$ **7.** 800   **8.** 40 mph   **9.** $\{x \mid x \leq -2\}$

10. $\{x \mid x < 0\}$    **11.** 6 hr   **12.** 8 l of 40% and 4 l of 70%   **13.** 3.5 hr   **14.** 40°, 60°, 80°   **15.** 16

16. $2x - y = 4$ 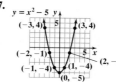   **17.** $y = x^2 - 5$ **18.** $y = -4x + 3$ **19.** $3x - 2y < -6$ **20.** $y \geq -1$

CHAPTER 5

Section 5.1

Check Point Exercises

1. a. solution   **b.** not a solution   **2.** $(1, 4)$ **3.** $(3, 3)$ **4.** inconsistent; no solution   **5.** infinitely many solutions

Exercise Set 5.1

1. solution   **3.** solution   **5.** not a solution   **7.** solution   **9.** not a solution   **11.** $(4, 2)$ **13.** $(-1, 2)$ **15.** $(3, 0)$ **17.** $(1, 0)$
19. $(-1, 4)$ **21.** $(2, 4)$ **23.** $(2, -1)$ **25.** no solution   **27.** $(-2, 6)$ **29.** infinitely many solutions   **31.** $(2, 3)$ **33.** no solution

35. infinitely many solutions   **37.** $(2, 4)$ **39.** no solution   **41.** no solution   **43.** $m = \dfrac{1}{2}; b = -3$ and -5; no solution

45. $m = -\dfrac{1}{2}$ and 3; $b = 4$; one solution   **47.** $m = 3; b = -6$; infinite number of solutions   **49.** $m = -3; b = 0$ and 1; no solution

51. a. Answers will vary. Example: (1996, 41); Mothers 30 years and older had 41 thousand births in 1996.
b. There are more births to mothers 30 years and older.   **61.** c   **67.** $(6, -1)$ **69.** $(3, 0)$ **71.** $(2, -1)$ **73.** $(-4, 4)$ **74.** -12
75. 6   **76.** 27

Section 5.2

Check Point Exercises

1. $(3, 2)$ **2.** $(1, -2)$ **3.** no solution   **4.** infinitely many solutions   **5.** $30; 400 units

Exercise Set 5.2

1. $(1, 3)$ **3.** $(5, 1)$ **5.** $(2, 1)$ **7.** $(-1, 3)$ **9.** $(4, 5)$ **11.** $\left(-\dfrac{2}{5}, -\dfrac{11}{5}\right)$ **13.** no solution   **15.** infinitely many solutions

17. $(0, 0)$ **19.** $\left(\dfrac{17}{7}, -\dfrac{8}{7}\right)$ **21.** no solution   **23.** $\left(\dfrac{43}{5}, -\dfrac{1}{5}\right)$ **25.** $(200, 700)$ **27.** $(7, 3)$ **29.** $(-1, -1)$ **31.** $(5, 4)$
33. $x + y = 81, y = x + 41$; 20 and 61   **35.** $x - y = 5, 4x = 6y$; 10 and 15   **37.** $x - y = 1, x + 2y = 7$; 2 and 3   **39.** $(2, 8)$
41. a. 6500 sold; 6200 supplied   **b.** $50; 6250 tickets   **43.** 2700 gal   **45.** 2032; about 0.6 deaths per 1000 live births or less than one death per
1000 live births   **53.** $x = 1, y = -3, z = 5$

55. $4x + 6y = 12$ **56.** 12   **57.** $-73, 0, \dfrac{3}{1}$

Section 5.3

Check Point Exercises

1. $(7, -2)$ **2.** $(6, 2)$ **3.** $(2, -1)$ **4.** $\left(\dfrac{60}{17}, -\dfrac{11}{17}\right)$ **5.** no solution   **6.** infinitely many solutions

Exercise Set 5.3

1. $(4, -7)$ **3.** $(3, 0)$ **5.** $(-3, 5)$ **7.** $(3, 1)$ **9.** $(2, 1)$ **11.** $(-2, 2)$ **13.** $(-7, -1)$ **15.** $(2, 0)$ **17.** $(1, -2)$ **19.** $(-1, 1)$

21. $(3, 1)$ **23.** $(-5, -2)$ **25.** $\left(\dfrac{11}{12}, -\dfrac{7}{6}\right)$ **27.** $\left(\dfrac{23}{16}, \dfrac{3}{8}\right)$ **29.** no solution   **31.** infinitely many solutions   **33.** no solution

35. $\left(\dfrac{1}{2}, -\dfrac{1}{2}\right)$ **37.** infinitely many solutions   **39.** $\left(\dfrac{1}{3}, 1\right)$ **41.** $(-10, 21)$ **43.** $(0, 1)$ **45.** $(2, -1)$ **47.** $(1, -3)$ **49.** $(4, 3)$

51. no solution   **53.** infinitely many solutions   **55.** $(3, 2)$ **57.** $(-1, 2)$ **59.** $(-1, 0)$ **61.** $(3, 1)$ **63.** dependent   **65.** 2002; 48% for and 48% against   **73.** $x = -a - b; y = -2a - b$ **77.** $(5, 1)$ **79.** 10   **80.** II   **81.** 26

Mid-Chapter 5 Check Point

1. $(2, 0)$ **2.** $(1, 1)$ **3.** no solution   **4.** $(5, 8)$ **5.** $(2, -1)$ **6.** $(2, 9)$ **7.** $(-2, -3)$ **8.** $(10, -1)$ **9.** no solution

10. $(-12, -1)$ **11.** $(-7, -23)$ **12.** $\left(\dfrac{16}{17}, -\dfrac{12}{17}\right)$ **13.** infinitely many solutions   **14.** $(7, 11)$ **15.** $(6, 10)$

Section 5.4

Check Point Exercises

1. A bustard weighs 46 lb; a condor weighs 27 lb.   **2.** A Quarter Pounder has 420 cal; a Whopper with cheese has 589 cal.
3. length: 100 ft; width: 80 ft   **4.** 17.5 yr; $24,250

Exercise Set 5.4

1. $x + y = 17, x - y = -3$; 7 and 10   **3.** $3x - y = -1, x + 2y = 23$; 3 and 10   **5.** 5.8 million lb of potato chips; 4.6 million lb of tortilla chips
7. pan pizza: 1120 calories; beef burrito: 430 calories   **9.** scrambled eggs: 366 mg; Double Beef Whopper: 175 mg   **11.** sweater: $12; shirt: $10
13. 44 ft by 20 ft   **15.** 90 ft by 70 ft   **17. a.** 300 min; $35   **b.** plan B; Answers will vary.   **19.** $600 of merchandise; $580
21. 26 yr; 2011; college grads: $1158; high school grads: $579   **23.** 2 servings of macaroni and 4 servings of broccoli   **25.** A: 100°; B: 40°; C: 40°
27. still water rate: 6 mph; current rate: 2 mph   **33.** 10 birds and 20 lions   **35.** There are 5 people downstairs and 7 people upstairs.

38. $2x - y < 4$

39. $y \geq x + 1$
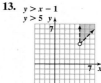

40. $x \geq 2$
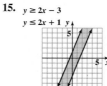

Section 5.5

Check Point Exercises

1.
$x + 2y > 4$
$2x - 3y \leq -6$

2.
$y \geq x + 2$
$x \geq 1$

3. a. Answers will vary.
b. Answers will vary.

Exercise Set 5.5

1.
$x + y \leq 4$
$x - y \leq 2$

3.
$2x - 4y \leq 8$
$x + y \geq -1$

5.
$x + 3y \leq 6$
$x - 2y \leq 4$

7.
$x - 2y > 4$
$2x + y \geq 6$

9.
$x + y > 1$
$x + y < 4$

11.
$y \geq 2x + 1$
$y \leq 4$

13.
$y > x - 1$
$y > 5$

15.
$y \geq 2x - 3$
$y \leq 2x + 1$

17.
$y > 2x - 3$
$y \leq -x + 6$

19.
$x + 2y \leq 4$
$y \geq x - 3$

21. $x \le 3$
$y \ge -2$
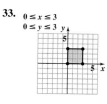

23. $x \ge 3$
$y < 2$
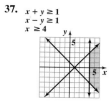

25. $x \ge 0$
$y \le 0$

27. $x \ge 0$
$y > 0$
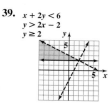

29. $x + y \le 5$
$x \ge 0$
$y \ge 0$

31. $4x - 3y > 12$
$x \ge 0$
$y \le 0$

33. $0 \le x \le 3$
$0 \le y \le 3$

35. $x - y \le 4$
$x + 2y \le 4$

37. $x + y \ge 1$
$x - y \ge 1$
$x \ge 4$

39. $x + 2y < 6$
$y > 2x - 2$
$y \ge 2$

41. $y \le -3x + 3$
$y \ge -x - 1$
$y < x + 7$

43. no solution $y \ge 2x + 2$
$y < 2x - 3$
$x \ge 2$

47. yes   **49.** about 140 to 190 lb

55. $x + y \ge 25,000$
$35x + 50y \ge 1,025,000$
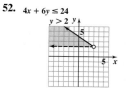

56. $-\dfrac{1}{4}$ **57.** $-\dfrac{11}{20}$ **58.** $y = x^2$

Review Exercises

1. solution **2.** not a solution **3.** no; $(-1, 3)$ does not satisfy $2x + y = -5$. **4.** $(4, -2)$ **5.** $(3, -2)$ **6.** $(2, 0)$ **7.** $(2, 1)$ **8.** $(4, -1)$
9. no solution **10.** infinitely many solutions **11.** no solution **12.** $(-2, -6)$ **13.** $(2, 5)$ **14.** no solution **15.** $(2, -1)$ **16.** $(5, 4)$

17. $(-2, -1)$ **18.** $(1, -4)$ **19.** $(20, -21)$ **20.** infinitely many solutions **21.** no solution **22.** $(4, 18)$ **23.** $\left(-1, -\dfrac{1}{2}\right)$ **24.** $12.50; 250 copies

25. $(2, 4)$ **26.** $(-1, -1)$ **27.** $(2, -1)$ **28.** $(3, 2)$ **29.** $(2, 1)$ **30.** $(0, 0)$ **31.** $\left(\dfrac{17}{7}, -\dfrac{15}{7}\right)$ **32.** no solution **33.** infinitely many solutions

34. $(4, -2)$ **35.** $(-8, -6)$ **36.** $(-4, 1)$ **37.** $\left(\dfrac{5}{2}, 3\right)$ **38.** $(3, 2)$ **39.** $\left(\dfrac{1}{2}, -2\right)$ **40.** no solution **41.** $(3, 7)$ **42.** Japan: 73.6 yr;
Switzerland: 72.8 yr **43.** gorilla: 485 lb; orangutan: 165 lb **44.** shrimp: 42 mg; scallops: 15 mg **45.** 9 ft by 5 ft **46.** 7 yd by 5 yd
47. room: $80; car: $60 **48.** 200 min; $25

49. $3x - y \le 6$
$x + y \ge 2$
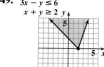

50. $x + y < 4$
$x - y < 4$

51. $y < 2x - 2$
$x \ge 3$

52. $4x + 6y \le 24$
$y > 2$

53. $x \le 3$
$y \ge -2$

54. $y \ge \dfrac{1}{2}x - 2$
$y \le \dfrac{1}{2}x + 1$
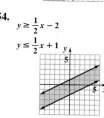

55. $x \le 0$
$y \ge 0$

Chapter 5 Test

1. solution **2.** not a solution **3.** $(2, 4)$ **4.** $(2, 4)$ **5.** $(1, -3)$ **6.** $(2, -3)$ **7.** no solution **8.** $(-1, 4)$
9. $(-4, 3)$ **10.** infinitely many solutions **11.** Mary: 2.6%; Patricia: 1.1% **12.** 12 yd by 5 yd **13.** 500 min; $40

14. $x - 3y > 6$
$2x + 4y \le 8$
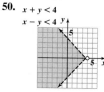

15. $y \ge 2x - 4$
$x < 2$

Cumulative Review Exercises (Chapters 1–5)

1. -36 **2.** $17x - 11$ **3.** -6 **4.** 20 **5.** $t = \dfrac{A - P}{Pr}$ **6.** $\{x | x > 2\}$;

7. **8.** **9.**

10. $(0, -2)$ **11.** $\left(\dfrac{3}{2}, -2\right)$ **12.** 1

13. $y - 6 = -4(x + 1)$; $y = -4x + 2$ **14.** 10 ft

15. pen: \$0.80; pad: \$1.20 **16.** $-93, 0, \dfrac{7}{1}, \sqrt{100}$ **17.** 20%

18. one computer; The percentage of households with one computer is increasing faster than the percentage for multiple computers.
19. 8 yr, 2005 **20.** 2012

CHAPTER 6

Section 6.1

Check Point Exercises

1. $5x^3 + 4x^2 - 8x - 20$ **2.** $5x^3 + 4x^2 - 8x - 20$ **3.** $7x^2 + 11x + 4$ **4.** $7x^3 + 3x^2 + 12x - 8$ **5.** $3y^3 - 10y^2 - 11y - 8$
6. approximately 4 per thousand; approximately (40, 4)

Exercise Set 6.1

1. binomial, 1 **3.** binomial, 3 **5.** monomial, 2 **7.** monomial, 0 **9.** trinomial, 2 **11.** trinomial, 4 **13.** binomial, 3
15. monomial, 23 **17.** $-8x + 13$ **19.** $12x^2 + 15x - 9$ **21.** $10x^2 - 12x$ **23.** $5x^2 - 3x + 13$ **25.** $4y^3 + 10y^2 + y - 2$
27. $3x^3 + 2x^2 - 9x + 7$ **29.** $-2y^3 + 4y^2 + 13y + 13$ **31.** $-3y^6 + 8y^4 + y^2$ **33.** $10x^3 + 1$ **35.** $-\dfrac{2}{5}x^4 + x^3 - \dfrac{1}{8}x^2$

37. $0.01x^5 + x^4 - 0.1x^3 + 0.3x + 0.33$ **39.** $11y^3 - 3y^2$ **41.** $-2x^2 - x + 1$ **43.** $-\dfrac{1}{4}x^4 - \dfrac{7}{15}x^3 - 0.3$ **45.** $-y^3 + 8y^2 - 3y - 14$
47. $-5x^3 - 6x^2 + x - 4$ **49.** $7x^4 - 2x^3 + 4x - 2$ **51.** $8x^2 + 7x - 5$ **53.** $9x^3 - 4.9x^2 + 11.1$ **55.** $-2x - 10$ **57.** $-5x^2 - 9x - 12$
59. $-5x^2 - x$ **61.** $-4x^2 - 4x - 6$ **63.** $-2y - 6$ **65.** $6y^3 + y^2 + 7y - 20$ **67.** $n^3 + 2$ **69.** $y^6 - y^3 - y^2 + y$
71. $26x^4 + 9x^2 + 6x$ **73.** $\dfrac{5}{7}x^3 - \dfrac{9}{20}x$ **75.** $4x + 6$ **77.** $10x^2 - 7$ **79.** $-4y^2 - 7y + 5$ **81.** $9x^3 + 11x^2 - 8$
83. $-y^3 + 8y^2 + y + 14$ **85.** $7x^4 - 2x^3 + 3x^2 - x + 2$ **87.** $0.05x^3 + 0.02x^2 + 1.02x$ **89.** $x^2 + 12x$ **91.** $y^2 - 19y + 16$
93. $2x^3 + 3x^2 + 7x - 5$ **95.** $-10y^3 + 2y^2 + y + 3$ **97.** 5 billion, 7.25 billion, 17 billion, 25.25 billion, and 5 billion; 3 days; 4 days
99. 2029 cigarettes; Answers will vary. **101. a.** 73% **b.** 72.8% **c.** 73.3%; Answers will vary. **103. a.** 42 human years
b. 41.8 human years **105.** 3 dog years **117.** $-3x^2 - x - 2$ **120.** -10 **121.** 5.6 **122.** -4

Section 6.2

Check Point Exercises

1. a. 2^6 or 64 **b.** x^{10} **c.** y^8 **d.** y^9 **2. a.** 3^{20} **b.** x^{90} **c.** $(-5)^{21}$ **3. a.** $16x^4$ **b.** $-64y^6$ **4. a.** $70x^3$ **b.** $-20x^9$
5. a. $3x^2 + 15x$ **b.** $30x^5 - 12x^3 + 18x^2$ **6. a.** $x^2 + 9x + 20$ **b.** $10x^2 - 29x - 21$ **7.** $5x^3 - 18x^2 + 7x + 6$
8. $6x^5 - 19x^4 + 22x^3 - 8x^2$

Exercise Set 6.2

1. x^{18} **3.** y^{12} **5.** x^{11} **7.** 7^{19} **9.** 6^{90} **11.** x^{45} **13.** $(-20)^9$ **15.** $8x^3$ **17.** $25x^2$ **19.** $16x^6$ **21.** $16y^{24}$ **23.** $-32x^{35}$
25. $14x^2$ **27.** $24x^3$ **29.** $-15y^7$ **31.** $\dfrac{1}{8}a^5$ **33.** $-48x^7$ **35.** $4x^2 + 12x$ **37.** $x^2 - 3x$ **39.** $2x^2 - 12x$ **41.** $-12y^2 - 20y$
43. $4x^3 + 8x^2$ **45.** $2y^4 + 6y^3$ **47.** $6y^4 - 8y^3 + 14y^2$ **49.** $6x^4 + 8x^3$ **51.** $-2x^3 - 10x^2 + 6x$ **53.** $12x^4 - 3x^3 + 15x^2$
55. $x^2 + 8x + 15$ **57.** $2x^2 + 9x + 4$ **59.** $x^2 - 2x - 15$ **61.** $x^2 - 2x - 99$ **63.** $2x^2 + 3x - 20$ **65.** $\dfrac{3}{16}x^2 + \dfrac{11}{4}x - 4$
67. $x^3 + 3x^2 + 5x + 3$ **69.** $y^3 - 6y^2 + 13y - 12$ **71.** $2a^3 - 9a^2 + 19a - 15$ **73.** $x^4 + 3x^3 + 5x^2 + 7x + 4$
75. $4x^4 - 4x^3 + 6x^2 - \dfrac{17}{2}x + 3$ **77.** $x^4 + x^3 + x^2 + 3x + 2$ **79.** $x^3 + 3x^2 - 37x + 24$ **81.** $2x^3 - 9x^2 + 27x - 27$
83. $2x^4 + 9x^3 + 6x^2 + 11x + 12$ **85.** $12z^4 - 14z^3 + 19z^2 - 22z + 8$ **87.** $21x^5 - 43x^4 + 38x^3 - 24x^2$
89. $4y^6 - 2y^5 - 6y^4 + 5y^3 - 5y^2 + 8y - 3$ **91.** $x^4 + 6x^3 - 11x^2 - 4x + 3$ **93.** $2x - 2$ **95.** $15x^5 + 42x^3 - 8x^2$
97. $2y^3$ **99.** $16y + 32$ **101.** $2x^2 + 7x - 15 \text{ ft}^2$ **103. a.** $(2x + 1)(x + 2)$ **b.** $2x^2 + 5x + 2$ **c.** $(2x + 1)(x + 2) = 2x^2 + 5x + 2$

115. $8x + 16$ **117.** $-8x^4$ **118.** $\{x | x < -1\}$ **119.** **120.** $-\dfrac{2}{3}$

Section 6.3

Check Point Exercises

1. $x^2 + 11x + 30$ **2.** $28x^2 - x - 15$ **3.** $6x^2 - 22x + 20$ **4. a.** $49y^2 - 64$ **b.** $16x^2 - 25$ **c.** $4a^6 - 9$ **5. a.** $x^2 + 20x + 100$
b. $25x^2 + 40x + 16$ **6. a.** $x^2 - 18x + 81$ **b.** $49x^2 - 42x + 9$

Exercise Set 6.3

1. $x^2 + 10x + 24$ **3.** $y^2 - 4y - 21$ **5.** $2x^2 + 7x - 15$ **7.** $4y^2 - y - 3$ **9.** $10x^2 - 9x - 9$ **11.** $12y^2 - 43y + 35$
13. $-15x^2 - 32x + 7$ **15.** $6y^2 - 28y + 30$ **17.** $15x^4 - 47x^2 + 28$ **19.** $-6x^2 + 17x - 10$ **21.** $x^3 + 5x^2 + 3x + 15$
23. $8x^5 + 40x^3 + 3x^2 + 15$ **25.** $x^2 - 9$ **27.** $9x^2 - 4$ **29.** $9r^2 - 16$ **31.** $9 - r^2$ **33.** $25 - 49x^2$ **35.** $4x^2 - \dfrac{1}{4}$ **37.** $y^4 - 1$
39. $r^6 - 4$ **41.** $1 - y^8$ **43.** $x^{20} - 25$ **45.** $x^2 + 4x + 4$ **47.** $4x^2 + 20x + 25$ **49.** $x^2 - 6x + 9$ **51.** $9y^2 - 24y + 16$
53. $16x^4 - 8x^2 + 1$ **55.** $49 - 28x + 4x^2$ **57.** $4x^2 + 2x + \dfrac{1}{4}$ **59.** $16y^2 - 2y + \dfrac{1}{16}$ **61.** $x^{16} + 6x^8 + 9$ **63.** $x^3 - 1$
65. $x^2 - 2x + 1$ **67.** $9y^2 - 49$ **69.** $12x^4 + 3x^3 + 27x^2$ **71.** $70y^2 + 2y - 12$ **73.** $x^4 + 2x^2 + 1$ **75.** $x^4 + 3x^2 + 2$
77. $x^4 - 16$ **79.** $4 - 12x^5 + 9x^{10}$ **81.** $\dfrac{3}{16}x^4 + 7x^2 - 96$ **83.** $x^2 + 2x + 1$ **85.** $4x^2 - 9$ **87.** $6x + 22$
89. $16x^4 - 72x^2 + 81$ **91.** $16x^4 - 1$ **93.** $x^3 + 6x^2 + 12x + 8$ **95.** $x^2 + 6x + 9 - y^2$ **97.** $(x + 1)(x + 2)$ yd^2 **99.** 56 yd^2; $(6, 56)$
101. $(x^2 + 4x + 4)$ in^2 **109.** $(x - 10)$ and $(x + 2)$ **111.** $x^2 + 2x$ **113.** Change $x^2 + 2x + 4$ to $x^2 + 4x + 4$.
115. Graphs coincide.   **116.** $(2, -1)$ **117.** $(1, 1)$ **118.**

Section 6.4

Check Point Exercises

1. -9 **2.** polynomial degree: 9;

Term	Coefficient	Degree
$8x^4y^5$	8	9
$-7x^3y^2$	-7	5
$-x^2y$	-1	3
$-5x$	-5	1
11	11	0

3. $2x^2y + 2xy - 4$ **4.** $3x^3 + 2x^2y + 5xy^2 - 10$ **5.** $60x^5y^5$ **6.** $60x^5y^7 - 12x^3y^3 + 18xy^2$
7. a. $21x^2 - 25xy + 6y^2$ **b.** $4x^2 + 16xy + 16y^2$ **8. a.** $36x^2y^4 - 25x^2$ **b.** $x^3 - y^3$

Exercise Set 6.4

1. 1   **3.** -47 **5.** -6 **7.** polynomial degree: 9;

Term	Coefficient	Degree
x^3y^2	1	5
$-5x^2y^7$	-5	9
$6y^2$	6	2
-3	-3	0

9. $7x^2y - 4xy$ **11.** $2x^2y + 13xy + 13$ **13.** $-11x^4y^2 - 11x^2y^2 + 2xy$ **15.** $-5x^3 + 8xy - 9y^2$ **17.** $x^4y^2 + 8x^3y + y - 6x$
19. $5x^3 + x^2y - xy^2 - 4y^3$ **21.** $-3x^2y^2 + xy^2 + 5y^2$ **23.** $8a^2b^4 + 3ab^2 + 8ab$ **25.** $-30x + 37y$ **27.** $40x^3y^2$ **29.** $-24x^5y^9$
31. $45x^2y + 18xy^2$ **33.** $50x^3y^2 - 15xy^3$ **35.** $28a^3b^5 + 8a^2b^3$ **37.** $-a^2b + ab^2 - b^3$ **39.** $7x^2 + 38xy + 15y^2$ **41.** $2x^2 + xy - 21y^2$
43. $15x^2y^2 + xy - 2$ **45.** $4x^2 + 12xy + 9y^2$ **47.** $x^2y^2 - 6xy + 9$ **49.** $x^4 + 2x^2y^2 + y^4$ **51.** $x^4 - 4x^2y^2 + 4y^4$ **53.** $9x^2 - y^2$
55. $a^2b^2 - 1$ **57.** $x^2 - y^4$ **59.** $9a^4b^2 - a^2$ **61.** $9x^2y^4 - 16y^2$ **63.** $a^3 - ab^2 + a^2b - b^3$ **65.** $x^3 + 4x^2y + 4xy^2 + y^3$
67. $x^3 - 4x^2y + 4xy^2 - y^3$ **69.** $x^2y^2 - a^2b^2$ **71.** $x^6y + x^4y + x^4 + 2x^2 + 1$ **73.** $x^4y^4 - 6x^2y^2 + 9$ **75.** $x^2 + 2xy + y^2 - 1$
77. $3x^2 + 8xy + 5y^2$ **79.** $2xy + y^2$ **81.** $x^{12}y^{12} - 2x^6y^6 + 1$ **83.** $x^4y^4 - 18x^2y^2 + 81$ **85.** $x^2 - y^2 - 2yz - z^2$
87. no; need 120 more board feet   **89.** 192 ft   **91.** 0 ft; The ball hits the ground.   **93.** 2.5 to 6 sec   **95.** $(2, 192)$
97. 2.5 sec; 196 ft   **101.** c   **103.** $5x^2 - 16y^2$ **105.** $W = \dfrac{2R - L}{3}$ **106.** 3.8   **107.** 105

Mid-Chapter 6 Check Point

1. $-55x^4y^6$ **2.** $6x^2y^3$ **3.** $12x^2 - x - 35$ **4.** $-x + 12$ **5.** $2x^3 - 11x^2 + 17x - 5$ **6.** $x^2 - x - 4$ **7.** $64x^2 - 48x + 9$ **8.** $70x^9$
9. $x^4 - 4$ **10.** $x^4 + 4x^2 + 4$ **11.** $18a^2 - 11ab - 10b^2$ **12.** $70x^5 - 14x^3 + 21x^2$ **13.** $5a^2b^3 + 2ab - b^2$ **14.** $18y^2 - 50$ **15.** $2x^3 - x^2 + x$
16. $10x^2 - 5xy - 3y^2$ **17.** $-4x^5 + 7x^4 - 10x + 23$ **18.** $x^3 + 27y^3$ **19.** $10x^7 - 5x^4 + 8x^3 - 4$ **20.** $y^2 - 12yz + 36z^2$ **21.** $-21x^2 + 7$

Section 6.5

Check Point Exercises

1. a. 5^8 **b.** x^7 **c.** y^{19} **2. a.** 1 **b.** 1 **c.** -1 **d.** 20 **e.** 1 **3. a.** $\dfrac{x^2}{25}$ **b.** $\dfrac{x^{12}}{8}$ **c.** $\dfrac{16a^{40}}{b^{12}}$

4. a. $-2x^8$ **b.** $\dfrac{1}{5}$ **c.** $3x^5y^3$ **5.** $-5x^7 + 2x^3 - 3x$ **6.** $5x^6 - \dfrac{7}{5}x + 2$ **7.** $3x^6y^4 - xy + 10$

Exercise Set 6.5

1. 3^{15} **3.** x^4 **5.** y^8 **7.** $5^3 \cdot 2^4$ **9.** $x^{75}y^{40}$ **11.** 1 **13.** 1 **15.** -1 **17.** 100 **19.** 1 **21.** 0 **23.** -2 **25.** $\dfrac{x^2}{9}$ **27.** $\dfrac{x^6}{64}$

29. $\dfrac{4x^6}{25}$ **31.** $-\dfrac{64}{27a^9}$ **33.** $-\dfrac{32a^{35}}{b^{20}}$ **35.** $\dfrac{x^8y^{12}}{16z^4}$ **37.** $3x^5$ **39.** $-2x^{20}$ **41.** $-\dfrac{1}{2}y^3$ **43.** $\dfrac{7}{5}y^{12}$ **45.** $6x^5y^4$ **47.** $-\dfrac{1}{2}x^{12}$ **49.** $\dfrac{9}{7}$

51. $-\dfrac{1}{10}x^8y^9z^4$ **53.** $5x^4 + x^3$ **55.** $2x^3 - x^2$ **57.** $y^6 - 9y + 1$ **59.** $-8x^2 + 5x$ **61.** $6x^3 + 2x^2 + 3x$ **63.** $3x^3 - 2x^2 + 10x$

65. $4x - 6$ **67.** $-6z^2 - 2z$ **69.** $4x^2 + 3x - 1$ **71.** $5x^4 - 3x^2 - x$ **73.** $-9x^3 + \dfrac{9}{2}x^2 - 10x + 5$ **75.** $4xy + 2x - 5y$

77. $-4x^5y^3 + 3xy + 2$ **79.** $4x^2 - x + 6$ **81.** $-xy^2$ **83.** $y + 5$ **85.** $3x^{12n} - 6x^{9n} + 2$ **87. a.** $\dfrac{6t^4 - 207t^3 + 2128t^2 - 6622t + 15{,}220}{28t^4 - 711t^3 + 5963t^2 - 1695t + 27{,}424}$

b. No; the divisor is not a monomial. **97.** $18x^8 - 27x^6 + 36x^4$ **99.** $\dfrac{3x^{14} - 6x^{12} - 9x^7}{-3x^7}$ **100.** 20.3 **101.** 0.875

102.

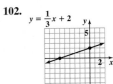

$y = \dfrac{1}{3}x + 2$

Section 6.6

Check Point Exercises

1. $x + 5$ **2.** $4x - 3 - \dfrac{3}{2x + 3}$ **3.** $x^2 + x + 1$

Exercise Set 6.6

1. $x + 4$ **3.** $2x + 5$ **5.** $x - 2$ **7.** $2y + 1$ **9.** $x - 5 + \dfrac{14}{x + 2}$ **11.** $y + 3 + \dfrac{4}{y + 2}$ **13.** $x^2 - 5x + 2$ **15.** $6y - 1$

17. $2a + 3$ **19.** $y^2 - y + 2$ **21.** $3x + 5 - \dfrac{5}{2x - 5}$ **23.** $x^2 + 2x + 8 + \dfrac{13}{x - 2}$ **25.** $2y^2 + y + 1 + \dfrac{6}{2y + 3}$

27. $2y^2 - 3y + 2 + \dfrac{1}{3y + 2}$ **29.** $9x^2 + 3x + 1$ **31.** $y^3 - 9y^2 + 27y - 27$ **33.** $2y + 4 + \dfrac{4}{2y - 1}$ **35.** $y^3 + y^2 - y - 1 + \dfrac{4}{y - 1}$

37. $4x - 3 + \dfrac{-7x + 7}{x^2 + 2}$ **39.** $x^2 + ax + a^2$ **41.** $2x^2 - 3x + 2$ **43.** $x^2 + 2x + 3$ **45.** $x^2 + 2x + 3$ units **47. a.** $\dfrac{30{,}000x^3 - 30{,}000}{x - 1}$

b. $30{,}000x^2 + 30{,}000x + 30{,}000$ **c.** \$94,575 **53.** b **55.** -3 **57.** Graphs coincide.

59. $2x + 3$ should be $2x + 23 + \dfrac{130}{x - 5}$. **61.** $x^2 - 2x + 3$ should be $x^2 + 2x + 3$. **62.**

$$2x - y \geq 4$$
$$x + y \leq -1$$

63. 1.2 **64.** -6

Section 6.7

Check Point Exercises

1. a. $\dfrac{1}{6^2} = \dfrac{1}{36}$ **b.** $\dfrac{1}{5^3} = \dfrac{1}{125}$ **c.** $\dfrac{1}{(-3)^4} = \dfrac{1}{81}$ **d.** $-\dfrac{1}{3^4} = -\dfrac{1}{81}$ **e.** $\dfrac{1}{8^1} = \dfrac{1}{8}$ **2. a.** $\dfrac{7^2}{2^3} = \dfrac{49}{8}$ **b.** $\dfrac{5^2}{4^2} = \dfrac{25}{16}$ **c.** $\dfrac{y^2}{7}$ **d.** $\dfrac{y^8}{x^1} = \dfrac{y^8}{x}$

3. $\dfrac{1}{x^{10}}$ **4. a.** $\dfrac{1}{x^8}$ **b.** $\dfrac{15}{x^6}$ **c.** $-\dfrac{2}{y^6}$ **5.** $\dfrac{36}{x^3}$ **6.** $\dfrac{1}{x^{20}}$ **7. a.** 7,400,000,000 **b.** 0.000003017 **8. a.** 7.41×10^9 **b.** 9.2×10^{-8}

9. a. 6×10^{10} **b.** 2.1×10^{11} **c.** 6.4×10^{-5} **10.** 5.2×10^5 mi

Exercise Set 6.7

1. $\dfrac{1}{8^2} = \dfrac{1}{64}$ **3.** $\dfrac{1}{5^3} = \dfrac{1}{125}$ **5.** $\dfrac{1}{(-6)^2} = \dfrac{1}{36}$ **7.** $-\dfrac{1}{6^2} = -\dfrac{1}{36}$ **9.** $\dfrac{1}{4^1} = \dfrac{1}{4}$ **11.** $\dfrac{1}{2^1} + \dfrac{1}{3^1} = \dfrac{1}{2} + \dfrac{1}{3} = \dfrac{5}{6}$ **13.** $3^2 = 9$ **15.** $(-3)^2 = 9$

17. $\dfrac{8^2}{2^3} = 8$ **19.** $\dfrac{4^2}{1^2} = 16$ **21.** $\dfrac{5^3}{3^3} = \dfrac{125}{27}$ **23.** $\dfrac{x^5}{6}$ **25.** $\dfrac{y^1}{x^8} = \dfrac{y}{x^8}$ **27.** $3 \cdot (-5)^3 = -375$ **29.** $\dfrac{1}{x^5}$ **31.** $\dfrac{8}{x^3}$ **33.** $\dfrac{1}{x^6}$ **35.** $\dfrac{1}{y^{99}}$

37. $\dfrac{3}{z^5}$ **39.** $-\dfrac{4}{x^4}$ **41.** $-\dfrac{1}{3a^3}$ **43.** $\dfrac{7}{5w^8}$ **45.** $\dfrac{1}{x^5}$ **47.** $\dfrac{1}{y^{11}}$ **49.** $\dfrac{16}{x^2}$ **51.** $216y^{17}$ **53.** $\dfrac{1}{x^6}$ **55.** $\dfrac{1}{16x^{12}}$ **57.** $\dfrac{x^2}{9}$ **59.** $-\dfrac{y^3}{8}$

61. $\dfrac{2x^6}{5}$ **63.** x^8 **65.** $16y^6$ **67.** $\dfrac{1}{y^2}$ **69.** $\dfrac{1}{y^{50}}$ **71.** $\dfrac{1}{a^{12}b^{15}}$ **73.** $\dfrac{a^8}{b^{24}}$ **75.** $\dfrac{4}{x^4}$ **77.** $\dfrac{y^9}{x^6}$ **79.** 870 **81.** 923,000 **83.** 3.4

85. 0.79 **87.** 0.0215 **89.** 0.000786 **91.** 3.24×10^4 **93.** 2.2×10^8 **95.** 7.13×10^2 **97.** 6.751×10^3 **99.** 2.7×10^{-3}

101. 2.02×10^{-5} **103.** 5×10^{-3} **105.** 3.14159×10^0 **107.** 6×10^5 **109.** 1.6×10^9 **111.** 3×10^4 **113.** 3×10^6 **115.** 3×10^{-6}

117. 9×10^4 **119.** 2.5×10^6 **121.** 1.25×10^8 **123.** 8.1×10^{-7} **125.** 2.5×10^{-7} **127.** 1 **129.** $\dfrac{y}{16x^8 z^6}$ **131.** $\dfrac{1}{x^{12}y^{16}z^{20}}$

133. $\dfrac{x^{18}y^6}{4}$ **135.** 2.5×10^{-3} **137.** 8×10^{-5} **139.** 9.2×10^3 **141.** 2.5×10^{-16} **143.** 6×10^8 **145.** 3.99×10^{11}

147. $\$4 \times 10^{11}$ **149.** $\$3.48 \times 10^{10}$ **151.** 1.25 sec **159.** b **167.** 225 deer **168.** 5 **169.** $0, \sqrt{16}$

Review Exercises

1. binomial, 4 **2.** trinomial, 2 **3.** monomial, 1 **4.** $8x^3 + 10x^2 - 20x - 4$ **5.** $13y^3 - 8y^2 + 7y - 5$ **6.** $11y^2 - 4y - 4$
7. $8x^4 - 5x^3 + 6$ **8.** $-14x^4 - 13x^2 + 16x$ **9.** $7y^4 - 5y^3 + 3y^2 - y - 4$ **10.** $3x^2 - 7x + 9$ **11.** $10x^3 - 9x^2 + 2x + 11$
12. 1451; 1451 per 100,000 is the death rate for men averaging 10 hr of sleep each night. **13.** x^{23} **14.** y^{14} **15.** x^{100}
16. $100y^2$ **17.** $-64x^{30}$ **18.** $50x^4$ **19.** $-36y^{11}$ **20.** $30x^{12}$ **21.** $21x^3 + 63x$ **22.** $20x^5 - 55x^4$ **23.** $-21y^4 + 9y^3 - 18y^2$
24. $16y^8 - 20y^7 + 2y^5$ **25.** $x^3 - 2x^2 - 13x + 6$ **26.** $12y^3 + y^2 - 21y + 10$ **27.** $3y^3 - 17y^2 + 41y - 35$
28. $8x^4 + 8x^3 - 18x^2 - 20x - 3$ **29.** $x^2 + 8x + 12$ **30.** $6y^2 - 7y - 5$ **31.** $4x^4 - 14x^2 + 6$ **32.** $25x^2 - 16$
33. $49 - 4y^2$ **34.** $y^4 - 1$ **35.** $x^2 + 6x + 9$ **36.** $9y^2 + 24y + 16$ **37.** $y^2 - 2y + 1$ **38.** $25y^2 - 20y + 4$
39. $x^4 + 8x^2 + 16$ **40.** $x^4 - 16$ **41.** $x^4 - x^2 - 20$ **42.** $x^2 + 7x + 12$ **43.** $x^2 + 50x + 600 \text{ yd}^2$ **44.** 28

45. polynomial degree: 5;

Term	Coefficient	Degree
$4x^2y$	4	3
$9x^3y^2$	9	5
$-17x^4$	-17	4
-12	-12	0

46. $-x^2 - 17xy + 5y^2$ **47.** $2x^3y^2 + x^2y - 6x^2 - 4$ **48.** $-35x^6y^9$ **49.** $15a^3b^5 - 20a^2b^3$ **50.** $3x^2 + 16xy - 35y^2$
51. $36x^2y^2 - 31xy + 3$ **52.** $9x^2 + 30xy + 25y^2$ **53.** $x^2y^2 - 14xy + 49$ **54.** $49x^2 - 16y^2$ **55.** $a^3 - b^3$ **56.** 6^{30} **57.** x^{15}
58. 1 **59.** -1 **60.** 400 **61.** $\dfrac{x^{12}}{8}$ **62.** $\dfrac{81}{16y^{24}}$ **63.** $-5y^6$ **64.** $8x^7y^3$ **65.** $3x^3 - 2x + 6$ **66.** $-6x^5 + 5x^4 + 8x^2$

67. $9x^2y - 3x - 6y$ **68.** $2x + 7$ **69.** $x^2 - 3x + 5$ **70.** $x^2 + 5x + 2 + \dfrac{7}{x - 7}$ **71.** $y^2 + 3y + 9$ **72.** $\dfrac{1}{7^2} = \dfrac{1}{49}$ **73.** $\dfrac{1}{(-4)^3} = -\dfrac{1}{64}$

74. $\dfrac{1}{2^1} + \dfrac{1}{4^1} = \dfrac{1}{2} + \dfrac{1}{4} = \dfrac{3}{4}$ **75.** $5^2 = 25$ **76.** $\dfrac{5^3}{2^3} = \dfrac{125}{8}$ **77.** $\dfrac{1}{x^6}$ **78.** $\dfrac{6}{y^2}$ **79.** $\dfrac{30}{x^5}$ **80.** x^8 **81.** $81y^2$ **82.** $\dfrac{1}{y^{19}}$

83. $\dfrac{x^3}{8}$ **84.** $\dfrac{1}{x^6}$ **85.** y^{20} **86.** 23,000 **87.** 0.00176 **88.** 0.9 **89.** 7.39×10^7 **90.** 6.2×10^{-4} **91.** 3.8×10^{-1} **92.** 3.8×10^0
93. 9×10^3 **94.** 5×10^4 **95.** 1.6×10^{-3} **96.** 1000 nanosec **97.** 1.26×10^{10} people

Chapter 6 Test

1. trinomial, 2 **2.** $13x^3 + x^2 - x - 24$ **3.** $5x^3 + 2x^2 + 2x - 9$ **4.** $-35x^{11}$ **5.** $48x^5 - 30x^3 - 12x^2$ **6.** $3x^3 - 10x^2 - 17x - 6$
7. $6y^2 - 13y - 63$ **8.** $49x^2 - 25$ **9.** $x^4 + 6x^2 + 9$ **10.** $25x^2 - 30x + 9$ **11.** 30 **12.** $2x^2y^3 + 3xy + 12y^2$
13. $12a^2 - 13ab - 35b^2$ **14.** $4x^2 + 12xy + 9y^2$ **15.** $-5x^{12}$ **16.** $3x^3 - 2x^2 + 5x$ **17.** $x^2 - 2x + 3 + \dfrac{1}{2x + 1}$
18. $\dfrac{1}{10^2} = \dfrac{1}{100}$ **19.** $4^3 = 64$ **20.** $-27x^6$ **21.** $\dfrac{4}{x^5}$ **22.** $-\dfrac{21}{x^6}$ **23.** $16y^4$ **24.** $\dfrac{x^8}{25}$ **25.** $\dfrac{1}{x^{15}}$ **26.** 0.00037
27. 7.6×10^6 **28.** 1.23×10^{-2} **29.** 2.1×10^8 **30.** $x^2 + 10x + 16$

Cumulative Review Exercises (Chapters 1–6)

1. $\dfrac{35}{9}$ **2.** -128 **3.** 15,050 ft **4.** $-\dfrac{2}{3}$ **5.** $-\dfrac{5}{3}$ **6.** 4 ft by 10 ft **7.** $\{x \mid x \geq 6\}$; **8.** 3 hr

9. 104 lb; 6 bags   **10.**   **11.**   **12.** $-\dfrac{6}{5}$; falling   **13.** $y + 1 = -2(x - 3); y = -2x + 5$ **14.** $(0, 5)$ **15.** $(3, -4)$

16. 500 min; \$40   **17.**   **18.** $3x^5 - 6x^3 + 9x + 2$

19. $x^2 + 2x + 3$ **20.** $\dfrac{81}{x^2}$

CHAPTER 7

Section 7.1

Check Point Exercises

1. a. $3x^2$ **b.** $4x^2$ **c.** x^2y **2.** $6(x^2 + 3)$ **3.** $5x^2(5 + 7x)$ **4.** $3x^3(5x^2 + 4x - 9)$ **5.** $2xy(4x^2y - 7x + 1)$
6. a. $(x + 1)(x^2 + 7)$ **b.** $(y + 4)(x - 7)$ **7.** $(x + 5)(x^2 + 2)$ **8.** $(y + 3)(x - 5)$

Exercise Set 7.1

1. Answers will vary; 3 examples are: $(2x)(4x^2)$, $(4x)(2x^2)$, and $(8x)(x^2)$.   **3.** Answers will vary; 3 examples are: $(-4x^3)(3x^2)$, $(2x^2)(-6x^3)$, and $(-3)(4x^5)$.   **5.** Answers will vary; 3 examples are: $(6x^2)(6x^2)$, $(-2x)(-18x^3)$, and $(4x^3)(9x)$.   **7.** 4   **9.** $4x$ **11.** $2x^3$ **13.** $3y$
15. xy **17.** $4x^4y^3$ **19.** $8(x + 1)$ **21.** $4(y - 1)$ **23.** $5(x + 6)$ **25.** $6(5x - 2)$ **27.** $x(x + 5)$ **29.** $6(3y^2 + 2)$
31. $7x^2(2x + 3)$ **33.** $y(13y - 25)$ **35.** $9y^4(1 + 3y^2)$ **37.** $4x^2(2 - x^2)$ **39.** $4(3y^2 + 4y - 2)$ **41.** $3x^2(3x^2 + 6x + 2)$
43. $50y^2(2y^3 - y + 2)$ **45.** $5x(2 - 4x + x^2)$ **47.** cannot be factored   **49.** $3xy(2x^2y + 3)$ **51.** $10xy(3xy^2 - y + 2)$
53. $8x^2y(4xy - 3x - 2)$ **55.** $(x + 5)(x + 3)$ **57.** $(x + 2)(x - 4)$ **59.** $(y + 6)(x - 7)$ **61.** $(x + y)(3x - 1)$
63. $(3x + 1)(4x + 1)$ **65.** $(5x + 4)(7x^2 + 1)$ **67.** $(x + 2)(x + 4)$ **69.** $(x - 5)(x + 3)$ **71.** $(x^2 + 5)(x - 2)$
73. $(x^2 + 2)(x - 1)$ **75.** $(y + 5)(x + 9)$ **77.** $(y - 1)(x + 5)$ **79.** $(x - 2y)(3x + 5y)$ **81.** $(3x - 2)(x^2 - 2)$ **83.** $(x - a)(x - b)$
85. $6x^2yz(4xy^2z^2 + 5y + 3z)$ **87.** $(x^3 - 4)(1 + 3y)$ **89.** $2x^2(x + 1)(2x^3 - 3x - 4)$ **91.** $(x - 1)(3x^4 + x^2 + 5)$
93. $36x^2 - 4\pi x^2; 4x^2(9 - \pi)$ **95. a.** 48 ft   **b.** $16x(4 - x)$ **c.** 48 ft; yes; no; Answers will vary.   **97.** $x^3 - 2$ **105.** d
109. $-3(x - 2)$ should be $-3(x + 2)$.   **111.** $x(x + 2) + 1$ should be $(x + 2)(x + 1)$.   **112.** $x^2 + 17x + 70$ **113.** $(-3, -2)$
114. $y - 2 = 1(x + 7)$, or $y - 5 = 1(x + 4); y = x + 9$

Section 7.2

Check Point Exercises

1. $(x + 2)(x + 3)$ **2.** $(x - 2)(x - 4)$ **3.** $(x + 5)(x - 2)$ **4.** $(y - 9)(y + 3)$ **5.** cannot factor over the integers; prime
6. $(x - 3y)(x - y)$ **7.** $2x(x - 4)(x + 7)$

Exercise Set 7.2

1. $(x + 6)(x + 1)$ **3.** $(x + 2)(x + 5)$ **5.** $(x + 1)(x + 10)$ **7.** $(x - 4)(x - 3)$ **9.** $(x - 6)(x - 6)$ **11.** $(y - 3)(y - 5)$
13. $(x + 5)(x - 2)$ **15.** $(y + 13)(y - 3)$ **17.** $(x - 5)(x + 3)$ **19.** $(x - 4)(x + 2)$ **21.** prime   **23.** $(y - 4)(y - 12)$ **25.** prime
27. $(w - 32)(w + 2)$ **29.** $(y - 5)(y - 13)$ **31.** $(r + 3)(r + 9)$ **33.** prime   **35.** $(x + 6y)(x + y)$ **37.** $(x - 3y)(x - 5y)$
39. $(x - 6y)(x + 3y)$ **41.** $(a - 15b)(a - 3b)$ **43.** $3(x + 2)(x + 3)$ **45.** $4(y - 2)(y + 1)$ **47.** $10(x - 10)(x + 6)$
49. $3(x - 2)(x - 9)$ **51.** $2r(r + 2)(r + 1)$ **53.** $4x(x + 6)(x - 3)$ **55.** $2r(r + 8)(r - 4)$ **57.** $y^2(y + 10)(y - 8)$
59. $x^2(x - 5)(x + 2)$ **61.** $2w^2(w - 16)(w + 3)$ **63.** $15x(y - 1)(y + 4)$ **65.** $x^3(x - y)(x + 4y)$ **67.** $2x^2(y - 15z)(y - z)$

69. $(a + b)(x + 5)(x - 4)$ **71.** $(x + 0.3)(x + 0.2)$ **73.** $\left(x - \dfrac{1}{5}\right)\left(x - \dfrac{1}{5}\right)$ **75.** $-(x + 8)(x - 5)$ **77. a.** $-16(t - 2)(t + 1)$

b. 0; yes; After 2 seconds, you hit the water.   **83.** c   **85.** 3, 4   **87.** $x(x + 1)(x + 2)$; the product of three consecutive integers
89. correctly factored   **91.** $(x + 1)(x - 1)$ should be $(x - 1)(x - 1)$.   **93.** $2x^2 - x - 6$ **94.** $9x^2 + 15x + 4$ **95.** 13

Section 7.3

Check Point Exercises

1. $(5x - 4)(x - 2)$ **2.** $(3x - 1)(2x + 7)$ **3.** $(3x - y)(x - 4y)$ **4.** $(3x + 5)(x - 2)$ **5.** $(2x - 1)(4x - 3)$ **6.** $y^2(5y + 3)(y + 2)$

Exercise Set 7.3

1. $(2x + 3)(x + 1)$ **3.** $(3x + 1)(x + 4)$ **5.** $(2x + 3)(x + 4)$ **7.** $(5y - 1)(y - 3)$ **9.** $(3y + 4)(y - 1)$ **11.** $(3x - 2)(x + 5)$
13. $(3x - 1)(x - 7)$ **15.** $(5y - 1)(y - 3)$ **17.** $(3x - 2)(x - 5)$ **19.** $(3w - 4)(2w - 1)$ **21.** $(8x + 1)(x + 4)$
23. $(5x - 2)(x + 7)$ **25.** $(7y - 3)(2y + 3)$ **27.** prime   **29.** $(5z - 3)(5z - 3)$ **31.** $(3y + 1)(5y - 2)$ **33.** prime
35. $(5y - 1)(2y + 9)$ **37.** $(4x + 1)(2x - 1)$ **39.** $(3y - 1)(3y - 2)$ **41.** $(5x + 8)(4x - 1)$ **43.** $(2x + y)(x + y)$
45. $(3x + 2y)(x + y)$ **47.** $(2x - 3y)(x - 3y)$ **49.** $(2x - 3y)(3x + 2y)$ **51.** $(3x - 2y)(5x + 7y)$ **53.** $(2a + 5b)(a + b)$
55. $(3a - 2b)(5a + 3b)$ **57.** $(3x - 4y)(4x - 3y)$ **59.** $2(2x + 3)(x + 5)$ **61.** $3(3x + 4)(x - 2)$ **63.** $2(2y - 5)(y + 3)$

65. $3(3y - 4)(y + 5)$ **67.** $x(3x + 1)(x + 1)$ **69.** $x(2x - 5)(x + 1)$ **71.** $3y(3y - 1)(y - 4)$ **73.** $5z(6z + 1)(2z + 1)$
75. $3x^2(5x - 3)(x - 2)$ **77.** $x^3(2x - 3)(5x - 1)$ **79.** $3(2x + 3y)(x - 2y)$ **81.** $2(2x - y)(3x + 4y)$ **83.** $2y(4x - 7)(x + 6)$
85. $2b(2a - 7b)(3a - b)$ **87.** $10(y + 1)(x + 1)(3x - 2)$ **89.** $-4y^4(8x + 3)(x - 1)$ **91. a.** $(2x + 1)(x - 3)$ **b.** $(2y + 3)(y - 2)$
93. $(x - 2)(3x - 2)(x - 1)$ **95. a.** $x^2 + 3x + 2$ **b.** $(x + 2)(x + 1)$ **c.** $x^2 + 3x + 2 = (x + 2)(x + 1)$ **101.** a   **103.** $5, 7, -5, -7$
105. $(2x^n + 1)(x^n - 4)$ **106.** $81x^2 - 100$ **107.** $16x^2 + 40xy + 25y^2$ **108.** $x^3 + 8$

Mid-Chapter 7 Check Point

1. $x^4(x + 1)$ **2.** $(x + 9)(x - 2)$ **3.** $x^2y(y^2 - y + 1)$ **4.** prime   **5.** $(7x - 1)(x - 3)$ **6.** $(x^2 + 3)(x + 5)$ **7.** $x(2x - 1)(x - 5)$
8. $(x - 4)(y - 7)$ **9.** $(x - 15y)(x - 2y)$ **10.** $(5x + 2)(5x - 7)$ **11.** $2(8x - 3)(x - 4)$ **12.** $(3x + 7y)(x + y)$

Section 7.4

Check Point Exercises

1. a. $(x + 9)(x - 9)$ **b.** $(6x + 5)(6x - 5)$ **2. a.** $(5 + 2x^5)(5 - 2x^5)$ **b.** $(10x + 3y)(10x - 3y)$ **3. a.** $2x(3x + 1)(3x - 1)$
b. $18(2 + x)(2 - x)$ **4.** $(9x^2 + 4)(3x + 2)(3x - 2)$ **5. a.** $(x + 7)^2$ **b.** $(x - 3)^2$ **c.** $(4x - 7)^2$ **6.** $(2x + 3y)^2$
7. $(x + 3)(x^2 - 3x + 9)$ **8.** $(1 - y)(1 + y + y^2)$ **9.** $(5x + 2)(25x^2 - 10x + 4)$

Exercise Set 7.4

1. $(x + 5)(x - 5)$ **3.** $(y + 1)(y - 1)$ **5.** $(2x + 3)(2x - 3)$ **7.** $(5 + x)(5 - x)$ **9.** $(1 + 7x)(1 - 7x)$ **11.** $(3 + 5y)(3 - 5y)$
13. $(x^2 + 3)(x^2 - 3)$ **15.** $(7y^2 + 4)(7y^2 - 4)$ **17.** $(x^5 + 3)(x^5 - 3)$ **19.** $(5x + 4y)(5x - 4y)$ **21.** $(x^2 + y^5)(x^2 - y^5)$
23. $(x^2 + 4)(x + 2)(x - 2)$ **25.** $(4x^2 + 9)(2x + 3)(2x - 3)$ **27.** $2(x + 3)(x - 3)$ **29.** $2x(x + 6)(x - 6)$ **31.** prime
33. $3x(x^2 + 9)$ **35.** $2(3 + y)(3 - y)$ **37.** $3y(y + 4)(y - 4)$ **39.** $2x(3x + 1)(3x - 1)$ **41.** $(x + 1)^2$ **43.** $(x - 7)^2$ **45.** $(x - 1)^2$
47. $(x + 11)^2$ **49.** $(2x + 1)^2$ **51.** $(5y - 1)^2$ **53.** prime   **55.** $(x + 7y)^2$ **57.** $(x - 6y)^2$ **59.** prime   **61.** $(4x - 5y)^2$
63. $3(2x - 1)^2$ **65.** $x(3x + 1)^2$ **67.** $2(y - 1)^2$ **69.** $2y(y + 7)^2$ **71.** $(x + 1)(x^2 - x + 1)$ **73.** $(x - 3)(x^2 + 3x + 9)$
75. $(2y - 1)(4y^2 + 2y + 1)$ **77.** $(3x + 2)(9x^2 - 6x + 4)$ **79.** $(xy - 4)(x^2y^2 + 4xy + 16)$ **81.** $y(3y + 2)(9y^2 - 6y + 4)$

83. $2(3 - 2y)(9 + 6y + 4y^2)$ **85.** $(4x + 3y)(16x^2 - 12xy + 9y^2)$ **87.** $(5x - 4y)(25x^2 + 20xy + 16y^2)$ **89.** $\left(5x + \dfrac{2}{7}\right)\left(5x - \dfrac{2}{7}\right)$

91. $y\left(y - \dfrac{1}{10}\right)\left(y^2 + \dfrac{y}{10} + \dfrac{1}{100}\right)$ **93.** $x(0.5 + x)(0.5 - x)$ **95.** $(x + 6)(x - 4)$ **97.** $(x - 3)(x + 1)^2$ **99.** $x^2 - 25 = (x + 5)(x - 5)$
101. $x^2 - 16 = (x + 4)(x - 4)$ **107.** b   **109.** $(x + y)(x - y + 3)$ **111.** $(2x^n + 3)^2$ **113.** $6, -6$
115. $(4x + 3)(4x - 3)$ should be $(2x + 3)(2x - 3)$.   **117.** $(4x - 1)^2$ should be $(2x - 1)^2$.   **119.** $80x^9y^{14}$ **120.** $-4x^2 + 3$ **121.** $2x + 5$

Section 7.5

Check Point Exercises

1. $5x^2(x + 3)(x - 3)$ **2.** $4(x - 6)(x + 2)$ **3.** $4x(x^2 + 4)(x + 2)(x - 2)$ **4.** $(x - 4)(x + 3)(x - 3)$ **5.** $3x(x - 5)^2$
6. $2x^2(x + 3)(x^2 - 3x + 9)$ **7.** $3y(x^2 + 4y^2)(x + 2y)(x - 2y)$ **8.** $3x(2x + 3y)^2$

Exercise Set 7.5

1. $5x(x + 2)(x - 2)$ **3.** $7x(x^2 + 1)$ **5.** $5(x - 3)(x + 2)$ **7.** $2(x^2 + 9)(x + 3)(x - 3)$ **9.** $(x + 2)(x + 3)(x - 3)$ **11.** $3x(x - 4)^2$
13. $2x^2(x + 1)(x^2 - x + 1)$ **15.** $2x(3x + 4)$ **17.** $2(y - 8)(y + 7)$ **19.** $7y^2(y + 1)^2$ **21.** prime   **23.** $2(4y + 1)(2y - 1)$ **25.** $r(r - 25)$
27. $(2w + 5)(2w - 1)$ **29.** $x(x + 2)(x - 2)$ **31.** prime   **33.** $(9y + 4)(y + 1)$ **35.** $(y + 2)(y + 2)(y - 2)$ **37.** $(4y + 3)^2$
39. $4y(y - 5)(y - 2)$ **41.** $y(y^2 + 9)(y + 3)(y - 3)$ **43.** $5a^2(2a + 3)(2a - 3)$ **45.** $(4y - 1)(3y - 2)$ **47.** $(3y + 8)(3y - 8)$
49. prime   **51.** $(2y + 3)(y + 5)(y - 5)$ **53.** $2r(r + 17)(r - 2)$ **55.** $2x^3(2x + 1)(2x - 1)$ **57.** $3(x^2 + 81)$ **59.** $x(x + 2)(x^2 - 2x + 4)$
61. $2y^2(y - 1)(y^2 + y + 1)$ **63.** $2x(3x + 4y)$ **65.** $(y - 7)(x + 3)$ **67.** $(x - 4y)(x + y)$ **69.** $12a^2(6ab^2 + 1 - 2a^2b^2)$
71. $3(a + 6b)(a + 3b)$ **73.** $3x^2y(4x + 1)(4x - 1)$ **75.** $b(3a + 2)(2a - 1)$ **77.** $7xy(x^2 + y^2)(x + y)(x - y)$ **79.** $2xy(5x - 2y)(x - y)$
81. $2b(x + 11)^2$ **83.** $(5a + 7b)(3a - 2b)$ **85.** $2xy(9x - 2y)(2x - 3y)$ **87.** $(y - x)(a + b)(a - b)$ **89.** $ax(3x + 7)(3x - 2)$
91. $y(9x^2 + y^2)(3x + y)(3x - y)$ **93.** $(x + 1)(5x - 6)(2x + 1)$ **95.** $(x^2 + 6)(6x^2 - 1)$ **97.** $(x - 7 + 2a)(x - 7 - 2a)$
99. $(x + 4 + 5a)(x + 4 - 5a)$ **101.** $y(y^2 + 1)(y^4 - y^2 + 1)$ **103.** $16(4 + t)(4 - t)$ **105.** $\pi b^2 - \pi a^2; \pi(b + a)(b - a)$ **109.** d
111. $5y^2(y - 1)(y + 2)(y - 2)$ **113.** $(x - 5)^2$ **115.** $(4x - 3)^2$ should be $(2x - 3)^2$.   **117.** correctly factored

119. $2(x + 5)(x^2 + 1)$ should be $2(x + 5)(x + 1)(x - 1)$.   **120.** $(3x + 4)(3x - 4)$ **121.**   **122.** $20°, 60°, 100°$

Section 7.6

Check Point Exercises

1. $-\dfrac{1}{2}$ and 4 **2.** 5 and 1 **3.** 0 and $\dfrac{1}{2}$ **4.** 5 **5.** $-\dfrac{5}{4}$ and $\dfrac{5}{4}$ **6.** -2 and 9 **7.** 1 sec and 2 sec; $(1, 192)$ and $(2, 192)$ **8.** length: 9 ft; width: 6 ft

Exercise Set 7.6

1. 0 and -7 **3.** 6 and -4 **5.** 9 and $-\dfrac{4}{5}$ **7.** 4 and $-\dfrac{9}{2}$ **9.** -5 and -3 **11.** -3 and 5 **13.** -3 and 7 **15.** -8 and -1

17. -4 and 0 **19.** 0 and 5 **21.** 0 and 4 **23.** 0 and $\dfrac{5}{2}$ **25.** $-\dfrac{5}{3}$ and 0 **27.** -2 **29.** 6 **31.** $\dfrac{3}{2}$ **33.** $-\dfrac{1}{2}$ and 4 **35.** -2 and $\dfrac{9}{5}$

37. -7 and 7 **39.** $-\dfrac{5}{2}$ and $\dfrac{5}{2}$ **41.** $-\dfrac{5}{9}$ and $\dfrac{5}{9}$ **43.** -3 and 7 **45.** $-\dfrac{5}{2}$ and $\dfrac{3}{2}$ **47.** -6 and 3 **49.** -2 and $-\dfrac{3}{2}$ **51.** 4 **53.** $-\dfrac{5}{2}$

55. $\dfrac{3}{8}$ **57.** $-3, -2,$ and 4 **59.** $0, -6,$ and 6 **61.** $-2, -1,$ and 0 **63.** 2 and 8 **65.** 4 and 5 **67.** 5 sec; Each tick represents one second.

69. 2 sec; (2, 276) **71.** $\dfrac{1}{2}$ sec and 4 sec **73.** 1995; (15, 1100) **75.** 10 yr **77.** (10, 7250) **79.** 10 teams **81.** length: 15 yd; width: 12 yd

83. base: 5 cm; height: 6 cm **85. a.** $4x^2 + 44x$ **b.** 3 ft **89.** d **91.** $-4, 1,$ and 4 **93.** 4 and 1 **95.** a **97.** b **99.** -3 and 2

101. 1 **104.** $y > -\dfrac{2}{3}x + 1$ **105.** $\dfrac{4}{x^6}$ **106.** -2

Review Exercises

1. $15(2x - 3)$ **2.** $4x(3x^2 + 4x - 100)$ **3.** $5x^2y(6x^2 + 3x + 1)$ **4.** $5(x + 3)$ **5.** $(7x^2 - 1)(x + y)$ **6.** $(x^2 + 2)(x + 3)$
7. $(x + 1)(y + 4)$ **8.** $(x^2 + 5)(x + 1)$ **9.** $(x - 2)(y + 4)$ **10.** $(x - 2)(x - 1)$ **11.** $(x - 5)(x + 4)$ **12.** $(x + 3)(x + 16)$
13. $(x - 4y)(x - 2y)$ **14.** prime **15.** $(x + 17y)(x - y)$ **16.** $3(x + 4)(x - 2)$ **17.** $3x(x - 11)(x - 1)$ **18.** $(x + 5)(3x + 2)$
19. $(y - 3)(5y - 2)$ **20.** $(2x + 5)(2x - 3)$ **21.** prime **22.** $2(2x + 3)(2x - 1)$ **23.** $x(2x - 9)(x + 8)$ **24.** $4y(3y + 1)(y + 2)$
25. $(2x - y)(x - 3y)$ **26.** $(5x + 4y)(x - 2y)$ **27.** $(2x + 1)(2x - 1)$ **28.** $(9 + 10y)(9 - 10y)$ **29.** $(5a + 7b)(5a - 7b)$
30. $(z^2 + 4)(z + 2)(z - 2)$ **31.** $2(x + 3)(x - 3)$ **32.** prime **33.** $x(3x + 1)(3x - 1)$ **34.** $2x(3y + 2)(3y - 2)$ **35.** $(x + 11)^2$
36. $(x - 8)^2$ **37.** $(3y + 8)^2$ **38.** $(4x - 5)^2$ **39.** prime **40.** $(6x + 5y)^2$ **41.** $(5x - 4y)^2$ **42.** $(x - 3)(x^2 + 3x + 9)$
43. $(4x + 1)(16x^2 - 4x + 1)$ **44.** $2(3x - 2y)(9x^2 + 6xy + 4y^2)$ **45.** $y(3x + 2)(9x^2 - 6x + 4)$ **46.** $(a + 3)(a - 3)$
47. $(a + 2b)(a - 2b)$ **48.** $A^2 + 2A + 1 = (A + 1)^2$ **49.** $x(x - 7)(x - 1)$ **50.** $(5y + 2)(2y + 1)$ **51.** $2(8 + y)(8 - y)$
52. $(3x + 1)^2$ **53.** $4x^3(5x^4 - 9)$ **54.** $(x - 3)^2(x + 3)$ **55.** prime **56.** $x(2x + 5)(x + 7)$ **57.** $3x(x - 5)^2$
58. $3x^2(x - 2)(x^2 + 2x + 4)$ **59.** $4y^2(y + 3)(y - 3)$ **60.** $5(x + 7)(x - 3)$ **61.** prime **62.** $2x^3(5x - 2)(x - 4)$
63. $(10y + 7)(10y - 7)$ **64.** $9x^4(x - 2)$ **65.** $(x^2 + 1)(x + 1)(x - 1)$ **66.** $2(y - 2)(y^2 + 2y + 4)$ **67.** $(x + 4)(x^2 - 4x + 16)$
68. $(3x - 2)(2x + 5)$ **69.** $3x^2(x + 2)(x - 2)$ **70.** $(x - 10)(x + 9)$ **71.** $(5x + 2y)(5x + 3y)$ **72.** $x(x + 5)(x^2 - 5x + 25)$
73. $2y(4y + 3)(4y + 1)$ **74.** $2(y - 4)^2$ **75.** $(x + 5y)(x - 7y)$ **76.** $(x + y)(x + 7)$ **77.** $(3x + 4y)^2$ **78.** $2x^2y(x + 1)(x - 1)$
79. $(10y + 7z)(10y - 7z)$ **80.** prime **81.** $3x^2y^2(x + 2y)(x - 2y)$ **82.** 0 and 12 **83.** 7 and $-\dfrac{9}{4}$ **84.** -7 and 2 **85.** -4 and 0
86. -8 and $\dfrac{1}{2}$ **87.** -4 and 8 **88.** -8 and 7 **89.** 7 **90.** $-\dfrac{10}{3}$ and $\dfrac{10}{3}$ **91.** -5 and -2 **92.** $\dfrac{1}{3}$ and 7 **93.** 2 sec
94. width: 5 ft; length: 8 ft **95.** 11 m by 11 m

Chapter 7 Test

1. $(x - 3)(x - 6)$ **2.** $(x - 7)^2$ **3.** $5y^2(3y - 1)(y - 2)$ **4.** $(x^2 + 3)(x + 2)$ **5.** $x(x - 9)$ **6.** $x(x + 7)(x - 1)$
7. $2(7x - 3)(x + 5)$ **8.** $(5x + 3)(5x - 3)$ **9.** $(x + 2)(x^2 - 2x + 4)$ **10.** $(x + 3)(x - 7)$ **11.** prime **12.** $3y(2y + 1)(y + 1)$
13. $4(y + 3)(y - 3)$ **14.** $4(2x + 3)^2$ **15.** $2(x^2 + 4)(x + 2)(x - 2)$ **16.** $(6x - 7)^2$ **17.** $(7x - 1)(x - 7)$ **18.** $(x^2 - 5)(x + 2)$
19. $3y(2y + 3)(2y - 5)$ **20.** $(y - 5)(y^2 + 5y + 25)$ **21.** $5(x - 3y)(x + 2y)$ **22.** -6 and 4 **23.** $-\dfrac{1}{3}$ and 2 **24.** -2 and 8
25. 0 and $\dfrac{7}{2}$ **26.** $-\dfrac{9}{4}$ and $\dfrac{9}{4}$ **27.** -1 and $\dfrac{6}{5}$ **28.** $x^2 - 4 = (x + 2)(x - 2)$ **29.** 6 sec **30.** width: 5 ft; length: 11 ft

Cumulative Review Exercises (Chapters 1–7)

1. -48 **2.** 0 **3.** 12 **4.** $\{x | x < -2\}$; **5.** $38°, 38°, 104°$ **6.** $150

7. **8.** $y + 4 = 5(x - 2)$, or $y - 1 = 5(x - 3)$; $y = 5x - 14$ **9.** **10.** $(3, -1)$ **11.** $\left(-\dfrac{2}{13}, \dfrac{23}{13}\right)$

12. $-\dfrac{13}{40}$ **13.** $2x^4 - x^3 + 3x + 9$

14. $6x^2 + 17xy - 45y^2$

15. $2x^2 + 5x - 3 - \dfrac{2}{3x - 5}$ **16.** 7.1×10^{-3} **17.** $(3x + 2)(x + 3)$ **18.** $y(y^2 + 4)(y + 2)(y - 2)$ **19.** $(2x + 3)^2$ **20.** width: 4 ft; length: 6 ft

CHAPTER 8

Section 8.1

Check Point Exercises

1. a. $x = 5$ **b.** $x = -7$ and $x = 4$ **2.** $\dfrac{x + 4}{3x}$ **3.** $\dfrac{x^2}{7}$ **4.** $\dfrac{x - 1}{x + 1}$ **5.** $-\dfrac{3x + 7}{4}$

Exercise Set 8.1

1. $x = 0$ **3.** $x = 8$ **5.** $x = 4$ **7.** $x = -9$ and $x = 2$ **9.** $x = \dfrac{17}{3}$ and $x = -3$ **11.** $x = -4$ and $x = 3$

13. defined for all real numbers   **15.** $y = -1$ and $y = \dfrac{3}{4}$ **17.** $y = -5$ and $y = 5$ **19.** defined for all real numbers   **21.** $2x$

23. $\dfrac{x - 3}{5}$ **25.** $\dfrac{x - 4}{2x}$ **27.** $\dfrac{1}{x - 3}$ **29.** $-\dfrac{5}{x - 3}$ **31.** 3 **33.** $\dfrac{1}{x - 5}$ **35.** $\dfrac{2}{3}$ **37.** $\dfrac{1}{x - 3}$ **39.** $\dfrac{4}{x - 2}$ **41.** $\dfrac{y - 1}{y + 9}$

43. $\dfrac{y - 3}{y - 2}$ **45.** cannot be simplified   **47.** $\dfrac{x + 6}{x - 6}$ **49.** $x^2 + 1$ **51.** $x^2 + 2x + 4$ **53.** $\dfrac{x - 4}{x + 4}$ **55.** cannot be simplified

57. cannot be simplified   **59.** -1 **61.** -1 **63.** cannot be simplified   **65.** -2 **67.** $-\dfrac{2}{x}$ **69.** $-x - 1$ **71.** $-y - 3$ **73.** $-\dfrac{1}{x}$

75. $\dfrac{x - y}{2x - y}$ **77.** $\dfrac{x - 6}{x^2 + 3x + 9}$ **79.** $\dfrac{3 + y}{3 - y}$ **81.** $\dfrac{y + 3}{x + 3}$ **83.** $\dfrac{2}{1 - 2x}$ **85. a.** It costs \$86.67 million to inoculate 40% of the population.
It costs \$520 million to inoculate 80% of the population. It costs \$1170 million to inoculate 90% of the population.   **b.** $x = 100$
c. The cost keeps rising; No amount of money will be enough to inoculate 100% of the population.   **87.** 400 mg   **89. a.** \$300   **b.** \$125

c. decrease   **91.** 1.5 mg per liter; $(3, 1.5)$ **93.** 20.8; no   **95. a.** $\dfrac{-0.4t + 14.2}{3.7t + 257.4}$ **b.** $0.04; 4000$ **c.** fairly well; Answers will vary.

105. $\dfrac{x^2 - x - 6}{x + 2}$ **107.** $2x^2 - 1$ should be $2x + 1$.   **110.** $\dfrac{3}{10}$ **111.** $\dfrac{1}{6}$ **112.** $(4, 2)$

Section 8.2

Check Point Exercises

1. $\dfrac{9x - 45}{2x + 8}$ **2.** $\dfrac{3}{8}$ **3.** $\dfrac{x + 2}{9}$ **4.** $-\dfrac{5(x + 1)}{7x(2x - 3)}$ **5.** $\dfrac{(x + 3)(x + 7)}{x - 4}$ **6.** $\dfrac{x + 3}{x - 5}$ **7.** $\dfrac{y + 1}{5y(y^2 + 1)}$

Exercise Set 8.2

1. $\dfrac{4x - 20}{9x + 27}$ **3.** $\dfrac{4x}{x + 5}$ **5.** $\dfrac{4}{5}$ **7.** $\dfrac{4}{9}$ **9.** 1 **11.** $\dfrac{x + 5}{x}$ **13.** $\dfrac{2}{y}$ **15.** $\dfrac{y + 2}{y + 4}$ **17.** $4(y + 3)$ **19.** $\dfrac{x - 1}{x + 2}$ **21.** $\dfrac{x^2 + 2x + 4}{3x}$

23. $\dfrac{x - 2}{x - 1}$ **25.** $-\dfrac{2}{x(x + 1)}$ **27.** $-\dfrac{y - 10}{y - 7}$ **29.** $(x - y)(x + y)$ **31.** $\dfrac{4(x + y)}{3(x - y)}$ **33.** $\dfrac{3x}{35}$ **35.** $\dfrac{1}{4}$ **37.** 10 **39.** $\dfrac{7}{9}$ **41.** $\dfrac{3}{4}$

43. $\dfrac{(x - 2)^2}{x}$ **45.** $\dfrac{(y - 4)(y^2 + 4)}{y - 1}$ **47.** $\dfrac{y}{3}$ **49.** $\dfrac{2(x + 3)}{3}$ **51.** $\dfrac{x - 5}{2}$ **53.** $\dfrac{y^2 + 1}{y^2}$ **55.** $\dfrac{y + 1}{y - 7}$ **57.** 6 **59.** $\dfrac{2}{3}$

61. $\dfrac{(x + y)^2}{32(x - y)^2}$ **63.** $\dfrac{y(x + y)}{(x + 1)^2}$ **65.** $\dfrac{y + 3}{y - 3}$ **67.** $\dfrac{5(x - 5)}{4}$ **69.** $\dfrac{(x + y)^2}{x - y}$ **71.** $\dfrac{2}{y^2 - by + b^2}$ **73.** $\dfrac{125x}{100 - x}$

79. numerator: x; denominator: $x - 4$ **81.** $(3x - y)(3x + y)$ **83.** correct answer   **85.** $2x - 1$ should be $2x + 1$.   **86.** $\{x \mid x > 18\}$

87. $3(x - 7)(x + 2)$ **88.** $\dfrac{1}{2}$ and -5

Section 8.3

Check Point Exercises

1. $x + 2$ **2.** $\dfrac{x - 5}{x + 5}$ **3. a.** $\dfrac{3x + 5}{x + 7}$ **b.** $3x - 4$ **4.** $\dfrac{3y + 2}{y - 4}$ **5.** $x + 3$ **6.** $\dfrac{-4x^2 + 12x}{x^2 - 2x - 9}$

Exercise Set 8.3

1. $\dfrac{9x}{13}$ **3.** $\dfrac{3x}{5}$ **5.** $\dfrac{x + 3}{2}$ **7.** $\dfrac{6}{x}$ **9.** $\dfrac{7}{3x}$ **11.** $\dfrac{9}{x + 3}$ **13.** $\dfrac{5x + 5}{x - 3}$ **15.** 2 **17.** $\dfrac{2y + 3}{y - 5}$ **19.** $\dfrac{7}{5y}$ **21.** $\dfrac{2}{x + 2}$ **23.** $\dfrac{x - 2}{x - 3}$

25. $\dfrac{3x - 4}{5x - 4}$ **27.** 1 **29.** 1 **31.** $\dfrac{x}{2x - 1}$ **33.** $-\dfrac{1}{y}$ **35.** $\dfrac{y + 3}{3y + 8}$ **37.** $\dfrac{3y + 2}{y - 3}$ **39.** $\dfrac{2}{x - 3}$ **41.** $\dfrac{3x + 7}{x - 6}$ **43.** $\dfrac{3x + 1}{3x - 4}$

45. $x + 2$ **47.** 0   **49.** $\dfrac{11}{x-1}$ **51.** $\dfrac{12}{x+3}$ **53.** $\dfrac{y+1}{y-1}$ **55.** $\dfrac{x-2}{x-7}$ **57.** $\dfrac{2x-4}{x^2-25}$ **59.** 1   **61.** $\dfrac{2}{x+y}$ **63.** $\dfrac{x-3}{x-1}$

65. $\dfrac{4b}{4b-3}$ **67.** $\dfrac{y}{y-5}$ **69.** $-\dfrac{1}{c+d}$ **71.** $\dfrac{3y^2+8y-5}{(y+1)(y-4)}$ **73. a.** $\dfrac{100W}{L}$ **b.** round   **75.** 10 m   **81.** d   **83.** $x-2$

85. $20x - 6$ **87.** $a + 12$ **89.** correct answer   **91.** $x + 4$ should be $x - 4$.   **92.** $\dfrac{31}{45}$ **93.** $(9x^2 + 1)(3x + 1)(3x - 1)$

94. $3x^2 - 7x - 5$

Section 8.4

Check Point Exercises

1. $30x^2$ **2.** $(x+3)(x-3)$ **3.** $7x(x+4)(x+4)$ or $7x(x+4)^2$ **4.** $\dfrac{9+14x}{30x^2}$ **5.** $\dfrac{6x+6}{(x+3)(x-3)}$ **6.** $-\dfrac{5}{x+5}$ **7.** $-\dfrac{5+y}{5y}$

8. $\dfrac{x-15}{(x+5)(x-5)}$

Exercise Set 8.4

1. $120x^2$ **3.** $30x^5$ **5.** $(x-3)(x+1)$ **7.** $7y(y+2)$ **9.** $(x+4)(x-4)$ **11.** $y(y+3)(y-3)$ **13.** $(y+1)(y-1)(y-1)$

15. $(x-5)(x+4)(2x-1)$ **17.** $\dfrac{3x+5}{x^2}$ **19.** $\dfrac{37}{18x}$ **21.** $\dfrac{8x+7}{2x^2}$ **23.** $\dfrac{6x+1}{x}$ **25.** $\dfrac{2+9x}{x}$ **27.** $\dfrac{x+1}{2}$ **29.** $\dfrac{7x-20}{x(x-5)}$

31. $\dfrac{5x+1}{(x-1)(x+2)}$ **33.** $\dfrac{11y+15}{4y(y+5)}$ **35.** $-\dfrac{7}{x+7}$ **37.** $\dfrac{3x-55}{(x+5)(x-5)}$ **39.** $\dfrac{x^2+6x}{(x-4)(x+4)}$ **41.** $\dfrac{y+12}{(y+3)(y-3)}$

43. $\dfrac{7x-10}{(x-1)(x-1)}$ **45.** $\dfrac{9y}{4(y-5)}$ **47.** $\dfrac{8y+16}{y(y+4)}$ **49.** $\dfrac{17}{(x-3)(x-4)}$ **51.** $\dfrac{7x-1}{(x+1)(x+1)(x-1)}$ **53.** $\dfrac{x^2-x}{(x+3)(x-2)(x+5)}$

55. $\dfrac{y^2+8y+4}{(y+4)(y+1)(y+1)}$ **57.** $\dfrac{2x^2-4x+34}{(x+3)(x-5)}$ **59.** $\dfrac{5-3y}{2y(y-1)}$ **61.** $-\dfrac{x^2}{(x+3)(x-3)}$ **63.** $\dfrac{2}{y+5}$ **65.** $\dfrac{4x-11}{x-3}$ **67.** $\dfrac{3}{y+1}$

69. $\dfrac{-x^2+7x+3}{(x-3)(x+2)}$ **71.** $\dfrac{x^2+2x-1}{(x+1)(x-1)(x+2)}$ **73.** $\dfrac{-y^2+8y+9}{15y^2}$ **75.** $\dfrac{x^2-2x-6}{3(x+2)(x-2)}$ **77.** $\dfrac{-y-2}{(y+1)(y-1)}$

79. $\dfrac{x+2xy-y}{xy}$ **81.** $\dfrac{5x+2y}{(x+y)(x-y)}$ **83.** $\dfrac{-x+6}{(x+2)(x-2)}$ **85.** $\dfrac{6x+5}{(x-5)(x+5)(x-6)}$ **87.** $\dfrac{9}{(x-3)(x^2+3x+9)}$

89. $\dfrac{3}{y-3}$ **91.** $-\dfrac{22y}{(x+3y)(x+y)(x-3y)}$ **93.** $\dfrac{D}{5}$ **95.** $\dfrac{D}{24}$ **97.** no   **99.** 5 yr   **101.** $\dfrac{4x^2+14x}{(x+3)(x+4)}$

105. $\dfrac{3}{x+5}$ should be $\dfrac{5+2x}{5x}$.   **109.** $\dfrac{y^3+y^2-10y-2}{(y-1)(y-1)(y+3)}$ **111.** $\dfrac{3x-1}{x^2(x-1)}$ **113.** $6x^2 - 11x - 35$

114. $3x - y < 3$

115. $y = x - 1$

Mid-Chapter 8 Check Point

1. $x = -2$ and $x = 4$ **2.** $\dfrac{x-2}{2x+1}$ **3.** $\dfrac{-3}{y-2}$ or $\dfrac{3}{2-y}$ **4.** $\dfrac{2}{w}$ **5.** $\dfrac{4}{x+4}$ **6.** $\dfrac{4}{(x-2)^2}$ **7.** $\dfrac{x+5}{x-2}$ **8.** $\dfrac{x+2}{x-4}$

9. $\dfrac{2x+1}{(x+2)(x+3)(x-1)}$ **10.** $\dfrac{9-x}{x-5}$ **11.** $-\dfrac{2y-1}{3y}$ or $\dfrac{1-2y}{3y}$ **12.** $\dfrac{-y^2}{(y+1)(y+2)}$ **13.** $\dfrac{w+1}{(w-3)(w+5)}$

14. $\dfrac{2z^2-3z+15}{(z+3)(z-3)(z+1)}$ **15.** $\dfrac{3z^2+5z+3}{(3z-1)^2}$ **16.** $\dfrac{3x-1}{(x+7)(x-3)}$ **17.** x **18.** $\dfrac{2x+1}{(x+1)(x-2)(x+3)}$ **19.** $\dfrac{5x-5y}{x}$ **20.** $\dfrac{x+3}{x+5}$

Section 8.5

Check Point Exercises

1. $\dfrac{11}{5}$ **2.** $\dfrac{2x-1}{2x+1}$ **3.** $y - x$ **4.** $\dfrac{11}{5}$ **5.** $\dfrac{2x-1}{2x+1}$ **6.** $y - x$

Exercise Set 8.5

1. $\dfrac{9}{10}$ **3.** $\dfrac{18}{23}$ **5.** $-\dfrac{4}{5}$ **7.** $\dfrac{3-4x}{3+4x}$ **9.** $\dfrac{7x-2}{5x+1}$ **11.** $\dfrac{2y+3}{y-7}$ **13.** $\dfrac{4-6y}{4+3y}$ **15.** $x-5$ **17.** $\dfrac{x}{x-1}$ **19.** $-\dfrac{1}{y}$

21. $\dfrac{xy+2}{x}$ **23.** $\dfrac{y+x}{x^2y^2}$ **25.** $\dfrac{x^2+y}{y(y+1)}$ **27.** $\dfrac{1}{y}$ **29.** $\dfrac{4-x}{5x-3}$ **31.** $\dfrac{2y}{y-3}$ **33.** $\dfrac{1}{x+3}$ **35.** $\dfrac{x^2-5x+3}{x^2-7x+2}$ **37.** $\dfrac{2y^2+3xy}{y^2+2x}$

39. $-\dfrac{6}{5}$ **41.** $\dfrac{1-x}{(x-3)(x+6)}$ **43.** $\dfrac{1}{y(y+5)}$ **45.** $\dfrac{1}{x-1}$ **47.** $\dfrac{x+1}{2x+1}$ **49.** $\dfrac{2r_1r_2}{r_1+r_2}; 34\dfrac{2}{7}$ mph **55.** It cubes x.

57. $\dfrac{5y+3}{(y-3)(y+1)}$ **59.** 2 should be $1+x$. **61.** $2x(x-5)^2$ **62.** -2 **63.** x^3+y^3

Section 8.6

Check Point Exercises

1. 4 **2.** 3 **3.** -3 and -2 **4.** 24 **5.** no solution **6.** 75%

Exercise Set 8.6

1. 12 **3.** 0 **5.** -2 **7.** 6 **9.** -2 **11.** 2 **13.** 15 **15.** 4 **17.** -6 and -1 **19.** -5 and 5 **21.** -3 and 3 **23.** -8 and 1

25. -1 and 16 **27.** 3 **29.** no solution **31.** $\dfrac{2}{3}$ and 4 **33.** no solution **35.** 1 **37.** $\dfrac{1}{5}$ **39.** -3 **41.** no solution

43. -6 and $-\dfrac{1}{2}$ **45.** -3 **47.** $\dfrac{-6x-18}{(x-5)(x+4)}$ **49.** 1 and 5 **51.** 3 **53.** 10,000 wheelchairs **55.** 50% **57.** 5 yr old

59. either 50 or approx 67 cases; (50, 350) or $\left(66\dfrac{2}{3}, 350\right)$ **61.** 10 more hits **69.** b **71.** $f_2 = \dfrac{-ff_1}{f-f_1}$ or $f_2 = \dfrac{ff_1}{f_1-f}$ **73.** no solution

75. 8 **77.** -3 and -2 **78.** $(x^3-3)(x+2)$ **79.** $-\dfrac{12}{x^8}$ **80.** $-20x+55$

Section 8.7

Check Point Exercises

1. 2 mph **2.** $2\dfrac{2}{3}$ hr or 2 hr 40 min **3.** 32 in. **4.** 32 yd **5. a.** $L = kN$ **b.** $L = 4N$ **c.** 68 in. **6. a.** $W = kL$ **b.** 12.5

c. $W = 12.5L$ **d.** 200 lb **7.** using $\dfrac{5}{12}$ for k, 137.5 lb per sq in. **8.** about 4.36 lb per sq in.

Exercise Set 8.7

1. Walking rate is 6 mph; car rate is 9 mph. **3.** Downhill rate is 10 mph; uphill rate is 6 mph. **5.** Water's current is 5 mph.
7. Walking rate is 3 mph; jogging rate is 6 mph. **9.** Still water rate is 10 mph. **11.** It will take about 8.6 min, which is enough time.
13. It will take about 171.4 hr, which is not enough time. **15.** It will take 2.4 hr, or 2 hr 24 min. **17.** 5 in. **19.** 6 m **21.** 16 in. **23.** 16 ft

25. $g = kh$ **27.** $w = \dfrac{k}{v}$ **29.** 20 **31.** 6000 **33.** 84 **35.** 25 **37. a.** $G = kW$ **b.** $G = 0.02W$ **c.** 1.04 in. **39.** $60

41. 2442 mph **43.** 0.5 hr or 30 min **45.** 6.4 lb per cm^2 **47.** 5000 pens **63.** 30 mph **65.** 7 hr **67.** 2.5 hr **69.** 7% and 8%

70. $(5x+9)(5x-9)$ **71.** 6 **72.**

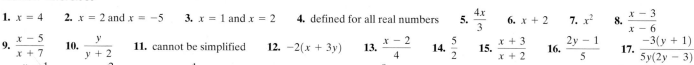

Review Exercises

1. $x = 4$ **2.** $x = 2$ and $x = -5$ **3.** $x = 1$ and $x = 2$ **4.** defined for all real numbers **5.** $\dfrac{4x}{3}$ **6.** $x+2$ **7.** x^2 **8.** $\dfrac{x-3}{x-6}$

9. $\dfrac{x-5}{x+7}$ **10.** $\dfrac{y}{y+2}$ **11.** cannot be simplified **12.** $-2(x+3y)$ **13.** $\dfrac{x-2}{4}$ **14.** $\dfrac{5}{2}$ **15.** $\dfrac{x+3}{x+2}$ **16.** $\dfrac{2y-1}{5}$ **17.** $\dfrac{-3(y+1)}{5y(2y-3)}$

18. $\dfrac{x-1}{4}$ **19.** $\dfrac{2}{x(x+1)}$ **20.** $\dfrac{1}{7(y+3)}$ **21.** $(y-3)(y-6)$ **22.** $\dfrac{8}{x-y}$ **23.** 4 **24.** 4 **25.** $3x-5$ **26.** $3y+4$

27. $\dfrac{4}{x-2}$ **28.** $\dfrac{2x+5}{x-3}$ **29.** $36x^3$ **30.** $x^2(x-1)^2$ **31.** $(x+3)(x+1)(x+7)$ **32.** $\dfrac{14x+15}{6x^2}$ **33.** $\dfrac{7x+2}{x(x+1)}$ **34.** $\dfrac{7x+25}{(x+3)^2}$

35. $\dfrac{3}{y-2}$ **36.** $-\dfrac{y}{y-1}$ **37.** $\dfrac{x^2+y^2}{xy}$ **38.** $\dfrac{3x^2-x}{(x+1)^2(x-1)}$ **39.** $\dfrac{5x^2-3x}{(x+1)(x-1)}$ **40.** $\dfrac{4}{(x+2)(x-2)(x-3)}$ **41.** $\dfrac{2x+13}{x+3}$

42. $\dfrac{y-4}{3y}$ **43.** $\dfrac{7}{2}$ **44.** $\dfrac{1}{x-1}$ **45.** $x+y$ **46.** $\dfrac{3}{x}$ **47.** $\dfrac{3x}{x-4}$ **48.** 12 **49.** -1 **50.** -6 and 1 **51.** no solution **52.** 5

53. -6 and 2   **54.** 5   **55.** -5 **56.** 3 yr   **57.** 30%   **58.** 4 mph   **59.** Slower car's rate is 20 mph; faster car's rate is 30 mph.   **60.** 4 hr

61. 5 ft   **62.** $12\frac{1}{2}$ ft   **63.** $154   **64.** 5 hr

Chapter 8 Test

1. $x = -9$ and $x = 4$ **2.** $\dfrac{x + 3}{x - 2}$ **3.** $\dfrac{4x}{x + 1}$ **4.** $\dfrac{x - 4}{2}$ **5.** $\dfrac{x}{x + 3}$ **6.** $\dfrac{2x + 6}{x + 1}$ **7.** $\dfrac{5}{(y - 1)(y - 3)}$ **8.** $2y$

9. $\dfrac{2}{y + 3}$ **10.** $\dfrac{x^2 + 2x + 15}{(x + 3)(x - 3)}$ **11.** $\dfrac{8x - 14}{(x + 2)(x - 1)(x - 3)}$ **12.** $\dfrac{-x - 1}{x - 3}$ **13.** $\dfrac{x + 2}{x - 1}$ **14.** $\dfrac{11}{(x - 3)(x - 4)}$ **15.** $\dfrac{4}{y + 4}$

16. $\dfrac{3x - 3y}{x}$ **17.** $\dfrac{5x + 5}{2x + 1}$ **18.** $\dfrac{y - x}{y}$ **19.** 6   **20.** -8 **21.** 0 and 2   **22.** Water's current is 2 mph.   **23.** 12 min

24. 3.2 in.   **25.** 52.5 amp

Cumulative Review Exercises (Chapters 1–8)

1. 2   **2.** $\{x | x < 2\}$ **3.** -6 and 3   **4.** -6 **5.** $(3, 3)$ **6.** $(-2, 2)$

7. $3x - 2y = 6$ **8.** $y > -2x + 3$ **9.** $y = -3$

10. -19 **11.** $8x^9$ **12.** $\dfrac{1 - 2x}{4x - 1}$ **13.** $(4x - 1)(x - 3)$ **14.** $(2x - 5)^2$ **15.** $3(x + 5)(x - 5)$ **16.** $-x^2 + 4x + 8$

17. $-2x^4 + 3x^2 - 1$ **18.** $\dfrac{3x^2 + 6x + 16}{(x - 2)(x + 3)}$ **19.** $1225 at 5% and $2775 at 9%   **20.** 17 in. and 51 in.

CHAPTER 9

Section 9.1

Check Point Exercises

1. a. 9   **b.** -3 **c.** $\dfrac{1}{5}$ **d.** 10   **e.** 14   **2.** The evaporation is $\dfrac{1}{2}$ in. on that day.   **3.** 50%; fairly well   **4. a.** 1   **b.** -3 **c.** $\dfrac{1}{5}$

5. a. 3   **b.** not a real number   **c.** -3 **d.** $-\dfrac{1}{2}$

Exercise Set 9.1

1. 6   **3.** -6 **5.** not a real number   **7.** $\dfrac{1}{3}$ **9.** $\dfrac{1}{10}$ **11.** $-\dfrac{1}{6}$ **13.** not a real number   **15.** 0.2   **17.** 5   **19.** 8   **21.** 13   **23.** 17

25. not a real number   **27.**

29. Answers will vary. example: Exercise 27 graph is shaped like $y = \sqrt{x}$, but shifted 1 unit to the right.

31. 2.646   **33.** 4.796   **35.** -8.062 **37.** 15.317   **39.** 7.658   **41.** 2.153   **43.** 2.828

45. not a real number   **47.** 4   **49.** -3 **51.** -2 **53.** $\dfrac{1}{5}$ **55.** -10 **57.** 1   **59.** 2

61. -2 **63.** not a real number   **65.** -1 **67.** not a real number   **69.** -4 **71.** 2

73. -2 **75.** $x \ge 0$ **77.** $x \ge 2$ **79.** $x \le 2$ **81.** all real numbers   **83.** $x \le 6$ **85.** 12 mph   **87.** 70 mph; He was speeding.

89. a. 36 cm   **b.** 44.7 cm   **c.** 46.9 cm   **d.** describes healthy children   **91. a.** $y = a\sqrt{x} + 261$ **b.** $y = 177\sqrt{x} + 261$

c. 821 thousand; Answers will vary.   **d.** 969 thousand.   **101.** 2   **103.** 0

105.

107.

108. $\{x | x < -8\}$

109. $\dfrac{x + 9}{x - 15}$

Section 9.2

Check Point Exercises

1. a. $\sqrt{30}$ **b.** $\sqrt{26xy}$ **c.** 5 **d.** $\sqrt{\dfrac{15}{22}}$ **2. a.** $2\sqrt{3}$ **b.** $2\sqrt{15}$ **c.** cannot be simplified   **3.** $2x^8\sqrt{10}$ **4.** $3x^4\sqrt{3x}$

5. $3x^6\sqrt{5x}$ **6. a.** $\dfrac{7}{5}$ **b.** $4x^2$ **7. a.** $2\sqrt[3]{5}$ **b.** $2\sqrt[5]{2}$ **c.** $\dfrac{5}{3}$

Exercise Set 9.2

1. $\sqrt{14}$ **3.** $\sqrt{15xy}$ **5.** 5 **7.** $\sqrt{\dfrac{10}{21}}$ **9.** $\sqrt{0.5xy}$ **11.** $\dfrac{1}{5}\sqrt{ab}$ **13.** $\sqrt{x}$ **15.** $5\sqrt{2}$ **17.** $3\sqrt{5}$ **19.** $10\sqrt{2}$ **21.** $5\sqrt{3x}$

23. $3\sqrt{x}$ **25.** cannot be simplified   **27.** y **29.** $8x$ **31.** $x\sqrt{11}$ **33.** $2x\sqrt{2}$ **35.** x^{10} **37.** $5y^5$ **39.** $2x^3\sqrt{5}$ **41.** $6y^{50}\sqrt{2}$

43. $x\sqrt{x}$ **45.** $x^3\sqrt{x}$ **47.** $y^8\sqrt{y}$ **49.** $5x^2\sqrt{x}$ **51.** $2x^8\sqrt{2x}$ **53.** $3y^9\sqrt{10y}$ **55.** $3\sqrt{5}$ **57.** $5\sqrt{2xy}$ **59.** $6x$ **61.** $3x\sqrt{5x}$

63. $5x^6\sqrt{3x}$ **65.** $\sqrt{21xy}$ **67.** $10xy\sqrt{2y}$ **69.** $\dfrac{7}{4}$ **71.** $\dfrac{\sqrt{3}}{2}$ **73.** $\dfrac{x}{6}$ **75.** $\dfrac{\sqrt{7}}{x^2}$ **77.** $\dfrac{6\sqrt{2}}{y^{10}}$ **79.** 3 **81.** $2\sqrt{2}$ **83.** $\sqrt{5}$

85. 4 **87.** $2x$ **89.** $5x\sqrt{2x}$ **91.** $2x^3\sqrt{10x}$ **93.** $2\sqrt[3]{2}$ **95.** $3\sqrt[3]{2}$ **97.** $2\sqrt[4]{2}$ **99.** 2 **101.** $3\sqrt[3]{2}$ **103.** $2\sqrt[4]{2}$ **105.** $\dfrac{3}{2}$

107. $\dfrac{\sqrt[3]{3}}{2}$ **109.** $3(x+4)\sqrt{10(x+4)}$ **111.** $x-3$ **113.** $2^{21}x^{52}y^6\sqrt{2y}$ **115.** $2x\sqrt[3]{3x^2}$ **117.** $20\sqrt{2}$ mph   **119.** $5\sqrt{3}$ sq ft

129. $\sqrt{25x^{14}}$ **133.** $4x\sqrt{2}$ should be $2x\sqrt{2}$.   **135.** $(6,-2)$ **136.** $\dfrac{3}{x-2}$ **137.** $2x(x-3)(x-5)$

Section 9.3

Check Point Exercises

1. a. $17\sqrt{13}$ **b.** $-19\sqrt{17x}$ **2. a.** $17\sqrt{3}$ **b.** $10\sqrt{2x}$ **3. a.** $\sqrt{10}+\sqrt{22}$ **b.** $11+6\sqrt{3}$ **c.** $4-4\sqrt{5}$ **4. a.** -2 **b.** 5

Exercise Set 9.3

1. $13\sqrt{3}$ **3.** $15\sqrt{6}$ **5.** $-5\sqrt{13}$ **7.** $15\sqrt{x}$ **9.** $-6\sqrt{y}$ **11.** $9\sqrt{10x}$ **13.** $6\sqrt{5y}$ **15.** $2\sqrt{5}$ **17.** $12\sqrt{2}$ **19.** $7\sqrt{7}$
21. 0 **23.** $3\sqrt{5}$ **25.** $\sqrt{2}$ **27.** $8\sqrt{2}$ **29.** $19\sqrt{3}$ **31.** $9\sqrt{3}-6\sqrt{2}$ **33.** $2\sqrt{5x}$ **35.** $9+5\sqrt{2}$ **37.** cannot be combined
39. $8\sqrt{5}+15\sqrt{3}$ **41.** $7\sqrt{6}+8\sqrt{5}$ **43.** $\sqrt{6}+\sqrt{10}$ **45.** $\sqrt{42}-\sqrt{70}$ **47.** $5\sqrt{3}+3$ **49.** $3\sqrt{2}-3$ **51.** $32+11\sqrt{2}$
53. $25-2\sqrt{5}$ **55.** $117-36\sqrt{7}$ **57.** $25-8\sqrt{10}$ **59.** $15+9\sqrt{2}$ **61.** $\sqrt{6}-6\sqrt{2}+\sqrt{3}-6$ **63.** 4 **65.** -5 **67.** -14
69. 2 **71.** -37 **73.** 7 **75.** $5+2\sqrt{6}$ **77.** $x-2\sqrt{10x}+10$ **79.** $9x\sqrt{3x}$ **81.** $8x^2y^2\sqrt{xy}$ **83.** $\sqrt[3]{2}$ **85.** $11x\sqrt[3]{4x}$
87. perimeter $=4(\sqrt{3}+\sqrt{5})=(4\sqrt{3}+4\sqrt{5})$ in.; area $=(\sqrt{3}+\sqrt{5})^2=(8+2\sqrt{15})$ sq in.
89. perimeter $=2(\sqrt{6}-1)+2(\sqrt{6}+1)=4\sqrt{6}$ in.; area $=(\sqrt{6}+1)(\sqrt{6}-1)=5$ sq in.

91. perimeter $=\sqrt{2}+\sqrt{2}+2=(2+2\sqrt{2})$ in.; area $=\dfrac{1}{2}(\sqrt{2})(\sqrt{2})=1$ sq in.   **93.** $3\sqrt{2}$; Answers will vary.   **103.** $11\sqrt{3}$

105. $(5+\sqrt{3})(5-\sqrt{3})$ **107.** correct   **109.** $x+1$ should be $x-2\sqrt{x}+1$.   **111.** $25x^2-9$

112. $x(8x+1)(8x-1)$ **113.**

$y=-\dfrac{1}{4}x+3$

Mid-Chapter 9 Check Point

1. $10\sqrt{3}$ **2.** $10\sqrt{6}$ **3.** $4x\sqrt{6x}$ **4.** $\dfrac{\sqrt[5]{4}}{2}$ **5.** $9\sqrt{3}$ **6.** 7 **7.** $2\sqrt{15}-30$ **8.** $-4x^{10}\sqrt{2x}$ **9.** $2x^3\sqrt{3x}$

10. $2\sqrt[3]{2}$ **11.** $-19\sqrt{10}$ **12.** $4-\sqrt{3}$ **13.** $2x\sqrt{2}$ **14.** $-2\sqrt[4]{2}$ **15.** $9+2\sqrt{14}$ **16.** $2\sqrt[3]{2}$ **17.** $6\sqrt{5}$

18. $\dfrac{\sqrt{10}}{4}$ **19.** $6+3\sqrt{10}$ **20.** 23

Section 9.4

Check Point Exercises

1. a. $\dfrac{5\sqrt{10}}{2}$ **b.** $\dfrac{\sqrt{14}}{7}$ **2. a.** $\dfrac{5\sqrt{2}}{2}$ **b.** $\dfrac{\sqrt{35x}}{10}$ **3.** $\dfrac{32-8\sqrt{5}}{11}$ **4.** $2\sqrt{7}+2\sqrt{3}$

Exercise Set 9.4

1. $\dfrac{\sqrt{10}}{10}$ **3.** $\sqrt{5}$ **5.** $\dfrac{\sqrt{6}}{3}$ **7.** $4\sqrt{7}$ **9.** $\dfrac{\sqrt{15}}{5}$ **11.** $\dfrac{\sqrt{21}}{3}$ **13.** $\dfrac{x\sqrt{3}}{3}$ **15.** $\dfrac{\sqrt{11x}}{x}$ **17.** $\dfrac{\sqrt{xy}}{y}$ **19.** $\dfrac{x^2\sqrt{2}}{2}$ **21.** $\dfrac{\sqrt{35}}{5}$

23. $\dfrac{\sqrt{42x}}{14}$ **25.** $\dfrac{\sqrt{5}}{10}$ **27.** $\dfrac{3\sqrt{2}}{2}$ **29.** $\dfrac{5\sqrt{3}}{2}$ **31.** $\dfrac{\sqrt{10}}{6}$ **33.** $\dfrac{\sqrt{2x}}{8}$ **35.** $\dfrac{\sqrt{5}}{15}$ **37.** $\dfrac{\sqrt{21}}{6}$ **39.** $2x\sqrt{2}$ **41.** $\dfrac{\sqrt{14y}}{4}$ **43.** $\dfrac{\sqrt{21x}}{6}$

45. $\dfrac{3\sqrt{5x}}{x}$ **47.** $\dfrac{5\sqrt{x}}{x^2}$ **49.** $\dfrac{3\sqrt{3y}}{y^2}$ **51.** $\dfrac{5x\sqrt{6y}}{6y^2}$ **53.** $\dfrac{4-\sqrt{3}}{13}$ **55.** $-6-3\sqrt{7}$ **57.** $8\sqrt{11}-24$ **59.** $9+3\sqrt{3}$

61. $2-\sqrt{2}$ **63.** $\dfrac{10+\sqrt{70}}{3}$ **65.** $2\sqrt{6}-2\sqrt{3}$ **67.** $\sqrt{5}+\sqrt{3}$ **69.** $\dfrac{8-2\sqrt{x}}{16-x}$ **71.** $\dfrac{6\sqrt{5}-4\sqrt{3}}{11}$ **73.** $\dfrac{7+2\sqrt{10}}{3}$

75. $\dfrac{3y^2\sqrt{2x}}{x}$ **77.** $\sqrt{x+2}+\sqrt{x}$ **79.** $\dfrac{5\sqrt{6}}{6}$ **81.** $2\sqrt{x+2}-h$ **83. a.** 18% **b.** $\dfrac{13\sqrt{x}+x}{5}$ **c.** 18%; yes; no; Answers will vary.

85. $\dfrac{\sqrt{5}+1}{2}\approx 1.62$ **89.** about 20%, 24%, and 28%; Answers will vary. **91.** $\dfrac{\sqrt[3]{4}}{2}$ **93.** 4 **95.** 5 and 1 **96.** $\dfrac{1}{8x^6}$ **97.** $\dfrac{x-3}{4(x+3)}$

Section 9.5

Check Point Exercises

1. 11 **2.** 4 **3.** no solution **4.** 6 **5.** 2012

Exercise Set 9.5

1. 25 **3.** 16 **5.** 7 **7.** 124 **9.** 7 **11.** 4 **13.** 2 **15.** 3 **17.** 24 **19.** no solution **21.** no solution **23.** 16
25. no solution **27.** 7 **29.** 2 **31.** -3 **33.** -6 **35.** 2 **37.** 1 and 3 **39.** no solution **41.** 12 **43.** no solution

45. 16 **47.** 8 **49.** $h=\dfrac{v^2}{2g}$ **51.** 1 **53.** 9 **55. a.** 17.4%; Answers will vary. **b.** 2010 **57.** 75 years old **59.** about 5 km

61. 400 lb **67.** b **69.** $x=2, y=2, z=2$ **71.** 4 **73.** 1 and 9 **76.** $7000 at 6% and $2000 at 4% **77.** $4.25

78. $(-1,-2)$

Section 9.6

Check Point Exercises

1. a. 5 **b.** 2 **c.** -3 **d.** -2 **2. a.** 81 **b.** 8 **c.** -8 **3. a.** $\dfrac{1}{5}$ **b.** $\dfrac{1}{4}$ **c.** $\dfrac{1}{16}$ **4.** $126.5 thousand

Exercise Set 9.6

1. 7 **3.** 11 **5.** 3 **7.** -5 **9.** 2 **11.** -2 **13.** $\dfrac{1}{3}$ **15.** $\dfrac{3}{4}$ **17.** 729 **19.** 25 **21.** 27 **23.** -8 **25.** $\dfrac{1}{3}$ **27.** $\dfrac{1}{5}$

29. $\dfrac{1}{2}$ **31.** 2 **33.** $\dfrac{1}{8}$ **35.** $\dfrac{1}{243}$ **37.** $\dfrac{1}{4}$ **39.** $\dfrac{5}{2}$ **41.** $\dfrac{5}{2}$ **43.** $\dfrac{1}{4}$ **45.** 17 **47.** 375 **49.** $x^{\frac{7}{12}}$ **51.** $\dfrac{1}{x^{\frac{2}{3}}}$ **53.** $x^{\frac{1}{6}}y^2$ **55.** $\dfrac{1}{x^7}$ **57.** 25 mph

59. 4.25% **61.** $V=194.8\sqrt[6]{t}$ and $E=339.6(\sqrt[3]{t})^2$ or $E=339.6\sqrt[3]{t^2}$ **69.** a **71.** $\dfrac{1}{5}$ **73. a.** 0, 94, 249, 1170, 1756, 2405, 3110

b. **c.** **75.** $(-2,0)$ **76.** $-3x+4y\le 12$, $x\ge 2$ **77.** $32x^2$

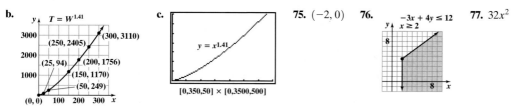

Review Exercises

1. 11 **2.** -11 **3.** not a real number **4.** $\dfrac{2}{5}$ **5.** -2 **6.** -3 **7.** 8.660 **8.** 19.824 **9.** 79.5 thousand people

10. Answers will vary. example: It's increasing, but the rate of increase is slowing down. **11.** 49 mi **12.** $3\sqrt{6}$

13. $12\sqrt{5}$ **14.** $3x\sqrt{7}$ **15.** $4x\sqrt{3x}$ **16.** x^4 **17.** $5x^4\sqrt{3x}$ **18.** $3x^{11}\sqrt{5x}$ **19.** $2\sqrt[3]{3}$ **20.** $\sqrt{77}$ **21.** 6 **22.** $5x\sqrt{2}$

23. $2x^2\sqrt{3x}$ **24.** $3\sqrt[3]{2}$ **25.** $\dfrac{\sqrt{15}}{4}$ **26.** $\dfrac{11}{2}$ **27.** $\dfrac{\sqrt{7x}}{5}$ **28.** $\dfrac{3\sqrt{2}}{x}$ **29.** 10 **30.** $4\sqrt{2}$ **31.** $6x^2\sqrt{2x}$ **32.** $\dfrac{\sqrt[3]{5}}{4}$

33. $\dfrac{2\sqrt[3]{5}}{3}$ **34.** $20\sqrt{5}$ **35.** $7\sqrt{2}$ **36.** $\sqrt{3}$ **37.** $17\sqrt{5}$ **38.** $24\sqrt{2}-8\sqrt{3}$ **39.** $6\sqrt{2}+7\sqrt{3}$ **40.** $5\sqrt{2}+2\sqrt{15}$ **41.** $3\sqrt{2}-6$

42. $92+19\sqrt{2}$ **43.** $-17+11\sqrt{7}$ **44.** $3\sqrt{2}-4\sqrt{3}+2\sqrt{6}-8$ **45.** -3 **46.** 6 **47.** $3+2\sqrt{2}$ **48.** $6\sqrt{5}$ **49.** $\dfrac{13\sqrt{2}}{10}$

50. $\dfrac{\sqrt{6}}{3}$ **51.** $\dfrac{\sqrt{6}}{4}$ **52.** $\dfrac{\sqrt{17x}}{x}$ **53.** $11\sqrt{5}-22$ **54.** $\dfrac{84+21\sqrt{3}}{13}$ **55.** $6\sqrt{5}-6\sqrt{3}$ **56.** $-1-2\sqrt{2}$ **57.** 13 **58.** 11 **59.** 5

60. 0 and 3   **61.** 8   **62.** no solution   **63.** no solution   **64.** 144 ft   **65.** 1250 ft   **66.** $\sqrt{16}=4$ **67.** $\sqrt[3]{125}=5$ **68.** $(\sqrt[3]{64})^2=16$

69. $\dfrac{1}{\sqrt{25}}=\dfrac{1}{5}$ **70.** $\dfrac{1}{\sqrt[3]{27}}=\dfrac{1}{3}$ **71.** $\dfrac{1}{(\sqrt[3]{-8})^4}=\dfrac{1}{16}$ **72.** approx 57 species

Chapter 9 Test

1. -8 **2.** 4 **3.** $4\sqrt{3}$ **4.** $6x\sqrt{2x}$ **5.** $x^{14}\sqrt{x}$ **6.** $\dfrac{5}{x}$ **7.** $\dfrac{5}{3}$ **8.** $\dfrac{\sqrt[3]{5}}{2}$ **9.** $2x\sqrt{10}$ **10.** $5\sqrt{2}$ **11.** $6\sqrt{xy}$ **12.** $2x^2\sqrt{5x}$

13. $11\sqrt{6}$ **14.** $6\sqrt{2}$ **15.** $\sqrt{30}+3$ **16.** $55+11\sqrt{5}$ **17.** 2 **18.** $16+6\sqrt{7}$ **19.** $\dfrac{4\sqrt{5}}{5}$ **20.** $\dfrac{20-5\sqrt{3}}{13}$ **21.** 12 **22.** 5

23. 1600 ft   **24.** $(\sqrt[3]{8})^2=4$ **25.** $\dfrac{1}{\sqrt{9}}=\dfrac{1}{3}$

Cumulative Review Exercises (Chapters 1–9)

1. -3 **2.** $\dfrac{3}{2}$ and -4 **3.** $(-3,-4)$ **4.** $\dfrac{9}{7}$ **5.** -5 **6.** 5 **7.** $-\dfrac{2}{x^4}$ **8.** $22\sqrt{3}$ **9.** $\dfrac{3}{x}$ **10.** $-\dfrac{2+x}{3x+1}$ **11.** 5

12. $(x-7)(x-11)$ **13.** $x(x+5)(x-5)$ **14.** $6x^2-7x+2$ **15.** $8x^3-27$ **16.** $\dfrac{1}{x-1}$ **17.** $\dfrac{5x-1}{4x-3}$ **18.** $-18\sqrt{3}$

19. $2x-y=4$ **20.** $y=-\dfrac{2}{3}x$ **21.** $x\ge-1$ **22.** $-\dfrac{8}{3}$ **23.** $y+3=5(x+2);\ y=5x+7$ **24.** 43 **25.** 954 deer

CHAPTER **10**

Section 10.1

Check Point Exercises

1. a. 6 and -6 **b.** $\sqrt{3}$ and $-\sqrt{3}$ **c.** $\dfrac{\sqrt{14}}{2}$ and $-\dfrac{\sqrt{14}}{2}$ **2.** 8 and -2 **3.** $2+\sqrt{7}$ and $2-\sqrt{7}$ **4.** 12 in.   **5.** 13 units

Exercise Set 10.1

1. 4 and -4 **3.** 9 and -9 **5.** $\sqrt{7}$ and $-\sqrt{7}$ **7.** $5\sqrt{2}$ and $-5\sqrt{2}$ **9.** 2 and -2 **11.** $\dfrac{7}{2}$ and $-\dfrac{7}{2}$ **13.** 5 and -5 **15.** $\dfrac{\sqrt{6}}{3}$ and $-\dfrac{\sqrt{6}}{3}$

17. $\dfrac{\sqrt{35}}{5}$ and $-\dfrac{\sqrt{35}}{5}$ **19.** 7 and -1 **21.** 6 and -16 **23.** $\dfrac{1}{3}$ and $-\dfrac{5}{3}$ **25.** $5+\sqrt{3}$ and $5-\sqrt{3}$ **27.** $-8+\sqrt{11}$ and $-8-\sqrt{11}$

29. $4+3\sqrt{2}$ and $4-3\sqrt{2}$ **31.** 2 and -6 **33.** 9 and -3 **35.** $5+\sqrt{2}$ and $5-\sqrt{2}$ **37.** $-1+\sqrt{5}$ and $-1-\sqrt{5}$ **39.** $7+2\sqrt{3}$ and $7-2\sqrt{3}$

41. 17 m   **43.** 39 m   **45.** 12 cm   **47.** $5\sqrt{7}$ m   **49.** $\sqrt{17}$ or 4.12 units   **51.** 17 units   **53.** $3\sqrt{5}$ or 6.71 units   **55.** $\sqrt{41}$ or 6.40 units

57. 16 units   **59.** -2 and 8   **61.** -3 and $\dfrac{5}{3}$ **63.** $r=\dfrac{\sqrt{A\pi}}{\pi}$ **65.** $d=\dfrac{\sqrt{kI}}{I}$ **67.** $2\sqrt{41}$ ft   **69.** $90\sqrt{2}$ ft   **71.** $\dfrac{25\sqrt{2}}{2}$ in.

73. 6 in.   **75.** 6 weeks   **77.** 5 sec   **79.** 8 m   **81.** 10 in.   **87.** c   **89.** -3 and 1   **91.** $2(2x+3)(3x-1)$

92. $\dfrac{x-3}{3(x-2)}$ **93.** -6

Section 10.2

Check Point Exercises

1. a. $x^2+10x+25=(x+5)^2$ **b.** $x^2-6x+9=(x-3)^2$ **c.** $x^2+3x+\dfrac{9}{4}=\left(x+\dfrac{3}{2}\right)^2$ **2. a.** 1 and -7

b. $5+\sqrt{7}$ and $5-\sqrt{7}$ **3.** $\dfrac{5\pm3\sqrt{3}}{2}$

Exercise Set 10.2

1. $x^2 + 10x + 25 = (x + 5)^2$ **3.** $x^2 - 2x + 1 = (x - 1)^2$ **5.** $x^2 + 5x + \dfrac{25}{4} = \left(x + \dfrac{5}{2}\right)^2$ **7.** $x^2 - 7x + \dfrac{49}{4} = \left(x - \dfrac{7}{2}\right)^2$

9. $x^2 + \dfrac{1}{2}x + \dfrac{1}{16} = \left(x + \dfrac{1}{4}\right)^2$ **11.** $x^2 - \dfrac{4}{3}x + \dfrac{4}{9} = \left(x - \dfrac{2}{3}\right)^2$ **13.** 1 and -5 **15.** 4 and 6   **17.** $1 + \sqrt{6}$ and $1 - \sqrt{6}$

19. $-2 + \sqrt{3}$ and $-2 - \sqrt{3}$ **21.** 7 and -4 **23.** $\dfrac{-3 \pm \sqrt{13}}{2}$ **25.** $\dfrac{7 \pm \sqrt{37}}{2}$ **27.** $\dfrac{1 \pm \sqrt{13}}{2}$ **29.** 1 and $\dfrac{1}{2}$ **31.** $\dfrac{-5 \pm \sqrt{3}}{2}$

33. $\dfrac{1 \pm \sqrt{13}}{4}$ **35.** $1 \pm \sqrt{7}$ **37.** $\dfrac{1 \pm \sqrt{29}}{2}$ **39.** $-5b$ and b **43.** d   **45.** $-\dfrac{1}{2} \pm \sqrt{\dfrac{1}{4} - c}$ or $\dfrac{-1 \pm \sqrt{1 - 4c}}{2}$

48. $\dfrac{11}{(x - 3)(x - 4)}$ **49.** $\dfrac{x}{3}$ **50.** 3

Section 10.3

Check Point Exercises

1. $\dfrac{1}{4}$ and $-\dfrac{1}{2}$ **2.** $3 \pm \sqrt{5}$ **3.** 26

Exercise Set 10.3

1. -2 and -3 **3.** $\dfrac{-5 \pm \sqrt{13}}{2}$ **5.** $-2 \pm \sqrt{10}$ **7.** $-2 \pm \sqrt{11}$ **9.** 6 and -3 **11.** $\dfrac{3}{2}$ and $-\dfrac{2}{3}$ **13.** $1 \pm \sqrt{11}$

15. $\dfrac{1 \pm \sqrt{57}}{2}$ **17.** $\dfrac{-3 \pm \sqrt{3}}{6}$ **19.** $\dfrac{2}{3}$ **21.** $\dfrac{1 \pm \sqrt{29}}{4}$ **23.** 1 and $-\dfrac{1}{2}$ **25.** 2 and $\dfrac{1}{5}$ **27.** $\pm 2\sqrt{5}$ **29.** $1 \pm \sqrt{2}$

31. $\dfrac{-11 \pm \sqrt{33}}{4}$ **33.** 0 and $\dfrac{8}{3}$ **35.** 2   **37.** 2 and -2 **39.** 0 and -9 **41.** 4 and $-\dfrac{2}{3}$ **43.** $\dfrac{2 \pm \sqrt{10}}{3}$ **45.** $\sqrt{3}$ and $-\sqrt{3}$

47. 0 and $-\dfrac{5}{2}$ **49.** $\sqrt{6}$ and $-\sqrt{6}$ **51.** $\dfrac{5 \pm \sqrt{73}}{2}$ **53.** about 3.8 sec   **55.** 10 ft   **57.** 2005   **59.** width: 4.7 m; length: 7.7 m

61. 2.3 and 3.3 ft   **63.** 9.8 yr   **71.** about 1.2 ft   **73.**   **74.** $\dfrac{1}{25}$ **75.** $9 - 3\sqrt{5}$ **76.** $x^3 - y^3$

Mid-Chapter 10 Check Point

1. $-\dfrac{8}{3}$ and 4   **2.** 0 and $\dfrac{1}{3}$ **3.** $1 \pm \sqrt{11}$ **4.** $4 \pm \sqrt{7}$ **5.** $-\dfrac{2}{3}$ and 1   **6.** $\dfrac{5 \pm \sqrt{7}}{6}$ **7.** -4 and 3   **8.** $-5 \pm 2\sqrt{10}$ **9.** -3 and 3

10. $\dfrac{-5 \pm \sqrt{17}}{4}$ **11.** $-\dfrac{1}{2}$ and 10   **12.** 4   **13.** -16 and 2   **14.** $2\sqrt{7}$ cm   **15.** 13 units   **16.** 12 in. and 16 in.

Section 10.4

Check Point Exercises

1. a. $4i$ **b.** $\sqrt{5}i$ **c.** $5\sqrt{2}i$ **2.** $-2 + 5i$ and $-2 - 5i$ **3.** $-3 \pm 2i$

Exercise Set 10.4

1. $6i$ **3.** $\sqrt{13}i$ **5.** $5\sqrt{2}i$ **7.** $2\sqrt{5}i$ **9.** $-2\sqrt{7}i$ **11.** $7 + 4i$ **13.** $10 + \sqrt{3}i$ **15.** $6 - 7\sqrt{2}i$ **17.** $3 \pm 3i$ **19.** $-7 \pm 8i$

21. $2 \pm \sqrt{7}i$ **23.** $-3 \pm 3\sqrt{2}i$ **25.** $-2 \pm i$ **27.** $3 \pm 2i$ **29.** $6 \pm 2i$ **31.** $5 \pm \sqrt{2}i$ **33.** $\dfrac{1 \pm \sqrt{14}i}{5}$ **35.** $\dfrac{2 \pm \sqrt{6}i}{2}$

37. $\pm\sqrt{5}i$ **39.** $1 \pm 3i$ **41.** $\dfrac{-1 \pm \sqrt{7}i}{2}$ **43.** $9 \pm \sqrt{19}i$; not real numbers; Applicant will not be hired.; Answers will vary.

51. b   **53.** The only numbers that satisfy the conditions are $1 \pm 2i$. These are not real numbers.

56. 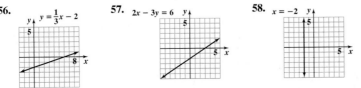   **57.** $2x - 3y = 6$ **58.** $x = -2$

Section 10.5

Check Point Exercises

1. a. upward

b.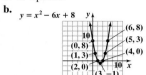

2. 2 and 4   **3.** 8   **4.** $(3, -1)$

5.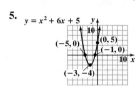

6.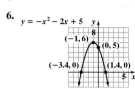

7. $(45, 190)$; The minimum number of accidents per 50 million miles driven is attributed to 45 year olds. That minimum number is 190.

Exercise Set 10.5

1. upward   **3.** downward   **5.** 1 and 3   **7.** 2 and 6   **9.** 1.2 and -3.2 **11.** 3   **13.** -12 **15.** -4 **17.** 0   **19.** $(2, -1)$ **21.** $(-1, -8)$

23. $(-3, -9)$ **25.** 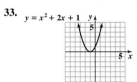   **27.**   **29.**

31. 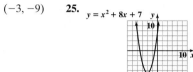   **33.**   **35.**

37. $(3, 2)$ **39.** $(-5, -4)$
41. $(1, -3)$ **43.** $(-2, 5)$ **45.** (h, k)
47. $(25, 13.5)$; The maximum amount that a redwood tree grows per year occurs when the annual rainfall is 25 in. This maximum growth is 13.5 in. per year.
49. 6.25 sec; 629 ft   **51.** length: 60 ft; width: 30 ft; 1800 sq ft

59. b   **61.** $(-2, 0)$ and $(2, 0)$ **65.** $(80, 1600)$; Answers will vary.

67. $(-4, 520)$; Answers will vary.   **69.** 2   **70.** $-\dfrac{8}{3}$ **71.** $(-2, 1)$

Section 10.6

Check Point Exercises

1. domain: {golf, lawn mowing, water skiing, hiking, bicycling}; range: $\{250, 325, 430, 720\}$ **2. a.** not a function   **b.** function
3. a. 23   **b.** -5 **c.** 3   **4. a.** 48   **b.** 3   **c.** 3   **5. a.** function   **b.** function   **c.** not a function
6. 17.014; In 1980, 17.014% of the U.S. population ages 25 or older had a college degree.

Exercise Set 10.6

1. function; domain: $\{1, 3, 5\}$; range: $\{2, 4, 5\}$ **3.** not a function; domain: $\{3, 4\}$; range: $\{4, 5\}$ **5.** function; domain: $\{-3, -2, -1, 0\}$; range: $\{-3, -2, -1, 0\}$ **7.** not a function; domain: $\{1\}$; range: $\{4, 5, 6\}$ **9. a.** 12   **b.** -1 **c.** 5   **11. a.** 70   **b.** -28 **c.** 0
13. a. 93   **b.** -7 **c.** -3 **15. a.** 10   **b.** -2 **c.** 0   **17. a.** 11   **b.** 27   **c.** 3   **19. a.** 5   **b.** 5   **c.** 5   **21. a.** 3   **b.** 7
23. a. 1   **b.** -1 **25.** function   **27.** function   **29.** not a function   **31.** function   **33.** $\{(-1, 1), (0, 3), (1, 5)\}$
35. $\{(-2, -6), (-1, -2), (0, 0), (1, 0), (2, -2)\}$ **37.** 6   **39.** $x + h$ **41. a.** {(U.S., 80%), (Japan, 64%), (France, 64%), (Germany, 61%), (England, 59%), (China, 47%)}   **b.** Yes; Each country corresponds to a unique percent.   **c.** {(80%, U.S.), (64%, Japan), (64%, France), (61%, Germany), (59%, England), (47%, China)}   **d.** No; 64% in the domain corresponds to two members of the range, Japan and France.
43. 186.6; At age 20, an American man's cholesterol level is 186.6, or approx 187.   **45.** 131; The ideal weight of a woman with a medium frame whose height is 64 in. is 131 lb.   **47.** 155; The ideal weight of a man with a medium frame whose height is 70 in. is 155 lb.
49. 39.1; 39.1% of 20-year-old Americans say they do volunteer work.   **57.** $f(x) = 3x + 7$ **59.** $V(x) = 22{,}500 - 3200x$; $V(3) = 12{,}900$;

After 3 years, the value of the car is \$12,900.   **60.** 3.97×10^{-3} **61.** $x^2 + 9x + 16 + \dfrac{35}{x - 2}$ **62.** $\left(\dfrac{4}{5}, \dfrac{9}{5}\right)$

Review Exercises

1. 8 and -8 **2.** $\sqrt{17}$ and $-\sqrt{17}$ **3.** $5\sqrt{3}$ and $-5\sqrt{3}$ **4.** 6 and 0   **5.** $-4 + \sqrt{5}$ and $-4 - \sqrt{5}$ **6.** $\dfrac{\sqrt{15}}{3}$ and $-\dfrac{\sqrt{15}}{3}$ **7.** 6 and 1

8. $-5 + 2\sqrt{3}$ and $-5 - 2\sqrt{3}$ **9.** 10 ft   **10.** $2\sqrt{13}$ in.   **11.** $2\sqrt{26}$ cm   **12.** 15 ft   **13.** 12 yd   **14.** 20 weeks   **15.** 2.5 sec

16. 5 units **17.** $2\sqrt{5}$ or 4.47 units **18.** $x^2 + 16x + 64 = (x + 8)^2$ **19.** $x^2 - 6x + 9 = (x - 3)^2$ **20.** $x^2 + 3x + \dfrac{9}{4} = \left(x + \dfrac{3}{2}\right)^2$

21. $x^2 - 5x + \dfrac{25}{4} = \left(x - \dfrac{5}{2}\right)^2$ **22.** 3 and 9 **23.** $3 \pm \sqrt{5}$ **24.** $\dfrac{6 \pm \sqrt{3}}{3}$ **25.** -3 and $\dfrac{1}{2}$ **26.** $1 \pm \sqrt{5}$ **27.** $\dfrac{9 \pm \sqrt{21}}{6}$

28. $\dfrac{1}{2}$ and 5 **29.** $\dfrac{10}{3}$ and -2 **30.** $\dfrac{7 \pm \sqrt{37}}{6}$ **31.** 3 and -3 **32.** $\dfrac{3 \pm \sqrt{5}}{2}$ **33.** 2005 **34.** about 8.8 sec **35.** $9i$

36. $\sqrt{23}i$ **37.** $4\sqrt{3}i$ **38.** $3 + 7i$ **39.** $\pm 10i$ **40.** $\pm 5i$ **41.** $\dfrac{-1 \pm 2\sqrt{2}i}{2}$ **42.** $2 \pm 3i$ **43.** $\dfrac{1 \pm \sqrt{23}i}{6}$

44. a. upward **b.** -1 and 7 **45. a.** downward **b.** 1 and -3 **46. a.** downward **b.** 2.2 and -0.2 **47. a.** upward **b.** 0 and 4

c. -7 **d.** $(3, -16)$ **c.** 3 **d.** $(-1, 4)$ **c.** 1 **d.** $(1, 4)$ **c.** 0 **d.** $(2, -4)$

e.
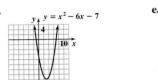

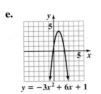

48. 3 sec; 147 ft **49.** function; domain: $\{2, 3, 5\}$; range: $\{7\}$ **50.** function; domain: $\{1, 2, 3\}$; range: $\{10, 500, \pi\}$
51. not a function; domain: $\{12, 14\}$; range: $\{13, 15, 19\}$ **52. a.** -19 **b.** 14 **c.** -4 **53. a.** 38 **b.** -4 **c.** 2 **54.** not a function
55. function **56.** function **57.** not a function **58.** 1000; In 2003, the average computer price was $1000.

Chapter 10 Test

1. 4 and -4 **2.** $3 + \sqrt{5}$ and $3 - \sqrt{5}$ **3.** $4\sqrt{5}$ yd **4.** $\sqrt{58}$ or 7.62 units **5.** $-2 \pm \sqrt{7}$ **6.** $\dfrac{-5 \pm \sqrt{13}}{6}$

7. 1 and $-\dfrac{4}{3}$ **8.** $\dfrac{5}{2}$ and $-\dfrac{7}{2}$ **9.** $\dfrac{3 \pm \sqrt{7}}{2}$ **10.** $\dfrac{1}{2}$ and -5 **11.** $11i$ **12.** $5\sqrt{3}i$ **13.** $\pm 6i$ **14.** $5 \pm 5i$ **15.** $1 \pm 2i$

16.
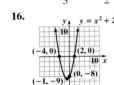
17.

18. 2 sec; 69 ft
19. 4.1 sec
20. function; domain: $\{1, 3, 5, 6\}$; range: $\{2, 4, 6\}$
21. not a function; domain: $\{2, 4, 6\}$; range: $\{1, 3, 5, 6\}$
22. 67 **23.** 17
24. function
25. not a function
26. 9; In 2000, 9 million people received food stamps.

Cumulative Review Exercises (Chapters 1–10)

1. 4 **2.** 10 **3.** $\{x \mid x \le 7\}$ **4.** $(-3, 4)$ **5.** $(3, 4)$ **6.** -8 **7.** -2 and -3 **8.** 9 **9.** $2 \pm 2\sqrt{5}$ **10.** $\dfrac{3 \pm \sqrt{3}}{3}$

11. $t = \dfrac{5r + 2}{A}$ **12.** $\dfrac{4}{x^9}$ **13.** 31 **14.** $10x^2 - 9x + 4$ **15.** $21x^2 - 23x - 20$ **16.** $25x^2 - 20x + 4$ **17.** $x^3 + y^3$ **18.** $\dfrac{x + 4}{3x^3}$

19. $\dfrac{-7x}{(x + 3)(x - 1)(x - 4)}$ **20.** $\dfrac{x}{5}$ **21.** $12\sqrt{5}$ **22.** $3x\sqrt{2}$ **23.** $\dfrac{2\sqrt{3}}{3}$ **24.** $6 + 2\sqrt{5}$ **25.** $(2x + 7)(2x - 7)$

26. $(x + 3)(x + 1)(x - 1)$ **27.** $2(x - 3)(x + 7)$ **28.** $x(x^2 + 4)(x + 2)(x - 2)$ **29.** $x(x - 5)^2$ **30.** $(x - 2)(x^2 + 2x + 4)$ **31.** $\dfrac{1}{4}$

32.
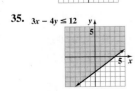
33.
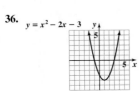
34.
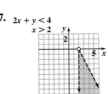

35.
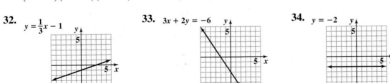
36.

37.

38. -2 **39.** $y - 2 = 2(x - 1)$ or $y - 6 = 2(x - 3)$; $y = 2x$ **40.** 43 **41.** \$320 **42.** length: 150 yd; width: 50 yd
43. \$12,500 at 7% and \$7500 at 9% **44.** 8 l of 40% and 4 l of 70% **45.** 7.25 hr or 7 hr, 15 min **46.** 2024 students **47.** 16 ft
48. $35°, 25°, 120°$ **49.** TV: \$350; stereo: \$370 **50.** length: 11 m; width: 5 m

APPLICATIONS INDEX

SUBJECT INDEX

PHOTO CREDITS

Robert F. Blitzer
Introductory Algebra for College Students 4e, Chapter Test Prep Video CD
0-13-185591-3
© 2006 Pearson Education, Inc.
Pearson Prentice Hall
Pearson Education, Inc.
Upper Saddle River, NJ 07458
Pearson Prentice Hall™ is a trademark of Pearson Education, Inc.

ESM Media Development
Higher Education Division
Pearson Education, Inc.
1 Lake Street
Upper Saddle River, NJ 07458
Should you have any questions concerning technical support, you may write to:
New Media Production
Higher Education Division
Pearson Education, Inc.
1 Lake Street
Upper Saddle River, NJ 07458
YOU ACKNOWLEDGE THAT YOU HAVE READ THIS AGREEMENT, UNDERSTAND IT, AND AGREE TO BE
BOUND BY ITS TERMS AND CONDITIONS. YOU FURTHER AGREE THAT IT IS THE COMPLETE AND
EXCLUSIVE STATEMENT OF THE AGREEMENT BETWEEN US THAT SUPERSEDES ANY PROPOSAL OR
PRIOR AGREEMENT, ORAL OR WRITTEN, AND ANY OTHER COMMUNICATIONS BETWEEN US RELATING
TO THE SUBJECT MATTER OF THIS AGREEMENT.

System Requirements
- Windows:
Pentium II 300 MHz processor
Windows 98, NT, 2000, ME, or XP
64 MB RAM (128 MB RAM required for Windows XP)
4.3 available hard drive space (optional—for minimum QuickTime installation)
800 x 600 resolution
8x or faster CD-ROM drive
QuickTime 6.x
Sound Card
- Macintosh:
Power PC G3 233 MHz or better
Mac OS 9.x or 10.x
64 MB RAM
10 MB available hard drive space for Mac OS 9, 19 MB on OS X (optional—if QuickTime installation is needed)
800 x 600 resolution
8x or faster CD-ROM drive
QuickTime 6.x

Support Information
If you are having problems with this software, call (800) 677-6337 between 8:00 a.m. and 8:00 p.m. EST, Monday through Friday,
and 5:00 p.m. through Midnight EST on Sundays. You can also get support by filling out the web form located at :
http://247.prenhall.com/mediaform
Our technical staff will need to know certain things about your system in order to help us solve your problems more quickly
and efficiently. If possible, please be at your computer when you call for support. You should have the following information
ready:
- Textbook ISBN
- CD-Rom/Diskette ISBN
- corresponding product and title
- computer make and model
- Operating System (Windows or Macintosh) and Version
- RAM available
- hard disk space available
- Sound card? Yes or No
- printer make and model
- network connection
- detailed description of the problem, including the exact wording of any error messages.
NOTE: Pearson does not support and/or assist with the following:
- third-party software (i.e. Microsoft including Microsoft Office Suite, Apple, Borland, etc.)
- homework assistance
- Textbooks and CD-Rom's purchased used are not supported and are non-replaceable. To purchase a new CD-Rom contact
  Pearson Individual Order Copies at 1-800-282-0693